69

Topics in Organometallic Chemistry

Aims and Scope

The series *Topics in Organometallic Chemistry* presents critical overviews of research results in organometallic chemistry. As our understanding of organometallic structure, properties and mechanisms increases, new ways are opened for the design of organometallic compounds and reactions tailored to the needs of such diverse areas as organic synthesis, medical research, biology and materials science. Thus the scope of coverage includes a broad range of topics of pure and applied organometallic chemistry, where new breakthroughs are being achieved that are of significance to a larger scientific audience.

The individual volumes of *Topics in Organometallic Chemistry* are thematic. Review articles are generally invited by the volume editors. All chapters from Topics in Organometallic Chemistry are published Online First with an individual DOI. In references, *Topics in Organometallic Chemistry* is abbreviated as Top Organomet Chem and cited as a journal.

More information about this series at http://www.springer.com/series/3418

Luis A. Oro • Carmen Claver
Editors

Iridium Catalysts for Organic Reactions

With contributions by

I. Cano · D. Carmona · J. G. de Vries · M. Diéguez ·
E. Fernández · F. J. Fernández-Alvarez · K. Fujita ·
M. Iglesias · R. Kanega · S. Kirchhecker · J. Margalef ·
L. M. Martínez-Prieto · L. A. Oro · O. Pàmies ·
V. Passarelli · M. Pilar Lamata · M. Reid · T. Shimbayashi ·
B. Spiegelberg · P. W. N. M. van Leeuwen

Springer

Editors
Luis A. Oro
Department of Inorganic Chemistry
University of Zaragoza-CSIC
Zaragoza, Spain

Carmen Claver
Department of Physical and Inorganic Chemistry
University Rovira i Virgili
Tarragona, Spain

ISSN 1436-6002 ISSN 1616-8534 (electronic)
Topics in Organometallic Chemistry
ISBN 978-3-030-69082-3 ISBN 978-3-030-69083-0 (eBook)
https://doi.org/10.1007/978-3-030-69083-0

This Springer imprint is published by the registered company Springer Nature Switzerland AG.
The registered company address is: Gewerbestrasse 11, 6330 Cham, Switzerland

Preface

Iridium catalysis, although not as established as rhodium catalysis, has seen considerable advances in recent years mainly due to its application in the synthesis of fine chemicals. Owing to the isoelectronic nature of iridium and rhodium, advances in either area have often been transferred and tested in the respective counterpart. The discovery by Schrock and Osborn that cationic rhodium phosphine complexes can be excellent homogeneous hydrogenation catalysts paved the way for Crabtree's cationic iridium catalyst, also offering the possibility of incorporating chiral ligands. From here, Pfaltz et al. and others revealed the potential of iridium catalysts first in the asymmetric hydrogenation of unsaturated organic bonds, including both imines and non-functionalized alkene bonds, and later extended this to other reactions that have proven to be key in the organic synthesis where iridium catalysts today play an indispensable role. Iridium has often been an alternative to the more studied rhodium catalyst as demonstrated in the production of acetic acid via the industrial carbonylation of methanol in the Cativa process. Furthermore, due to the stability of iridium-carbon and iridium-hydrogen bonds, iridium has also played an important role as a catalytic model because it allows the isolation and characterization of catalytic intermediates, which are difficult to detect in rhodium catalysis.

The first collection of modern applications of iridium organometallic complexes in this series was published in 2011 entitled "Iridium Catalysis" edited by Pher Andersson. Many of the iridium-catalyzed reactions presented in that volume have shown remarkable advances in the last decade and new applications to obtain building blocks in organic synthesis have been developed. This new volume of *Topics in Organometallic Chemistry* entitled "Iridium Catalysts for Organic Reactions" provides updates on previous topics and summarizes recent progress in iridium catalysis. The most relevant advances in iridium-catalyzed organic processes are considered throughout this volume including new developments in the synthesis of organosilanes, asymmetric hydrogenation, transfer hydrogenation, and hydroboration. The rapid development of iridium-catalyzed dehydrogenative reactions is described demonstrating that many of these catalytic systems have become important tools for organic synthesis. Due to the importance and interest in biomass conversion, a chapter is also dedicated to recent work on the use of homogeneous iridium catalysts in this field.

Despite the progress made in the last decade, there is still a challenge for iridium nanoparticles to be applied in the fine chemical industry. Therefore, the use of iridium nanoparticles as catalysts for hydrogenation reactions is reviewed including recent contributions presenting nanoparticles immobilized on supports, confined nanoparticles, nanoparticles stabilized by ionic liquids and polymers, and generated in situ without stabilizing agents. Another interesting area of development presented in this volume is the approach to CO_2 valorization using homogeneous and electrocatalytic iridium catalysts. Due to the high catalytic performance of several iridium catalysts in CO_2 transformation, a chapter is dedicated to the recent progress in the field of iridium-catalyzed reduction of CO_2 by using hydrogen and/or hydrosilanes as reducing agents. Another chapter focuses on homogeneous iridium catalysts for the electroreduction of CO_2, providing a mechanistic insight into the design of catalysts to efficiently transform CO_2 into new products using electricity.

In summary, the interest of the volume lies in the presentation of the advances and applications in iridium catalysis through a variety of catalytic processes that offer different approaches of efficiency and sustainability. The content of this volume is intended for researchers, graduate students, and synthetic chemists at all levels in academia and industry.

We would like to thank all the authors for their great contributions. We are also grateful to Dr. Charlotte Hollingworth and Ms. Alamelu Damodharan (Springer Nature) for their continuous assistance to complete this volume.

Zaragoza, Spain Luis A. Oro
Tarragona, Spain Carmen Claver
December 2020

Contents

Top Organomet Chem (2021) 69: 1–66
https://doi.org/10.1007/3418_2020_56

Published online: 4 August 2020

Iridium-Catalyzed Dehydrogenative Reactions

Takuya Shimbayashi and Ken-ichi Fujita

Contents

Abstract This chapter summarizes the recent advances in homogeneous iridium complex-catalyzed dehydrogenative reactions, covering the dehydrogenation of alkanes, heterocyclic compounds, alcohols, and formic acid. Both simple reactions

---

The original version of this chapter was revised. A correction to this chapter can be found at https://doi.org/10.1007/3418_2020_73

T. Shimbayashi and K. Fujita (✉)
Graduate School of Human and Environmental Studies, Kyoto University, Kyoto, Japan
e-mail: shimbayashi.takuya.5z@kyoto-u.ac.jp; fujita.kenichi.6a@kyoto-u.ac.jp

affording unsaturated products and relatively sophisticated organic transformations triggered by substrate dehydrogenation are discussed, and many of the listed catalytic systems are revealed to have become important tools for organic synthesis.

Keywords Cooperative catalysis · Dehydrogenation · Functional ligand · Hydrogen production · Hydrogen storage · Organic synthesis · Pincer complex

1 Introduction

Catalytic chemistry of iridium for organic synthesis is entering a maturity phase. A wide variety of reactions using iridium catalysts is continuously being developed, some of which are indispensable to organic synthesis.

The relative stability of Ir–H and Ir–C bonds allows the isolation of organoiridium complexes and the investigation of their structural and fundamental properties, as exemplified by the work of Vaska et al., who uncovered the basics of the fundamental reactions (e.g., oxidative addition) involved in organometallic chemistry [1]. However, the catalytic chemistry of Ir has advanced at a considerably slower pace than that of Rh and Pd. For example, $[Ir(cod)(Py)(PCy_3)]^+$ (cod: 1,5-cyclooctadiene; Py: pyridine; Cy: cyclohexyl), which shows a very high activity for alkene hydrogenation, was reported by Crabtree et al. only in 1977 [2, 3]. This discovery triggered the development of Ir catalytic chemistry and the discovery of other high-performance Ir catalysts, as exemplified by the use of $[Ir(cod)(phox)]^+$ (phox: phosphinooxazoline) for asymmetric hydrogenation, which was reported by Pfaltz et al. [4].

Although the application scope of Ir catalysts was initially restricted to the hydrogenation of unsaturated organic compounds, it subsequently expanded to include allylic substitution [5–14] and C–H borylation [15–18] and even the industrial carbonylation of methanol to afford acetic acid (the Cativa process) [19]. In addition, many other reactions catalyzed exclusively by Ir have been developed [20–22].

The ability of Ir to catalyze hydrogenation reactions suggests that this metal should also promote the reverse reaction, i.e., dehydrogenation. Indeed, the last three decades have witnessed the rapid development of Ir-catalyzed dehydrogenative reactions.

This chapter lists the applications of homogeneous Ir catalysts for the dehydrogenation of hydrogen acceptor-free organic molecules, demonstrating that many of these catalytic systems have become important organic synthesis tools.

2 Dehydrogenation of Alkanes

Alkenes are one of the most synthetically and industrially important classes of compounds, featuring high synthetic utility due to the presence of reactive π-bonds, whereas the reactivity of the parent alkanes is low because of their strong

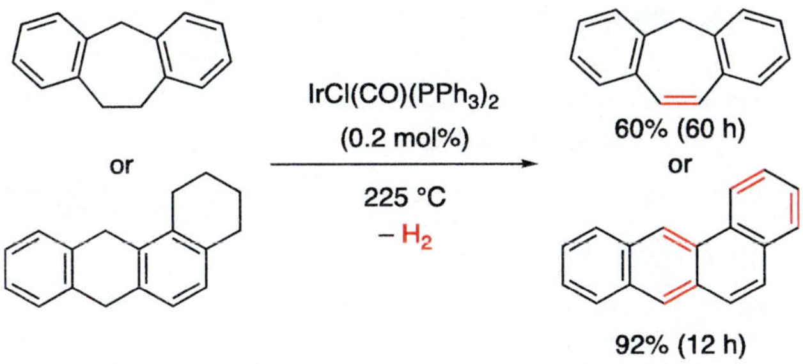

Scheme 1 Ir-catalyzed dehydrogenation of hydrocarbons containing aromatic moieties

and inert C–H σ-bonds. Considering the wide availability of alkanes, which are produced on an industrial scale by hydrotreatment of crude oil, dehydrogenation of alkanes is the most straightforward and atom-economical route to alkenes. Compared to the well-documented reverse hydrogenation, which can be carried out even under ambient conditions, the dehydrogenation of alkanes to alkenes requires high temperatures to overcome the enthalpic barrier of going from two C–H σ-bonds to one H–H σ-bond and a C–C π-bond. Despite this challenge, catalytic dehydrogenation of alkanes has attracted much attention because of its fascinating nature.

The Ir complex-catalyzed homogeneous dehydrogenation of alkyl functionalities was first reported in 1970 [23]. In the presence of a catalytic amount of Vaska's complex ($IrCl(CO)(PPh_3)_2$), the benzylic C–H bonds of 10,11-dihydro-5*H*-dibenzo [a,d]cycloheptene at 10- and 11-positions were cleaved to afford 5*H*-dibenzo[a,d] cycloheptene (yield = 60% after 60 h at 225°C) and H_2 (Scheme 1). 1,2,3,4,7,12-Hexahydrobenz[a]anthracene underwent dehydrogenation much more smoothly, affording benz[a]anthracene in 92% yield after 12 h. In this case, the intrinsically energetically unfavorable dehydrogenation was facilitated by the aromatization of the starting material.

The dehydrogenation of simple alkanes mediated by Ir complexes under stoichiometric conditions was pioneered by the Crabtree's group [24], which subsequently also investigated the catalytic dehydrogenation of cycloalkanes and linear alkanes under thermal or photochemical conditions [25]. Ir complex **1** exhibited catalytic activity for the transfer dehydrogenation of cycloalkanes and *n*-hexane at 150°C in the presence of *t*-butylethylene as a hydrogen acceptor; however, no thermal dehydrogenation occurred in the absence of *t*-butylethylene in a closed reaction system (Scheme 2a). In the presence of Ir complex **2**, cyclooctane could be dehydrogenated to cyclooctene at 25°C upon irradiation with UV light (254 nm) from a low-pressure Hg lamp in the absence of a hydrogen acceptor, with a turnover number (TON) of 8 observed after 7 days (Scheme 2b). The use of reflux conditions for dehydrogenation allowed the constant removal of evolved H_2 and thus improved reaction efficiency by displacing the equilibrium to the right [26]. Crabtree et al. reported that the Ir-catalyzed acceptorless dehydrogenation of cyclooctane to cyclooctene proceeded much more smoothly under reflux conditions than under closed-system conditions, with TONs of up to 35.8 achieved for Ir complex

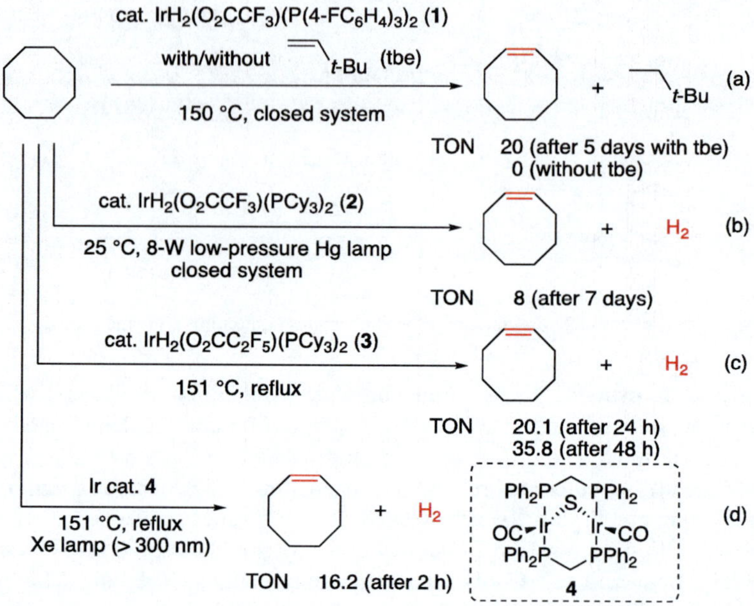

Scheme 2 Ir-catalyzed dehydrogenation of cyclooctane. (**a**) Thermal dehydrogenation with/without *t*-butylethylene. (**b**) Photochemical dehydrogenation at 25°C. (**c**) Thermal dehydrogenation under reflux conditions. (**d**) Photochemical dehydrogenation under reflux conditions

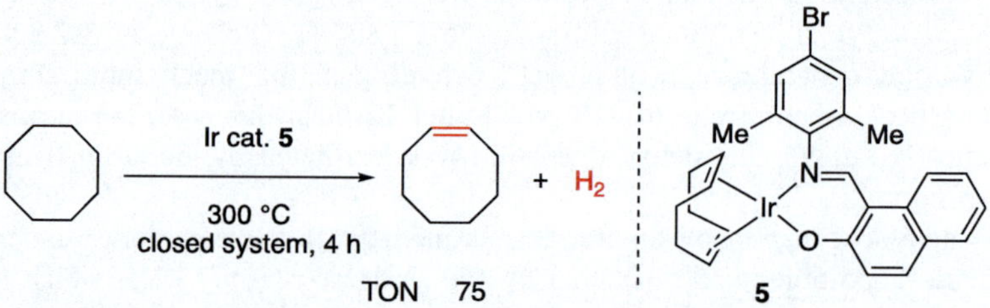

Scheme 3 Ir-catalyzed dehydrogenation of cyclooctane with relatively high turnover in closed system

3 (Scheme 2c) [27]. Saito et al. achieved a TON of 16.2 in 2 h for the photodehydrogenation (irradiation of Xe lamp (2 kW) through cut-off filter lower than 300 nm of cyclooctane under reflux conditions catalyzed by a dinuclear Ir complex **4** [28] (Scheme 2d). Alt et al. studied the catalytic activity of various [Ir(I)(cod)LL] complexes (LL = *N,N*- or *N,O*-chelating ligand, amine, or phosphine) for the dehydrogenation of cyclooctane at 300°C in a closed system, showing that TONs of up to 75 were obtained after 4 h for complex **5** (Scheme 3) [29, 30].

As another example, a CpIr(I) (Cp = cyclopentadienyl) complex was investigated for the photocatalytic dehydrogenation of cyclohexane (neat or in fluorinated solvent) [31]. Based on the known ability of the CpIr(CO) complex derived from [CpIr

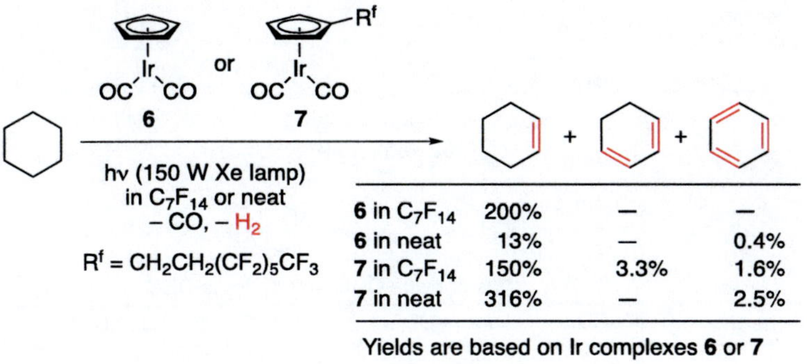

Scheme 4 Photocatalytic dehydrogenation of cyclohexane promoted by Cp-based Ir complexes

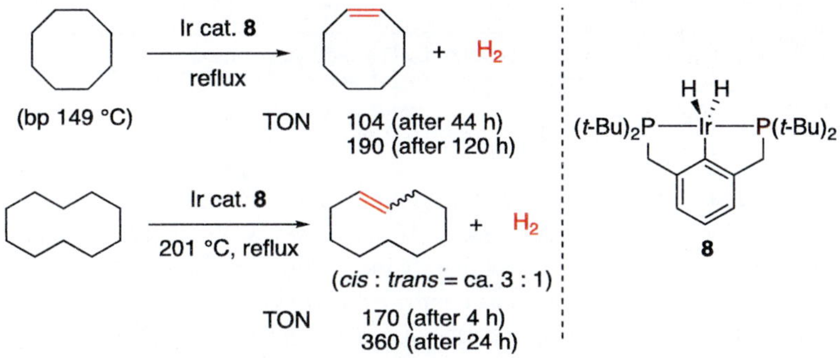

Scheme 5 Pincer Ir complex-catalyzed dehydrogenation of cycloalkanes with high TONs

$(CO)_2$] (**6**) to activate alkane C–H bonds [32], a fluoroalkyl-substituted CpIr complex **7** and the parent complex **6** were used for the photocatalytic dehydrogenation of cyclohexane (neat or in perfluoromethylcyclohexane (C_7F_{14})) to cyclohexene, cyclohexadiene, and benzene, with TONs ranging between 1.5 and 3.2 (Scheme 4). Fluorous solvents or fluoroalkyl substituents on the Cp ring increased the stability of Ir complexes under irradiation conditions.

The thermodynamic durability of Ir complexes under catalytic conditions is of key importance for the development of highly active alkane dehydrogenation catalysts. The thermodynamic parameters of dehydrogenation necessitate the use of relatively high temperatures, which may cause the decomposition of catalytically active species and thus result in low TONs. The first highly active Ir catalyst (TON >100) featured a rigid pincer ligand scaffold and was developed by Jensen and Goldman's group [33]. PCP pincer Ir complex **8**, which exhibited long-term stability at 200°C and superior catalytic activity for the transfer dehydrogenation of alkanes [34–36], promoted the dehydrogenation of cyclooctane to cyclooctene under reflux conditions. The initial turnover frequency (TOF) was 11 h^{-1}, and TONs of 104 and 190 were reached after 44 h and 120 h, respectively (Scheme 5). When the above catalyst was applied to the dehydrogenation of cyclodecane under reflux at a higher temperature of 201°C, the rate of dehydrogenation increased (TONs of 170 and

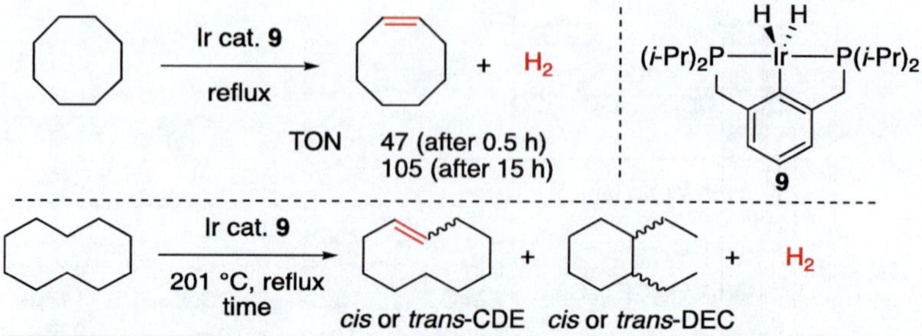

time (h)	*cis*-CDE	*trans*-CDE	*cis*-DEC	*trans*-DEC
1	378	82	16	10
4	619	145	38	59
8	692	163	42	79
20	700	163	43	81

TONs are shown. CDE = cyclodecene. DEC = diethylcyclohexane

Scheme 6 Pincer Ir complex-catalyzed dehydrogenation of cycloalkanes

360 were obtained after 4 and 24 h, respectively), and cyclodecene was produced as a 3:1 mixture of *cis*- and *trans*-isomers. The decrease of the dehydrogenation rate with time was ascribed to the concomitantly increasing content of the alkene product and not to catalyst decomposition as confirmed by the fact that the addition of 10% cyclooctene in cyclooctane completely stopped the dehydrogenation reaction.

The promising performance of the above PCP pincer complex for catalytic alkane dehydrogenation has inspired the development of other catalysts based on pincer Ir complexes, e.g., an Ir complex **9** bearing less sterically hindered bis(di-isopropyl) phosphino groups promoted the dehydrogenation of cyclooctane under reflux conditions and achieved an initial TOF of >94 h^{-1} (Scheme 6) [37]. This complex also smoothly catalyzed the dehydrogenation of refluxing cyclodecane, achieving a TON of 460 after 1 h and affording cyclodecene with a 4.6:1 *cis*:*trans* selectivity. As the reaction time increased to 20 h, TON increased to 863, and small amounts of isomerized products, *cis*- and *trans*-diethylcyclohexane, were observed. Additionally, the complex **9** catalyzed the dehydrogenation of refluxing *n*-undecane (196°C), affording a mixture of undecenes with TONs of 42.3 and 44.1 after 1 h and 45 h, respectively.

Krogh-Jespersen and Goldman et al. reported the mechanistic study of PCP pincer Ir-catalyzed dehydrogenation of alkanes based on density functional theory (DFT) calculations [38]. Two possible reaction pathways were compared, namely the dissociative pathway, in which the reaction proceeds via the liberation of H_2 from a dihydride complex followed by the oxidative addition of a cycloalkane (Scheme 7, right part), and the associative pathway, in which a cycloalkane directly interacts with the dihydride complex to give an alkyl complex intermediate via σ-bond metathesis followed by the liberation of H_2 (Scheme 7, left part). Based on the theoretical and experimental results, the latter pathway was excluded, and the

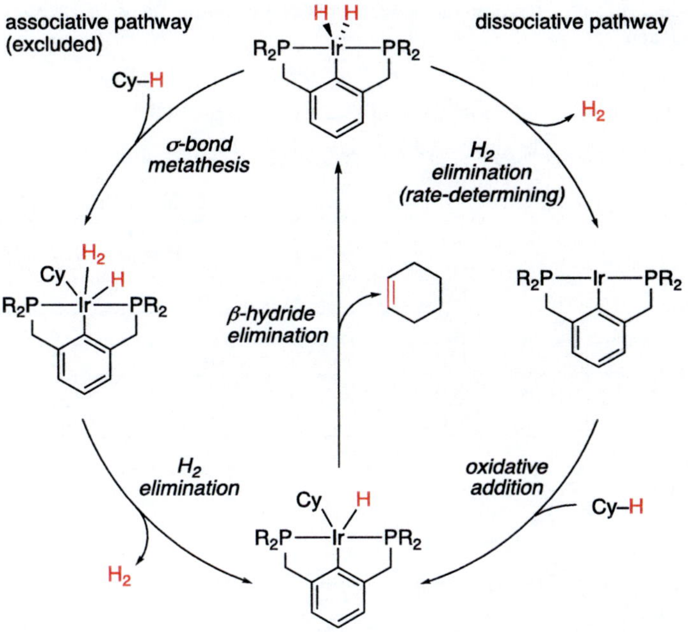

Scheme 7 Reaction mechanism for the activation of cyclohexane as a model substrate

elimination of H_2 to afford a $14e^-$ Ir complex was proposed to be the rate-determining step of the dissociative pathway. The oxidative addition of an alkane to the $14e^-$ Ir intermediate was expected to be much easier than the prior step and was followed by β-hydride elimination from the alkyl iridium intermediate to complete the catalytic cycle, forming a cycloalkene, and regenerating the dihydride complex.

The same group investigated the effects of introducing a methoxy group at the *p*-position of the pincer aromatic scaffold on catalytic dehydrogenation performance [39]. In the acceptorless dehydrogenation of cyclodecane to cyclodecene, methoxy-substituted complex **10** exhibited higher catalytic activity than the parent catalyst **8** (Table 1). An Ir complex **11** bearing a methoxy-substituted pincer ligand with isopropyl groups on phosphorus atoms was superior to other catalysts achieving TONs of 357 and 3,050 after 1 h and 78 h, respectively. In the case of diisopropylphosphine-based pincer complex **11**, a tetrahydride complex was used as a catalyst precursor because of the difficulty of dihydride complex isolation. No significant difference in catalytic performance was observed between the dihydride complex and the tetrahydride precursor. The catalytic activity of these complexes for the dehydrogenation of *n*-undecane was also investigated, and a TON of 108 was achieved with **10** after 17 h.

The effects of substituting a linker CH_2 moiety for an oxygen atom and changing the steric properties of the phosphino group of the PCP pincer ligand on catalytic activity for the acceptorless dehydrogenation of alkanes were also investigated

Table 1 Comparison of the catalytic activities (TONs) of pincer Ir complexes for acceptorless dehydrogenation

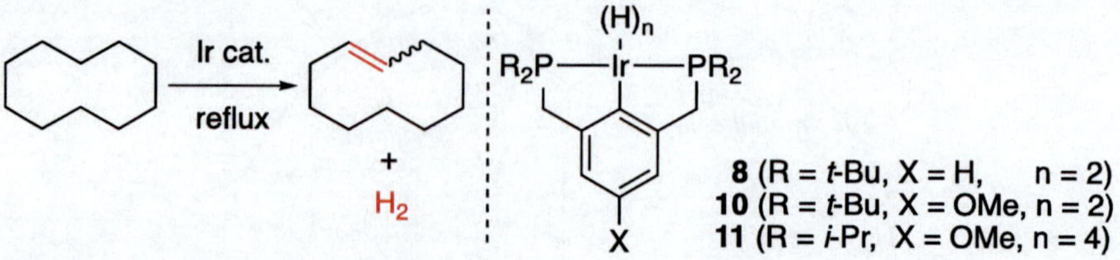

Time (h)	8	10	11
1	60	158	357
4	170	430	714
24	360	820	2,120
48	360	820	2,970
78	–	–	3,050

TONs are shown

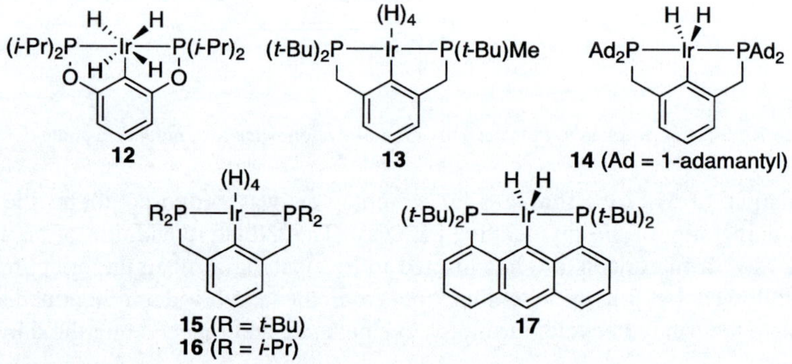

Fig. 1 PCP pincer Ir complexes used for the acceptorless dehydrogenation of alkanes

(Fig. 1) [40–43]. For example, when *O*-linked PCP pincer Ir complex **12** was used to catalyze the acceptorless dehydrogenation of *n*-undecane under reflux conditions, conversion to the corresponding alkene was as low as 1.8%, probably because of the inhibitory effect of the product [40]. Recently, an *O*-linked PCP pincer Ir complex supported on silica via covalent bonds with the ligand backbone was applied to continuous-flow alkane dehydrogenation at >300°C, achieving a TON of 3,500 for the dehydrogenation of butane to butene [44].

Ir complex **13**, bearing a *t*-butylmethylphosphino group and a di-*t*-butylphosphino group on its PCP pincer scaffold and therefore featuring reduced steric crowding (compared to that of the parent tetra-*t*-butyl-substituted complex **8**), exhibited higher performance for the catalytic dehydrogenation of cyclodecane to cyclodecene than complexes **15** and **16** in terms of both the initial (in the first 1 h)

Table 2 Comparison of the catalytic performances of Ir complexes **13**, **15**, and **16** for the acceptorless dehydrogenation of cyclodecane

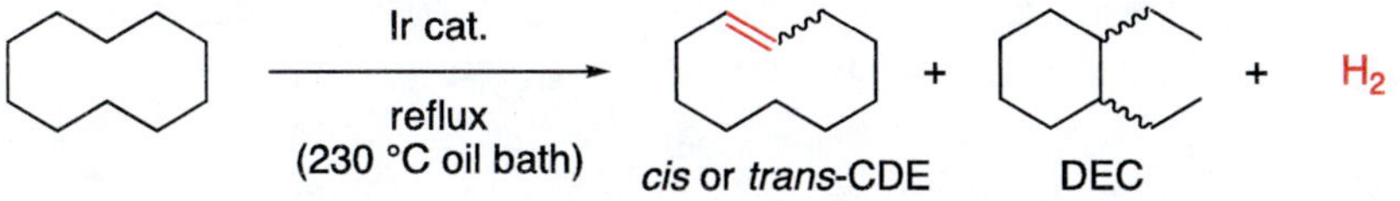

Ir cat.	Time (h)	*cis*-CDE	*trans*-CDE	DEC	Total
13	1	388	65	36	489
	9	807	163	119	1,089
15	1	49	49	3	101
	9	325	72	11	408
16	1	153	31	12	196
	9	410	86	96	592

TONs are shown
CDE cyclodecene, *DEC* diethylcyclohexane

Table 3 Comparison of the catalytic activities of Ir complexes **8**, **14**, and **16** for the acceptorless dehydrogenation of cyclodecane

Total cyclodecene (mM)			
Time (h)	Ir cat. **14**	Ir cat. **8**	Ir cat. **16**
1	74	102	136
3	179	218	274
5	251	240	–
24	509	267	364
48	526	294	–
72	534	305	366
96	543	314	–

Conditions: [catalyst] = 1.0 mM, cyclodecane volume = 1.5 mL. Oil bath temperature = 230°C

reaction rate and total TON after 9 h (Table 2) [41]. Complex **13** could also catalyze the dehydrogenation of *n*-undecane with slightly higher efficiency than **15** and **16**.

The structurally more rigid 1-adamantyl-substituted PCP pincer complex **14** was more stable than complexes **8** and **16** during the catalytic dehydrogenation of cyclodecane [42]. Although the initial rate of **14**-catalyzed cyclodecane dehydrogenation was lower than that observed for *t*-butyl- or isopropyl-substituted complexes **8** and **16** at reaction times of <3 h, the total TON exceeded those of **8** and **16** after 24 h (Table 3). A similar trend was also observed for the dehydrogenation of *n*-dodecane (bp 216°C), in which case a TON of 71 was reached after 72 h.

Anthracene-based PCP pincer Ir complex **17** features a much more rigid backbone than *m*-xylene-based complexes and should therefore exhibit a higher thermal stability [43]. This complex catalyzed the acceptorless dehydrogenation of cyclododecane (bp 244°C) to a mixture of *cis*- and *trans*-cyclododecene (1:2.4) even at 250°C, achieving an initial TOF of 40 h^{-1} and a TON of 136 after 148 h. At 230°C, the catalytic performance decreased to an initial TOF of 30 h^{-1} and a TON of 126 after 192 h. Thus, the rigidness and steric bulk of **17** effectively increased

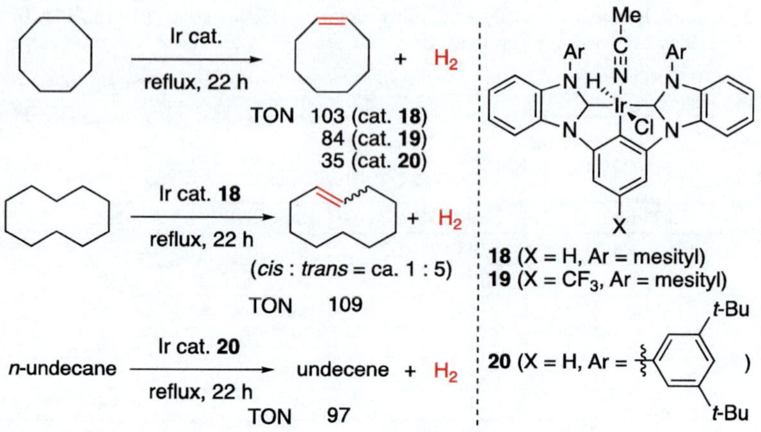

Scheme 8 CCC pincer Ir complex-catalyzed dehydrogenation of alkanes

thermal stability, albeit the catalytic activity at lower temperature was not as high as that of **8**.

Besides that of PCP pincer complexes, the catalytic activity of CCC pincer Ir complexes **18–20** was also investigated [45, 46]. CCC pincer ligands bearing two *N*-heterocyclic carbene (NHC)-derived coordinating moieties with mesityl or 3,5-di-*t*-butylphenyl groups on nitrogen atoms effectively catalyzed the acceptorless dehydrogenation of cyclic and linear alkanes. A TON of 103 after 22 h was achieved for the dehydrogenation of refluxing cyclooctane catalyzed by Ir complex **18** (Scheme 8). Cyclodecane also underwent acceptorless dehydrogenation under reflux conditions in the presence of **18**, in which case a TON of 109 and a 1:5 mixture of *cis*- and *trans*-cyclodecene were obtained after 22 h. In the case of *n*-undecane dehydrogenation, Ir complex **20** exhibited the best catalytic performance, achieving a TON of 97 after 22 h.

Recently, Celik and Goldman's group has developed a tandem catalytic system consisting of a PCP pincer Ir complex and a zeolite for the intramolecular dehydrogenative cyclization of *n*-pentylbenzene leading to 1-methyl-1,2,3,4-tetrahydronaphthalene and other carbocyclic compounds [47]. The PCP pincer Ir complex catalyzed the acceptorless dehydrogenation of *n*-pentylbenzene (bp 205°C), affording a mixture of alkene products under reflux conditions (Scheme 9a). H-SSZ-25 zeolite catalyzed the intramolecular alkylation of in situ generated alkene-tethered benzenes, affording 1,2,3,4-tetrahydronaphthalene and indane derivatives together with products of further dehydrogenation, e.g., naphthalene derivatives. For a tandem system composed of PCP pincer Ir complex **21** and H-SSZ-25, cyclized products were obtained in a yield of 94% (based on the amount of evolved H_2) with a TON of 6,860 after 24 h (Scheme 9b).

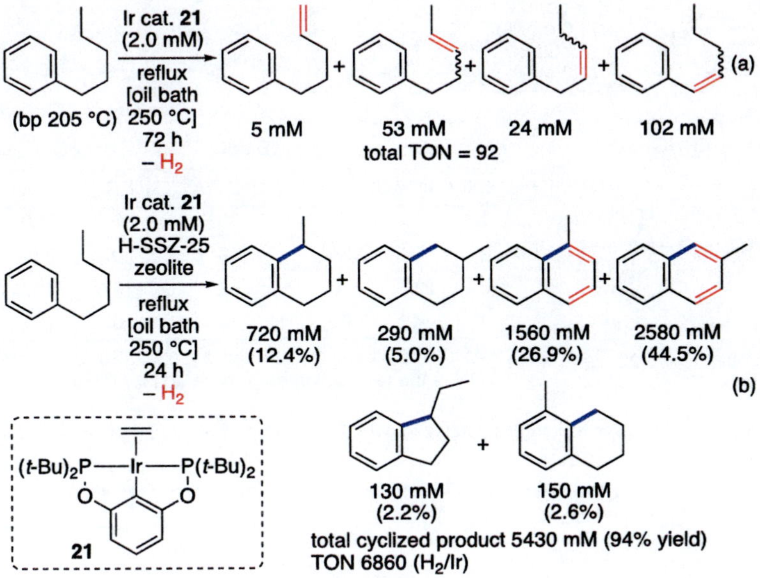

Scheme 9 Tandem catalytic system for the dehydrogenative cyclization of *n*-pentylbenzene. (**a**) Ir-catalyzed dehydrogenation of *n*-pentylbenzene. (**b**) Tandem system for dehydrogenation followed by cyclization catalyzed by Ir complex and zeolite

3 Dehydrogenation of Heterocyclic Compounds

Compared to the dehydrogenation of cycloalkanes, that of heterocyclic compounds containing at least one heteroatom (e.g., O, N, P, S) in the ring is much more facile because of the decreased reaction endothermicity [48]. Hence, heteroatom incorporation allows acceptorless dehydrogenative transformations to be carried out at lower temperatures, e.g., under homogeneous catalysis conditions, most *N*-heterocycles undergo dehydrogenation to afford aromatized products below 200°C. In this context, the design of transition metal complexes for catalytic dehydrogenation is not necessarily restricted by their thermal durability, which leaves enough room for efficiency improvement from a kinetic viewpoint.

3.1 Dehydrogenation of N*-Heterocyclic Compounds*

N-Heterocyclic structures are ubiquitous in indispensable natural and industrial products including biologically active compounds such as pharmaceuticals and agrochemicals, fragrances, dyes, pigments, and other functional materials [49, 50]. Hence, the development of efficient method for the transformations of heterocyclic compounds has been pursued for a long time since the evolvement of synthetic organic chemistry. The dehydrogenation of saturated heterocyclic

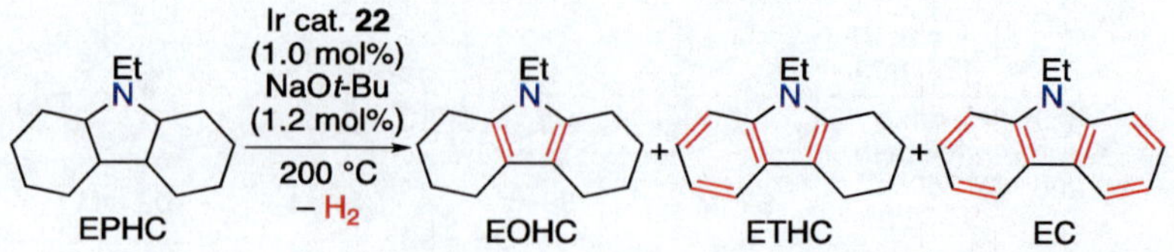

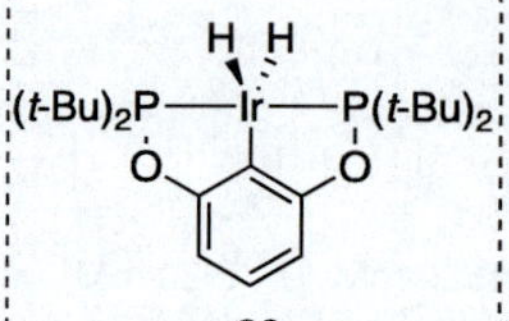

22

entry	time (h)	EOHC (%)	ETHC (%)	EC (%)	TON
1	9	70	30	0	260
2	24	60	40	0	280
3	216	0	82	18	436
4[a]	48	90	10	0	110

[a] At 150 °C, Ir cat. **22** (2.0 mol%), NaO*t*-Bu (2.4 mol%)
TON is the ratio between generated H_2 / Ir cat. **22**.

Scheme 10 PCP pincer Ir complex-catalyzed dehydrogenation of *N*-ethyl perhydrocarbazole

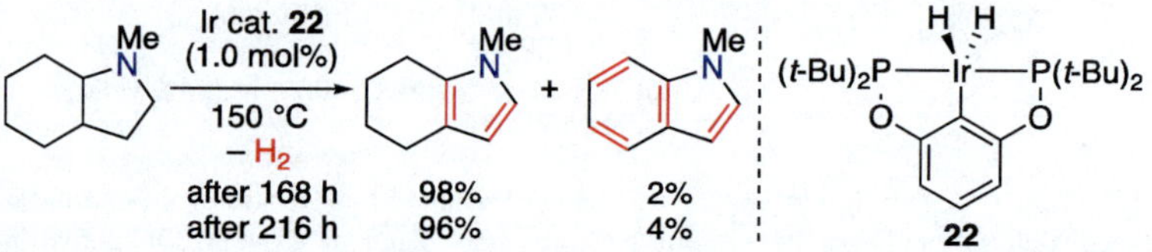

22

Scheme 11 Dehydrogenation of *N*-methyl perhydroindole

compounds offers direct access to the related unsaturated heterocycles with variable unsaturation degrees, finally affording heteroaromatics via perdehydrogenation [51]. Furthermore, *N*-heterocyclic compounds have recently attracted much attention as promising organic hydrogen carriers for reversible hydrogen storage [52–54].

In addition to promoting the dehydrogenation of cycloalkanes, PCP pincer Ir complexes also catalyze the dehydrogenation of *N*-heterocycles. In 2009, Jensen et al. reported the catalytic dehydrogenation of *N*-ethyl perhydrocarbazole (EPHC) to *N*-ethyl octahydrocarbazole (EOHC), *N*-ethyl tetrahydrocarbazole (ETHC), and *N*-ethyl carbazole (EC) (Scheme 10) [55]. In the presence of PCP pincer Ir complex **22** (1.0 mol%) and NaO*t*-Bu (1.2 mol%) at 200°C, EPHC was fully consumed to afford EOHC and ETHC in yields of 70 and 30%, respectively, and a TON (molar ratio between the generated H_2 and **22**) of 260 was obtained. After 24 h, the product ratio changed to 60% EOHC/40% ETHC, and a further extension of the reaction time to 216 h afforded an 82% ETHC/18% EC mixture. At 150°C in the presence of 2.0 mol% **22**, the full conversion of EPHC was reached in 48 h to afford EOHC (90%) and ETHC (10%).

The same group also reported the dehydrogenation of other *N*-heterocyclic compounds catalyzed by the above PCP pincer Ir complex and its analogues [56–59]. The dehydrogenation of *N*-methyl perhydroindole at 150°C in the presence of **22** (1.0 mol%) afforded *N*-methyl tetrahydroindole (98%) and *N*-methyl indole (2%) after 168 h (Scheme 11) [56]. The further dehydrogenation of *N*-methyl tetrahydroindole to *N*-methyl indole was very sluggish, proceeding in only 4%

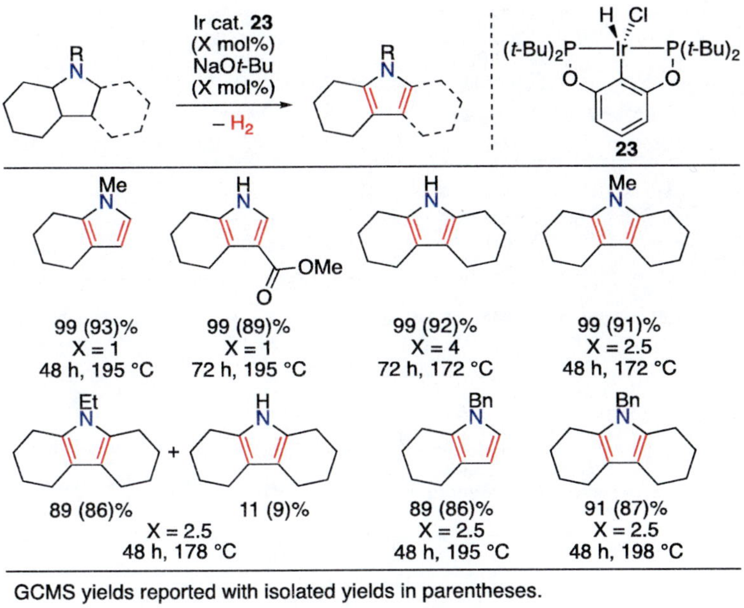

Scheme 12 Selective dehydrogenation of perhydroindoles and perhydrocarbazoles

yield after 216 h. This low catalytic performance for the perdehydrogenation of perhydroindole is not desirable from the standpoint of hydrogen storage, in which case maximal hydrogen weight capacity is preferable; however, the selective dehydrogenation of heterocyclic ring moieties can be applied to organic synthesis. The selective dehydrogenation of perhydroindole and perhydrocarbazole was achieved by employing PCP pincer Ir complex **23** under neat conditions at 172–198°C (Scheme 12) [57]. The methoxycarbonyl group was tolerated, and the corresponding tetrahydroindole was obtained in quantitative yield. Regarding the scope of substituents on the nitrogen atom, *N*-H, *N*-methyl, *N*-ethyl, and *N*-benzyl substrates were applicable. In the case of *N*-ethyl carbazole, a small amount of *N*-H octahydrocarbazole was formed via cleavage of the ethyl group. The catalytic performance of Ir complex **24** bearing an AsCAs pincer ligand as an arsenic congener of the PCP pincer ligand was also investigated, although the activity was only half of that of **23** (Scheme 13) [58].

Kinetic studies on the complex **23**-catalyzed dehydrogenation of pyrrolidine-based *N*-heterocycles including *N*-ethyl perhydrocarbazole, *N*-methyl perhydroindole, and *N*-butyl pyrrolidine revealed that the steric constraints of these *N*-heterocycles had a larger influence on the reaction rate than the C–H bond activation barrier [59].

Yamaguchi and Fujita et al. reported the catalytic dehydrogenation of 1,2,3,4-tetrahydroquinolines in the presence of a Cp*Ir (Cp* = 1,2,3,4,5-pentamethylcyclopentadienyl) complex bearing a functional 2-pyridonate ligand

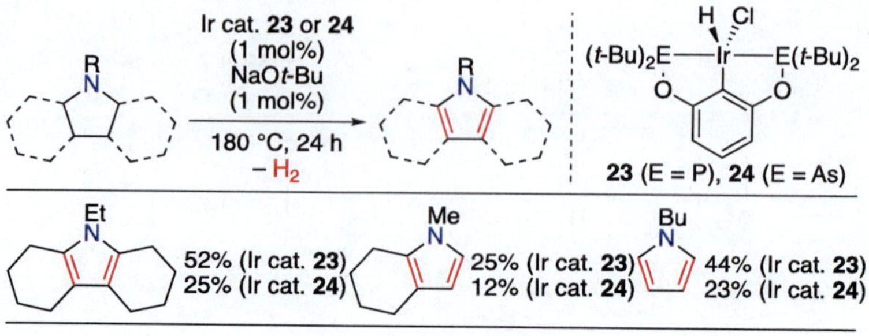

Scheme 13 Comparison of the catalytic activities of PCP and AsCAs pincer Ir complexes

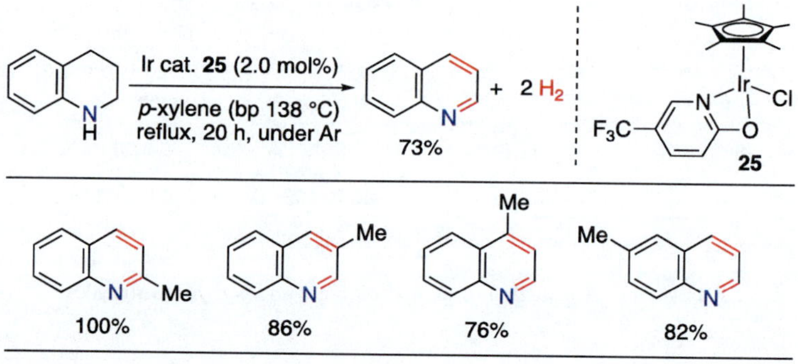

Scheme 14 Dehydrogenation of 1,2,3,4-tetrahydroquinoline catalyzed by Cp*Ir pyridonate **25**

[60]. The screening of 2-pyridonate ligands with various substituents showed that 5-trifluoromethyl-2-pyridonato Ir complex **25** exhibited the best catalytic performance for the dehydrogenation of 1,2,3,4-tetrahydroquinoline in refluxing *p*-xylene (bp 138°C), with quinoline obtained in 73% yield (Scheme 14). The incorporation of a methyl group into the 1,2,3,4-tetrahydroquinoline significantly affected the efficiency of catalytic dehydrogenation. In particular, 2-methyl-substituted 1,2,3,4-tetrahydroquinoline showed the highest reactivity in this dehydrogenation and was quantitatively converted to 2-methylquinoline. In general, the dehydrogenation enthalpy of substituted *N*-heterocyclic compounds is lower than that of non-substituted ones [61]. Notably, **25** also catalyzed the reverse hydrogenation of 2-methylquinoline to 2-methyl-1,2,3,4-tetrahydroquinoline. Moreover, a reversible and repetitive dehydrogenation/hydrogenation sequence well suited for reversible hydrogen storage was also possible.

Later, the same group developed a more efficient catalyst for the dehydrogenation of 2-methyl-1,2,3,4-tetrahydroquinoline to 2-methylquinoline [62]. Cp*Ir bipyridonate complexes **26** and **27** promoted this dehydrogenation at a reduced catalyst loading (1.0 mol%) to quantitatively afford the quinoline product, whereas catalyst **25** could not achieve full conversion (Scheme 15a). Catalysts **26** and **27** also promoted the dehydrogenation of 2,6-dimethyldecahydro-1,5-naphthyridine to

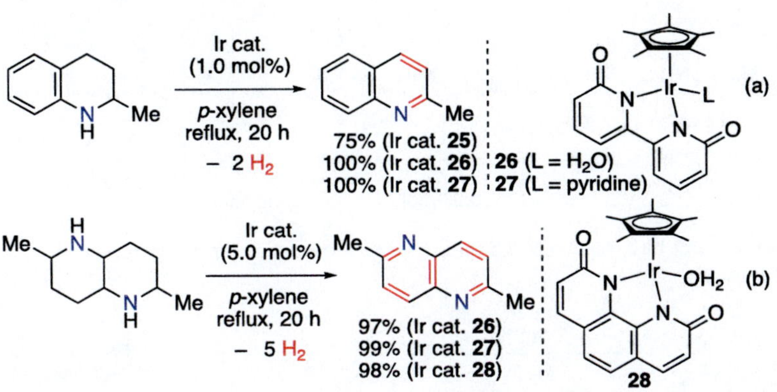

Scheme 15 Dehydrogenation of fused bicyclic compounds. (**a**) Dehydrogenation of 2-methyl-1,2,3,4-tetrahydroquinoline. (**b**) Dehydrogenation of 2,6-dimethyldecahydro-1,5-naphthyridine

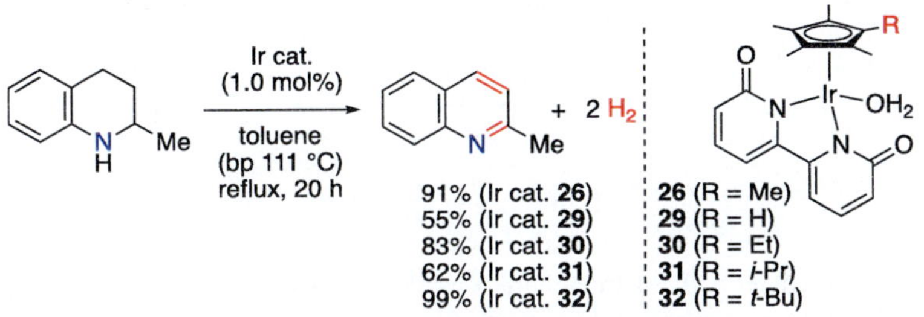

Scheme 16 Dehydrogenation of 2-methyl-1,2,3,4-tetrahydroquinoline catalyzed by Ir bipyridonate complexes bearing various Cp ligands

2,6-dimethyl-1,5-naphthyridine with a release of 5 equiv. of H_2 (Scheme 15b). Cp*Ir complex **28** with a structurally more rigid 1,10-phenanthroline-2,9-dione ligand was also effective for this dehydrogenation and allowed the quantitative reverse perhydrogenation of 2,6-dimethyl-1,5-naphthyridine. Moreover, the successive perhydrogenation and dehydrogenation sequence in the presence of **28** was concluded to hold great promise for hydrogen storage applications.

Very recently, Fujita et al. investigated the effect of Cp ligand substituents on Ir bipyridonate complex-catalyzed dehydrogenation reactions by replacing one methyl group of Cp* with hydrogen or various alkyl (e.g., ethyl, isopropyl, and *t*-butyl) groups [63]. Among the series of modified Cp* complexes **29–32**, the 1-*t*-butyl-2,3,4,5-tetramethyl CpIr complex **32** exhibited the best performance for the catalytic dehydrogenation of 2-methyl-1,2,3,4-tetrahydroquinoline in refluxing toluene (Scheme 16).

In addition to that of fused bicyclic compounds, Fujita et al. reported the dehydrogenation of 2,5-dimethylpiperazine to 2,5-dimethylpyrazine catalyzed by Cp*Ir bipyridonate complex **26** [64], showing that quantitative dehydrogenation was achieved in the presence of 2.0 mol% **26** in refluxing *p*-xylene after 20 h

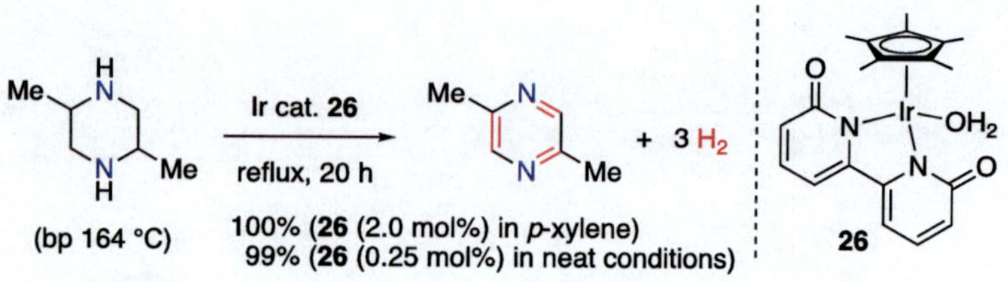

Scheme 17 Dehydrogenation of 2,5-dimethylpiperazine to 2,5-dimethylpyrazine

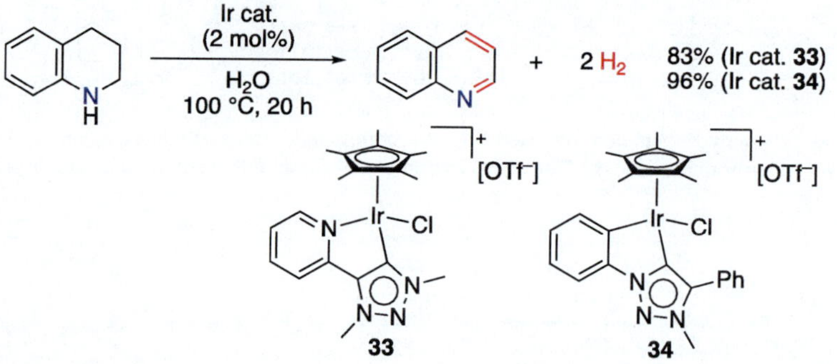

Scheme 18 Catalytic dehydrogenation of 1,2,3,4-tetrahydroquinoline in water

(Scheme 17). Under solvent-free conditions, the catalyst loading could be reduced to 0.25 mol% without significant loss of efficiency in refluxing 2,5-dimethylpiperazine. The reverse hydrogenation proceeded smoothly in the presence of **26** and free 6,6′-dihydroxy-2,2′-bipyridine in a *p*-xylene/H_2O mixed solvent. Moreover, dehydrogenation could also be performed by employing the reaction mixture used for hydrogenation under reflux conditions, and a reversible and repetitive hydrogenation/dehydrogenation sequence was successfully carried out for at least four cycles without any loss of catalytic performance.

Recently, Cp*Ir-catalyzed dehydrogenations of 1,2,3,4-tetrahydroquinolines and related *N*-heterocycles in refluxing water were reported. Albrecht et al. revealed the good ability of Cp*Ir complexes **33** and **34** bearing a bidentate abnormal NHC ligand to catalyze the dehydrogenation of 1,2,3,4-tetrahydroquinoline in refluxing water (Scheme 18) [65]. The same complexes could also promote the reverse hydrogenation reaction in water. Fischmeister et al. employed Cp*Ir complex **35** with a bispyridylamine-based ligand for the dehydrogenation of 1,2,3,4-tetrahydroquinoline derivatives in water under reflux conditions (Scheme 19) [66]. 1,2,3,4-Tetrahydroquinoxaline and 1,2,3,4-tetrahydroisoquinoline were also applicable to this dehydrogenation reaction. The reverse hydrogenation could also be performed in water with the same catalyst **35**.

Dehydrogenation of *N*-heterocycles catalyzed by Cp*Ir complexes under milder reaction conditions was developed by Xiao et al. for the efficient synthesis of various *N*-heteroarenes [67]. Ketimine-based iridacycle complex **36** exhibited high catalytic

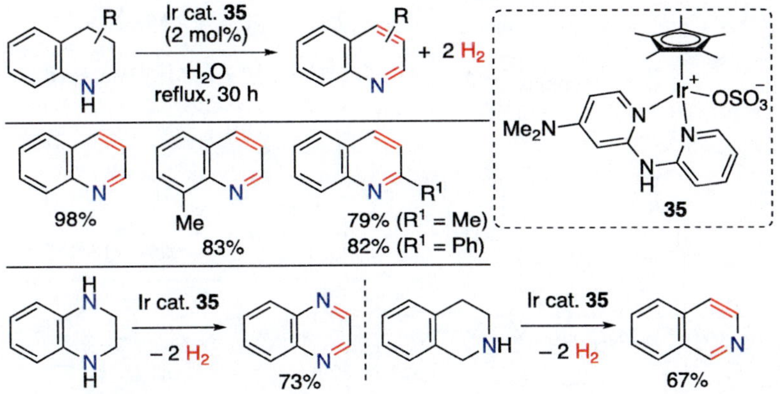

Scheme 19 Dehydrogenation of bicyclic *N*-heterocycles in water catalyzed by Ir complex **35**

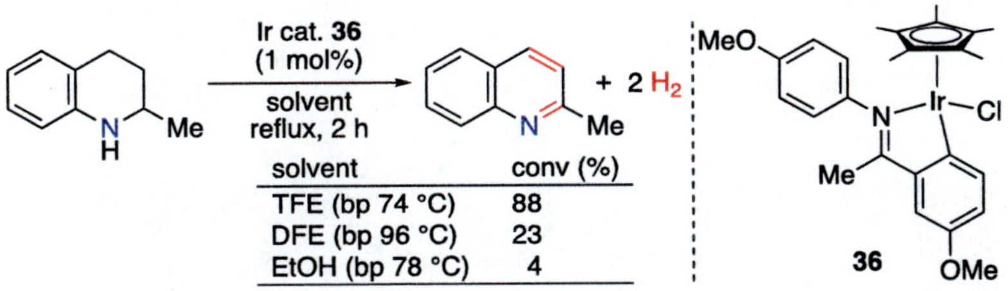

solvent	conv (%)
TFE (bp 74 °C)	88
DFE (bp 96 °C)	23
EtOH (bp 78 °C)	4

Scheme 20 Effect of solvent on the dehydrogenation of 2-methyl-1,2,3,4-tetrahydroquinoline

performance for the dehydrogenation of 2-methyl-1,2,3,4-tetrahydroquinoline in refluxing 2,2,2-trifluoroethanol (TFE: bp 74°C), achieving an 88% conversion after 2 h (Scheme 20). TFE was essential for efficient catalysis, probably because it facilitated H_2 formation by protonating the Ir hydride intermediate. In aprotic solvents, almost no conversion of the starting material was observed, while significantly diminished dehydrogenation yields were obtained in the less acidic 2,2-difluoroethanol (DFE) or ethanol. This dehydrogenation system was applicable to various *N*-heterocyclic compounds such as 1,2,3,4-tetrahydroquinolines, 9,10-dihydroacridine, 1,2,3,4,9,10-hexahydroacridine, 1,2,3,4-tetrahydroisoquinolines, 3,4-dihydroisoquinolines, 1,2,3,4-tetrahydro-β-carbolines, indolines, and 1,2,3,4-tetrahydroquinoxalines (Scheme 21). Later, a pyrene-tethered iridacycle immobilized onto multiwalled carbon nanotubes was prepared as a reusable catalyst for the dehydrogenation of indoline to indole in aqueous media [68]. Notably, **36** could also promote the hydrogenation of 1-methylisoquinoline to 1-methyl-1,2,3,4-tetrahydroisoquinoline.

Apart from pincer or Cp* complexes exhibiting catalytic activity for the dehydrogenation of *N*-heterocyclic compounds, other complexes were also investigated by Crabtree et al. [69]. The pre-activation of Ir complex **37** by treatment with an equimolar amount of PPh_3 at 100°C and H_2 bubbling led to the formation of the

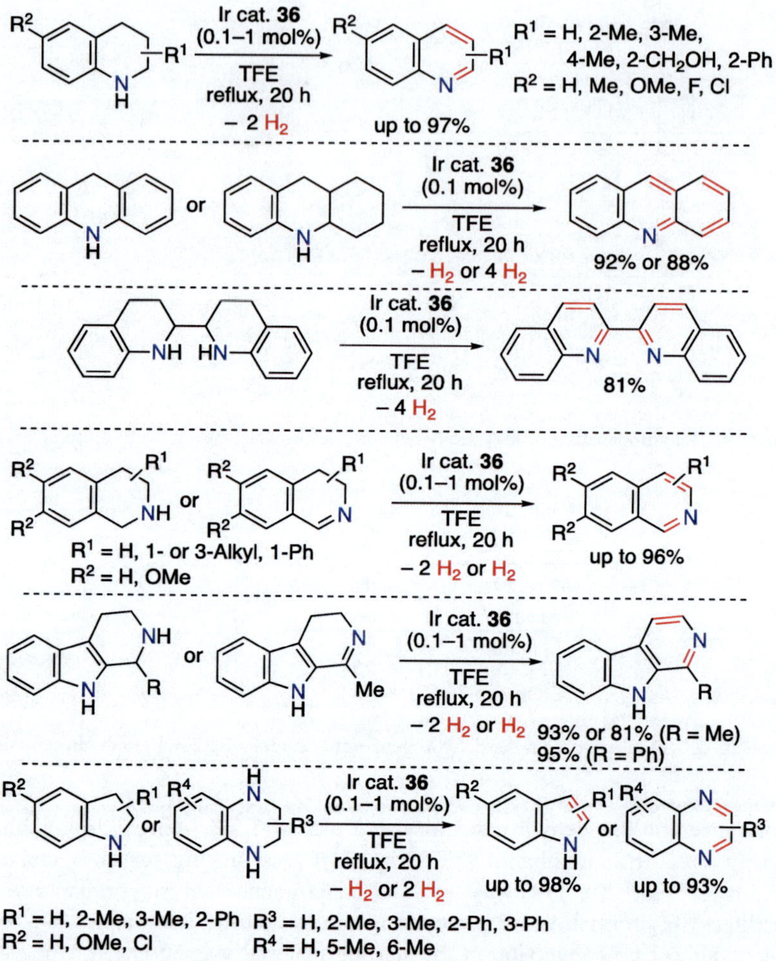

Scheme 21 Dehydrogenation of various *N*-heterocycles catalyzed by Ir complex **36**

proposed active catalyst **38**, which promoted the dehydrogenation of 2-methyl-1,2,3,4-tetrahydroquinoline (0.04 M) under reflux conditions at 135°C in chlorobenzene or at 145°C in *p*-xylene, affording 2-methylquinoline in 85 or 95% yield, respectively, after 72 h (Scheme 22). Catalyst **39** promoted the same dehydrogenation at a higher concentration (1.1 M) to afford 2-methylquinoline in 85% yield after 24 h. Both **38** and **39** also catalyzed the reverse hydrogenation.

Crabtree's catalyst (**40**) was employed for the dehydrogenation of 2-methylindoline to 2-methylindole in a liquid–liquid biphasic system [70]. The dehydrogenation proceeded at 130°C in the presence of **40** (0.5 mol%), PPh_3 (2 mol %), and $[PPh_4][NTf_2]$ (20 mol%) in Bu_2O as solvent (Scheme 23). PPh_3 was believed to stabilize the cationic Ir catalyst, which was immobilized in the molten

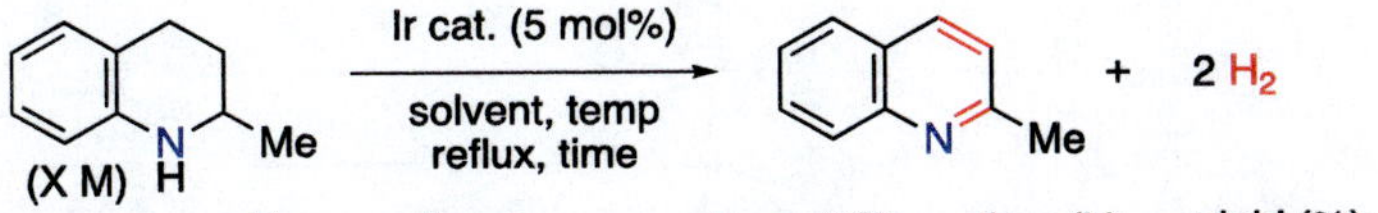

entry	Ir cat.	X	solvent	temp (°C)	time (h)	yield (%)
1[a]	**37**	0.04	chlorobenzene	135	72	85
2[a]	**37**	0.04	*p*-xylene	145	72	~95
3	**39**	1.1	chlorobenzene	145	24	85

[a] Pre-activation of **37** by treatment with PPh_3 and H_2.

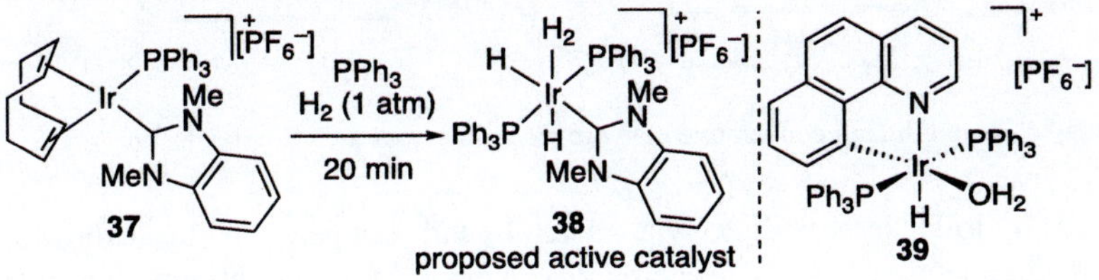

Scheme 22 Dehydrogenation of 2-methyl-1,2,3,4-tetrahydroquinoline by Ir complexes **37** and **39**

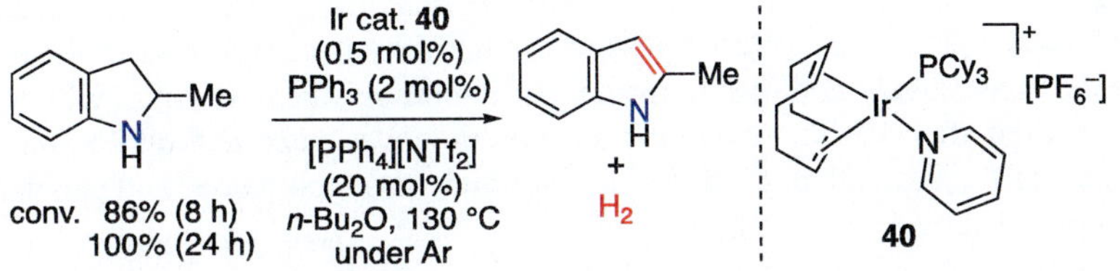

Scheme 23 Dehydrogenation of 2-methylindoline in a liquid–liquid biphasic system

salt phase and could therefore be retrieved and reused by simple separation of the liquid–liquid phase. The catalytic performance was sufficiently maintained in the second run of the dehydrogenation reaction.

3.2 Dehydrogenation of Other Heterocyclic Compounds

The dehydrogenation of cyclic compounds other than *N*-heterocycles remains underexplored, although the expansion of Ir catalysis scope is of great interest.

Jensen et al. examined the catalytic performance of PCP pincer Ir complex **22** for the dehydrogenation of perhydrodibenzofuran [56]; however, after 24 h at 200°C, the conversion to octahydrodibenzofuran (accompanied by the release of 2 equiv. of H_2) was only 15%.

Nozaki et al. developed a new type Ir complex promoting the dehydrogenation of cyclic ketones, esters, amides, and ethers [71]. According to the working hypothesis, a hydroxycyclopentadienyl ligand previously utilized for Ru catalysis in the well-known Shvo's complex [72, 73] coordinated to the Ir center allows H_2 liberation

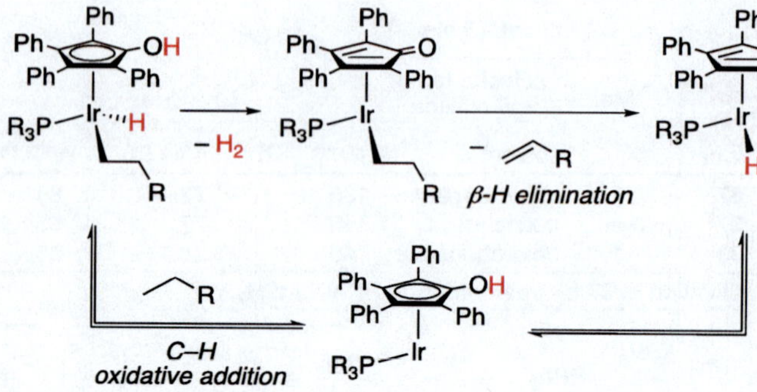

Scheme 24 Metal–ligand cooperative dehydrogenation across a C–C bond

from a hydride intermediate via metal–ligand cooperation to afford an Ir–cyclopentadienone complex (Scheme 24). After β-hydrogen elimination, intramolecular hydrogen transfer from the Ir center to oxygen generates a coordinatively unsaturated intermediate that can undergo oxidative addition of C–H bond. A TON of 11 after 29 h was achieved for the acceptorless dehydrogenation of cyclodecane to cyclodecene at 200°C in a closed system catalyzed by Ir complex **41** (Scheme 25), while the dehydrogenation of α-tetralone to 1-naphthol was more rapid, with a TON of 19 reached after 30 min. When the reaction time was extended to 29 h, the TON increased to 97, although byproducts such as tetrahydronaphthalene and naphthalene were also observed. The dehydrogenation of other cyclic compounds was also investigated, and a TON of 66 was observed for dihydrobenzofuran.

Later, the same group found that a simple $[Cp^*IrCl_2]_2$ (**42**) exhibited catalytic activity comparable or superior to that of the hydroxycyclopentadienyl complex **41** for the dehydrogenation of these cyclic compounds [74]. In cyclodecane as an almost inert solvent, catalyst **42** (0.33 mol%Ir) promoted the dehydrogenation of tetralone, achieving a conversion of 59% after 20 h at 200°C to afford 1-naphthol in 47% yield, whereas catalyst **41** achieved a conversion of only 15% (Scheme 26). An increase in catalyst loading to 1.0 mol%Ir improved the yield of 1-naphthol to 71%. For the dehydrogenation of other cyclic compounds, the performance of **42** was mostly better or comparable to that of **41** except for the case of 1-isochromanone. Albeit the mechanism was unclear, the results of the mercury test implied that the active species was present in the homogeneous phase.

3.3 N-Heteroarene Synthesis via Combination of Ring Construction–Dehydrogenation

The incorporation of *N*-heterocyclic moiety dehydrogenation into the synthesis of *N*-heteroarenes involving a cyclization step was reported by several groups. Such processes provide direct access to *N*-heteroarenes from acyclic starting materials

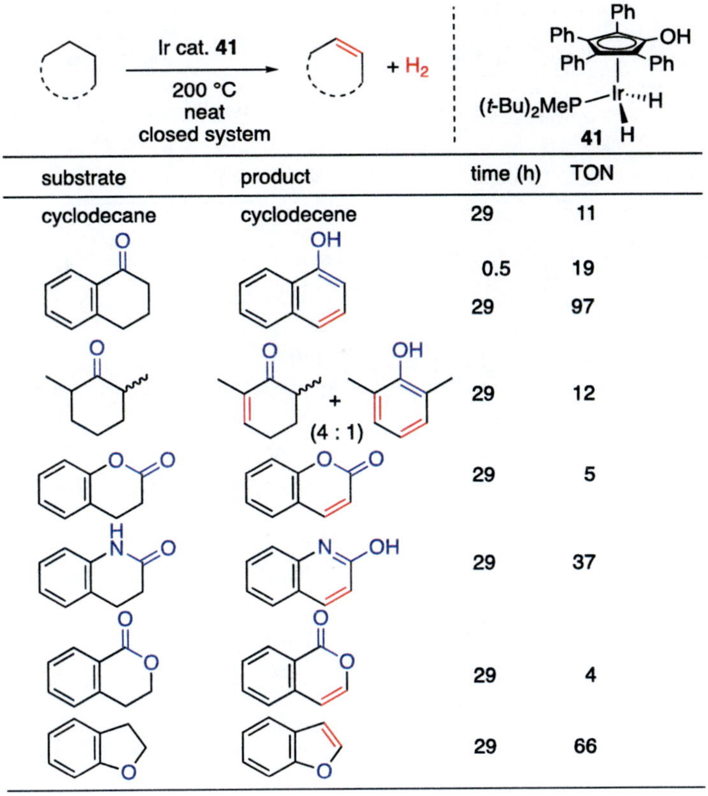

Scheme 25 Dehydrogenation of various cyclic compounds

without the use of external oxidants and are therefore of high synthetic utility from the viewpoint of green chemistry.

Yamamoto et al. reported the dehydrogenation of intermediate dihydropyrroles to pyrroles in cascade Ir-catalyzed intramolecular ene-yne cyclizations/Diels-Alder cycloadditions starting from 1,6-enynes prepared by Cu-catalyzed three-component coupling of allylamines and glyoxylate esters with terminal alkynes (Scheme 27) [75, 76]. In the presence of $[IrCl(cod)]_2$ (3 mol%) and acetic acid (12 mol%) in refluxing toluene, 1,6-enynes underwent cycloisomerization [77] to afford 1,3-diene intermediates that were trapped with dienophiles such as maleimide, maleic anhydride, or 1,4-naphthoquinone to afford polycyclic 3-pyrroline intermediates. Further dehydrogenation probably proceeded via C–H activation at the C-2 position of the pyrroline moiety by the Ir catalyst to finally afford polycyclic pyrrole-2-carboxylates. This reaction could be performed under microwave conditions, although a slight decrease in product yield occurred in some cases. In the case of the microwave-assisted reaction of a 5-alkyl-1,6-enyne, the pre-dehydrogenation (pre-aromatized) intermediate (3-pyrroline) could be isolated (Scheme 28). The

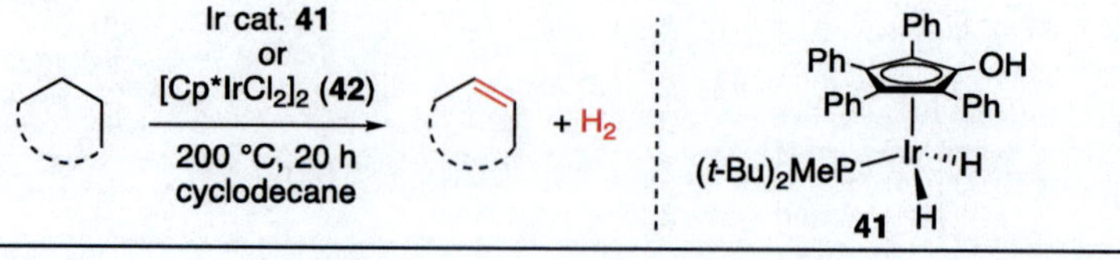

substrate	product	cat. (mol%Ir)	conv. (%)	yield (%)
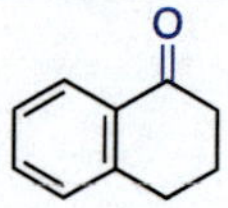	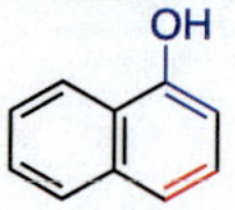	**41** (0.33) **42** (0.33) **42** (1.0)	15 59 92	13 47 71
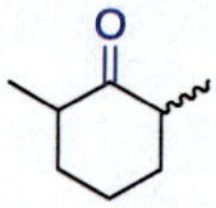	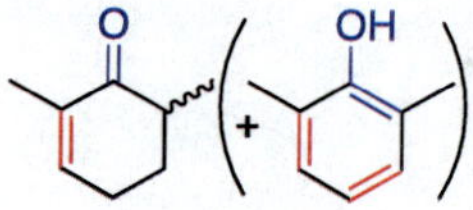	**41** (0.34) **42** (0.34)	8.5 9.3	1.9 (0.1) 3.7 (4.8)
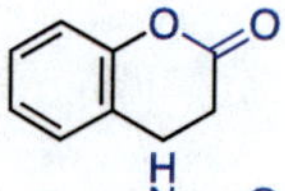	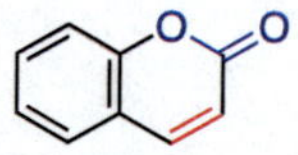	**41** (0.30) **42** (0.30)	3.1 21	0.9 6.0
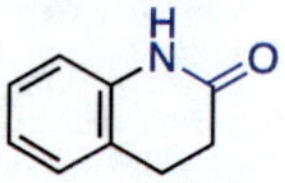	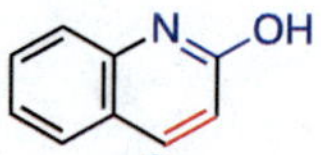	**41** (0.37) **42** (0.37)	12 34	0.2 25
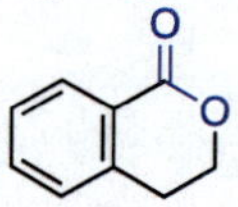	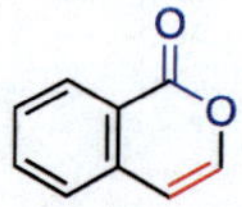	**41** (0.30) **42** (0.30)	13 4.4	0.9 0.2
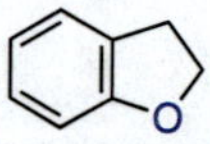	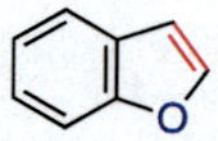	**41** (0.30) **42** (0.30)	68 56	3.9 3.8

Scheme 26 Comparison of the catalytic activities of **41** and **42** for the dehydrogenation of cyclic compounds

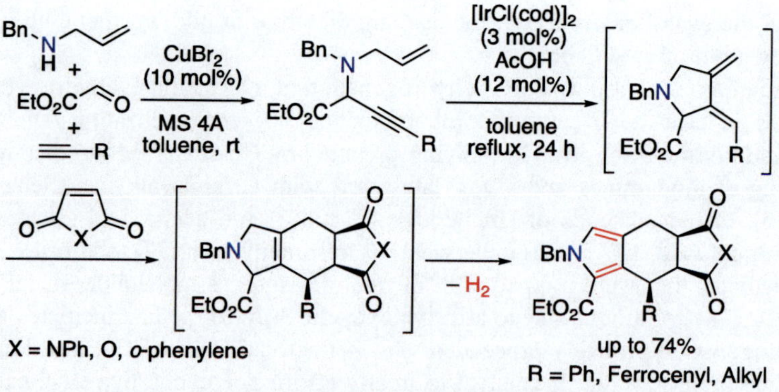

Scheme 27 Synthesis of polycyclic pyrroles via a cascade reaction involving Ir-catalyzed dehydrogenation

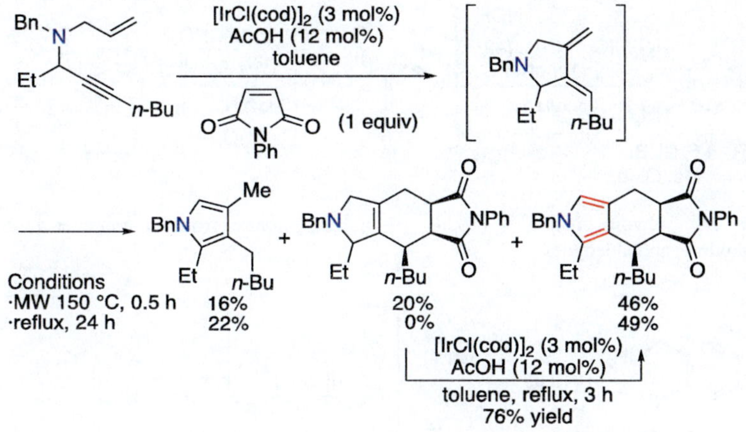

Scheme 28 Intermediacy of 3-pyrroline and following dehydrogenation

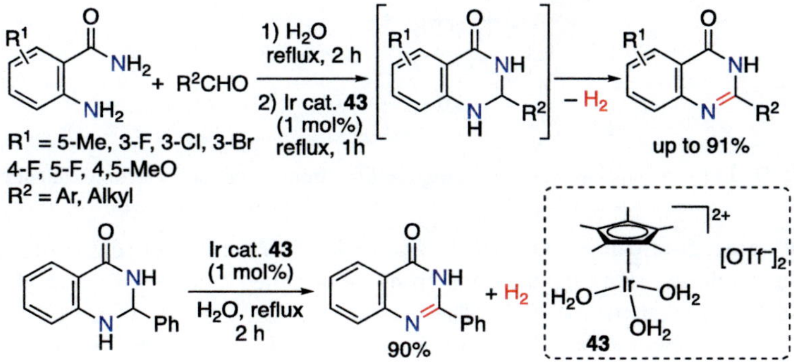

Scheme 29 Dehydrogenative synthesis of quinazolinones from *o*-aminobenzamide and aldehydes

intermediacy of this compound was confirmed by the reaction of the isolated 3-pyrroline intermediate in the presence of [IrCl(cod)]$_2$ and acetic acid in refluxing toluene for 3 h, which furnished the desired pyrrole in 76% yield. Although this step did not necessarily require a hydrogen acceptor, the added *N*-phenylmaleimide significantly accelerated the reaction and was converted into *N*-phenylsuccinimide.

Li et al. reported the dehydrogenative synthesis of quinazolinones from *o*-aminobenzamides and aldehydes catalyzed by a water-soluble Ir complex in aqueous media [78]. Specifically, the dehydrative condensation of *o*-aminobenzamide with aldehydes was carried out in refluxing water for 2 h to afford 2,3-dihydroquinazolin-4(1*H*)-ones, which was followed by the addition of [Cp*Ir(OH$_2$)$_3$][OTf]$_2$ (**43**) (1 mol%) and further reflux for 1 h to afford quinazolinones in good yields (Scheme 29). The dehydrogenation of isolated 2,3-dihydroquinazolin-4(1*H*)-one in refluxing water was confirmed to yield quinazolinones almost as equal efficiently as the one-pot process.

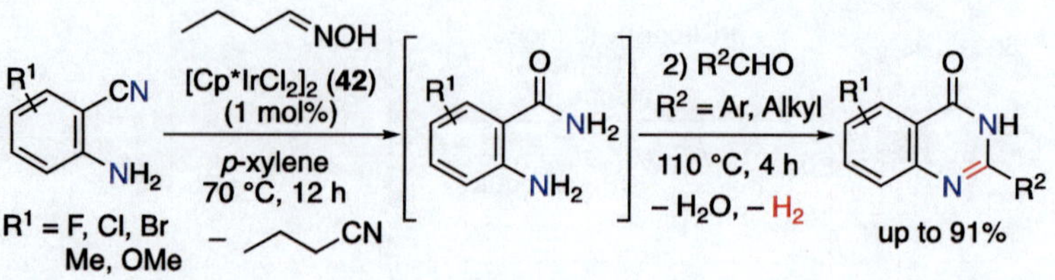

Scheme 30 Dehydrogenative synthesis of quinazolinones from *o*-aminobenzonitrile, *n*-butylaldoxime, and aldehydes

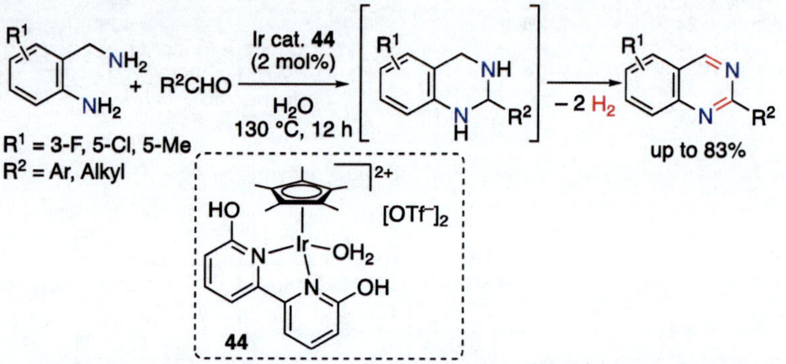

Scheme 31 Dehydrogenative synthesis of quinazolines from *o*-aminobenzylamine and aldehydes

The same group also reported a $[Cp^*IrCl_2]_2$ (**42**)-catalyzed dehydrogenative one-pot synthesis of quinazolinones from *o*-aminobenzonitrile, *n*-butylaldoxime, and aldehydes (Scheme 30) [79]. The reaction proceeded in *p*-xylene at 70°C for 12 h via the hydration of *o*-aminobenzonitrile to *o*-aminobenzamide by *n*-butylaldoxime accompanied by the release of *n*-butyronitrile. Subsequently, the aldehyde was added, and the mixture was further heated at 110°C for 4 h to afford quinazolinones according to a mechanism similar to that described previously [78].

The use of *o*-aminobenzylamines instead of *o*-aminobenzamides allowed for the dehydrogenative synthesis of quinazolines in the presence of Cp*Ir complexes [80], with the best catalytic performance in aqueous media observed for the Cp*Ir 6,6′-dihydroxy-2,2′-bipyridine complex **44** (Scheme 31). The reaction probably proceeded through the formation of 1,2,3,4-tetrahydroquinazolines via the dehydrative condensation of *o*-aminobenzylamines with aldehydes followed by two-fold dehydrogenation catalyzed by **44** to afford quinazolines.

Cross dehydrogenative coupling, in which two C–H bonds are cleaved to form a new C–C bond, is a powerful synthetic method for the construction of C–C bond because of no necessity of pre-functionalization; however, most transformations of this type require oxidants or hydrogen acceptors [81–84]. Yan and Xiao et al. reported an Ir-catalyzed intramolecular dehydrogenative C–C coupling of *N*-*o*-acylphenyl tetrahydroisoquinoline in the absence of hydrogen acceptors [85]. In refluxing acetic acid, $[IrCl(cod)]_2$ (2.5 mol%) and 1,2-bis(diphenylphosphino)ethane

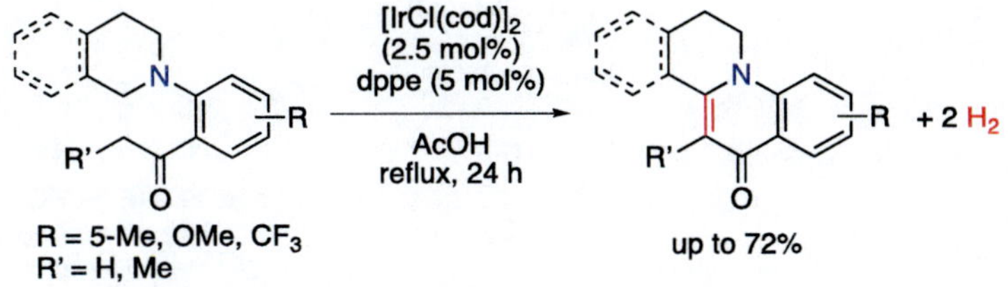

Scheme 32 Dehydrogenative intramolecular C–C coupling of *N*-*o*-acylphenyl tetrahydroisoquinoline in the absence of hydrogen acceptors

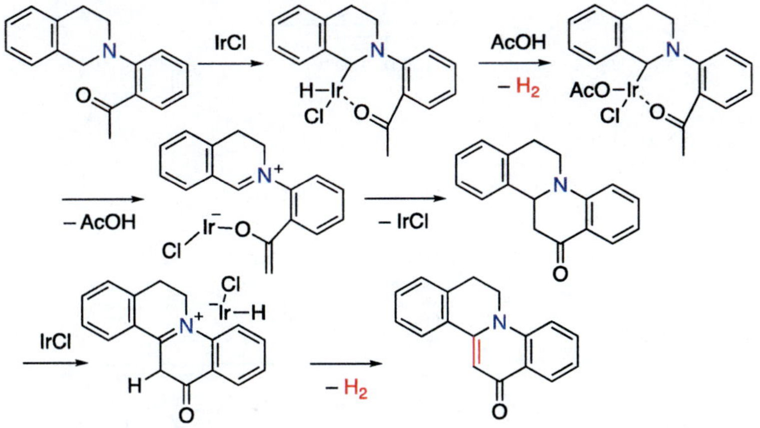

Scheme 33 Proposed mechanism of Ir-catalyzed intramolecular dehydrogenative C–C coupling of *N*-*o*-acylphenyl tetrahydroisoquinoline in the absence of hydrogen acceptors

(dppe, 5 mol%) promoted the intramolecular two-fold dehydrogenative coupling of two C(sp^3)–H bonds to afford a C=C bond (Scheme 32). According to the proposed mechanism (Scheme 33), the oxidative addition of a benzylic C–H bond adjacent to the nitrogen atom to an Ir(I) species occurs, and the resulting Ir–H intermediate is protonated by acetic acid to liberate H_2. The release of acetic acid from the produced Ir acetate along with deprotonation at the acetyl group affords an Ir enolate and an iminium moiety via Ir–C bond cleavage. The intramolecular nucleophilic addition of the enolate to the iminium moiety affords a cyclized intermediate that was detected in the reaction mixture at short reaction times. Finally, Ir-catalyzed hydride abstraction followed by elimination of a second molecule of H_2 completes the catalytic cycle.

The same group also reported an intramolecular dehydrogenative C–N coupling in refluxing TFE catalyzed by [Cp*IrCl$_2$]$_2$ (**42**) in the presence of 2-hydroxypyridine [86]. Although this ligand was not necessarily required for the reaction progress, it significantly accelerated the reaction. 2-Aminobenzamide, 2-aminobenzenesulfonamide, and 2-aminobenzylamine derivatives could be catalytically converted into the corresponding nitrogen-containing polycyclic products with

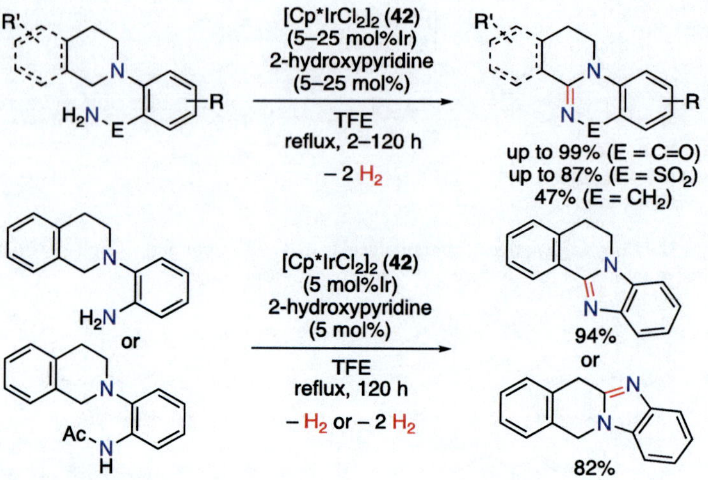

Scheme 34 Dehydrogenative intramolecular C–N coupling

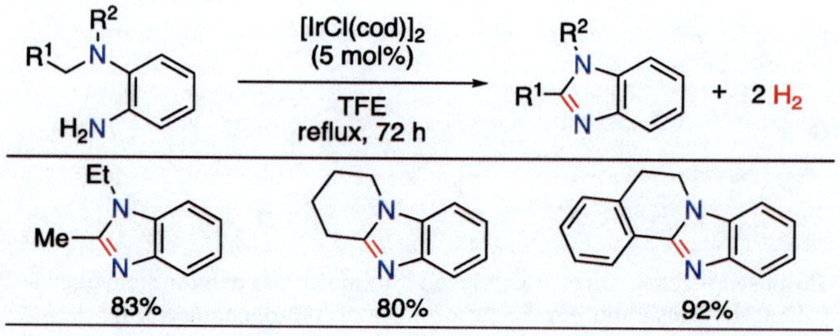

Scheme 35 Dehydrogenative synthesis of benzimidazoles via intramolecular C–N coupling

liberation of H_2 (Scheme 34). The reaction of 2-[3,4-dihydroisoquinolin-2(1*H*)-yl] aniline derivatives selectively afforded tetracyclic benzimidazoles according to the substituent on the nitrogen atom. Later, the same group expanded the scope of the above synthesis by employing [IrCl(cod)]$_2$ in refluxing TFE [87] and showed that simple benzimidazole, tricyclic benzimidazole, and tetracyclic benzimidazole derivatives can be prepared under these conditions (Scheme 35).

4 Dehydrogenation of Alcohols

The oxidation of alcohols to carbonyl compounds such as aldehydes, ketones, esters, and carboxylic acids is one of the most fundamental reactions in organic chemistry and is typically accomplished using stoichiometric amounts of oxidants. However, the catalytic dehydrogenation of alcohols accompanied by H_2 evolution in the absence of oxidants is a better alternative from the viewpoint of environmental friendliness. In this section, catalytic systems for the dehydrogenative oxidation of alcohols and related applications in organic synthesis are surveyed.

4.1 *Dehydrogenative Oxidation of Alcohols to Aldehydes and Ketones*

The Ir complex-catalyzed dehydrogenation of alcohols was pioneered by Lu et al. in 1987 [88], who successfully converted secondary alcohols to ketones in hexamethyldisiloxane as a solvent at 100°C in the presence of an Ir hydride complex **45** as a catalyst (Scheme 36). The reaction was accompanied by the evolution of H_2, which was detected by gas chromatography.

Another pioneering work on the Ir-catalyzed dehydrogenation of alcohols was reported by Saito et al. [89, 90], who found that *trans*-$[IrCl_2(SnCl_3)_4]^{3-}$ (**46**) effectively promoted the conversion of 2-propanol to acetone with a concomitant release of H_2. However, only the reaction of 2-propanol was reported.

In 2007, Fujita and Yamaguchi et al. reported a catalytic system for the dehydrogenative oxidation of secondary alcohols based on cooperativity between Ir and a functional ligand (Scheme 37) [91], revealing that the introduction of 2-hydroxypyridine as a functional ligand was of key importance for high catalytic activity.

Dehydrogenation catalyzed by **47** proceeds as illustrated in Scheme 38. The active species bearing a functional ligand in pyridonate form acts as a proton acceptor in the process of the activation of alcohol. Dehydrogenation via protonolysis of the Ir hydride complex bearing a functional ligand in hydroxypyridine form is another important step of the catalytic cycle.

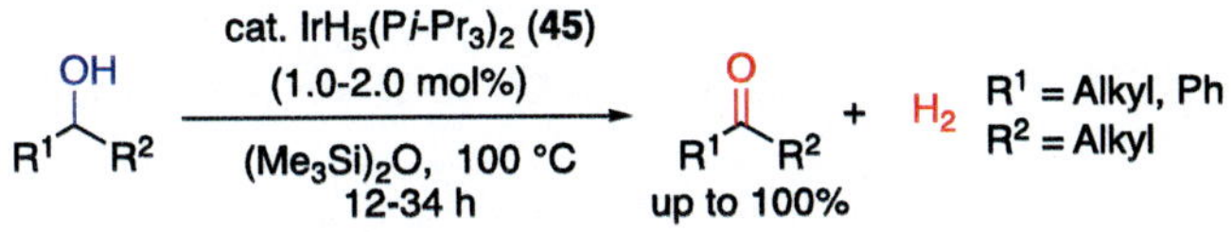

Scheme 36 Pioneering study on Ir complex-catalyzed dehydrogenation of alcohols

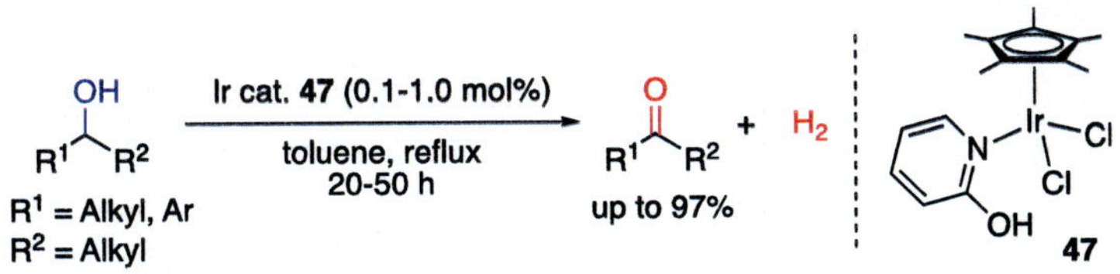

Scheme 37 Dehydrogenation of secondary alcohols catalyzed by Ir complex **47**

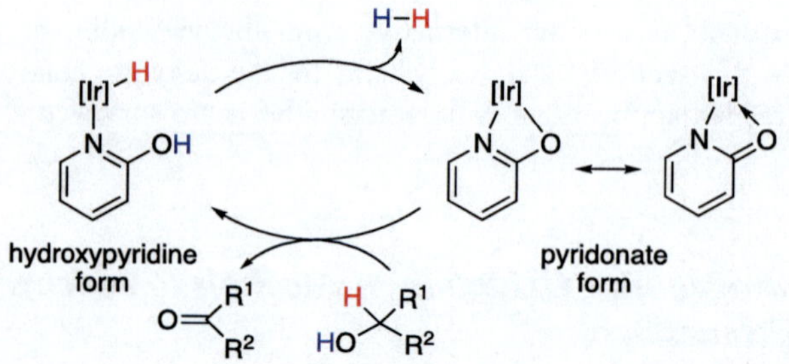

Scheme 38 Mechanism for dehydrogenation of alcohols catalyzed by Ir complex **47**

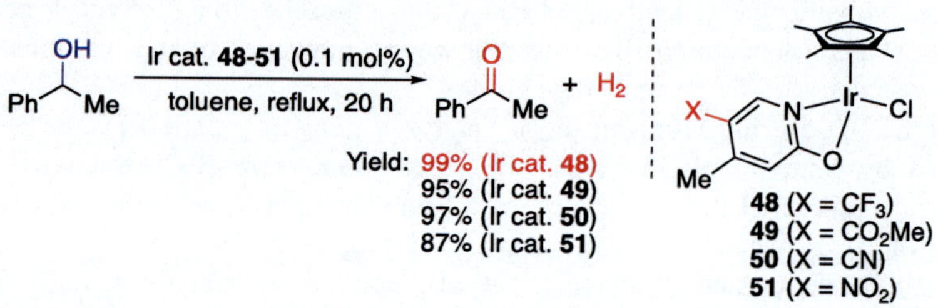

Scheme 39 Effect of substituent on the functional pyridonate ligand on the dehydrogenation of 1-phenylethanol

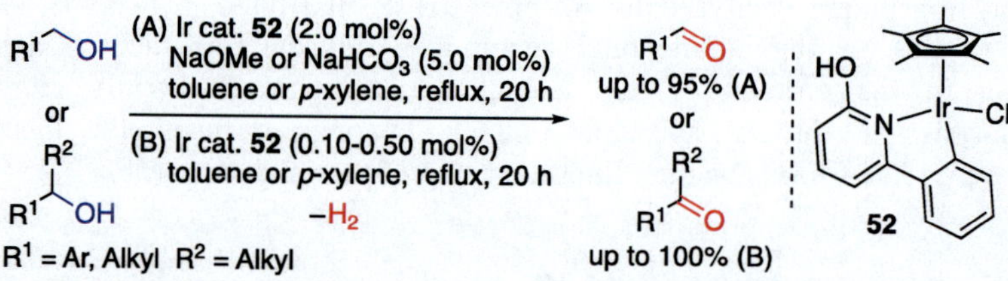

Scheme 40 Dehydrogenation of primary and secondary alcohols catalyzed by Ir complex **52**

Later, the same group reported the effect of substituents in the functional ligand [92]. As shown in Scheme 39, the catalytic activity of a series of Ir complexes **48–51** for the dehydrogenation of 1-phenylethanol was examined, and the highest activity was observed for the complex **48** with a 2-pyridonate ligand bearing CF_3 group at 5-position.

A system for the Ir-catalyzed dehydrogenation of both primary and secondary alcohols to afford aldehydes and ketones, respectively, was first developed by Fujita and Yamaguchi et al. The introduction of a *C,N*-chelate ligand based on 2-hydroxy-6-phenylpyridine improved the stability of catalyst **52**, and allowed the dehydrogenative oxidation of primary alcohols to aldehydes to be performed in the presence of only catalytic amounts of base (Scheme 40) [93], while the reactions

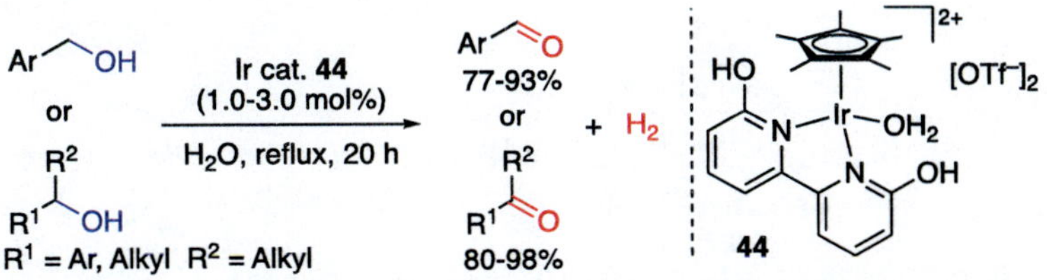

Scheme 41 Dehydrogenation of alcohols in water catalyzed by the water-soluble Ir complex **44**

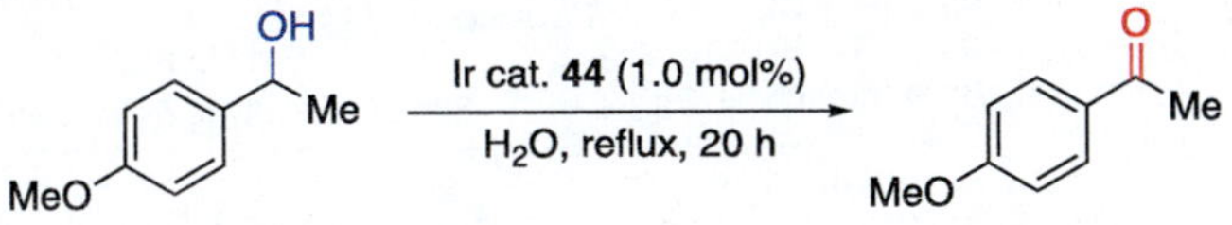

reuse	1	2	3	4	5	6	7	8
yield (%)	98	97	96	95	95	96	95	94

Scheme 42 Reusability of catalyst **44**

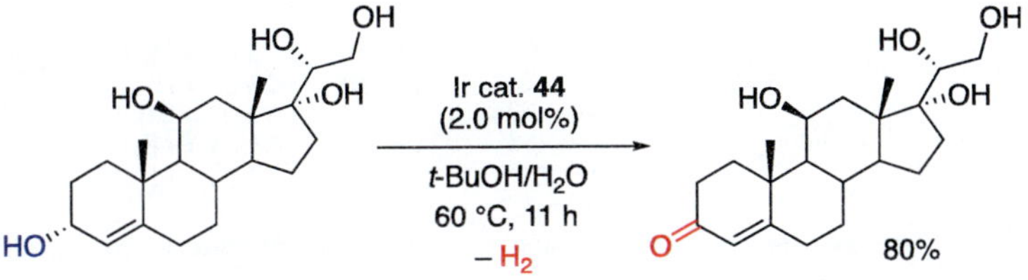

Scheme 43 Selective oxidation of one of the alcoholic moieties in a glucocorticoid derivative

of secondary alcohols proceeded well under base-free conditions at a lower loading of **52**.

The same group reported a water-soluble and reusable Ir catalyst **44** with 6,6′-dihydroxy-2,2′-bipyridine as a functional ligand (Scheme 41) [94]. This complex was stable in water and air and highly soluble in water because of its dicationic character, which enabled the dehydrogenation of alcohols in aqueous media.

Another merit of the above system is catalyst reusability. Owing to the high solubility of **44** in water, the separation of the organic product from the aqueous solution of the catalyst was easily accomplished, and the recovered catalyst could be reused at least eight times without marked activity loss (Scheme 42). Later, the above authors reported the immobilization of **44** onto mesoporous organosilica to afford a heterogeneous catalyst for alcohol dehydrogenation [95].

Do et al. applied the catalyst **44** to the interconversion of glucocorticoids [96], selectively oxidizing one of the alcoholic moieties in glucocorticoid derivatives (Scheme 43). The reverse hydrogenation could also be accomplished using the same catalyst.

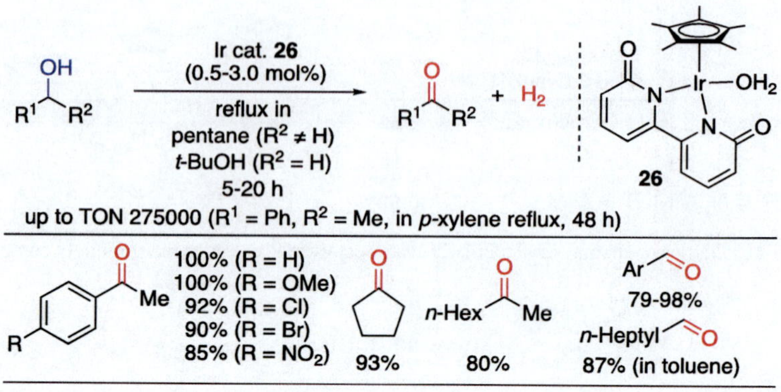

Scheme 44 Dehydrogenation of alcohols at low temperature catalyzed by complex **26**

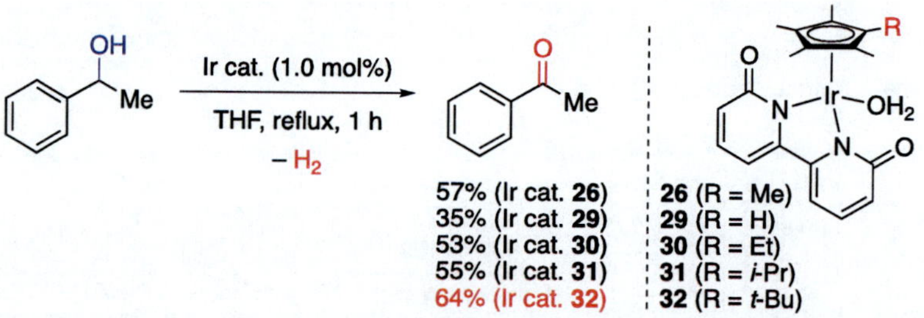

Scheme 45 Dehydrogenation of 1-phenylethanol catalyzed by Ir bipyridonate complexes bearing various Cp ligands

In 2012, Fujita and Yamaguchi et al. reported a neutral Ir catalyst **26** with a 6,6′-bipyridonate functional ligand (Scheme 44) [97], revealing that this catalyst allows the dehydrogenation of secondary alcohols to ketones under reflux in pentane (bp: 36°C) and the conversion of primary alcohols to aldehydes under reflux in *t*-butyl alcohol (bp: 82°C). A high TON of 275,000 was achieved when the reaction was performed under reflux in *p*-xylene (bp: 138°C). Studies on the mechanism of catalysis by **26** based on theoretical computations were also reported [98].

The same group disclosed the effect of substituents on the Cp ligand (Scheme 45) [63]. Among Ir complexes **26** and **29–32**, complex **32** with a *t*-butyl group exhibited the highest activity for the dehydrogenative oxidation of 1-phenylethanol to acetophenone.

A dicationic water-soluble Ir catalyst **53** with a bidentate functional ligand that consisting of 2-hydroxypyridine and NHC components developed by Fujita and Yamaguchi et al. [99] showed high activity for the dehydrogenative oxidation of secondary alcohols in aqueous media (Scheme 46). Compared to the water-soluble catalyst **44**, the catalyst **53** exhibited a three-fold higher activity.

Gelman et al. reported the design and synthesis of an Ir complex **54** with a PCP pincer ligand based on dibenzobarrelene structure [100], demonstrating that this

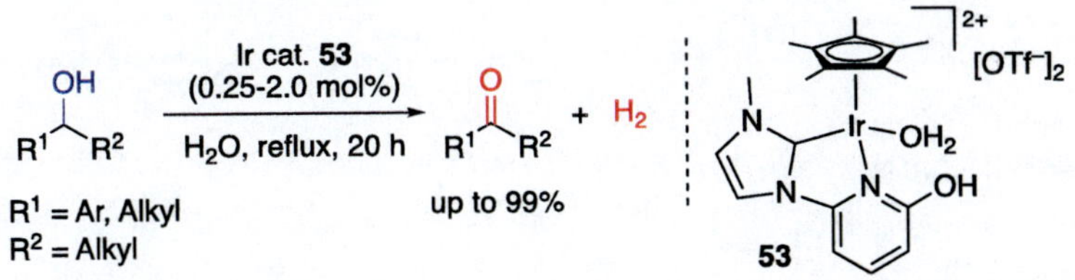

Scheme 46 Dehydrogenation of alcohols catalyzed by an Ir complex **53** bearing a functional NHC-hydroxypyridine ligand

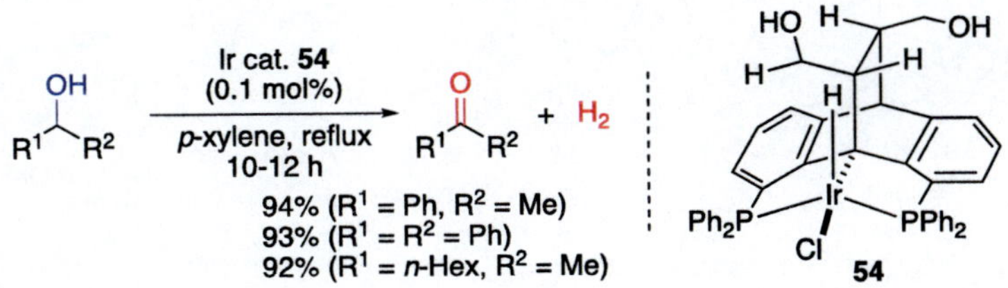

Scheme 47 Dehydrogenation of secondary alcohols catalyzed by an Ir complex **54** bearing a functional PCP pincer ligand

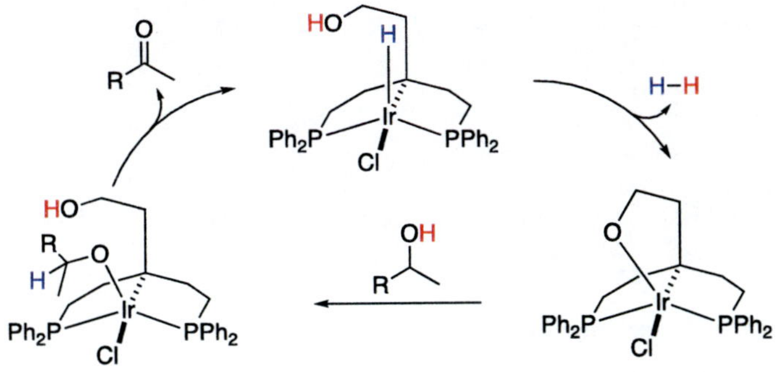

Scheme 48 Mechanism for dehydrogenation of alcohols catalyzed by Ir complex **54**

complex showed high catalytic activity for the dehydrogenative oxidation of secondary alcohols (Scheme 47).

The mechanism for the dehydrogenation of alcohols catalyzed by **54** is illustrated in Scheme 48. The tethered hydroxy group in the ligand plays an important role in the hydrogen release process. The same group also reported the heterogenization of the catalyst **54** within a silica sol-gel matrix [101].

In addition to the catalyst mentioned above, numerous other Ir complexes (Fig. 2) were shown to exhibit catalytic activity for the dehydrogenative oxidation of alcohols to ketones and/or aldehydes [102–111].

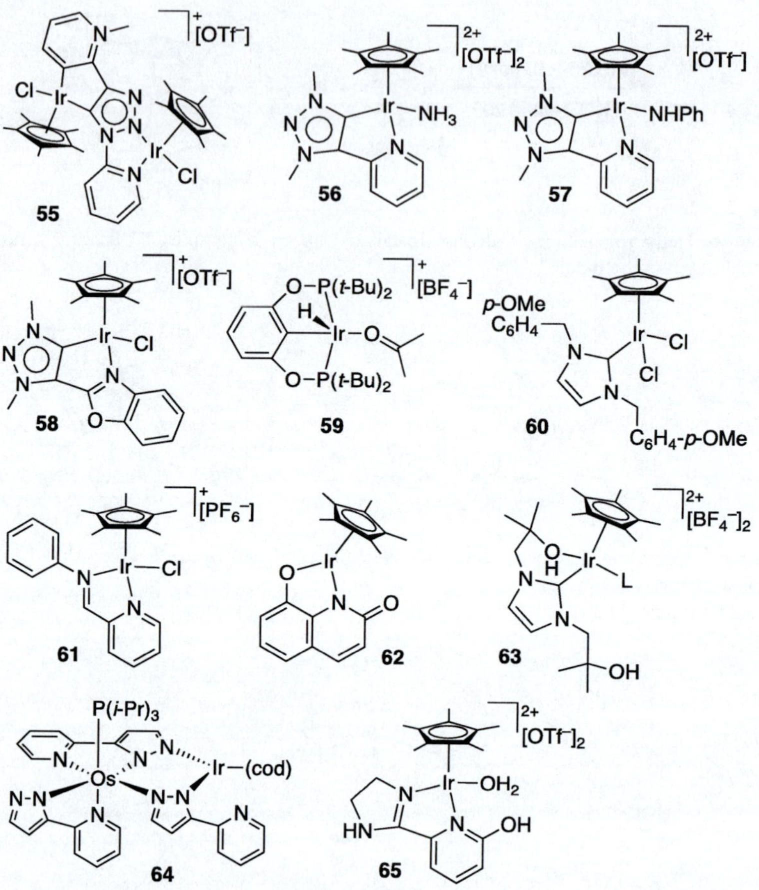

Fig. 2 Ir catalysts for the dehydrogenation of alcohols

4.2 Dehydrogenative Oxidation of Primary Alcohols to Esters, Carboxylic Acids, and Acetals

Catalyst **54**, mentioned in the preceding section, effectively promoted the dehydrogenative esterification of two molecules of primary benzylic alcohols in the presence of catalytic amounts of Cs_2CO_3 (Scheme 49) [100]. This esterification is believed to proceed through several steps involving dehydrogenation, hemiacetalyzation, and dehydrogenation. Moreover, the above catalyst also promoted the lactonization of diols.

Esteruelas et al. employed an Ir complex **64** with an Os-based metalloligand to realize a similar conversion of primary alcohols to esters [110] and identified the key intermediate of this transformation (Scheme 50, square brackets). The selectivity for

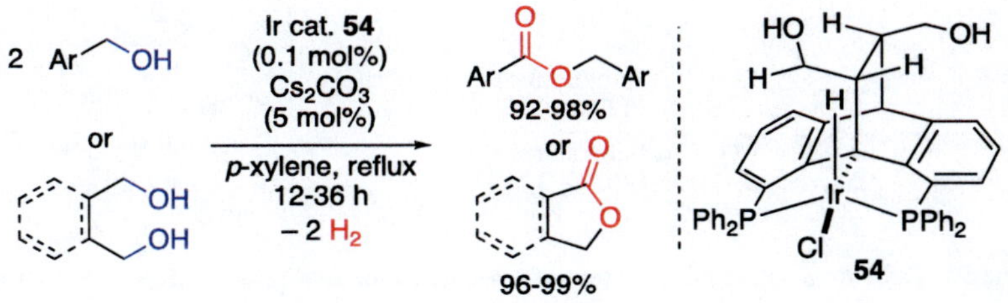

Scheme 49 Esterification and lactonization catalyzed by **54**

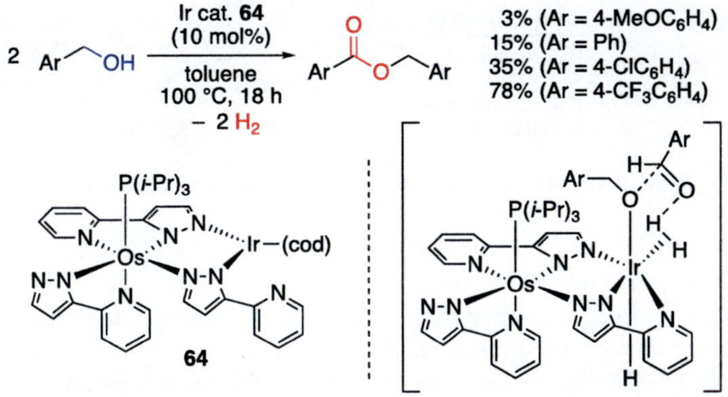

Scheme 50 Esterification of primary alcohols catalyzed by **64**

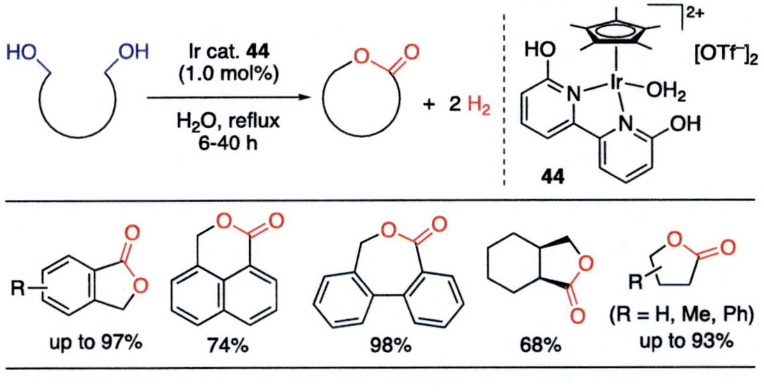

Scheme 51 Lactonization of diols in water catalyzed by Ir complex **44**

esters over aldehydes was improved by introducing an electron-withdrawing group at the *para*-positions of benzylic alcohols.

Dehydrogenative lactonization of diols in aqueous media catalyzed by a water-soluble Ir complex **44** was developed by Fujita and Yamaguchi et al. (Scheme 51)

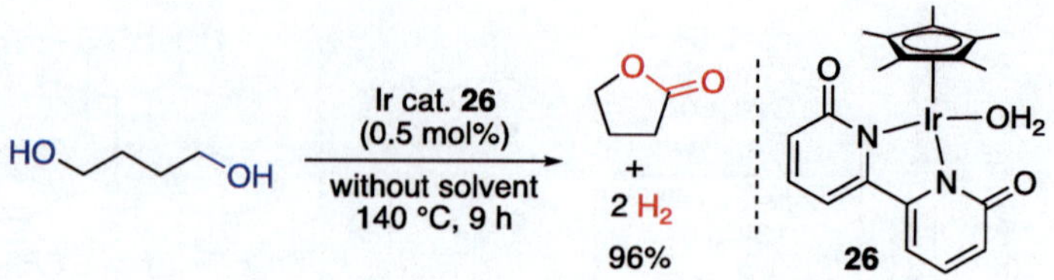

Scheme 52 Dehydrogenative lactonization of 1,4-butanediol to γ-butyrolactone

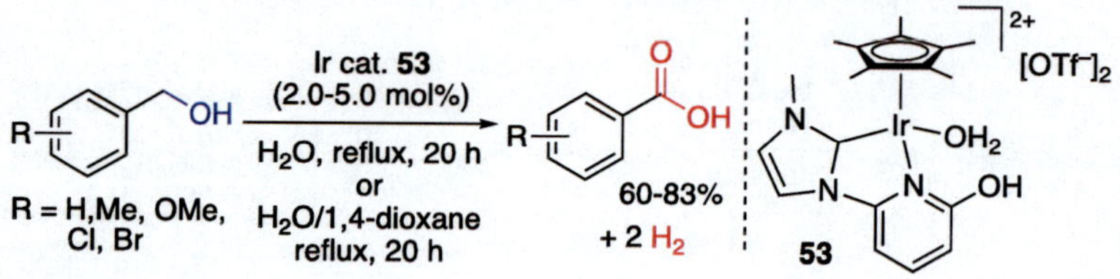

Scheme 53 Conversion of primary benzyl alcohols to benzoic acids in water catalyzed by **53**

[112]. Wide variety of diols was applicable for this catalytic system, selectively giving lactones in good to high yields.

The same group also reported the synthesis of γ-butyrolactone in excellent yield via the dehydrogenative lactonization of 1,4-butanediol catalyzed by Ir complex **26** (Scheme 52) [113] under neat conditions at 140°C. The reverse hydrogenation could be accomplished using the same catalyst, and the reversible catalytic interconversion of 1,4-butanediol to γ-butyrolactone was therefore regarded as a promising prototype of a hydrogen storage system.

Fujita and Yamaguchi et al. reported the dehydrogenative oxidation of primary benzylic alcohols to benzoic acid derivatives under aqueous reflux conditions catalyzed by a water-soluble Ir complex **53** (Scheme 53) [99], suggesting that this transformation involves the dehydrogenation of the primary alcohol to an aldehyde, the hydration of this aldehyde to afford a *gem*-diol, and the dehydrogenation of the *gem*-diol to afford a carboxylic acid.

A versatile catalytic system for the dehydrogenation of primary alcohols to carboxylic acids under basic aqueous conditions in the presence of an Ir catalyst **66** with a bipyridonate ligand bearing *N,N*-dimethylamino substituents was reported by Fujita et al. [114] (Scheme 54). The carboxylic acids were obtained after acidic work-up, and the developed methodology was successfully applied to the quantitative synthesis of acetic acid starting from an ethanol–water solution.

Williams et al. have recently developed an efficient catalytic system for the dehydrogenative oxidation of primary alcohols to carboxylic acids [115]. Various types of alcohols (R = alkyl, aryl, and heteroaryl) were converted to the corresponding carboxylic acids in the presence of Ir complex **67** as a catalyst after acidic work-up (Scheme 55).

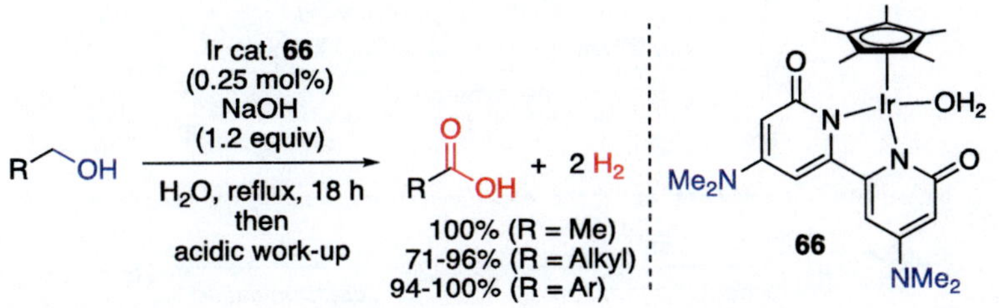

Scheme 54 Dehydrogenative oxidation of aqueous solutions of primary alcohols to carboxylic acids catalyzed by Ir complex **66**

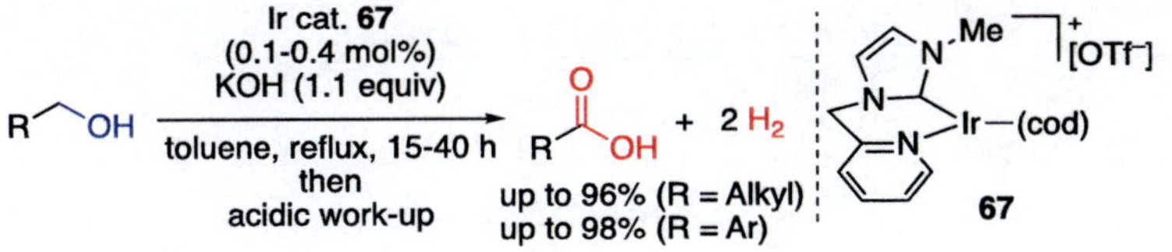

Scheme 55 Dehydrogenative oxidation of primary alcohols to carboxylic acids catalyzed by the Ir complex **67**

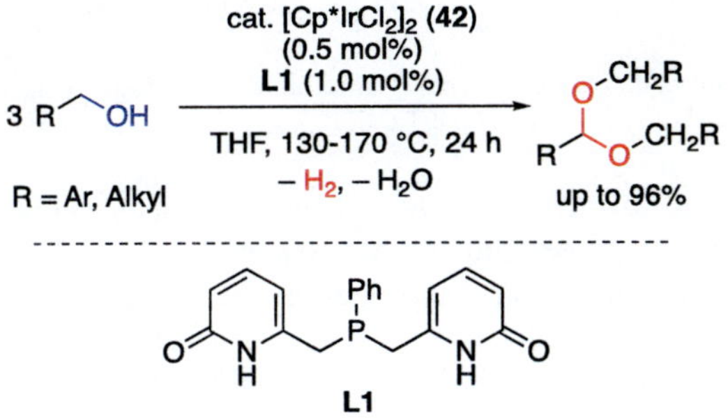

Scheme 56 Acetalization of three molecules of primary alcohols

A system for the dehydrogenative acetalization of three molecules of primary alcohols promoted by $[Cp^*IrCl_2]_2$ (**42**) and bis(pyridone)phosphine ligand **L1** was developed by Achard et al. [116] (Scheme 56).

4.3 Dehydrogenative Decarbonylation of Primary Alcohols

Madsen et al. developed a system for the dehydrogenative decarbonylation of primary alcohols accompanied by the release of syngas (H_2 and CO) [117], showing that this transformation is effectively promoted by $[IrCl(coe)_2]_2$ in the presence of *rac*-BINAP (Scheme 57). The reaction is believed to proceed through the dehydrogenation of the primary alcohol to the corresponding aldehyde followed by its

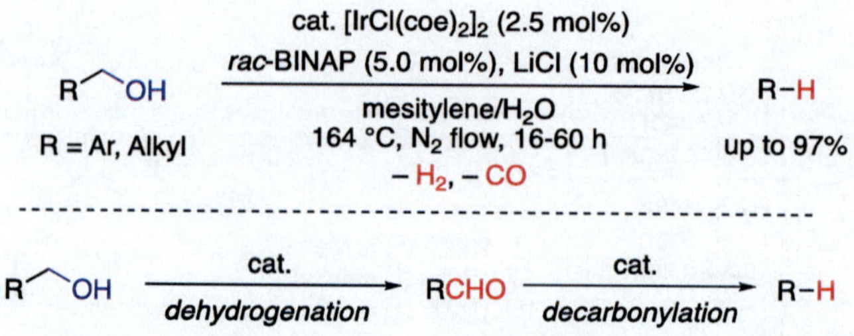

Scheme 57 Dehydrogenative decarbonylation of primary alcohols releasing syngas

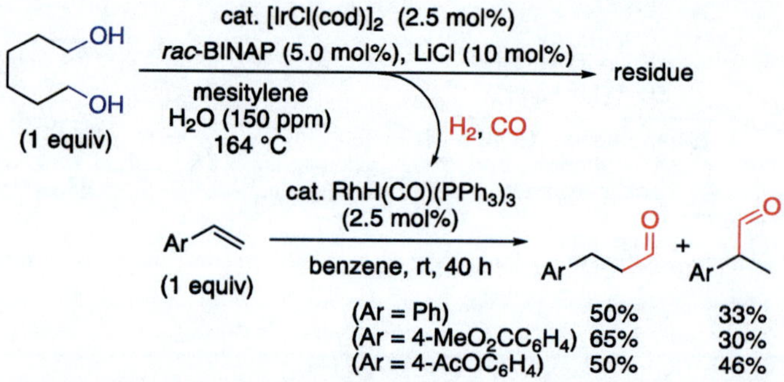

Scheme 58 Hydroformylation of styrene accomplished using syngas released by the dehydrogenative decarbonylation of hexane-1,6-diol

decarbonylation to afford an alkane. Theoretical mechanistic studies on this catalysis were also performed by Andersson and Madsen et al. [118].

The same group later reported the further application of the above dehydrogenative decarbonylation [119]. As illustrated in Scheme 58, syngas released during the dehydrogenative decarbonylation of hexane-1,6-diol catalyzed by $[IrCl(cod)]_2$ and *rac*-BINAP was successfully used for the hydroformylation of styrene.

The same paper also reported the reductive carbonylation of *p*-bromoanisole to afford *p*-anisaldehyde (Scheme 59) [119].

4.4 Dehydrogenative Cross-Coupling Reactions Involving Alcohols

Gelman et al. reported a catalytic system for the dehydrogenative cross-coupling of primary and secondary alcohols [120]. As shown in Scheme 60, Ir complex **54** effectively catalyzed this reaction to afford alkylated ketones. The above coupling is believed to proceed through dehydrogenation and hydrogen transfer processes involving the dehydrogenation of alcohols to ketones and aldehydes, base-mediated

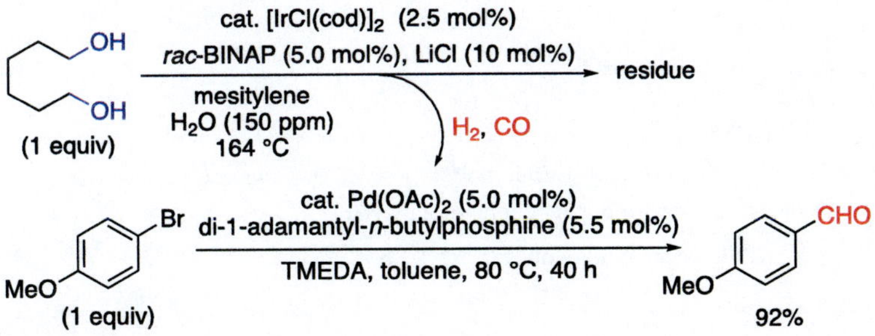

Scheme 59 Reductive carbonylation of *p*-bromoanisole to afford *p*-anisaldehyde using syngas released by the dehydrogenative decarbonylation of hexane-1,6-diol

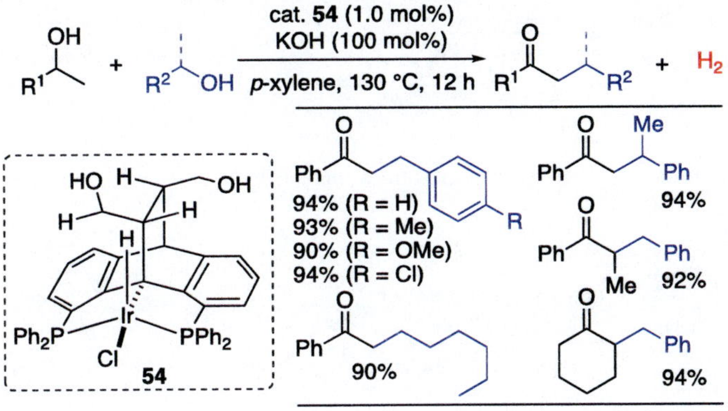

Scheme 60 Dehydrogenative cross-coupling of primary and secondary alcohols catalyzed by **54**

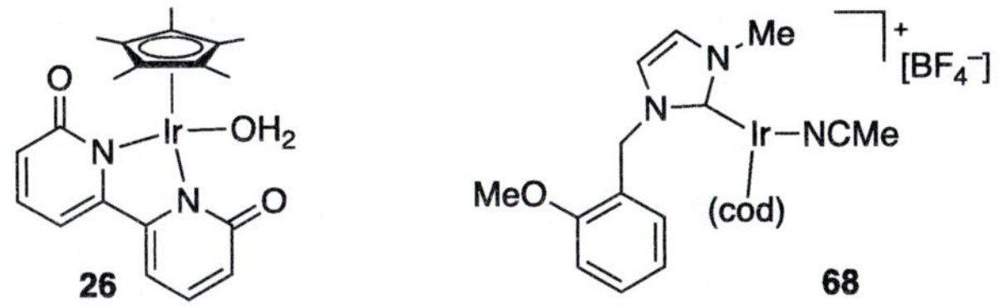

Fig. 3 Ir catalysts for the dehydrogenative coupling of two alcohols

cross aldol condensation, and the transfer hydrogenation of the aldol condensation product to afford an alkylated ketone.

Closely related couplings of two alcohols catalyzed by the Ir complexes illustrated in Fig. 3 have also been reported [121, 122].

Donohoe et al. developed a catalytic system for the chemo- and regioselective synthesis of acylcyclohexenes from pentamethylacetophenone and diols [123], revealing that an Ir catalyst composed of $[IrCl(cod)]_2$ and di-1-adamantyl-*n*-butylphosphine showed high catalytic activity for this transformation (Scheme 61), which is believed to proceed through a sequence of diol dehydrogenation, aldol condensation, and [1,5]-hydride shift.

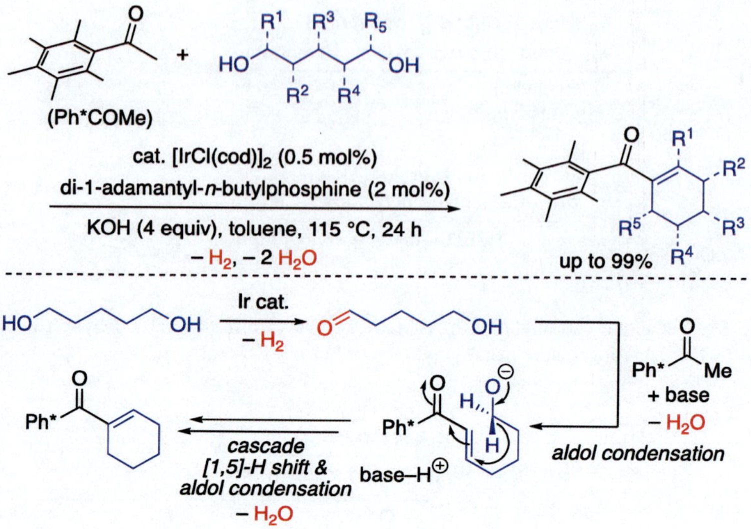

Scheme 61 Synthesis of acylcyclohexenes from pentamethylacetophenone and diols

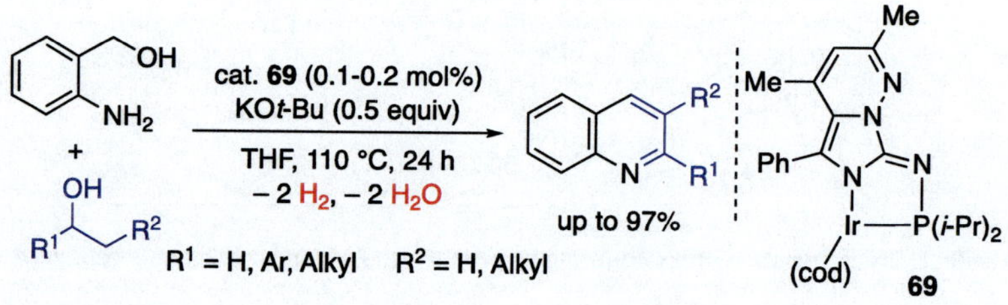

Scheme 62 Synthesis of quinoline derivatives by the dehydrogenative coupling of 2-aminobenzyl alcohol and alcohols

4.5 *Synthesis of* N*-Heterocycles Based on Alcohol Dehydrogenation*

Kempe et al. developed a catalytic system for the dehydrogenative coupling of 2-aminobenzyl alcohol and alcohols to afford quinoline derivatives [124], revealing that Ir complex **69** with a bidentate *P*,*N*-chelate ligand shows high catalytic activity for this reaction (Scheme 62). The above transformation is believed to proceed via the dehydrogenative oxidation of the alcoholic moieties of the starting materials followed by condensation accompanied by the release of H_2 (2 equiv.) and water (2 equiv.).

Li et al. reported a similar system for the synthesis of quinoline derivatives via the coupling of 2-aminobenzyl alcohol and ketones in water [125] catalyzed by the water-soluble Ir complex **44** (Scheme 63). The first step of this reaction is believed to

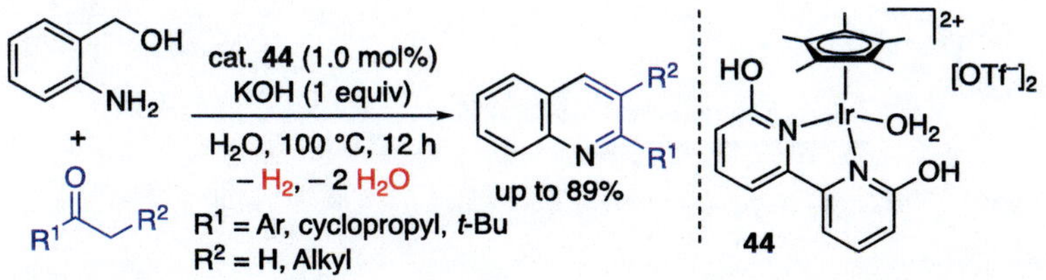

Scheme 63 Synthesis of quinoline derivatives by the dehydrogenative coupling of 2-aminobenzyl alcohol and ketones

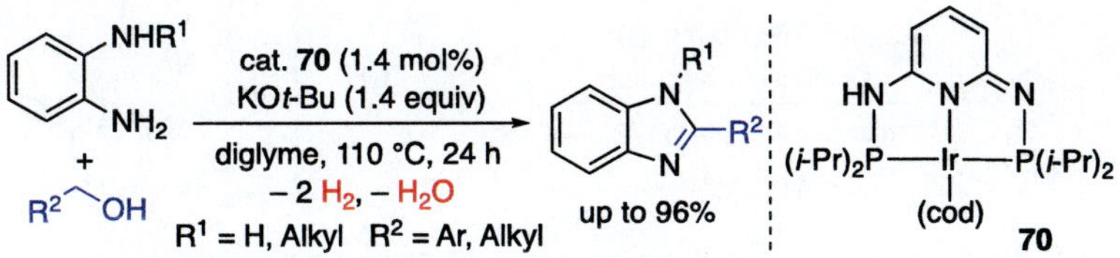

Scheme 64 Synthesis of benzimidazoles from benzene-1,2-diamines and primary alcohols

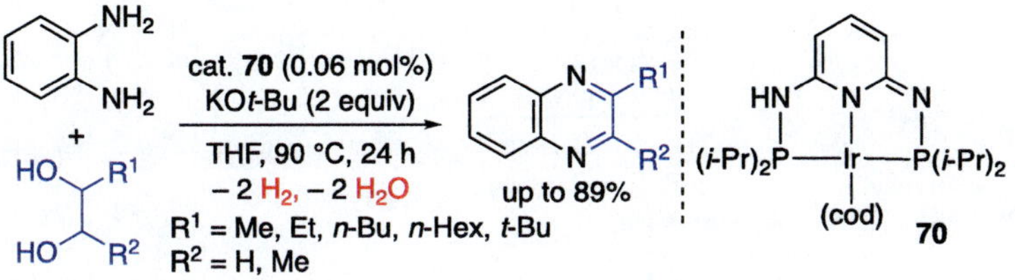

Scheme 65 Synthesis of quinoxalines from benzene-1,2-diamines and 1,2-diols

be the dehydrogenative oxidation of the alcoholic moiety of 2-aminobenzyl alcohol. A similar reaction promoted by an Ir-NHC catalyst in an organic solvent was also reported by Gülcemal et al. [126].

Catalytic synthesis of benzimidazoles from benzene-1,2-diamines and primary alcohols triggered by the dehydrogenation of alcohols was reported by Kempe et al. [127]. The iridium complex **70** with a tridentate *P∧N∧P* ligand [bis (diisopropylphosphino)pyridine-2,6-diamine] exhibited catalytic activity in diglyme, affording benzimidazoles accompanied by the evolution of two equivalents of H_2 (Scheme 64).

The same paper [127] also described the reactions of 1,2-phenylenediamine with 1,2-diols promoted by Ir catalyst **70** (Scheme 65). In these reactions, the formation of quinoxalines was accompanied by the release of two equivalents of H_2.

Kundu et al. synthesized quinazolines by reacting 2-aminobenzylamine with aromatic primary alcohols in water in the presence of a dicationic Ir complex **65** with a 2-hydroxypyridine-based *N,N*-chelate ligand [128] (Scheme 66), showing that the reaction is accompanied by the evolution of three equivalents of H_2. The same paper also demonstrated reactions of 1,2-phenylenediamine with 1,2-diols

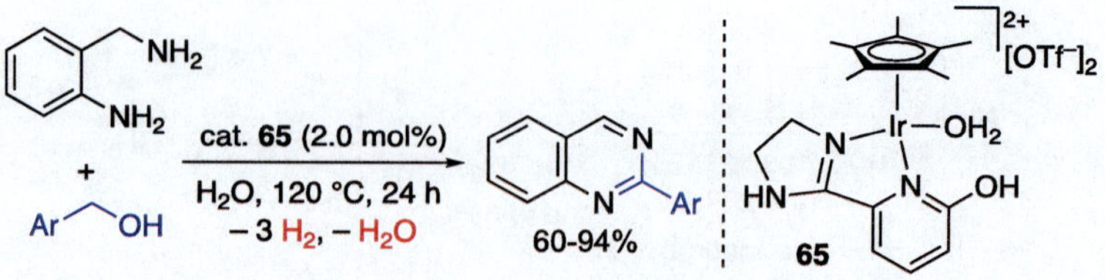

Scheme 66 Synthesis of quinazolines from 2-aminobenzylamine and aromatic primary alcohols

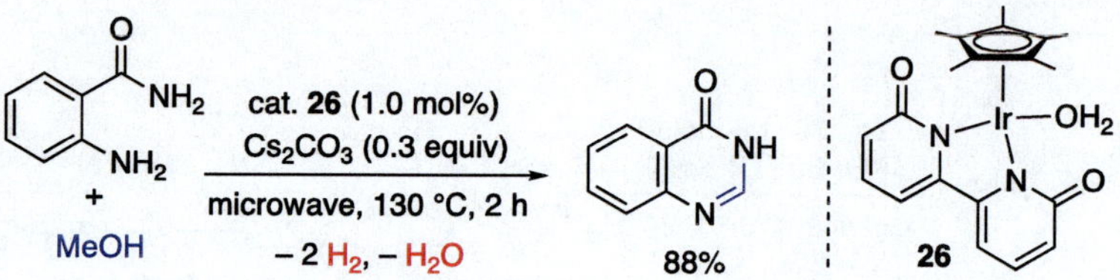

Scheme 67 Synthesis of quinazolinones from 2-aminobenzamide and methanol

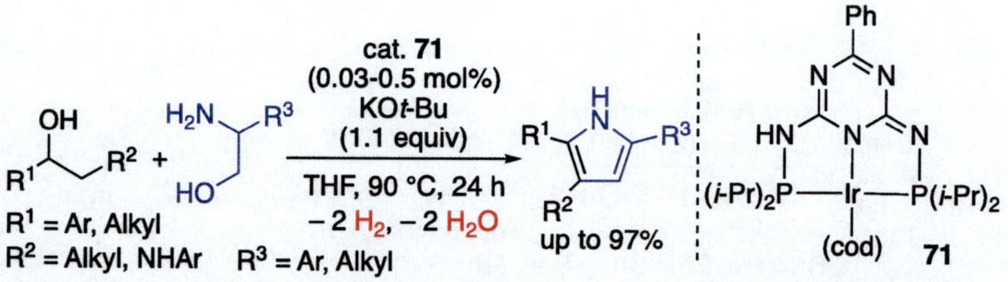

Scheme 68 Synthesis of pyrroles from secondary alcohols and 2-aminoalcohols

leading to quinoxalines and those of 1,2-phenylenediamine with aromatic primary alcohols leading to benzimidazoles.

Li et al. developed a catalytic system for the synthesis of quinazolinones starting from 2-aminobenzamide and methanol [129], showing that this reaction proceeds in the presence of Ir catalyst **26** under microwave heating conditions and is accompanied by the release of two equivalents of H_2 (Scheme 67). The initial step of this reaction is believed to be the catalytic dehydrogenation of methanol to formaldehyde, which then reacts with 2-aminobenzamide.

Kempe et al. developed an effective catalytic system for the synthesis of pyrroles from secondary alcohols and 2-aminoalcohols [130]. In the presence of Ir catalyst **71**, 2,5-disubstituted and 2,3,5-trisubstituted pyrroles were obtained along with two equivalents of H_2 (Scheme 68). The initial step of this reaction is believed to be the dehydrogenative oxidation of the secondary alcohol to a ketone, which then undergoes condensation with the amino moiety of the aminoalcohol. It should be noted that in this case, the initial dehydrogenation of the secondary alcohol triggered successive C–N and C–C bond formation. Later, the same group reported the

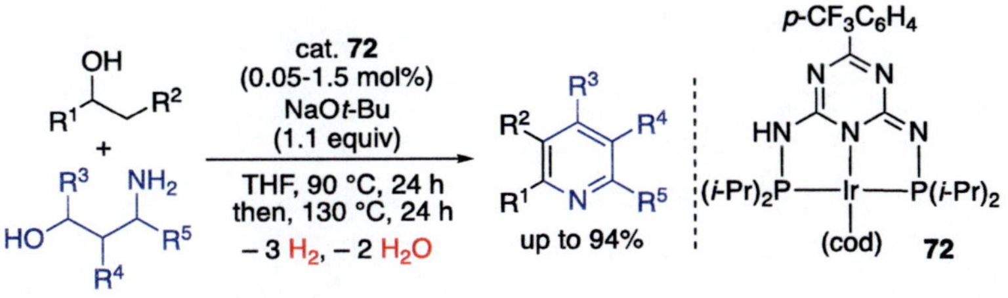

Scheme 69 Synthesis of pyridines from primary or secondary alcohols and 3-aminoalcohols

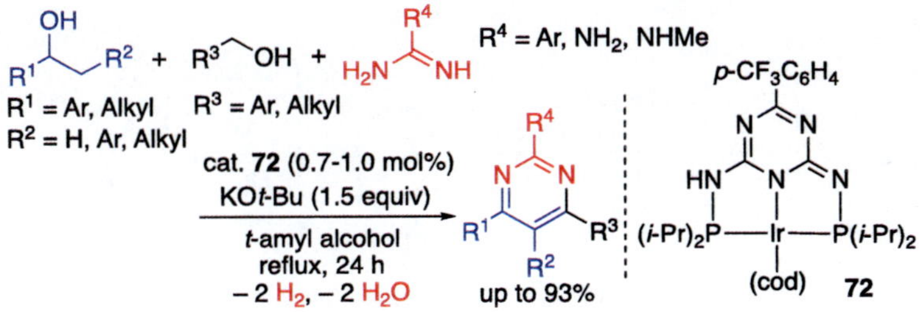

Scheme 70 Synthesis of pyrimidine derivatives by a multi-component dehydrogenative reaction

synthesis of tetrahydro-1*H*-indoles by a similar type of reaction using Ir catalyst **71** [131].

Kempe et al. reported the synthesis of substituted pyridines by the reaction of primary and secondary alcohols with 3-aminoalcohol derivatives [132], revealing that pyridines with various substitution patterns (2,4-, 2,5-, 2,6-, and 3,5-disubstituted) could be synthesized in the presence of Ir catalyst **72** (Scheme 69).

An effective synthesis of pyrimidines by a multi-component reaction was reported by Kempe et al. [133]. As illustrated in Scheme 70, various pyrimidine derivatives were synthesized starting from secondary alcohols, primary alcohols, and amidines in the presence of Ir catalyst **72** with the release of two equivalents of H_2. The same paper also demonstrated the synthesis of tetra-substituted pyrimidines via a consecutive four-component reaction.

4.6 Other Reactions Based on Alcohol Dehydrogenation

H_2 production from methanol–water mixtures in the presence of a homogeneous catalyst under mild conditions has become a hot topic in the catalytic chemistry of transition metal complexes [134]. Hence, Ir complex-catalyzed H_2 production systems have attracted much attention, e.g., Ir complexes **73** [135, 136] and **74** [137] are promising homogeneous catalysts for efficient H_2 production below 100°C (Scheme 71). These reactions are believed to proceed through the dehydrogenation of methanol to formaldehyde, the hydration of formaldehyde to a *gem*-diol, the

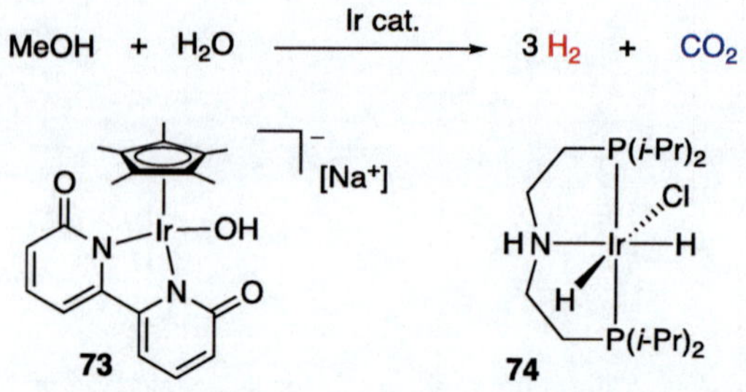

Scheme 71 H_2 production under mild conditions from methanol–water mixtures

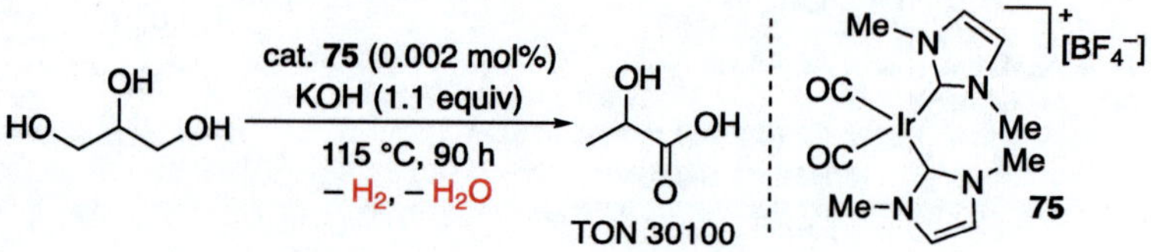

Scheme 72 Conversion of glycerol to lactic acid catalyzed by Ir complex **75**

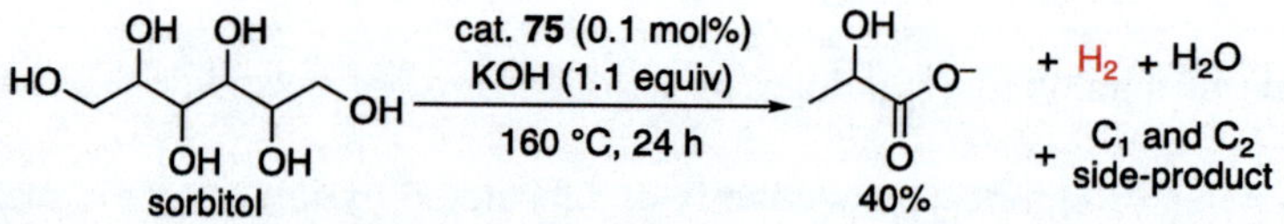

Scheme 73 Dehydrogenative conversion of sugar alcohols to lactic acid

dehydrogenation of the *gem*-diol to formic acid, and the dehydrogenation of formic acid to afford CO_2.

Crabtree et al. developed an efficient catalytic system for the conversion of glycerol to lactic acid [138] in the presence of a cationic bis-NHC Ir complex **75** (Scheme 72), revealing that a very high TON of 30,100 was achieved. It should be noted that this reaction could be conducted under mild solvent-free conditions in air. Williams et al. also reported a similar catalytic system [139].

Crabtree et al. reported the dehydrogenative conversion of sugar alcohols such as sorbitol, mannitol, and xylitol into lactic acid [140] catalyzed by Ir complex **75** (Scheme 73). As the above substrates can be easily obtained from biomass, this catalytic method holds great promise for the production of both carbon building blocks and hydrogen as an energy carrier.

Apart from the above studies, the dehydrogenation of glucose in water catalyzed by an Ir complex with an NHC ligand was reported by García and Mata et al. [141].

Fu et al. developed a catalytic system for the chemoselective dehydrogenation of an alcoholic moiety in a lignin model compound in the presence of Ir catalyst **26** [142] (Scheme 74). Additionally, hydrogen evolution from native lignin was accomplished using the same Ir catalyst.

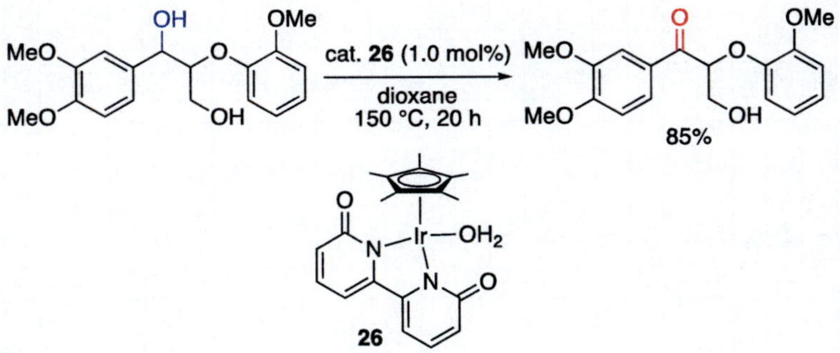

Scheme 74 Dehydrogenation of an alcoholic moiety in lignin model compound

5 Dehydrogenation of Formic Acid

Formic acid is one of the best studied compounds in the field of dehydrogenation catalysis. Early attempts of formic acid dehydrogenation under the conditions of homogeneous catalysis were reported in the 1960s–1970s, shedding light onto the potentially high catalytic performance of Ir complexes [143–145]. This dehydrogenation is a thermodynamically favored process, producing an equimolar mixture of H_2 and CO_2 (Scheme 75) [146]. The reverse hydrogenation can be accomplished under pressurized conditions in the presence of bases. Therefore, formic acid has recently attracted considerable attention as an organic hydrogen carrier for hydrogen storage applications [147–149]. The liquid nature of formic acid under ambient conditions as well as its low toxicity and good gravimetric hydrogen capacity (4.4 wt%) makes this compound a suitable hydrogen carrier.

The catalytic dehydrogenation reactions of formic acid in the presence of Ir complexes under homogeneous conditions are summarized in Table 4, and the structures of Ir complexes utilized for this purpose are depicted in Fig. 4. Most catalytic systems employ aqueous conditions in the temperature range of ambient-100°C. Basic additives such as formates and Et_3N are utilized in some systems to enhance catalytic performance by pH control. Various Cp*Ir complexes such as **35** and **76–101** with chelating ancillary ligands have been developed and shown to exhibit excellent catalytic activities, achieving TONs of up to ten million and initial TOFs of three million [150–174]. Recently, the incorporation of nitrogen-based coordination sites into polymer scaffolds has been implemented to render Cp*Ir complexes suitable for heterogeneous catalysis [175, 176]. Other types of Ir complexes **102–109** bearing phosphine or NHC ligands have also been investigated [177–184].

$$HCO_2H\,(l) \longrightarrow H_2\,(g) + CO_2\,(g) \qquad \Delta H° = +31.2\ \text{kJ/mol} \quad \Delta G° = -32.9\ \text{kJ/mol}$$

Scheme 75 Dehydrogenation of formic acid affording H_2 and CO_2

Table 4 Ir-catalyzed dehydrogenation of formic acid

Entry	Catalyst	Solvent	Additives	Temp (°C)	Time	TON	TOF (h^{-1})	Ref
1	$[IrH_3(PPh_3)_3]$	CH_3CO_2H	–	Reflux	1 h	8,889[a]	8,889	[143]
2	**76**	H_2O	–	90	N.A.	N.A.	14,000	[150]
3[b]	**77**	H_2O	–	23	30 h	>500	>17[a]	[151]
4	**78**	H_2O	–	90	4.5 h	30,000	6,667[a]	[152]
5	**79**	H_2O	–	60	24 h	33,000	1,375[a]	[153]
6	**80**	H_2O	–	60	2,600 h	5,000,000	1,923[a]	[154]
7	**81**	H_2O	HCO_2Na	80	5 min	3,292[a]	39,500	[152]
8	**82**	H_2O	HCO_2Na	80	12 h	308,000	25,667[a]	[155]
9	**83**	H_2O	–	80	14 h	2,400,000	171,429[a]	[156]
10	**84**	H_2O	HCO_2Na	90	N.A.	N.A.	144,000	[157]
11	**85**	H_2O	–	90	10 min	47,000	282,000[a]	[156]
12	**86**	H_2O	–	100	4.8 h	90,000	18,750[a]	[158]
13	**87**	H_2O	–	80	10 min	5,667[a]	34,000	[159]
14	**88**	H_2O	–	50	363 h	2,000,000	5,510[a]	[160]
15	**89**	H_2O	–	60	2.5 h	40,000	16,000[a]	[161]
16	**90**	H_2O	HCO_2Na	100	10 min	53,666[a]	322,000	[162]
17	**91**	H_2O	–	60	580 h	2,050,000	3,534[a]	[162]
18	**92**	H_2O	–	70	35 days	10,000,000	11,905[a]	[163]
19	**35**	H_2O	–	80	10 days	710,000	2,958[a]	[164]
20	**93**	–	Et_3N	40	2 h	6,680[a]	3,340	[165]
21	**94**	H_2O	HCO_2Na	90	2 min	3,333[a]	100,000	[166]
22	**95**	H_2O	HCO_2Na	50	100 h	1,000,000	10,000[a]	[167, 168]
23	**96**	H_2O	HCO_2K	25	N.A.	14,400	N.A.	[169]
24	**97**	H_2O	HCO_2Na	100	1 h	2,826	2,826	[170]
25	**98**	H_2O	–	70	175 h	5,020,000	28,686[a]	[171]
26	**99**	DME/H_2O	–	35	1 h	2,570	2,570[a]	[172]

27	**100**	H_2O	–	90	N.A.	N.A.	3,278	[173]
28	**101**	DME/H_2O	–	60	80 min	15,400	11,550[a]	[174]
29	**102**	t-BuOH	Et_3N	80	4 h	5,000	1,250[a]	[177]
30	**103**	H_2O	HCO_2Na	100	12 min	59,600[a]	298,000	[178]
31	**104**	Dioxane	–	85	N.A.	N.A.	3,271	[179]
32	**105**[c]	H_2O	HCO_2Na	80	≦1 h	N.A.	15,110	[180]
33	**106**	–	HCO_2Na	90	4 months	2,160,000	N.A.	[181]
34	**107**	DME	HCO_2Na	70	N.A.	498,190	N.A.	[182]
35	**108**	–	HCO_2Na, H_2O	90	5 h	8,030	1,606[a]	[183]
36	**109**	H_2O	HCO_2Na	80	1 min	80[a]	4,790	[184]

N.A. not available, *DME* 1,2-dimethoxyethane. bp of $CH_3CO_2H = 118°C$

[a]Calculated as TON = TOF × time

[b]Dehydrogenation of sodium formate under irradiation with 400–500 nm light

[c]Tris(*m*-sulfophenyl)phosphine trisodium salt was added as a ligand

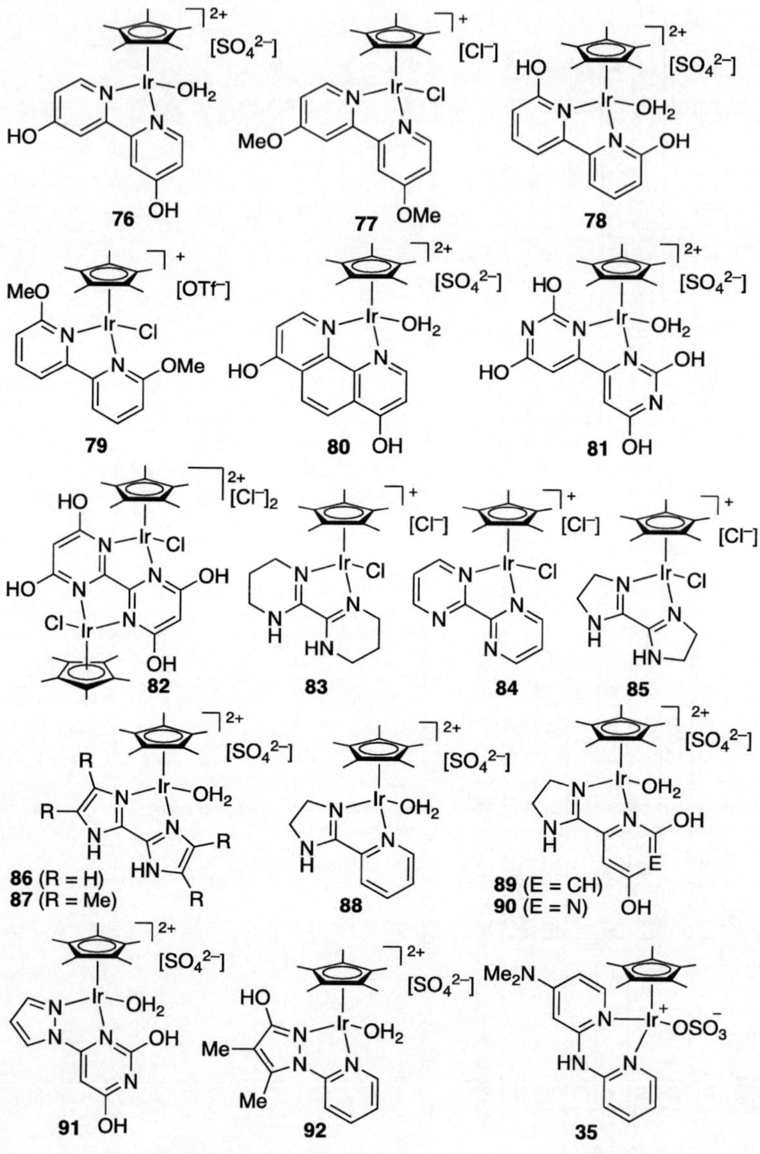

Fig. 4 Ir catalysts for dehydrogenation of formic acid

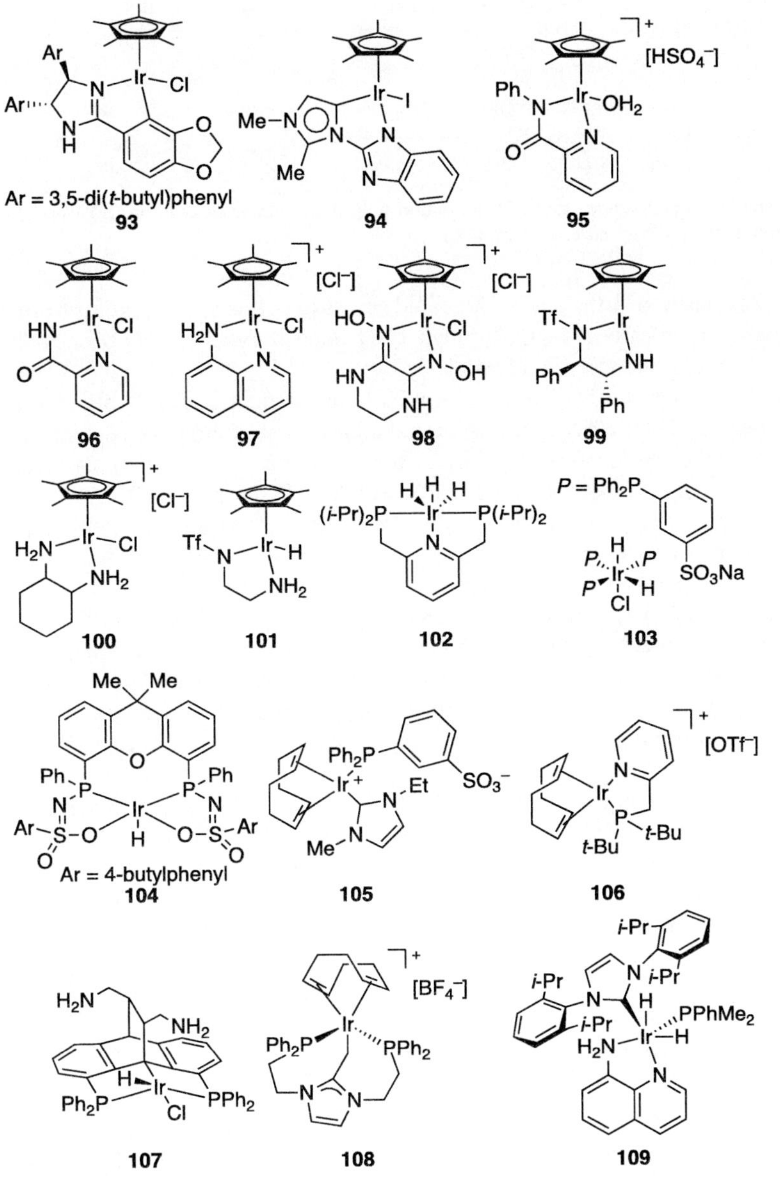

Fig. 4 (continued)

6 Dehydrogenative Borylation

Organoboronate esters are versatile building blocks for organic synthesis because of their ease of handling stemming from adequate stability toward air and moisture, low toxicity, and high functional group compatibility in the Suzuki-Miyaura cross-

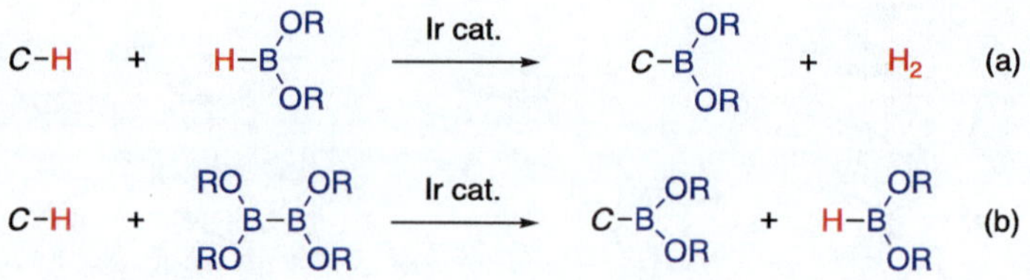

Scheme 76 Stoichiometry of C–H borylation with hydroborane or diboron. (a) Borylation with hydroborane. (b) Borylation with diboron

coupling reaction [185]. Therefore, rapid access to boronate compounds from readily available hydrocarbons, e.g., through C–H bond borylation, has been intensively investigated in the last few decades.

C–H borylation of hydrocarbons with hydroboronates affords organoboronates together with H_2 as a byproduct and can therefore be performed as a dehydrogenative transformation in the absence of hydrogen acceptors (Scheme 76a). Computational studies on the dissociation energies of B–H and C–H bonds imply that this process is thermodynamically admissible [186, 187]. In fact, numerous dehydrogenative borylations of hydrocarbons utilizing hydroboranes as a boron source have been achieved in the presence of various transition metal compounds, including Ir complexes as some of the most reliable and powerful catalysts [18, 188]. Diboron compounds with B–B bonds have also been broadly used for C–H borylation reactions that afford an organoboron product and a hydroborane byproduct in the primary catalytic cycle (Scheme 76b). In some cases, the hydroborane byproduct can also act as a second boron source for the borylation reaction with the release of H_2, which allows C–H borylation to be carried out in a dehydrogenative manner using 0.5 equiv. of the diboron reagent relative to the hydrocarbon substrate. However, from the standpoint of dehydrogenative transformation, C–H borylations employing hydroboranes are featured in this section for the sake of a brief summary. For comprehensive details of C–H borylations with both diboron and hydroborane compounds, the reader should refer to recent well-organized reviews [189–191].

6.1 Aryl and Alkyl Group Borylation

Ir-catalyzed C–H borylation with hydroboranes was first demonstrated by Smith's group [192], who showed that the reaction of benzene (solvent) with pinacolborane (HBpin) in the presence of a Cp*Ir catalyst **110** afforded phenylboronate in moderate yield and released H_2 (Scheme 77). Shortly after this report, the same group examined the regioselectivity of this catalytic system for substituted arenes [193]. The isomer distributions for the reaction of mono-substituted arenes except anisole were almost statistically controlled, corresponding to 1:2 *para*:*meta* mixtures, whereas a 1:4 ratio was observed for anisole (Scheme 78). The borylation of 1,3-disubstituted arenes exclusively occurred at 5-position for both *m*-xylene and 1,3-di(trifluoromethyl)benzene, which implied that regioselectivity was

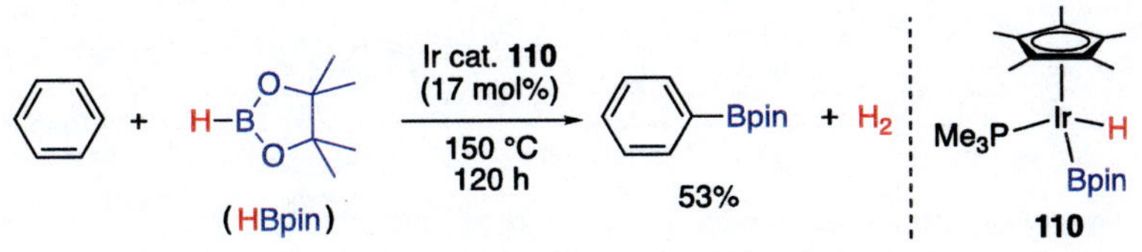

Scheme 77 C–H borylation of benzene catalyzed by Ir complex **110**

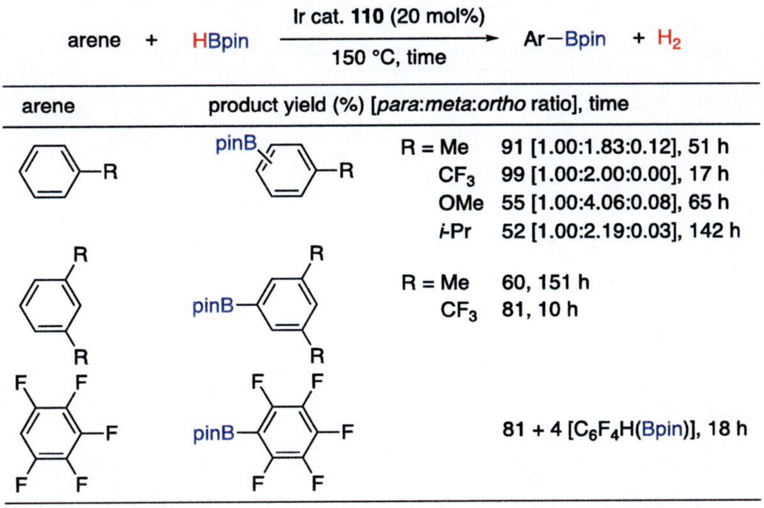

Scheme 78 C–H borylation of various arenes catalyzed by Ir complex **110**

predominantly controlled by steric factors. Additionally, this method could be applied to pentafluorobenzene to produce pentafluorophenyl boronate and a small amount of the defluoroborylated product.

A much more efficient catalytic system for the dehydrogenative C–H borylation at a lower Ir catalyst loading was explored by the same group using appropriate phosphine ligands and Ir precatalysts [15]. (η^5-Indenyl)Ir(cod) (**111**) or (η^6-mesitylene)Ir(Bpin)$_3$ (**112**) (2 mol%) in combination with PMe_3 ([P]:[Ir] ratio = 2:1) promoted the borylation of benzene at 150°C in high yields (Scheme 79). The use of bisphosphine ligands such as dppe and 1,1-bis(dimethylphosphino)ethane (dmpe) instead of PMe_3 combined with precatalyst **111** at the same [P]:[Ir] ratio of 2:1 was also efficient, with TONs of up to 4,500 achieved for a catalyst loading of 0.02 mol %. The borylation of other aromatic compounds bearing halo, methoxy, and methoxycarbonyl functionalities was also smoothly catalyzed by Ir–bisphosphine complexes. A putative mechanism involving an Ir(III)/Ir(V) cycle was proposed, in which the oxidative addition of an arene C–H bond to an Ir(III)–Bpin intermediate is followed by the reductive elimination of Ar–Bpin and the successive H_2-releasing regeneration of Ir(III)–Bpin by a reaction with HBpin (Scheme 80).

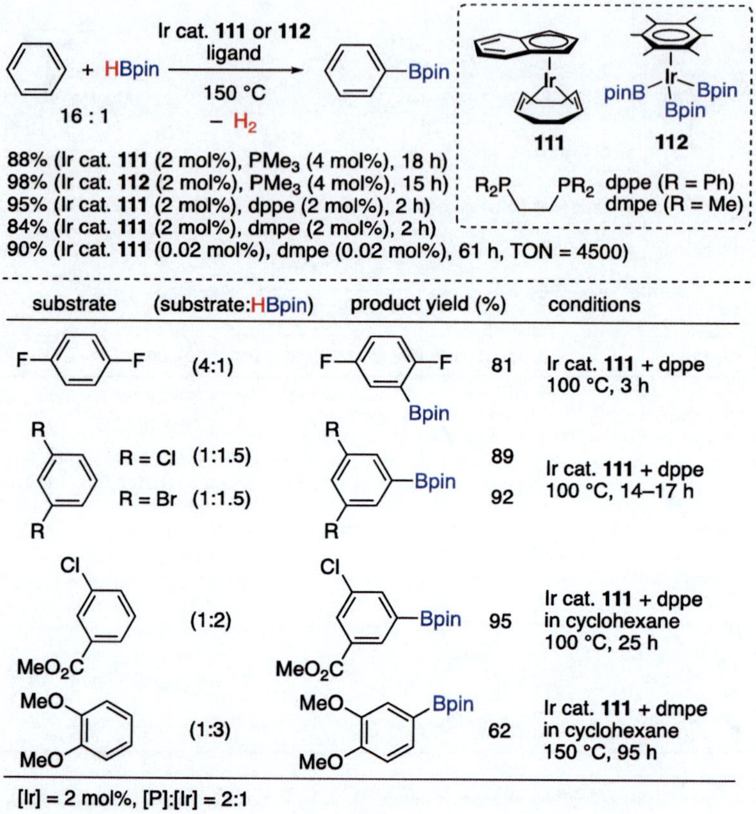

Scheme 79 Ir-phosphine complex-catalyzed dehydrogenative borylation of arenes

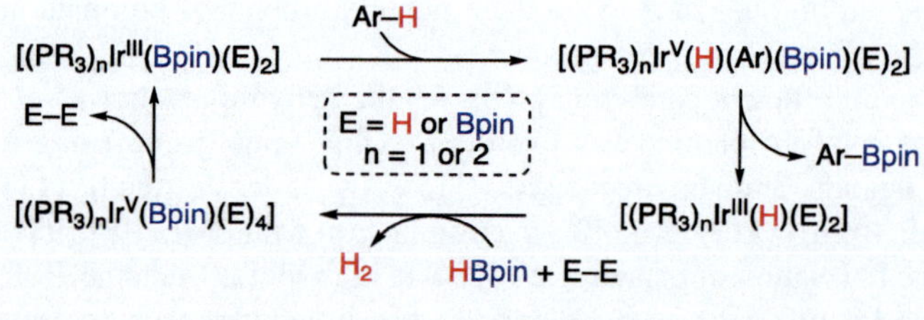

Scheme 80 Proposed mechanism of Ir-catalyzed dehydrogenative borylation of arenes

Independently from the Smith's work, the group of Hartwig, Miyaura, and Ishiyama discovered the excellent catalytic performance of an Ir-dtbpy (dtbpy = 4,4′-di-*t*-butyl-2,2′-bipyridine) complex for arene C–H borylation with B_2pin_2 under mild conditions even at room temperature in 2002 [16, 194, 195]. The catalytic system could fully utilize the boryl unit of B_2pin_2 to afford ArBpin with the release of H_2. Shortly afterwards, the same group demonstrated the Ir-dtbpy

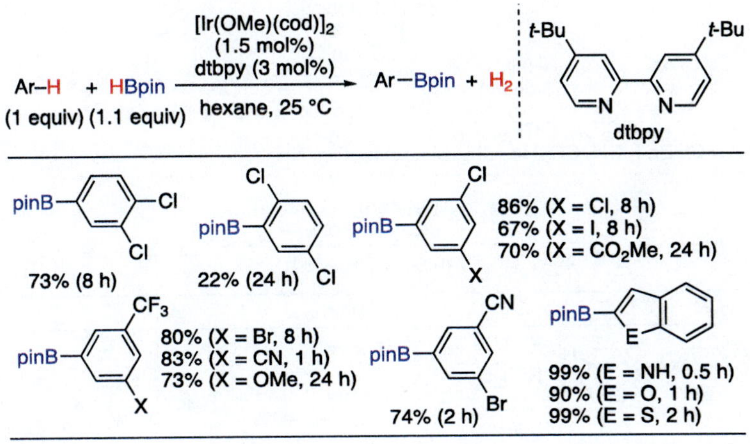

Scheme 81 Ir-dtbpy complex-catalyzed dehydrogenative borylation of aromatic compounds

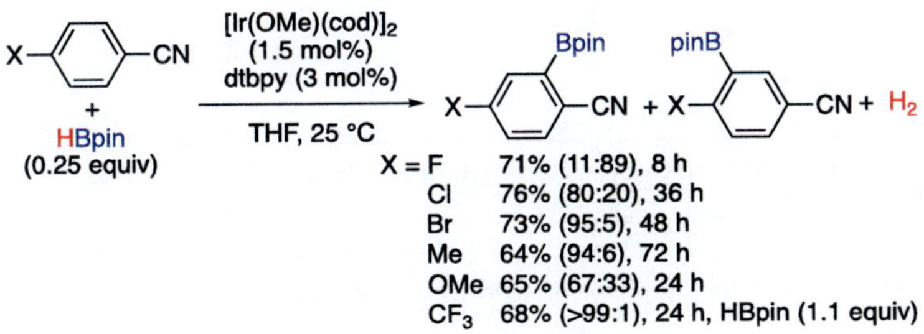

Scheme 82 Regioselectivities of borylation of 4-substituted benzonitriles

complex-catalyzed dehydrogenative borylation of arenes with HBpin (Scheme 81) [196]. In hexane as a solvent, the borylation of various arenes proceeded at 25°C in the presence of $[Ir(OMe)(cod)]_2$ (1.5 mol%) and dtbpy (3 mol%), selectively occurring at 2-position for indole, benzofuran, and benzothiophene. After these pioneering reports, dtbpy and related bipyridine- or phenanthroline-type ligand-based catalytic systems have been intensively exploited for the dehydrogenative C–H borylation of various classes of aromatic compounds with HBpin [197–212].

The regioselectivity of substituted arene borylation is predominantly controlled by steric factors [189]. Hence, the choice of an appropriately substituted arene substrate enables regioselective borylation. Borylation of various 4-substituted benzonitriles was examined by the Smith's group, proving good regioselectivities in the case of Br-, Me-, and CF_3-substituted substrates to afford borylation products at *ortho*-position to the cyano group (Scheme 82) [198].

Regioselective borylation reactions of di- or tri-substituted arenes were reported by the Chotana's group. The incorporation of the boryl group at the sterically least hindered position was achieved for various substrates including fluoroalkoxyarenes,

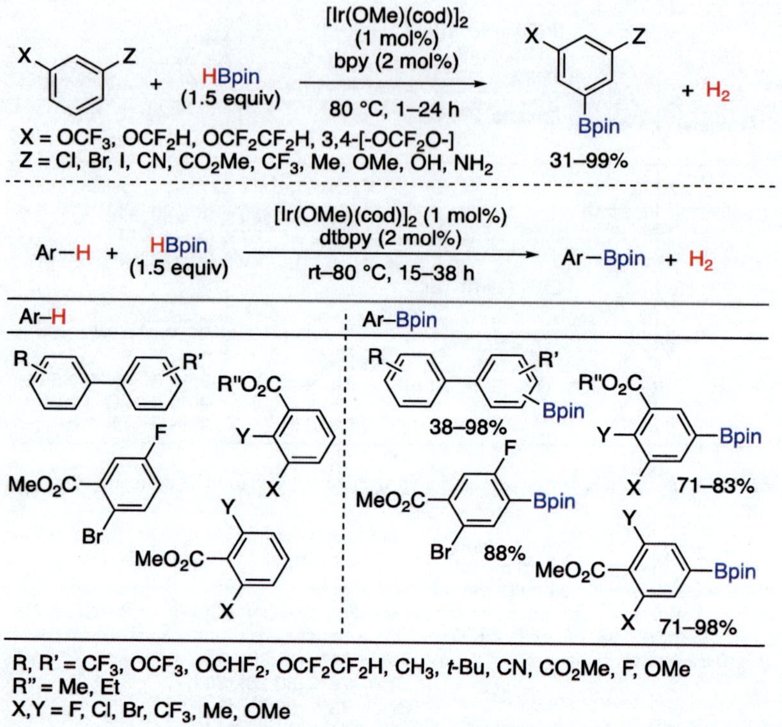

Scheme 83 Regioselective borylation of multi-substituted arenes

biaryls, and benzoate esters using Ir-bpy or Ir-dtbpy catalysts (Scheme 83) [204, 209, 210].

In contrast to the borylation regioselectivity of simple substituted arenes, that of heteroarenes is mainly dominated by their electronic character. Typical five-membered heterocycles such as pyrroles, indoles, thiophenes, benzothiophenes, and benzofurans undergo borylation at C-2 [197, 200]. However, specific substituents on such heterocycles can alter borylation regioselectivity, e.g., C-2-substituted indoles underwent borylation at C-7 in the presence of an Ir-dtbpy catalyst (Scheme 84) [199]. The reaction of a non-substituted simple indole with 2.2 equiv. of HBpin afforded a 2,7-diborylated product in high yield. The nitrogen atom of indole was proposed to act as a potential directing group for the Ir catalyst, causing C–H activation at C-7. 2,7-Diborylated indole derivatives could be selectively converted to 7-borylated indoles via acid-mediated or Pd-catalyzed protodeboronation at C-2; hence, direct one-pot syntheses of 7-borylated indoles could be achieved by an Ir-catalyzed double borylation/protodeboronation sequence [203, 213]. The attachment of a *t*-butoxycarbonyl (Boc) group to the nitrogen atom of pyrroles or indoles dramatically switched borylation regioselectivity, resulting in the exclusive formation of C-3-borylated products (Scheme 85) [201]. 2,5-Disubstituted thiophenes also selectively underwent C-3 borylation [200].

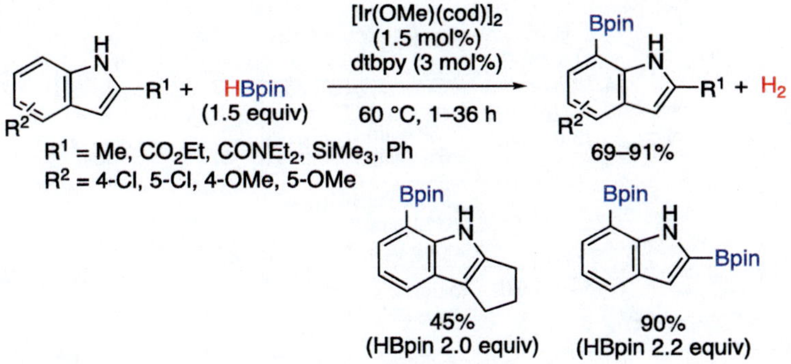

Scheme 84 Borylation of indoles at C-7 and C-2 positions

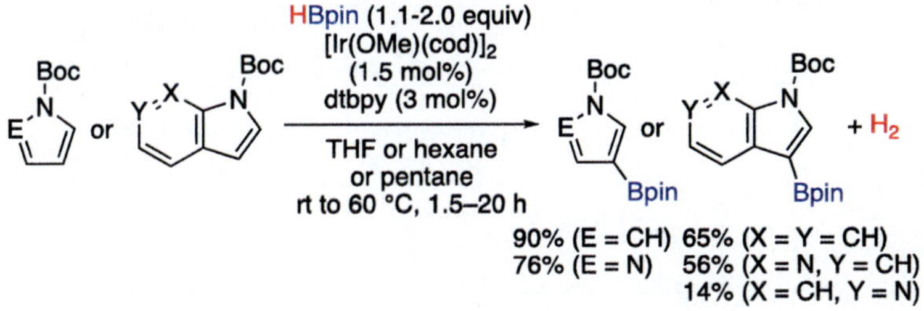

Scheme 85 Borylation of *N*-Boc-protected heterocycles

Directing group-controlled regioselective borylation reactions with HBpin have been reported by a few groups. Suginome et al. reported an efficient route to 1,2-diborylarenes via boronyl group-directed *ortho* C–H borylation catalyzed by $[Ir(OMe)(cod)]_2$ without an ancillary ligand (Scheme 86a) [214]. The use of pyrazolylaniline-modified phenylboronic acids (Ph-Bpza) resulted in successful borylation at *ortho* positions with both B_2pin_2 and HBpin as boron sources. Moreover, the Bpza group could be selectively converted to Bpin, OH, and Ar groups with retention of the newly introduced Bpin group.

Phosphine-directed borylation catalyzed by $[Ir(cod)_2][BF_4]$ with 3,4,7,8-tetramethyl-1,10-phenanthroline (tmphen) ligand was demonstrated by Clark et al. (Scheme 86b) [211]. The borylated products were isolated as forms of phosphine-borane by adding $BH_3{\cdot}THF$ after the Ir-catalyzed reaction. Benzylic phosphines, indole-based phosphines, and a ferrocenyl phosphine were applicable to this dehydrogenative borylation.

Xu et al. reported pyridine-directed dehydrogenative borylation of benzyl-2-pyridines proceeding at room temperature and affording intramolecularly coordinated pyridine-boronate adducts (Scheme 86c) [215]. Borylation at the *ortho* positions of benzylic moieties was smoothly catalyzed by $[Ir(OMe)(cod)]_2$ without any supporting ligands.

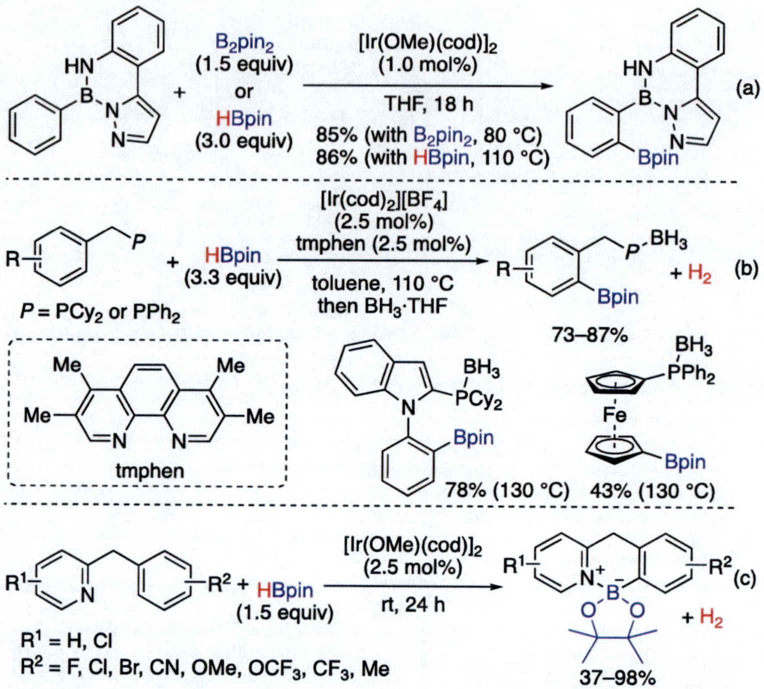

Scheme 86 Directed dehydrogenative borylation. (**a**) Bpza group-directed borylation. (**b**) Phosphinomethyl group-directed borylation. (**c**) Pyridylmethyl group-directed borylation

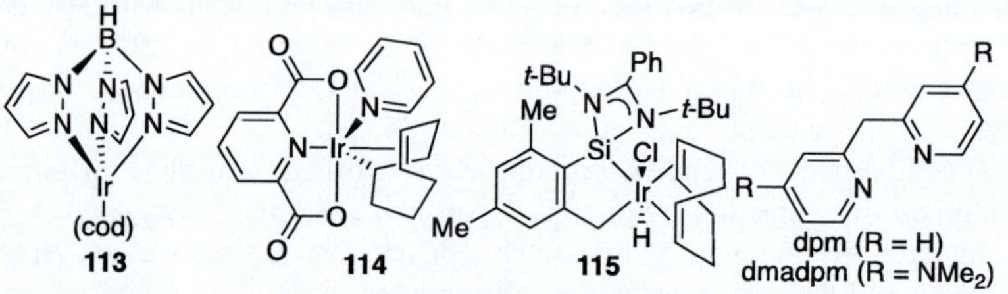

Fig. 5 Ligands and complexes for Ir-catalyzed dehydrogenative borylation of arenes

Other ligands and their Ir complexes used for the catalytic dehydrogenative borylation of arenes are summarized in Fig. 5. Complexes **113–115** exhibited typical regioselectivities in the borylation of mono- and di-substituted arenes [216–218]. Maleczka and Smith et al. reported that Ir-dpm and -dmadpm complexes exhibited high regioselectivities for the borylation of 1,3- or 1,4-disubstituted arenes with small substituents such as fluorine [219].

Apart from arene borylation, the dehydrogenative borylation of alkyl groups remains rare. Yamaguchi and Itami et al. reported the dehydrogenative borylation of cyclopropanes using a catalytic system of $[Ir(OMe)(cod)]_2$ with tmphen ligand [207], which had been proven to be effective for the borylation of the same substrate

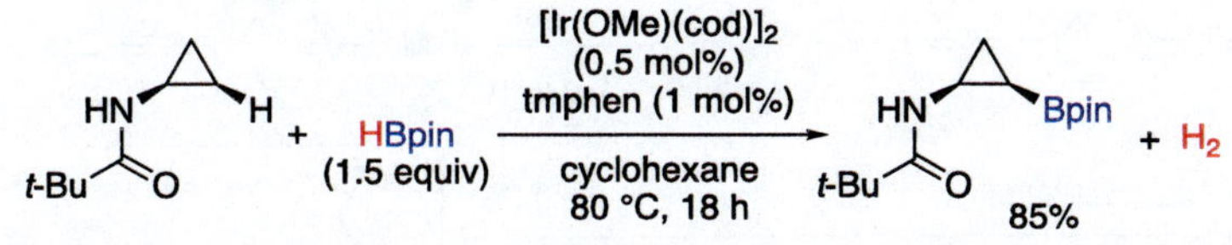

Scheme 87 Dehydrogenative borylation of a cyclopropane

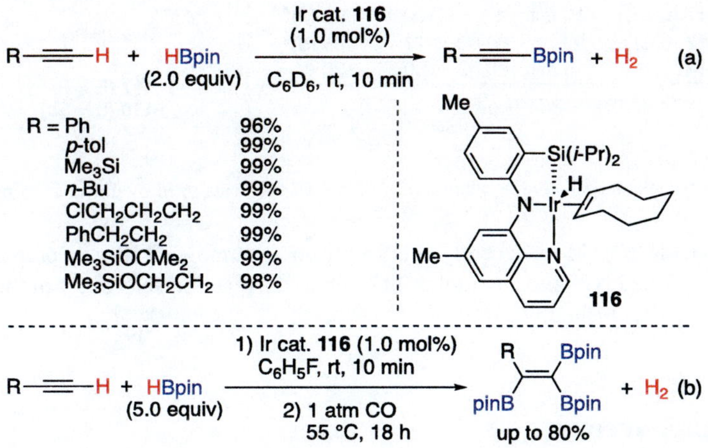

Scheme 88 (**a**) SiNN pincer Ir complex-catalyzed dehydrogenative borylation of terminal alkynes. (**b**) Triborylation of terminal alkyne via terminal alkyne borylation followed by diboration of borylalkyne

with B_2pin_2 by the Hartwig's group [220]. The borylation of a BocNH-substituted cyclopropane proceeded *cis*-selectively in cyclohexane at 70°C (Scheme 87).

6.2 *Terminal Alkyne Borylation*

The dehydrogenative borylation of C(sp)–H bonds in terminal alkynes has recently been investigated by the Ozerov's group. In particular, a SiNN pincer Ir complex **116** exhibited excellent catalytic performance at room temperature, accomplishing borylation within 10 min in benzene-d_6 (Scheme 88a) [221]. Both aromatic and aliphatic alkynes were quantitatively converted to the corresponding boronate esters. Later, the same group discovered an efficient one-pot route to triboryl alkenes based on dehydrogenative terminal alkyne borylation followed by diboration in a CO atmosphere (Scheme 88b) [222].

The same group also screened various pincer Ir complexes as catalysts for the dehydrogenative borylation of terminal alkynes [223]. The catalytic performance of PNP pincer Ir complex **117** was superior to that of **116** at low catalyst loadings, although a small amount of a hydrogenated byproduct was produced in the former

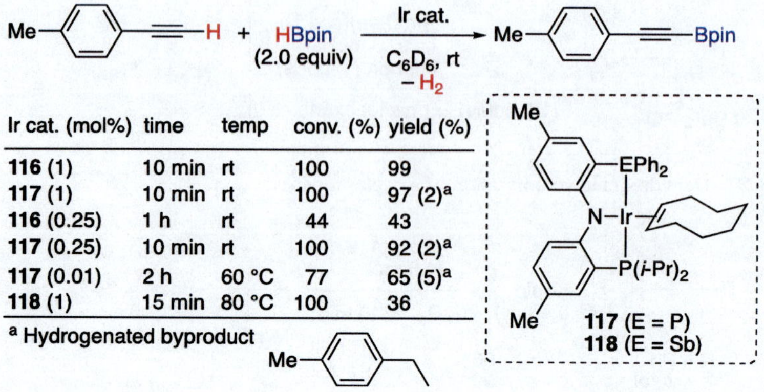

Ir cat. (mol%)	time	temp	conv. (%)	yield (%)
116 (1)	10 min	rt	100	99
117 (1)	10 min	rt	100	97 (2)[a]
116 (0.25)	1 h	rt	44	43
117 (0.25)	10 min	rt	100	92 (2)[a]
117 (0.01)	2 h	60 °C	77	65 (5)[a]
118 (1)	15 min	80 °C	100	36

[a] Hydrogenated byproduct

Scheme 89 Dehydrogenative borylation of terminal alkynes catalyzed by pincer Ir complexes

case (Scheme 89). More recently, PNSb pincer Ir complex **118** was prepared as a congener of the PNP pincer complex **117**, albeit the reaction selectivity of the former was greatly diminished [224].

7 Miscellaneous

Compared to the well-studied dehydrogenative transformation of alcohols, the Ir-catalyzed dehydrogenation of amines has been underexplored. Albrecht et al. described the catalytic dehydrogenative homocoupling of amines affording imines in the presence of a Cp*Ir triazolylidene complex **119** (Scheme 90) [225]. Complex **119** features a pendant pyridyl moiety not coordinated to the metal center and therefore capable of interacting with the N–H moiety of an amine substrate through hydrogen bonding. Catalytic reactions of benzylic amines in the presence of **119** at 150°C in 1,2-dichlorobenzene (1,2-DCB) performed in a closed system afforded a mixture of imines (dehydrogenated product) and secondary and tertiary amines. Imines were formed via the dehydrogenation of primary amines to N–H imines followed by amine exchange with the second molecule of the primary amine and the release of ammonia. The secondary and tertiary amines were probably produced via the hydrogenation of the in situ formed imines. Later, the same type of reaction catalyzed by an Ir-indazolyl-pyridinyl-triazole pincer complex supported by hydrotalcite was reported by Zhang and Wang et al. [226]. Catalytic transformation of primary amines to nitriles in the presence of hydrogen acceptor by employing iridium complex **22** as a catalyst was reported by Brookhart et al. [227].

C–H silylation with hydrosilane as a silyl source can be potentially carried out in a dehydrogenative manner; however, most of the reported Ir-catalyzed silylation reactions required stoichiometric amounts of hydrogen acceptors such as norbornene or *t*-butylethylene [228–230], in contrast to the C–H borylation reactions summarized in the preceding section. Only limited examples of C–Si bond formation through dehydrogenative silylation have been reported.

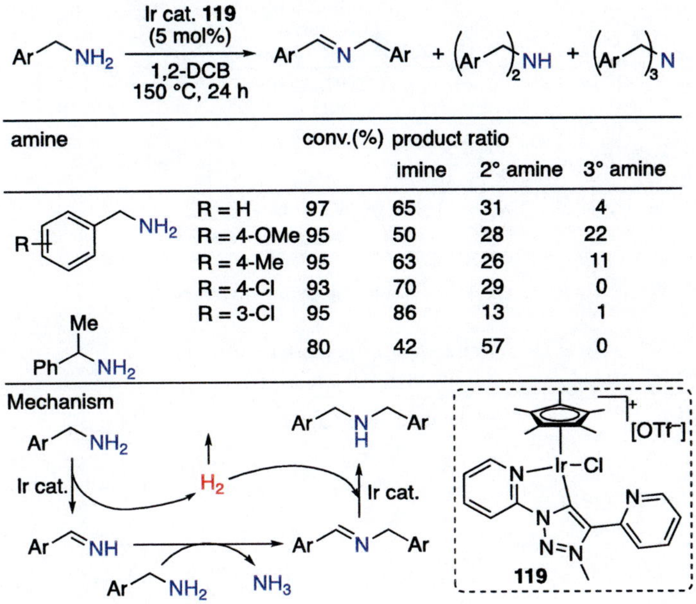

Scheme 90 Dehydrogenative homocoupling of amines

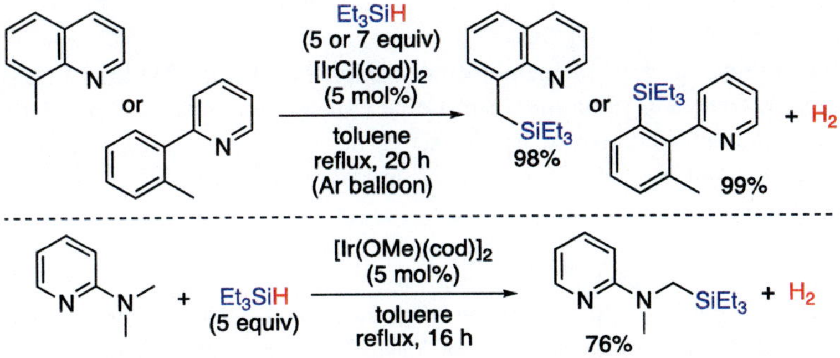

Scheme 91 Dehydrogenative silylation of C(sp^3)–H and C(sp^2)–H bonds in *N*-heteroarenes

Mita and Sato et al. presented the dehydrogenative silylation of 8-methylquinoline and 2-(*o*-tolyl)pyridine catalyzed by $[IrCl(cod)]_2$ in refluxing toluene (Scheme 91) [231]. The silylation of 8-methylquinoline occurred at the methyl group in an almost quantitative yield and was directed by the quinoline scaffold. The silylation of 2-(*o*-tolyl)pyridine afforded an *ortho*-silylated product in quantitative yield. Later, the same group reported the dehydrogenative silylation of 2-*N*,*N*′-dimethylaminopyridine at the methyl group [232].

Ishiyama and Miyaura et al. developed a direct dehydrogenative silylation of simple arenes with 1-hydrosilatrane catalyzed by an Ir-2,9-dmphen

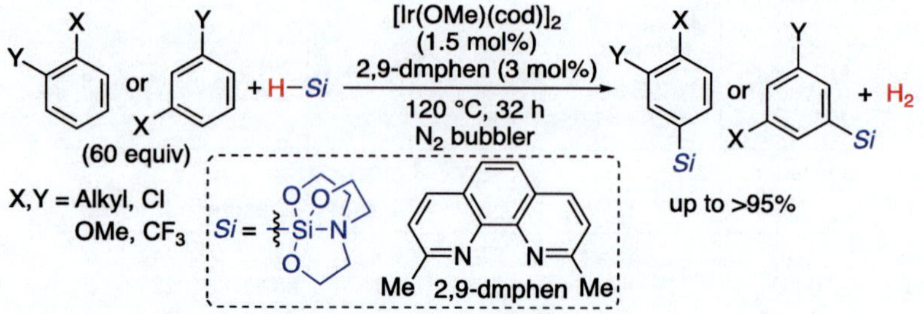

Scheme 92 Direct dehydrogenative silylation employing 1-hydrosilatrane

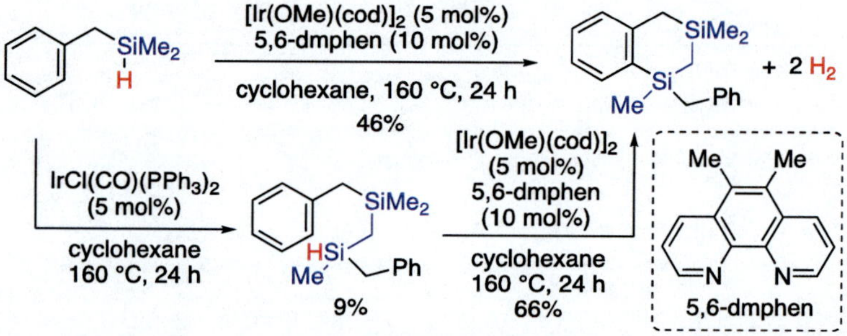

Scheme 93 Dehydrogenative dimerization of benzyldimethylsilane

(dmphen = dimethyl-1,10-phenanthroline) complex (Scheme 92) [233]. In this case, the regioselectivities of silylation were similar to those of Ir-catalyzed borylation and were predominantly controlled by the steric factors of arene substrates. 1,2- and 1,3-Disubstituted arenes underwent silylation at 4- and 5-positions in yields of up to >95%.

Murai and Takai et al. reported the dehydrogenative dimerization of benzyldimethylsilane catalyzed by an Ir-5,6-dmphen complex and affording 1,2,3,4-tetrahydrobenzo[*d*][1,3]disiline (Scheme 93) [234]. The reaction of benzyldimethylsilane in the presence of Vaska's complex allowed the isolation of the pre-cyclized intermediate that could be converted to the final cyclized product under the parent catalytic conditions. Hence, this catalytic reaction is thought to proceed via dehydrogenative dimerization followed by dehydrogenative intramolecular cyclization.

A similar dehydrogenative intramolecular cyclization involving C(sp^2)–Si bond formation was performed in the presence of Ir-bpy-based complexes immobilized into metal-organic-frameworks [235].

Dehydrogenative coupling of hydrosilanes with protic reagents such as alcohols, carboxylic acids, and amines has been recently investigated. Three component coupling of amines with hydrosilanes and CO_2 leading to silyl carbamate was reported by Fernández-Alvarez and Oro et al., by employing Ir complex **120** with a facially coordinated bis(pyridine-2-yloxy)methylsilyl ligand (Scheme 94)

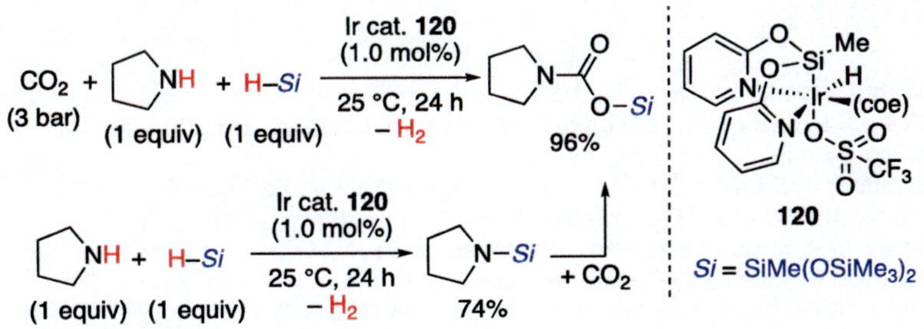

Scheme 94 Dehydrogenative coupling of an amine with a hydrosilane and CO_2

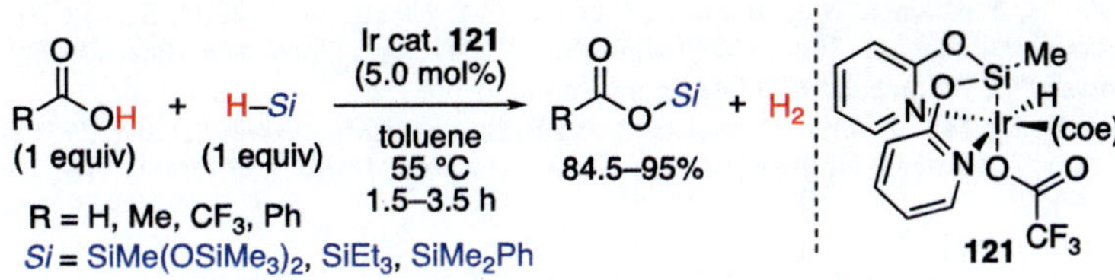

Scheme 95 Dehydrogenative coupling of carboxylic acids with hydrosilanes

[236]. The reaction of the amine with the hydrosilane was smoothly catalyzed by Ir complex **120**, giving silylamine with liberation of H_2. The condensation of the freshly prepared silylamine with CO_2 afforded a silyl carbamate, as a well-known process [237]. The same group also described the dehydrogenative coupling of hydrosilanes with carboxylic acids catalyzed by a similar Ir complex **121** (Scheme 95) [238].

Mata et al. reported that a Cp*Ir-NHC complex bearing a pyrenyl group and immobilized onto reduced graphene oxide catalyzed the fast dehydrogenative coupling of hydrosilanes with alcohols at low temperature and thus could potentially be applied in hydrogen storage systems [239]. The relatively high stability under ambient conditions, good reactivity for H_2 evolution, and ease of handling make hydrosilanes promising hydrogen carriers.

8 Summary

This chapter reviewed Ir-catalyzed dehydrogenation and related reactions, demonstrating the use of Ir catalysts for the dehydrogenation of alkanes, nitrogen-containing heterocycles, alcohols, formic acid, and other substrates and showcasing the utility of Ir-catalyzed dehydrogenation for organic synthesis. In addition, numerous reactions such as the silylation and borylation of C–H bonds in organic molecules accompanied by the evolution of H_2 have attracted much attention to the catalytic chemistry of Ir. As this chemistry will undoubtedly continue to make great progress, it deserves constant attention.

References

1. Vaska L (1968) Acc Chem Res 1:335
2. Crabtree RH, Felkin H, Morris GE (1977) J Organomet Chem 141:205
3. Crabtree R (1979) Acc Chem Res 12:331
4. Lightfoot A, Schnider P, Pfaltz A (1998) Angew Chem Int Ed 37:2897
5. Janssen JP, Helmchen G (1997) Tetrahedron Lett 38:8025
6. Takeuchi R, Kashio M (1997) Angew Chem Int Ed 36:263
7. Takeuchi R, Kashio M (1998) J Am Chem Soc 120:8647
8. Fuji K, Kinoshita N, Kawabata T, Tanaka K (1999) Chem Commun:2289
9. Takeuchi R, Shiga N (1999) Org Lett 1:265
10. Ohmura T, Hartwig JF (2002) J Am Chem Soc 124:15164
11. López F, Ohmura T, Hartwig JF (2003) J Am Chem Soc 125:3426
12. Miyabe H, Yoshida K, Matsumura A, Yamauchi M, Takemoto Y (2003) Synlett 2003:0567
13. Fischer C, Defieber C, Suzuki T, Carreira EM (2004) J Am Chem Soc 126:1628
14. Lipowsky G, Helmchen G (2004) Chem Commun:116
15. Cho J-Y, Tse MK, Holmes D, Maleczka Jr RE, Smith III MR (2002) Science 295:305
16. Ishiyama T, Takagi J, Ishida K, Miyaura N, Anastasi NR, Hartwig JF (2002) J Am Chem Soc 124:390
17. Ishiyama T, Miyaura N (2004) Chem Rec 3:271
18. Mkhalid IAI, Barnard JH, Marder TB, Murphy JM, Hartwig JF (2010) Chem Rev 110:890
19. Sunley GJ, Watson DJ (2000) Catal Today 58:293
20. Chatani N, Ikeda S, Ohe K, Murai S (1992) J Am Chem Soc 114:9710
21. Murakami M, Itami K, Ubukata M, Tsuji I, Ito Y (1998) J Org Chem 63:4
22. Okimoto Y, Sakaguchi S, Ishii Y (2002) J Am Chem Soc 124:1590
23. Blum J, Biger S (1970) Tetrahedron Lett 11:1825
24. Crabtree RH, Mihelcic JM, Quirk JM (1979) J Am Chem Soc 101:7738
25. Burk MJ, Crabtree RH (1987) J Am Chem Soc 109:8025
26. Fujii T, Saito Y (1990) J Chem Soc Chem Commun 757
27. Aoki T, Crabtree RH (1993) Organometallics 12:294
28. Itagaki H, Einaga H, Saito Y (1993) Chem Lett:2097
29. Taubmann S, Alt HG (2008) J Organomet Chem 693:1808
30. Taubmann S, Alt HG (2008) J Mol Catal A Chem 284:134
31. Rábay B, Braun T, Falkenhagen JP (2013) Dalton Trans 42:8058
32. Hoyano JK, Graham WAG (1982) J Am Chem Soc 104:3723
33. Xu WW, Rosini GP, Gupta M, Jensen CM, Kaska WC, Krogh-Jespersen K, Goldman AS (1997) Chem Commun:2273
34. Gupta M, Hagen C, Flesher RJ, Kaska WC, Jensen CM (1996) Chem Commun:2083
35. Gupta M, Hagen C, Kaska WC, Cramer RE, Jensen CM (1997) J Am Chem Soc 119:840
36. Gupta M, Kaska WC, Jensen CM (1997) Chem Commun:461
37. Liu FC, Goldman AS (1999) Chem Commun:655
38. Krogh-Jespersen K, Czerw M, Summa N, Renkema KB, Achord PD, Goldman AS (2002) J Am Chem Soc 124:11404
39. Zhu K, Achord PD, Zhang X, Krogh-Jespersen K, Goldman AS (2004) J Am Chem Soc 126:13044
40. Morales-Morales D, Redon R, Yung C, Jensen CM (2004) Inorg Chim Acta 357:2953
41. Kundu S, Choliy Y, Zhuo G, Ahuja R, Emge TJ, Warmuth R, Brookhart M, Krogh-Jespersen-K, Goldman AS (2009) Organometallics 28:5432
42. Punji B, Emge TJ, Goldman AS (2010) Organometallics 29:2702
43. Haenel MW, Oevers S, Angermund K, Kaska WC, Fan HJ, Hall MB (2001) Angew Chem Int Ed 40:3596
44. Sheludko B, Cunningham MT, Goldman AS, Celik FE (2018) ACS Catal 8:7828
45. Chianese AR, Mo A, Lampland NL, Swartz RL, Bremer PT (2010) Organometallics 29:3019

46. Chianese AR, Drance MJ, Jensen KH, McCollom SP, Yusufova N, Shaner SE, Shopov DY, Tendler JA (2014) Organometallics 33:457
47. Dinh LV, Li B, Kumar A, Schinski W, Field KD, Kuperman A, Celik FE, Goldman AS (2016) ACS Catal 6:2836
48. Crabtree RH (2008) Energy Environ Sci 1:134
49. Katritzky AR (2010) Handbook of heterocyclic chemistry.3rd edn. Elsevier, Amsterdam
50. Quin LD, Tyrell JA (2010) Fundamentals of heterocyclic chemistry: importance in nature and in the synthesis of pharmaceuticals. Wiley, Hoboken
51. Hati S, Holzgrabe U, Sen S (2017) Beilstein J Org Chem 13:1670
52. Gianotti E, Taillades-Jacquin M, Rozière J, Jones DJ (2018) ACS Catal 8:4660
53. Wei ZZ, Shao FJ, Wang JG (2019) Chin J Catal 40:980
54. Shimbayashi T, Fujita K (2020) Tetrahedron 76:130946
55. Wang ZH, Tonks I, Belli J, Jensen CM (2009) J Organomet Chem 694:2854
56. Wang Z, Belli J, Jensen CM (2011) Faraday Discuss 151:297
57. Brayton DF, Jensen CM (2014) Chem Commun 50:5987
58. Brayton DF, Beaumont PR, Fukushima EY, Sartain HT, Morales-Morales D, Jensen CM (2014) Organometallics 33:5198
59. Brayton DF, Jensen CM (2015) Int J Hydrog Energy 40:16266
60. Yamaguchi R, Ikeda C, Takahashi Y, Fujita K (2009) J Am Chem Soc 131:8410
61. Clot E, Eisenstein O, Crabtree RH (2007) Chem Commun:2231
62. Fujita K, Tanaka Y, Kobayashi M, Yamaguchi R (2014) J Am Chem Soc 136:4829
63. Jeong J, Shimbayashi T, Fujita K (2019) Catalysts 9:846
64. Fujita K, Wada T, Shiraishi T (2017) Angew Chem Int Ed 56:10886
65. Vivancos Á, Beller M, Albrecht M (2018) ACS Catal 8:17
66. Wang S, Huang H, Bruneau C, Fischmeister C (2019) ChemSusChem 12:2350
67. Wu J, Talwar D, Johnston S, Yan M, Xiao J (2013) Angew Chem Int Ed 52:6983
68. Liu H, Chen JG, Wang C, Liu ZT, Li Y, Liu ZW, Xiao JL, Lu J (2017) Ind Eng Chem Res 56:11413
69. Manas MG, Sharninghausen LS, Lin E, Crabtree RH (2015) J Organomet Chem 792:184
70. Sogaard A, Scheuermeyer M, Bosmann A, Wasserscheid P, Riisager A (2019) Chem Commun 55:2046
71. Kusumoto S, Akiyama M, Nozaki K (2013) J Am Chem Soc 135:18726
72. Shvo Y, Czarkie D, Rahamim Y, Chodosh DF (1986) J Am Chem Soc 108:7400
73. Conley BL, Pennington-Boggio MK, Boz E, Williams TJ (2010) Chem Rev 110:2294
74. Ando H, Kusumoto S, Wu W, Nozaki K (2017) Organometallics 36:2317
75. Yamamoto Y, Hayashi H, Saigoku T, Nishiyama H (2005) J Am Chem Soc 127:10804
76. Yamamoto Y, Hayashi H (2007) Tetrahedron 63:10149
77. Chatani N, Inoue H, Morimoto T, Muto T, Murai S (2001) J Org Chem 66:4433
78. Li F, Lu L, Ma J (2015) Org Chem Front 2:1589
79. Zhao W, Liu PC, Li F (2016) ChemCatChem 8:1523
80. Fan H, Zhang W, Zhao W, Li F (2017) ChemistrySelect 2:5735
81. Li CJ (2009) Acc Chem Res 42:335
82. Yeung CS, Dong VM (2011) Chem Rev 111:1215
83. Varun BV, Dhineshkumar J, Bettadapur KR, Siddaraju Y, Alagiri K, Prabhu KR (2017) Tetrahedron Lett 58:803
84. Huang CY, Kang H, Li J, Li CJ (2019) J Org Chem 84:12705
85. Nie S-Z, Sun X, Wei W-T, Zhang X-J, Yan M, Xiao J-L (2013) Org Lett 15:2394
86. Sun X, Hu Y, Nie S-Z, Yan Y-Y, Zhang X-J, Yan M (2013) Adv Synth Catal 355:2179
87. Sun X, Lv XH, Ye LM, Hu Y, Chen YY, Zhang XJ, Yan M (2015) Org Biomol Chem 13:7381
88. Lin YR, Ma DW, Lu XY (1987) Tetrahedron Lett 28:3115
89. Matsubara T, Saito Y, Yamakawa T, Shinoda S (1991) J Mol Catal 66:171
90. Matsubara T, Saito Y, Yamakawa T, Shinoda S (1991) J Mol Catal 67:175
91. Fujita K, Tanino N, Yamaguchi R (2007) Org Lett 9:109

92. Yamaguchi R, Kobayashi D, Shimizu M, Fujita K (2017) J Organomet Chem 843:14
93. Fujita K, Yoshida T, Imori Y, Yamaguchi R (2011) Org Lett 13:2278
94. Kawahara R, Fujita K, Yamaguchi R (2012) J Am Chem Soc 134:3643
95. Shimizu M, Michikawa K, Maegawa Y, Inagaki S, Fujita K (2020) ACS Appl Nano Mater 3:2527
96. Ngo AH, Adams MJ, Do LH (2014) Organometallics 33:6742
97. Kawahara R, Fujita K, Yamaguchi R (2012) Angew Chem Int Ed 51:12790
98. Zeng G, Sakaki S, Fujita K, Sano H, Yamaguchi R (2014) ACS Catal 4:1010
99. Fujita K, Tamura R, Tanaka Y, Yoshida M, Onoda M, Yamaguchi R (2017) ACS Catal 7:7226
100. Musa S, Shaposhnikov I, Cohen S, Gelman D (2011) Angew Chem Int Ed 50:3533
101. Oded K, Musa S, Gelman D, Blum J (2012) Catal Commun 20:68
102. Valencia M, Muller-Bunz H, Gossage RA, Albrecht M (2016) Chem Commun 52:3344
103. Vivancos Á, Petronilho A, Cardoso J, Müller-Bunz H, Albrecht M (2018) Dalton Trans 47:74
104. Mazloomi Z, Pretorius R, Pamies O, Albrecht M, Dieguez M (2017) Inorg Chem 56:11282
105. Polukeev AV, Petrovskii PV, Peregudov AS, Ezernitskaya MG, Koridze AA (2013) Organometallics 32:1000
106. Gulcemal S, Gulcemal D, Whitehead GF, Xiao J (2016) Chem Eur J 22:10513
107. Thangavel S, Boopathi S, Mahadevaiah N, Kolandaivel P, Pansuriya PB, Friedrich HB (2016) J Mol Catal A Chem 423:160
108. Toyomura K, Fujita K (2017) Chem Lett 46:808
109. González Miera G, Martínez-Castro E, Martín-Matute B (2018) Organometallics 37:636
110. Alabau RG, Esteruelas MA, Martínez A, Oliván M, Oñate E (2018) Organometallics 37:2732
111. Yoshida M, Wang H, Shimbayashi T, Fujita K (2018) Catalysts 8:312
112. Fujita K, Ito W, Yamaguchi R (2014) ChemCatChem 6:109
113. Onoda M, Nagano Y, Fujita K (2019) Int J Hydrog Energy 44:28514
114. Kuwahara M, Nishioka M, Yoshida M, Fujita K (2018) ChemCatChem 10:3636
115. Cherepakhin V, Williams TJ (2018) ACS Catal 8:3754
116. Sahoo AR, Jiang F, Bruneau C, Sharma GVM, Suresh S, Achard M (2016) RSC Adv 6:100554
117. Olsen EPK, Madsen R (2012) Chem Eur J 18:16023
118. Olsen EPK, Singh T, Harris P, Andersson PG, Madsen R (2015) J Am Chem Soc 137:834
119. Christensen SH, Olsen EPK, Rosenbaum J, Madsen R (2015) Org Biomol Chem 13:938
120. Musa S, Ackermann L, Gelman D (2013) Adv Synth Catal 355:3077
121. Wang R, Ma J, Li F (2015) J Org Chem 80:10769
122. Jiménez MV, Fernández-Tornos J, Modrego FJ, Pérez-Torrente JJ, Oro LA (2015) Chem Eur J 21:17877
123. Smith LB, Armstrong RJ, Matheau-Raven D, Donohoe TJ (2020) J Am Chem Soc 142:2514
124. Ruch S, Irrgang T, Kempe R (2014) Chem Eur J 20:13279
125. Wang R, Fan H, Zhao W, Li F (2016) Org Lett 18:3558
126. Genç S, Arslan B, Gülcemal S, Günnaz S, Çetinkaya B, Gülcemal D (2019) J Org Chem 84:6286
127. Hille T, Irrgang T, Kempe R (2014) Chem Eur J 20:5569
128. Chakrabarti K, Maji M, Kundu S (2019) Green Chem 21:1999
129. Li F, Lu L, Liu P (2016) Org Lett 18:2580
130. Michlik S, Kempe R (2013) Nat Chem 5:140
131. Forberg K, Kempe (2019) Inorganics 7:97
132. Michlik S, Kempe R (2013) Angew Chem Int Ed 52:6326
133. Deibl N, Ament K, Kempe R (2015) J Am Chem Soc 137:12804
134. Campos J, Sharninghausen LS, Manas MG, Crabtree RH (2015) Inorg Chem 54:5079
135. Fujita K, Kawahara R, Aikawa T, Yamaguchi R (2015) Angew Chem Int Ed 54:9057
136. Enomoto A, Kajita S, Fujita K (2019) Chem Lett 48:106
137. Prichatz C, Alberico E, Baumann W, Junge H, Beller M (2017) ChemCatChem 9:1891
138. Sharninghausen LS, Campos J, Manas MG, Crabtree RH (2014) Nat Commun 5:5084

139. Lu ZY, Demianets I, Hamze R, Terrile NJ, Williams TJ (2016) ACS Catal 6:2014
140. Manas MG, Campos J, Sharninghausen LS, Lin E, Crabtree RH (2015) Green Chem 17:594
141. Borja P, Vicent C, Baya M, García H, Mata JA (2018) Green Chem 20:4094
142. Zhu R, Wang B, Cui M, Deng J, Li X, Ma Y, Fu Y (2016) Green Chem 18:2029
143. Coffey RS (1967) Chem Commun:923
144. Forster D, Beck GR (1971) J Chem Soc D:1072
145. Yurtchenko EN, Anikeenko NP (1975) React Kinet Catal Lett 2:65
146. Loges B, Boddien A, Gärtner F, Junge H, Beller M (2010) Top Catal 53:902
147. Johnson TC, Morris DJ, Wills M (2010) Chem Soc Rev 39:81
148. Iglesias M, Oro LA (2018) Eur J Inorg Chem 2018:2125
149. Sordakis K, Tang C, Vogt LK, Junge H, Dyson PJ, Beller M, Laurenczy G (2018) Chem Rev 118:372
150. Himeda Y (2009) Green Chem 11:2018
151. Barrett SM, Slattery SA, Miller AJM (2015) ACS Catal 5:6320
152. Wang W-H, Xu S, Manaka Y, Suna Y, Kambayashi H, Muckerman JT, Fujita E, Himeda Y (2014) ChemSusChem 7:1976
153. Siek S, Burks DB, Gerlach DL, Liang G, Tesh JM, Thompson CR, Qu F, Shankwitz JE, Vasquez RM, Chambers N, Szulczewski GJ, Grotjahn DB, Webster CE, Papish ET (2017) Organometallics 36:1091
154. Iguchi M, Himeda Y, Manaka Y, Kawanami H (2016) ChemSusChem 9:2749
155. Hull JF, Himeda Y, Wang WH, Hashiguchi B, Periana R, Szalda DJ, Muckerman JT, Fujita E (2012) Nat Chem 4:383
156. Wang Z, Lu SM, Li J, Wang J, Li C (2015) Chem Eur J 21:12592
157. Du Y, Shen YB, Zhan YL, Ning FD, Yan LM, Zhou XC (2017) Chin Chem Lett 28:1746
158. Manaka Y, Onishi N, Iguchi M, Kawanami H, Himeda Y (2017) J Jpn Petrol Inst 60:53
159. Manaka Y, Wang WH, Suna Y, Kambayashi H, Muckerman JT, Fujita E, Himeda Y (2014) Cat Sci Technol 4:34
160. Onishi N, Ertem MZ, Xu S, Tsurusaki A, Manaka Y, Muckerman JT, Fujita E, Himeda Y (2016) Cat Sci Technol 6:988
161. Wang L, Onishi N, Murata K, Hirose T, Muckerman JT, Fujita E, Himeda Y (2017) ChemSusChem 10:1071
162. Wang WH, Ertem MZ, Xu SA, Onishi N, Manaka Y, Suna Y, Kambayashi H, Muckerman JT, Fujita E, Himeda Y (2015) ACS Catal 5:5496
163. Onishi N, Kanega R, Fujita E, Himeda Y (2019) Adv Synth Catal 361:289
164. Wang S, Huang H, Roisnel T, Bruneau C, Fischmeister C (2019) ChemSusChem 12:179
165. Barnard JH, Wang C, Berry NG, Xiao JL (2013) Chem Sci 4:1234
166. Semwal S, Kumar A, Choudhury J (2018) Cat Sci Technol 8:6137
167. Kanega R, Onishi N, Wang L, Murata K, Muckerman JT, Fujita E, Himeda Y (2018) Chem Eur J 24:18389
168. Kanega R, Ertem MZ, Onishi N, Szalda DJ, Fujita E, Himeda Y (2020) Organometallics 39:1519
169. Menendez Rodriguez G, Domestici C, Bucci A, Valentini M, Zuccaccia C, Macchioni A (2018) Eur J Inorg Chem 2018:2247
170. Fidalgo J, Ruiz-Castañeda M, García-Herbosa G, Carbayo A, Jalón FA, Rodríguez AM, Manzano BR, Espino G (2018) Inorg Chem 57:14186
171. Lu S-M, Wang Z, Wang J, Li J, Li C (2018) Green Chem 20:1835
172. Matsunami A, Kayaki Y, Ikariya T (2015) Chem Eur J 21:13513
173. Fink C, Laurenczy G (2017) Dalton Trans 46:1670
174. Matsunami A, Kuwata S, Kayaki Y (2017) ACS Catal 7:4479
175. Broicher C, Foit SR, Rose M, Hausoul PJC, Palkovits R (2017) ACS Catal 7:8413
176. Shen Y, Zhan Y, Bai C, Ning F, Wang H, Wei J, Lv G, Zhou X (2020) Sustain Energy Fuels 4:2519

177. Tanaka R, Yamashita M, Chung LW, Morokuma K, Nozaki K (2011) Organometallics 30:6742
178. Papp G, Ölveti G, Horváth H, Kathó Á, Joó F (2016) Dalton Trans 45:14516
179. Oldenhof S, de Bruin B, Lutz M, Siegler MA, Patureau FW, van der Vlugt JI, Reek JNH (2013) Chem Eur J 19:11507
180. Horvath H, Papp G, Szabolcsi R, Katho A, Joo F (2015) ChemSusChem 8:3036
181. Celaje JJA, Lu Z, Kedzie EA, Terrile NJ, Lo JN, Williams TJ (2016) Nat Commun 7:11308
182. Cohen S, Borin V, Schapiro I, Musa S, De-Botton S, Belkova NV, Gelman D (2017) ACS Catal 7:8139
183. Iturmendi A, Iglesias M, Munarriz J, Polo V, Passarelli V, Pérez-Torrente JJ, Oro LA (2018) Green Chem 20:4875
184. Iturmendi A, Rubio-Pérez L, Pérez-Torrente JJ, Iglesias M, Oro LA (2018) Organometallics 37:3611
185. Hall DG (ed) (2011) Boronic acids: preparation and applications in organic synthesis, medicine and materials.2nd edn. Wiley-VCH Verlag Gmbh, Weinheim
186. Rablen PR, Hartwig JF, Nolan SP (1994) J Am Chem Soc 116:4121
187. Rablen PR, Hartwig JF (1996) J Am Chem Soc 118:4648
188. Ishiyama T, Miyaura N (2006) Pure Appl Chem 78:1369
189. Hartwig JF (2011) Chem Soc Rev 40:1992
190. Hartwig JF (2012) Acc Chem Res 45:864
191. Xu L, Wang G, Zhang S, Wang H, Wang L, Liu L, Jiao J, Li P (2017) Tetrahedron 73:7123
192. Iverson CN, Smith III MR (1999) J Am Chem Soc 121:7696
193. Cho J-Y, Iverson CN, Smith III MR (2000) J Am Chem Soc 122:12868
194. Takagi J, Sato K, Hartwig JF, Ishiyama T, Miyaura N (2002) Tetrahedron Lett 43:5649
195. Ishiyama T, Takagi J, Yonekawa Y, Hartwig JF, Miyaura N (2003) Adv Synth Catal 345:1103
196. Ishiyama T, Nobuta Y, Hartwig JF, Miyaura N (2003) Chem Commun:2924
197. Mertins K, Zapf A, Beller M (2004) J Mol Catal A Chem 207:21
198. Chotana GA, Rak MA, Smith III MR (2005) J Am Chem Soc 127:10539
199. Paul S, Chotana GA, Holmes D, Reichle RC, Maleczka Jr RE, Smith III MR (2006) J Am Chem Soc 128:15552
200. Chotana GA, Kallepalli VA, Maleczka Jr RE, Smith III MR (2008) Tetrahedron 64:6103
201. Kallepalli VA, Shi F, Paul S, Onyeozili EN, Maleczka Jr RE, Smith III MR (2009) J Org Chem 74:9199
202. Preshlock SM, Ghaffari B, Maligres PE, Krska SW, Maleczka Jr RE, Smith III MR (2013) J Am Chem Soc 135:7572
203. Loach RP, Fenton OS, Amaike K, Siegel DS, Ozkal E, Movassaghi M (2014) J Org Chem 79:11254
204. Batool F, Parveen S, Emwas AH, Sioud S, Gao X, Munawar MA, Chotana GA (2015) Org Lett 17:4256
205. Kato S, Serizawa Y, Sakamaki D, Seki S, Miyake Y, Shinokubo H (2015) Chem Commun 51:16944
206. Maegawa Y, Inagaki S (2015) Dalton Trans 44:13007
207. Miyamura S, Araki M, Suzuki T, Yamaguchi J, Itami K (2015) Angew Chem Int Ed 54:846
208. Li HL, Kanai M, Kuninobu Y (2017) Org Lett 19:5944
209. Asghar S, Shahzadi T, Alazmi M, Gao X, Emwas A-H, Saleem RSZ, Batool F, Chotana GA (2018) Synthesis 50:2211
210. Shahzadi T, Saleem R, Chotana GA (2018) Synthesis 50:4336
211. Wright SE, Richardson-Solorzano S, Stewart TN, Miller CD, Morris KC, Daley CJA, Clark TB (2019) Angew Chem Int Ed 58:2834
212. Kato S, Akahori S, Serizawa Y, Lin X, Yamauchi M, Yagai S, Sakurai T, Matsuda W, Seki S, Shinokubo H, Miyake Y (2020) J Org Chem 85:62
213. Kallepalli VA, Gore KA, Shi F, Sanchez L, Chotana GA, Miller SL, Maleczka Jr RE, Smith III MR (2015) J Org Chem 80:8341

214. Yamamoto T, Ishibashi A, Suginome M (2017) Org Lett 19:886
215. Yang Y, Gao Q, Xu S (2019) Adv Synth Catal 361:858
216. Murata M, Odajima H, Watanabe S, Masuda Y (2006) Bull Chem Soc Jpn 79:1980
217. Nguyen DH, Perez-Torrente JJ, Lomba L, Jimenez MV, Lahoz FJ, Oro LA (2011) Dalton Trans 40:8429
218. Cabeza JA, García-Álvarez P, González-Álvarez L (2017) Chem Commun 53:10275
219. Miller SL, Chotana GA, Fritz JA, Chattopadhyay B, Maleczka Jr RE, Smith III MR (2019) Org Lett 21:6388
220. Liskey CW, Hartwig JF (2013) J Am Chem Soc 135:3375
221. Lee CI, Zhou J, Ozerov OV (2013) J Am Chem Soc 135:3560
222. Lee C-I, Shih W-C, Zhou J, Reibenspies JH, Ozerov OV (2015) Angew Chem Int Ed 54:14003
223. Lee CI, DeMott JC, Pell CJ, Christopher A, Zhou J, Bhuvanesh N, Ozerov OV (2015) Chem Sci 6:6572
224. Kosanovich AJ, Jordan AM, Bhuvanesh N, Ozerov OV (2018) Dalton Trans 47:11619
225. Valencia M, Pereira A, Müller-Bunz H, Belderraín TR, Pérez PJ, Albrecht M (2017) Chem Eur J 23:8901
226. Ge C, Sang X, Yao W, Zhang L, Wang D (2018) Green Chem 20:1805
227. Bernskoetter WH, Brookhart M (2008) Organometallics 27:2036
228. Hartwig JF, Romero EA (2019) Tetrahedron 75:4059
229. Cheng C, Hartwig JF (2015) Chem Rev 115:8946
230. Xu Z, Xu L-W, Huang W-S, Zhang J (2015) Synthesis 47:3645
231. Mita T, Michigami K, Sato Y (2012) Org Lett 14:3462
232. Mita T, Michigami K, Sato Y (2013) Chem Asian J 8:2970
233. Ishiyama T, Saiki T, Kishida E, Sasaki I, Ito H, Miyaura N (2013) Org Biomol Chem 11:8162
234. Murai M, Takeuchi Y, Takai K (2017) Chem Lett 46:1044
235. Manna K, Zhang T, Greene FX, Lin W (2015) J Am Chem Soc 137:2665
236. Julián A, Polo V, Jaseer EA, Fernández-Alvarez FJ, Oro LA (2015) ChemCatChem 7:3895
237. Kraushaar K, Schmidt D, Schwerzer A, Kroke E (2014) Adv Inorg Chem 66:117
238. Julián A, Garcés K, Lalrempuia R, Jaseer EA, García-Orduña P, Fernández-Alvarez FJ, Lahoz FJ, Oro LA (2018) ChemCatChem 10:1027
239. Ventura-Espinosa D, Sabater S, Carretero-Cerdán A, Baya M, Mata JA (2018) ACS Catal 8:2558

Top Organomet Chem (2021) 69: 67–152
https://doi.org/10.1007/3418_2020_59

Published online: 24 December 2020

Recent Advances in Iridium-Catalysed Transfer Hydrogenation Reactions

M. Pilar Lamata, Vincenzo Passarelli, and Daniel Carmona

Contents

Abstract This review focuses on the contributions of the last 5 years to the application of iridium complexes as homogeneous catalysts in transfer

---

M. Pilar Lamata and D. Carmona (✉)
Instituto de Síntesis Química y Catálisis Homogénea, CSIC-Universidad de Zaragoza, Zaragoza, Spain
e-mail: plamata@unizar.es; dcarmona@unizar.es

V. Passarelli
Instituto de Síntesis Química y Catálisis Homogénea, CSIC-Universidad de Zaragoza, Zaragoza, Spain

Centro Universitario de la Defensa, Zaragoza, Spain
e-mail: passarel@unizar.es

hydrogenation (TH) reactions. The reduction of carbonyls, imines, alkenes and alkynes is considered. The TH of unsaturated alkene-carbonyl substrates and heterocycles is particularly studied. Recent results on the reduction of CO_2 are also included. Special attention is paid to THs performed in aqueous medium as well as to the development of TH in biological media. The employ of biomass-derived products as reagents or solvents in TH transformations is also reviewed. Finally, the proposed mechanisms for TH reactions are revised.

Keywords Aldehydes · Alkenes · Alkynes · Biological transfer hydrogenation · Carbene complexes · Half-sandwich complexes · Ketones · Mechanisms · *N*-Heterocycles · Pincer complexes · Sustainability · *α,β*-Unsaturated substrates

1 Introduction

Transfer hydrogenation (TH) reaction refers to the addition of hydrogen to an unsaturated molecule from a sacrificial hydrogen donor other than H_2 usually with the aid of a catalyst (Scheme 1).

The use of readily available, inexpensive and easy to handle hydrogen donors avoids the risks associated with hazardous pressurized hydrogen, does not require special experimental setups and allows the selection of the most appropriate donor.

The first antecedent of this reaction is the Meerwein-Ponndorf-Verley reduction of alhehydes and ketones with alcohols. In 1925, Meerwein and Schmidt performed the reduction of aldehydes with ethanol in the presence of aluminium ethoxide [1]. In the same year, Verley reported the reduction of butyraldehyde by geraniol using the same catalyst [2]. One year later, Ponndorf published the reduction of ketones by secondary alcohols catalysed by aluminium isopropoxide [3].

An important step in the evolution of the TH reaction involved the incorporation in the 1960s of transition metal compounds as catalysts. As the first examples, Mitchell, Henbest and co-workers reported the reduction of cyclohexanones and *α,β*-unsaturated ketones by TH from 2-propanol by using iridium compounds as catalysts [4, 5]. Shortly after, Sasson and Blum showed that $[RuCl_2(PPh_3)_3]$ promotes the selective TH of C=C bonds in *α,β*-unsaturated carbonyl compounds by using primary and secondary alcohols as hydrogen donors [6, 7]. Aromatic aldehydes may also be used as hydrogen donors for this reaction which can be also catalysed by $[RhCl(PPh_3)_3]$ or $[IrCl(CO)(PPh_3)_2]$, although in lower yield [8].

An important breakthrough came some years later when Bäckwall and Chowdhury discovered that the TH of both aliphatic and aromatic ketones with

Scheme 1 Catalysed transfer hydrogenation reaction

$$A + DH_2 \xrightarrow{\text{catalyst}} AH_2 + D$$

A = Unsaturated molecule; DH_2 = Hydrogen donor

2-propanol catalysed by $[RuCl_2(PPh_3)_3]$ was significantly accelerated upon adding a catalytic amount of NaOH [9]. Furthermore, in 1995, Noyori and co-workers reported that η^6-arene complexes of ruthenium containing the ligand (*1S,2S*)-(*N-p*-toluenesulfonyl)-1,2-diphenylethylenediamine [(*S,S*)-TsDPEN] efficiently catalysed the TH of aromatic ketones with 2-propanol affording high enantiomeric ratios (e.r.) [10]. The same group discovered that the TH reaction took place through a new bifunctional mechanism that involved cooperation between the metal and an NH moiety of the ligand [11].

These two important findings fuelled an extraordinary flourishing of the TH reaction which still continues nowadays. On the one hand, the scope of the substrates to be hydrogenated and that of the hydrogen donors to be used have been extended, asymmetric versions have been developed and water has been incorporated as one of the solvents to be taken into account. On the other hand, metals and the class of compounds employed as catalysts have expanded notably. So, along with complexes of Ru, Rh and Ir, it has been shown that compounds of Ni [12, 13], Pd [14, 15], Re [16], Os [17, 18], Pt [19, 20], Au [21, 22], Mo [23, 24], Zn [25, 26], Zr [27, 28] and Bi [29] also efficiently catalyse the TH of a broad variety of unsaturated substrates. In particular, in recent years, more abundant, cheap and healthy transition metal such as Fe [12, 30–39], Co [12, 40, 41] and Mn [42–55] have attracted increasing attention.

Iridium compounds have played a key role in the development of TH, allowing for the achievement of important milestones. Among the most important advances propelled by iridium compounds, it is worth mentioning the TH of linear imines. Back in 1996, Noyori's group reported that ruthenium complexes containing (*S,S*)-TsDPEN as a ligand could catalyse the asymmetric TH of cyclic imines with outstanding enantioselectivities [56]. However, it took until 2010 to have efficient catalysts for the TH of acyclic imines. In that year, Xiao and co-workers reported that the metallacycle complex resulting from the reaction of $[Cp^*IrCl_2]_2$ with the imine 4-methoxy-*N*-(1-phenylethylidene)aniline (complex **A**) was highly active for the TH of several acyclic imines using formic acid as a hydrogen source [57]. Since then a series of related iridacycles have been prepared and have shown excellent activity in reducing acyclic imines in a variety of reactions.

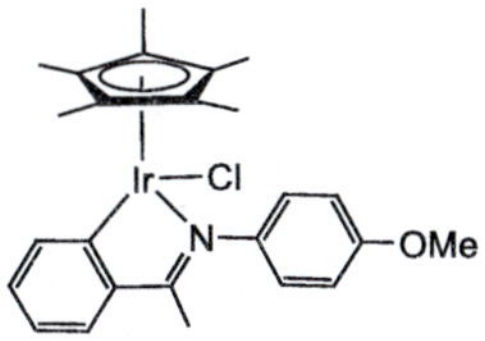

A

Another relevant example of the remarkable features of TH iridium catalysts comes from the employment of intracellular NADH as a hydrogen donor in order to promote anticancer activity. As we will see in what follows, half-sandwich iridium compounds are among the most efficient in catalysing NADH oxidation. In particular, compound **B** catalyses the reaction of dioxygen and NADH yielding H_2O_2, a reactive oxygen species that causes cellular oxidative stress and shows anticancer activity higher than cisplatin [58].

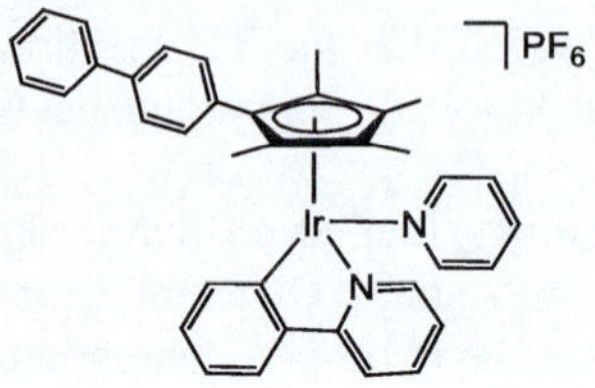

B

Although the first good results in the reduction of *N*-heterocyclic compounds were obtained with rhodium-based catalysts [59], the use of iridium compounds was gaining more and more importance in the reduction of this type of substrates. As we will see below, pyridines, quinolines, isoquinolines, dehydroisoquinolines, quinoxalines, indoles and phenanthrolines are effectively reduced for a variety of iridium complexes bearing *N*-donor ligands.

In the present report, we summarise the progress made thanks to complexes based on iridium in TH reactions. The chapter covers the work published in this area in the 5-year period from 2015 to January 2020. Previous contributions have been included in the excellent review by Astruc and Wang published in 2015 [60].

The work is structured in sections that classify the results obtained according to the type of ligand. Thus, in the first section, we consider the pentamethylcyclopentadienyl ligand that under η^5 coordination originates semisandwich compounds. This ample section contains subsections specifically dedicated to TH of CO_2, TH in water or to the fascinating biological TH. Subsequent sections are devoted to the study of catalysts containing carbene or pincer ligands. In parallel, relevant topics such as TH of *α,β*-unsaturated alkene-carbonyl substrates, TH of *N*-heterocycles or TH and sustainability are treated in separate sections. To finish, in the last section, the mechanistic aspects of the process under study are discussed.

It is worth noting that Schemes 3, 4, 5, and 6 list the aldehydes (Scheme 3), ketones (Scheme 4), imines (Scheme 5) and alkenes (Scheme 6) that have been successfully tested in TH reactions. Within each scheme, the substrates are grouped into families, e. g. aliphatic aldehydes and aromatic aldehydes in Scheme 3, 1,2-disubstituted alkynes and terminal alkynes in Scheme 6 and so on.

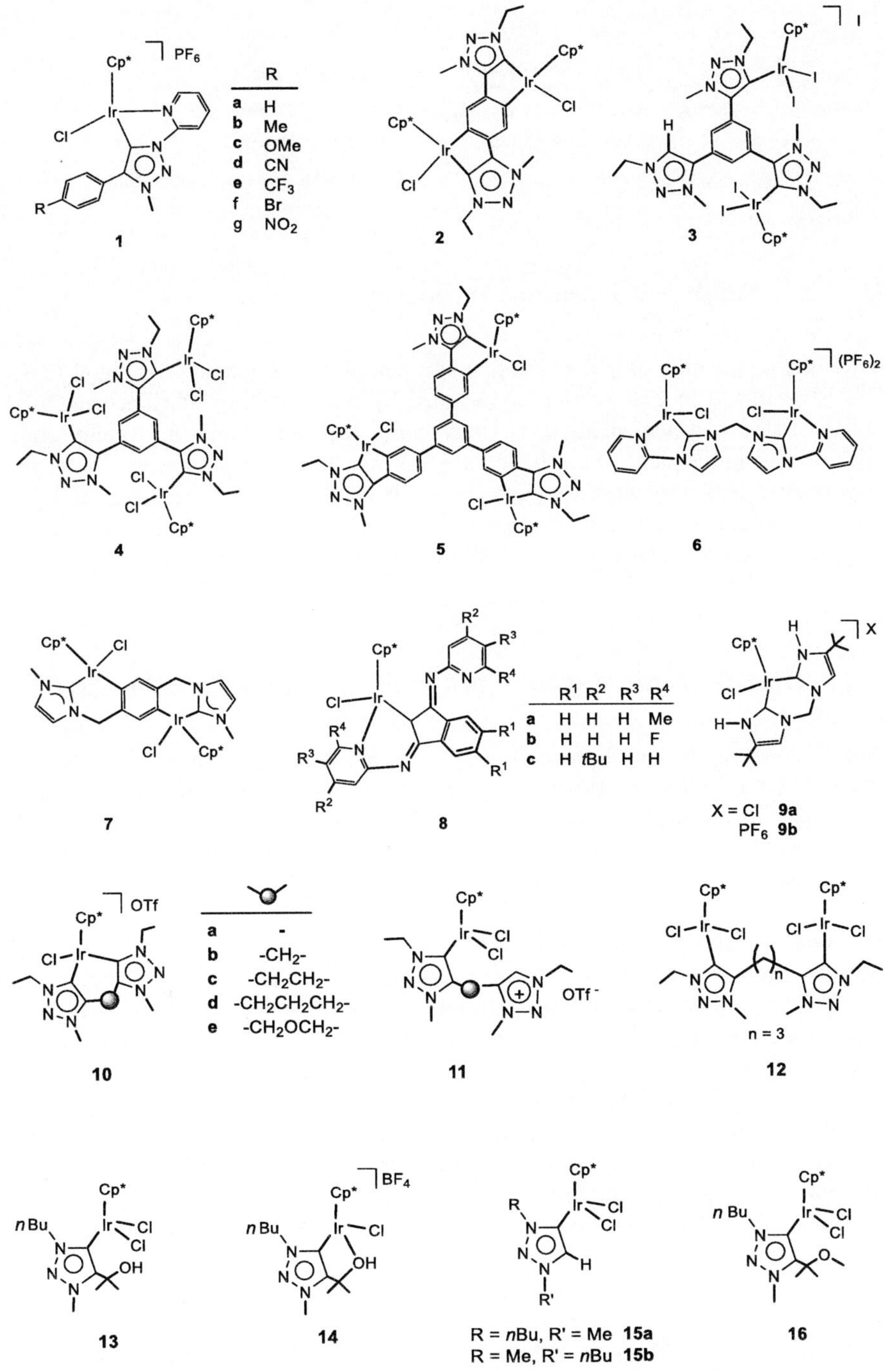

Scheme 2 Half-sandwich iridium catalyst precursors (**1–18**)

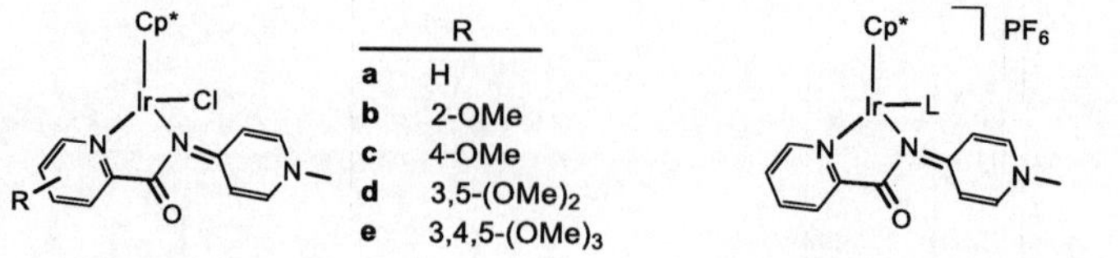

17

L = MeCN (**18a**), DMSO (**18b**)

Scheme 2 (continued)

2 Half-Sandwich Iridium Complexes

After the application of η^6-arene ruthenium complexes containing the chiral (*S,S*)-TsDPEN ligand in TH reactions developed by Noyori, Tani, Mashima and Abe demonstrated that pentamethylcyclopentadienyl (Cp*) rhodium and iridium complexes bearing the same chiral ligand are also excellent catalyst precursors for the asymmetric reduction of ketonic substrates [61]. Since then, a number of Cp*-iridium complexes have been successfully applied as TH catalysts, and the class of hydrogenated substrates has been widely extended.

2.1 *Transfer Hydrogenation of C=O, C=N and C=C Bonds*

Transfer hydrogenation of aldehydes, ketones and imines with molecular metallic catalysts is one of the most powerful and versatile tool to access alcohols and amines. Asymmetric versions complete the potential of this methodology. Although much less studied, alkenes can also be hydrogenated by TH. TH in water of the double bonds indicated in the Sect. 2.1 will be treated in Sect. 2.3 of this work.

Sarkar's group reported that mononuclear **1** [62, 63], dinuclear **2, 3** [64, 65] and trinuclear **4, 5** [65, 66] Cp*Ir complexes containing multidentate mesoionic carbene ligands (Scheme 2) promoted the reduction of benzaldehyde (Scheme 3), acetophenone, benzophenone and cyclohexanone (Scheme 4). 2-Propanol was employed as both a hydrogen source and a solvent and KOH as a promoter. At 100°C, benzaldehyde was quantitatively reduced to benzyl alcohol after 3 h of reaction, at 0.01 mol% of precatalyst loading. Under the same conditions, lower conversions to the corresponding alcohol (55–92%) were achieved for the tested ketones. Mononuclear iridium complexes **1** bearing electron-withdrawing groups at the carbene ligands are catalysts superior to the related complexes with electron-donating substituents. For the homologous ruthenium and osmium complexes, the influence of the substituents was opposite [62, 63].

Transfer hydrogenation of *N*-benzylideneaniline (Scheme 5) and of the olefins cyclooctene (**D30**), *trans-β*-methylstyrene (**D10**) and trans-stilbene (**D11**) (Scheme 6) with 2-propanol was also tested with precatalysts **1a**, **1c** and **1e**. At 100°C, olefins **D10**, **D11**, **D30** were reduced at 1.0 mol% of catalyst precursor

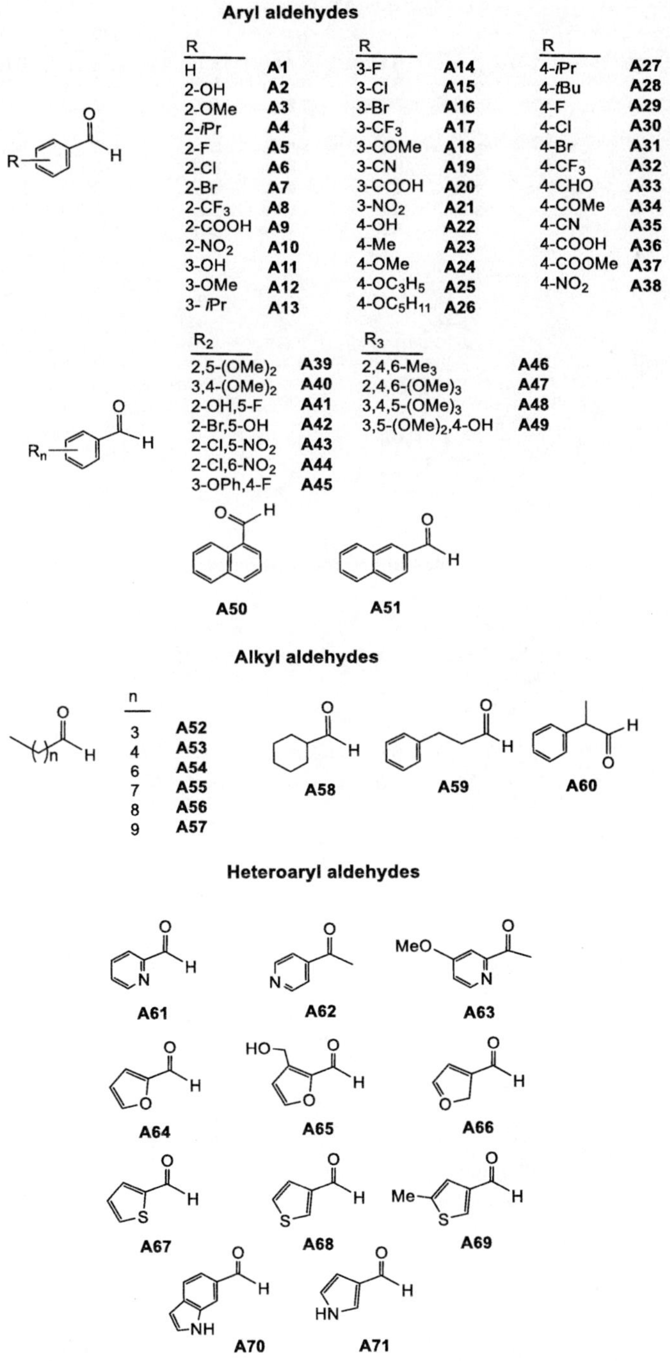

Scheme 3 Aldehydes

Methyl aryl ketones

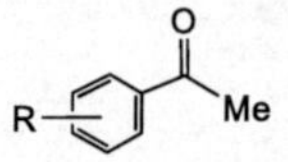

R		R		R	
H	**B1**	3-OMe	**B10**	4-SMe	**B19**
2-Me	**B2**	3-Cl	**B11**	4-F	**B20**
2-OMe	**B3**	3-Br	**B12**	4-Cl	**B21**
2-F	**B4**	3-CF_3	**B13**	4-Br	**B22**
2-Cl	**B5**	3-NO_2	**B14**	4-CF_3	**B23**
2-Br	**B6**	4-Me	**B15**	4-CN	**B24**
2-CF_3	**B7**	4-*t*Bu	**B16**	4-COOH	**B25**
3-Me	**B8**	4-OH	**B17**	4-COOMe	**B26**
3-OH	**B9**	4-OMe	**B18**	4-NO_2	**B27**

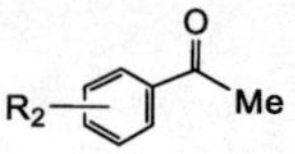

R_2		R_2	
3,4-Me_2	**B28**	2,4-Cl_2	**B31**
3,5-Me_2	**B29**	2-Cl,4-F	**B32**
2,4-$(OMe)_2$	**B30**	2,5-F_2	**B33**
		3,5-$(CF_3)_2$	**B34**

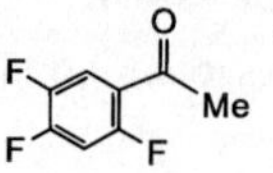
B35

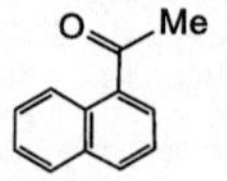
B36

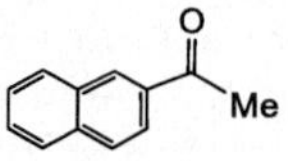
B37

Methyl heteroaryl ketones

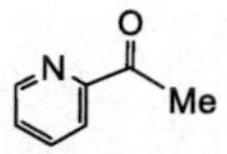
B38

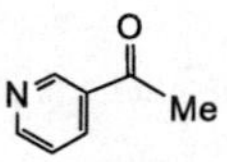
B39

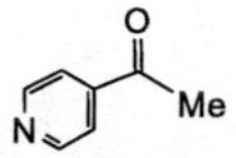
B40

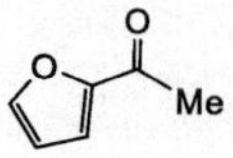
B41

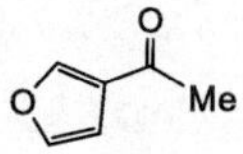
B42

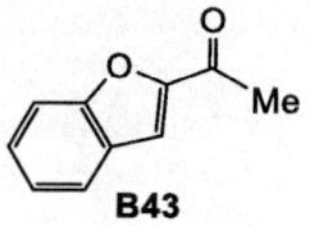
B43

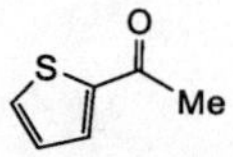
B44

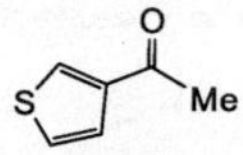
B45

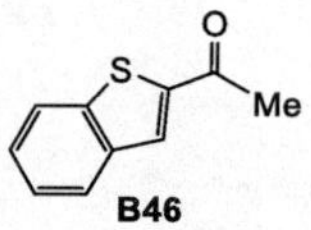
B46

Alkyl aryl ketones

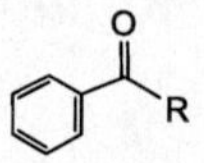

R		R		R	
Et	**B47**	CH_2OH	**B53**	Bn	**B59**
*n*Pr	**B48**	CH_2OCOMe	**B54**	CPh_3	**B60**
*i*Pr	**B49**	CH_2Cl	**B55**	COOEt	**B61**
*t*Bu	**B50**	$CHCl_2$	**B56**	CH_2COOEt	**B62**
*n*Hex	**B51**	CH_2F	**B57**	CH_2CN	**B63**
Cy	**B52**	CF_3	**B58**		

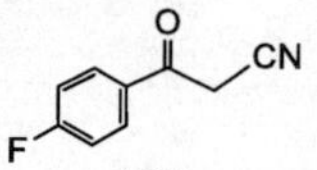
B64

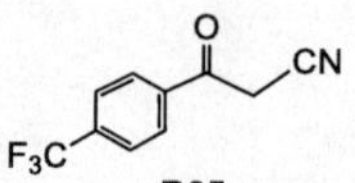
B65

Scheme 4 Ketones

Diaryl ketones

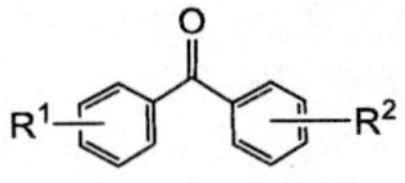

R^1	R^2	
H	H	**B66**
H	4-Br	**B67**
2-Me	H	**B68**
2-F	H	**B69**
4-Me	4-Me	**B70**
4-OMe	H	**B71**
4-CF_3	H	**B72**

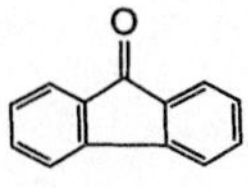

B73

Aryl heteroaryl ketones

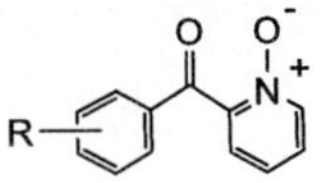

R	
H	**B74**
3-Me	**B75**
3-OMe	**B76**
3-CF_3	**B77**
4-Me	**B78**
4-F	**B79**
4-Cl	**B80**
4-Br	**B81**
4-CF_3	**B82**
3,5-Me_2	**B83**
3,5-$(OMe)_2$	**B84**
3,5-$(CF_3)_2$	**B85**

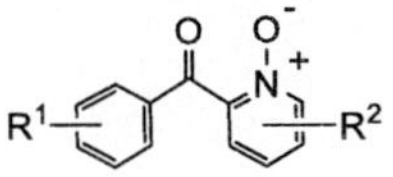

R^1	R^2	
H	3-Me	**B86**
H	4-Me	**B87**
3,5-$(CF_3)_2$	3-Me	**B88**
3,5-$(CF_3)_2$	4-Me	**B89**

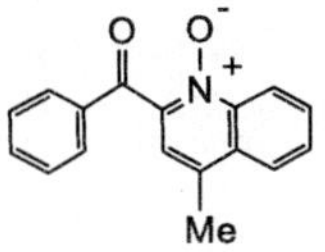

B90

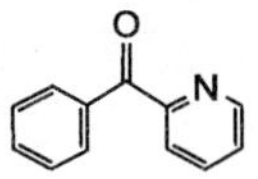

B91

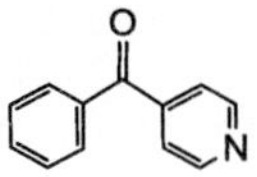

B92

Methyl alkyl ketones

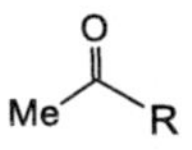

R	
Me	**B93**
Et	**B94**
*n*Pr	**B95**
*i*Pr	**B96**
*n*Bu	**B97**
*i*Bu	**B98**
*t*Bu	**B99**
*n*Pent	**B100**
*n*Hex	**B101**
*n*Hep	**B102**
*n*Oct	**B103**
Cy	**B104**
CH_2CH_2OH	**B105**
CH_2OPh	**B106**
$(CH_2)Ph$	**B107**
CH_2COOEt	**B108**

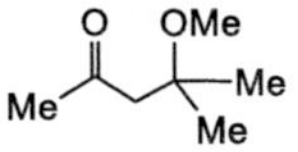

B109

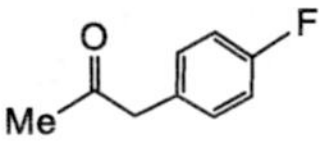

B110

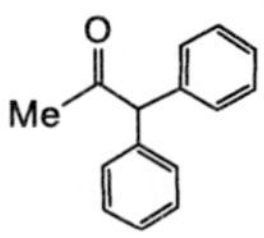

B111

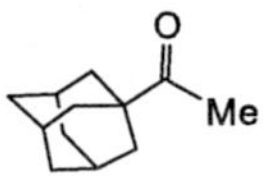

B112

Scheme 4 (continued)

Dialkyl ketones

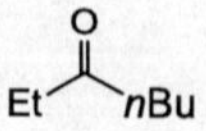

B113

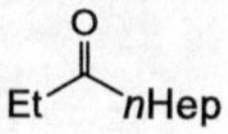

B114

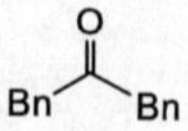

B115

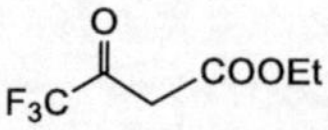

B116

Cyclic ketones

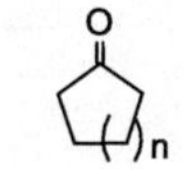

n	
1	**B117**
2	**B118**
3	**B119**
4	**B120**

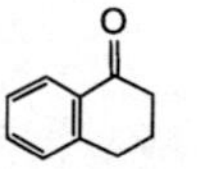

B121

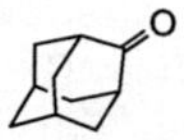

B122

Alkynyl ketones

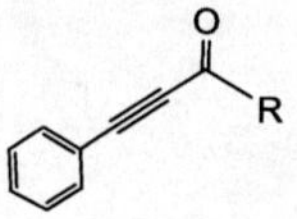

R		R	
Me	**B123**	CF_3	**B126**
Et	**B124**	$(CH_2)_2COOEt$	**B127**
*i*Pr	**B125**	$(CH_2)_3COOEt$	**B128**

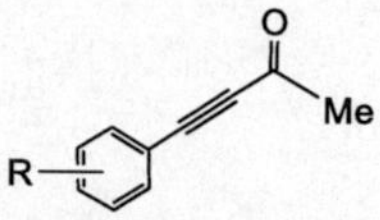

R		R	
2-OMe	**B129**	4-Me	**B133**
2-Cl	**B130**	4-OMe	**B134**
3-Me	**B131**	4-Cl	**B135**
3-Cl	**B132**	4-Br	**B136**

α-Ketophosphonates

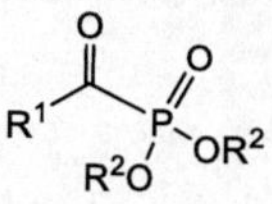

R^1	R^2		R^1	R^2	
Ph	Me	**B137**	4-(OMe)C_6H_4	*i*Pr	**B143**
Ph	Et	**B138**	4-FC_6H_4	*i*Pr	**B144**
Ph	*i*Pr	**B139**	4-ClC_6H_4	*i*Pr	**B145**
2-BrC_6H_4	*i*Pr	**B140**	4-BrC_6H_4	*i*Pr	**B146**
3-BrC_6H_4	*i*Pr	**B141**	Bn	*i*Pr	**B147**
4-MeC_6H_4	*i*Pr	**B142**	$(CH_2)_2Ph$	*i*Pr	**B148**

Scheme 4 (continued)

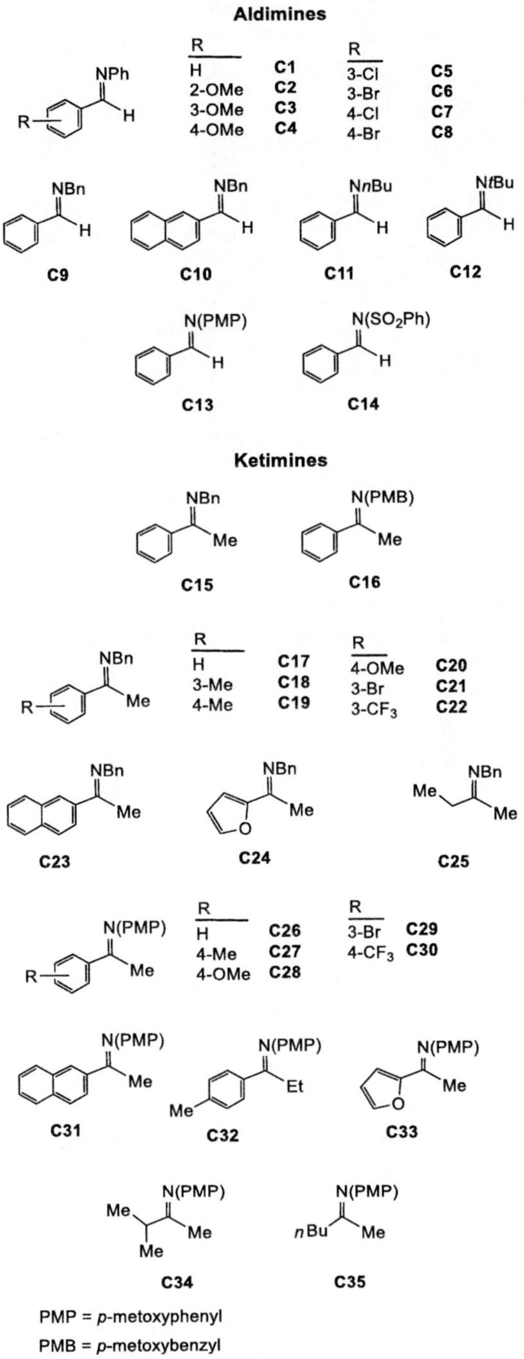

Scheme 5 Imines

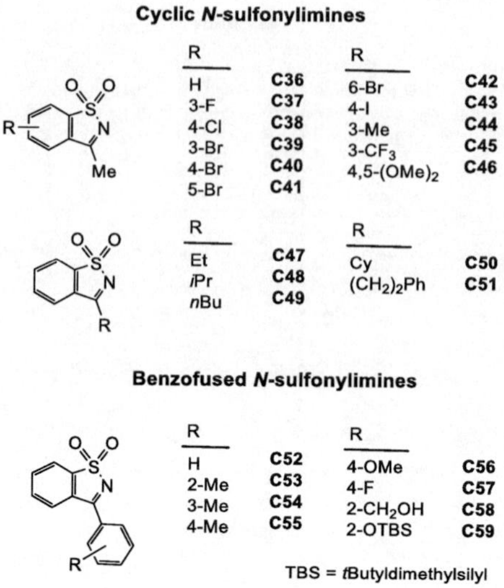

Scheme 5 (continued)

loading. Conversions up to 80% were achieved after 24 h of treatment. With 1.0 mol% precatalyst loading, the same compounds converted *N*-benzylideneaniline into *N*-benzylaniline (17–50%) after 18 h of treatment at 100°C [63].

The dinuclear half-sandwich iridium complexes **6** and **7** having a bridging bis-*N*-heterocyclic carbene ligand catalysed the TH of methyl aryl (**B1**, **B10**), diaryl (**B66**), dialkyl (**B102**, **B113**) and cyclic (**B117**, **B118**) ketones (Scheme 4). Reactions were carried out in refluxing 2-propanol, under basic conditions (*i*PrONa, 3 mol%) and at 0.5 mol% of catalyst loading. After several hours, conversions in the range 5–100% were achieved, the cationic complex **6** being more active than the neutral *ortho*-metallated complex **7** [67].

Half-sandwich iridium complexes **8a-c** bearing di- and tridentate bis(2-pyridylimino)isoindolato ligands are catalytically active in the TH of ketones and imines. In refluxing 2-propanol, with 0.5 equiv. of KOH and at 1 or 5 mol% of precatalyst loading, ketones **B1**, **B18**, **B27**, **B37**, **B66** and **B118** (Scheme 4) were converted to the corresponding hydrogenated species in good yield. Under similar conditions, TH of the imines **C1** and **C14** (Scheme 5) afforded the corresponding amines in 53 and 16% yield, respectively, after 4 h of reaction [68].

Reduction of acetophenone was carried out at 95°C in 2-propanol and KOH as a base. At 2 mol% loading of the precatalysts **9a** and **9b** conversions of 96 and 98%, respectively, was achieved after 24 h of reaction [69].

The half-sandwich complexes **10–12** (Scheme 2) with monodentate, bidentate chelating or bidentate bridging ditriazolylidine ligands containing alkyl or ether linkers of different lengths between the triazolium heterocycles were investigated as catalyst precursors for the TH of benzophenone. At a substrate/base/catalyst molar ratio of 100:10:1, in refluxing 2-propanol, yields in the range 30–98% were obtained

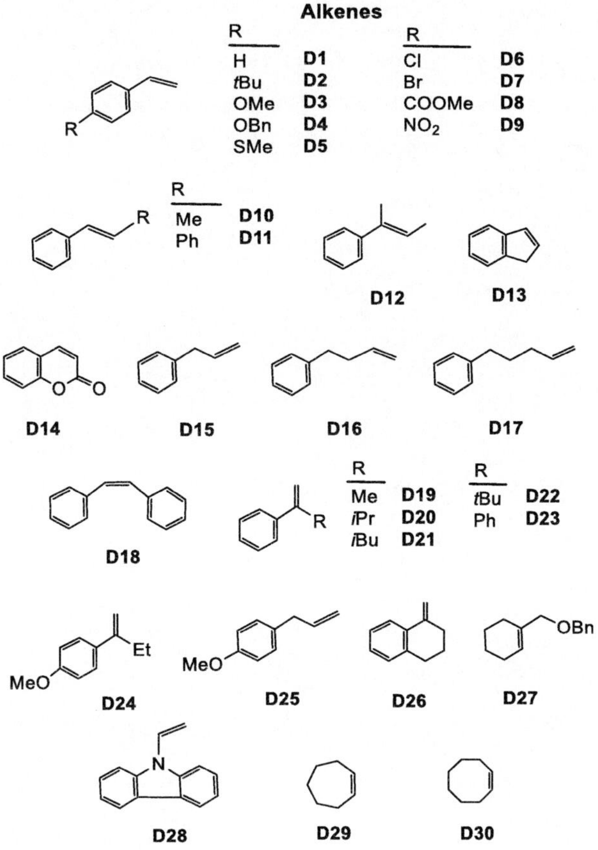

Scheme 6 Alkenes and alkynes

after 4 h of reaction. Activity generally increases with the length of the linker. Bimetallic complex **12** is among the least active species [70].

The catalytic activity of complexes **13–16** was evaluated in TH under standard conditions, i.e. refluxing 2-propanol as the sacrificial hydrogen source and KOH (10 mol%) and at 2 mol% of catalyst loading. Benzophenone was taken as the model substrate. In general, good conversions (35–97%) were obtained after 8 h of treatment [71]. The presence of the hydroxyl functionality enhances the reactivity of **13** and **14** with respect to that of the unfunctionalised triazolylidene ligand in **15**. It has been suggested [71] that, in basic medium, a metal-bound alkoxide could be formed. Such alkoxide would increase the electron density at the iridium centre and enhance the activity in TH. Alternatively, the alkoxy group may facilitate the dihydrogen abstraction from 2-propanol through an outer sphere mechanism.

A set of aryl-substituted pyridylideneamide (PYA) compounds has been used as ligands in combination with the Cp*Ir moiety affording complexes **17** and **18**

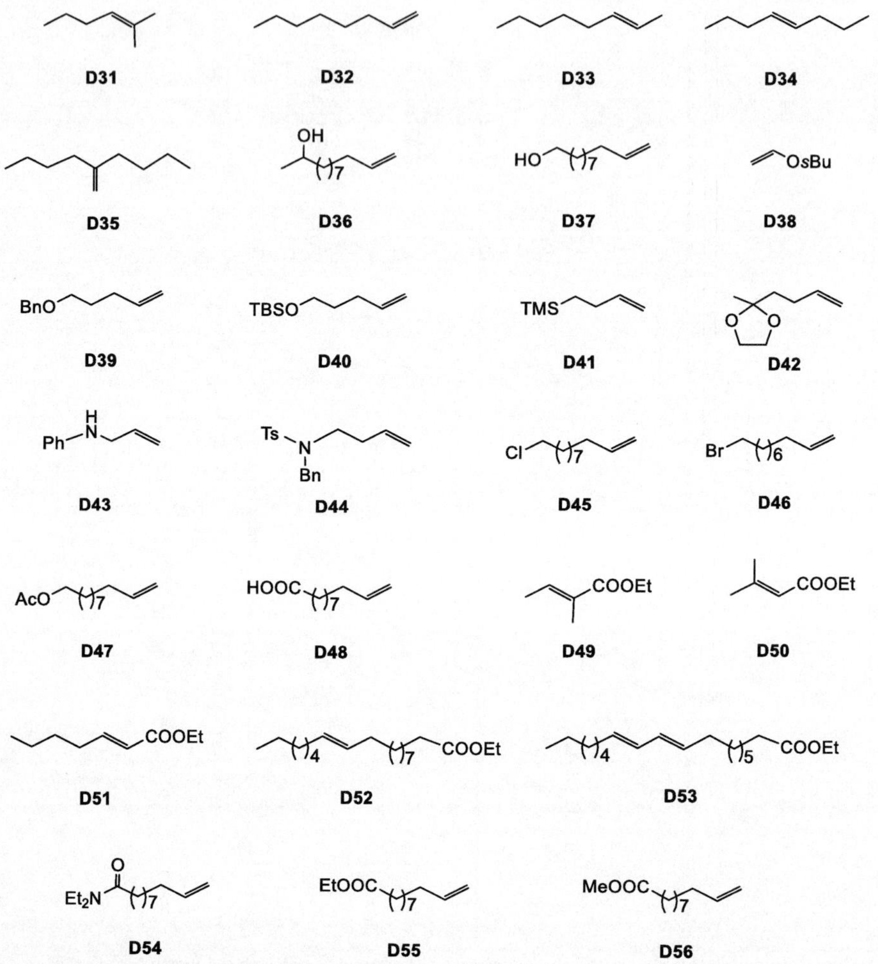

TBSO = Tris(tertbutoxy)silanoxy
TMS = Trimethylsilyl
Ts = Tosyl

Scheme 6 (continued)

[72]. Up to three methoxy substituents have been incorporated to the phenyl ring of the *C,N*-bidentate chelating PYA (complexes **17b-e**, Scheme 2). Under optimised conditions, these complexes catalyse the TH of benzophenone using 2-propanol as a hydrogen source. Incorporation of one methoxy substituent to the phenyl ring of the PYA ligand markedly improved the catalytic activity from 27% conversion for **17a** to 99 and 96% for **17b** and **17c**, respectively, at 4 h of reaction. Incorporation of a second methoxy group further enhances the catalytic rate (93% conversion for

1,2-Disubstituted alkynes

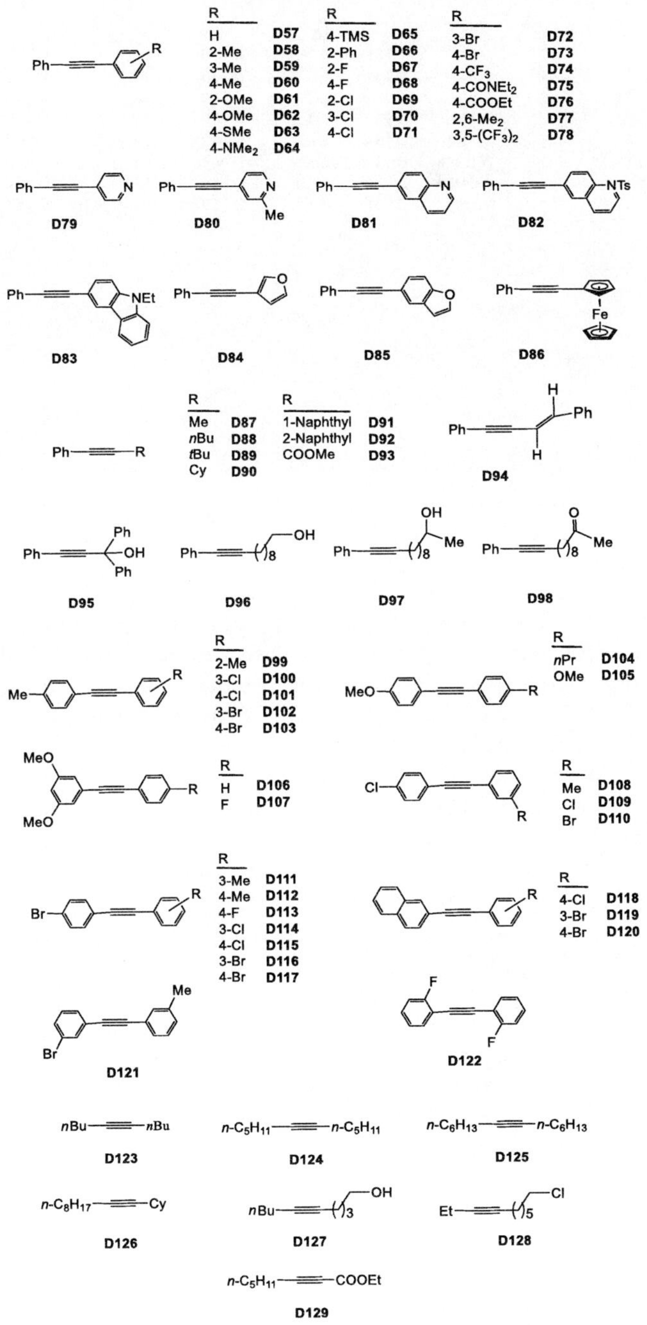

Scheme 6 (continued)

Terminal alkynes

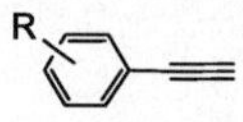

R		R		R	
H	**D130**	4-Ph	**D138**	3-Br	**D146**
2-Me	**D131**	2-F	**D139**	4-Br	**D147**
3-Me	**D132**	3-F	**D140**	$3\text{-}CF_3$	**D148**
4-Me	**D133**	4-F	**D141**	$4\text{-}CF_3$	**D149**
2-OMe	**D134**	2-Cl	**D142**	4-CN	**D150**
3-OMe	**D135**	3-Cl	**D143**	$3\text{-}NO_2$	**D151**
4-OMe	**D136**	4-Cl	**D144**	$4\text{-}NO_2$	**D152**
4-*t*Bu	**D137**	2-Br	**D145**	$3,5\text{-}(CF_3)_2$	**D153**

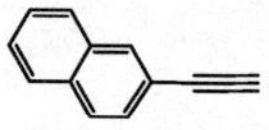

D154

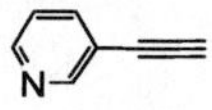

D155

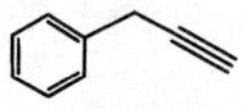

D156

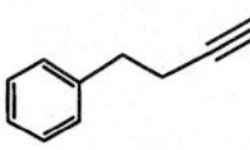

D157

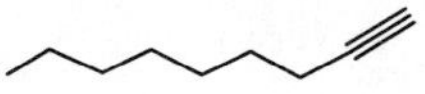

D158

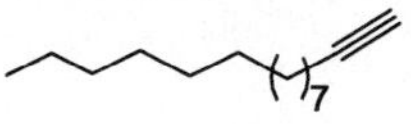

D159

Scheme 6 (continued)

complex **17d**, after 2 h) probably due to the increase of the electron density at the iridium centre. However, in the presence of a third methoxy group, complex **17e**, the conversion drops even below that of the unsubstituted complex **17a** (30% conversion, at 24 h of reaction). The relative low activity of **17e** has been tentatively attributed to steric reasons [72].

Complex **17d** has been tested as the catalyst precursor in the TH of ketones **B1**, **B18**, **B20**, **B38-B40** and **B118** (Scheme 4). Under standard conditions good conversions were achieved in all cases. Electron-withdrawing substituents on the aromatic ring (**B20**, **B38-B40**) increased the reduction rate, and electro-donating substituents (**B18**) had the opposite effect.

Imines were also investigated as substrates for TH under the standard conditions. Complexes **17a–17e** catalysed the hydrogenation of *N*-benzylideneaniline to the corresponding amine in 80–94% conversion after 4 h of treatment. With these substrates, the different degree of substitution at the phenyl ring of the PYA ligand does not affect significantly the catalytic activity. Complex **17b** hydrogenates aldimines **C4**, **C8**, **C9**, **C12** and **C14** (Scheme 5) in yields up to 90%, after 24 h of reaction. Notably, ketimine **C15** was not converted at all, and it was suggested that substitution at the *α* carbon may hamper or even prevent substrate binding due to steric constraints.

The iridium complex **17e** was investigated in the TH of aldehydes. Under standard conditions, i.e. refluxing 2-propanol, 1 mol% catalyst loading and 10 mol% KOH, essentially full conversion of benzaldehyde was achieved within 5 min; however the yield of benzyl alcohol was only 50%, and benzoic acid and benzoate were also obtained. When this reaction was carried out without a base, aldehyde conversion was slower (85%, 24 h), but it was completely selective to benzyl alcohol. In the absence of a base, other aryl aldehydes such as **A24**, **A29** and **A35** were also selectively converted to the corresponding alcohol. Both electron-donating and electron-withdrawing substituents at the phenyl group of the PYA ligand slow down the catalysis. In contrast to the behaviour of ketones, 2-pyridinecarboxaldehyde and the aliphatic aldehyde **A54** (Scheme 3) were not hydrogenated in the absence of a base. Probably, under base-free conditions, coordination of the pyridine nitrogen to the iridium centre is favoured thus preventing the catalytic TH [72].

Chiral amidoiridium complexes **19–21** (Scheme 7) catalyse the asymmetric TH of acetophenone in 2-propanol, in the absence of a base, with a substrate/catalyst (S/C) ratio of 200 at 30°C [73]. The mononuclear amido complexes **20** and **21** reached almost complete conversion within 30 min. The catalytic activity of the dinuclear complex **19** is remarkably lower. Catalytic rates can be related to the relative ability of complexes **19–21** to dehydrogenate 2-propanol. The bridging complex **19** is inert in 2-propanol, but the mononuclear complexes **20** and **21** were completely converted into diastereomeric mixtures of the amino complexes **A** and **B**, respectively, by addition of excess of 2-propanol (Scheme 8). The (*S*)-product was obtained in all cases with an e.r. of about 80/20.

The $C_{carborane}$-cyclometallated complex **22** and the $B_{carborane}$-cyclometallated complexes **23** and **24** reduce acetophenone to 2-phenylethanol, in the presence of *t*BuOH, with 2-propanol as a hydrogen source and at a catalyst loading of 0.5–1.0 mol%. Conversions greater than 90% were achieved after 1 h of reaction. Cyclometallation of the carborane moiety enhances catalysis greatly compared to the non-cyclometallated counterparts. The B-cyclometallated complex **23** is more active than the corresponding C-cyclometallated **22** [74].

Cp*Ir(III) complexes (**25–27**) bearing imidazolium ion-tethered TsDPEN ligands (Scheme 7) are efficient catalysts for asymmetric TH of α-ketophosphonates **B137-B148** (Scheme 4) in water. At RT, moderate yields (44–78%, 4–8 h of reaction) and good to excellent e.r.'s (up to >99.5/0.5) were obtained by using 1 mol% of $[Cp^*IrCl_2]_2$, 2 mol% of the ligand and HCOONa as hydrogen donor [75].

Half-sandwich iridium complexes bearing pyridinesulfonamide ligands (**28–34**, Scheme 7) were assessed as precatalysts for TH of methyl aryl (**B1**, **B3**, **B10**, **B15**, **B18**, **B22**, **B27**) diaryl (**B66**), methyl alkyl (**B96**, **B103**) and cyclic (**B117-B119**) ketones (Scheme 4) [76]. In general, good conversions are obtained after 3 h of reaction, in refluxing 2-propanol, at 1 mol% of catalyst loading and in the absence of a base. With electro-donating groups, on the 4-substituted acetophenone substrate, the observed conversion after 6 h drops in comparison with substrates with electron-withdrawing substituents. On the other hand, precatalysts bearing electron-rich substituents on the pyridinesulfonamide ligand (complexes **28**, **31** and **33**) exhibited the highest conversion of 4-nitroacetophenone to the corresponding alcohol, while precatalysts **29**, **32** and **34**, which possess electron-withdrawing substituents on the ligand, afforded only moderate conversions. Under the same conditions, trials conducted using precatalyst **28** reduced benzophenone (82% conversion, 24 h

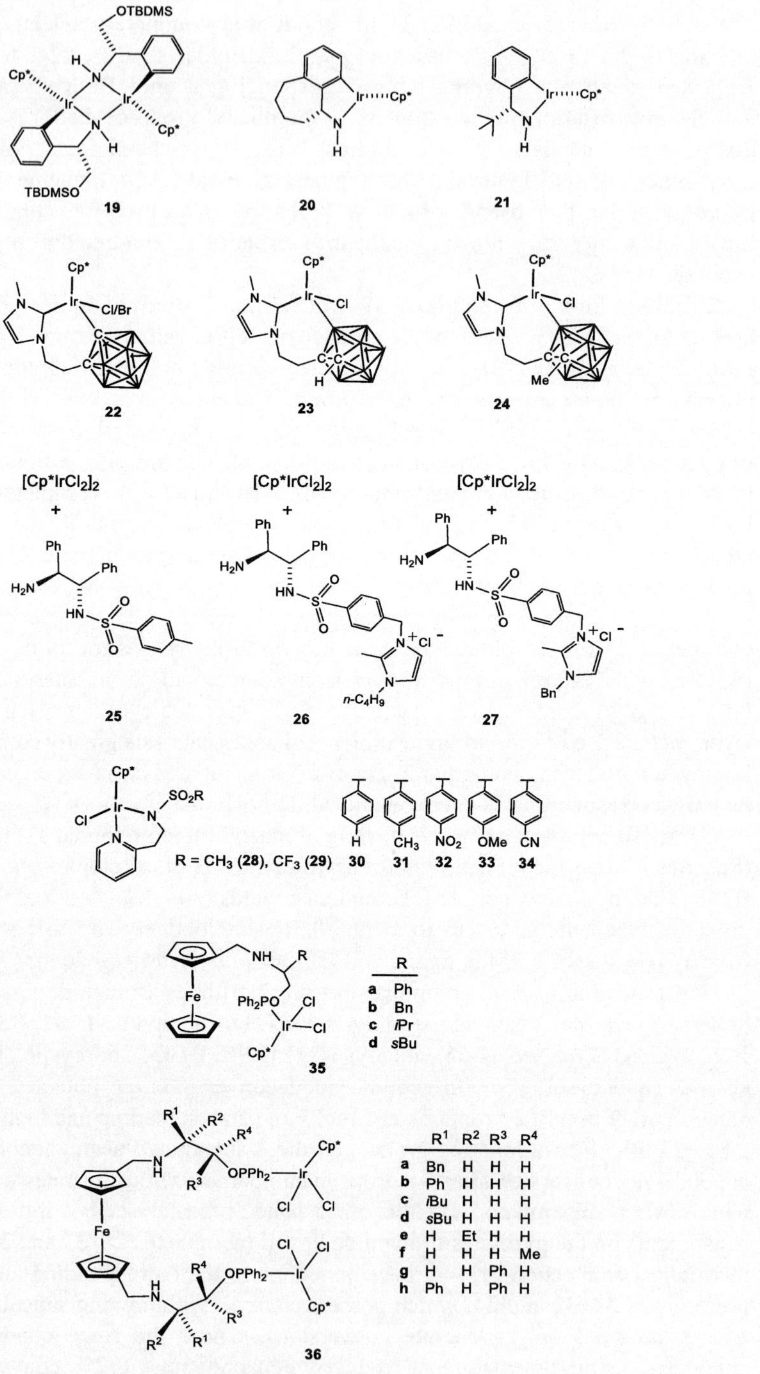

Scheme 7 Half-sandwich iridium catalyst precursors (**19–36**)

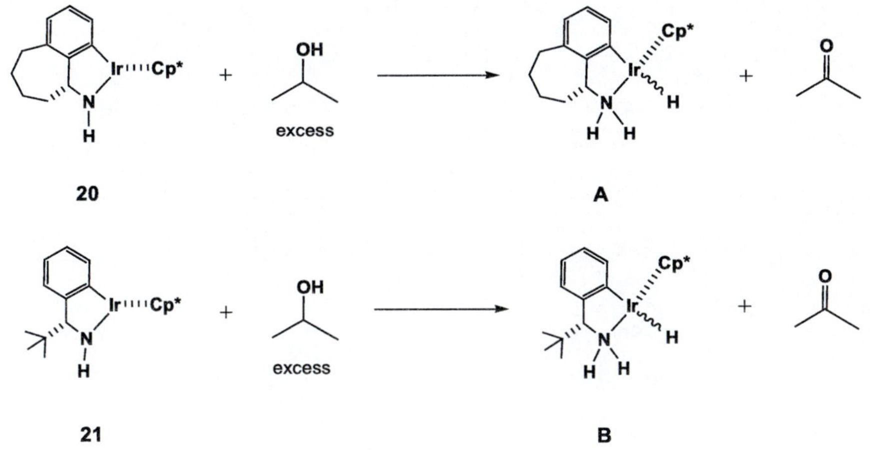

Scheme 8 Dehydrogenation of 2-propanol by the amido complexes **20** and **21**

reaction), dialkyl ketones **B96** (52%, 24 h) and **B103** (65%, 3 h), and cycloaliphatic ketones **B117** (85%, 3 h), **B118** (99%, 3 h) and **B119** (63%, 3 h) [76].

The same iridium pyridinesulfonamide complexes **28–34** are active for the TH of a variety of aryl (**A1**, **A3**, **A4**, **A12**, **A13**, **A22-A24**, **A27**, **A30**, **A32**, **A38**, **A43**), alkyl (**A56**) and heterocyclic (**A64**, **A65**) aldehydes (Scheme 3) [77]. Reductions occur with moderate to high conversions (39–100%), under base-free conditions and at high rates. Thus, for example, benzaldehyde derivatives are hydrogenated within 30 min in 2-propanol at 85°C, at 1 mol% of catalyst loading. As observed for substituted acetophenones, the combination of electron-donating groups on the ligand of the precatalysts with substrates possessing electron-withdrawing moieties entailed the highest rate of conversion to the alcohol. Decylaldehyde was quantitatively reduced in the presence of complex **30** in 12 h, under the above-mentioned conditions. The heterocyclic substrates 2-furfural **A64** and 5-hydroxymethylfurfural **A65** were selectively reduced to the corresponding alcohol in 95 and 100% conversion, respectively, in 30 min. A metal-ligand cooperative mechanism has been proposed for the catalytic TH reaction, the trihydride bridged dimer $[Cp^*Ir(\mu\text{-}H)_3IrCp^*]^+$ being the resting state [77].

Methyl aryl (**B1**, **B3**, **B18**, **B20-B22**, **B27**, **B36**), methyl alkyl (**B94**, **B96-B98**, **B101**, **B107**) and alkyl aryl (**B51**) ketones (Scheme 4) are reduced using chiral Cp*Ir compounds **35** and **36** (Scheme 7) containing C_2-symmetric ferrocenyl bis (phosphinite) ligands [78–81]. At 82°C, using 2-propanol as hydrogen source and solvent, with KOH (5 mol%) as a base and at 1 mol% of catalyst loading, almost quantitative conversions and moderate-to-good e.r.'s (from 68/32 to 99.5/0.5) were obtained, after 0.25–8 h of reaction.

Chiral cyclometallated 1-naphthylethaneamine iridium complexes featuring up to three stereogenic centres displayed excellent activity for asymmetric TH, albeit with modest enantioselectivities [82]. Complex **37** (Scheme 9), at 2 mol% catalyst

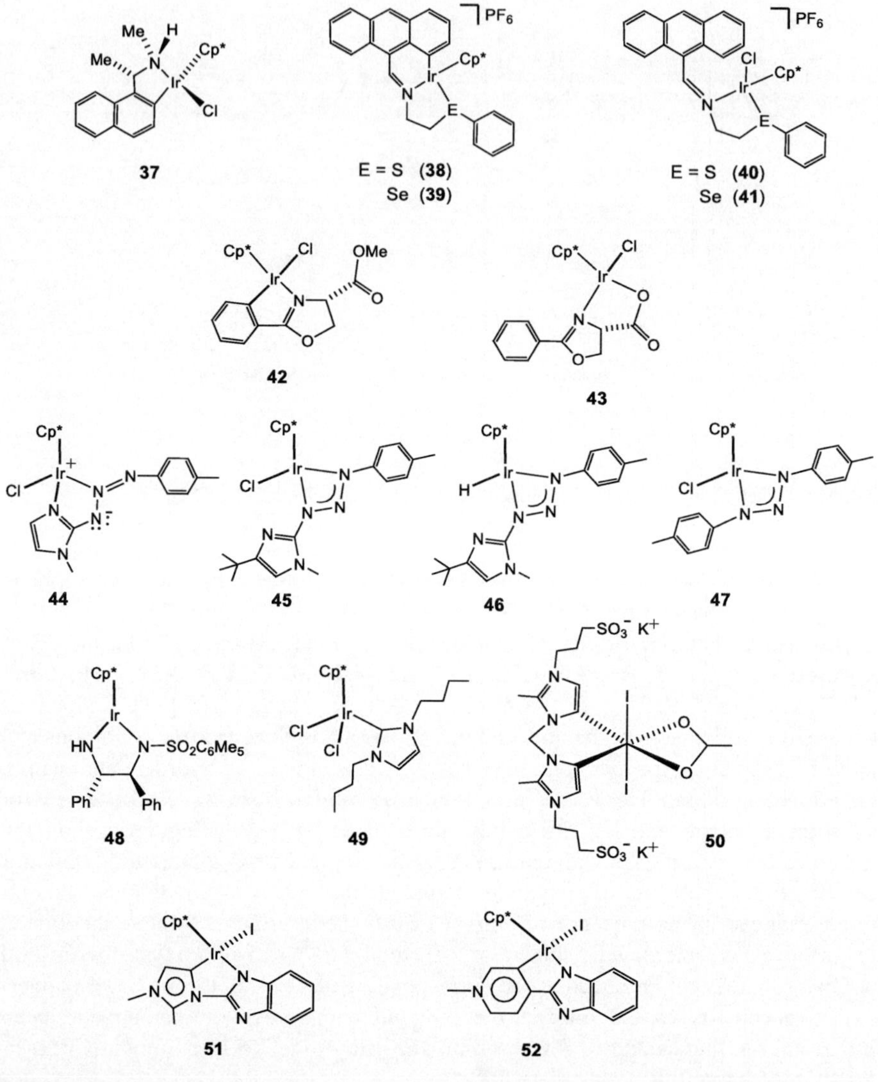

Scheme 9 Half-sandwich iridium catalyst precursors (**37–52**)

loading, with acetophenone as the substrate, 2-propanol as the reducing agent and in the presence of *t*BuOK (5 mol%) afforded 2-phenylethanol in 99% conversion, after 15 min of treatment at 20°C, with an e.r. of 76/24. Working at −30°C, a conversion of 97%, with 84.5/15.5 e.r., was achieved after 1 h of reaction.

Cp*Ir(III) complexes with chalcogenated Schiff bases of anthracene-9-carbaldehyde **38–41** (Scheme 9) were tested for the TH of aldehydes **A1**, **A23**, **A24**, **A30**, **A31**, **A38** and **A61** (Scheme 3) and ketones **B1**, **B15**, **B21**, **B47 B101** and

B117 (Scheme 4) under base-free conditions [83]. At 80°C, in 2-propanol, at catalyst loading of 0.1–0.5 mol%, conversions in the 66–98% range were obtained, after 1–6 h of reaction. For iridacycles **38** and **39**, a mechanism involving the loss of the Cp* ring was proposed [83].

The iridium dimer $[Cp^*IrCl_2]_2$ reacts with methyl (*S*)-2-phenyl-4,5-dihydrooxazole-4-carboxylate selectively giving a *N,C*- (**42**) or a *N,O*-chelated (**43**) complex (Scheme 9) depending on the reaction conditions [84]. Interestingly, the *N,O*-chelated complex **43** is much more active and selective in the asymmetric TH of ketones than its *N,C*-chelated isomer **42**, using mixtures of formic acid/amine as a hydrogen source, in CH_2Cl_2. Furthermore, the sense of asymmetric induction is different for each isomer. The favoured configuration of the alcohol is *S* for the *N,O*-chelated complex **43** and *R* for *N,C*-chelated analogue **42**. Thus, for example, in the presence of 1 mol% of complexes **42** or **43**, 4-nitroacetophenone could be reduced by using an azeotropic mixture of formic acid/triethylamine, in CH_2Cl_2 at RT. However, while with precatalyst **43** 100% of the corresponding alcohol was obtained after 2 h of reaction, with an 86.5/13.5 of e.r. in favour of the *S* enantiomer, with the analogous precatalyst **42**, only 75% of conversion was measured after 15 h of treatment with a 48/52 e.r. in favour of the *R* enantiomer. Regarding the formic acid/amine mixture, besides NEt_3, also Et_2NH, Cy_2NH, $i PrNH_2$ and $t BuNH_2$ have been tested. Using the catalyst precursor **43**, the highest e.r. (99/1) for the TH of ketone **B27** was obtained with a formic acid/$i PrNH_2$, 2/1 mixture. No reasons have been proposed for this behaviour.

The *N,O*-chelated complex **43** catalyses the TH of a variety of methyl aryl (**B1-B6**, **B14**, **B15**, **B17**, **B18**, **B22**, **B27**, **B30-B33**, **B35**) and alkyl aryl ketones (**B47**, **B49**, **B55**, Scheme 4) using a formic acid/$i PrNH_2$, 2/1 mixture. Excellent e.r.'s, between 95/5 and 99.5/0.5, were obtained with yields ranging from 7 to 98%, when the reactions were carried out overnight, at RT in CH_2Cl_2. The lowest activity corresponds to acetophenones that bear highly electro-donating substituents (e.g. **B17**, 7% yield) or 2-substituted acetophenones (e.g. **B2**, 18% yield) [84].

Half-sandwich iridium complexes **44–47** (Scheme 9) containing a triazenide group as a ligand have been tested as precatalysts in the TH of acetophenone, in 2-propanol as both hydrogen donor and solvent, at 70°C [85]. Reactions were performed in the presence of KOH as a base with a catalyst loading of 2 mol%. The chloride compounds **44** and **45** need 42 and 62 h of reaction, respectively, to achieve, in both cases, 95% yield. However, if the hydride complex **46**, derived from **45**, is used as the catalyst, a yield of 83% was obtained just after 4 h and in the absence of a base. Complex **47**, without imidazole substituents on the triazenido moiety, also reduced acetophenone without a base (98% yield in 86 h) [85].

The chiral iridium complex **48** (Scheme 9) in cooperation with the chiral phosphoric acid depicted in Scheme 10 efficiently catalysed the asymmetric TH of *N*-benzyl **C15-C25**, *N-p*-methoxybenzyl **C16**, *N-p*-methoxyphenyl **C26-C31** and **C33**, dialkyl substituted **C34**, **C35** and aryl ethyl **C32** imines [86]. Unexpectedly, the efficiency and enantioselectivity of the TH can be tuned by the use of different alcohols as the hydrogen donor. In general, benzylic alcohols provided high

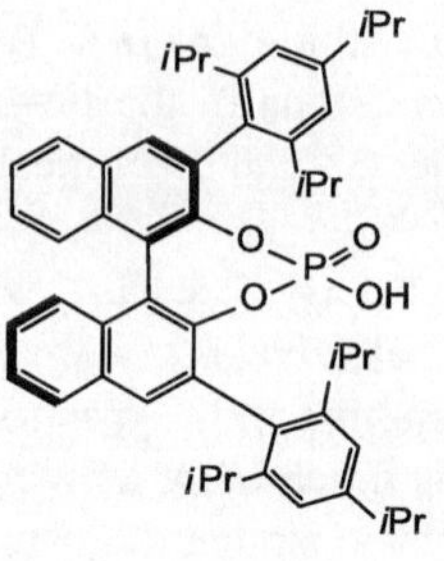

Scheme 10 Chiral phosphoric acid used as additive in the TH of imines

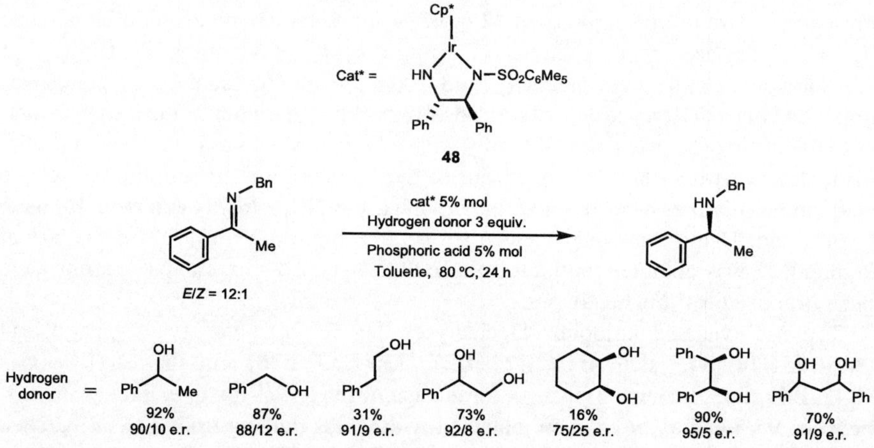

Scheme 11 Screening of hydrogen donors for the TH of imines

efficiency, and 1,2-diphenylethane-1,2-diol gave the highest enantioselectivity (Scheme 11). Treatment of the imines in toluene, at 80°C, over 24 h, with 3 equivalents of this alcohol, in the presence of 5 mol% of complex **48** and 5 mol% of phosphoric acid afforded the corresponding amines with yields from 60 to 90% and e.r.'s from 87/13 to 98/2. Based on experimental studies and DFT calculations, a mechanism involving an iridium alkoxide as the reducing agent was proposed [86].

2.2 *Transfer Hydrogenation of CO_2*

With the aim of reducing the concentration of CO_2 in the atmosphere, methods to convert it into valuable chemicals have been the subject of intense investigation in recent years [87]. Among them, catalytic methods for CO_2 direct hydrogenation to one-carbon molecules such as formic acid or methanol using homogeneous metallic

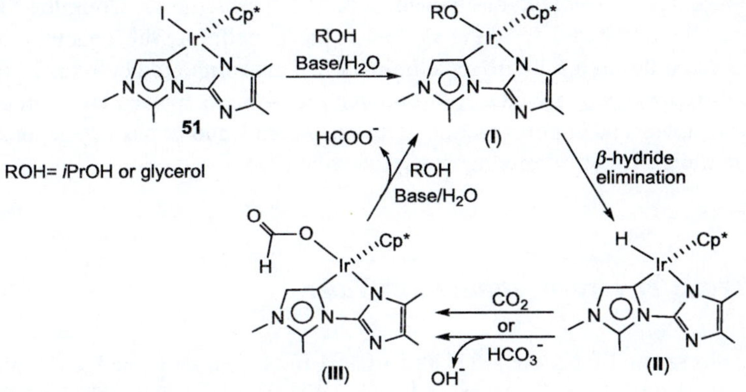

Scheme 12 Proposed catalytic cycle for the TH of CO_2

catalysts are, probably, the most intensively studied ones (for recent reviews, see Refs. [88–91]). However, the protocols reported so far require, in general, harsh conditions of temperature and pressure. TH approaches could try to avoid this problem and allow for the use of renewable hydrogen sources with inherent advantages from a perspective of sustainability. However, reports on TH of CO_2 are scarce. Most of them involve half-sandwich ruthenium or iridium complexes bearing strongly electron-donating *N*-heterocyclic carbene ligands. In 2010, the group of Peris reported the formation of formic acid from CO_2 in the presence of 2-propanol as the hydrogen donor. With the iridium catalyst precursor **49** under 50 bar of CO_2, in a 0.5 M KOH solution, a TON of 150 was obtained after 72 h at 110°C [92]. The same group showed that the sulfonate substituted bis-abnormal carbene complex **50** afforded a TON of 2,700 under 50 bar of CO_2, in a 0.5 M KOH solution, after 75 h at 200°C [93].

More recently, Choudhury's group studied the application of the half-sandwich iridium complexes **51** and **52** in the catalytic conversion of CO_2 to formates via TH using glycerol, a renewable biomass derivative, as a hydrogen source [94]. Complex **51** performs better than complex **52** under all the tested conditions. Notably, the catalytic system works at ambient-pressure of CO_2, in water as the solvent. Glycerol showed to be superior to methanol or 2-propanol as the hydrogen donor. Under optimised reaction conditions (0.15 μmol of complex **51**, 1.0 M aqueous K_2CO_3 as a base, at 150°C), a TON of 149 was achieved after 1 h of reaction. Control experiments showed that (i) formate was not formed from the carbonate used as the base; (ii) under standard conditions, formation of hydrogenocarbonate was observed in the reaction mixture; (iii) hydrogenocarbonate also renders formate at lower rate than CO_2, in parallel and simultaneously; (iv) hydride **II** (Scheme 12) was stoichiometrically prepared by treating complex **51** with glycerol in acetonitrile or with 2-propanol in water, in both cases, in the presence of K_2CO_3; and (v) this hydride reacts with aqueous hydrogenocarbonate affording formate.

All these data suggest the mechanism depicted in Scheme 12. Complex **51** forms the alkoxy intermediate **I** from the alcohol in basic medium. Subsequently, hydride **II** is generated through *β*-hydrogen elimination. This compound delivers its hydride to the substrates, CO_2 or HCO_3^-, to produce the formato species **III**. Complex **III** releases formate and, upon reaction with an alcohol and a base, regenerates the alkoxy complex **I** that restarts the catalytic cycle [94].

2.3 *Transfer Hydrogenation in Water*

The development of aqueous TH reactions offers economic and environmental benefits because water is cheap and non-toxic. Research into TH reactions in water started in the 1980s. Sasson, Blum and co-workers reported on the aqueous-organic biphasic reduction of C=C and C=O bonds with phosphine ruthenium and rhodium complexes in the presence of formate salts as hydrogen donor [95–97]. Joó's group reported on the hydrogenation of unsaturated bonds by using water soluble rhodium and ruthenium complexes bearing sulfonated phosphine ligands [98–100]. Thereafter, great progress was made in the TH of C=C and C=O and also C=N bonds. The most widely employed catalysts are half-sandwich compounds, Cp*Ir complexes playing a remarkable role [101–108]. We collect here new contributions in this area that have appeared in the last 5 years.

Iridium complexes containing imidazolium ion-tethered TsDPEN ligands (Scheme 13) catalyse the TH of a range of *α*-ketophosphonates (**B137-B148**, Scheme 4) in water, employing HCOONa as the hydrogen donor. The products were obtained in moderate to good isolated yield (44–78%) with good to excellent e.r.'s (up to >99.5/0.5) after 4–8 h of treatment at RT [75].

Cationic half-sandwich iridium complexes bearing pyridine-2-yl-methyl aniline ligands (**53**, Scheme 14) were investigated for catalytic TH of acetophenone in water

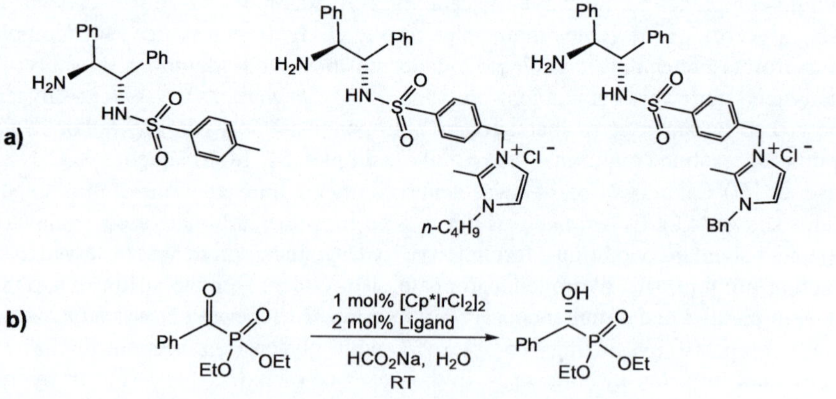

Scheme 13 TsDPEN and imidazolium ion-tethered TsDPEN ligands and phosphonate reduction

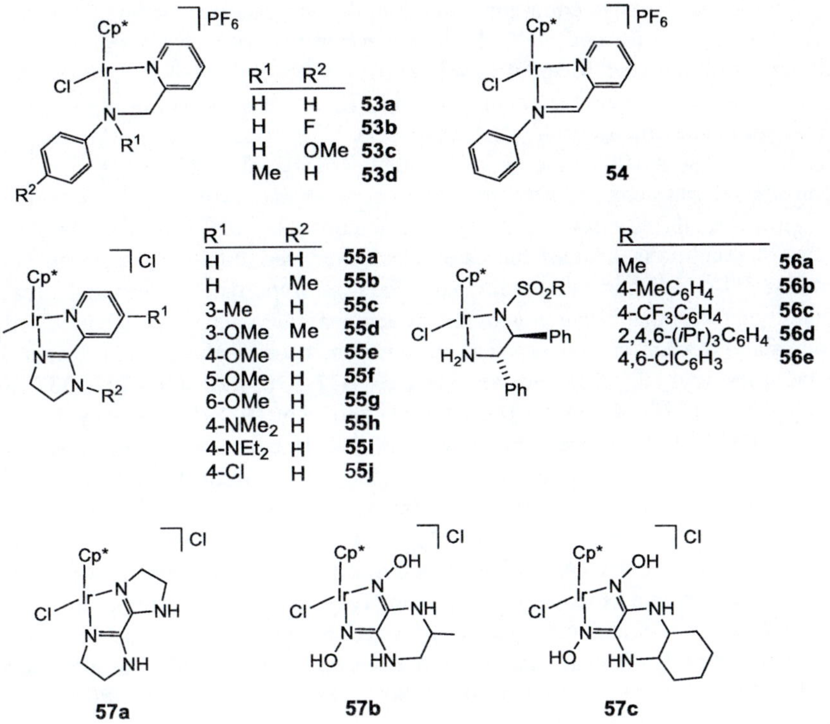

Scheme 14 Half-sandwich iridium catalyst precursors (**53–57**)

as a solvent. A HCOOH/HCOONa mixture was used as a hydrogen source. The best results were obtained with a HCOOH/HCOONa, 7.5/3.5 M ratio, in 2 mL of H_2O (pH = 2.60). At 60°C, with a catalyst loading of 0.5 mol%, 95–96% of conversion was achieved after 4–6 h of reaction [109].

When complex **53d** with a protected amine group (N–Me) was tested for the TH of acetophenone, moderate catalytic activity was observed (52% conv, 24 h). This result indicates that for complexes **53** the preferred mechanism is the Noyori's bifunctional mechanism.

The catalytic activity of the related imine compound **54** (Scheme 14) is lower than that of the corresponding amine compound **53a**. To achieve 96% conversion, 24 h of reaction was necessary, under the optimised conditions [109].

In 2017, the Tang's group reported on the reduction of a range of aldehydes in water in the presence of formic acid as the hydrogen source using half-sandwich iridium complexes **55** containing imidazolyl-pyridine ligands (Scheme 14). It is remarkable the good tolerance for a wide variety of functional groups such as alkene (**A25**) and alkyloxy (**A26**), halogens (**A16, A30, A31, A44, A45**), phenols (**A2, A41, A42**), ketones (**A34**), carboxylic acids (**A20, A36**), cyano (**A19**) or nitro (**A38, A44**).

Under the optimal conditions (0.02 mol% of catalyst loading, 4 equiv. of HCOOH, 2 mL of water, 80°C) high conversion (91–99%) was obtained for most of the substrates tried, in about 15 min [110].

The catalytic process is chemoselective. Thus, for example, only the aldehyde group of the ketonic substrate **A34** was reduced. The system also works at a catalyst loading as low as 0.005 mmol %, and, notably, a TOF value of 73,800 h^{-1}, at 49% of conversion, was achieved for 4-methoxybenzaldehyde [110].

However, the group of Tang and Xu demonstrated that properly adjusting the reaction conditions, ketones can be efficiently reduced employing the same type of imidazolyl-pyridine iridium complexes. The most important changes in the reaction conditions were that the amount of formic acid was increased up to 12 equiv. and that the catalytic reactions were carried out under inert atmosphere [111]. Under these conditions, aryl alkyl (**B1**, **B3-B5**, **B9**, **B10**, **B12-B15**, **B20-B26**, **B27**, **B37**, **B47**), methyl alkyl (**B100**, **B104**, **B107**), methyl heterocyclic (**B38**, **B41**) and cyclic (**B118**, **B119**, **B121**) ketones were hydrogenated. Yields between 57 and >99% were obtained after 4–12 h of reaction. The reduction rates are dependent on the pH values of the reaction media, conversions dramatically decreasing when pH increases. The standard S/C ratio employed was 10,000 but catalysis also occurs at an S/C ratio of 100,000. A TOF of 26,000 h^{-1}, at 52% of conversion, was achieved in the reduction of acetophenone, employing catalyst **55a** [111]. Furthermore, a wide range of ketones containing a variety of functional groups such as aryloxy (**B106**), halogens (**B57**, **B64**), cyano (**B63–65**) or ester (**B54**, **B61**, **B108**) was also reduced. Typically, conversions higher than 90% were obtained, at 80°C, within 3–12 h [111].

The Fu's and Zhou's group reported on the asymmetric TH of non-*ortho*-substituted-2-pyridylketone *N*-oxides (Scheme 4) using chiral diamine-derived iridium complexes (**56**) (Scheme 14). A H_2O/*i*PrOH mixture and sodium formate were employed as the solvent and the hydrogen source, respectively. The *N*-oxide function was removed by treating the resulting alcohols with Zn/NH_4Cl [112].

After screening, by examining the reduction of 2-(4-chlorobenzoyl)pyridine *N*-oxide (**B80**), complex **56a** was selected as the best catalyst, in terms of yield and enantioselectivity. Under the optimised conditions (5 mol% of catalyst loading, 10 equiv. of HCOONa, H_2O/*i*PrOH, v/v:1/1, RT, 24 h), a wide variety of 2-pyridyl ketone *N*-oxides with different functionalities and electronic properties (**B74-B90**) were reduced. Yields from 56 to 90% and e.r.'s from 78.7/26.3 to 99.1/0.9 were obtained [112].

Structurally diverse aldehydes and ketones were reduced, in water, with formic acid as a hydrogen source, by using half-sandwich iridium complexes **57** as catalyst precursors (Scheme 14). Under aerobic conditions, the reduction of benzaldehyde was completed with catalyst **57a** within 9 min using an S/C ratio of 2,000. At 5,000 and 10,000 S/C ratios, reduction of benzaldehyde was not observed. However, under nitrogen atmosphere, in degassed water, good results were obtained at S/C ratios as high as 20,000 and 10,000 for aldehydes (including benzaldehyde) and ketones, respectively.

Catalyst **57a** proved to be more active than its congeners **57b** and **57c**. With complex **57a** as the catalyst and at the standard conditions of a S/C ratio of 20,000, 5 equiv. of HCOOH, at 90°C and under N_2 atmosphere, a series of diverse aldehydes (**A1**, **A2**, **A16**, **A19**, **A23**, **A24**, **A28**, **A30-A32**, **A34**, **A36-A38**, **A42**, **A44**, **A46**, **A47**, **A50**, **A55**, **A59**, **A63**, **A64**, **A67**, Scheme 3) were reduced in excellent isolated yield (90–99%), after 2 h of reaction [113].

Catalysts **57** also actively reduce ketones. Like for Tang's imidazolyl-pyridine iridium catalysts, to obtain good results, an excess of formic acid (15 equiv) greater than that for TH of aldehydes (5 equiv) has to be used. Higher excess of formic acid provides more acidic conditions to activate carbonyl groups and dissolves ketones. In fact, ketones that were not reduced in the presence of 5 equiv. of formic acid were easily reduced by using 15 equiv. of this acid.

A variety of substituted acetophenones with electron-donating or withdrawing group was reduced to the corresponding alcohols. Halogen (**B20-B22**), nitro (**B27**) and cyano (**B24**) groups were compatible as well as heterocyclic substituents (**B38**, **B41**). Methyl alkyl ketones (**B100**, **B107**), cyclohexanone, cycloheptanone or ketones containing acidic functional groups (**B25**) were also reduced.

Again, complex **57a** was the catalyst of choice. With this complex, yields from 78 to 99% were obtained after 12 h of reaction, with an S/C ratio of 10,000, at 90°C and under N_2 atmosphere [113].

2.4 *Biological Transfer Hydrogenation*

Artificial metalloenzymes (ArMs) are hybrid catalysts that result from combining an abiotic metal cofactor with a protein with the aim of taking advantage of the features of both metal-based and enzymatic catalysis [114, 115]. In the field of artificial transfer hydrogenases, many efforts have been devoted to the development of imine reductases. Since 2005, the Ward's group has been using Noyori-type piano stool metallic cofactors combined with either wild-type streptavidin (Sav WT) or streptavidin mutants as host proteins to develop artificial asymmetric transfer hydrogenases. The metallic cofactor was completed by bonding the Noyori's diamine ligand to biotin which acts as an anchor to the protein [116–118] (Scheme 15). The intricate network of interactions in the biological scaffold creates a chiral second coordination sphere environment around the catalytic metal site responsible for enantioselection. Given the prevalence of the chiral 1,2,3,4-tetrahydroquinoline in natural alkaloids and synthetic drugs [119], the salsolidine precursor imine 1-methyl-6,7-dimethoxy-3,4-dihydroquinoline was usually employed as a model substrate (Scheme 15). 3-(*N*-morpholino)propanesulfonic acid (MOPS) buffer and sodium formate were usually employed as the reaction medium and the hydrogen source, respectively, [120–122] (Scheme 15).

Based on the biotin-streptavidin technology developed by the Ward's group, Rimoldi et al. have investigated the catalytic activity of the iridium complexes

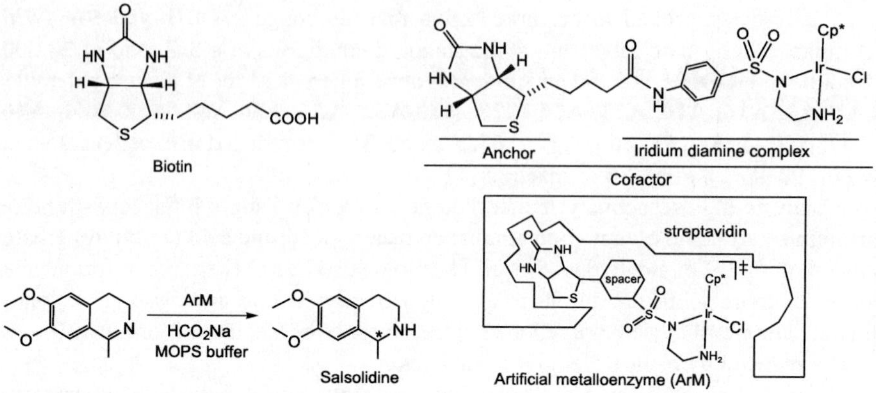

Scheme 15 Transfer hydrogenation of imines catalysed by artificial metalloenzymes

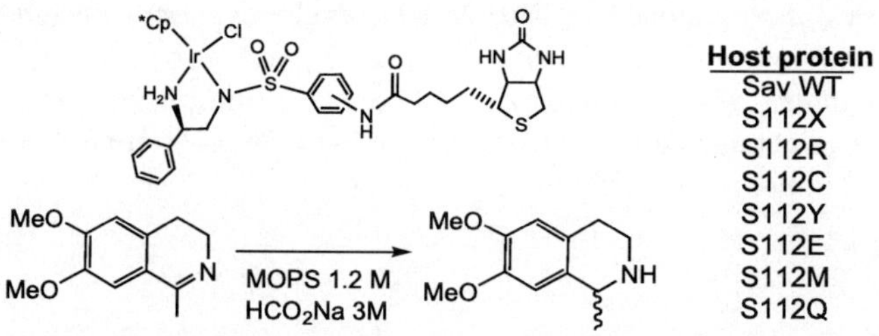

Scheme 16 Artificial metalloenzymes for the TH of the salsolidine precursor developed by Rimoldi et al. [123]

depicted in Scheme 16 employing wild-type Sav and the Sav mutants indicated in the same Scheme.

Reactions were carried out at 30–40°C, by using 1 mol% of iridium complex, 0.33 mol% of tetrameric wild-type Sav or Sav mutants, in 1 mL of MOPS buffer at pH 6.5–7.0 and in the presence of HCOONa (3 M) as the hydride source. When the transition metal compound is embedded in the host protein, a decrease in activity was observed accompanied by an improvement of stereoselectivity. The best results (60% conversion, 82.5/17.5 e.r., 24 h) were achieved with the S112C mutant in combination with the complex bearing the biotin anchored to the para position in the ligand (Scheme 16) [123].

Rimoldi's group has also investigated half-sandwich iridium complexes with the biotin moiety anchored to the Cp* ring for the TH of the imine precursor of salsolidine. Sav WT and different Sav mutants at position S112 or K121 and a double mutation at positions S112 and K121 or L124 were investigated as the host proteins (Scheme 17).

Reactions were carried out at 30°C, in 0.6 M MOPS buffer with 1 mol% catalyst loading and 3 M HCOONa as a hydrogen source. Poor conversion and e.r. were obtained, the best results being achieved for the chiral amino hydroquinoline ligand

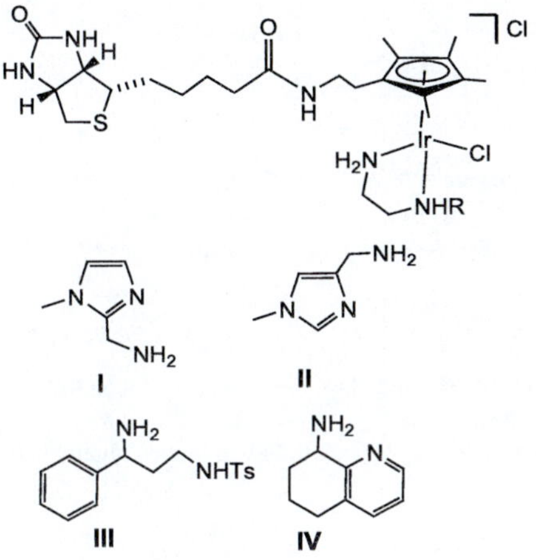

Host protein	
Sav WT	S112K
S112Q	S112W
S112R	K121R
S112M	K121M
S112C	K121H
S112H	K121F
S112A	K121A
S112E	S112Y + K121R
S112T	S112K + L124E
S112Y	S112C + K121H

Scheme 17 Artificial metalloenzymes for the TH of imines [124]

IV (5–77% conversion, 46/54–16/84 e.r., 18 h). In all cases, both conversion and e.r. decrease in the presence of host protein with respect to the values obtained by using metallic cofactors alone. The *S* enantiomer of the hydroquinoline ligand **IV** always gives preferentially *S*-salsolidine, and the *R* enantiomer of the ligand always gives preferentially *R*-salsolidine indicating that the chiral environment of the host protein affected the chirality of the product to a lesser extent [124].

Ward, Maréchal and co-workers have applied directed evolution to an artificial transfer hydrogenase to improve its catalytic activity and selectivity for the reduction of cyclic imines. The introduction of the cofactor depicted in Scheme 18 within Sav isoforms affords asymmetric transfer hydrogenases that can be optimised by directed evolution protocols [125, 126]. Two mutants were identified (see Scheme 18) that increased the reaction rate for the reduction of the cyclic imine shown in Scheme 18, and, under the indicated conditions, enantiomeric rates of 97.5/2.5 (*R*) and 7.5/92.5 (*S*) were achieved [119].

A biotinylated iridium half-sandwich complex was incorporated into streptavidin mutant S112A to generate an ArM (Scheme 19). Selective deuteration of nicotinamide adenine dinucleotide (NAD^+) mediated by this ArM, employing deuterated sodium formate as a deuterium source, generated deuterated NAD^2H in high diastereomeric excess [127].

ArM based on Sav variants and biotinylated iridium cofactor enable the regeneration of various synthetic NADH mimics (mNADH) from NAD^+, with formic acid as a hydrogen source. The involved TH can be coupled with ene reductase-catalysed reduction of *α*,*β*-unsaturated compounds. Scheme 20 shows the iridium cofactor and NAD^+ mimics as well as the *α*,*β*-unsaturated substrates which are hydrogenated. Sav 112A and Sav 112 K mutants were used as the host protein, and the ene reductase of the Old Enzyme family from *Thermus scotuductus* was selected as an mNADH-accepting enzyme. Scheme 20 also shows a particular case of the two coupled processes [128].

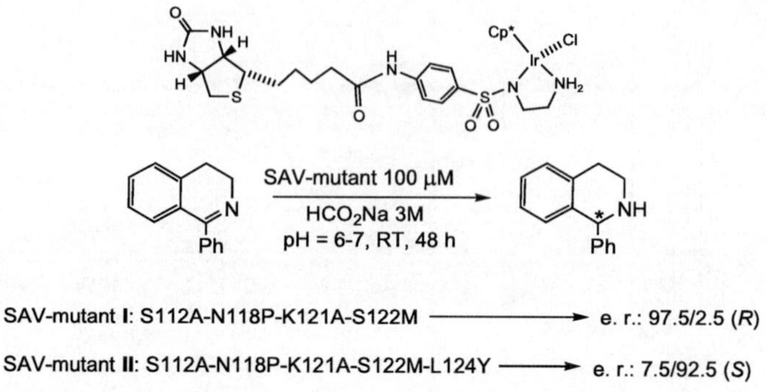

Scheme 18 Observed enantiodivergence with two selected Sav mutants obtained by directed evolution

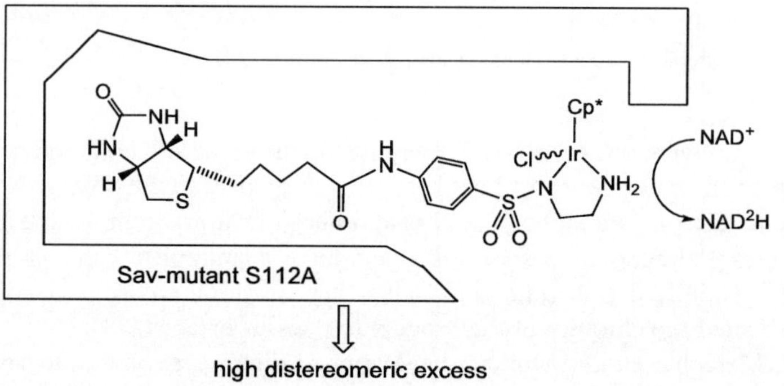

Scheme 19 Diastereoselective reduction of NAD^+

Biological reducing agent nicotinamide adenine dinucleotide phosphate (NAD(P)H) serves as an efficient hydrogen source for the reduction of cyclic imines. An ArM formed by combining a biotinylated Cp*Ir compound with either wild-type Sav or Sav mutants was used as the catalyst. To regenerate the consumed NAD(P)H, glucose dehydrogenase was incorporated into the process, and, to increase the enantiomeric excess, a monoamine oxidase and a catalase were also integrated into the system. The resulting four-enzyme system catalyses the reduction of 1-methyl-3,4-dihydroisoquinoline with NAD(P)H as a hydrogen source. Quantitative conversion and perfect enantioselectivity, in favour of the (*R*)-enantiomer, were achieved by combining the iridium cofactor, shown in Scheme 21, with the Sav K121R mutant after 24 h of treatment at 37°C [129].

Human carbonic anhydrase II (hCA II) offers an attractive scaffold for the assembly of ArMs using arylsulfonamide as a high-affinity anchor [130]. An X-ray structure determination shows that arylsulfonamide can interact with the Zn ion which lies at the bottom of a deep hydrophobic funnel-shaped cavity of the protein [131]. To anchor to hCA II, the arylsulfonamide bearing iridium complexes

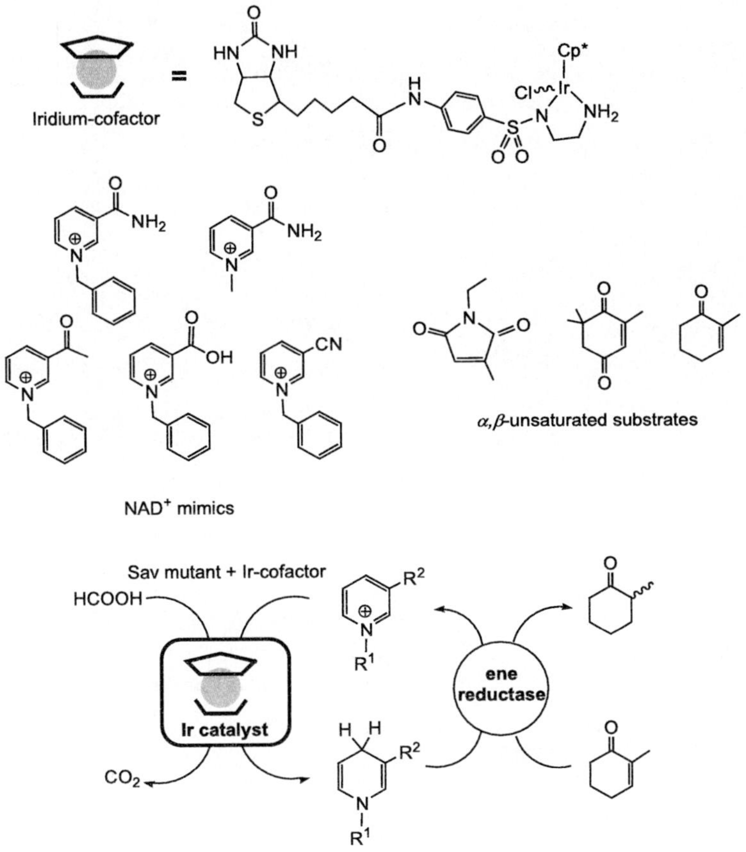

Scheme 20 Reduction of a NAD$^+$ mimics by the Sav 112 K mutant combined with the iridium shown cofactor (top). Reduction is coupled with the enzymatic hydrogenation of an α,β-unsaturated substrate (bottom)

[(η^5-Cp*)Ir(picolylamine)Cl] and [(η^5-Cp^{propyl})Ir(picolylamine)Cl] (Cp^{propyl} = penta-*n*-propylcyclopentadienyl) depicted in Scheme 22 were prepared. Application of the computational design software Rosetta [132] to the combination [(η^5-Cp*)Ir(pico)Cl]/hCA II WT afforded four hCA II variants that significantly increased affinity for the catalysts. The combination of designed hCA II mutations with the η^5-Cp^{propyl} complex renders productive (TON up to 100) and highly selective (up to 98/2 e.r.) ArMs for the reduction of cyclic imines, under the conditions indicated in Scheme 22 [133].

A half-sandwich iridium complex was assembled to a periplasmic-binding protein such as CeuE, an iron siderophore of *Campylobacter jejuni*, by using an azotochelin siderophore as binding anchor. The remaining two coordination sites around the iron centre were occupied by solvent molecules in the free cofactor. Nonetheless, it was hypothesised that those sites are used to bind Y288 and H227 of

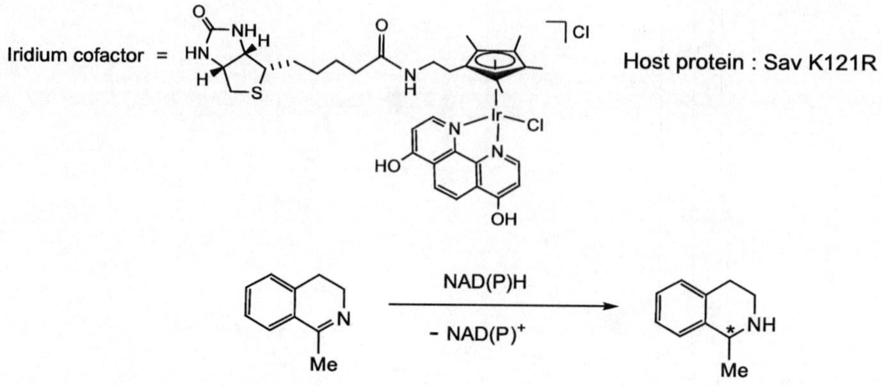

Scheme 21 Transfer hydrogenation employing NAD(P)H as a hydrogen source

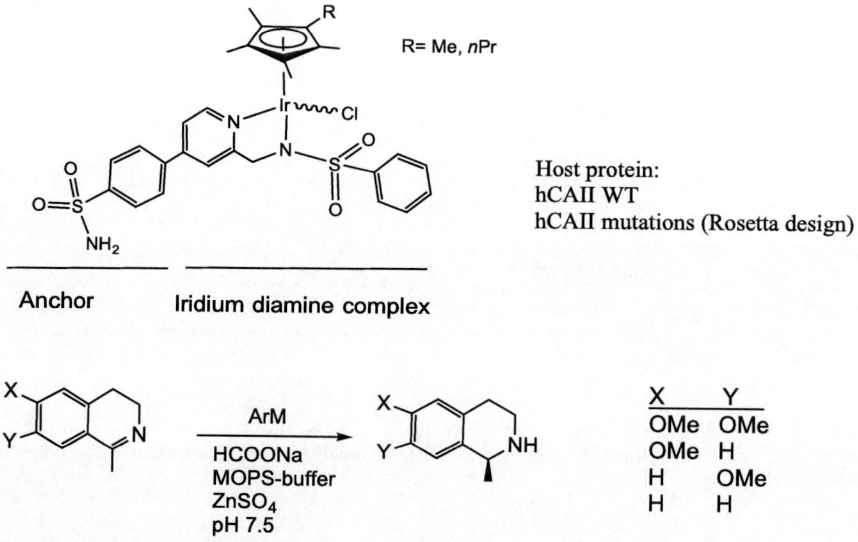

Scheme 22 TH of isoquinolines mediated by ArMs. The Ir-cofactors are shown at the top, and the host proteins employed are hCAII WT or mutants computationally designed

CeuE. The resulting ArM slowly reduces salsolidine precursor (TON 400, 24 h) with an e.r. of 67.7/32.3. CeuE mutant H227A increases TON (400, 2 h) but at the expense of a substantially lowered e.r. (51.7/48.4) [134] (Scheme 23).

NADH and NAD(P)H play important roles as cofactors in numerous biocatalysed processes [135]. In 2012, the groups of Sadler [136] and Fukuzumi [137] independently showed that half-sandwich iridium complexes can utilise NADH as a hydride donor to generate an iridium-hydride complex. The hydrido complex is able to transfer the hydride moiety to organic substrates. Regeneration of the iridium-

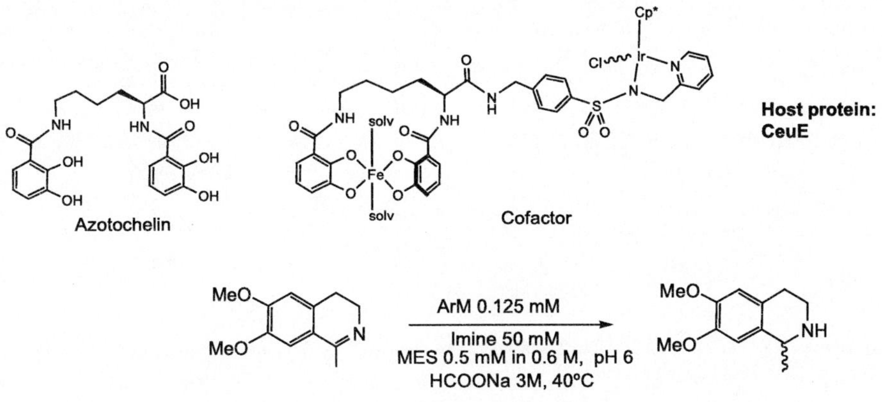

Scheme 23 TH of 1-methyl-6,7-dihydroxy-3,4-dihydroisoquinoline by the ArM formed by combination of the Ir-cofactor with the host protein shown at the top

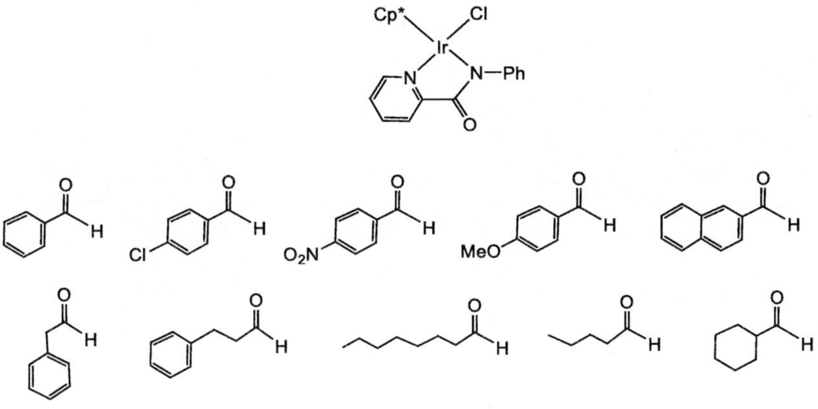

Scheme 24 Iridium catalyst and aldehydes that are hydrogenated. NADH was employed as a hydrogen donor

hydride from NADH completes the TH catalytic cycle. In particular, the hydride can be transferred to molecular oxygen increasing the levels of hydrogen peroxide and reactive oxygen species (ROS). Oxidative stress caused by generation of ROS is an effective method of killing cancer cells [138].

The half-sandwich iridium complex depicted in Scheme 24 bearing an *N*-phenyl-2-pyridinecarboxamidate ligand catalysed the reduction of aldehydes to the corresponding alcohols in *t*-BuOH/phosphate-buffered saline 2/8 at 37°C, employing NADH as a hydrogen donor. After 24 h of treatment, conversions of 30 to 90% were obtained, at a catalyst loading of 2 mol%. Neither acetophenone nor 4-heptanone was reduced, and the electron-deficient 4-nitroacetophenone was converted to the corresponding alcohol but in only 11% yield [139].

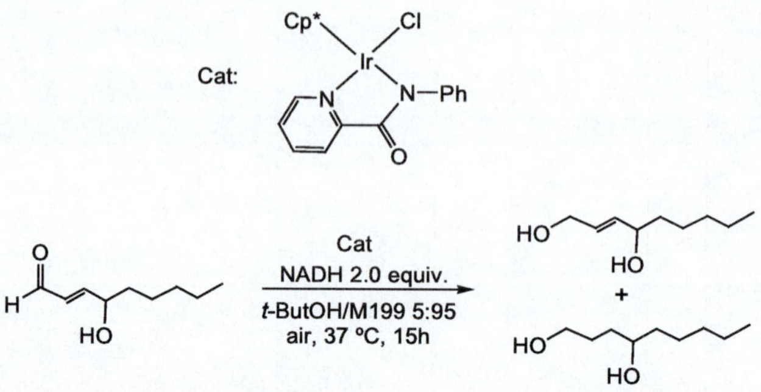

Scheme 25 TH of 4-hydroxynon-2-enal pyridine catalysed by the carboxamidate iridium complex shown. NADH was the H-donor

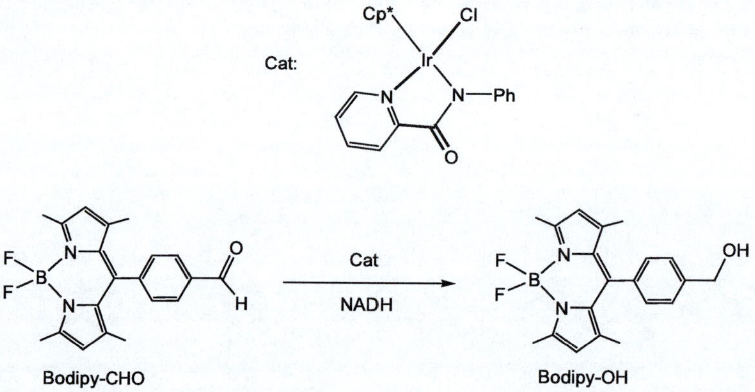

Scheme 26 Reduction of BODIPY-CHO

Aldehydes derived from lipid peroxidation, such as 4-hydroxynon-2-enal, are implicated in various diseases, and, therefore, reduction of these aldehydes to the corresponding alcohol may lead to detoxification. The same pyridinecarboxamidate iridium complex reduced 4-hydroxynon-2-enal with NADH acting as a hydrogen source. Under the conditions shown in Scheme 25, 88% of 4-hydroxynon-2-en-1-ol and 6% of 4-hydroxynonan-1-ol were obtained. Increasing the reaction time led to higher ratios of the saturated dialcohol suggesting that the C=O group is first hydrogenated. The putative hydride intermediate [Cp*IrH(pyridinecarboxamidate)] was detected and characterised by NMR spectroscopy [139].

Do's group showed that the reduction of aldehydes catalysed by the pyridinecarboxamidate iridium complex above-mentioned can take place inside living cells [140]. This complex catalyses the reduction of BODIPY-CHO to BODIPY-OH (Scheme 26) inside of NIH-3T3 mouse embryo fibroblast cells, using NADH as the hydrogen donor. The fluorescence that the BODIPY-OH compound develops within the cytoplasm of the cell was employed to monitor the TH reaction. When the generation of NADH was artificially reduced, the cells did

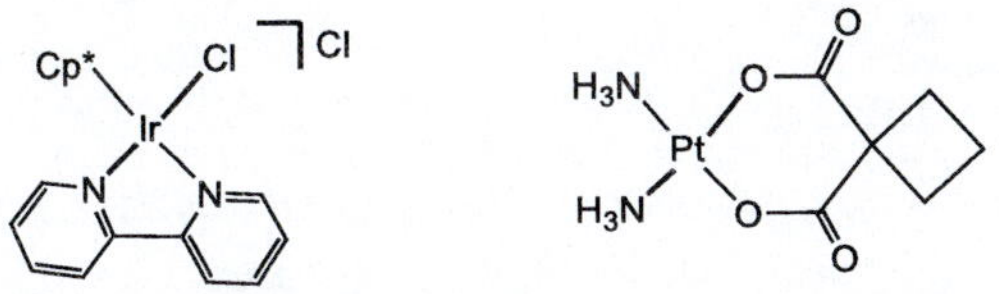

Scheme 27 Bipyridine iridium complex and carboplatin

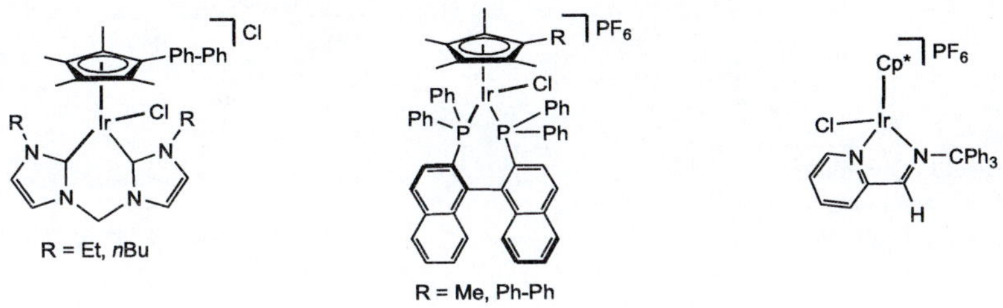

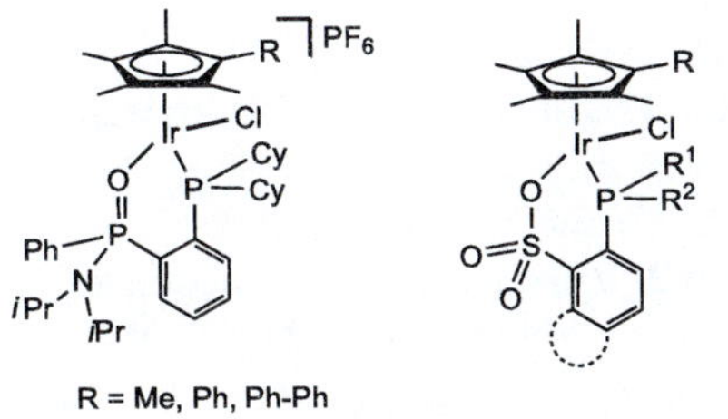

R	R1	R2	◌
Me	Cy	Cy	H
Ph	Cy	Cy	H
Ph-Ph	Cy	Cy	H
Me	Ph	Ph	H
Me	Ph	2-[2,6-$(MeO)_2C_6H_3$]C_6H_4	H
Me	o-$OMeC_6H_4$	o-$OMeC_6H_4$	H
Me	Cy	Cy	C_4H_4

Scheme 28 Iridium complexes with anticancer activity

not show an increase in fluorescence indicating that NADH acts as an endogenous hydride source [140].

Chemosensitiser agents are compounds that added to drug cocktails make cells more susceptible to the effects of the active drug [141]. Do's group has shown that the biocompatible bipyridine Cp*Ir complex presented in Scheme 27 is an excellent metallochemosensitiser in combination with carboplatin. Treatment of Y79 eye/retina cancer cells with this complex and carboplatin led to a NAD^+/NADH ratio 2.2 times greater than that in untreated cells. In contrast, after a similar treatment, the non-cancerous ARPE-19 eye/retina cells did not show significant differences in the NAD^+/NADH ratio, relative to untreated cells. Most probably, NADH acts as a hydrogen donor to the iridium complex. The plausible iridium-hydride $[Cp^*IrH(2,2'\text{-bipyridine})]^+$ was independently synthesised and isolated as a hexafluorophosphate salt. Notably, this hydrido complex dissolved in a mixture of methanol/phosphate-buffered saline 1/1 and when exposed to the air yields H_2O_2 in up to 40% yield [142].

The half-sandwich iridium complexes shown in Scheme 28 catalyse the oxidation of NADH to NAD^+ and lead to the generation of ROS [143–147]. These iridium complexes display promising anticancer activities toward Hela human cervical

Scheme 29 Reduction of imines with a half-sandwich iridium catalyst with NADH and an NADH mimic as a hydrogen donor

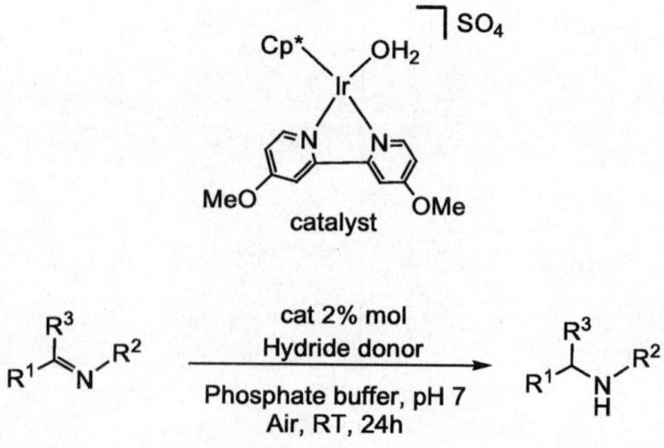

Scheme 30 Quinoline-based half-sandwich organoiridium complex

[143–146] and A549 [144–146] cancer cells. The increase of ROS appears to contribute to the anticancer activity.

The cationic Cp*Ir complex depicted in Scheme 29 catalyses the TH of a set of linear and cyclic imines in water. NADH or a *N*-benzyl-1,4-dihydronicotinamide (an NADH mimic) can be used as the hydrogen source. NAD(P)H and HCOONa can also be used as the hydrogen donors. Moderate to excellent yield (58–99%) was obtained under the conditions indicated in Scheme 29 [148].

The iridium complex presented in Scheme 30 facilitates the formation of NADH from NAD^+ in the presence of formate. Intracellular co-administration of the iridium complex with sodium formate enhances the antiplasmodial activity in the chloroquine-resistant strain of *Plasmodium falciparum*. This result indicates that TH reactions could be studied in terms of application to infectious diseases such as malaria [149].

Rimoldi's group has shown that antibiotic vancomycin is able to coordinate Cp*Ir(III) moieties, although its coordination mode remains uncertain. The resulting system was applied to the asymmetric TH of cyclic imines. Different cyclic imines were reduced in aqueous media, under mild reaction conditions, affording the corresponding amine, with moderate to appreciable e.r.'s. In particular, for imine **II** (Scheme 31), a conversion of 35% with 80.5/19.5 e.r. was achieved, after 18 h of treatment at 25°C, in 2-(*N*-morpholino)ethanesulfonic acid (MES) buffer 1.2 M pH 5, with 4 mol% of vancomycin and 1 mol% of $[(\eta^5\text{-Cp*})IrCl_2]_2$, in the presence of HCOONa 3 M as hydrogen source [150].

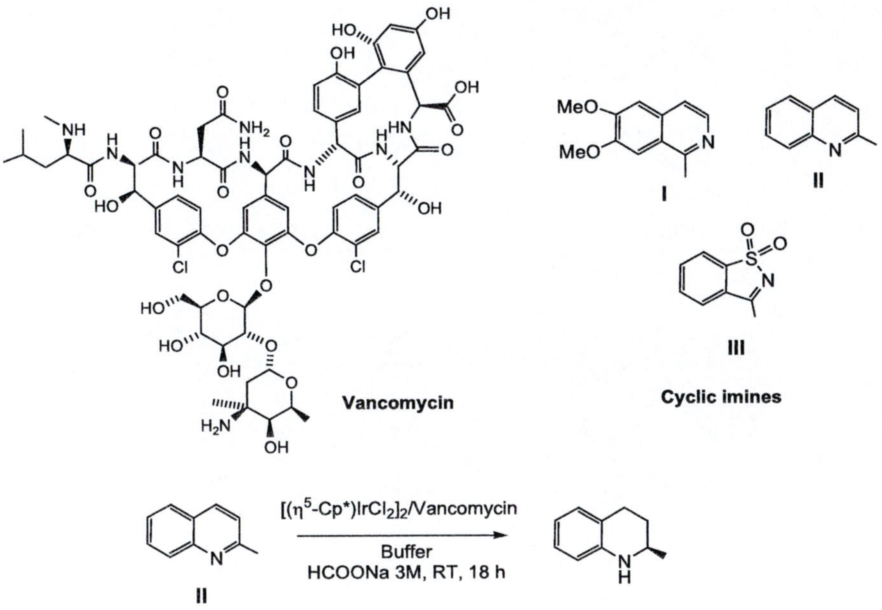

Scheme 31 Structure of vancomycin and application of the $[(\eta^5\text{-Cp*})IrCl_2]_2$/vancomycin system to the TH of cyclic imines

3 Carbene Iridium Complexes

The use of *N*-heterocyclic carbenes as ancillary ligands in catalysis and, in particular, in TH has experienced a remarkable increase in the last years. In this section we will discuss the contributions of the last 5 years in iridium complexes containing carbene ligands that have been applied as TH catalysts. Note that the half-sandwich iridium complexes bearing carbene ligands have been considered in Sect. 2.

Crabtree's group explored the application of the iridium bis(*N*-heterocyclic carbene) complexes **58** and **59** (Scheme 32) to the TH of ketones and imines using MeOH as a hydrogen source.

The best results were achieved with catalyst **59a**. Several aromatic ketones were reduced using 5 mol% of **59a** and 1–5 equiv. of KOH versus substrate. Reactions were carried out under microwave irradiation (120°C) because this mode of heating greatly improved yields compared to conventional heating at the same temperature. Acetophenones (**B1, B18, B23**) gave generally poorer yields (28–70%, 5 h reaction) than benzophenones (**B66, B71, B72**, 61 – >95%, 5 h) due to competing methylation of the α-CH_3 group [151].

Under the same reaction conditions, several imines were reduced using complex **59a** as the catalyst precursor. *N*-benzylideneaniline was reduced to the corresponding amine (> 95% yield); however, imines with $R^2 \neq$ Ph (Scheme 33) underwent both reduction and *N*-methylation. When $R^1 =$ H, a double methylation at nitrogen was also observed [151].

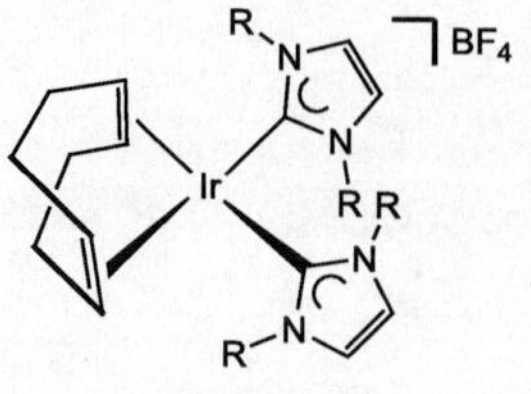

R = Me (**58a**), Et (**58b**), *n*Bu (**58c**)

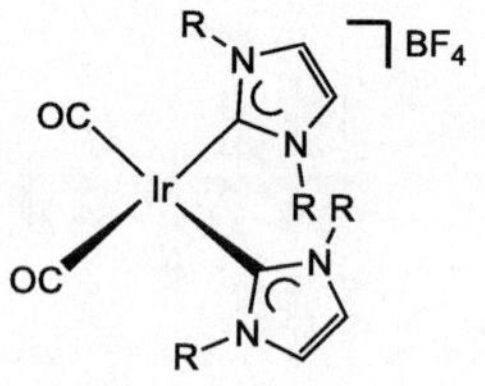

R = Me (**59a**), Et (**59b**), *n*Bu (**59c**)

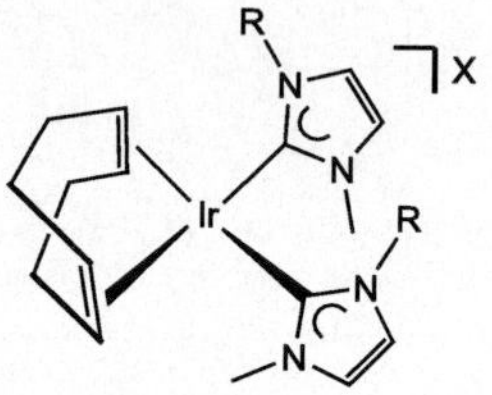

R =

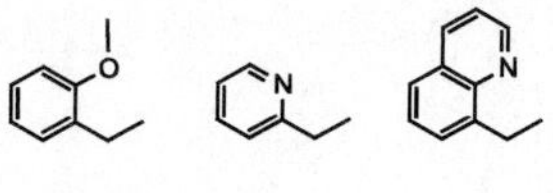

X =	Br **60a**	Br **60b**	Br **60c**
X =	PF$_6$ **60d**	PF$_6$ **60e**	

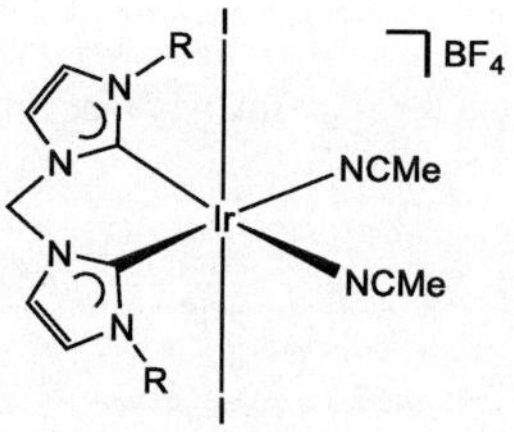

R = Me (**61a**), *n*Bu (**61b**), CH$_2$-CH$_2$OPh (**61c**)

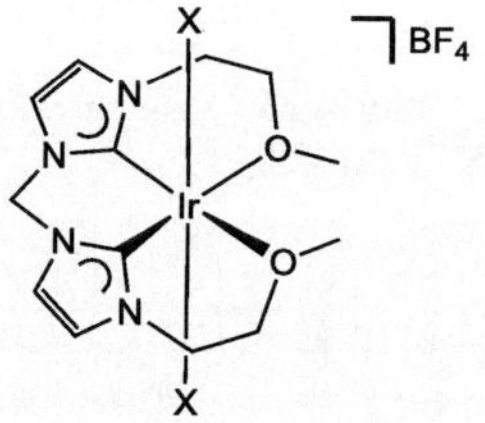

X = I (**61d**), CF$_3$-COO (**61e**)

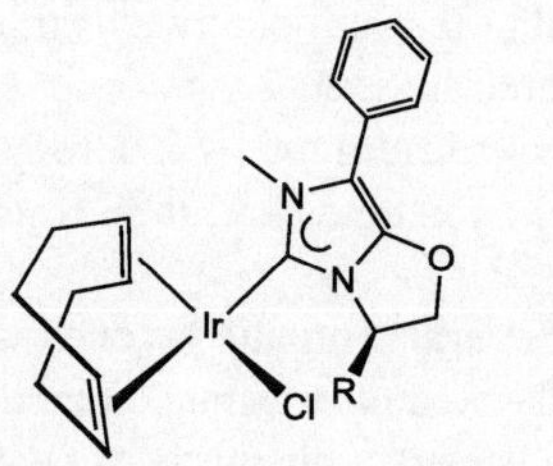

R = *s*Bu (**62a**), *i*Bu (**62b**), *i*Pr (**62c**)

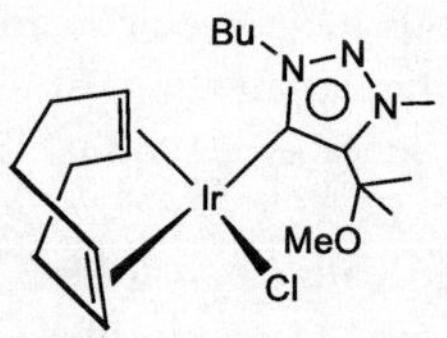

63

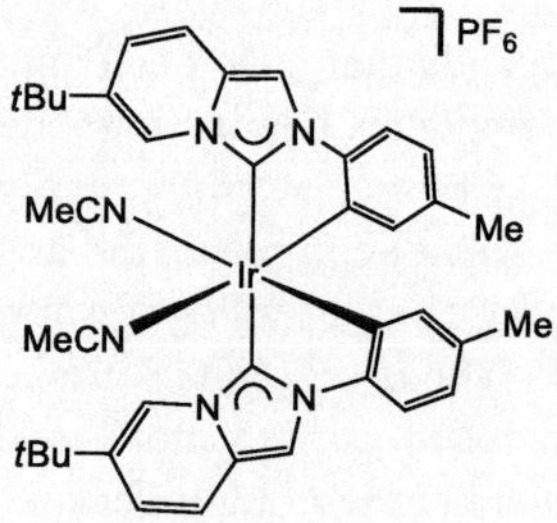

64

Scheme 32 Carbene iridium catalysts (**58–64**)

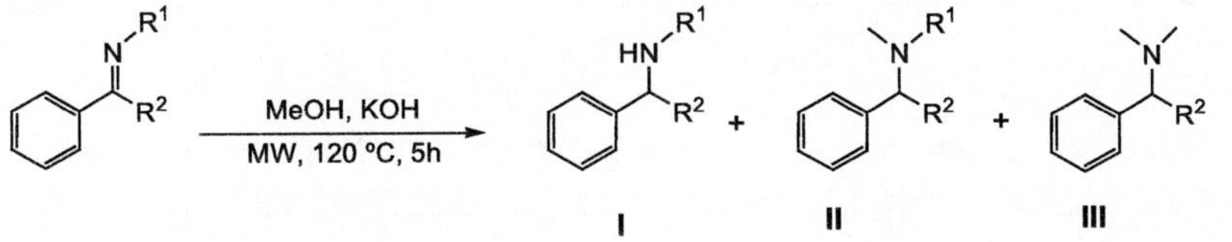

R^1	R^2	Yield (%) I	Yield (%) II	Yield (%) III
Ph	H	>95	>1	--
Bn	H	15	70	--
Me	H	<1	63	--
*n*Bu	H	<1	50	--
H	Ph	20	34	20

Scheme 33 Reduction of imines by catalyst **59a**

Cationic iridium complexes containing two *N*- or *O*-functionalised *N*-heterocyclic carbene ligands (**60**, Scheme 32) reduce cyclohexanone using 2-propanol as a hydrogen source and KOH as a base. Under the optimised reaction conditions (0.1 mol% catalyst, 5 equiv. of KOH, 80°C), good conversions were obtained after 6–8 h of reaction. Neither the wingtip fragment of the carbenes nor the anion (Br^- or PF_6^-) has significant influence on the catalytic activity [152].

Iridium(III) complexes featuring methylene bridged bis-*N*-heterocyclic carbene ligands **61** catalyse the reduction of a variety of ketones and imines in 2-propanol as a hydrogen donor, *t*BuOK as a base (5 equiv. versus substrate) and at catalyst loading of 1 mol%. Cyclohexanone and methyl aryl ketones **B1**, **B3**, **B15**, **B18** and **B20** were efficiently reduced. Conversions from 77 to 99% were obtained in 0.33–23 h of reaction with TOF of up to 615 h^{-1} at 50% conversion. A remarkable enhancement of the activity in the reduction of both ketones and imines was observed for carbenes with *N*-substituents where an oxygen atom is present [153]. *N*-benzylideneanilines **C1–C4** were also efficiently hydrogenated. Conversions from 38 to 96% were obtained after 2–24 h of treatment [153].

Iridium complexes containing *N*-heterocyclic carbene ligands derived from a chiral fused bicyclic scaffold (**62**, Scheme 32) have been explored in asymmetric TH of ketones. Under the optimised reaction conditions (0.5 mol% of catalyst **62c**, *i*PrOH, 8 equiv. of *t*BuOK, 75°C), a variety of aryl ketones (**B1, B2, B15, B16, B19, B20-B22, B27, B28, B49, B50, B52**, Scheme 4) were reduced. Moderate to high yields (18–95%) and low e.r.'s, from 51.5/48.5 to 62/38 were obtained. Iridium complex **62c** performs better than its rhodium counterpart in terms of both yield and enantioselectivity for most of the tried ketones [154].

An iridium(I) compound with a triazolylidene ligand that contains a pendant methyl group (**63**, Scheme 32) efficiently catalyses the TH of ketones, aldehydes, imines, allylic alcohol and olefins [155].

After screening it was established that the optimised reaction conditions are 0.5 mol% of catalyst and 5 mol% of *i*PrONa, in refluxing 2-propanol. Under these conditions, ketones with different steric and electronic properties (**B1, B3, B10, B13,**

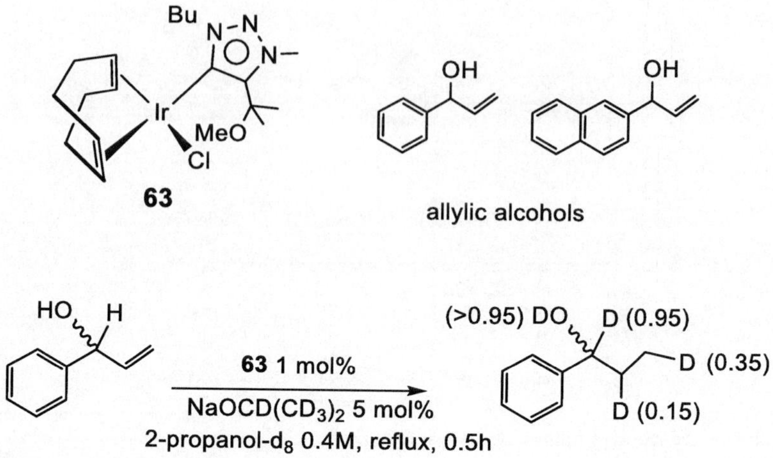

Scheme 34 Deuterium-labelling experiments in the reduction of allylic alcohols by complex **63**

B15, **B18**, **B23**, **B47**, **B52**, **B66**, **B118**) including heterocyclic ketones (**B38-B42**, **B45**) were reduced in high yield (81–99%) with TOFs in the 260–1,800 h^{-1} range, measured after 5 min of reaction. As usual, conversion increases both when the steric demand decreases and when the electro-withdrawing character of ketone substituents increases.

Several aldehydes (**A1**, **A3**, **A12**, **A17**, **A24**, **A32**, **A64**, **A68**) were also reduced by catalyst **63** under the same reaction conditions. The activity follows the same trend as that observed for the reduction of ketones, although, in general, when using complex **63**, aldehydes are reduced significantly faster than ketones.

Complex **63** was also applied to the TH of the imines *N*-benzylideneaniline and *N*-2-methyl benzylideneaniline, under the optimised reaction conditions above indicated. The TOF (770 h^{-1}) measured for the aldimine was significantly greater than that observed for the ketimine (30 h^{-1}), which was rationalised in terms of the higher steric demand of the ketimine.

Full reduction of the allylic alcohols depicted in Scheme 34 to the corresponding saturated alcohols was observed within 1 h. Deuterium-labelling experiments showed that the formed alcohol contained deuterium not only at the double bond but also at the allylic position, which indicates a competing isomerisation pathway.

Complex **63** induces the efficient reduction of a range of olefins (**D1**, **D10-D12**, **D15**, **D19**, **D21**, **D22**, **D29**, **D30**, **D32**, **D33**, Scheme 6) under mild reaction conditions (0.5 mol% catalyst, 5 mol% *i*PrONa, 4–6 h, refluxing 2-propanol, quantitative conversion). In general, the catalytic activity was insensitive to the olefin substitution pattern as well as to the geometry of the double bond, and activities were similar for linear mono- and di-substituted olefins.

To assess the relative rate of olefin hydrogenation versus alkene double bond isomerisation, deuteration-labelling experiments were performed using *trans-β*-methylstyrene and allyl benzene as the substrates (Scheme 35). Deuterium incorporates at both olefinic and allylic positions with almost equal ratios for both substrates.

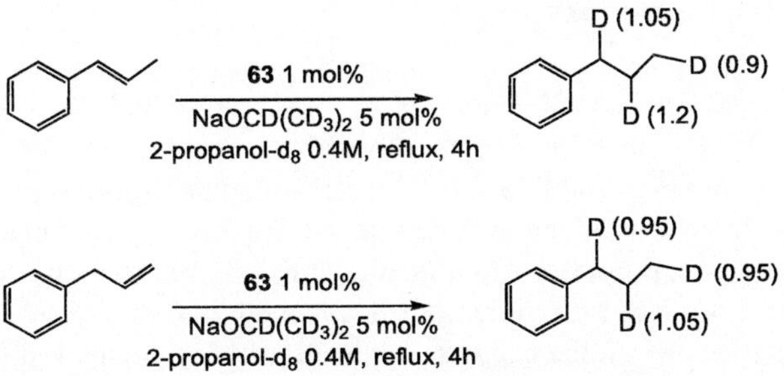

Scheme 35 Deuterium-labelling experiments in the reduction of olefins by complex **63**

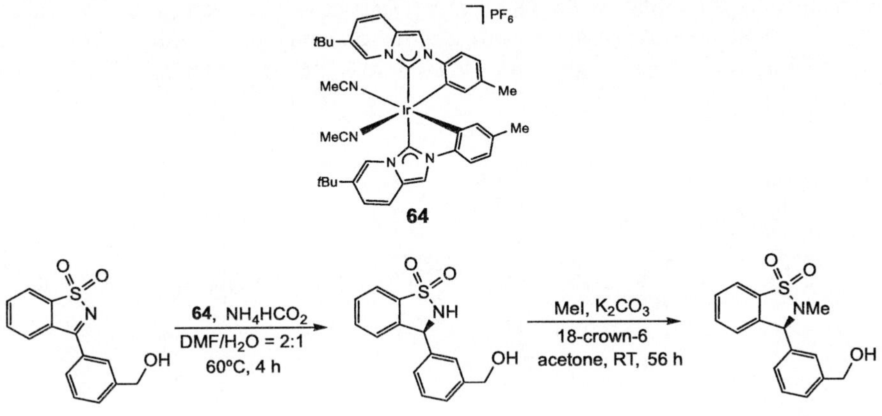

64

C58

sultam I

Scheme 36 Enantioselective synthesis of sultam **I**

This result suggests that the reductions proceed via a tandem isomerisation/TH reaction and that isomerisation is faster than reduction [155].

A cyclometalating *N*-heterocyclic carbene iridium complex featuring metal-centred chirality (**64**, Scheme 32) is an efficient catalyst for the asymmetric TH of a wide variety of cyclic *N*-sulfonylimines (**C36-C51**) and benzofused *N*-sulfonylimines (**C52-C59**) (Scheme 5). Enantioselective reduction of *N*-sulfonylimines yields chiral sultams, which are present in a range of biologically active compounds [156]. Yields in the range 82–99% and e.r.'s greater than 97/3 were obtained after 3–9 h of treatment, using HCO_2NH_4 (9 equiv) as a hydrogen source, at a catalyst loading of 0.05 or 0.2 mol%, in DMF/H_2O (2/1 mixture) at 60°C. Other common hydrogen donors that do not contain ammonium such as HCO_2Na or HCO_2H/NEt_3, 5/2, led to poorer results. A tentative mechanism including NH_3 containing intermediates was proposed [157].

A half-gram-scale synthesis of bioactive sultam **I** (Scheme 36) with anti-HIV activity was developed based on asymmetric TH of *N*-sulfonylimine **C58** followed by an *N*-methylation step [157].

4 Pincer Iridium Complexes

The term "pincer" refers to ligands that generally coordinate in a *mer* tridentate configuration [158]. This tridentate coordination mode results in strong binding to the metal centre and high stability for the metal-tridentate ligand unit. Additionally, pincer ligands have a modular nature that allows for a high tunability of their electronic and steric properties. Within the last two decades, the use of pincer compounds for chemical transformations has increased greatly [158–161]. A number of reviews focusing on the catalytic applications of this type of complexes have been published in the recent years [162–170]. In particular, some pincer iridium complexes have been found to be effective for the TH of unsaturated substrates [171–175]. Here, we summarise the progress in this area since 2015.

With only 0.1 mol% of the chiral *P,N,N'*-pincer catalyst **65a** (Scheme 37), at 40°C, several aromatic ketones with electron-donating or electron-withdrawing substituents in 2–3- and 4-positions (**B1-B3**, **B5**, **B6**, **B8**, **B10-B12**, **B15**, **B18**,

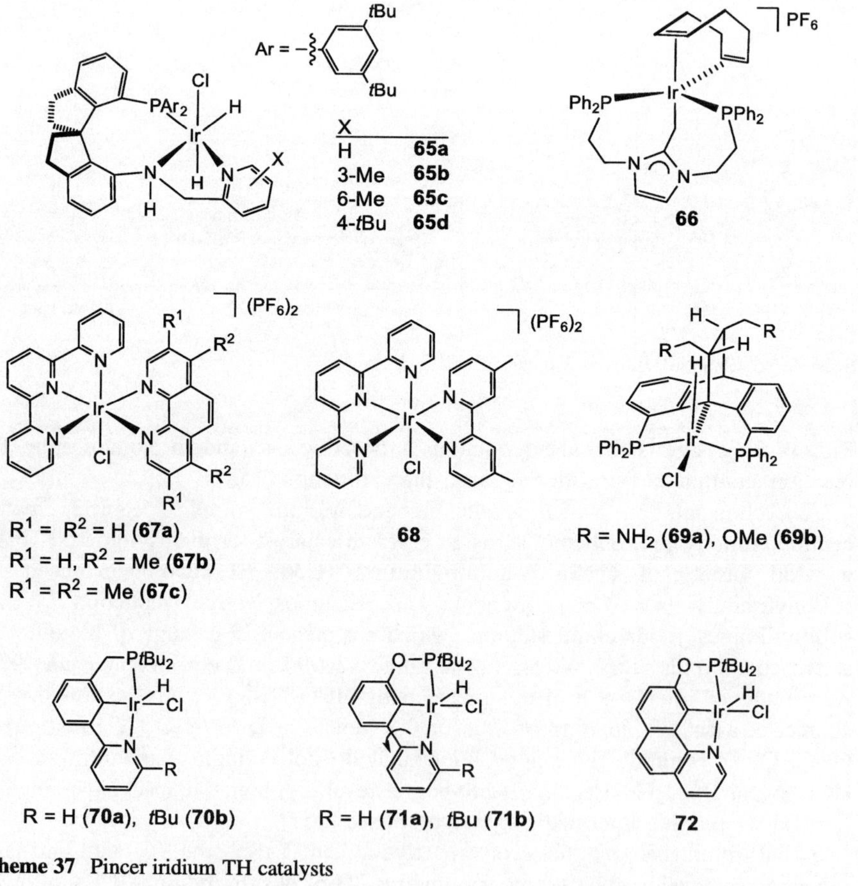

Scheme 37 Pincer iridium TH catalysts

B21, **B22**, **B34**, **B37**, **B47**) as well as selected methyl alkyl (**B99**, **B104**) and the methyl pyridinyl ketone **B39** were reduced. Potassium *t*-butoxide (4 equiv. versus substrate) was employed as a base but also in the presence of other bases such as *t*BuONa, KOH and K_2CO_3, acetophenone was hydrogenated to 2-phenylethanol with comparable activity and selectivity. Isolated yields from 87 to 99% and e.r.'s from 90/10 to 99/1 were obtained within 10 h. Ethanol was used as a hydrogen source. Using 2-propanol, instead, gave rise to significantly lower e.r.'s (99/1, EtOH versus 92/8, *i*PrOH) [176].

A practical and sustainable method for the preparation of optically active propargyl alcohols was developed by the Zhou's group [177]. The tridentate spiro-pyridine-aminophosphine iridium complexes **65** depicted in Scheme 37 catalyse the chemoselective reduction of a series of alkynyl ketones to the corresponding propargyl alcohols. Both the unsubstituted complex **65a** and the methyl-substituted derivatives **65b** and **c** were more active and enantioselective than compound **65d** having a *t*Bu substituent. The optimised reaction conditions were 1 mol% of catalyst **65b**, HCO_2Na (2 equiv), 60°C and ethanol as solvent. Under these conditions, high yield (86–99%) and high e.r.'s (93/7–98.5/1.5) were obtained within 4–48 h. Alkynyl ketones containing electron-withdrawing (**B130**, **B132**, **B135**, **B136**) and electron-donating (**B123-B125**, **B129**, **B131**, **B133**, **B134**) or additional ester (**B127**, **B128**) groups as well as the trifluoromethyl ketone **B126** were efficiently and chemoselectively reduced [177].

Some remarkable features of the process are:

1. No base is required. Indeed, in the presence of *t*BuOK, 4-phenylbut-3-yn-2-one (**B123**) only gave the Michael addition byproduct formed by the ethoxide addition to the carbon-carbon triple bond of **B123.**
2. The TH of **B123** did not occur using HCO_2H/NEt_3 instead of HCO_2Na as a hydrogen donor.
3. Other alkali metal formates such as HCO_2Li, HCO_2K and HCO_2Cs can also be used as hydrogen sources although HCO_2Li gave lower reaction rate and conversion.
4. Ethanol is the best solvent; indeed, the reaction in MeOH or *i*PrOH gave low conversion and low enantioselectivity.
5. Under the optimised conditions but using HCO_2Cs as a hydrogen donor, the TH of the trifluoro alkynyl ketone **B126** mediated by **65b** was monitored by in situ IR spectroscopy.

These results indicate that the formate salt and EtOH served as the hydride and proton sources, respectively, in the TH reaction (Scheme 38) [177].

The iridium complex **66** (Scheme 37) containing a functionalised *N*-heterocyclic olefin (Scheme 39) acting as a tridentate ligand has been applied as a catalyst for the reduction of ketones, benzaldehyde and imines [178].

The reaction conditions entailed the use of 0.1 mol% of catalyst, *i*PrOH as a solvent and a hydrogen donor, *t*BuOK (5 equiv) as a base and working at 80°C. Under these conditions, cyclohexanone, substituted acetophenones (**B1**, **B3**, **B15**, **B18**, **B20**), dialkyl (**B114**) and diphenyl (**B66**) ketones were efficiently reduced. For example, the TH of cyclohexanone to cyclohexanol was completed in 7 min with a

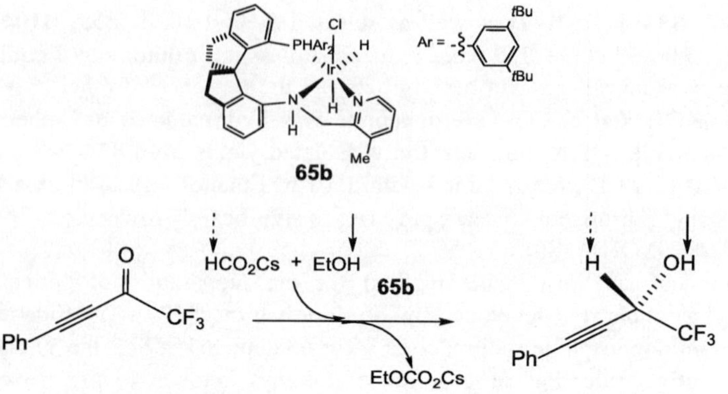

Scheme 38 Transfer hydrogenation of ketone **B126** with HCO_2Cs in ethanol

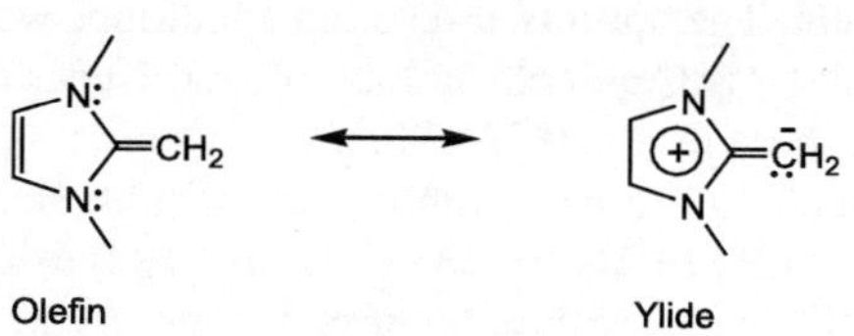

Scheme 39 Resonance structures of *N*-heterocyclic olefins

TOF value of 2,222 h^{-1}, at 10% conversion [178]. No relationship was observed between the substitution of the acetophenones and the catalytic activity. The reduction rate of aliphatic ketones is higher than that of their aromatic counterparts. Remarkably, the TH of benzaldehyde to benzyl alcohol was completed in 40 s.

Catalyst **66** showed good activity in the TH of imines. *N*-benzylideneaniline was transformed into *N*-benzylaniline, using 1 mol% of catalyst, in 45 min with a TOF value of 1,118 h^{-1} at 10% conversion. Substituted imines 3-methoxy-*N*-benzylideneaniline and 4-methoxy-*N*-benzylideneaniline were also converted into their corresponding amines, under the same reaction conditions, featuring TOFs of 548 h^{-1} and 110 h^{-1}, respectively [178].

Polypyridyl iridium(III) compounds **67** and **68** (Scheme 37) were tested as catalysts for the TH of a range of aromatic aldehydes, in aqueous ethanol, with HCO_2Na as a hydride source and using microwave-assisted heating. The best results were obtained with catalyst **67a**. Catalytic reactions were performed at 100°C, at a catalyst loading of 0.2 mol%, with 4.5 equiv. of HCO_2Na, in an EtOH/H_2O mixture, 70/30 (v/v). Under these reaction conditions, yields from 68 to 99% were obtained within 5–90 min. Reaction tolerates a wide range of substituents including halogens (**A5-A7**, **A14-A16**, **A29-A31**), phenols (**A11**, **A22**, **A49**), alkoxy (**A3**, **A12**, **A24**, **A40**, **A48**, **A49**), ketones (**A18**, **A34**), carboxylic acids (**A9**, **A20**, **A36**), cyano (**A19**, **A35**) and nitro (**A21**) groups, as well as heteroarenes (**A66**, **A68-A71**). Under the same conditions, catalysts **67a** and **67b** (Scheme 37) also reduced alkyl aldehydes **A54**, **A56**, **A58–A60** in 82–97% yield within 25–90 min [179].

Pincer compounds **69** (Scheme 37), equipped with different functional groups in the secondary coordination sphere, were employed as the catalysts in the chemoselective TH of nitroarenes (Scheme 43) to anilines. Reactions were carried out in dimethoxyethane, at 60–80°C with a catalyst loading of 1 mol%, with 2 equiv. of $HCOOH/NEt_3$ azeotrope as a hydrogen source. Although reduction of nitroarenes to amines is often accompanied by incomplete reduction to hydroxylamines or azocompounds, almost complete conversion of nitrobenzene into aniline (99%) with excellent selectivity (99%) was achieved with catalyst **69a**. The OMe-modified catalyst **69b** gave a more modest conversion but with the same selectivity [180].

Methyl- (**E4**), amino- (**E9**) and chloro- (**E10-E12**) substituted nitrobenzenes were fully converted into the corresponding anilines. Cyano- (**E18**), amido- (**E16**), ester- (**E17**) and keto- (**E15**) substituted nitrobenzenes as well as 2-nitrostyrene displayed moderate chemoselectivity owing to partial reduction of the functional group present in the substrates [180].

In contrast with the well-established TH of aldehydes, ketones and imines, TH reactions of nonactivated C-C multiple bonds have been much less studied. Huang's group has recently demonstrated that a series of *N,C,P*-pincer iridium(III) complexes (**70–72**, Scheme 37) efficiently reduced a wide variety of alkenes and alkynes [181, 182] (Scheme 6). Notably, ethanol was employed as a solvent and as a hydrogen donor. Ethanol is an abundant, sustainable and environmentally benign source of hydrogen. However, it has been seldom applied to TH reactions [38, 176, 177, 179, 183–184] because acetaldehyde, its dehydrogenation product, readily undergoes metal-mediated decarbonylation leading to catalytically inactive metal carbonyl species. In Huang's system EtOAc was the only detectable byproduct. Probably, the acetaldehyde resulting from EtOH dehydrogenation reacts with another molecule of EtOH to yield hemiacetal, which is further dehydrogenated to EtOAc thus eliminating the possibility of catalyst poisoning by the acetaldehyde [181].

Complex **72** gave the best results. Thus, for example, upon activation with *t*BuONa, this complex (1 mol%) completely reduced both cyclooctene and 1-octene at 60°C within 20 min. Under these conditions, other nonactivated (**D17, D25, D31, D33, D36, D37, D39-D48, D52-D56**), aryl (**D1, D5-D7, D11, D13, D18, D23, D24**) and electron-rich (**D28, D38**) alkenes were efficiently reduced. Yields from 65 to >99% were obtained, within 0.5–12 h [181].

Complex **72** was also active for the full reduction of alkynes. Nonactivated aliphatic alkynes, 6-dodecyne and alkynes bearing Cl (**D128**) or OH (**D127**) functionalities were reduced (94–97% yield, 1.5 h) to the corresponding alkanes using 1 mol% of **72**, at 80°C. Ethyl 2-octynoate and internal alkynes (**D57, D58, D88, D90**) gave the reduction products in >95% isolated yield, at 2 mol% of catalyst loading. Reduction was always complete due the good activity of catalyst **72** for alkene TH [181].

However, complex **72** can be used for the semi-hydrogenation of a range of internal alkynes [182]. Ethanol was also employed as a solvent, and *E*-alkenes were stereoselectively obtained. Transformations can be stopped at the alkene step

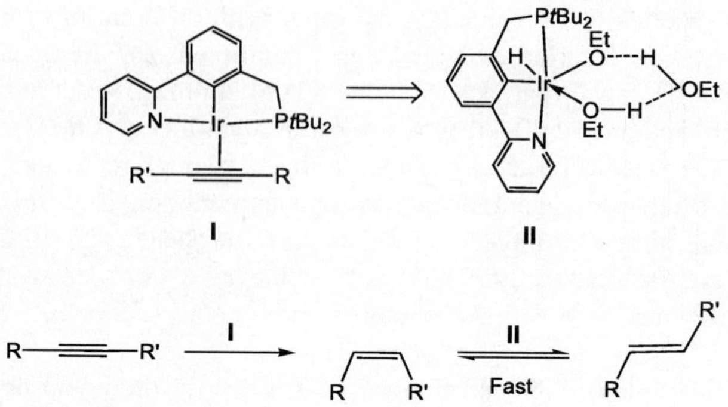

Scheme 40 Selective TH of alkynes to *E*-alkenes with EtOH

because the catalyst itself was used as a colour indicator for endpoint detection. Catalysis proceeds through the sequence alkyne → *Z*-alkene → *E*-alkene, the second step being faster than the first. The green iridium(I) π-alkyne complex **I** and the yellow iridium(III) hydride **II** act as the resting states for the first and second step, respectively (Scheme 40).

The full consumption of alkyne produces a sharp colour change of the solution from green to yellow due to the shift of the resting state from **I** to **II**. Quenching the reaction at this point allows for a high *E*-selectivity and minimizes over-reduction.

A broad variety of internal alkynes, including diaryl (**D57-D60**, **D62-D68**, **D71**, **D73-D77**, **D91**, **D92**, **D122**), aryl alkyl (**D88**, **D90**, **D95-D98**) and dialkyl (**D124**, **D126-D128**) alkynes were transformed in the corresponding *E*-alkenes. Functional groups such as amine (**D64**, **D83**), aryl halides (**D67**, **D68**, **D71**, **D73**), amide (**D75**), ester (**D76**) and ferrocenyl (**D86**) were tolerated. Alkynes containing various *O*- or *N*-heteroaromatic rings including furane (**D84**), benzofuran (**D85**), *N*-ethylcarbazole (**D83**), *N*-tosylindole (**D82**) and quinoline (**D81**) were also reduced. Generally, yields greater than 95% were achieved within 2.5–292 min.

Several alkynes containing biologically relevant skeletons can also be semi-hydrogenated to the corresponding *E*-alkene using catalyst **72** in EtOH. Scheme 41 shows some of the *E*-alkenes obtained [182].

5 Other Iridium Complexes

Mixtures of $[IrCl(COD)]_2$ (COD = 1,5-cyclooctadiene) and a variety of ligands (Scheme 42) have been tested as catalysts for the TH of a range of nitroarenes (Scheme 43).

Reactions were carried out under nitrogen atmosphere, in *i*PrOH, at 83°C, with 1 mol% iridium dimer and 2 mol% of the corresponding ligand. Monodentate phosphines displayed low catalytic activity. Bidentate phosphines and nitrogen

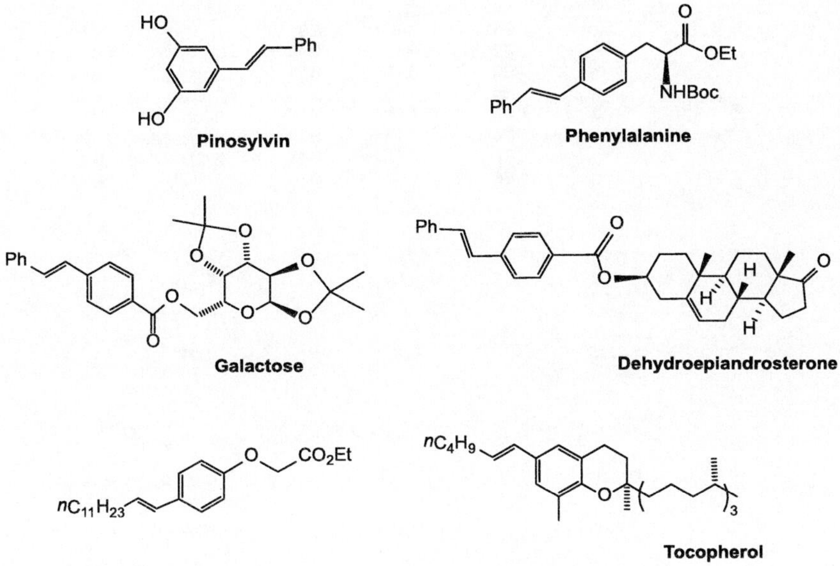

Scheme 41 Alkenes containing biologically active skeletons

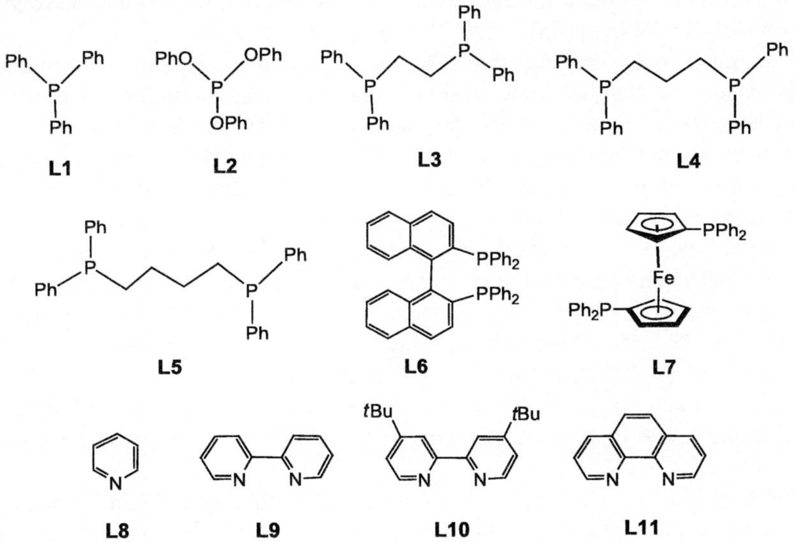

Scheme 42 Phosphorus and nitrogen donor ligands employed in combination with $[IrCl(COD)]_2$ in the TH of ketones

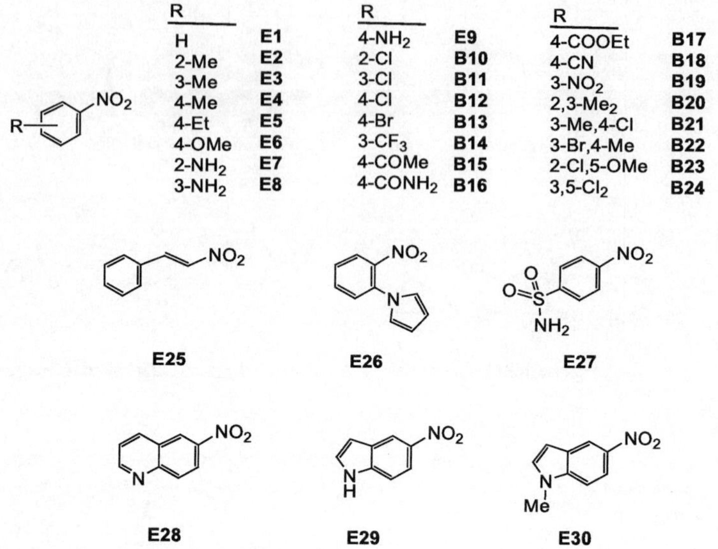

Scheme 43 Nitroarenes

donor ligands gave full conversion of nitroarenes but variable aniline yields (53–98%). The bidentate nitrogen ligand 1,10-phenanthroline (**L11**) showed the highest yield in aniline [185].

A monitoring experiment and a series of controlled experiments employing nitrobenzene as the substrate showed that the transformation proceeds via phenylhydroxylamine and azobenzene intermediates. However, after 15 h of treatment under the reaction conditions above indicated, using $[IrCl(COD)]_2$/**L11** as the catalyst, a series of nitroarenes (**E1-E3**, **E5-E14**, **E19-E24**, **E26-E30**, Scheme 43) were completely reduced to the corresponding anilines in 70–98% yield.

A phosphine bisbenzothienyl iridium(III) complex (**73**, Scheme 44) was applied to the TH of ketones. Under the best conditions found, namely, 2 mol% of **73**, 1.5 equiv. of *t*BuOH, toluene as the reaction solvent, at 110°C, several methyl aryl ketones (**B1**, **B2**, **B10**, **B156**, **B20**, **B27**, Scheme 4) were hydrogenated. Yields of the corresponding alcohol up to 68% were obtained [186].

A range of ketones was reduced to the corresponding alcohols by $[IrCl(COD)]_2$/chiral amine ligand mixtures. The amino ligands employed are collected in Scheme 45.

The best reaction conditions were as follows: a 1/2 metal/ligand ratio, a 1/1 mixture of formic acid and sodium formate (pH 3.5 at the beginning of the reaction) and water/methanol (1/1) as a solvent. After 1 h of treatment at RT under argon atmosphere, the reducing agent was added and the solution heated at 70°C. Under these conditions, moderate to high isolated yields (41–87%) and e.r.'s (57.5/42.5–99.5/0.5) were obtained for the reduction of a range of methyl aryl ketones (**B1**, **B2**, **B8**, **B12**, **B18**, **B21-B25**, **B27**, **B34**) [187].

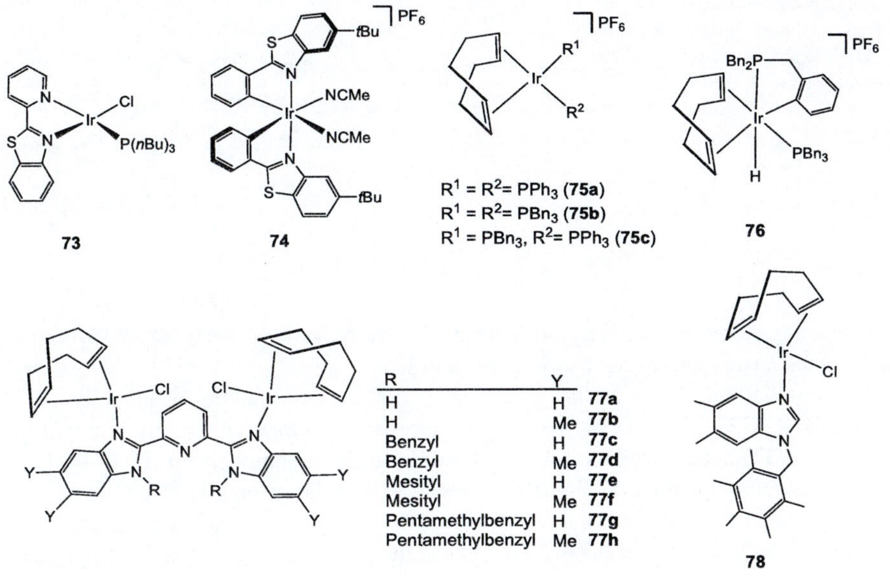

Scheme 44 Other iridium complexes employed as TH catalysts (**73–78**)

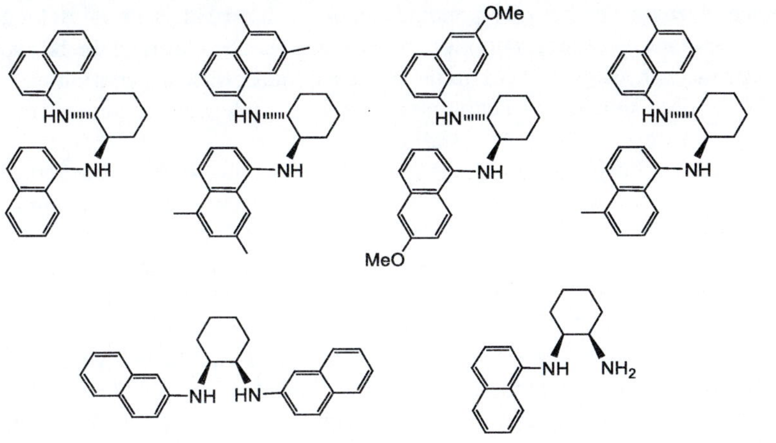

Scheme 45 Chiral amino ligands employed in combination with $[IrCl(COD)]_2$ in the TH of ketones

A bis-cyclometallated iridium(III) complex (**74**, Scheme 44) with metal-centred chirality catalyses the enantioselective TH of ketones in the presence of an additional pyrazole ligand. Reactions were carried out in THF/H_2O 1/1 (v/v) mixtures, at 40 or 60°C, with $HCOONH_4$ (9 equiv) as the hydrogen source. The best results were obtained by adding to the catalyst precursor **74** (0.2 mol%) 10 equiv. of 5-(4-methoxyphenyl)-3-methyl-1*H*-pyrazole. Under these reaction conditions, a

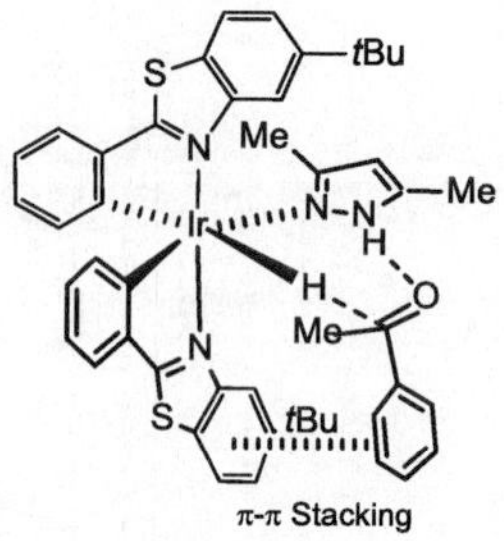

Scheme 46 Proposed transition state for the TH of ketones catalysed by complex **74**

wide range of ketones were hydrogenated. Generally, yields greater than 90% with e.r.'s > 98/2 were achieved, within 8–36 h [188].

Acetophenones with electro-donating or electro-withdrawing groups (**B1**, **B2**, **B15**, **B18**, **B21**, **B29**, **B34**, Scheme 4), aromatic ketones containing a naphthyl moiety (**B37**) heteroaromatic ring (**B41-B46**), larger aliphatic groups (**B47-B49**) or an additional ester functionality (**B62**) as well as a cyclic ketone **B121** provided both high yield and good e.r. Diaryl ketones **B68** and **B69** also afforded satisfactory results. Dialkyl ketones **B107** and **B110** gave the corresponding alcohol in high yield (> 90%) but with low e.r. (< 30%). However, ketone **B111** with a bulky alkyl substituent gave the desired alcohol with 93% yield and 97/3 e.r. The catalytic system also works at a lower catalyst loading. Thus, 2-acetyl benzothiophene was completely reduced to the corresponding alcohol within 108 h at a catalyst loading as low as 0.005 mol%. The reaction was proposed to proceed through the bifunctional Noyori's mechanism [189, 190]. In the proposed transition state, an iridium-hydride and the N-H functionality of a coordinated pyrazole molecule interact with the C=O bond of the ketone (Scheme 46) [188].

The iridium phosphine complexes **75** and **76** (Scheme 44) are active in the TH of ketones. The best results in conversion to the alcohol were obtained using catalyst **75c**. Reactions were carried out in *i*PrOH, at 80°C, with a catalyst loading of 0.5 mol% and KOH (5 mol%) as a base. A series of methyl ketones (**B1**, **B18**, **B27**), one example of an ethyl ketone (**B47**), cyclic ketones (**B118**, **B122**) or the bulky ketones **B112** and **B122** were tested in the reaction. Generally, moderate to high yields were obtained (69 – > 99%), but the ethyl ketone **B47** and the bulky ketones **B112** and **B122** were converted to the corresponding alcohol in low yield (< 36%) [191].

Mixtures of the dimer $[IrCl(COD)]_2$ and chiral ferrocenyl alcohols (Scheme 47) have been applied to the TH of alkyl aryl (**B1**, **B47**) and phenyl heteroaryl **B91** and **B92** ketones.

The reaction conditions were as follows: a 1/1 $[IrCl(COD)]_2$/ligand ratio in *i*PrOH as a solvent and a hydrogen source, at 25 or 50°C and in the presence of KOH (2 equiv) as a base. Moderate yield (35–89%) and e.r. (52/48–91/9) were obtained [192].

The iridium complexes **77** and **78** (Scheme 44) reduced ketones under TH conditions. Typically, reactions were performed at 0.25 mol% of catalyst loading, in the presence of KOH (40 equiv) as a base and at 82°C. Alkyl aryl ketones **B1**, **B5**,

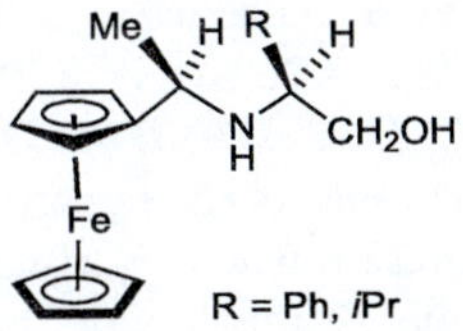

Scheme 47 Chiral ferrocenyl amino alcohols employed in combination with $[IrCl(COD)]_2$ in the TH of ketones

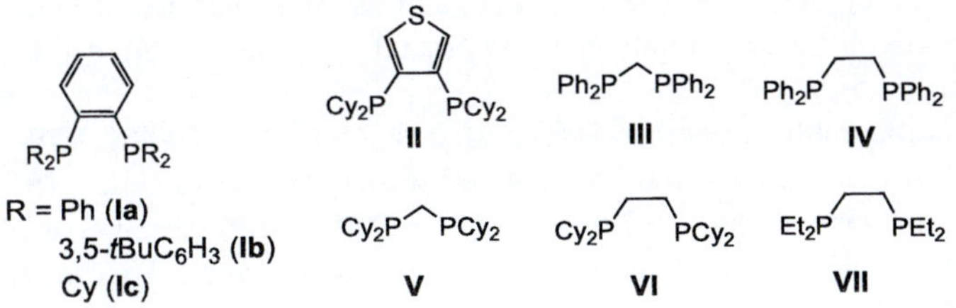

Scheme 48 Diphosphines employed in combination with $[IrCl(COD)]_2$ in the TH of alkenes and alkynes

B6, **B18**, **B21**, **B22**, **B28**, and **B47**; methyl alkyl ketones **B93–B95**, **B97**, and **B109** and naphthyl aldehyde were reduced. It was found that bimetallic complexes **77** were more active than monometallic complex **78**. Yields higher than 90% within 15–30 min were obtained [193].

Sawamura's group reported on the reduction of C=C bonds of alkenes with 1,4-dioxane as a solvent and a two-hydrogen donor, in the presence of $[IrCl(COD)]_2$ and a diphosphine ligand (Scheme 48). Reactions were performed at 120–130°C at a catalyst loading of 1–4 mol%.

The reactivity of the TH was enhanced by using bulky and electron-donating diphosphines such as 1,2-bis(dicyclohexylphosphino)ethane **VI**. Styrene derivatives **D2-D4**, **D8**, **D9**, **D13**, **D20**, **D23** and **D26** (Scheme 6), exocyclic (**D26**) and cyclic (**D13**) olefinic bonds as well as aliphatic alkenes (**D16**, **D32**, **D35**) including 1,1-disubstituted terminal (**D35**) and disubstituted (**D27**, **D34**) or trisubstituted (**D30**) internal alkenes were efficiently hydrogenated by $[IrCl(COD)]_2$/diphosphine **VI** mixtures. Tetrasubstituted alkenes such as 2,3-dimethyl-1*H*-indene and tetraphenylethylene were not hydrogenated even at higher reaction temperature (140°C).

Alkynes also underwent TH reaction with the $[IrCl(COD)]_2$/diphosphine **VI** catalyst using 1,4-dioxane as a hydrogen donor. Thus, diphenylacetylene was converted into 1,2-diphenylethane through double TH with 4 mol% of catalyst loading, at 140°C after 20 h of treatment [194].

Transfer hydrogenation of internal and terminal alkynes to alkenes can be achieved with the $[IrCl(COD)]_2$/DPPE (DPPE = 1,2-bis(diphenylphosphino)ethane) using EtOH as a hydrogen donor [195, 196].

Internal alkynes were reduced to *E*-alkenes in THF, at 120°C using 2.5 mol% of $[IrCl(COD)]_2$, 20 mol% of DPPE and 20 equiv. of EtOH. Under these conditions, a variety of diaryl alkynes (**D57**, **D61**, **D70-D74**, **D78**, **D100-D102**, **D104-D117**, **D121**, Scheme 6) were converted into *E*-alkenes in 73–92% yield, with *E*/*Z* molar ratios from 4/1 to >99/1, within 22 h under N_2 atmosphere. Notably, the *E*/*Z* ratio can be inverted in favour of the *Z* isomer working under the conditions above indicated but adding to the reaction medium 2 equiv. of COD. Thus, yields between 71 and 95% with *Z*/*E* molar ratios from 4/1 to >99/1 were achieved in the TH of the diaryl alkynes **D57**, **D61**, **D67**, **D69**, **D78**, **D100**, **D101** **D103-D105**, **D107**, **D109** and **D118** (Scheme 6). The observed change in the alkene selectivity was attributed to the increase of the steric hindrance at the metal centre generated by the addition of the diolefin [195].

Under comparable conditions (THF, 70°C, 2.5 mol% of $[IrCl(COD)]_2$, 20 mol% of DPPE, 2 equiv. of EtOH) terminal alkynes were reduced to alkenes [196]. A series of activated (**D130-D155**) and unactivated (**D156-D159**) terminal alkynes were reduced in 67–94% yield within 24 or 48 h, respectively. The reaction of the nonactivated alkynes was carried out at 100°C.

6 Transfer Hydrogenation of *α,β*-Unsaturated (and Non-conjugated) Alkene-Carbonyl Substrates

Transfer hydrogenation protocols have also been applied to the reduction of *α,β*-unsaturated carbonyl compounds [197]. In principle, both partial hydrogenation to aliphatic ketones or allylic alcohols and complete reduction to saturated alcohols are possible (Scheme 49). The energy barriers for the reduction of C=O and C=C conjugated bonds are often similar. For this reason, mixtures of reduction products are sometimes obtained, and one of the major goals in the reduction of unsaturated carbonyl compounds is to achieve high chemoselectivity. As it will be seen herein, the pH value of the catalytic medium plays a key role in the chemoselectivity in TH of unsaturated alkene-carbonyl substrates.

Iridium compounds are among the most used for *α,β*-unsaturated carbonyl compounds reduction. In this section we present the new contributions made in this field during the last 5 years.

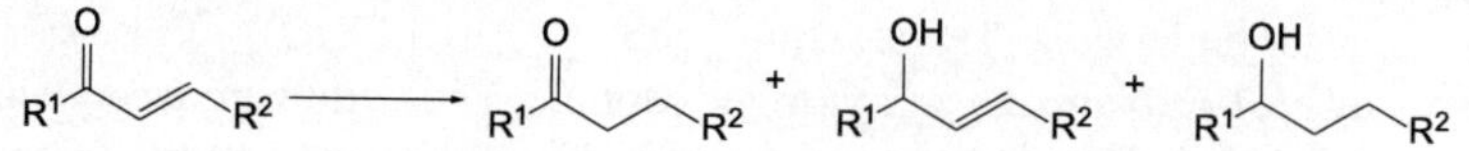

Scheme 49 Partial and complete reduction of *α,β*-unsaturated carbonyl compounds

6.1 Transfer Hydrogenation of α,β-Unsaturated Aldehydes

A variety of α,β-unsaturated aldehydes (Scheme 50) have been reduced employing iridium compounds as the catalysts (Scheme 51).

Compound **57a** catalyses the chemoselective reduction of cinnamaldehyde to the corresponding alcohol in 90% yield, in water, using HCOOH (5 equiv) as a hydrogen donor [113].

Under neutral conditions, complex **79** is an efficient catalyst for chemoselective TH of unsaturated aldehydes using *i*PrOH as a solvent and a hydrogen donor [198]. Cinnamaldehydes **F1**, **F2**, **F4**, **F11-F13**, **F22** and **F23**, 3-(9-anthryl) acrylaldehyde, 3-(2-furyl)acrylaldehyde and aliphatic α,β-unsaturated aldehydes **F29** and **F30** were reduced in 80–97% yield. A metal-ligand bifunctional mechanism was proposed for the catalytic TH reaction [198].

The polypyridyl iridium(III) **67a** in EtOH/H_2O, at 100°C, at 0.2 mol% of catalyst loading, in the presence of 4.5 equiv. of HCOONa, reduced cinnamaldehyde. A 1/1 mixture of the allylic and saturated alcohol was obtained. However, using 9 equiv. of HCOONa (initial pH of the reaction, 9) full reduction to the saturated alcohol was achieved. Similar results were obtained for α-methyl cinnamaldehyde and 4-methoxy cinnamaldehyde. Alternatively, when HCOOH was used to maintain acidic the reaction medium, unsaturated alcohol was predominantly obtained [179].

The Luo's group investigated the TH of a wide variety of α,β-unsaturated aldehydes, including cinnamaldehydes **F1**, **F3**, **F6- F11** and **F13-F23**, furan-substituted aldehyde **F27** and aliphatic aldehydes **F31-F35**, mediated by catalyst **55 h**. All the reactions were carried out at 0.1 mol% of catalyst loading, in H_2O at 80°C. Nonetheless, a set of reactions was performed using HCOONa (5 equiv) as a hydrogen donor, and another set was carried out with HCOOH (5 equiv). Full reduction to the saturated alcohol was achieved with HCOONa, and semi-reduction to the unsaturated alcohol was obtained with the acid. Typically, yields higher than 90% were achieved, in both cases [199]. Control experiments showed that 3-phenylpropanal was reduced to the saturated alcohol under the same conditions, whereas the C=C bond of cinnamyl alcohol was not reduced by complex **55h**, when using HCOONa as a hydrogen donor. These results indicated that the formation of the fully reduced product starts with the reduction of the C=C bond to give the saturated aldehyde intermediate and follows with the reduction of the C=O bond to give the final product [199].

6.2 Transfer Hydrogenation of α,β-Unsaturated Ketones

A range of α,β-unsaturated ketones (Scheme 52) have been reduced employing iridium catalysts (Scheme 53).

The dimer $[Cp^*IrCl_2]_2$ chemoselectively reduced α,β-unsaturated ketones such as (*E*)-chalcones **G1-G3**, **G6**, **G8-G17**, **G21**, and **G28**, heterocyclic α-enones **G29** and

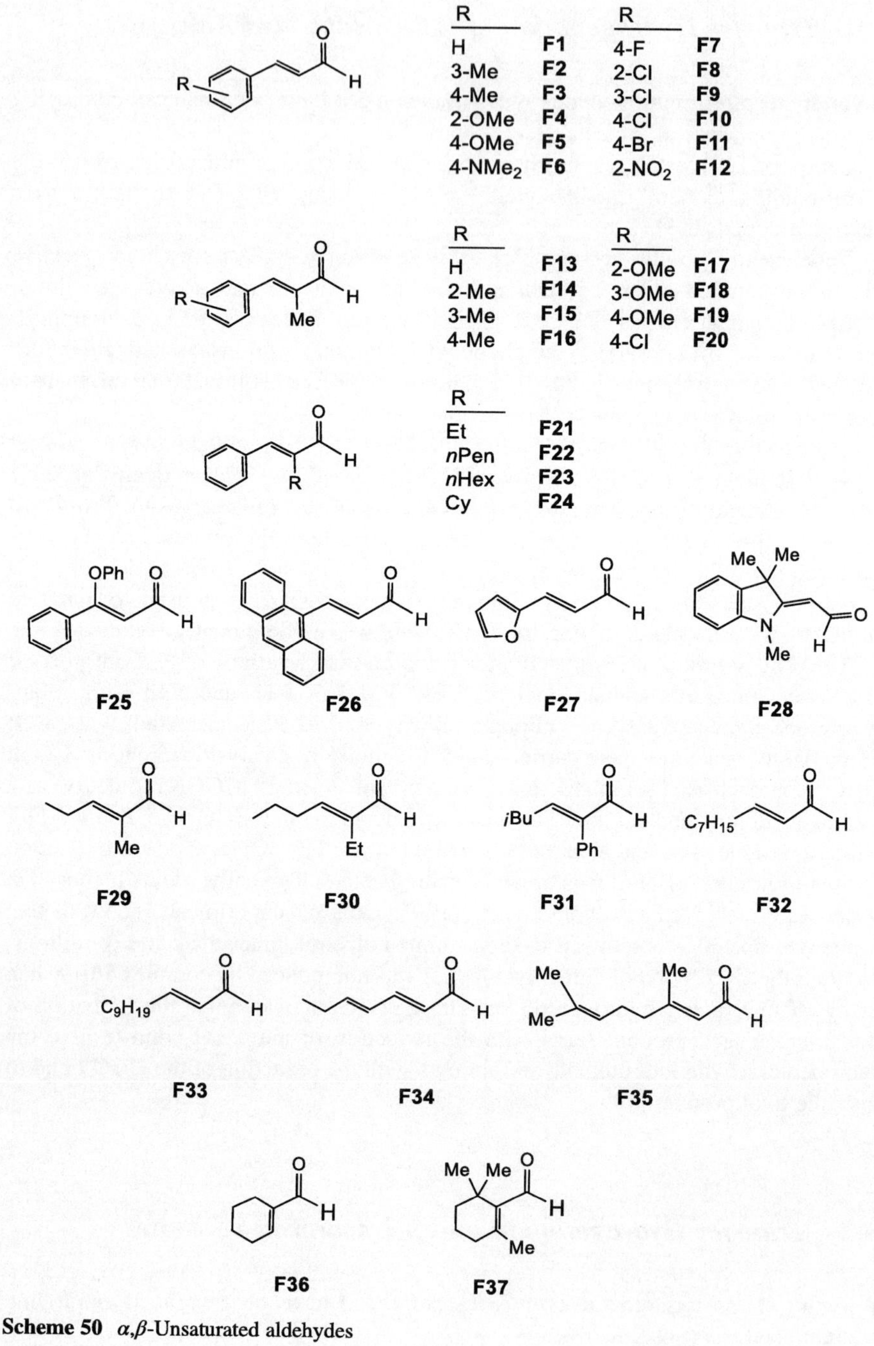

Scheme 50 α,β-Unsaturated aldehydes

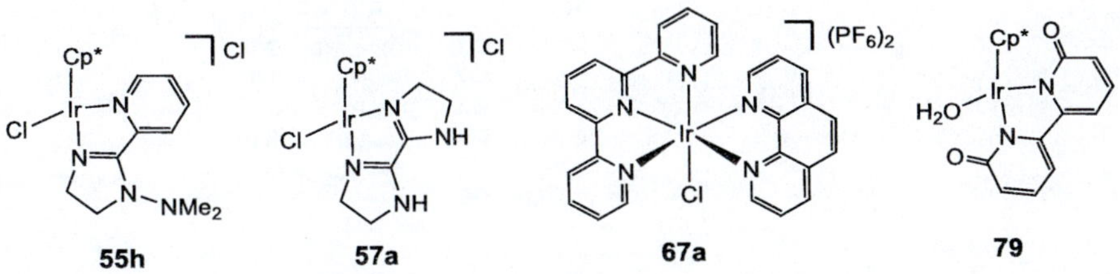

Scheme 51 Catalysts for the reduction of α,β-unsaturated aldehydes

G31, activated α,β-unsaturated ketones **G32** and **G40** and cyclic enones **G49** and **G50**. All the reactions were performed in *i*PrOH, at 85°C, using 1 mol% of catalyst loading. When K_2CO_3 (5 mol%) was employed as a base, the saturated ketone was obtained in high yield (85–98%). However, when in the reduction of the (*E*)-chalcones **G1–G3**, **G6**, **G8–G14** and **G28**, KOH (50 mol%) was employed as a base, the corresponding saturated alcohols were obtained in 83–91% yield [200].

The iridium(I) complex with triazolylidene ligands **63** catalysed the reduction of α,β-unsaturated ketones **G1**, **G3** and **G50**. Full reduction to the corresponding saturated alcohol was observed within 1 h when the substrates were treated with 0.5 mol% of the catalyst, in refluxing *i*PrOH with *i*PrONa (5 mol%) as a base. Yields higher than 94% were achieved. From deuterium-labelling experiments, it was suggested that the reaction proceeds via a tandem isomerisation/TH reactions [155].

In *i*PrOH, at 80°C, using KOH (10 mol%) as a base, the bis(phosphine) complex **75c** (0.5 mol%) reduced 2-cyclohexenone to the saturated ketone cyclohexanone as the main product (63% yield). The unsaturated alcohol and full reduced cyclohexanol were obtained in 6% and 31% yield, respectively. Under similar conditions, the TH of carvone gave the product of the hydrogenation of the conjugate olefin and that of the hydrogenation of both conjugated olefin and carbonyl group. Hydrogenation of the isolated C=C bond was not observed [191].

The outcome of the reduction of 2-cyclohexenone with catalyst **55a** (S/C = 10,000, HCOOH 12 equiv., water, 80°C, N_2 atmosphere) was the saturated alcohol cyclohexanol. Further studies showed that under the reaction conditions cyclohexanone could be completely reduced to cyclohexanol, while cyclo-2-en-1-ol could not. This result indicated that the TH of 2-cyclohexanone should first occur at the C=C bond and then at the C=O bond. Experiments on 4-methylpent-3-en-2-one gave similar results [111].

The combination of the dimer $[IrCl(COD)]_2$ (4 mol%) with the ligand 1,2-bis (dicyclohexylphosphino)ethane (4 mol%) promoted the chemoselective TH of α,β-unsaturated ketones such as chalcone, benzylideneacetone and its derivatives **G32–G45**, conjugated enoates **G46** and **G47**, enamide **G48** and cyclic enone **G51**. 1,4-dioxane was employed as a solvent. Chemoselectivity toward C=C hydrogenation was found for all these enones. Yields from 93 to 99% were obtained after 10 h of reaction at 130°C [194].

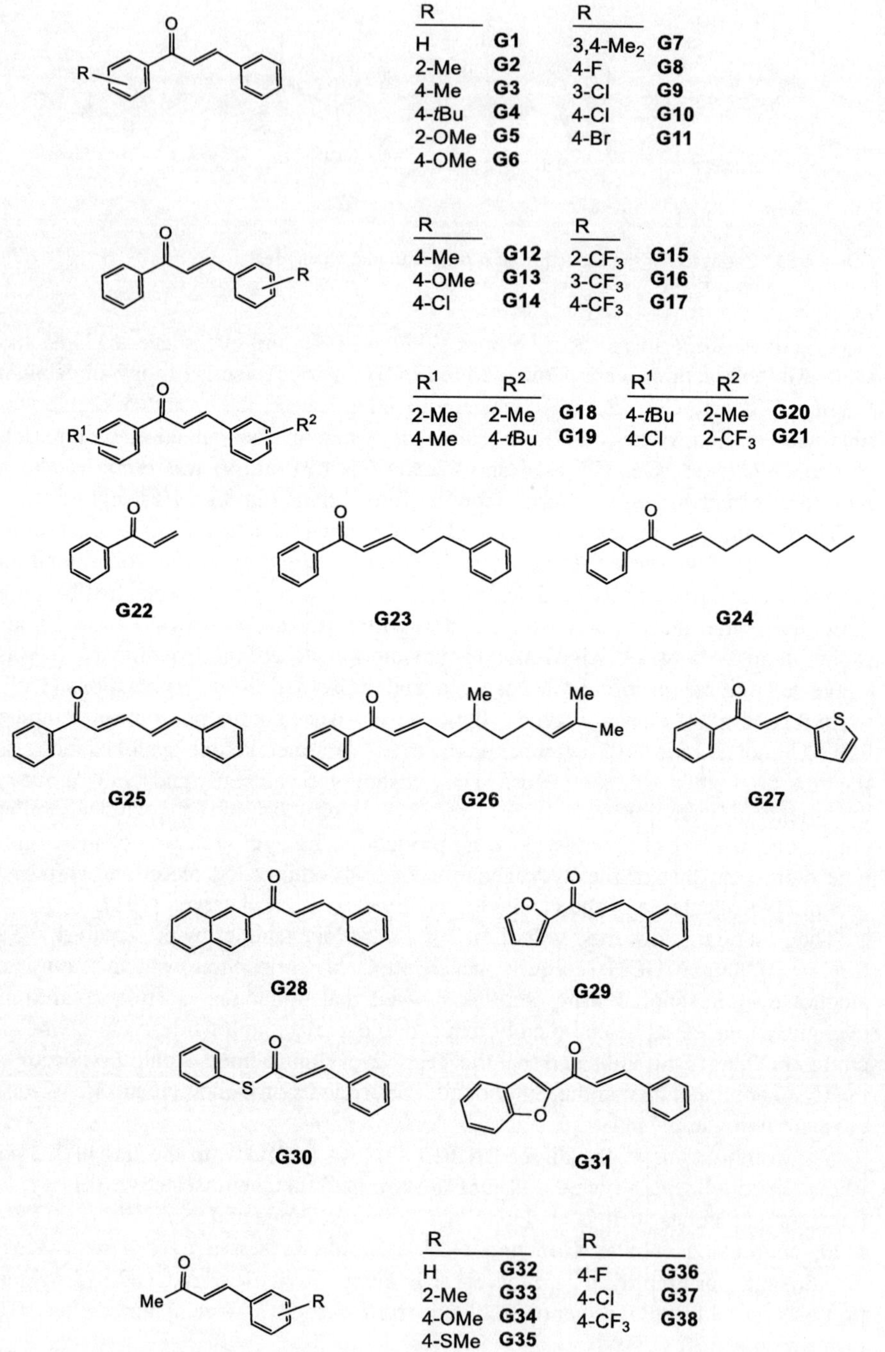

Scheme 52 α,β-Unsaturated ketones

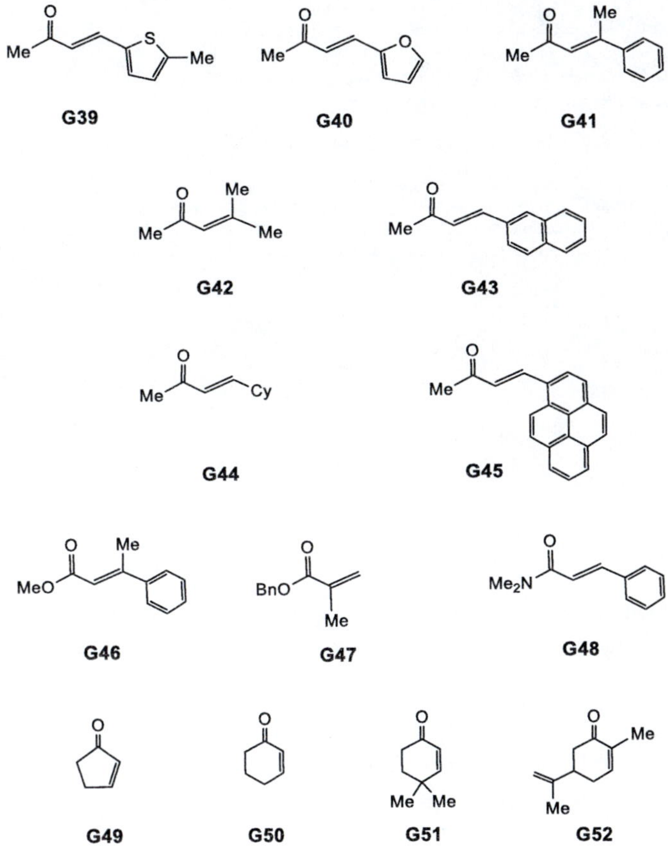

Scheme 52 (continued)

The iridium complex **80** bearing a tricyclic bisoxazoline-fused imidazole-derived *N*-heterocyclic carbene ligand (Scheme 53) catalysed the asymmetric TH of (*E*)-chalcones **G1**, **G2**, **G4**, **G5**, **G7**, **G8**, **G10**, **G11**, and **G18–G20**, as well as that of the heterocyclic *α*,*β*-unsaturated ketones **G27** and **G31**. In *i*PrOH, at 75°C, at a catalyst loading of 1 mol%, using NaOH (8 mol%) as a base, moderate to good yields (36–91%) in the corresponding full hydrogenated alcohol, were obtained within 3 h. Probably, the existence of an equilibrium between conformational isomers in the iridium catalyst **80** at the temperature of the reaction, originates a poor enantioface discrimination during the reaction resulting in poor e.r.'s (52.5/47.5–62.5/37.5). From detailed mass spectrometric studies, a mechanism consisting of two catalytic cycles working in tandem was proposed [201].

Hydrogenation of the (*E*)-chalcone mediated by the bis(carbene)catalyst **9a** (Scheme 2) was performed in *i*PrOH, at 95°C using KOH (10 mol%) as a base. The catalytic reaction was selective for the C=C bond and required only 30 min to

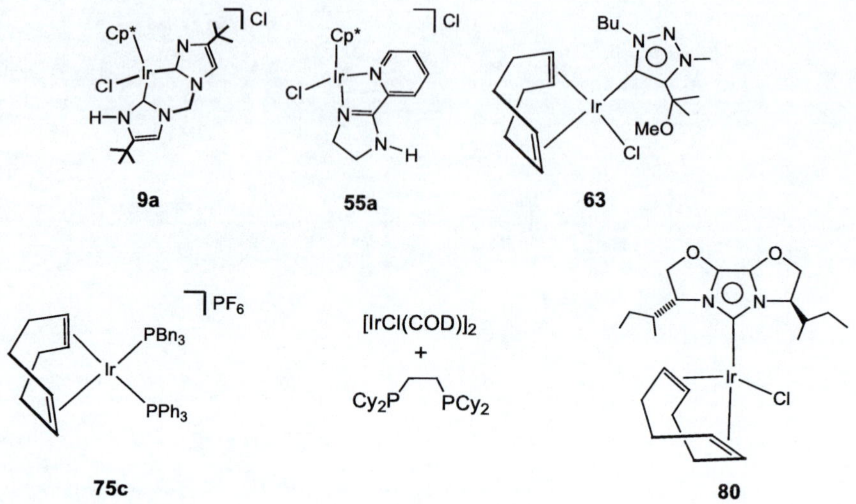

Scheme 53 Catalysts for the reduction of *α*,*β*-unsaturated ketones

give 97% yield. After this time, maintaining the reaction conditions, hydrogenation of the C=O groups gave the saturated alcohol in 63% yield within 4 h [69].

6.3 Transfer Hydrogenation of Non-conjugated Unsaturated Aldehydes and Ketones

A handful of non-conjugated unsaturated aldehydes and ketones (Scheme 54) have been subjected to TH protocols based on iridium catalysts (Scheme 55).

In *i*PrOH, under a N_2 atmosphere, non-conjugated unsaturated aldehydes 3-cyclohexene-1-carboxaldehyde and 5-norbornadiene-2-carboxaldehyde were chemoselectively converted into the corresponding unsaturated alcohols in 93 and 97% yield using the half-sandwich bipyridonate complex **79** (2 mol%) [198].

The non-conjugated ketone 5-hexen-2-one was chemoselectively transformed into 5-hexen-2-ol in 6 h with the dinuclear di(*N*-heterocyclic carbene) iridium complex **6** (0.5 mol% of catalyst loading, *i*PrOH, *i*PrONa (3 mol%), 82°C) in 96% yield. Under the same conditions, 86% yield was obtained with catalyst **7** within 10 h [67].

The same ketone and the related non-conjugated ketone 6-methyl-5-hepten-2-one were hydrogenated to the corresponding unsaturated alcohol by the bis(*N*-heterocyclic carbene) complex **9a** (2 mol% of catalyst loading, *i*PrOH, KOH (10 mol%), 95°C). Yields of 83 and 96% were obtained within 6 h. Notably, the same catalyst under similar conditions hydrogenated selectively the C=C bond of the conjugated *α*,*β*-unsaturated ketone (*E*)-chalcone [67].

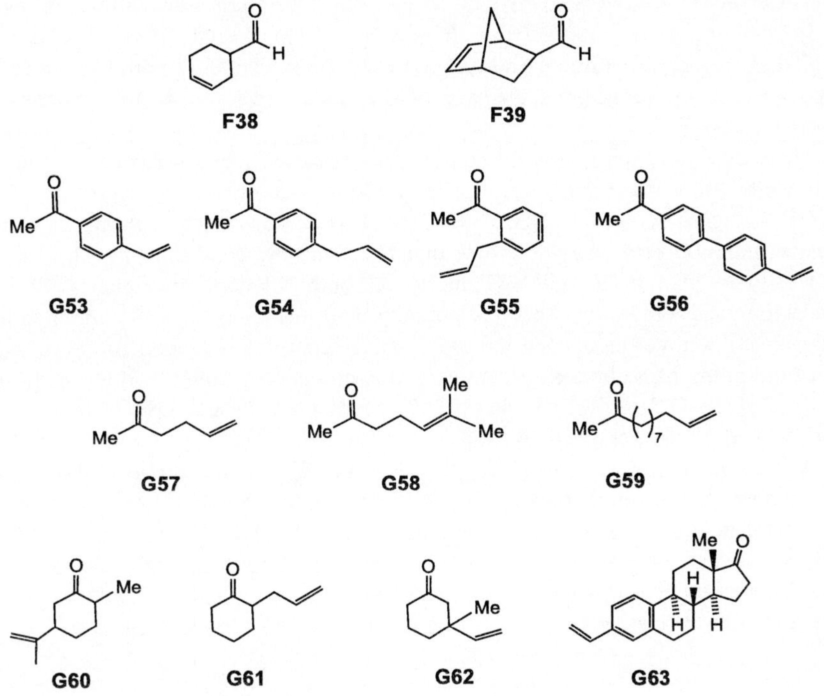

Scheme 54 Non-conjugated unsaturated aldehydes and ketones

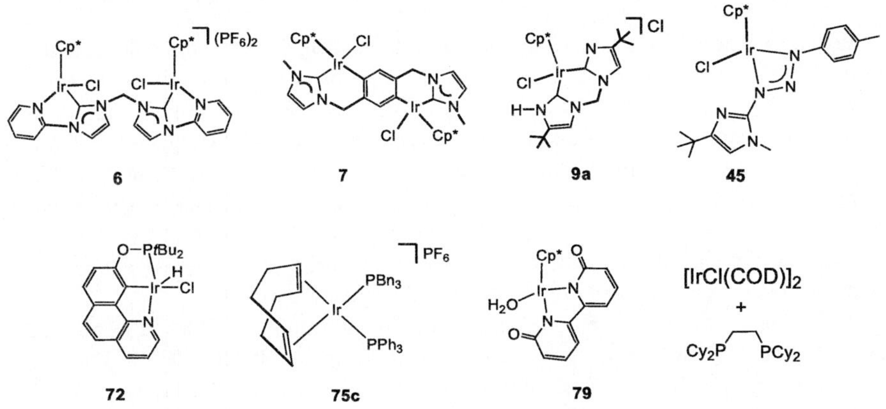

Scheme 55 Catalysts for the reduction of non-conjugated unsaturated aldehydes and ketones

The catalytic behaviour of the half-sandwich iridium triazenide complex **45** in the hydrogenation of non-conjugated 5-alken-2-ones depends on the presence of a base. If KOH is present, then the carbonyl group will be selectively hydrogenated in the first place. However, in the absence of the base, the C=C bond is preferably hydrogenated [85].

The carbonyl group of the dihydrocarvone **G60** was chemoselectively reduced with the bis(phosphine) iridium(I) complex **75c**. At 0.5 mol% of catalyst loading, in *i*PrOH, using 10 equiv. of KOH, at 80°C, 90% yield was obtained within 24 h. The isolated olefin moiety was not hydrogenated under these conditions [191].

However, the isolated C=C bond of both 11-dodecen-2-one **G59** and 4-acetylstyrene **G53** was chemoselectively hydrogenated to the corresponding unsaturated ketone using the pincer iridium complex **72**, containing a rigid benzoquinoline backbone, as a catalyst. At 1 mol% of catalyst loading, in EtOH, using *t*BuONa (2,2 mol%) as a base, 95% yields for **G59** and **G53** were obtained after 18 and 30 min, respectively [181].

A mixture of the dimer $[IrCl(COD)]_2$ (1 mol%) and the diphosphine 1,2-bis (dicyclohexylphosphino)ethane (4 mol%) promoted the highly chemoselective TH of the isolated alkene group of unsaturated ketones [194]. In 1,4-dioxane, at 130°C, terminal alkene groups bearing an acetophenone (**G53-G56**) or cyclohexanone (**G61**) underwent selective reduction of the alkene moiety. The isolated C=C bond of the estrone derivative **G63** was also chemoselectively reduced [194].

7 Transfer Hydrogenation of *N*-Heterocycles

Reduced *N*-heterocycles are frequently found in drugs, agrochemicals and dyes [120, 202], and, therefore, having efficient methods for their synthesis is highly desirable. Hydrogenation of *N*-heterocycles is the most obvious route to that type of compounds, and heterogeneous [203] and homogeneous [204, 205] catalytic hydrogenation has been the most widely used methodology. Despite the advantages of TH in terms of sustainability, simplicity of application and safety, this approach has been much less investigated for the reduction of *N*-heterocycles [206–211]. Here we present recent contributions in this area. An example of selective reduction of benzofurans by TH protocols is also included in this Section.

The iridacycle complex **81** (Scheme 56) reduced quinolines **H1–H20** and **H22–H28** (Scheme 57) to tetrahydroquinolines in an aqueous HCOOH/HCOONa solution, using 0.1 mol% of catalyst. The solution pH is critical for the catalytic activity pH 4.5 giving the best yield. Typically, isolated yields higher than 90% were obtained within 14 h [212].

Complex **81** also catalyses the reduction of activated isoquinolines and pyridines by quaternisation. Thus, isoquinolinium **H29–H34** and pyridinium **H35–H44** (Scheme 58) were hydrogenated in 24–36 h, under the same conditions but at refluxing temperature. Again, isolated yields higher than 90% were achieved [212].

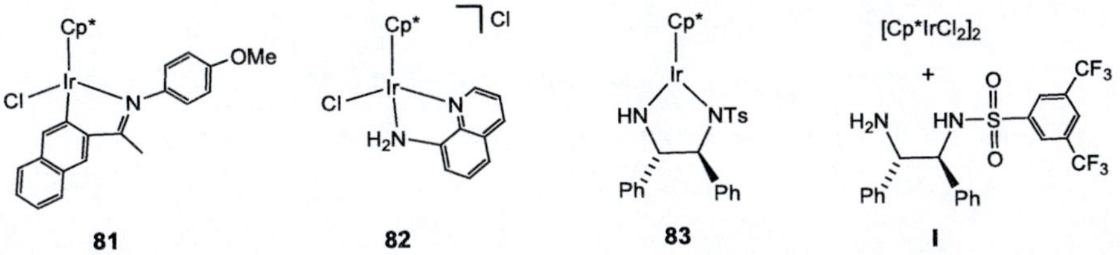

Scheme 56 Iridium complexes for the reduction of *N*-heterocycles

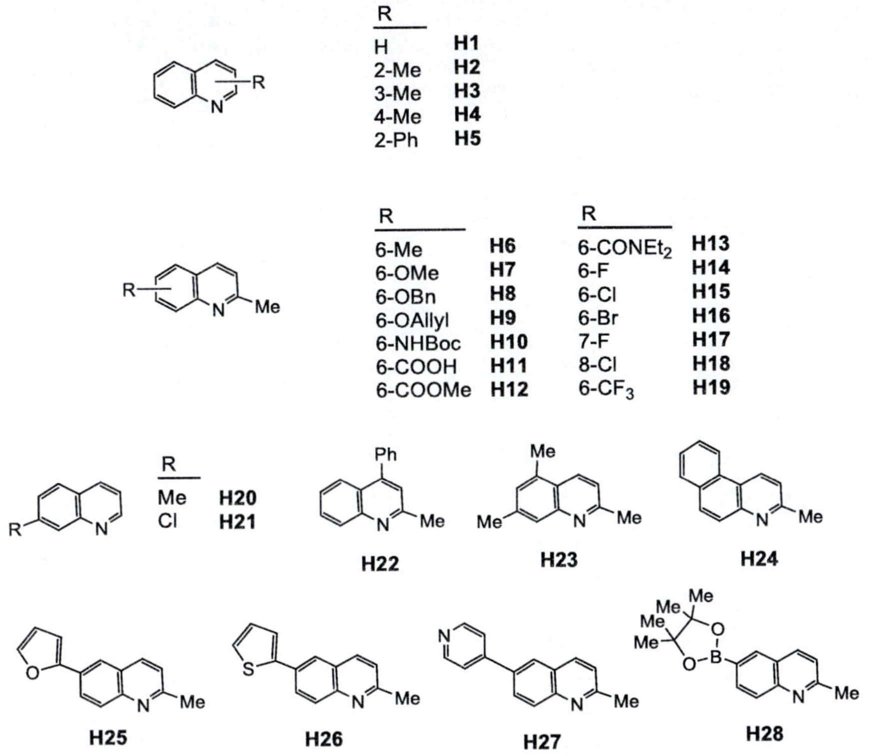

Scheme 57 Quinolines

The protocol of reduction of complex **81** was also efficiently applied to indoles **H63–H69** (Scheme 59) as well as to a range of diverse heterocycles **H58** (Scheme 60), **H72** (Scheme 61) and **H90–H92** (Scheme 62). These substrates were all reduced with excellent yield although quinoxaline **H90** was isolated as its mono *N*-formyl derivative [212].

2-Methylquinoline was quantitatively reduced to 2-methyl-1,2,3,4-tetrahydroquinoline within 24 h at room temperature using 0.1 mol% loading of

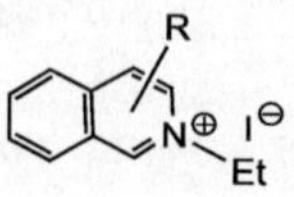

R	
H	**H29**
1-Me	**H30**
3-Me	**H31**
6-Me	**H32**
6-Br	**H33**

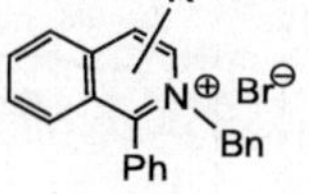

H34

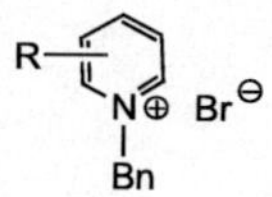

R		R	
2-Ph	**H35**	3-COOEt	**H40**
2-C(Ph)$_2$OH	**H36**	4-Bn	**H41**
2-CH$_2$NHCbz	**H37**	4-CF$_3$	**H42**
6-CH$_2$NHBoc	**H38**	4-COOEt	**H43**
6-*p*-C$_6$H$_4$(OMe)	**H39**	3,4-Me$_2$	**H44**

Scheme 58 Isoquinolinium and pyridinium

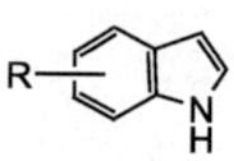

R	
H	**H63**
5-OMe	**H64**
5-Br	**H65**

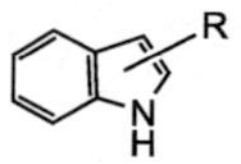

R	
2-Me	**H66**
3-Me	**H67**
2-Ph	**H68**

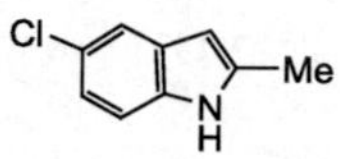

H69

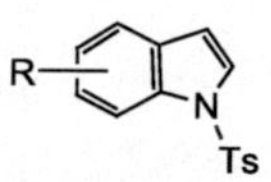

H70

Scheme 59 Indoles

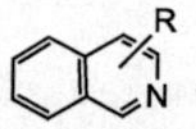

R	
H	**H45**
4-CO$_2$Et	**H46**
3-Ph,4-CO$_2$Et	**H47**

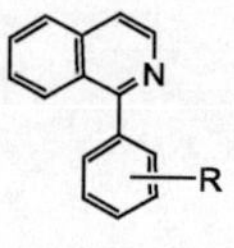

R		R	
H	**H48**	3-CF$_3$	**H53**
2-Me	**H49**	4-CF$_3$	**H54**
4-Me	**H50**	4-COOMe	**H55**
4-OMe	**H51**	4-NO$_2$	**H56**
4-Br	**H52**	3,4,5-(OMe)$_3$	**H57**

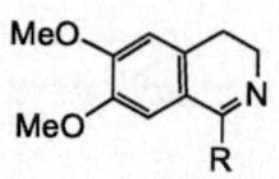

R		R	
Me	**H58**	4-(OMe)C$_6$H$_4$CH$_2$	**H61**
Ph	**H59**	3,4-(OMe)$_2$C$_6$H$_3$CH$_2$	**H62**
2,3,4,5-(OMe)$_4$C$_6$H	**H60**		

Scheme 60 Isoquinolines and dehydroquinolines

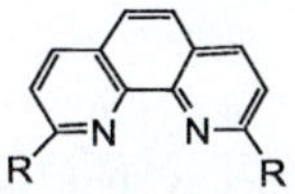

R		R	
H	**H71**	*n*Bu	**H76**
Me	**H72**	*i*Bu	**H77**
Et	**H73**	*t*Bu	**H78**
*n*Pr	**H74**	*n*-C_6H_{13}	**H79**
*i*Pr	**H75**	Ph	**H80**

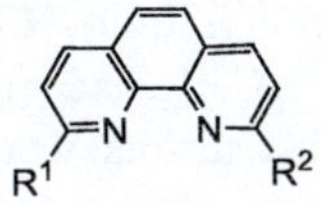

R^1	R^2		R^1	R^2	
Et	H	**H81**	*t*Bu	H	**H86**
*n*Pr	H	**H82**	*n*-C_6H_{13}	H	**H87**
*i*Pr	H	**H83**	Ph	H	**H88**
*n*Bu	H	**H84**	*i*Bu	Me	**H89**
*i*Bu	H	**H85**			

Scheme 61 Phenanthrolines

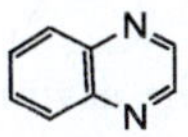

H90

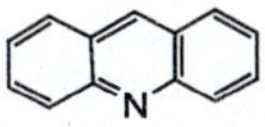

H91

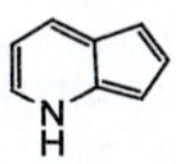

H92

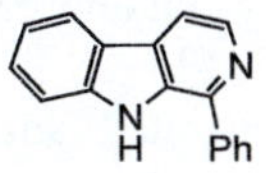

H93

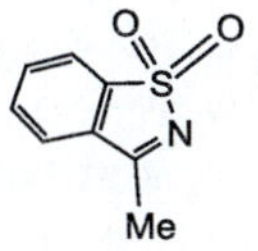

H94

Scheme 62 Other *N*-heterocycles

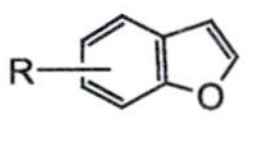

R	
H	**H95**
6-Me	**H96**
5-*t*Bu	**H97**
5-OMe	**H98**

Scheme 63 Benzofurans

the iridium complex **82** (Scheme 56) and aqueous HCOOH/HCOONa as a solvent and a hydrogen source. The same result was obtained using $[Cp^*IrCl_2]_2$ with the ligand 8-aminoquinoline as the catalyst precursor. Application of the same conditions to unsubstituted quinoline led to the hydrogenation of the pyridinic ring along with the introduction of a formyl group on the nitrogen atom in 33% yield [213].

The *N,C,P* pincer iridium complex **72** (Scheme 55), at 0.25 mol% catalyst loading, selectively reduced benzofurans **H95-H98** (Scheme 63) at 60°C in EtOH as a solvent and a hydrogen donor, using *t*BuONa (1 mol%) as a base. Dihydrobenzofurans were isolated in ≥95 yield [181]. *N*-Tosylindole, quinoline **H1** and its derivatives **H20** and **H21** (Scheme 57) were also reduced to the

Scheme 64 Chiral diamine (*S*)-CAMPY

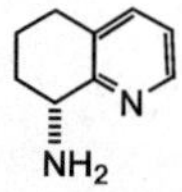

corresponding indoline and tetrahydroquinolines, respectively. Isoquinoline exhibited low reactivity (27% isolated yield) [181].

Asymmetric TH of isoquinolines **H46** and **H47**; dehydroquinolines **H58**, **H61** and **H62** (Scheme 60); quinoline **H2** (Scheme 57); and sulfonyl imine **H94** (Scheme 62) was accomplished by using a combination of $[Cp^{*}IrCl_2]_2$ and chiral cyclic diamine ligands as the catalyst precursor [214]. The best results were obtained employing the cyclic diamine (*S*)-CAMPY (Scheme 64) as a ligand, in MES or MOPS buffer (1.2 M, pH 6–8), in the presence of HCOONa (6 M) at 20°C. Yields up to 99% and e.r.'s from 65/35 to 88/12 were obtained. For the disubstituted isoquinoline **H47** (Scheme 60), the *syn* diastereomers were obtained in >99% d.r. with 86/14 e.r [214].

The Vilhanová's group developed a protocol for the asymmetric TH of 1-aryl-substituted dihydroquinolines **H48–H57**, **H59** and **H60** (Scheme 60) using anhydrous phosphoric acid as an additive and the iridium complex **83** (Scheme 56) as the catalyst. Reactions were performed in *i*PrOH at 30°C with 1 mol% of catalyst and using a 1/1 $HCOOH/NEt_3$ mixture as the hydrogen donor. Isolated yields up to 92% and e.r's from 75/25 to 93/7 were obtained [215].

Reduction of a wide variety of 2-substituted and 2,9-disubstituted 1,10-phenathrolines to exclusively give 1,2,3,4-tetrahydro-1,10-phenanthrolines was achieved employing the dimer $[Cp^{*}IrCl_2]_2$ or combinations of $[Cp^{*}IrCl_2]_2$/chiral diamine ligand as a catalyst. The products were obtained in high yields using HCOOH as the hydrogen source. When the dimer $[Cp^{*}IrCl_2]_2$ in combination with the chiral diamine **I** (Scheme 56) was employed as a catalyst precursor, e.r.'s up to 99.5/0.5 were achieved [216].

8 Transfer Hydrogenation and Sustainability

The design and development of efficient chemical processes that meet the requirements of the green chemistry principles [217] remain a global challenge. Nowadays, the vast majority of chemicals are derived from fossil resources which are limited and non-renewable. Sustainability has become an imperative issue, and renewables are destined to increasingly replace fossil chemicals. Biomass is the major renewable feedstock on the planet, and, consequently, biomass-derived chemicals are promising alternatives to replace them.

Levulinic acid (LA) is one of such chemicals that, in turn, can be converted to higher value compounds such as γ-valerolactone (GVL), 1,4-pentanediol or 2-methyl tetrahydrofuran. A number of homogeneous catalysts based on Ru, Ir, Pd or Fe have been applied to the transformation of LA to GVL, an important green fuel

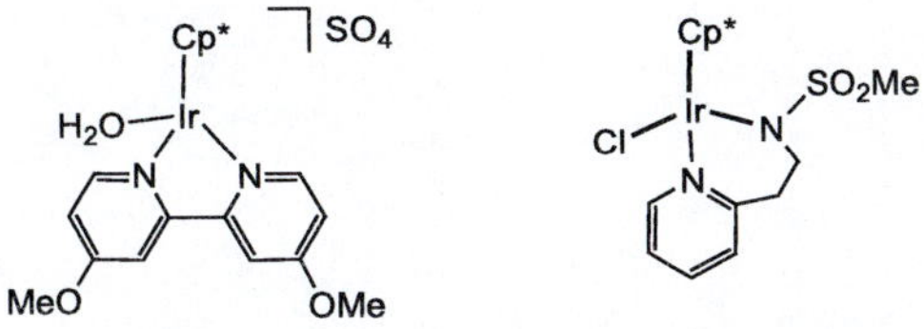

84 **28**

Scheme 65 Iridium catalysts for the TH of levulinic acid

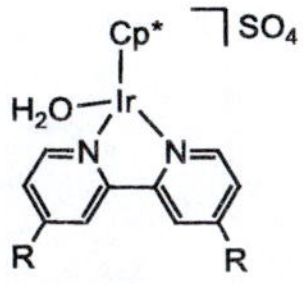

R = H (**85a**), CH_2Ph (**85b**), $CH(CH_3)Ph$ (**85c**)

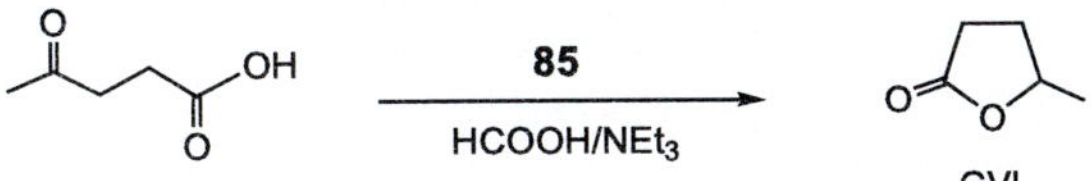

LA GVL

Scheme 66 Transfer hydrogenation of levulinic acid

additive, solvent and fine chemical intermediate [218]. However, reports dealing with homogeneous TH of LA are scarce. In particular, the generation of GLV from aqueous mixtures of LA and HCOOH catalysed by the half-sandwich iridium complex **84** (Scheme 65) has been reported [219].

More recently, Fischmeister and co-workers reported on the solvent free reduction of LA to GVL (Scheme 66) mediated by the dipyridylamine iridium complexes **85**. Reactions were performed at 120–150°C, employing mixtures of $HCOOH/NEt_3$ as a hydrogen donor. Yields from 22 to 98% were obtained within 16 h. Complex **85c** with the bulkiest ligand was found to be the less active catalyst [220].

The half-sandwich complex **28** (Scheme 7) catalysed the transformation of LA to GVL. Under standard TH conditions (1 mol% of catalyst, *i*PrOH, 85°C), 69% yield was obtained within 12 h [76].

Furfural is one of the organic compounds readily available from nonedible biomass (corncobs, oat hulls, bagasse, etc.) [221]. Furfural is mainly used as a raw material for the synthesis of a multitude of important nonpetroleum-derived chemicals such as furfuryl alcohol, methyltetrahydrofuran and furan [222]. In this line, Grushin and van Leeuween et al. reported a highly efficient chemo- and stereoselective synthesis of chiral hydrofuroins via asymmetric TH of racemic furoin that is prepared from furfural (Scheme 67) using the half-sandwich iridium complex [Cp*IrCl((*S*,*S*)-TsDPEN)] as catalyst precursor [223].

On the other hand, solvents often account for most of the mass wasted in syntheses and processes [224]. The utilisation of biogenetic alcohols such as glycerol as a H-donor solvent in catalytic TH reactions is an interesting contribution to the field of sustainable chemistry. Glycerol is a non-toxic and non-flammable

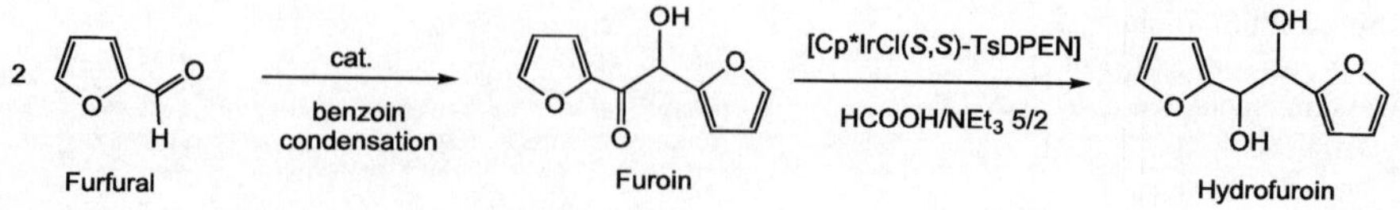

Scheme 67 Two steps synthesis of hydrofuroin from furfural

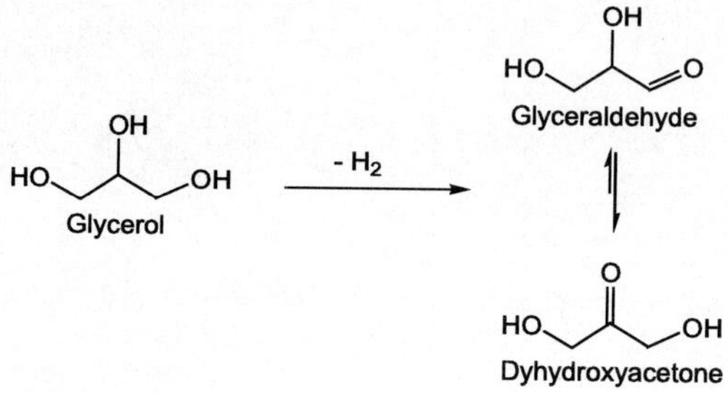

Scheme 68 Dehydrogenation of glycerol

byproduct from biodiesel processing. Its physical and solvation properties [225] make glycerol a promising candidate to be employed as an environmentally friendly reaction medium for synthetic chemistry [226]. Furthermore, glycerol has advantages, for example, over ethanol, another H-donor also produced from biomass. Dihydroxyacetone, the thermodynamic product of the dehydrogenation of glycerol, has a low tendency to decarbonylation. Although glyceraldehyde can also be formed in glycerol dehydrogenation, it would tautomerize to dihydroxyacetone under the usual catalytic conditions (Scheme 68). However, dehydrogenation of ethanol renders acetaldehyde which is expected to have a much greater tendency than dihydroxyacetone to deactivate the catalyst by decarbonylation.

The last decade has witnessed an increasing use of glycerol as a solvent and a reducing agent in homogeneous TH reactions. In fact, carbon dioxide, carbonyl compounds, olefins, nitroarenes and carboxylic acids have been efficiently hydrogenated in glycerol using Ir, Ru, Pd, Rh and Mo complexes as the catalysts [227, 228].

Recently, it has been reported the reduction of aldehydes, ketones, imines and, to a lesser extent, olefins mediated by iridium carbene complexes **86** and **50** (Scheme 69) with glycerol as a H-donor solvent. Imines were more readily reduced than carbonyl. This trend is opposite to that encountered in TH from *i*PrOH. TH was thought to proceed through a monohydride mechanism [229].

The iridium precatalyst **51** (Scheme 69) bearing an abnormal carbene ligand converts CO_2 to formate salt at a TOF value of 90 h^{-1} in 12 h of reaction at 150°C, using glycerol as a hydrogen source [94].

Nishina reported on the $[Cp^*IrCl_2]_2$-catalysed reduction of carbonyl groups using biogenetic alcohols such as glycerol, monosaccharides and polysaccharides as a

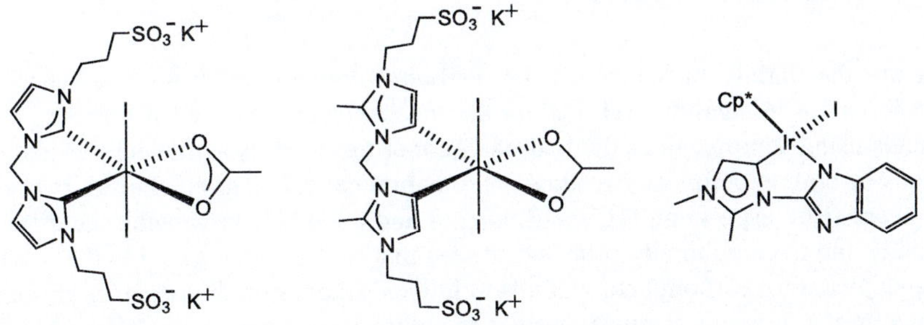

86 **50** **51**

Scheme 69 Iridium catalysts used in TH from biomass-derived H-donors

hydrogen source. This system did not require any base [230]. 2-Naphthaldehyde was reduced in 1,4-dioxane/water 1/1 (v/v) mixtures, at 85°C, at 5.0 mol% of catalyst loading using glycerol (1 equiv., 24 h, 87% yield) or glucose (1 equiv., 24 h, 95% yield) as a hydrogen source. Monosaccharides galactose and xylose; disaccharides lactose, sucrose and maltose; and the trisaccharide raffinose can also be employed as a hydrogen source. Under the same conditions, yields from 79 to 92% were obtained [230] with these biogenetic H-donors.

Aryl aldehydes bearing different substituents (**A7**, **A23**, **A31**, **A32**, **A35**), heterocyclic aldehydes such as 4-pyridine carboxaldehyde and 2-thiophene carboxaldehyde, alkyl aldehyde **A56** (Scheme 3) and diphenylketone were converted to the corresponding alcohols in yields ranging from 61 to 83%, using 1 equiv. of glucose as a hydrogen source [230].

Various aldehydes and ketones were efficiently converted to the corresponding alcohols with 2 equiv. of glucose as a H-donor, in the presence of 0.1–1 mol% of the bipyridonate iridium complex **79** (Scheme 51). Reactions were carried out at 100°C, using water or *N*,*N*-dimethylacetamide as a solvent and Na_2CO_3 as a base. Aryl aldehydes (**A1**, **A10**, **A21**, **A23**, **A24**, **A28**, **A30**, **A31**, **A35**, **A37**, **A50**, **A51**), alkyl aryl (**B1**, **B3**, **B5**, **B8**, **B11**, **B18**, **B21-B24**, **B26**, **B27**, **B47**, **B58**), methyl alkyl (**B107**) and cyclic (**B118**, **B120**) ketones were reduced to the corresponding alcohols. Isolated yields from 54 to 91% were obtained [231].

To determine which of the glucose OH groups functioned as a hydrogen donor, the reaction of acetophenone was conducted using α-glucopyranoside, in which only the hydroxyl group at the C1 position was methylated, or, alternatively, using 2,3,4,6-tetra-*O*-methyl α-glucopyranose, in which all the hydroxyl groups except that at C1 are methylated. The TH reaction did not proceed with the former but gave 97% yield in the alcohol with the latter. These results suggested that TH of the CO group of the carbonyl substrates implied the hydroxyl group at the C1 position glucose [231].

9 Mechanistic Aspects

Despite the variety of substrates, i.e. carbonyl compounds, imines, alkenes and alkynes, used in catalytic TH, mainly the mechanism of the TH of carbonyl compounds using 2-propanol as the hydrogen donor has been systematically studied by means of both experimental methods and theoretical calculations. For an overview of mechanistic aspects of TH, see Refs. 228 and 232–234. In general, the different basicity and nucleophilicity of the substrates makes it difficult to apply the reaction map depicted for carbonyl compounds to imines, alkenes or alkynes. In addition, as shown in the previous sections, besides the ubiquitous 2-propanol, different hydrogen donors have been successfully used as well, e.g. formic acid, 1,4-dioxane, glycerol, ethanol and diols.

As far as TH of carbonyl compounds using 2-propanol is concerned, overall, the formal transfer of a hydride and a hydrogen ion from the donor to the substrate takes place.

Three scenarios have been depicted so far:

1. In an inner sphere mechanism, a metal alkoxide is the entry species of the catalytic cycle (Scheme 70), eventually undergoing a β-hydrogen elimination rendering a metal monohydride complex. The hydride moiety is transferred by means of the insertion of the substrate into the metal hydride bond, whereas the protonolysis of the resulting alkoxide intermediate enables the transfer of the hydrogen ion.

   Alternatively a metal dihydride species has been proposed as the active species along the catalytic cycle. In this case both hydride moieties come from the hydrogen donor thanks to the oxidative addition of the O-H bond followed by the β-hydrogen elimination in the resulting alkoxide (Scheme 70). In this case, the hydrogen atoms of the metal dihydride moiety transfer to the substrate in a sequential way by means of the insertion of the substrate into the metal hydride bond and the following reductive elimination of the hydrogenated substrate from the hydride alkoxide intermediate.

   As a result, in the monohydride route, the C-H hydrogen of the donor exclusively turns into the C-H hydrogen of the product, while in the dihydride route, a scrambling of the C-H and O-H hydrogen atoms of the donor takes place.
2. The hydride moiety and the hydrogen ion are transferred to the substrate in an outer sphere mechanism implying both a metal hydride moiety and a protic end of one ligand at the metal centre (Scheme 71). This metal-ligand bifunctional mechanism was initially described by Noyori and co-workers [235] for ruthenium catalysts and has been later observed in a number of structurally related transition metal complexes as well as in metal-ligand platforms in which the ligand act as a non-innocent fragment [236]. Notably recent reviews on this mechanism have raised concerns about the concertedness of the H^+/H^- transfer to the substrate and the genuine non-innocence of the ancillary ligand [236–238].

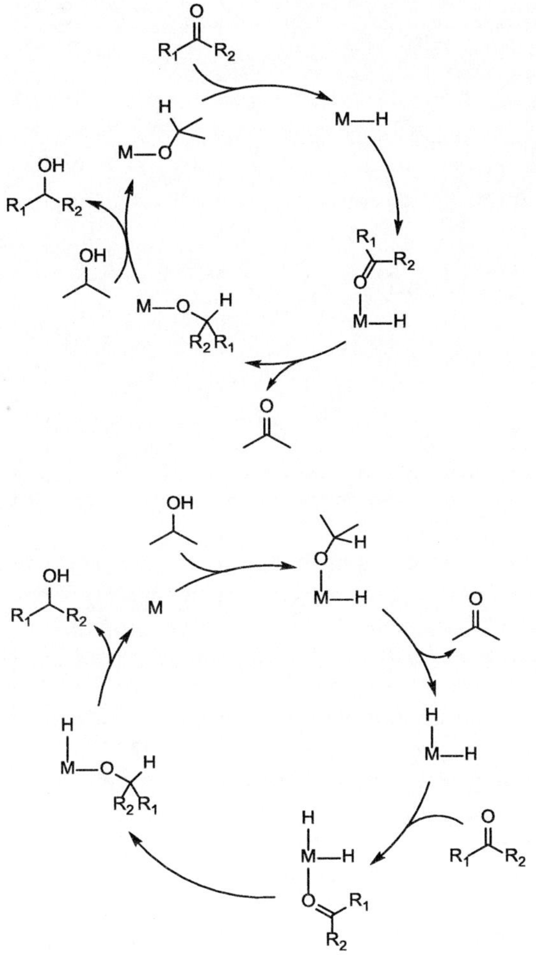

Scheme 70 Metal monohydride (*top*) and metal dihydride routes (*bottom*) for the TH of carbonyl compounds using 2-propanol as the hydrogen donor

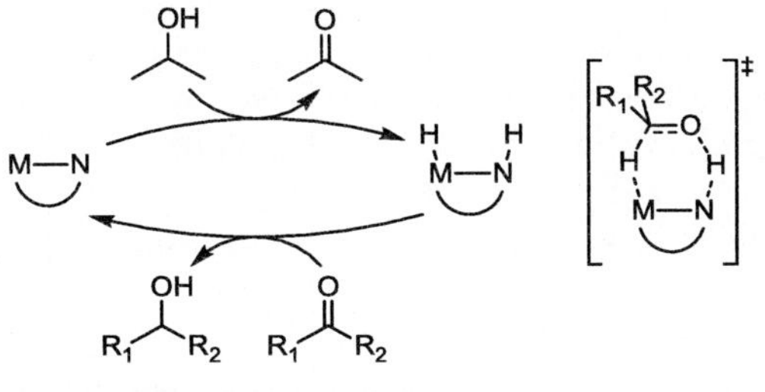

Scheme 71 Metal-ligand bifunctional mechanism for the TH of carbonyl compounds using 2-propanol as the hydrogen donor (*left*) and the transition state for the concerted transfer of the two hydrogen atoms from the metal-ligand platform to the substrate (*right*)

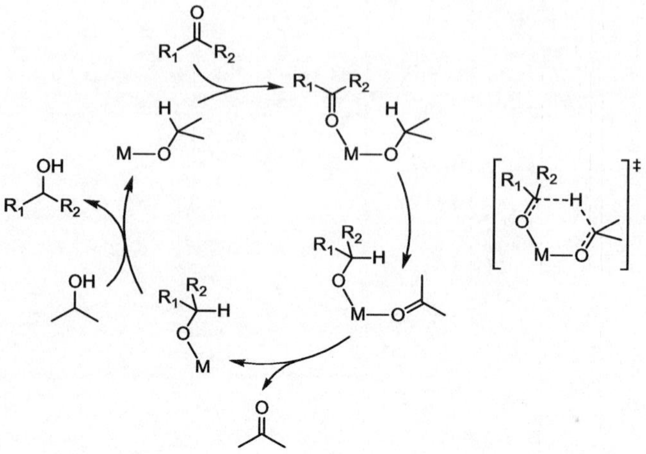

Scheme 72 Meerwein-Ponndorf-Verley mechanism of the TH of carbonyl compounds using 2-propanol as the hydrogen donor (*left*) and the transition state of the direct hydrogen transfer (*right*)

It is worth mentioning that metal-ligand bifunctional catalysts are very competent in the asymmetric transfer hydrogenation (ATH) of either carbonyl compounds or imines. Nonetheless, it has been proposed that the origin of enantioselection is different for carbonyl compounds or imines. In the case of carbonyl compounds, it relies on the formation of weak interactions between ancillary ligands and substituents at the substrate, while in the case of imines, an ionic outer sphere mechanism could be operative [237].

3. In an inner sphere mechanism, the hydride moiety is transferred directly from the donor to the substrate, and the protonolysis of the resulting alkoxide intermediate enables the transfer of the hydrogen ion (Scheme 72). This route is known as the Meerwein-Ponndorf-Verley mechanism and was first reported for the TH reaction catalysed by aluminium 2-propoxide. Nonetheless it has also been observed when transition metal complexes have been used as the catalysts.

As a milestone in the DFT-based elucidation of the mechanism of iridium catalysed TH of carbonyl compounds, back in 2003 the pioneering DFT study by Meijer [239] explored the viability of both the hydride route (inner/outer sphere) and the direct transfer when using iridium(I) catalysts of general formula Ir(COD)(L) (HL = 2-amino ethanol, 2-aminoethylmethylsulfide). The study concluded that the direct transfer exhibits the lowest activation barrier and consequently is the most accessible path (Scheme 73). Further, the hemilabile character of the amino ligand proved to be decisive given that the dissociation of either the SCH_3 or the OH moiety takes place generating the necessary coordination vacant for the coordination of the substrates to the iridium(I) centre.

Within the time frame covered herein, namely, from early 2015 to early 2020, a selection of mechanistic proposals, backed by experimental data and/or theoretical calculations, will be presented in the following in order to highlight different aspects

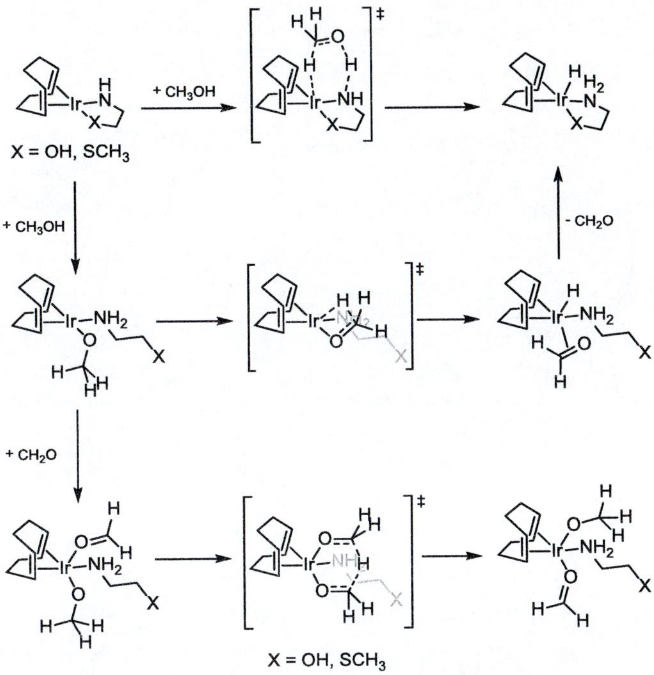

Scheme 73 Intermediates and transition states for the TH of carbonyl compounds catalysed by Ir(COD)(HL) (HL = 2-amino ethanol, 2-aminoethylmethylsulfide) [239]

of TH (or ATH) of different C = X (X = C, N, O) and C ≡ C bonds catalysed by iridium complexes using either alcohols or formic acid as the hydrogen donor.

In 2015 Xiao [212] explored the reaction mechanism of the TH of the C=N bond in *N*-heterocycles using κ^2*C,N*-iridacycle catalysts and formic acid as the hydrogen source. Using 2-methylquinoline as the model substrate, stoichiometric assays as well as isotopic-labelling experiments allowed shedding light on the main steps of the TH reaction (Scheme 74). Indeed, the authors clearly demonstrated that the rate-limiting step is the hydrogen transfer from the metal hydride to the substrate. Furthermore, the occurrence of both the 1,2-addition (Scheme 74, *bottom*) and the 1,4-addition of hydrogen (Scheme 74, *top*) to the H-heterocycle was proved.

Also, Xiao and Catlow [240] have described the remarkable activity of κ^2*C,N*-iridacycles in reducing imino groups by TH with formic acid. In this case, based on a combination of kinetic measurements, crystallographic studies and DFT calculations, the metal hydride route depicted in Scheme 75 was proposed, pointing out that the rate-limiting step is the hydride formation rather than the hydrogen transfer to the protonated substrate.

Stirling [241] reported the ATH of imines using formic acid as the hydrogen source (in the presence of NEt_3) using an iridium CATHy catalyst (Scheme 76). On one hand, under the experimental conditions, the imine is expected to be protonated rendering the iminium ion, and, on the other one, the observed KIEs as well as the

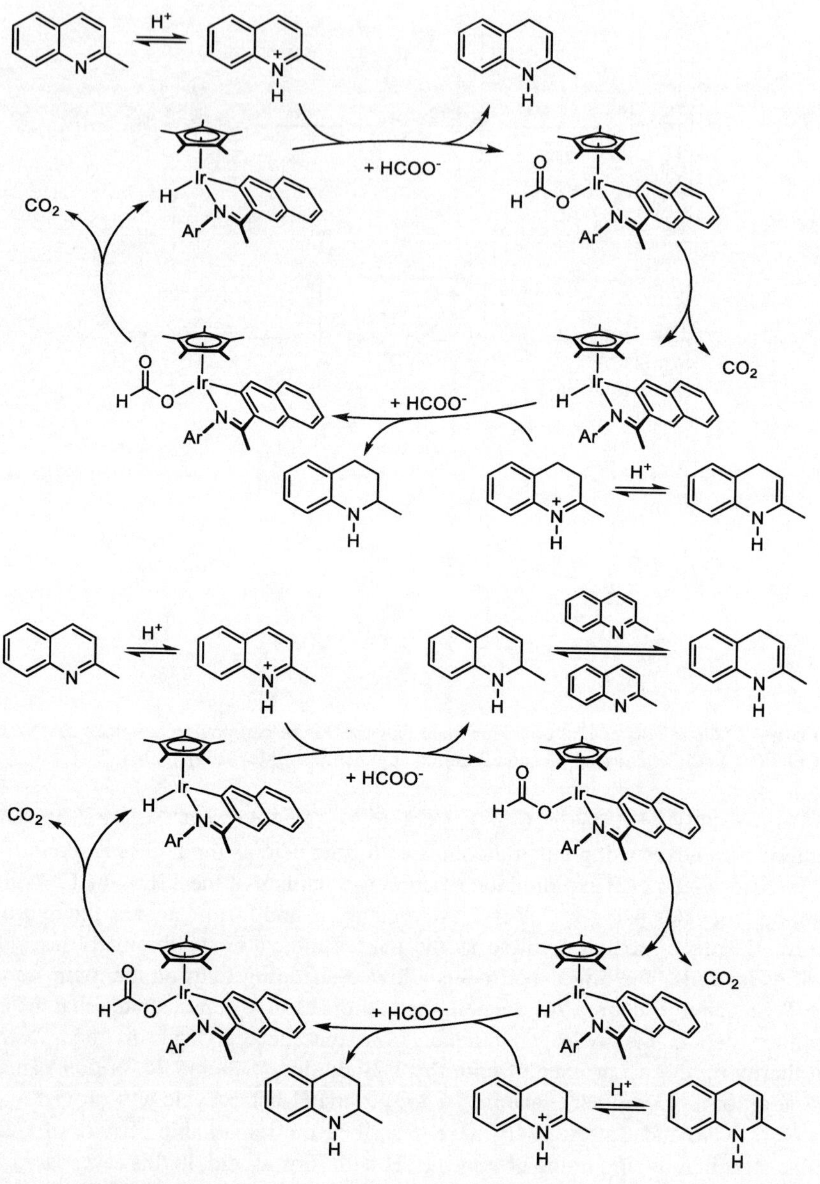

Scheme 74 Catalytic cycle proposed by Xiao [212] for the TH of N-heterocycles: 1,4-addition (*top*), 1,2-addition (*bottom*)

rate profiles point out that the order of the transfer hydrogenation kinetics of the iminium ion depends on the enantioface of the iminium implied in the hydride transfer, being first order when the *R* enantiomer is formed and zero order for the *S* enantiomer. On this background, one catalytic species could be operative for the

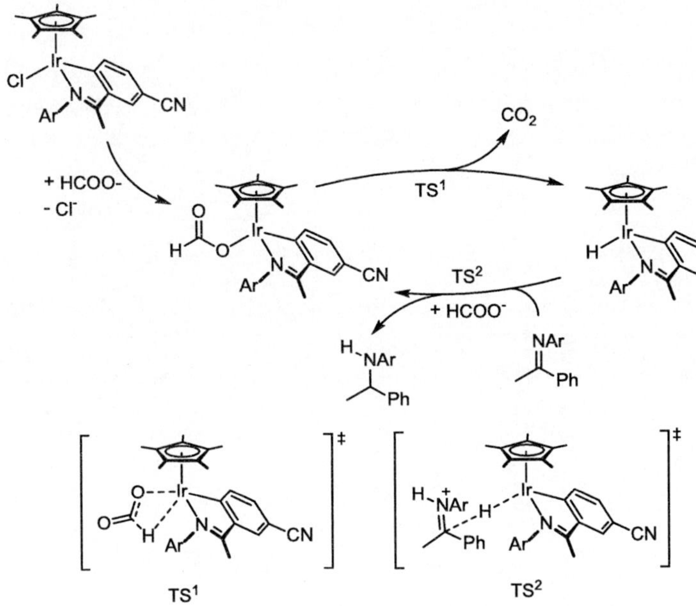

Scheme 75 Catalytic cycle and transition states proposed by Xiao [240] for the TH of imine catalysed by iridacycles using formic acid as the hydrogen donor

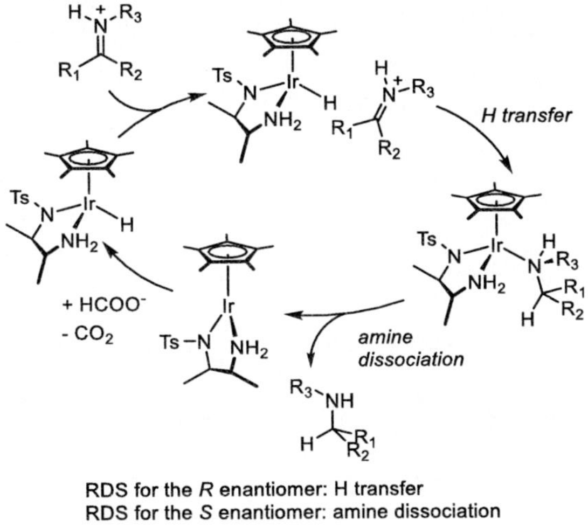

Scheme 76 Catalytic cycle proposed by Stirling [241] for the TH of imines catalysed by an iridium CATHy catalyst. The rate-determining steps (RDS) for the formation of the *R* and *S* enantiomers are indicated

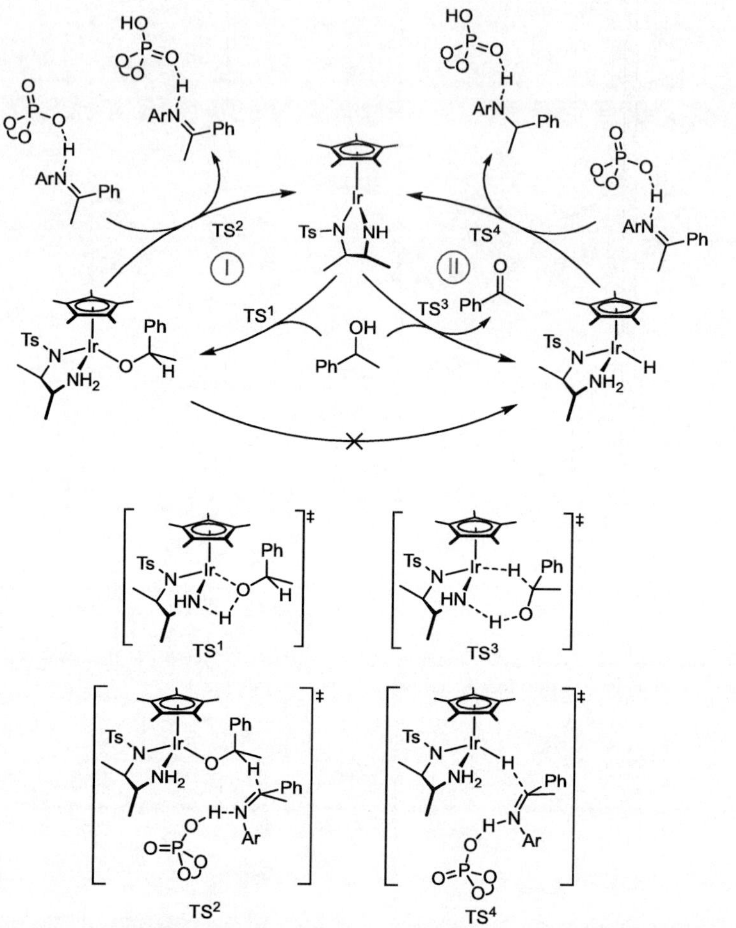

Scheme 77 Reaction paths and transition states for the ATH of imines proposed by Zhao and Lan [86]. Cycle I: iridium-alkoxide path; cycle II: iridium-hydride path

formation of the two enantiomers, but there should exist different rate-limiting steps for each one, namely, the slow dissociation of the *S* enantiomer, leading to a cero order kinetics, and the slow hydride transfer when the *R* enantiomer forms, leading to a first order kinetics. Nonetheless, Stirling does not rule out the possibility that two different catalytic species could form, each one leading to a different enantiomer with its own activity.

Zhao and Lan [86] have investigated the ATH of imines catalysed by a chiral Cp*Ir platform using different alcohols as the hydrogen donor, focusing, among other things, on the influence of the hydrogen donor. In this study, the experimental data support that the iridium alkoxide pathway (direct transfer) prevails over the classical hydride pathway and that the external phosphoric acid cocatalyst is necessary in order to activate the substrate (Scheme 77). As a confirmation, extensive DFT

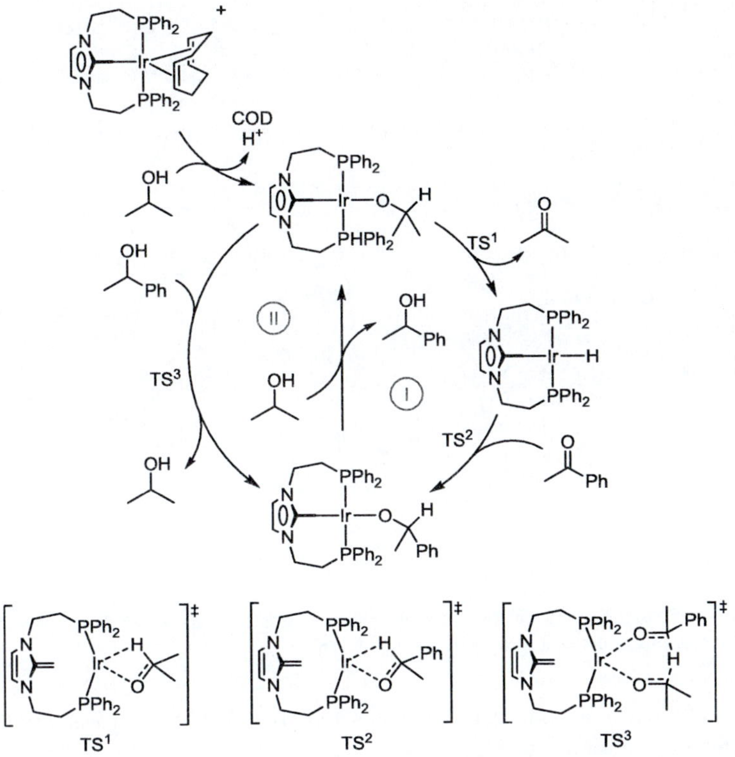

Scheme 78 Reaction paths and transition states for the TH of carbonyl compounds proposed by Iglesias and Oro [178]. Cycle I, monohydride route; cycle II, direct transfer

calculations nicely confirm that the activation barrier for the classical bifunctional mechanism is higher than that calculated for the outer sphere hydrogen transfer from the alkoxide moiety to the iminium ion (stabilised by a hydrogen-bonded phosphate anion). In addition, the hydride route is discarded due to the significantly higher barrier calculated for the formation of the hydride via either β-hydrogen elimination in the alkoxide intermediate or transfer from the hydrogen source.

Iglesias and Oro [178] have explored the TH of both carbonyl compounds and imines using 2-propanol as the hydrogen donor, catalysed by novel iridium complexes containing an *N*-heterocyclic olefin (NHO) as the ligand. DFT calculations indicate that the inner sphere metal hydride route is operative (Scheme 78, cycle I). It is worth a mention that the proposed catalytic cycle only implies iridium(I) complexes as the active species. In addition the NHO ligand plays a crucial role switching between the coordination modes κ^3C,P,P' and κ^2P,P', thus allowing the β-hydrogen elimination in the alkoxide intermediate and the insertion of the substrate into the metal hydride bond. For the sake of comparison, the energy profile of the inner sphere direct hydrogen transfer has also been computed

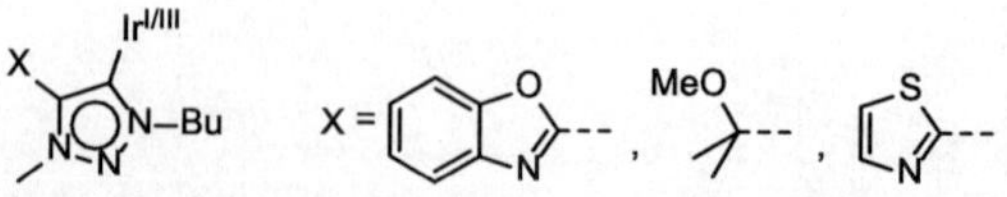

Scheme 79 Triazolylidene iridium complexes used as the catalysts for the TH of carbonyl groups, imines and alkenes [155]

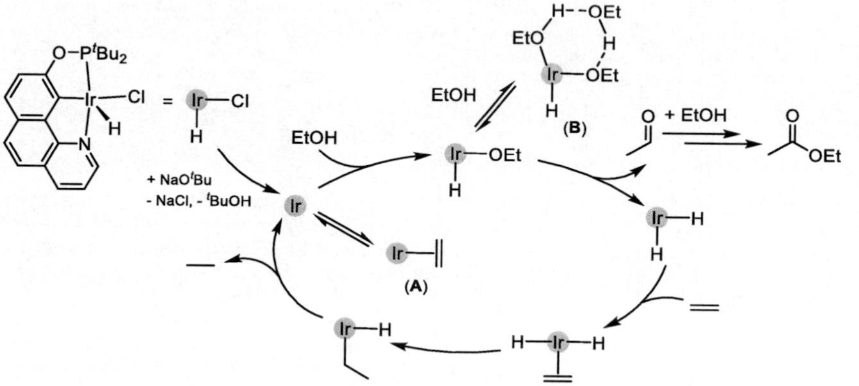

Scheme 80 Catalytic cycle for the TH of alkenes using ethanol as the hydrogen donor and an iridium *P,C,N* pincer catalyst, proposed by Huang [181]

(Scheme 78, cycle II), thus ruling out that this route could be operative under the experimental conditions.

The TH of carbonyl compounds, imines and alkenes with a family of triazolylidene iridium complexes (Scheme 79) has been studied by Pámies, Albrecht and Diéguez [155]. A combination of isotope-labelling experiments, KIE measurements and Hammet parameter correlations reveals that the reaction takes place via a monohydride route and that the turn-over limiting step is the hydride transfer from the metal to the substrate.

More recently, in 2018 Huang [181] described a family of *N,C,P* pincer iridium catalysts for the TH of alkenes, alkynes and heteroarenes using ethanol as the hydrogen source. As for the TH of alkene, a thorough experimental and computational study shed light on the operating mechanism. A metal dihydride route is proposed and remarkably the nature of the alkene determines the resting state of the cycle. Indeed, strongly bound alkenes give rise to a square planar iridium(I) complex (Scheme 80, **A**), whereas weakly bound alkenes lead to an unprecedented iridium (III) hydride alkoxide as the resting state (Scheme 80, **B**). Both DFT data and crystallographic studies reveal that this resting state contains two-hydrogen-bonded ethanol molecules forming a six-member iridacycle.

In 2019, in a related study, Huang [182] delved into the application of *P,C,N*-pincer iridium(III) complexes to the catalytic TH of alkynes to *E*-alkenes using ethanol as the hydrogen source. Remarkably the authors make the most of a finely balanced catalytic cycle in order to gain control over the selective hydrogenation of

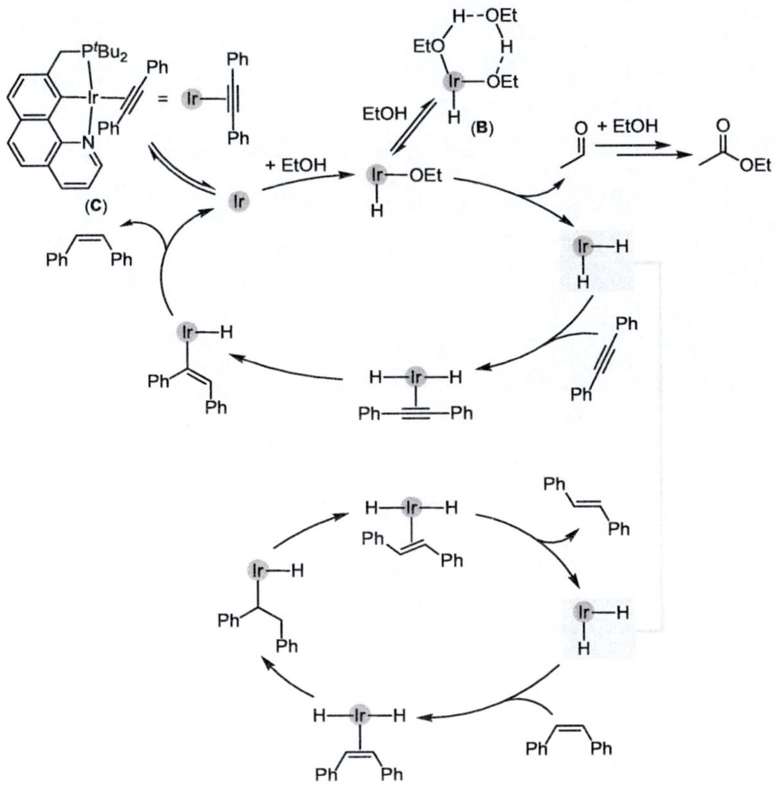

Scheme 81 Catalytic cycles, proposed by Huang [182], for the TH of alkynes to *Z*-alkene (*top*) and *Z-E* isomerization (*bottom*) catalysed by an iridium *P,C,N* pincer complex using ethanol as the hydrogen donor

alkynes to *E*-alkenes. Crystallographic studies, KIE measurements and kinetic studies reveal that a metal dihydride mechanism operates. As a matter of fact, the dihydride intermediate is responsible for both the TH of alkyne to *Z*-alkene and the following (fast) *Z-E* isomerization of the formed alkene. In addition, two crucial resting states have been proposed for the formation of the *Z*-alkene, namely, the iridium(I) alkyne complex **C** (Scheme 81) and the iridium(III) hydride alkoxide derivative **B** stabilised by two-hydrogen-bonded ethanol molecules (Scheme 81b). It is worth a mention that these two resting states differ in their colour and their relative stability in the presence of the alkyne. Indeed, as long as the alkyne is present, **C** forms, and the reaction mixture has a green colour. On the other hand, when the full conversion of the alkyne is reached, the solution turns yellow as a result of the formation of **B**.

In 2019, Sawamura and Iwai [194] have reported the use of 1,4-dioxane as the hydrogen source in the TH of alkenes and alkyne catalysed by an in situ-generated diphosphino iridium(I) catalyst. Isotopic-labelling tests and KIE measurements point to a dihydride route (Scheme 82).

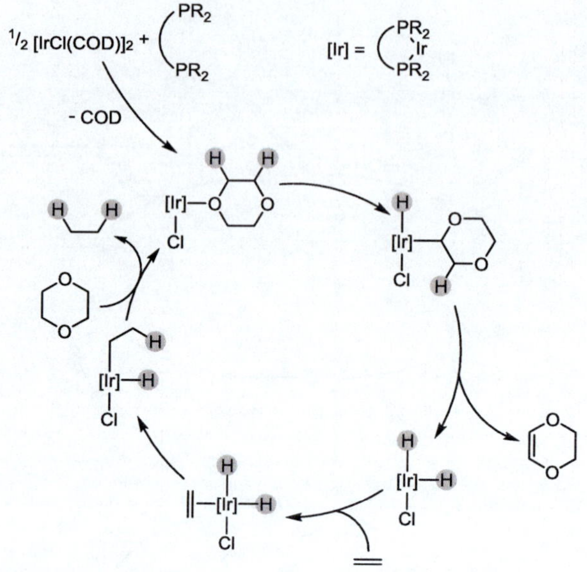

Scheme 82 Catalytic cycle for the TH of alkenes using 1,4-dioxane as the hydrogen donor with an in situ generated iridium(I) catalyst [194]

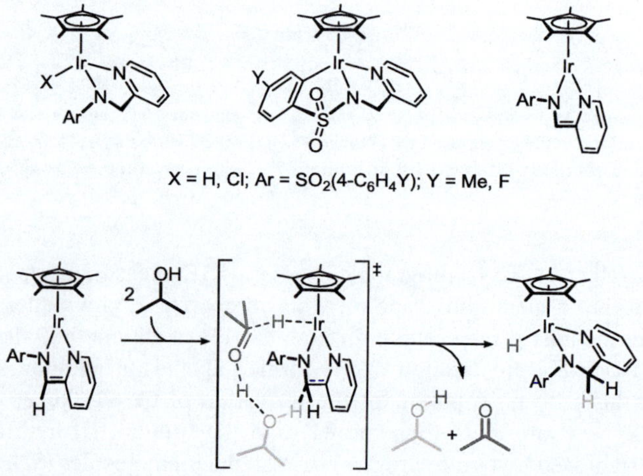

Scheme 83 Cp*Ir catalysts used by Williams [242] for the TH of carbonyl compounds using 2-propanol as the hydrogen donor (*top*) and outer sphere metal-ligand bifunctional reaction sequence (*bottom*)

Very recently Williams [242] described the TH of ketones catalysed by a family of Cp*Ir derivatives using 2-propanol as the hydrogen donor (Scheme 83). Isotopic-labelling experiments, KIE measurements and NMR kinetic studies allow the authors to propose the occurrence of a single kinetically relevant step in which a concerted proton-hydride abstraction takes place in an associative event implying two molecules of donor. Further synthetic studies and DFT calculations pointed out that the active species should imply the pyridinyl ligand acting as a non-innocent platform.

10 Conclusions

Although the first examples of TH date back to the first quarter of the twentieth century, the systematic development of this methodology started in the early 1970s. Despite the inherent advantage that the use of a sacrificial hydrogen donor presents over the employ of hazardous molecular hydrogen, the good results that were obtained in hydrogenation placed TH as a non-competitive variant with respect to H_2-hydrogenation for the reduction of unsaturated substrates.

Two fundamental milestones changed the landscape of the TH in the 1990s. In 1991, Bäckwall, based on mechanistic proposals, introduced the use of bases as an additive in TH reactions. Increases of several orders of magnitude in reaction rate supported this innovation. In the second half of the 1990s, Noyori presented a new catalytic system that meant a conceptual change in the mechanism of TH reactions and allowed almost perfect enantioselectivities to be achieved. These events gave rise to an explosive and fruitful development of the area in which HT has become a powerful and efficient alternative in the field of the reduction of unsaturated species. Iridium catalysts have played a fundamental role in this process since, after ruthenium catalysts, they have been the most widely used.

Most of the catalytic processes, including TH reactions, are carried out in organic media. HT reactions in water can make the process cheaper and environmentally more friendly. In this regard, some iridium catalysts show a high tolerance toward water and acids. Therefore, they are appealing candidates to develop TH reactions in water as a solvent and with formic acid as a hydrogen source.

Along with the search for more robust and efficient catalysts, the development of environmentally benign catalytic systems is currently another key point. Iridium catalysts have been shown to be compatible with the use of biomass-derived compounds, such as glycerol or ethanol, as the solvents and the hydrogen donors in TH reactions. Likewise, iridium-based catalysts have been successfully applied in the transformation of this type of compounds into high added-value products.

Half-sandwich iridium compounds are among the most widely used as the metallic part of metalloenzymes that have been successfully applied in TH reactions. Artificial metalloenzymes have the unique characteristic of being able to be improved both from a chemical and a genetic point of view. This feature provides them with immense room for improvement in order to optimise them and obtain catalysts with the desired capabilities. On the other hand, new anticancer strategies associated with the use of NADH or NAD(P)H as hydrogen donor in TH reactions mediated by iridium complexes have also been developed.

Nowadays, the field of HT has a solid and proven mechanistic basis that allows for the design, development and optimization of new catalytic systems to tackle new issues, fulfilling efficiency and sustainability criteria, with a high possibility of success.

In summary, TH affords an effective alternative for the reduction of unsaturated substrates, and the last 5 years has witnessed great efforts to develop iridium catalysts for this purpose. In particular, iridium catalysts have substantially contributed to the development of green and sustainable as well as new anticancer strategies associated with TH processes. Without a doubt, in the coming years, in this rapidly evolving area, there will be interesting findings for both academia and industry.

References

1. Meervein H, Schmidt R (1925) Liebigs Ann Chem 444:221–238
2. Verley A (1925) Bull Soc Chim Fr 37:537–542
3. Ponndorf W (1926) Angew Chem 39:138–143
4. Addadh YMY, Henbest HB, Husbands J, Mitchell TRB (1964) Proc Chem Soc Lond:361
5. Trocha-Grimshaw J, Henbest HB (1967) Chem Commun:544
6. Sasson Y, Blum J (1971) Tetrahedron Lett 12:2167–2170
7. Sasson Y, Blum J (1975) J Org Chem 40:1887–1896
8. Blum J, Sasson Y, Iflah S (1972) Tetrahedron Lett 13:1015–1018
9. Chowdhury RL, Bäckvall JE (1991) J Chem Soc Chem Commun:1063–1064
10. Hashiguchi S, Fujii A, Takehara J, Ikariya T, Noyori R (1995) J Am Chem Soc 117:7562–7563
11. Noyori R, Yamakawa M, Hashiguchi S (2001) J Org Chem 66:7931–7944
12. Li YY, Yu SL, Shen WY, Gao JX (2015) Acc Chem Res 48:2587–2598
13. Magubane MN, Nyamato GS, Ojwach SO, Munro OQ (2016) RSC Adv 6:65205–65221
14. Dubey P, Gupta S, Singh AK (2019) Organometallics 38:944–961
15. Shaikh MN (2019) RSC Adv 9:28199–28206
16. Foubelo F, Nájera C, Yus M (2015) Tetrahedron Asymmetry 26:769–790
17. Chelucci G, Baldino S, Baratta W (2015) Acc Chem Res 48:363–379
18. Gichumbi JM, Friedrich HB (2018) 866:123–143
19. Wang F, Lin Xu L, Sun C, Yu L, Xu Q (2018) Appl Organometal Chem 32:e4505
20. Tang Z, Liu P, Cao H, Bals S, Heeres HJ, Pescarmona PP (2019) ACS Catal 9:9953–9963
21. Vilhanová B, van Bokhoven JA, Ranocchiaric M (2017) Adv Synth Catal 359:677–686
22. Zhao J, Li Q, Zhuang S, Song Y, Morris DJ, Zhou M, Wu Z, Zhang P, Jin R (2018) J Phys Chem Lett 9:7173–7179
23. Hou SF, Chen JY, Xue M, Jia M, Zhai X, Liao RZ, Tung CH, Wang W (2020) ACS Catal 10:380–390
24. Wu W, Seki T, Walker KL, Waymouth RM (2018) Organometallics 37:1428–1431
25. Ma F, Chen J, Yang F, Shinde MV, Zhoua Y, Fan B (2017) Org Biomol Chem 15:2359–2362
26. Xu Y, Long J, Zhao W, Li H, Yang S (2019) Front Chem 7:590
27. Mishra AA, Bhanage BM (2019) New J Chem 43:6160–6167
28. Cabanillas M, Franco A, Lázaro N, Balu AM, Luque R, Pineda A (2019) Mol Catal 477:110522
29. Wang F, Planas O, Cornella J (2019) J Am Chem Soc 141:4235–4240
30. Bolm C (2009) Nat Chem 1:420–420
31. Bullock RM (2013) Science 342:1054–1055
32. Zuo WW, Lough AJ, Li YF, Morris RH (2013) Science 342:1080–1083

33. Morris RH (2015) Acc Chem Res 48:1494–1502
34. Zuo W, Morris RH (2015) Nat Protoc 10:241–257
35. Demmans KZ, Seo CSG, Lough AJ, Morris RH (2017) Chem Sci 8:6531–6541
36. Morris RH (2018) Dalton Trans 47:10809–10826
37. De Luca L, Passera A, Mezzetti A (2019) J Am Chem Soc 141:2545–2255
38. Farrar-Tobar RA, Wozniak B, Savini A, Hinze S, Tin S, de Vries JG (2019) Angew Chem Int Ed 58:1129–1133
39. Liang Q, Song D (2020) Chem Soc Rev 49:1209–1232
40. Jiang BL, Ma SS, Wang ML, Liu DS, Xu BH, Zhang SJ (2019) ChemCatChem 11:1701–1706
41. Gao Y, Zhang X, Laishram RD, Chen J, Li K, Zhang K, Zeng G, Fan B (2019) Adv Synth Catal 361:3991–3997
42. Bruneau-Voisine A, Wang D, Dorcet V, Roisnel T, Darcel C, Sortais J-B (2017) Org Lett 19:3656–3659
43. Demmans KZ, Olson ME, Morris RH (2018) Organometallics 37:4608–4618
44. Zhou Y-P, Mo Z, Luecke M-P, Driess M (2018) Chem Eur J 24:4780–4784
45. Wei D, Bruneau-Voisine A, Dubois M, Bastin S, Sortais JB (2019) ChemCatChem 11:5256–5259
46. Schneekönig J, Junge K, Beller M (2019) Synlett 30:503–507
47. Putten RV, Benschop J, Munck VJ, Weber M, Müller C, Filonenko GA, Pidko EA (2019) ChemCatChem 11:5232–5235
48. Dubey A, Rahaman SMW, Fayzullin RR, Khusnutdinova JR (2019) ChemCatChem 11:3844–3852
49. Garduño JA, Flores-Alamo M, García JJ (2019) ChemCatChem 11:5330–5338
50. Nikita V, Shvydkiy NV, Vyhivskyi O, Nelyubina YV, Perekalin DS (2019) ChemCatChem 11:1602–1605
51. Ganguli K, Shee S, Panja D, Kundu S (2019) Dalton Trans 48:7358–7366
52. Vigneswaran V, MacMillan SN, Lacy DC (2019) Organometallics 38:4387–4391
53. Zhang C, Hu B, Chen D, Xia H (2019) Organometallics 38:3218–3226
54. Zhou L, Liu D, Lan H, Wang X, Zhao C, Ke Z, Hou C (2020) Cat Sci Technol 10:169–179
55. Passera A, Mezzetti A (2020) Angew Chem Int Ed 59:187–191
56. Uematsu N, Fujii A, Hashiguchi S, Ikariya T, Noyori R (1996) J Am Chem Soc 118:4916–4917
57. Wang C, Pettman A, Bacsa J, Xiao J (2010) Angew Chem Int Ed 49:7548–7552
58. Liu Z, Romero-Canelón I, Qamar B, Hearn JM, Habtemariam A, Barry NPE, Pizarro AM, Clarkson GJ, Sadler PJ (2014) Angew Chem Int Ed 53:3941–3946
59. Wang C, Li C, Wu X, Pettman A, Xiao J (2009) Angew Chem Int Ed 48:6524–6528
60. Wang D, Astruc D (2015) Chem Rev 115:6621–6686
61. Mashima K, Abe T, Tani K (1998) Chem Lett:1199–1200
62. Bolje A, Hohloch S, van der Meer M, Kosmrlj J, Sarkar B (2015) Chem Eur J 21:6756–6764
63. Bolje A, Hohloch S, Kosmrlj J, Sarkar B (2016) Dalton Trans 45:15983–15993
64. Maity R, Hohloch S, Su C-Y, van der Meer M, Sarkar B (2014) Chem Eur J 20:9952–9961
65. Maity R, van der Meer M, Hohloch S, Sarkar B (2015) Organometallics 34:3090–3096
66. Maity R, Mekic A, van der Meer M, Verma A, Sarkar B (2015) Chem Commun 51:15106–15109
67. Volpe A, Baldino S, Tubaro C, Baratta W, Basato M, Graiff C (2016) Eur J Inorg Chem:247–251
68. Müller AL, Bleith T, Roth T, Wadepohl H, Gade LH (2015) Organometallics 34:2326–2342
69. Gómez-López JL, Chávez D, Parra-Hake M, Royappa AT, Rheingold AL, Grotjahn DB, Miranda-Soto V (2016) Organometallics 35:3148–3153
70. Vivancos Á, Albrecht M (2017) Organometallics 36:1580–1590
71. Pretorius R, Mazloomi Z, Albrecht M (2017) J Organomet Chem 845:196–205
72. Navarro M, Smith CA, Albrecht M (2017) Inorg Chem 56:11688–11701
73. Sato Y, Kayaki Y, Ikariya T (2016) Chem Asian J 11:2924–2931

74. Holmes J, Pask CM, Willans CE (2016) Dalton Trans 45:15818–15827
75. Sun M, Campbell J, Kang G, Wang H, Ni B (2016) J Organomet Chem 810:12–14
76. Ruff A, Kirby C, Chan BC, O'Connor AR (2016) Organometallics 35:327–335
77. Townsend TM, Kirby C, Ruff A, O'Connor AR (2017) J Organomet Chem 843:7–13
78. Meriç N, Aydemir M (2016) J Organomet Chem 819:120–128
79. Ak B, Aydemir M, Durap F, Meriç N, Elma D, Baysal A (2015) Tetrahedron Asymmetry 26:1307–1313
80. Durap F, Karakas DE¸ Ak B, Baysal A, Aydemir M (2016) J Organometal Chem 818:92–97
81. Baysal A, Karakas DE, Meriç N, Ak B, Aydemir M, Durap F (2017) Trans Met Chem 42:365–372
82. Chen HJ, Teo RHX, Li Y, Pullarkat SA, Leung P-H (2018) Organometallics 37:99–106
83. Dubey P, Gupta S, Singh AK (2108) Dalton Trans 47:3764–3774
84. Zhou G, Aboo AH, Robertson CM, Liu R, Li Z, Luzyanin K, Berry NG, Chen W, Xiao J (2018) ACS Catal 8:8020–8026
85. Camarena-Díaz JP, Iglesias AL, Chávez D, Aguirre G, Grotjahn DB, Rheingold AL, Parra-Hake M, Miranda-Soto V (2019) Organometallics 38:844–851
86. Pan H-J, Zhang Y, Shan C, Yu Z, Lan Y, Zhao Y (2016) Angew Chem Int Ed 55:9615–9619
87. Wang W-H, Himeda Y, Muckerman JT, Manbeck GF, Fujita E (2015) Chem Rev 115:12936–12973
88. Sordakis K, Tang C, Vogt LK, Junge H, Dyson PJ, Beller M, Laurenczy G (2018) Chem Rev 118:372–433
89. Bernskotter WH, Hazari N (2017) Acc Chem Res 50:1049–1058
90. Klankermayer J, Wesselbaum S, Beydoun K, Leitner W (2016) Angew Chem Int Ed 55:7296–7343
91. Goeppert A, Czaun M, Jones J-P, Prakash GKS, Olah GA (2014) Chem Soc Rev 43:7995–8048
92. Sanz S, Benítez M, Peris E (2010) Organometallics 29:275–277
93. Azua A, Sanz S, Peris E (2011) Chem Eur J 17:3963–3967
94. Kumar A, Semwal S, Choudhury J (2019) ACS Catal 9:2164–2168
95. Bar R, Sasson Y (1981) Tetrahedron Lett 22:1709–1710
96. Bar R, Sasson Y, Blum J (1984) J Mol Catal 26:327–332
97. Bar R, Bar LK, Sasson Y, Blum J (1985) J Mol Catal 33:161–177
98. Joó F, Csuhai E, Quinn PJ, Vigh L (1988) J Mol Catal 49:L1–L5
99. Joó F, Bényei A (1989) J Organomet Chem 363:C19–C21
100. Bényei A, Joó F (1990) J Mol Catal 58:151–163
101. Wu X, Xiao J (2007) Chem Commun:2449–2466
102. Wang C, Wu X, Xiao J (2008) Chem Asian J 3:1750–1770
103. Wu X, Wang C, Xiao J (2010) Platin Met Rev 54:3–19
104. Robertson A, Matsumoto T, Ogo S (2011) Dalton Trans 40:10304–10310
105. Wei Y, Wub X, Wanga C, Xiao J (2015) Catal Today 247:104–115
106. He Y-M, Fan Q-H (2015) ChemCatChem 7:398–400
107. Wu X, Wang C, Xiao J (2016) Chem Rec 16:2772–2786
108. Wang C, Xiao J (2017) Chem Commun 53:3399–3411
109. Thangavel S, Friedrich HB, Omondi B (2017) J Mol Caltal A Chem 429:27–42
110. Yang Z, Zhu Z, Luo R, Qiu X, Liu J-T, Yang J-K, Tang W (2017) Green Chem 19:3296–3301
111. Liu J-T, Yang S, Tang W, Yang Z, Xu J (2018) Green Chem 20:2118–2124
112. Liu Q, Wang C, Zhou H, Wang B, Lv J, Cao L, Fu Y (2018) Org Lett 20:971–974
113. Yang Z, Cheng W, Li Z (2018) Catal Commun 117:38–42
114. Davis HJ, Ward TR (2019) ACS Cent Sci 5:1120–1136
115. Schwizer F, Okamoto Y, Heinisch T, Gu Y, Pellizzoni MM, Lebrun V, Reuter R, Köhler V, Lewis JC, Ward TR (2018) Chem Rev 118:142–231
116. Letondor C, Humbert N, Ward TR (2005) Proc Natl Acad Sci U S A 102:4683–4687

117. Dürrenberger M, Heinisch T, Wilson YM, Rossel T, Nogueira E, Knörr L, Mutschler A, Kersten K, Zimbron MJ, Pierron J, Schirmer T, Ward TR (2011) Angew Chem Int Ed 50:3026–3029
118. Hestericov M, Heinisch T, Alonso-Cotchico L, Maréchal J-D, Vidossich P, Ward TR (2018) Angew Chem Int Ed 57:1863–1868
119. Scott JD, Williams RM (2002) Chem Rev 102:1669–1730
120. Ward TR (2011) Acc Chem Res 44:47–57
121. Heinisch T, Ward TR (2016) Acc Chem Res 49:1711–1721
122. Liang AD, Serrano-Plana J, Peterson RL, Ward TR (2019) Acc Chem Res 52:585–595
123. Pellizzoni M, Facchetti G, Gandolfi R, Fusé M, Contini A, Rimoldi I (2016) ChemCatChem 8:1665–1670
124. Facchetti G, Rimoldi I (2018) New J Chem 42:18773–18776
125. Reetz MT, Wu S, Zheng H, Prasad S (2010) Pure Appl Chem 82:1575–1584
126. Denard CA, Ren H, Zhao H (2015) Curr Opin Chem Biol 25:55–64
127. Quinto T, Häussinger D, Köhler V, Ward TR (2015) Org Biomol Chem 13:357–360
128. Okamoto Y, Köhler V, Paul CE, Hollmann F, Ward TR (2016) ACS Catal 6:3553–3557
129. Okamoto Y, Köhler V, Ward TR (2016) J Am Chem Soc 138:5781–5784
130. Monnard FW, Nogueira ES, Heinisch T, Schirmerb T, Ward TR (2013) Chem Sci 4:3269–3274
131. Monnard FW, Heinisch T, Nogueira ES, Schirmerb T, Ward TR (2011) Chem Commun 47:8238–8240
132. Tinberg CE, Khare SD, Dou J, Doyle L, Nelson JW, Schena A, Jankowski W, Kalodimos CG, Johnsson K, Stoddard BL, Baker D (2013) Nature 501:212–218
133. Heinisch T, Pellizzoni M, Dürrenberger M, Tinberg CE, Köhler V, Klehr J, Häussinger D, Baker D, Ward TR (2018) J Am Chem Soc 137:10414–10419
134. Raines DJ, Clarke JE, Blagova EV, Dodson EJ, Wilson KS, Duhme-Klair A-K (2018) Nat Catal 1:680–688
135. Lin S-J, Guarente L (2003) Curr Opin Cell Biol 15:241–246
136. Betanzos-Lara S, Liu Z, Habtemariam A, Pizarro AM, Qamar B, Sadler PJ (2012) Angew Chem Int Ed 51:3897–3900
137. Maenaka Y, Suenobu T, Fukuzumi S (2014) J Am Chem Soc 134:367–374
138. Soldevila-Barreda JJ, Romero-Canelón I, Habtemariam A, Sadler PJ (2015) Nat Commun 6:6582–6590
139. Ngo AH, Ibáñez M, Do LH (2016) ACS Catal 6:2637–2641
140. Bose S, Ngo AH, Do LH (2017) J Am Chem Soc 139:8792–8795
141. Jang B, Kwon H, Katila P, Lee SJ, Lee H (2016) Adv Drug Deliv Rev 98:113–133
142. Yang L, Bose S, Ngo AH, Do LH (2017) ChemMedChem 12:292–299
143. Wang C, Liu J, Tian Z, Tian M, Tian L, Zhao W, Liu Z (2017) Dalton Trans 46:6870–6883
144. Li JJ, Tian M, Tian Z, Zhang S, Yan C, Shao C, Liu Z (2018) Inorg Chem 57:1705–1716
145. Li JJ, Guo L, Tian Z, Zhang S, Xu Z, Han Y, Li R, Li Y, Liu Z (2018) Inorg Chem 57:13552–13563
146. Du Q, Guo L, Tian M, Ge X, Yang Y, Jian X, Xu Z, Tian Z, Liu Z (2018) Organometallics 37:2880–2889
147. Du Q, Yang Y, Guo L, Tian M, Ge X, Tian Z, Zhao L, Xu Z, Li J, Liu Z (2019) Dyes Pigments 162:821–830
148. Soetens M, Drouet F, Riant O (2017) ChemCatChem 9:929–933
149. Stringer T, Melis DR, Smith GS (2019) Dalton Trans:13143–13148
150. Facchetti G, Pellegrino S, Bucci R, Nava D, Gandolfi R, Christodoulou MS, Rimoldi I (2019) Molecules 2:2771–2779
151. Campos J, Sharninghausen LS, Manas MG, Crabtree RH (2015) Inorg Chem 54:5079–5084

152. Jiménez MV, Fernández-Tornos J, Pérez-Torrente JJ, Modrego FJ, García-Orduña P, Oro LA (2015) Organometallics 34:926–940
153. García N, Jaseer EA, Munarriz J, Sanz Miguel PJ, Polo V, Iglesias M, Oro LA (2015) Eur J Inorg Chem:4388–4395
154. Ramasamy B, Gangwar MK, Ghosh P (2017) Eur J Inorg Chem:3253–3268
155. Mazloomi Z, Pretorius R, Pàmies O, Albrecht M, Diéguez M (2017) Inorg Chem 57:11282–11298
156. Wang H, Jiang T, Xu M-H (2013) J Am Chem Soc 135:971–974
157. Li Y, Lei M, Yuan W, Meggers E, Lei Gong L (2017) Chem Commun 53:8089–8092
158. Albrecht M, van Koten G (2001) Angew Chem Int Ed 40:3650–3781
159. Peris E, Crabtree RH (2004) Coord Chem Rev 248:2239–2246
160. Leis W, Mayera HA, Kaska WC (2008) Coord Cherm Rev 252:1787–1797
161. Morales-Morales D (2008) Mini Rev Org Chem 5:141–152
162. van der Boom ME, Milstein D (2003) Chem Rev 103:1759–1792
163. Benito-Garagorri D, Kirchner K (2008) Acc Chem Res 41:201–213
164. Choi J, MacArthur AHR, Brookhart M, Goldman AS (2011) Chem Rev 111:1761–1779
165. Selander N, Szabó KJ (2011) Chem Rev 111:2048–2076
166. Gunanathan C, Milstein D (2011) Acc Chem Res 44:588–602
167. Ito J-I, Nishiyama H (2011) Tetrahedron Lett 55:3133–3145
168. Gunanathan C, Milstein D (2014) Chem Rev 114:12024–12087
169. Younus HA, Ahmad N, Su W, Verpoort F (2014) Coord Chem Rev 276:112–152
170. Werkmeister S, Neumann J, Junge K, Beller M (2015) Chem Eur J 21:12226–12250
171. Clarke ZE, Maragh PT, Dasgupta TP, Gusev DG, Lough AJ, Abdur-Rashid K (2006) Organometallics 25:4113–4117
172. Paredes P, Díez J, Gamasa P (2008) Organometallics 27:2597–2607
173. Bi S, Xie Q, Zhao X, Zhao Y, Kong X (2008) J Organomet Chem 693:633–638
174. Azerraf C, Gelman D (2008) Chem Eur J 14:10364–10368
175. Li Y-H, Zhang Y, Ding X-H (2011) Inorg Chem Commun 14:1306–1310
176. Liu W-P, Yuan M-L, Yang X-H, Li K, Xie J-H, Zhou Q-L (2015) Chem Commun 51:6123–6125
177. Zhang Y-M, Yuan M-L, Liu W-P, Xie J-H, Zhou Q-L (2018) Org Lett 20:4486–4489
178. Iturmendi A, García N, Jaseer EA, Munárriz J, Sanz Miguel PJ, Polo V, Iglesias M, Oro LA (2016) Dalton Trans 45:12835–12845
179. Pandrala M, Resendez A, Malhotra SV (2019) J Catal 378:283–288
180. Cohen S, Bilyachenko AN, Gelman D (2019) Eur J Inorg Chem:3203–3209
181. Wang Y, Huang Z, Leng X, Zhu H, Liu G, Huang Z (2018) J Am Chem Soc 140:4417–4429
182. Wang Y, Huang Z, Huang Z (2019) Nat Catal 2:529–536
183. Zweifel T, Naubron JV, Büttner T, Ott T, Grützmacher H (2008) Angew Chem Int Ed 47:3245–3249
184. Reddy AS, Swarmy KCK (2017) Angew Chem Int Ed 56:6984–6988
185. Chen S, Lu G, Cai C (2015) New J Chem 39:5360–5365
186. Wang D, Zhao K, Yang S, Ding Y (2015) Z Anorg Allg Chem 641:400–404
187. El-Asaad B, Guicheret B, Métay E, Karamé I, Marc Lemaire M (2016) J Mol Catal A Chem 411:196–202
188. Tian C, Gong L, Meggers E (2016) Chem Commun 52:4207–4210
189. Ikariya K, Murata K, Noyori R (2006) Org Biomol Chem 4:393–406
190. Ikariya K, Blacker AJ (2007) Acc Chem Res 40:1300–1308
191. Landaeta VR, Salazar-La Rosa AD, Rodríguez-Lugo RE (2018) Inorg Chim Acta 470:303–311
192. Zhang YM, Li PQ, Liu P (2018) Russ J Coord Chem 44:688–692
193. Günnaz S, Gökçe AG, Türkmen H (2018) Dalton Trans 47:17317–17328
194. Zhang D, Iwai T, Sawamura M (2019) Org Lett 21:5867–5872
195. Yang J, Wang C, Sun Y, Man X, Li J, Sun F (2019) Chem Commun 55:1903–1906

196. Wang C, Gong S, Liang Z, Sun Y, Cheng R, Yang B, Liu Y, Jinfei Yang J, Sun F (2019) ACS Omega 4:16045–16051
197. Farrar-Tobar RA, Tin S, de Vries JG (2018) Selective transfer hydrogenation of α,-β-unsaturated carbonyl compounds. In: Dixneuf P, Soulé JF (eds) Organometallics for green catalysis. Topics in organometallic chemistry, vol 63. Springer, Cham
198. Wang R, Tang Y, Xu M, Meng C, Li F (2018) J Org Chem 83:2274–2281
199. Luo N, Liao J, Ouyang L, Wen H, Liu J, Tang W, Luo R (2019) Organometallics 38:3025–3011
200. Chen S-J, Lu G-P, Chun Cai C (2015) RSC Adv 5:13208–13211
201. Ramasamy B, Prakasham AP, Gangwar MK, Ghosh P (2019) Chem Select 4:357–365
202. Chrzanowska M, Rozwadowska M (2004) Chem Rev 104:3341–3370
203. Augustine RL (1995) Heterogeneous catalysis for the synthetic chemist. CRC Press, Boca Raton
204. Zhou Y-G (2007) Acc Chem Res 40:1357–1366
205. Wang D-S, Chen Q-A, Lu S-M, Zhou Y-G (2012) Chem Rev 112:2557–2590
206. Matsunami A, Kayaki Y (2018) Tetrahedron Lett 59:504–513
207. Wu J, Tang W, Pettman A, Xiao J (2013) Adv Synth Catal 355:35–40
208. Wu J, Wang C, Tang W, Pettman A, Xiao J (2012) Chem Eur J 18:9525–9529
209. Voutchkova AM, Gnanamgari D, Jakobsche CE, Butler C, Miller SJ, Parr J, Crabtree RH (2008) J Organomet Chem 693:1815–1821
210. Frediani P, Rosi L, Cetarini L, Frediani M (2006) Inorg Chim Acta 359:2650–2657
211. Fujita K-I, Kitatsuji C, Furukawa S, Yamaguchi R (2004) Tetrahedron Lett 45:3215–3217
212. Talwar D, Li HY, Durham E, Xiao J (2015) Chem Eur J 21:5370–5379
213. Fidalgo J, Ruiz-Castañeda M, García-Herbosa G, Carbayo A, Jalón FA, Rodríguez AM, Manzano BR, Espino G (2018) Inorg Chem 57:14186–14198
214. Facchetti G, Bucci R, Fusè M, Rimoldi I (2018) Chem Select 3:8797–8800
215. Vilhanová BV, Budinská A, Václavík J, Matoušek V, Kuzma M, Červený L (2017) Eur J Org Chem:5131–5134
216. Xu C, Zhang L, Dong C, Xu J, Pan Y, Li Y, Zhang H, Li H, Yu Z, Xu L (2016) Adv Synth Catal 358:567–572
217. Anastas P, Eghbali N (2010) Chem Soc Rev 39:301–312
218. Omoruyi U, Page S, Hallett J, Miller PW (2016) ChemSusChem 9:2037–2047
219. Deng J, Wang Y, Pan T, Xu Q, Guo Q-X, Fu Y (2013) ChemSusChem 6:1163–1167
220. Wang S, Dorcet V, Roisnel T, Bruneau C, Fischmeister C (2017) Organometallics 36:708–713
221. Xing R, Qi W, Huber GW (2011) Energy Environ Sci 4:2193–2205
222. Karinen R, Kati Vilonen K, Niemel M (2011) ChemSusChem 4:1002–1016
223. Kabro A, Escudero-Adán EC, Grushin VV, Leeuwen PWNM (2012) Org Lett 14:4014–4017
224. Constable DJC, Jiménez-González C, Henderson RK (2007) Org Process Res Dev 11:133–137
225. Gu Y, Jérôme F (2010) Green Chem 12:1127–1138
226. Wolfson A, Dlugy C, Tavor D (2011) Trends Org Chem 15:41–50
227. Díaz-Álvarez AE, Cadierno V (2013) Appl Sci 3:55–69
228. Crabtree RH (2019) ACS Sustain Chem Eng 7:15845–15853
229. Azua A, Finn M, Yi H, Dantas AB, Voutchkova-Kostal A (2017) ACS Sustain Chem Eng 5:3963–3972
230. Nishina Y (2019) Inorganics 7:114
231. Yoshida M, Hirahata R, Inoue T, Shimbayashi T, Fujita K-I (2019) Catalysts 9:503
232. Mwansa JM, Page MI (2020) Cat Sci Technol 10:590–612
233. Cheng-Xing C, Chen H, Li SJ, Zhang T, Qu LB, Lan Y (2020) Coord Chem Rev 412:213251
234. Samec JSM, Bäckvall JE, Andersson PG, Brandt P (2006) Chem Soc Rev 35:237–248
235. Noyori R, Ohkuma T (2001) Angew Chem Int Ed 40:40–73, and references therein
236. Dub PA, Gordon JC (2016) Dalton Trans 45:6756–6781

237. Dub PA, Gordon JC (2017) ACS Catal 7:6635–6655
238. Dub PA, Gordon JC (2018) Nat Rev Chem 2:396–408
239. Handgraaf JW, Reek JNH, Meijer EJ (2003) Organometallics 22:3150–3157
240. Chen HYT, Wang C, Wu X, Jiang X, Catlow CRA, Xiao J (2015) Chem Eur J 21:16564–16577
241. Stirling MJ, Sweeney G, MacRory K, Blacker AJ, Page MI (2016) Org Biomol Chem 14:3614–3622
242. Demianets I, Cherepakhin V, Maertens A, Lauridsen PJ, Sharada SM, Williams TJ (2020) Polyhedron 182:114508

Top Organomet Chem (2021) 69: 153–206
https://doi.org/10.1007/3418_2020_64

Published online: 11 December 2020

Iridium-Catalyzed Asymmetric Hydrogenation

Jèssica Margalef, Oscar Pàmies, and Montserrat Diéguez

Contents

Abstract In this chapter, we describe the development in homogeneous Ir-catalyzed asymmetric hydrogenation with particular emphasis on the achievements made during the last 10 years. We also present their application to the synthesis of complex molecules. The first section deals with the hydrogenation of unfunctionalized olefins or with poorly coordinative groups. The second section includes the advances made in the hydrogenation of functionalized olefins. The last two sections cover the hydrogenation of imines and ketones, respectively.

J. Margalef, O. Pàmies, and M. Diéguez (✉)
Departament de Química Física i Inorgànica, Universitat Rovira i Virgili, Tarragona, Spain
e-mail: montserrat.dieguez@urv.cat

Keywords (Un)functionalized olefins · Asymmetric hydrogenation · Catalysis · Imines · Iridium · Ketones

1 Introduction

Metal-catalyzed asymmetric hydrogenation (AH) offers some of the most sustainable and straightforward reactions for producing pharmaceuticals, flavors, fragrances, agrochemicals, and fine chemicals due to its perfect atom economy and operational simplicity [1–5]. It is estimated that around 10% of all chemical steps in the synthesis of these compounds are hydrogenations. Despite the extensive research dedicated to the asymmetric hydrogenation and the important progress reached, some issues still need to be solved. Most catalysts only work with a limited number of substrates, and each type of substrates needs a specific catalyst for optimal enantioselectivity. For example, the asymmetric hydrogenation of functionalized alkenes is mostly carried out by Ru- and Rh-diphosphine catalysts (see, e.g., [6–9]), while the asymmetric hydrogenation of unfunctionalized olefins or with poorly coordinative groups is mainly carried out with Ir-P,N catalysts (for reviews, see [10–15]). A broad substrate scope is desirable to reduce the time dedicated to ligand/catalyst design and preparation. A desired additional condition is that the catalyst family should be synthesized from available starting materials and be easy to handle.

The number and types of functionalized substrates have been remarkably expanded, and their use is commonplace, as illustrated in the commercial production of the Parkinson's L-DOPA drug [16, 17], the broad-spectrum antibiotics levofloxacin [18] and sitagliptin [19], and the pesticide (*S*)-metolachlor [20]. The success of catalysts relies on the ability of the substrate to form a metal chelate involving the double bond and a donor atom. Although the reduction of functionalized olefins has been thoroughly studied for decades, there are some substrate types that are still a challenge. Among them it can be found the cyclic β-enamides, which have recently attracted attention because their hydrogenation products are found in many pharmaceutically and biologically active products. Two representative examples are rotigotine, used to treat Parkinson's disease, and alnespirone, a selective 5-HT1A receptor with antidepressant and anxiolytic properties [21–24]. In the last decade, it has been found that Ir-containing catalysts [25–27] can be used in their reduction with results that surpass the most studied Rh- and Ru-catalysts [28–38]. Other challenging substrates where Ir-catalysts have shown to be superior or complementary to the Rh-/Ru-catalysts are unsaturated carboxylic acids and nitroolefins, among others (see Sects. 3–5).

The absence of a coordinative group in the olefins makes their hydrogenation a great challenge. So, compared to the AH of functionalized olefins, the reduction of unfunctionalized alkenes or with poor coordinative groups is much less mature [10–15]. The best catalysts have two characteristics in common: (1) they mainly contain P,N-ligands [39–41] and (2) their optimal structure is highly dependent on the geometry and substitution pattern of the olefin [10–15]. The consequence is that for each particular olefin type, a different ligand family needs to be developed. It is also important to notice that catalysts have been developed in different grades for each olefin substitution pattern. The most successful cases have been reported for trisubstituted olefins and, to a less extent, for disubstituted. The asymmetric hydrogenation of tetrasubstituted unfunctionalized substrates is still underdeveloped.

In this book chapter, we describe the development in Ir-catalyzed asymmetric hydrogenation with particular emphasis on the achievements made during the last 10 years. We also present their applications to the synthesis of complex molecules. Most of the work has been devoted to the hydrogenation of nonfunctionalized olefins or with poorly coordinative groups (Sect. 2), with significant advances in both substrate scope and mechanistic studies. However, also notable advances have been made in improving the catalytic performance in the reduction of relevant and more challenging functionalized substrates, such as unsaturated carboxylic acids, nitroolefins, cyclic β-enamides, imines, etc. (see Sects. 3–5).

2 Ir-Catalyzed Asymmetric Hydrogenation of Unfunctionalized Olefins or with Poorly Coordinative Groups

Among the most challenging substrates to date are the unfunctionalized olefins or olefins with poorly coordinative groups [10–15]. A breakthrough in the hydrogenation of this type of substrates came in 1997 when Pfaltz et al. used phosphine-oxazoline PHOX ligands **L1** (Fig. 1) to design [Ir(**L1**)(cod)]PF_6 (cod = 1,5-cyclooctadiene), a chiral analogue of Crabtree's catalyst ([Ir(py)(PCy_3)(cod)]PF_6) that enantioselectively hydrogenated imines [42]. Although this catalyst also hydrogenated unfunctionalized olefins highly enantioselectively, it was unstable to the reaction conditions. Pfaltz and co-workers overcame this problem by changing the catalyst anion to $[(3,5\text{-}(F_3C)_2\text{-}C_6H_3)_4B]^-$($[BAr_F]^-$). The result was [Ir(**L1**)(cod)] BAr_F (Fig. 1), an active, enantioselective, and stable catalyst library for olefin hydrogenation. Despite this success, its scope was limited to mainly *E*-trisubstituted olefins [43]. It was also seen that the optimal catalyst was highly dependent on the geometry and substitution pattern of the olefin. This triggered the search for new catalysts that would reach a wider substrate scope.

In this respect, Pfaltz group continued to develop new versions of the PHOX complexes, modifying the ligand backbone, with the discovery of very efficient ligand libraries [44–52]. Successive work incorporated pyridine and quinoline rings instead of the oxazoline, which allowed the successful reduction of challenging purely alkyl-substituted substrates in high ee [53–55]. A notable application was the total synthesis of γ-tocopherol as a single diastereoisomer in 98% ee, controlling two stereocenters in one reductive step (see below) [54]. The various developed ligands also enabled the reduction of various type of substrates, such as allylic alcohols,

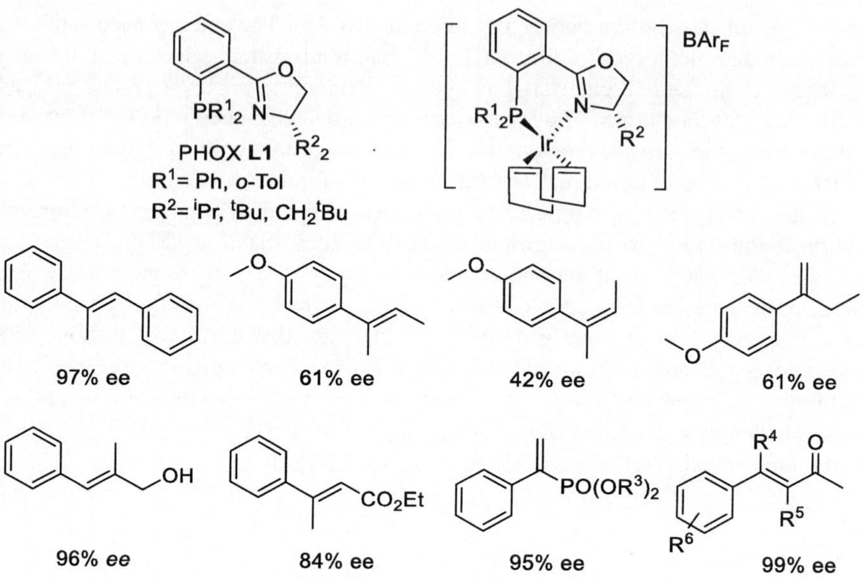

Fig. 1 Selected Ir-catalyzed asymmetric hydrogenation results with [Ir(**L1**)(cod)]BAr$_F$

α,β-unsaturated esters [56], furan derivatives [57], boronic esters [58], and tetrasubstituted olefins [59].

Other groups also provided new successful ligand libraries, by modifying the chiral backbone and by replacing either the P-group by a N-heterocyclic carbene moiety or the oxazoline moiety by other N-donor groups (such as oxazole, thiazole, and imidazole) and by O- and S-donor groups [14, 15, 60, 61]. All these modifications allow to further extend the substrate scope.

Concerning mechanistic aspects, although the mechanism of olefin hydrogenation by Rh-catalysts is well understood, the mechanism when Ir-catalysts are used has not been fully determined until recently. In this context, computational and experimental research with P,N- and C,N- ligands have shown that the hydrogenation of minimally functionalized olefins proceeds via an Ir^{III}/Ir^{V} migratory-insertion/reductive-elimination catalytic cycle (Fig. 2) [62–67]. Very recently, Pfaltz's group, based on mechanistic studies under hydrogenation conditions, was able to detect the Ir(III) dihydride alkene intermediates responsible for the catalytic performance for the first time [68]. They found that, similar to the classical Halpern mechanism for asymmetric hydrogenation with Rh-catalysts, the minor intermediate, which is less stable, is converted to the major product enantiomer.

In the next sections, we collect the catalytic results on the asymmetric hydrogenation of unfunctionalized olefins or with poorly coordinative groups.

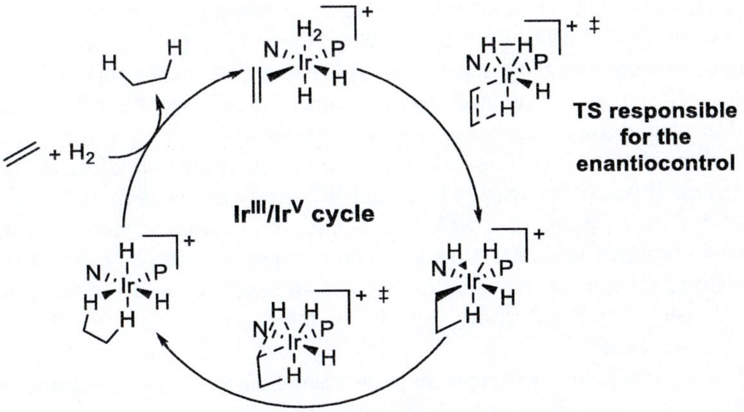

Fig. 2 Ir^{III}/Ir^{V} catalytic cycle for the hydrogenation of minimally functionalized olefins

2.1 *Di- and Trisubstituted Unfunctionalized Olefins or with Poorly Coordinative Groups*

Aryl/alkyl trisubstituted alkenes have become the model substrates for evaluating the efficiency of new catalytic systems. In general, the hydrogenation of 1,2-diarylalkenes (i.e., *trans* α-methylstilbene) proceeded with higher enantioselectivities than monoarylated ones (such as *E*-2-(4-methoxyphenyl)-2-butene) for which only a limited number of catalysts provided high enantioselectivities [10–15]. The geometry of the olefin also affects the catalytic performance. *Z*-Trisubstituted olefins are usually hydrogenated less enantioselectively than the related *E*-trisubstituted olefins. The lower enantioselectivities can be mainly attributed to a *Z*/*E* isomerization process to form the more stable *E*-alkene, which gives the opposite enantiomer of the hydrogenated product [10–15]. *Z*-2-(4-Methoxyphenyl)-2-butene and dihydronaphthalenes (i.e., 7-methoxy-4-methyl-1,2-dihydronaphthalene) are frequently used to study the ligand scope in the hydrogenation of *Z*-alkenes. Dihydronaphthalenes have recently received much attention because the corresponding chiral tetraline motif is found in numerous natural products [69]. Trialkyl substituted alkenes have been much less studied. This is due in part to the difficulty in developing methods for ee determination and also the lack of an aryl group that could direct the reaction via π-stacking interaction between the substrate and the chiral catalyst. The best results have been reported in the reduction of 1-methoxy-4-(3-methyl-pent-3-enyl)-benzene (ees up to 95%) [54].

Nowadays, Ir-catalysts have also been able to reduce olefins with a variety of relevant poorly coordinative groups such as α,β-unsaturated esters, ketones, and lactames and vinyl boronates, among others [10–15]. The effective hydrogenation of such a range of olefins is of great importance since their reduced products are key structural chiral units found in many high-value chemicals (e.g., α- and β-chiral ketones and carboxylic acid derivatives are ubiquitous in natural products, fragrances, agrochemicals, and drugs). Substrate scope has also been extended to 1,1-diaryl or 1,1,2-triaryl substituted substrates (i.e., 1-(1,2-diphenyl-vinyl)-3,5-dimethyl-benzene) and more recently to 1,4- and cyclic dienes (i.e., 1,5-dimethyl-cyclohexa-1,4-diene), linear and cyclic sulfones, and alkyl fluorides, which are present in several important drugs and natural products.

Unlike trisubstituted olefins, a large range of 1,1-disubstituted olefins have not been successfully asymmetrically hydrogenated until very recently [10–15]. This is because the catalyst has the added difficulty of controlling not only the face selectivity coordination (only two substituents compared with the three of trisubstituted olefins, Scheme 1a) but also the isomerization of the olefins to form the more stable *E*-trisubstituted substrates, which are hydrogenated to form the opposite enantiomer (Scheme 1b).

Next we compile the most representative catalytic results in the hydrogenation of di- and trisubstituted olefins organized by the type of ligands.

2.1.1 Phosphine-Oxazoline Ligands

Inspired by the work of Pfaltz et al. with PHOX ligands, many other phosphine-oxazoline ligands have been developed. Künding, Pfaltz et al. reported a modification in the oxazoline moiety with the phosphine-benzoxazine analogues **L2** (Fig. 3, R = ^{t}Bu, ^{i}Pr) [46]. The enantioselectivities were lower than those recorded with PHOX ligands. The presence of a bulky substituent, a ^{t}Bu, at the oxazine group provided good enantioselectivities for *E*-trisubstituted olefins (ees up to 89%), except for trisubstituted allylic alcohols, but low for *Z*-trisubstituted olefins, 1,1′-di- and tetrasubstituted olefins.

The rest of the new developments in the ligand design were based on modifications of the ligand backbone. Ligands **L3** (Fig. 3, R^1 = Ph, *o*-Tol and R^2 = Me, ^{t}Bu, 1-Ad, CPh_3), developed by Burgess et al., were applied in the hydrogenation of several aryl-alkyl alkenes [70]. These ligands proved to be superior to the PHOX ligands in the hydrogenation of *Z*-trisubstituted alkenes, while ees for *E*-trisubstituted alkenes were lower. The best enantioselectivities for *Z*-olefins were obtained

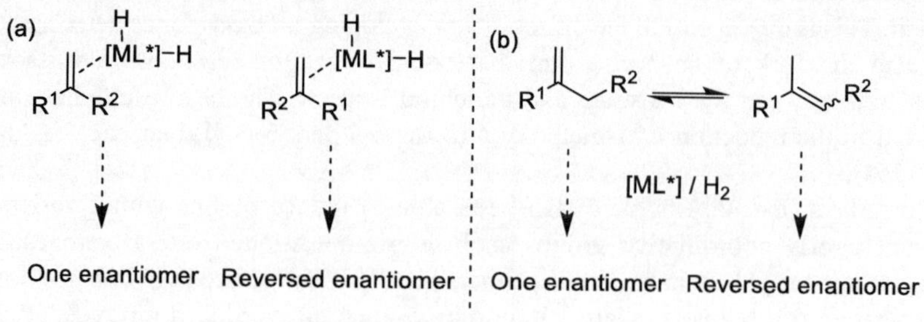

Scheme 1 Proposed reasons for the low enantioselectivities in the reduction of 1,1′-disubstituted olefins

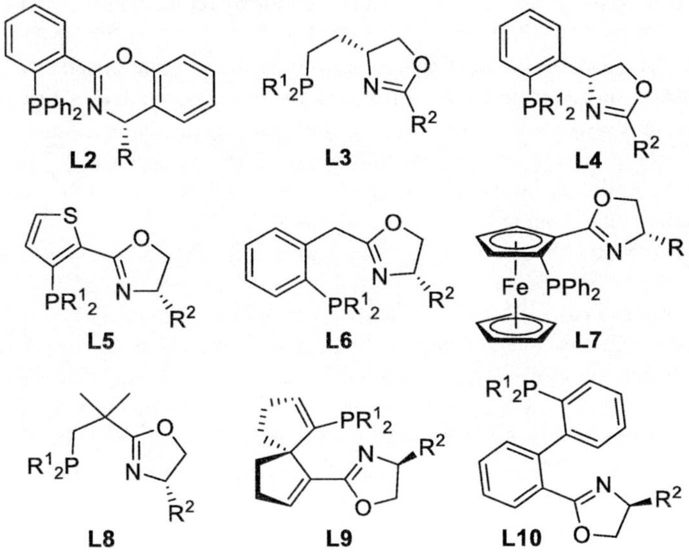

Fig. 3 Selected phosphine-oxazoline ligand libraries developed for the Ir-catalyzed asymmetric hydrogenation of di- and trisubstituted olefins

with a ^{t}Bu group at the oxazoline and a diphenylphosphanyl group, while for *E*-olefins a bis(*o*-tolyl)phosphanyl group was needed (ees up to 80%). A further modification of ligands **L3** was to introduce again the *ortho*-phenylene motif of the PHOX ligands. New ligands **L4** (Fig. 3, R^1 = Ph, Cy and R^2 = ^{t}Bu, 1-Ad, $CHPh_2$, 3,5-${}^{t}Bu_2$-C_6H_3) provided excellent results in the reduction of trans-α-methylstilbene derivatives and trisubstituted α,β-unsaturated esters (ees up to 99%) [71]. Again bulky groups in oxazoline and phosphine moieties were needed (R^1 = Cy; R^2 = ^{t}Bu).

Later, Cozzi's group developed ligands **L5**, in which the phenyl ring of the PHOX ligands was replaced by a thiophene group (Fig. 3, R^1 = Ph, *o*-Tol, Cy and R^2 = ^{i}Pr, ^{t}Bu) [72]. This modification also led to high enantioselectivities but only in the hydrogenation of *trans*-α-methylstilbene (ees up to 99%). Hou et al. developed phosphine-oxazoline ligands **L6** in which the flat *ortho*-phenylene group in the PHOX ligands was replaced by a benzyl group (Fig. 3, R^1 = Ph, *o*-Tol, *p*-Tol and R^2 = Me, ^{i}Pr, ^{t}Bu) [73, 74]. These ligands allow to extend the type of substrates successfully hydrogenated. High enantioselectivities were achieved with *E*-trisubstituted aryl-alkyl alkenes, allylic alcohols, and α,β-unsaturated esters and ketones (ees up to 98%). The best enantioselectivities were obtained with an ^{i}Pr oxazoline group and a diphenylphosphanyl functionality.

Another modification was to introduce a ferrocenyl group (ligands **L7**, Fig. 3, R^2 = Me, ^{i}Pr, ^{t}Bu, Ph, Bn). The best results were obtained with the ligand that contains a small methyl substituent in the oxazoline group that proved to be superior than PHOX in the *Z*-substrates (89% ee), while ees for *E*-alkenes were lower (ees up to 89%) [75].

Pfaltz et al. also further modified PHOX ligands by replacing the *ortho*-phenylene tether by a branched alkyl chain (ligands **L8**; Fig. 3, R^1 = Ph, *o*-Tol, Xyl and R^2 = ^{i}Pr, ^{t}Bu, Bn) [69]. These ligands provided higher enantioselectivities in the hydrogenation of trisubstituted *E*- and *Z*-aryl alkenes than the PHOX ligands (ees up to 98%). The best results were achieved with the ligand that contains bulky substituents at both phosphine and oxazoline groups (R^1 = Xyl and R^2 = *t*Bu). The authors showed its applicability with the synthesis of (*R*)-7-demethyl-2-methoxycalamenene, an antitumor natural product.

The spirocyclic phosphine-oxazoline ligands **L9** (Fig. 3, R^1 = *o*-Tol, Ph and R^2 = Ph, Bn) were also successfully used in the hydrogenation of α,β-unsaturated Weinreb amides [76] and α,α′-bis(2-hydroxyarylidene) ketones [77].

Afterward, Zhang et al. developed phosphine-oxazoline ligands **L10** with a biphenyl backbone (Fig. 3, R^1 = Ph, 3,5-^{t}Bu$_2$-C$_6$H$_3$, 3,5-^{t}Bu$_2$-4-MeO-C$_6$H$_2$ and R^2 = ^{i}Pr, ^{t}Bu, Ph, Me) which successfully hydrogenated exocyclic α,β-unsaturated carbonyl compounds (including ketones, lactones, and lactams) [78], 3-substituted 2,5-dihydropyrroles [79], and 2,5-dihydrothiophene 1,1-dioxides [79].

2.1.2 Aminophosphine-Oxazoline Ligands

Some aminophosphine-oxazoline ligands have also showed comparable high efficiency than phosphine-oxazolines in the reduction of unfunctionalized olefins or with poorly coordinative groups. In this context, Pfaltz et al. modified the PHOX ligands by replacing the *ortho*-phenylene group by a pyrrole group leading to ligands **L11** (Fig. 4, R^1 = Ph, *o*-Tol, Cy and R^2 = ^{i}Pr, ^{t}Bu) [80]. Enantiomeric excesses surpassed those previously obtained with the PHOX ligands, with ligands bearing a bulky *tert*-butyl oxazoline substituent and either an *ortho*-tolyl or cyclohexyl P-group. Nevertheless, the enantioselectivities for *Z*-trisubstituted olefins were not above 80% ee. Then, Gilbertson et al. developed the proline-based aminophosphine-oxazoline ligands **L12** (Fig. 4, R^1 = Ph, *o*-Tol and R^2 = ^{i}Pr, ^{t}Bu), related to previous ligands **L11**; however, they provided lower enantioselectivities [81]. The best result was obtained with the ligand bearing a bulky *tert*-butyl oxazoline substituent.

Andersson et al. developed ligands **L13** and **L14** (Fig. 4, **L13**; R^1 = Ph, *o*-Tol, Cy; R^2 = H, ^{t}Bu, Ph and R^3 = H, Ph; and **L14**; R^1 = Ph; R^2 = H, ^{i}Pr, Ph and R^3 = H, ^{i}Pr, Ph) [82–91]. Ligands **L13**, which are based on a rigid bicyclic backbone, provided higher enantioselectivities than ligands **L14**, with a more flexible backbone. Ir/**L13** catalyst (with R^1 = R^2 = R^3 = Ph) afforded, for first time, high enantioselectivities in the hydrogenation of enol phosphinates [84, 85], vinylsilanes [86], fluorinated olefins [87], vinyl boronates [88], α,β-unsaturated acyclic esters [89], α,β-unsaturated lactones [90] and γ,γ-disubstituted and β,γ-disubstituted allylic alcohols [91] (Fig. 5).

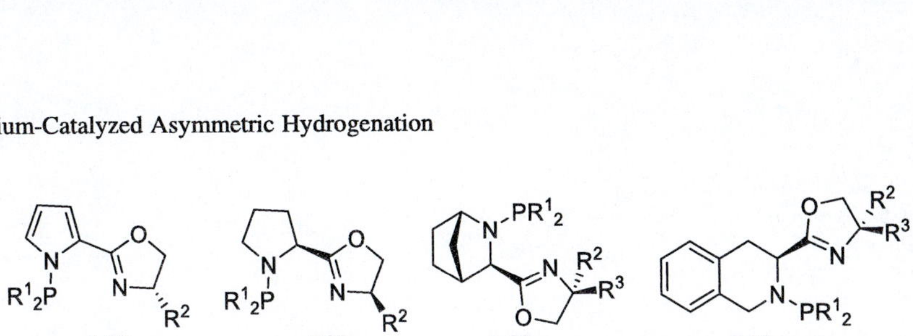

Fig. 4 Selected aminophosphine-oxazoline ligand libraries developed for the hydrogenation of di- and trisubstituted olefins

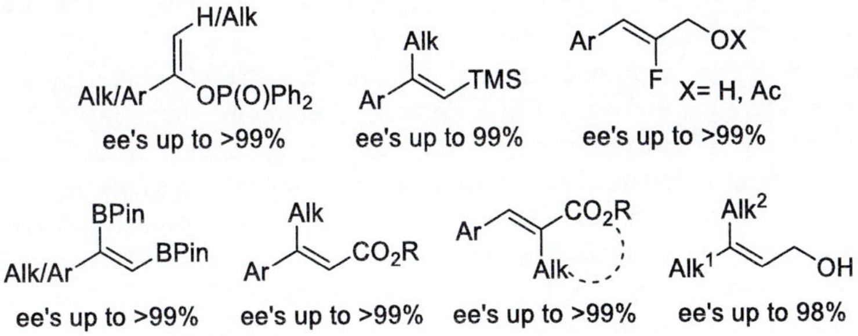

Fig. 5 Representative hydrogenation results with Ir/**L13** catalyst

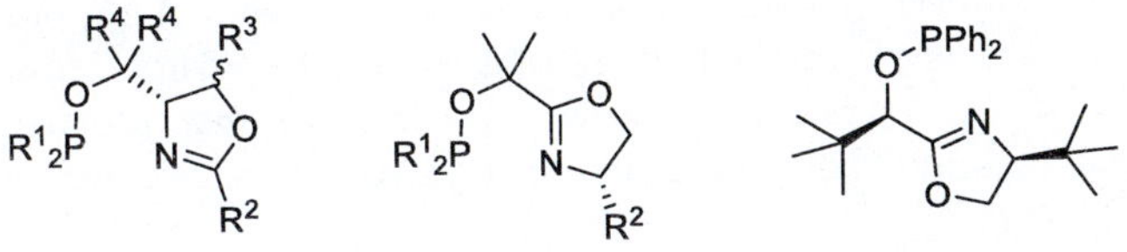

Fig. 6 Selected phosphinite-oxazoline ligand libraries developed for hydrogenation of di- and trisubstituted olefins

2.1.3 Phosphinite-Oxazoline Ligands

It can be highlighted the family of phosphinite-oxazoline ligands **L15** (Fig. 6, R^1 = Ph, *o*-Tol, Cy; R^2 = ^{t}Bu, Ph, ferrocenyl, 2-Naph; R^3 = H, Me, 3,5-Me_2-C_6H_3 and R^4 = Me, ^{i}Pr, ^{t}Bu, Bn), developed by Pfaltz et al. [47, 49, 92–94]. Ligands **L15** constitute one of the most privileged ligands for this process. They provided excellent enantioselectivities in the reduction of a broad range of both *E*- and *Z*-trisubstituted olefins, including α,β-unsaturated esters and for the first time, in a limited range of more challenging terminal olefins as well as in the reduction of 1,1′-disubstituted enamines (ees up to 99%, Fig. 7) [92–95]. More recently, Ir-catalysts containing ligands **L15** have also been successfully applied in the reduction of α,β-unsaturated nitriles (ees up to 98%, Fig. 7) [94]. The best enantioselectivities were achieved with ligands containing a methyl substituent at R^3, a benzyl substituent at R^4, and a phenyl at R^1. However, the appropriate substituent at the oxazoline and the configuration of the carbon of R^3 depend on

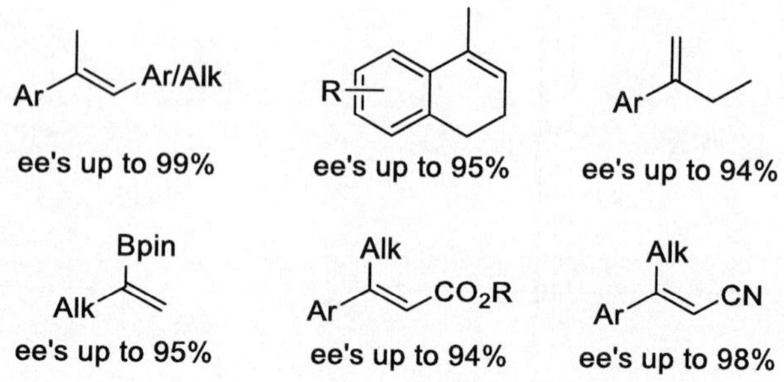

Fig. 7 Representative catalytic hydrogenation results with Ir/**L15** catalyst

the substrate to be hydrogenated. For *E*-trisubstituted olefins, ees are best with ligands containing a Ph or a 3,5-Me_2-Ph and a *S*-configuration, while for *Z*-olefins the highest enantioselectivities were achieved using ligands with Ph and a *R*-configuration. In addition, these catalysts work efficiently in propylene carbonate as an environmentally friendly solvent, and this allowed the Ir-catalysts to be reused maintaining the excellent enantioselectivities [96, 97]. Based on ligands **L15**, Pfaltz group developed ligands **L16** (Fig. 6, R^1 = Ph, *o*-Tol and R^2 = ^{i}Pr, ^{t}Bu) where the alkyl chain is bonded in the C-2 instead of the C-4 of the oxazoline moiety, which shifts the chirality from the alkyl chain to the oxazoline moiety [50, 75]. The scope of these ligands is narrower than with the privileged phosphinite-oxazoline ligands **L15**; however, they are complementary. Ligands **L16** provided high enantioselectivities for allylic alcohols and alkenes with heteroaromatic substituents.

Kazmeier et al. synthetized ligands **L17** (Fig. 6) that provided excellent enantioselectivities for linear and cyclic α,β-unsaturated ketones (ees up to >99%) [95].

2.1.4 Phosphite-Oxazoline Ligands

Phosphite-containing ligands have shown to be particularly useful for asymmetric catalysis. They have a greater resistance to oxidation than phosphines and phosphinites, they are easily synthesized from readily available chiral alcohols, and their modular constructions are easy (for reviews, see [98–100]). Despite this, it was not until 1999 that a publication reported their use in the reduction of unfunctionalized olefins using Ir-TADDOL-based phosphite-oxazoline catalysts (**L18**, Fig. 8). However, their substrate scope and enantioselectivities were lower than their related Ir-phosphinite/phosphine-oxazoline catalysts. Additionally, the use of high pressures (100 bars) and high catalyst loadings (4 mol%) was required to obtain full conversions [51].

Our group contributed to the asymmetric Ir-catalyzed hydrogenation of unfunctionalized olefins with a new series of air-stable ligands that are applicable to a wide variety of substrates (di- and trisubstituted). The key was the introduction of a flexible biaryl phosphite group in the ligand. We first developed the pyranoside phosphite-oxazoline ligand library **L19** (Fig. 8, R = Me, ^{i}Pr, ^{t}Bu, Ph, Bn) synthetized

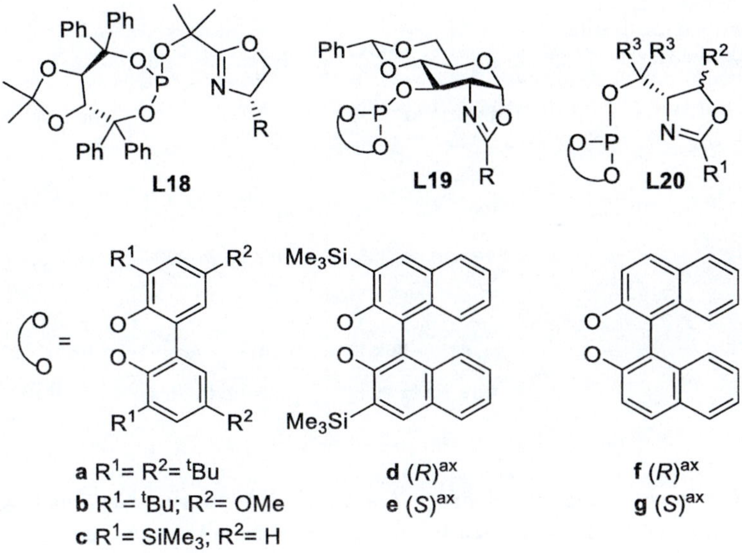

Fig. 8 Selected phosphite-oxazoline ligand libraries developed for the Ir-catalyzed asymmetric hydrogenation of di- and trisubstituted olefins

from D-glucosamine, an inexpensive natural feedstock, that contains several biaryl phosphite groups [67, 101, 102]. It was found that for enantioselectivities to be high, the presence of bulky substituents in the biaryl phosphite group and less sterically demanding substituents in the oxazoline moiety was required. Thus, it was possible to identify two general ligands (**L19c** and **L19e** with R = Ph) that provided high enantioselectivities. For comparative purposes, the related phosphinite-oxazoline analogues were also tested, but with lower success [102]. With ligands **L19c** and **L19e**, high enantioselectivities and activities (ees up to >99%) in many trisubstituted olefins (25 examples, Fig. 9), even in the reduction of the more challenging Z-isomers, could be reached and triarylsubstituted substrates, which provide an easy entry point to diarylmethine chiral centers that are present in several important drugs and natural products [103–106]. High enantioselectivities could also be achieved in the reduction of many trisubstituted substrates with poorly coordinative groups, such as α,β-unsaturated esters and ketones, vinylsilane, allylic alcohol, and acetates. Also, it should be noted the excellent enantioselectivities obtained in the hydrogenation of vinyl boronates (ees ranging from 92% to >99%). Their hydrogenation provides chiral borane compounds, which are useful building blocks in organic synthesis because the C-B bond can be readily converted to C-O, C-N, and C-C bonds with retention of the chirality. Even more remarkable were the high enantioselectivities obtained for the first time in the reduction of a broad range of 1,1′-disubstituted olefins (19 examples, Fig. 9). It was found that the Ir/**19e** system was robust against variations in the electronic nature of the substrate aryl substituents (ees up to 99%). Also high levels of enantioselectivity were obtained in the reduction of

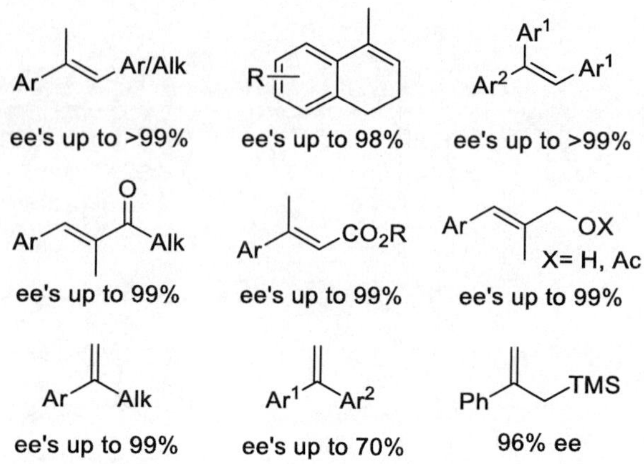

Fig. 9 Representative hydrogenation results with Ir/**L19c,e** catalyst

heteroaromatic terminal olefins (ees up to 99%). Nevertheless, the enantioselectivities were affected by the nature of the alkyl chain and diminished in 1,1′-diaryl alkenes due to an isomerization process.

With the aim of understanding the catalytic performance of Ir/**L19** catalysts, a DFT computational study was performed in collaboration with Norrby et al. [67]. It was found that the preferred reaction path is an Ir(III/V) cycle with migratory insertion of a hydride as the selectivity-determining step. In addition, the effect of the ligand parameters could be rationalized by using a simple quadrant model (Fig. 10), where the phenyl oxazoline's substituent occupies the upper left quadrant and one of the aryls of the biaryl phosphite moiety partly blocks the lower right quadrant (Fig. 10a). The other two quadrants are free. The calculated structure had a chiral pocket that fits perfectly *E*-olefins (Fig. 10b). This quadrant model also explains the change of ligand (from **L19e** to **L19c**) to obtain a high enantioselectivity in *Z*-olefins (Fig. 10c). Ligand **L19c** has bulky substituents in the *para* position, which increases the dihedral angle of the biaryl group and results in lower occupancy of the lower right quadrant than with ligand **L19e**. Therefore, the substituent in the biphenyl group can tune the occupancy of the lower right quadrant, and therefore *Z*-alkenes can also be successfully hydrogenated (Fig. 10c). The same explanation accounts for the triaryl- and disubstituted substrates. In conclusion, the DFT studies confirm that the flexibility of the biaryl phosphite group is a crucial parameter in the achievement of high enantioselectivities for substrates with different geometries and steric requirements.

Following this contribution comes the developments of new biaryl phosphite-oxazoline ligand libraries with the aim to increase even further the range of substrates successfully hydrogenated [27, 107–110]. Among them, we can highlight the application of phosphite-oxazoline ligands (**L20**, Fig. 8) [107, 109], which were inspired in one of the best families developed for this transformation (previous ligands **L15**, Fig. 6) by replacing the phosphinite groups by several biaryl phosphite moieties. Selecting the ligand parameters' high enantioselectivities has been reported

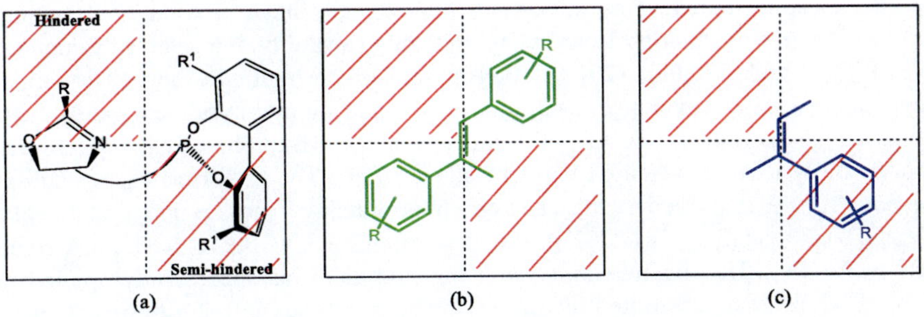

Fig. 10 Quadrant diagram describing the enantioselective substrate-ligand interactions

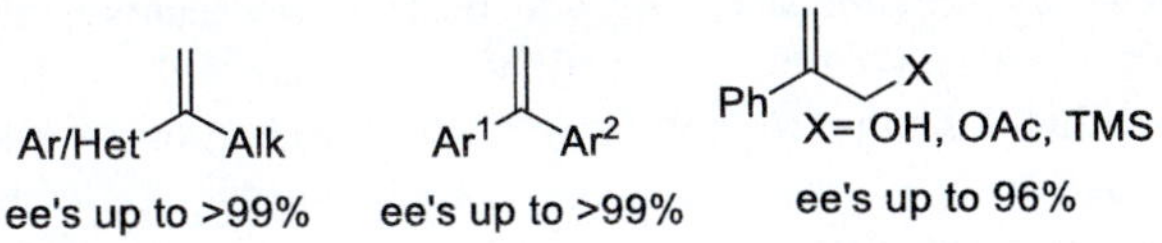

Fig. 11 Representative results achieved with Ir/**L20** catalysts in the hydrogenation of 1,1-′-disubstituted substrates

for trisubstituted olefins. Whereas Ir/**L20g** catalyst provided the best enantioselectivities for linear and cyclic olefins and a α,β-unsaturated ester, the best ees for the more demanding *Z*-isomers and allylic alcohol and acetate were obtained with Ir/**20a** catalyst. This high catalytic performance was also extended to the hydrogenation of the more challenging 1,1′-disubstituted olefins (29 compounds, Fig. 11), surpassing the previous family **L19** and becoming one of the best catalysts for the reduction of this type of substrates. High enantioselectivities were achieved in a broad range of aryl-alkyl (ees up to >99%), even with substrates bearing decreasingly sterically alkyl substituents, and heteroaromatic-alkyl (ees up to >99%) olefins. These catalyst precursors also tolerate very well the presence of neighboring polar groups. High enantioselectivities were achieved in the reduction of allylic alcohols and an allylic silane. Interestingly, the reaction showed no loss of enantioselectivity when dichloromethane was replaced by propylene carbonate. In addition, the use of propylene carbonate allowed the catalysts to be recycled up to five times by a simple two-phase extraction maintaining the excellent enantioselectivities [107].

2.1.5 Phosphorus-Other Nitrogen Donor Ligands

In the recent years, the research has also focused on the design of ligands containing more robust groups than oxazolines. A collection of the most representative phosphorus-other nitrogen donor ligands will be next presented.

As an alternative to P-oxazoline ligands, pyridine-containing ligands have attracted interest due to the robustness and the easy incorporation of pyridine group. Despite this, few pyridine-containing ligands have provided outstanding results in terms of enantioselectivity and substrate versatility. For selected examples, see Fig. 12 [53, 57, 111–123]. Among them, we can highlight the first pyridine-containing ligand developed by Pfaltz et al. (phosphinite-pyridine ligands **L21**; Fig. 12, R^1 = Ph, *o*-Tol, Cy, ^{t}Bu and R^2 = Me, ^{t}Bu, Ph, CPh_3), which was successfully used in a limited range of alkenes [57]. The performance was subsequently further improved by the same group introducing a more rigid chiral bicyclic ligand backbone (ligands **L22**, Fig. 12, R^1 = Ph, *o*-Tol, Cy, ^{t}Bu; R^2 = H, Ph, Me and R^3 = H, Me). This ligand family with high rigidity was successfully applied in several kinds of trisubstituted olefins, including purely alkyl trisubstituted alkenes, furans, and benzofurans as well as trisubstituted pinacol derivatives, α,β-unsaturated lactones, and N-protected indoles [57, 114, 116, 117]. The enantioselectivity was highest with a Ph substituent at the R^2 and bulky substituents at the phosphinite moiety (^{t}Bu or *o*-Tol). To obtain excellent enantiocontrol in the reduction of 7-methoxy-4-methyl-1,2-dihydronaphthalene, the introduction of a large aryl substituent at R^2 (2,4,6-tri-Me-Ph) was needed. Its applicability was demonstrated in the reduction of γ-tocotrienyl acetate to obtain γ-tocopherol, a principal component of vitamin E [118], resulting in enantioselectivity >98% for the *RRR* enantiomer. Another synthetic application can be found in the diastereo- and enantioselective hydrogenation of farnesol stereoisomers. By changing the bond's geometry, these catalysts give access to the four stereoisomers of the product in high selectivity.

To benefit from the advantages of phosphite and pyridine moieties, our group replaced in ligands **L21** the phosphinite moieties by several biaryl phosphite groups increasing even further the substrate scope (Fig. 12; ligands **L23**, R^1 = H, Me, Br, Ph and R^2 = Me, ^{t}Bu, Ph) [122]. Excellent enantioselectivities (ees up to 99%) were obtained in a wide range of *E*- and *Z*-trisubstituted alkenes, including more demanding triarylsubstituted olefins and dihydronaphthalenes, and also terminal disubstituted olefins and alkenes containing neighboring polar groups.

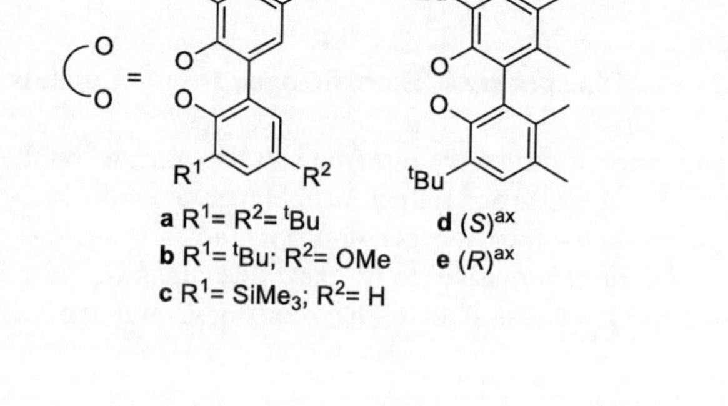

Fig. 12 Most representative P-pyridine ligands for Ir-catalyzed asymmetric hydrogenation

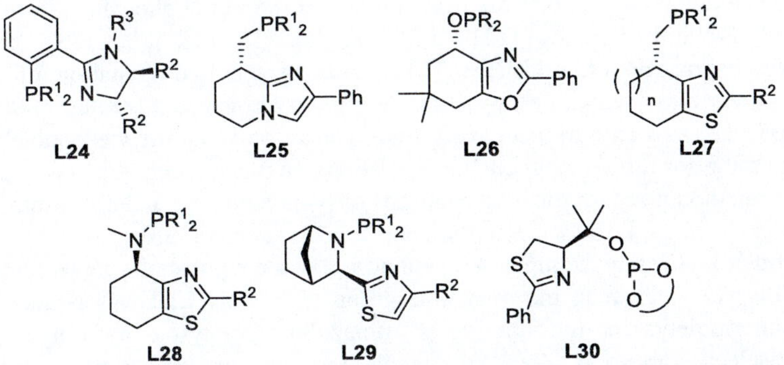

Fig. 13 Most representative P-other N-donor ligands for Ir-catalyzed asymmetric hydrogenation

Another interesting change in the nitrogen donor group is the replacement of the oxazoline by imidazole, oxazole, thiazole, thiazoline, and sulfoximine [124] groups. For selected examples, see Fig. 13. The first application was reported by Pfaltz et al. with the phosphine-imidazole ligands **L24** (Fig. 13, R^1 = Ph, *o*-Tol; R^2 = ^{i}Pr, *t*Bu and R^3 = ^{i}Pr, tBu, Cy, Ph, Bn, *p*-Tol) [48]. One advantage of the imidazoline group over the oxazoline is the possibility to introduce a new substituent R^3 at the nitrogen that could serve as a linker to attach the ligand to a solid support. Ligands **L24** provided better enantioselectivities in the hydrogenation of *Z*-trisubstituted olefins (ees up to 88%) than PHOX ligands (ees up to 42%). The best results were achieved with ligands containing bulky substituents at both R^1 and R^2 positions, while the substituent at R^3 had to be optimized for each substrate. Andersson group also developed the phosphine-imidazole ligands **L25** (Fig. 13, R^1 = Ph, *o*-Tol, 3,5-diMe-Ph) that gave high enantioselectivities for *E*-aryl/alkyl trisubstituted olefins (ees up to 98%) [125, 126] and cyclic dienes (ee's up to >99% for the *trans* isomer) [127–129], but was only moderate in the reduction of *Z*-olefins (ees up to 72%) [126]. Interestingly, the hydrogenation of dienes was also found to be regioselective, and by controlling the reaction conditions, selective hydrogenation of one of the two trisubstituted olefins was achieved. In addition, trisubstituted olefins were selectively hydrogenated in the presence of tetrasubstituted olefins. Thus, enantioselectivities were best with ligand containing a bisphenylphosphanyl group except for the reduction of *trans*-α-methylstilbene for which a bis-(*o*-tolyl) phosphanyl group was needed. More recently, they also showed its applicability in the enantioconvergent formal deoxygenation of racemic alcohols (Fig. 13). This methodology was successfully used in the total synthesis of antidepressant sertraline and σ_2 receptor PB [75] [35j].

Several classes of other P,N-ligands have been developed by Andersson et al. (Fig. 13, ligands **L26-L29**; R^1 = Ph, o-Tol … and R^2 = Ph, ^{t}Bu …). The investigation of different bicyclic heteroaromatic rings led to highly enantioselective iridium catalysts containing oxazoles [69] and thiazoles [70] (Fig. 13). These catalysts perform excellently on the typically tested trisubstituted nonfunctionalized olefins and also allow extending the substrate scope to vinyl allylsilanes [86], fluorinated olefins [87, 126, 130], vinyl boronates [88], enol phosphinates [84], and *E*- and *Z*-chiral sulfones [131, 132], allylic alcohols [90, 133], and the monohydrogenation of 1,4-dienes [134]. Among the hydrogenation of dienes, many purely alkyl-substituted were successfully hydrogenated (ees up to 99%). In addition, they are able to selectively hydrogenate only one of the double bonds, leaving room for further synthetic manipulations. In this respect, Andersson's group use this methodology for the total synthesis of (-)-juvabione, a natural sesquiterpene exhibiting juvenile hormone activity using Ir/**L28** catalyst [135].

Another interesting example of ligand design was the phosphite-thiazoline ligand **L30** (Fig. 13), in which the oxazoline group in ligands **L20** was replaced by a thiazoline moiety. The introduction of a thiazoline moiety has not only provided enantioselectivities up to >99% for a range of α,β-unsaturated ketones, vinylsilane, and trifluoromethyl olefins but also has increased the enantioselectivities of *Z*-trisubstituted olefins while maintaining the excellent enantioselectivities for a range of *E*-trisubstituted and 1,1-disubstituted minimally functionalized olefins [136].

2.1.6 Carbene-Nitrogen Ligands

Another type of effective catalysts is the Ir/carbene-nitrogen complexes. An important advantage of N-heterocyclic carbene (NHC) catalysts compared to their phosphine analogues concerns their better tolerance for acid-sensitive substrates. In 2001, Burgess' group reported for the first time that NHC-oxazoline-based Ir-catalysts (ligand **L31**, Fig. 14, R^1 = 2,6-^{i}Pr$_2$-Ph and R^2 = 1-Ad) can also be applied in the hydrogenation of unfunctionalized olefins with results comparable to the commonly used Ir-P,N catalysts [137, 138]. These catalysts afforded high enantioselectivities (up to 98% ee) in a limited group of unfunctionalized olefins, mainly trisubstituted, and for the more challenging disubstituted olefins, only one example was reported with low enantioselectivity. Since then, a few more carbene-N ligands have been developed but with less success [139–143], except for the family of Ir-NHC-pyridine catalysts [144] developed by Pfaltz's group (with ligands **L32**, Fig. 14, R = 2,6-diisopropylbenzene) that showed similar enantioselectivities to the Burgess ones. So, high enantioselectivities (>90% ee) were observed, even for *Z*-trisubstituted (94% ee) and endocyclic substrates (96% ee).

2.1.7 Application of P-O/S Ligands

In contrast to other catalytic processes and to the Rh-/Ru-hydrogenation, for the reduction of unfunctionalized olefins, the possibility of changing the nature of the N-donor atom in the ligand design of heterodonor ligands was not contemplated until recently. In 2011, Pfaltz successfully reported the application of proline-based P,O ligands **L33** in the asymmetric hydrogenation of trisubstituted alkenes (Fig. 15, R^1 = Ph, ^{t}Bu, Cy, *o*-Tol and R^2 = ^{t}Bu, 1-Ad, CPh_3, 1Ad-NH, MesNH, CPh_3NH) [145–147]. Phosphines bearing either a bulky amide or urea groups at the pyrrolidine

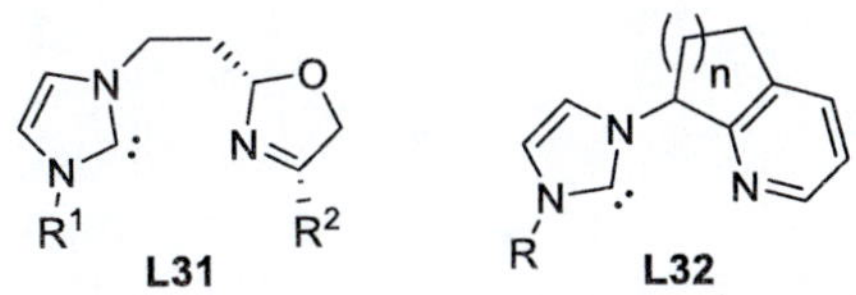

Fig. 14 Selected P-carbene ligands for Ir-catalyzed hydrogenation of olefins

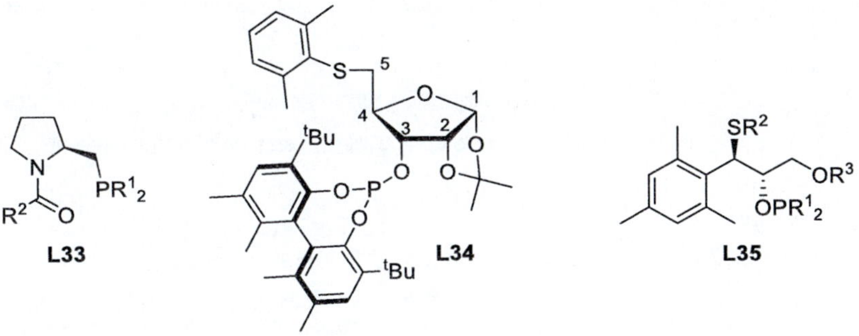

Fig. 15 Selected P-O/S ligands for the Ir-catalyzed hydrogenation of olefins

N-atom formed efficient Ir-catalysts for the asymmetric hydrogenation of several minimally functionalized olefins (ees up to 99%).

At the same time, our group reported the application of a highly modular furanoside phosphite-thioether ligand library (ligands **L34**, Fig. 15) [148, 149]. By selecting the ligand components in these furanoside-based ligands (position of the thioether group at either C-5 or C-3 of the furanoside backbone, the configuration of C-3, the thioether substituent, and the substituents/configuration in the biaryl phosphite moiety), we found that the best enantioselectivities were obtained using ligands with a 5-deoxy-ribofuranoside backbone **L34**. Excellent enantioselectivities were obtained (ees up to 99%) in the reduction of a range of trisubstituted alkenes, including relevant examples with poorly coordinative groups (such as α,-β-unsaturated esters and vinyl boronates). The results are comparable to the best ones reported in the literature except for the hydrogenation of 1,1′-disubstituted aryl/alkyl olefins. For this substrate class, our results indicated that enantioselectivity is dependent on the nature of the alkyl substrate substituent and much less affected by the electronic nature of the aryl ring. This has been attributed to an isomerization process that was supported by the fact that the hydrogenation of substrates bearing a *tert*-butyl group, for which isomerization cannot occur, provides high levels of enantioselectivity (ees up to 98%). We also studied the effect on catalytic performance of introducing either phosphinite or phosphine moieties with lower success.

Since then, several other P-thioether ligands have been developed [150–152]. From them, we can highlight the modular phosphite/phosphinite-thioether ligand library **L35** (Fig. 15; R^1 = Ph, Tol, Cy, Mes; R^2 = Ph, 2,6-Me_2-Ph, 4-MeOPh, 2-Naph, ^{t}Bu, Ad, Cy; R^3 = Me, Tr, Mes). In a simple three-step procedure, several ligand parameters were easily tuned to maximize the enantioselectivities for each substrate

Fig. 16 Representative hydrogenation results with Ir/**L35** catalysts

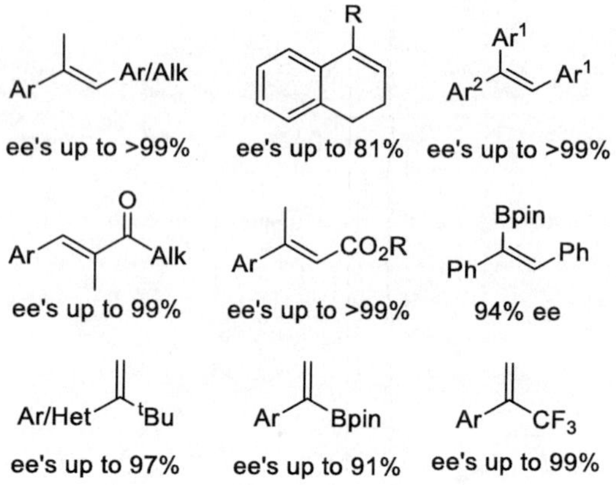

(ees up to 99% in 43 hydrogenated products, Fig. 16) [153]. In contrast to the furanoside-based ligands mentioned above (**L34**), the best enantioselectivities were obtained with the phosphinite-S ligands, while results achieved with the phosphite-S analogues were less optimal. The crystal structures of the Ir-catalyst precursors indicate an equatorial disposition of the thioether group for the phosphite-based ligands, while an axial disposition of the thioether group is found in the analogues phosphinite ligands. The modularity of the ligands together with DFT studies were crucial to find which ligand parameters could be modified to generate more selective catalysts. In this respect, the use of a bulky mesityl group instead of a phenyl group in the ligand backbone improved enantioselectivity. With catalyst Ir/**L35**, with R^1 = Tol, R^2 = 2,6-Me_2-Ph, and R^3 = Me, excellent enantioselectivities (ees up to >99%) were recorded for many trisubstituted olefins, including olefins with relevant neighboring polar groups such as α,β-unsaturated esters, ketones, vinyl boronates, and allylic alcohols (Fig. 16). High enantioselectivities were also achieved in the hydrogenation of 1,1′-disubstituted alkenes. Excellent enantioselectivities were also maintained by using propylene carbonate as an environmentally benign solvent, which allowed the Ir-catalyst to be reused up to three times. DFT studies also confirmed that the preferred reaction path is an Ir^{III}/Ir^{V} cycle where the selectivity-determining step is the migratory insertion of a hydride. DFT results also allowed the formulation of a quadrant model which explains the effect of the ligand parameters on selectivities. In this quadrant model, the thioether substituent occupies the upper left quadrant, and one of the P-substituents partly occupies the lower right quadrant, while the other two quadrants are free. This explains the high enantioselectivities obtained with the DFT-optimized guided design of thioether-phosphinite ligands in the reductions of (*E*)-olefins. In the case of the analogous phosphite-thioether ligands, the upper left quadrant is not enough blocked due to the equatorial disposition of the thioether group, which explains that they provided lower enantioselectivities than the related phosphinites.

2.2 *Tetrasubstituted Unfunctionalized Olefins or with Poorly Coordinative Groups*

Despite the advances during the last 10 years in the asymmetric hydrogenation of unfunctionalized olefins with the development of ligand libraries that allowed a significant increase in the range of substrates that can been successfully hydrogenated, the reduction of tetrasubstituted olefins remains a challenge. The range of such substrates that can be efficiently hydrogenated is still narrow [154].

In 1999, Buchwald's group reported the first successful asymmetric hydrogenation of tetrasubstituted unfunctionalized olefins [155]. Although high enantioselectivities were achieved for substituted indenes using the zirconocene catalyst **1** (Fig. 17; ee's in the range 52–99%), the high catalyst loading (8 mol%), the high H_2 pressure (typically >110 bar), and the low stability of the catalyst hampered their broad use. Much more recently, Zhang's group reported a Rh-catalyst **2** (Fig. 17), containing a P-stereogenic diphosphine ligand synthesized in nine steps that provided 85–95% ees in the reduction of some indenes [156]. But it still required high catalyst loading (10 mol%), 60 °C, and longer reaction times (4 days). Again, Pfaltz's group made an important breakthrough in this field. In 2007, they found that the stability and/or the harsh reaction condition issues of the Zr/Rh-catalysts can be overcome with Ir/P,N catalysts. Another important finding was that the optimum ligand structures for tri- and tetrasubstituted olefins differed strongly [59]. Using Ir-catalysts **3** (Fig. 17), containing ligands that form a

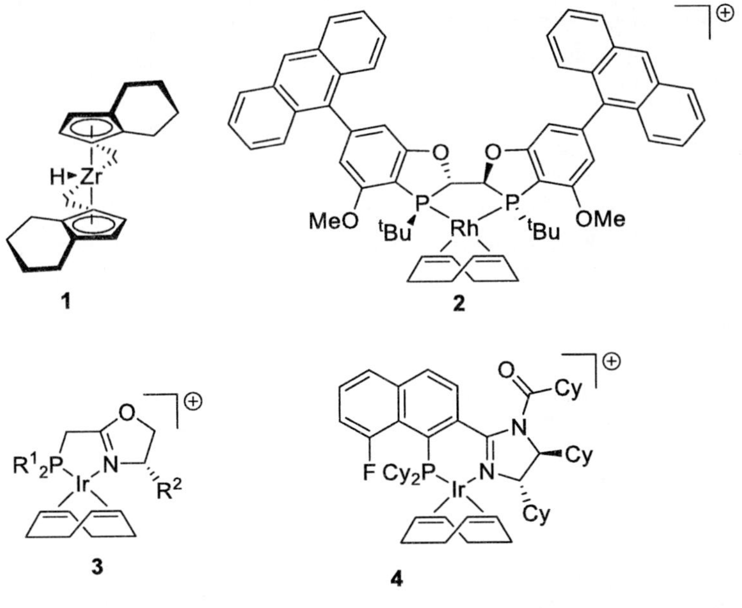

Fig. 17 Representative catalysts for the Ir-catalyzed hydrogenation of tetrasubstituted olefins

5-membered chelate ring, a wide range of indenes were hydrogenated with ees in the range of 94–96%, under milder reaction conditions and low catalyst loading (typically 1–2 mol%). Nevertheless, ees diminished for noncyclic olefins and for 1,2-dihydro-napthalenes (ees between 89 and 97% and up to 77%, respectively). It should be mentioned that these catalysts provided low enantioselectivities in the hydrogenation of trisubstituted olefins.

This finding prompted the interest in the design of new specific ligands for Ir-hydrogenation of unfunctionalized tetrasubstituted olefins. In 2013, Busacca's group found that the Ir-catalyst **4** could hydrogenate two cyclic substrates with ees up to 96% at low catalyst loading. An inconvenience was that low temperature (0 °C) was required [157].

Several more successful reports have appeared in the last few years. Notably, Andersson's group showed the efficient asymmetric hydrogenation of the challenging acyclic tetrasubstituted olefins. They successfully hydrogenated a broad range of tetrasubstituted vinyl fluorides using one of their privileged Ir-P,N catalysts for the reduction of trisubstituted olefins, with a small modification in the P-group (ligand **L29** with R^1 = *o*-EtPh; R^2 = ^{i}Pr, see above Fig. 13) [158]. The challenge of these substrates is that the catalyst must not only control de face selectivity but also avoid the side defluorination reaction. Advantageously, the reaction proceeded smoothly without defluorination in high diastereo- and enantioselectivities. Various aromatic, aliphatic, and heterocyclic systems with a variety of functional groups were efficiently hydrogenated. The successful asymmetric hydrogenation of these substrates opens up a direct, atom-efficient path to synthesize chiral fluorine molecules with two contiguous stereogenic centers. However, the catalyst was not successful for other classes of cyclic and acyclic tetrasubstituted olefins and also required the use of high H_2 pressure (20–100 bar).

Our group in collaboration with Riera's group reported the application of an Ir/P-stereogenic aminophosphine-oxazoline catalyst library **L36** (Fig. 18, R = Ph, ^{i}Pr, ^{t}Bu), with a simple, modular architecture, in the asymmetric hydrogenation of a broad range of different types of unfunctionalized tetrasubstituted olefins [159]. Improving previous results reported until now, the same family of catalysts is able to efficiently reduce indenes and the challenging 1,2-dihydro-napthalene derivatives (ees up to 96%) and also a broad range of the elusive acyclic olefins with enantioselectivities up to 99% under mild reaction conditions. Moreover, the excellent catalytic performance is maintained for a range of aryl and alkyl vinyl fluorides (drs > 99% and ees up to 98%), where two vicinal stereogenic centers are created.

Then, by substitution of the aminophosphine group by simple readily available phosphinite groups (Ir/**L37** catalyst, R^1 = Ph, *o*-Tol, Cy; R^2 = Ph, ^{i}Pr, ^{t}Bu, Fig. 18), we could also efficiently reduce many unfunctionalized tetrasubstituted olefins (ees up to 98%) under mild reaction conditions [27]. It should be noted that the more rigid the tetrasubstituted olefin is, the less bulky phosphinite moieties are required to reach the maximum enantioselectivity. For the more rigid cyclic indene

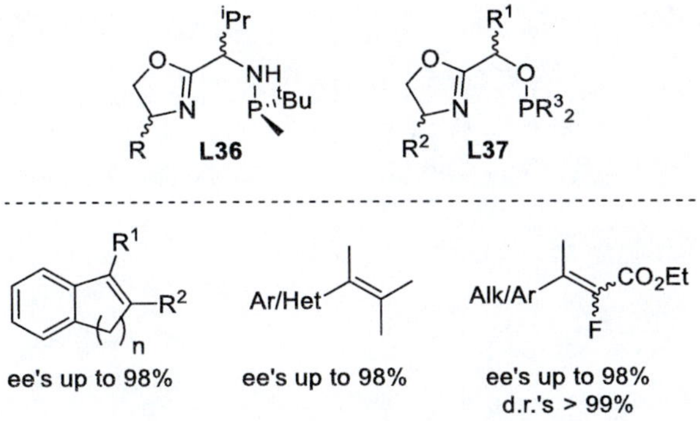

Fig. 18 Selected catalytic results obtained with Ir/**L36**–**L37** catalysts in the reduction of tetrasubstituted olefins

derivatives, the best catalytic performance is therefore reached with the phosphinite-based ligand with Ph phosphinite substituent, while for the less rigid cyclic substrate, the phosphinite ligand with a bulkier *o*-tolyl group is needed. Finally, the even less rigid acyclic substrates require the ligand with the bulkiest cyclohexyl phosphinite group. Even more interestingly, maintaining the same skeleton of the ligand by simply changing the phosphinite functionality by the right phosphite group (ligands **L37a-c**, Fig. 19), we could also efficiently reduce many unfunctionalized tri- and disubstituted olefins (ees up to 98%, Fig. 19). In summary, from a common simple skeleton, the correct choice of either phosphite or phosphinite groups gives for the first time ligands that are suitable for di-, tri-, and tetrasubstituted unfunctionalized substrates and also for cyclic β-enamides (62 examples, with ees up to 99%).

A notable last contribution is the identification of an Ir-catalyst that is able to successfully hydrogenate a very broad range of diverse acyclic unfunctionalized tetrasubstituted olefins (around 30 examples) [160]. A first parallel screening of a set of 34 different Ir-catalysts found previous Ir-catalyst **3** (Fig. 17, R^1 = *o*-Tol; R^2 = ^{i}Pr) to be the best candidate. A subsequent optimization of its structure (phosphine and oxazoline substituent) identified **3** with R^1 = Cy and R^2 = 3,5-bis-^{t}Bu-Ph as the optimal catalyst.

3 Ir-Catalyzed Asymmetric Hydrogenation of Olefins Containing Coordinating Groups

As already mentioned, the hydrogenation of olefins with strongly coordinating groups has been predominantly performed using Rh- or Ru-catalysts bearing chiral diphosphine ligands. They still constitute the optimal choice for the synthesis of optically active α-amino acids and many pharmaceutically relevant compounds. Nowadays, excellent enantioselectivities can be achieved for N-acyl α-dehydroamino acid derivatives, enamides and acrylates or itaconates, among

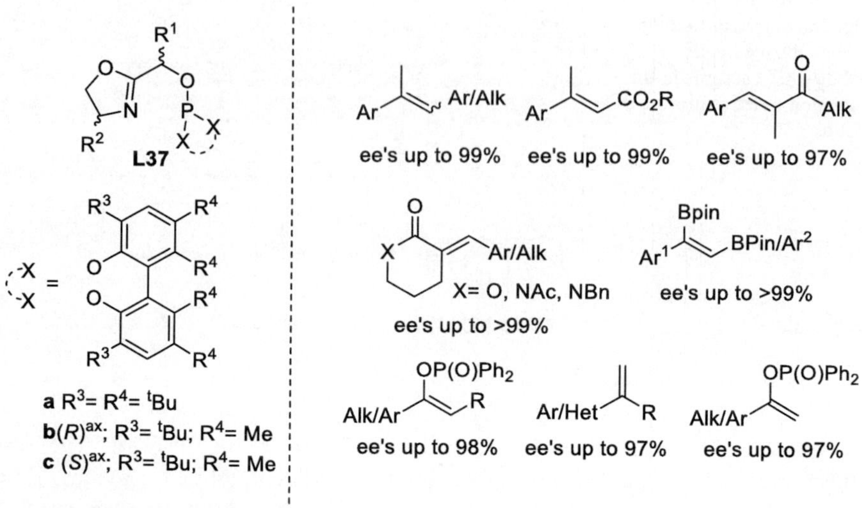

Fig. 19 Selected results obtained with phosphite-oxazoline of ligands **L37a–c** in the Ir-catalyzed hydrogenation of di- and trisubstituted olefins

others [6–9]. However, in the last decade, Ir-based catalysts appear as a good alternative in the reduction of challenging functionalized olefins, providing higher catalytic performance than the Rh- and Ru-catalysts. In this respect, we next show the improved catalytic performance in the reduction of carboxylic acids and nitroolefins using Ir-catalyst. Chiral carboxylic acids are important intermediates for the preparation of biologically active compounds ([161], for a review, see [162]; for selected examples, see [163–166]). On the other hand, enantiomerically pure nitroalkanes can be easily converted to other versatile building blocks, such as amines, aldehydes, carboxylic acids, nitrile oxides, and denitrated compounds [167, 168].

3.1 *Ir-Catalyzed Asymmetric Hydrogenation of Carboxylic Acids*

Although only few Ir-catalysts have been studied for this transformation, they have allowed to overcome the limitations of most studied Rh- and Ru-catalysts. Indeed, most of the reported Rh- and Ru-catalysts showed a scope limited to acrylic and cinnamic acids, and especially with Ru-catalysts, high pressures and catalyst loadings are usually needed [6–9]. Scrivanti and co-workers explored for the first time the Ir-PHOX catalyst in the hydrogenation of 2-phenethylacrylic acid, recording enantioselectivities not higher than 81% ee [169]. Later, Burgess applied a chiral N-heterocyclic carbene-oxazoline **L31** (Fig. 14) in the hydrogenation of tiglic acid with only 55% ee [170]. The inefficiency of these iridium catalysts was partly attributed to their tendency to aggregate into inactive trimers under a hydrogen atmosphere [171].

Zhou et al. showed for the first time that Ir-catalysts, the spiro phosphine-oxazoline (SIPHOX) Ir-catalysts (Fig. 20a) [172], could efficiently hydrogenate unsaturated carboxylic acids with the presence of a base [173]. The addition of a base results in the formation of a carboxylate anion, which act as a strong coordinating group. Under mild reaction conditions, excellent yields (90–97%) and enantioselectivities (96–>99% ee) could be achieved for a broad range of α-aryloxy- and α-alkyloxy-substituted α,β-unsaturated acids, with TONs up to 10,000 (Fig. 20a). It was found that the best catalysts contained a ligand with a bulky P-aryl group (Ar = 3,5-$^{t}Bu_2Ph$), which was the best choice for most of the substrates studied afterward. The hydrogenation protocol was efficiently used for the preparation of α-benzyloxy-carboxylic acid, a key intermediate in the syntheses of the rhinovirus protease inhibitor rupintrivir (Fig. 20b) [174]. In contrast to previous Ir-catalysts, the rigidity and bulkiness of the spiro scaffold on SIPHOX ligands seemed to prevent the Ir-catalysts to trimerize under hydrogenation conditions. The authors also found that while Ir-SIPHOX catalysts were not effective for the hydrogenation of α,β-unsaturated esters [175], the Ir-PHOX analogue did provide excellent enantioselectivities [11]. Thus, both catalyst types have complementary substrate scope.

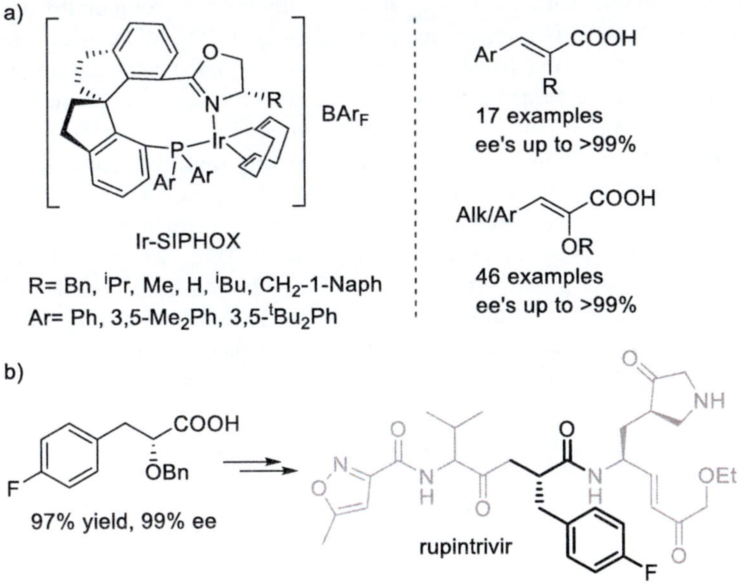

Fig. 20 **(a)** Ir-SIPHOX catalysts and their application in the asymmetric hydrogenation of α-alkyl, α-aryloxy-, and α-alkyloxy-substituted α,β-unsaturated carboxylic acids. **(b)** Synthesis of rupintrivir, a rhinovirus protease inhibitor

With the use of Ir-SIPHOX catalysts, Zhou and co-workers have largely unblocked the amount of unsaturated carboxylic acid derivatives that can be hydrogenated enantioselectively [175, 176] and for which Rh- and Ru-catalysts showed no success. Thus, they further expanded its application to the reduction of other trisubstituted α,β-unsaturated carboxylic acids, such as α-aryl- and α-oxymethyl-substituted cinnamic acids (Fig. 21). The hydrogenation of both types of substrates were used as a key step for the total synthesis of two natural products. Thus, the enantioselective reduction of α-arylcinnamic acid (Ar^1 = 2,4-$(MeO)_2$Ph, Ar^2 = 4-MeOPh) [177] and α-oxymethylcinnamic acid (Ar = 4-MeOPh, R = 3,4-(OCH_2O) Ph) was used for preparing (*S*)-equol and (*S*)-(+)-homoisoflavone [178]. Similarly, a range of N- and O-heterocycles of different ring sizes could be also hydrogenated with enantioselectivities ranging from 89 to 99% ee (Fig. 21). The methodology allowed the direct preparation of (*R*)-nipecotic acid and (*R*)-tiagabine in excellent yields and enantioselectivities [179]. Again, with a ligand containing a 3,5-^{t}Bu_2Ph substituent in the phosphine, a range of tetrasubstituted acrylic acids with α-aryl, α-alkyl, α-aryloxy, or α-alkyloxy substituents were reduced in high enantioselectivities (90–99%) (Fig. 21). It should be noted that some of the hydrogenated products are key intermediates of chiral drugs, such as mibefradil and fenvalerate [180].

The excellent results of spiro phosphine-oxazoline ligands (SIPHOX) are not only limited to α,β-unsaturated acids. The authors also explored the reduction of several β,γ-unsaturated acids, which gives access to molecules with a chiral center at the γ-position. A range of 4-alkyl-4-aryl-3-butenoic acids could be hydrogenated in up to 97% ee with a ligand containing an α-naphthylmethyl group on the oxazoline ring and a 3,5-Me_2Ph as a phosphine substituent (Fig. 22). With this asymmetric hydrogenation as the key step, the concise total syntheses of the natural products (*R*)-aristelegone-A, (*R*)-curcumene, and (*R*)-xanthorrhizol were accomplished [181]. It should be noted that β,γ-unsaturated ester (*E*)-methyl 4-phenylpent-3-enoate was inert under hydrogenation conditions, thus indicating that the functional carboxy group is crucial for the reaction by acting as a directing group. The use of a carboxy directing group was also extended to the hydrogenation of terminal 1,1-dialkyl, 1,1-diaryl and 1-aryl-1-alkyl γ,δ-unsaturated acids (Fig. 22, ees up to 99%) [182, 183]. This strategy is particularly useful for the reduction of 1,1-diaryl and 1,1-dialkylethenes, in which most of the catalysts fail in differentiating the *Re*- and *Si*-faces due to the similarity on size of both substituents of the olefin. In addition, it was shown that the directing carboxy group on these substrates can be subsequently

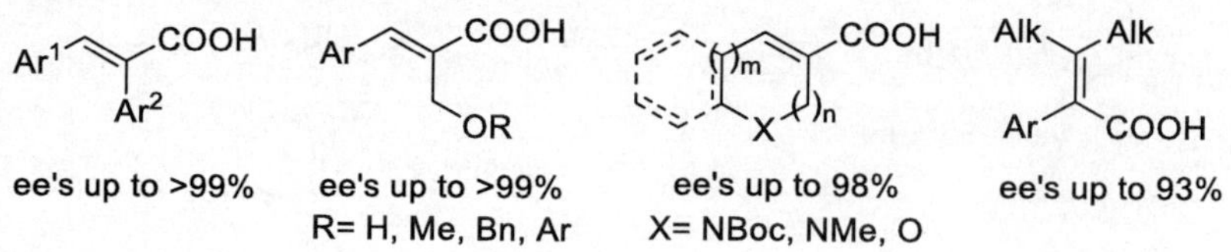

Fig. 21 Ir-catalyzed asymmetric hydrogenation of trisubstituted α,β-unsaturated carboxylic acids and acrylic acids with SIPHOX ligands

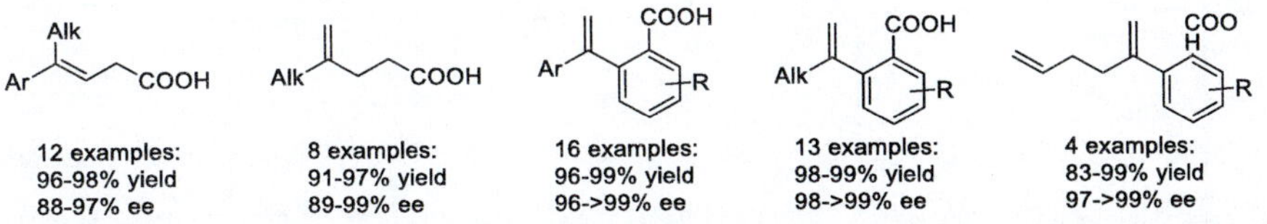

Fig. 22 Ir-catalyzed asymmetric hydrogenation of β,γ-unsaturated acids and γ,δ-unsaturated acids using SIPHOX ligands

removed or easily transformed to other useful functional groups if desired [182]. A range of α-alkyl-α-aryl terminal olefins were also reduced in excellent enantioselectivities (98–>99%; Fig. 22), yielding valuable compounds with a chiral benzylmethyl center. The developed hydrogenation process was also used as a key step for preparing (*S*)-curcudiol and (*S*)-curcumene in excellent enantioselectivities and overall yields [183]. Finally, the authors further confirmed the role of the carboxylate moiety as a directing group by showing that when no free carboxylic acid was present or no basic conditions were used, the reaction didn't proceed [182]. Moreover, it was found that for substrates having an extra C=C double bond in the alkyl side chain, the presence of the carboxy directing group makes the reaction chemoselective toward the α-alkyl-α-aryl double bond (Fig. 22), even when the additional double bond was placed in the terminal position [183].

As found with Rh- and Ru-catalysts, the Ir-SIPHOX catalysts showed unsatisfactory results for the hydrogenation of 2-substituted α-arylacrylic acids. To overcome this limitation, Zhou and co-workers developed a new series of spiro P, N-ligands (SpiroBAP), with a benzylamino moiety instead of the oxazoline group (Fig. 23a). The new generation of ligands exhibited extremely high reaction rates (TOFs up to 6,000 h^{-1}) and excellent enantioselectivities (94–98% ee) in the reduction of α-aryl and α-alkyl acrylic acids to the corresponding chiral carboxylic acids, including ibuprofen, naproxen, and flurbiprofen, which are widely used nonsteroidal anti-inflammatory drugs (Fig. 23a). As for SIPHOX ligands, ligands with a bulky P-aryl group (Ar = 3,5-^{t}Bu$_2$Ph) gave the best catalytic results [184].

Zhou et al. have also recently developed a neutral version of spiro-based Ir-catalysts (Ir-SpiroCAP, Fig. 23b), by replacing the oxazoline moiety on SIPHOX ligands by an anionic carboxy group. The resulting Ir-complexes do not require the use of a tetrakis[3,5-bis(tri-fluoromethyl)phenyl]borate (BAr_F^-) counterion, which is necessary for stabilizing chiral cationic Crabtree-type catalysts, while remaining highly stable for a long time in air. These new generations of catalysts exhibited an unprecedented high enantioselectivity (up to >99% ee) in the hydrogenation of the challenging 3-alkyl-3-methylenepropionic acids (Fig. 23b). To demonstrate its potential application in organic synthesis, the synthesis of (*S*)-14-methyloctadec-1-en, a female sex pheromone of the peach leaf miner moth (*Lyonetia clerkella*), was carried out. The new catalysts were also effective (ees up to 99.4%) in the reduction of other α-methyl cinnamic acid, tiglic acid, and α-substituted acrylic acids, among others [176].

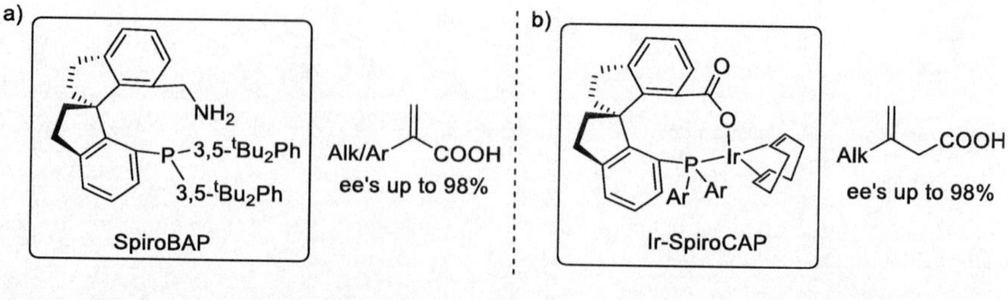

Fig. 23 Ir-catalyzed asymmetric hydrogenation of (**a**) α-substituted acrylic acids using a SpiroBAP ligand and (**b**) 3-alkyl-3-methylenepropionic acids with Ir/SpiroCAP catalysts

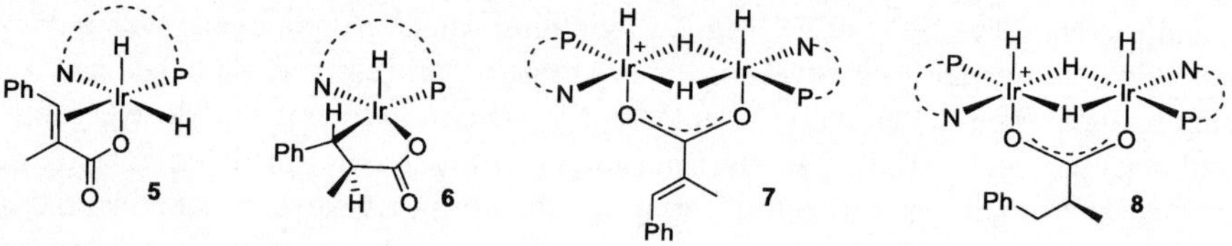

Fig. 24 Structures of isolated intermediates **5–8** in the model reaction of sodium (*E*)-2-methyl-3-phenyl acrylate with the Ir-SIPHOX catalyst (Ar = 3,5-^{t}Bu$_2$Ph, R = H)

The authors also performed a mechanistic study including DFT studies that strongly supported an Ir(III)/Ir(V) cycle. The high stability of the chiral spiro-iridium catalysts under reaction conditions [172] allowed the trapping of the active intermediates and facilitated the mechanistic study [185]. To mimic the basic conditions used in the hydrogenation reactions, the authors used sodium (*E*)-2-methyl-3-phenyl acrylate as a model substrate. The isolation of the monohydride intermediate **5** (Fig. 24), resulting from the migratory insertion of **6** (Fig. 24), was key to understand the mechanism. Dimeric species **7** and **8** (Fig. 24), which are off-cycle species, were also isolated and characterized by X-ray diffraction. Both dinuclear intermediates have the carboxy group acting as a bridge of the two Ir-centers. The isolation of intermediates **5–8** confirms the coordination of carboxy group to Ir when the reaction is performed under basic conditions. It should be noted that in contrast to Ir-hydrogenation of unfunctionalized olefins, the Ir-dihydride olefin complex **5** can undergo migratory insertion in the absence of H_2 [68]. This mechanistic divergence could indicate that the mechanism of Ir-catalyzed hydrogenation of alkenes may vary depending on the type of substrate and/or catalyst.

Two other Ir-catalysts have been studied for this transformation. Already in 2010, Ding et al. tested the spiro-based P,N-ligands **L9** (Fig. 3) in the reduction of α-aryl-β-substituted acrylic acids. Enantioselectivities up to 96% ee were achieved, leading to the production of a series of biologically interesting carboxylic acids, such as those containing a β-tetrahydro-2H-pyran-4-yl moiety [186]. These ligands were also applied in the hydrogenation of (*E*)-2-(hydroxymethyl)-3-arylacrylic acids in good-to-high enantioselectivities (Fig. 25a) [187].

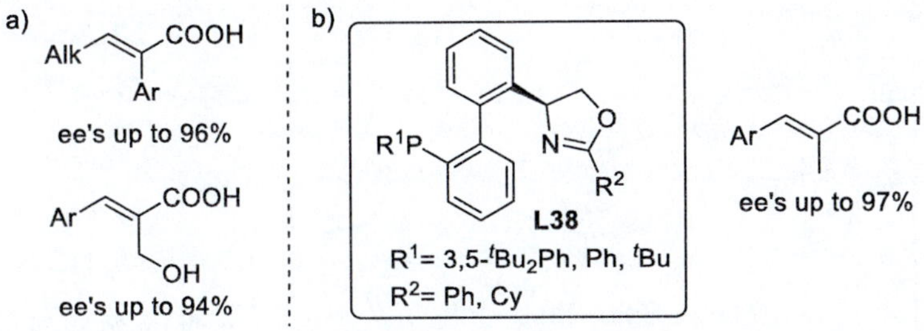

Fig. 25 Ir-catalyzed asymmetric hydrogenation of (**a**) α-aryl-β-substituted acrylic acids and α-hydroxymethyl cinnamic acids using ligand **L9** and (**b**) α-methyl cinnamic acids with **L38** ligand

Recently Zhang et al. reported the application of the phosphine-oxazoline ligands **L38** (Fig. 6b), with a biphenyl moiety, which has the advantage of a simple synthesis from the readily available (*S*)-(+)-2-phenylglycinol. Ir/**L38** (Ar = 3,5-^{t}Bu$_2$Ph, R = Ph) provided high yields and enantioselectivities in the widely studied α-methyl cinnamic acids (Fig. 25b, up to 97% ee, 98% yield, 2.000 TON) [78].

3.2 *Ir-Catalyzed Asymmetric Hydrogenation of Nitroolefins*

Despite the synthetic utility of reduced chiral nitroalkanes, the direct asymmetric hydrogenation of these types of substrates was not achieved until quite recently, by using diphosphine-based Rh-catalysts [188–190]. Although these catalysts were quite efficient in the hydrogenation of β,β-disubstituted nitroolefins, they were sensitive to the steric hindrance of the substrate. Hou and co-workers developed an Ir-(*R*,*R*)-f-SpiroPhos complex for the reduction of β-acylamino nitroolefins (Fig. 26). This newly developed Ir-catalysts allowed the preparation of a range of β-amino nitroalkanes in high yields and excellent optical purities (up to >99% ee), including substrates with *ortho*-substituted phenyl groups in the β-position [191]. The substrates studied contained a NH-acyl group which by chelation could facilitate the enantioselective hydrogenation. However, the authors showed later that Ir-(*R*,*R*)-f-SpiroPhos-catalyst was also highly enantioselective without the presence of this additional chelating group. Thus, excellent enantioselectivities (up to 98% ee, Fig. 26) were also achieved in the reduction of β,β-disubstituted nitroalkenes, including nitroalkenes with an *ortho*-substituted phenyl ring and thus, overcoming the limitations of the catalytic systems developed by Zhang and co-workers [192]. After this, the groups of Zhang and Zhou have also explored the enantioselective Ir-catalyzed reduction of nitroolefins without an extra chelating group, obtaining also high enantioselectivities [193, 194].

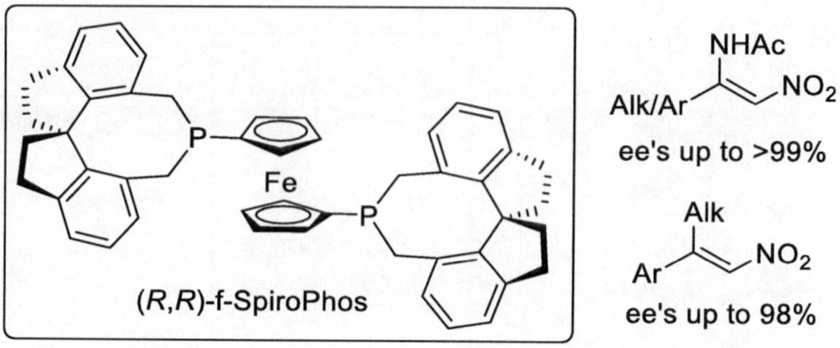

Fig. 26 Ir-catalyzed hydrogenation of β-acylamino nitroolefins and β,β-disubstituted nitroolefins using (*R*,*R*)-f-SpiroPhos ligand

3.3 Ir-Catalyzed Asymmetric Hydrogenation of Enamines, Enamides, and Allylic Amines

Ir-catalysts have also been used in the hydrogenation of amino-functionalized alkenes, albeit to a lesser extent than other alkenes [111, 195–199]. Nevertheless, they proved to be useful in the asymmetric hydrogenation of very attractive and challenging substrates, such as enamide esters, cyclic β-enamides, etc. In this context, the asymmetric hydrogenation of β-enamine esters is a straightforward way to prepare enantiopure β-amino acids and their derivatives. However, the methodology has been largely limited to the involvement of an N-acyl group that assists the reaction by chelation to the metal and facilitates to achieve high reactivity and enantioselectivity. The direct hydrogenation of unprotected enamine esters would be a more high-atom economical approach to obtain these valuable building blocks. Some Ir-catalysts modified with chiral monophosphoroamidites or diphosphine ligands have been applied in the hydrogenation of some unprotected NH- and N-aryl enamine esters with good-to-high enantioselectivities (ees up to 97%) (see, for instance, [200, 201]). Recently, Dong and Zhang et al. disclosed the highly effective asymmetric hydrogenation of tetrasubstituted α-fluoro-β-enamino esters using bisphosphine-thiourea ZhaoPhos ligand (Fig. 27) [202]. A series of valuable chiral α-fluoro-β-amino esters containing two adjacent tertiary stereocenters were afforded with high yields and excellent diastereo- and enantioselectivities (drs up to >25:1, ees up to >99% ee, and TON values up to 8,600). Importantly, no defluorinated by-product was detected.

Ir-catalysts have also shown to be very useful for the enantioselective hydrogenation of cyclic β-aryl enamides. The reduction of this type of substrates constitutes a direct route to 2-aminotetralines and 3-aminochromanes, which are key structural units found in numerous therapeutic agents and biologically active natural products [21–24]. However, their hydrogenation has provided unsatisfactory results, and only few Rh- and Ru-catalysts have been successful [28–38]. In 2012, a neutral Ir-complex, with a sulfonimidamido-based phosphoramidite (SIAPhos) ligand,

Fig. 27 Hydrogenation of tetrasubstituted α-fluoro-β-enamino esters catalyzed by Ir-ZhaoPhos catalyst

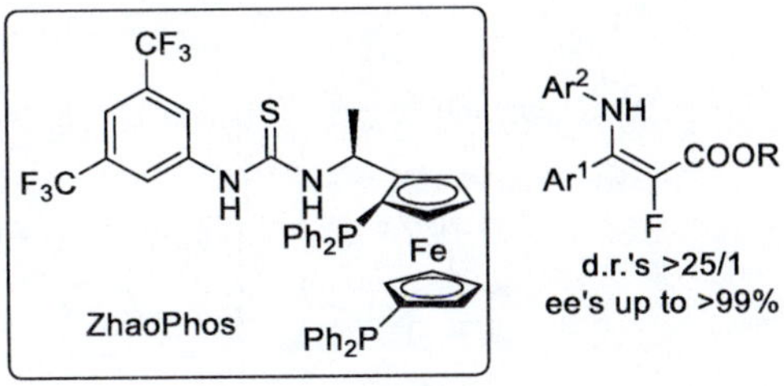

Fig. 28 Ir-catalyzed hydrogenation of β-aryl cyclic enamides using SIAPhos ligand

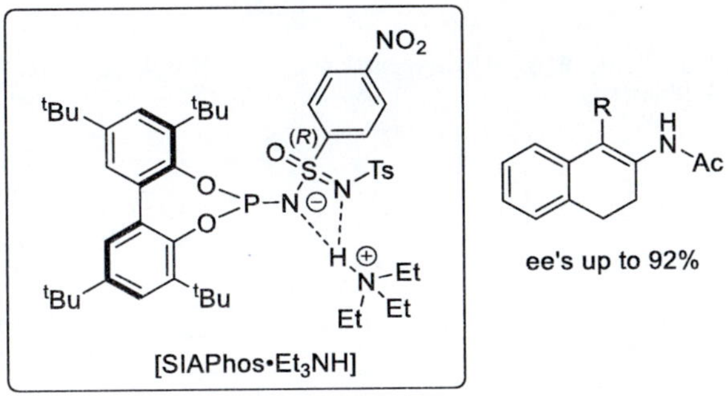

catalyzed the reduction of three cyclic β-enamides with promising enantioselectivities (up to 92% ee; Fig. 28). However, this catalyst showed low activities (55–81% conversion after 18 h at rt. and at $P_{H2} = 50$ bar) [203].

It has not been until very recently that Ir-catalysts have shown their high efficiency in the reduction of this type of challenging substrates. In 2016, two reports appeared demonstrating the potential of Ir-catalysts modified with P,N-ligands and showing that Ir-P,N catalysts can be also efficient in the reduction of alkenes bearing metal-coordinating groups [25, 26]. Riera and Verdaguer et al. found that bulky P-stereogenic phosphine-oxazoline ligands **L36** (Fig. 18) provided the highest selectivity ever reported for the reduction of cyclic enamides derived from α- and β-tetralones (Fig. 29a, ees up to 99% at only 3 bars of H_2), surpassing the results obtained with the widely studied Rh- and Ru-catalysts [25]. In the same year, Diéguez and co-workers also reported the successful application of PHOX-derived phosphite ligands **L39** (Fig. 29b) to the hydrogenation of cyclic β-enamides [26]. A range of 2-aminotetralines and 3-aminochromanes were obtained in high yields and excellent enantioselectivities (ees up to 99%, Fig. 29b). The hydrogenation of cyclic α-enamides also proceeded in high enantioselectivities (ees up to 96%). In addition, the reactions could be carried out in environmentally friendly solvents, propylene carbonate, with no loss of selectivity.

The already mentioned ligands **L37** (Fig. 18) that were successful for disubstituted, trisubstituted, and tetrasubstituted unfunctionalized olefins also provided excellent enantioselectivities in the hydrogenation of various cyclic β-enamides (6 examples, ees up to 99%) [27]. Moreover, both enantiomers of the reduced products could be accessed with the correct choice of the phosphite-based ligand.

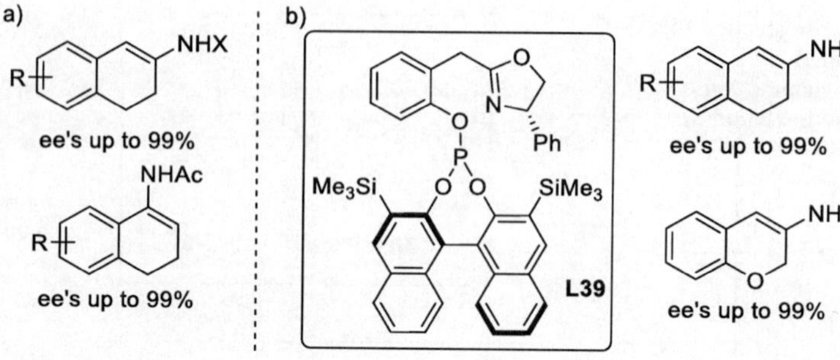

Fig. 29 Ir-catalyzed hydrogenation of α- and β-aryl cyclic enamides using **(a)** (S,R,R_P)-**L36** and **(b)** **L39** ligands

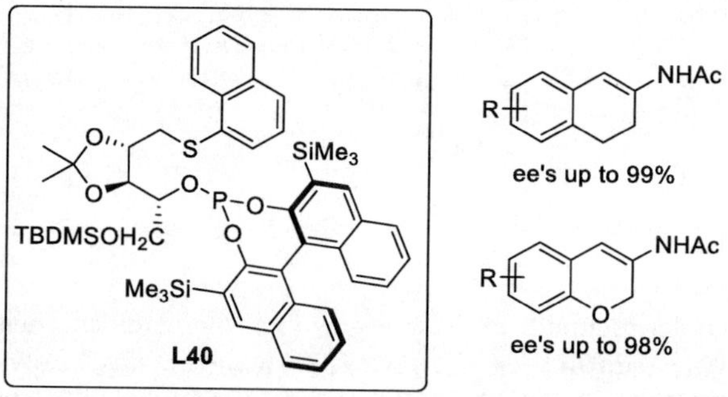

Fig. 30 Ir-catalyzed hydrogenation of β-aryl cyclic enamides using P,S-ligand **L40**

Finally, it has been recently shown that Ir-phosphite-thioether catalysts can also successfully catalyze the hydrogenation of cyclic β-enamides. The sugar-derived phosphite-thioether ligands (**L40**, Fig. 30) provided a range of 2-aminotetralines and 3-aminochromanes with excellent enantioselectivities (Fig. 30, ees up to 99%) [204]. Interestingly, both enantiomers of the hydrogenated products were obtained by simply switching from Rh to Ir. Moreover, low hydrogen pressure (10 bar) and environmentally friendly propylene carbonate could be used, with no loss of selectivity.

Very recently, it has been reported that Ir-catalysts can also successfully catalyze the hydrogenation of N-sulfonyl allyl amines [205] and aryl allyl phthalimides [206]. The hydrogenation of both types of substrates is another way to produce valuable chiral amines, such as β-aryl propanamines, which are important precursors for the synthesis of several pharmaceutical drugs (see, for instance, [207–211]). The commercially available threonine-derived phosphinite (UbaPHOX) iridium catalysts were found to be the best candidates for the hydrogenation of several N-sulfonyl allyl amines (Fig. 31a). A range of β-methyl amines were afforded with good to excellent ees of up to 94%. The synthetic potential of this methodology was shown with the synthesis of the biologically active compounds (R)-lorcaserin and LY-404187.

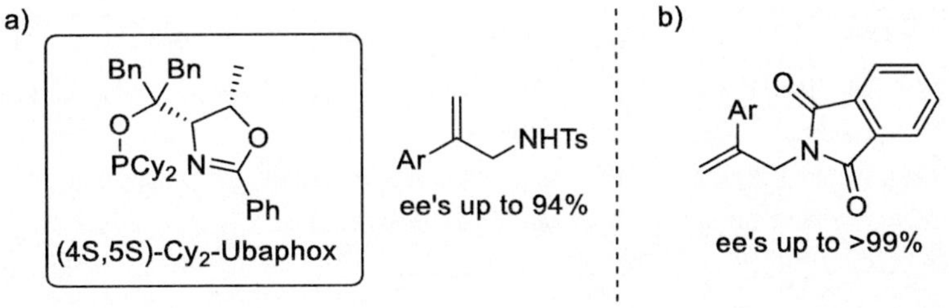

Fig. 31 Ir-catalyzed hydrogenation of (**a**) N-sulfonyl allyl amines using UbaPHOX ligand and (**b**) 2-aryl allyl phthalimides using **L36** ligand

Concerning the reduction of aryl allyl phthalimides, the previously mentioned Ir-**L36** catalysts were the best choice. Various enantioenriched β-aryl-β-methyl amines were yielded with excellent enantiomeric excess values (up to >99% ee) and using a low catalyst loading (1 mol%) and low hydrogen pressure (1 bar H_2) (Fig. 31b). The utility of the methodology was exemplified with the formal synthesis of (*R*)-lorcaserin, OTS514, and enantiomerically enriched 3-methyl indolines.

4 Asymmetric Ir-Catalyzed Hydrogenation of Imines

The asymmetric hydrogenation of imines is a high-atom economical way to prepare chiral amines. Despite all the advances made, their hydrogenation still remains a challenging task, and most of the reported catalysts present low reactivity and enantioselectivity, harsh reaction conditions, and narrow substrate scope. The reason is probably due to instability of certain imines; coordination of substrates, which can take place through both the nitrogen donor atom and the double bond; and *E/Z* imine interconversion in the case of acyclic imines [212, 213]. The first catalysts developed for this reaction were based mostly on Rh-diphosphine catalysts, and they showed only moderate enantioselectivities (60–70%) [214]. Later, some Ir-, Ru-, and Ti-catalysts also appeared improving enantioselectivities. Among the early applications, it should be highlighted the iridium-Xyliphos catalyst, which led to the large-scale production of the amine herbicide (S)-metolachlor (Fig. 32) [215]. Later Pfaltz's PHOX ligand (**L1**) [42] and Zhang's (*S,S*)-f-binaphane ligand [216] also exhibited excellent results for catalytic enantioselective hydrogenation of imines. This inspired several groups to explore the asymmetric hydrogenation of imines, and to date, Ir-complexes are among the most efficient catalysts for this transformation [212, 213].

Both cyclic and acyclic imines provide useful chiral amines, but usually, specific catalysts are required for each type of substrates. The reason is that acyclic imines might exist as Z/E mixtures, while cyclic imines usually have a fixed configuration imposed by the cycle. For the asymmetric hydrogenation of acyclic imines, in general, two types of iridium precursors are used, neutral and cationic complexes.

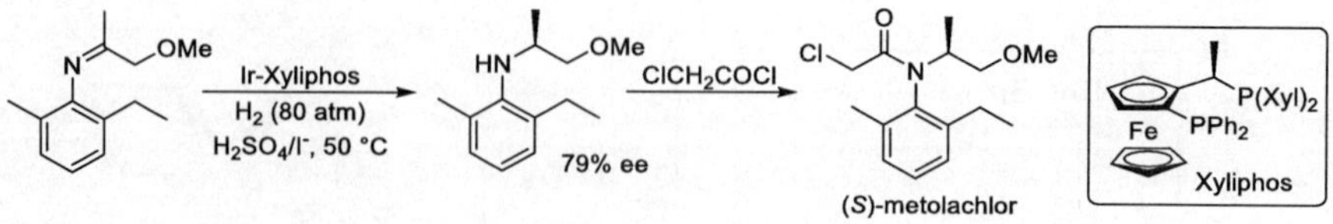

Fig. 32 Early synthesis of metolachlor using Ir-Xyliphos catalyst

Neutral complexes are generated from [{Ir(μ-Cl)(cod)}$_2$] precursor and are usually combined with diphosphine, monophosphoramidite, or phosphoramidite-N ligands. Some additives are also usually added, such as I_2 and KI. Cationic complexes are prepared from [Ir(cod)$_2$]X catalyst precursor and P-oxazoline ligands. Although cheaper counterions (X^-) were studied [217], usually BAr_F has been the most used counterion.

4.1 Asymmetric Hydrogenation Using Cationic Catalyst Precursors

Phosphine-oxazoline PHOX ligands **L1** (Fig. 1) and **L8** (Fig. 3) are among the most efficient ligands for the enantioselective hydrogenation of acyclic N-aryl imines. Excellent activities and enantioselectivities (up to 96% ee; Fig. 33) were obtained by using low catalyst loadings (0.1–0.5 mol%) at −20 °C and 5–50 bar hydrogen pressure in the reduction of aryl/alkyl N-aryl ketimines (Fig. 33a) [218]. Other related P,N-ligands, such as **L6** (Fig. 3) and **L16** (Fig. 6), have also been applied albeit with moderate success [73, 122]. A few years later, the success of PHOX-based catalytic systems was extended to dialkyl ketimines [219]. A key to this improvement was the addition of the appropriate imine as additive. Mechanistic investigation revealed that the active species is a cyclometalated complex **9** (Fig. 33b). Thus, a range of N-aryl dialkyl imines were hydrogenated in high enantioselectivities (up to 92% ee; Fig. 33a). Nevertheless, N-alkyl aliphatic imines were only hydrogenated in moderate ees (up to 77%; Fig. 33a).

Recently, cyclometalated complex **10** containing the P-stereogenic-oxazoline ligand **L36** proved to be successful in the asymmetry of a wide range of N-alkyl amines, including the more challenging N-methylated ones (ees up to 94%; Fig. 34) [220]. Interestingly, high ees (up to 96%) were also achieved in the reduction of N-aryl ketimines (Fig. 34) [221].

Cationic iridium complexes based on the already mentioned **L9** ligands (Fig. 3) were found to be highly efficient in the hydrogenation of a broad range of aryl alkyl N-benzyl imines (ees up to 93%). Importantly, even higher enantioselectivities were obtained for various exocyclic N-alkyl imines (ees up to 98%) [222]. The protocol was successfully employed in the synthesis of the antidepressant chiral drug sertraline [223]. High enantioselectivities in the reduction of some

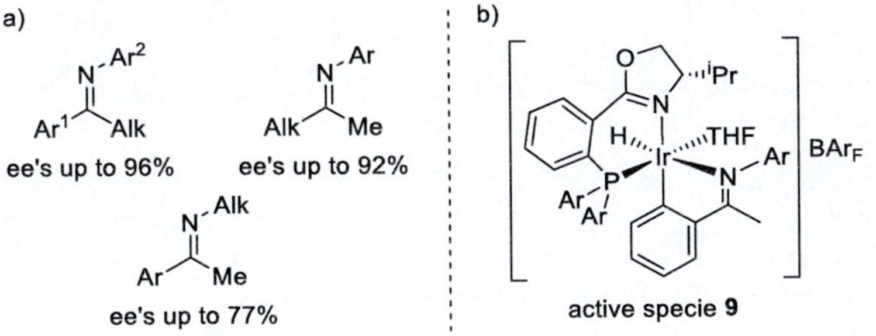

Fig. 33 (**a**) Ir-catalyzed asymmetric hydrogenation of N-aryl and N-alkyl imines. (**b**) Cyclometalated active species **9**

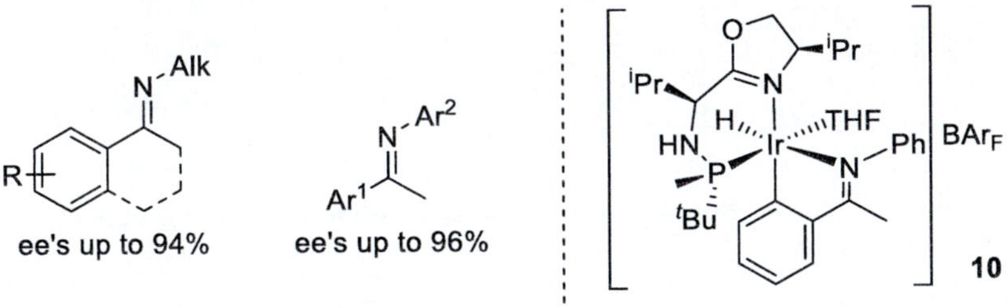

Fig. 34 Ir-catalyzed asymmetric hydrogenation of N-aryl and N-alkyl imines using Ir-**L36** catalytic system

exocyclic N-aryl-dihydronaphthalene imines were also achieved using P-chiral dihydrobenzooxaphosphole-oxazoline LalithPhos ligand (*S*)-2-((2*R*,3*R*)-3-(*tert*-butyl)-4-methoxy-2,3-dihydrobenzo[*d*][1,3]oxaphosphol-2-yl)-4-phenyl-4,5-dihydrooxazole (ees up to 99%) [224].

4.2 Asymmetric Hydrogenation Using Neutral Catalyst Precursors

Neutral catalyst modified with Xyliphos and (*R*,*R*)-f-binaphane ligands early proved to be suited for the AH of sterically hindered N-aryl imines [215, 216]. More recently, Hu et al. improved previous results with the use of phosphine-phosphoroamidite ligand **L41** (Fig. 35) [225]. A range of sterically hindered N-aryl imines were therefore hydrogenated featuring high ees (up to 99%) and turnover numbers (up to 100,000; Fig. 35). Later, the same group disclosed that the steric effect of substituents on the *o*-positions of the binaphthyl showed a significant influence on the enantioselectivity and found that **L42** (Fig. 35) was also highly enantioselective in the Ir-hydrogenation of sterically hindered N-aryl imines (ees up to 98%) [226]. The utility of this methodology was demonstrated in

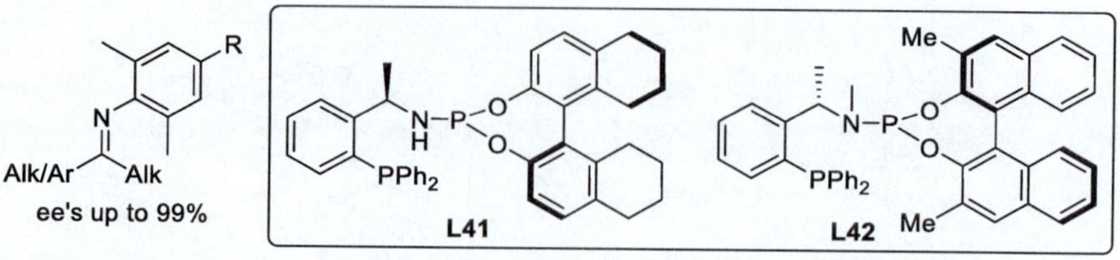

Fig. 35 Ir-catalyzed asymmetric hydrogenation of sterically hindered N-aryl imines using phosphite-phosphoroamidite ligands **L41** and **L42**

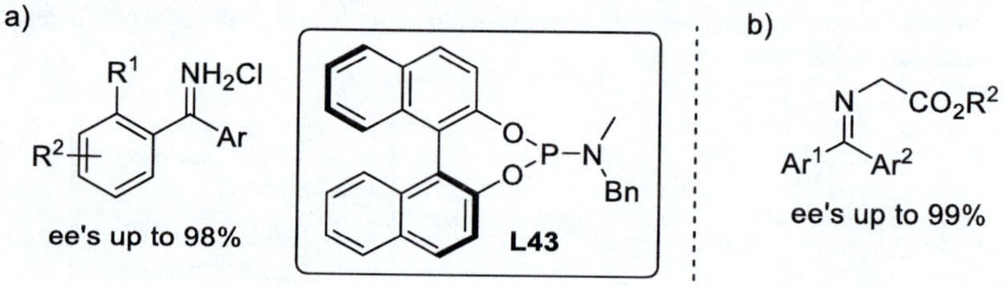

Fig. 36 Ir-catalyzed asymmetric hydrogenation of (**a**) diaryl iminium salts using Ir/**L43** catalysts and (**b**) N-alkylester-substituted diarylimines using Ir/(*R*,*R*)-f-SpiroPhos catalyst

the synthesis of the chiral herbicide (*S*)-metolachlor and the chiral fungicide (*R*)-metalaxyl.

Neutral Ir-catalysts have also shown excellent enantioselectivities in the asymmetric hydrogenation of diarylmethanimines, whose hydrogenation products are found in numerous biologically active compounds of pharmaceutical relevance [227–233]. In this context, two catalytic systems should be highlighted. Ir/**L43** catalytic system provided high enantioselectivities in the reduction of benzophenone N-H iminium salts with one of the aryl groups *ortho*-substituted (ees up to 98%; Fig. 36a) [234]. More recently, (*R*,*R*)-f-SpiroPhos ligand (Fig. 26) allowed the enantioselective hydrogenation of N-alkylester-substituted diarylimines under mild reaction conditions (ees up to >99%; Fig. 36b) [235]. A feature of this catalyst was that the presence of an *ortho*-substituent in one of the aryl groups was not required to achieve high enantioselectivities.

Hu et al. have also disclosed the Ir-hydrogenation of α-imino esters. A range of optically active α-aryl glycines were synthesized using Ir-**L44** catalytic system (ees up to 96%; Fig. 37a) [236]. Very recently a high-throughput experimentation (HTE) at AstraZeneca enabled the identification of highly enantioselective catalytic systems for the hydrogenations of N-alkyl α-aryl ketimines containing a furyl moiety [237]. After an extensive screening of Ru-, Rh-, and Ir- catalysts, Ir-catalytic system bearing the (*S*,*S*)-f-binaphane ligand was found to be the most enantioselective (ees up to 90%; Fig. 37b).

Neutral complexes generated from $[Ir(I)(COD)Cl]_2$ and activated by addition of halogen-based oxidants such as iodine are among the most successful systems for the asymmetric hydrogenation of cyclic amines. Among the cyclic imines studied, the

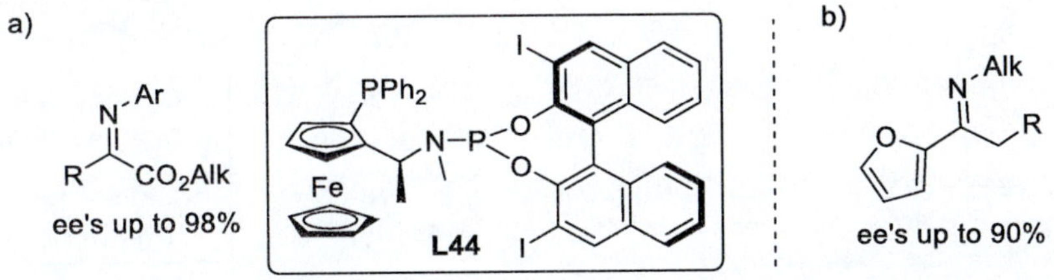

Fig. 37 Ir-catalyzed asymmetric hydrogenation of (**a**) α-imino esters using Ir/**L44** catalysts and (**b**) furyl-based N-alkyl α-aryl ketimines using Ir/(*S*,*S*)-f-binaphane catalyst

hydrogenation of isoquinolines and 3,4-dihydroisoquinolines is the most desired since they provide a straightforward synthetic route toward valuable chiral compounds with a 1,2,3,4-tetrahydroisoquinoline motif, which is present in several natural alkaloids and pharmaceutical molecules. However, the strong coordination ability of isoquinolines and their enhanced stability due to their aromaticity make them less reactive toward hydrogen. In this context, Zhou et al. developed the first example of highly enantioselective hydrogenation of quinoline derivatives with a (*R*)-MeO-BIPHEP [6,6′-dimethoxy-2,2′-bis(diphenylphosphino)-1,1′-biphenyl] as the ligand [238–240]. However, these systems were restricted only to quinolines. Importantly, the protocol was later expanded to isoquinolines, which is a very challenging class of substrates, but stoichiometric amounts of chloroformate as a substrate activator were needed [241]. In 2012, a more efficient Ir(COD)Cl]$_2$/(*R*)-SynPhos catalytic system was disclosed, which used 1-bromo-3-chloro-5,5-dimethylhydantoin (BDCMH) as a catalyst activator, and therefore, only catalytic amounts of activator were required [242]. A range of chiral 3,4-disubstituted tetrahydroisoquinoline derivatives were obtained with ee values as high as 96% (Fig. 38). The scope of this catalytic system was extended to other aromatic imines, such as polycyclic nitrogen-containing heteroaromatics pyrrolo/indolo[1,2-a] quinoxalines and phenanthridines [243], sulfur-containing dibenzo[b,f][1,4] thiazepines [244], activated N-benzyl-pyridinium bromides [245], and fluorinated isoquinoline derivatives [246] (Fig. 38). Finally, the authors also reported the use of (*R*)-SynPhos ligand in the deracemization of secondary and tertiary amines with a tetrahydroisoquinoline core [247]. The process consisted in a redox N-bromosuccinimide oxidation of the amines and the subsequent Ir-catalyzed asymmetric hydrogenation. A range of chiral 1-substituted 1,2,3,4-tetrahydroisoquinolines were generated with up to 98% ee in 93% yield.

Other diphosphine-based catalysts have also been successfully used in the asymmetric hydrogenation of aromatic iminium salts. Thus, for instance, a range of 1- and 3-substituted isoquinolinium salts [248], 2,6-disubstituted pyridinium hydrochloride (3·HCl) [249], and pyrrolo[1,2-a]pyrazines [250] were successfully hydrogenated using (*S*,*S*,*R*^{ax})-C3*-TunePhos ligand (ees up to 96%, Fig. 39). Ir/(*R*,*Sp*)-Josiphos catalyst also provided excellent ees in the hydrogenation of a range of pyrazinium salts (ees up to 96%, Fig. 39) [251]. It should be noticed that recently, an efficient catalytic system for the hydrogenation of pyrrolo[1,2-a]pyrazines without the need

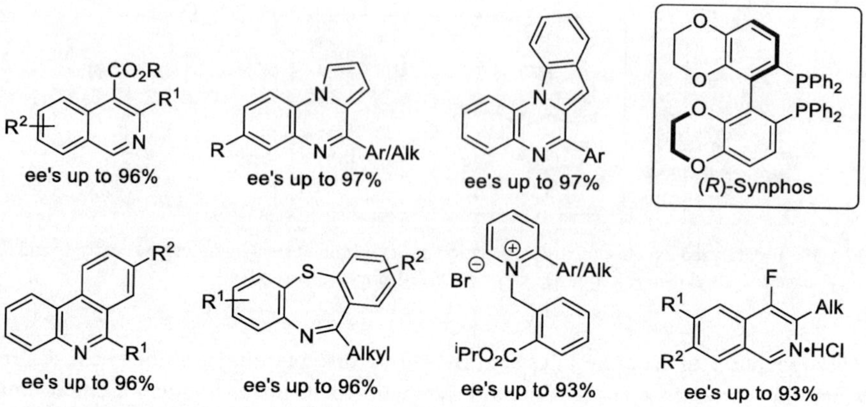

Fig. 38 Iridium-catalyzed enantioselective hydrogenation of cyclic imines using Ir/(*R*)-SynPhos catalyst

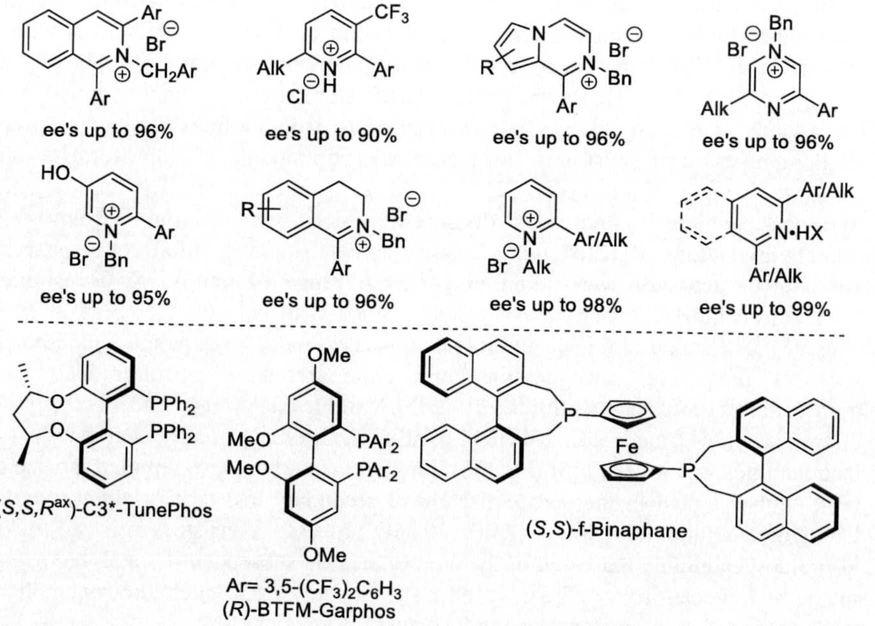

Fig. 39 Representative examples of iminium salts successfully hydrogenated

of formation of the corresponding salts or the addition of any additive has been reported using Ir/(*R*)-BTFM-Garphos catalyst (ees up to 96%, Fig. 39) [252]. A highly enantioselective hydrogenation of heteroaromatics bearing a hydroxyl group, 3-hydroxypyridinium salts, has been successfully developed using Ir/(*S,S*)-f-binaphane catalyst, providing a direct access to *trans* 6-substituted piperidin-3-ols with up to 95% ee [253]. Another interesting example can be found in the successful hydrogenation of iminium salts of N-alkyl tetrahydroisoquinolines (ees up to 96%, Fig. 39) [254] and of N-alkyl-2-arylpyridium salts (ees up to 98%, Fig. 39) [255] using SegPhos-type ligands. Recently, a new strategy has been developed, in which pyridinium and isoquinolinium salts are generated in situ by employing halogenide trichloroisocyanuric acid as a traceless activation reagent. Mechanistic studies indicated that hydrogen halide generated in situ acted as an activator. This method allowed the Ir/(*R*)-SegPhos-mediated hydrogenation of a range of isoquinolines and pyridines in excellent yields and enantioselectivities (up to 99% ee, Fig. 39) while avoiding tedious steps of installation and removal of the activating groups [256]. Finally, it should be mentioned that not only Ir/diphosphine catalysts are able to catalyze the reduction of iminium salts. Thus, it should be highlighted the recent works of Qu's group in the use of dihydrobenzooxaphosphole-pyridine ligand MeO-BoQPhos for the asymmetric hydrogenation of 2-alkyl-pyridinium salts including examples containing an α-heteroaryl substituent [257, 258].

Similarly, the successful hydrogenation of 6-membered ring cyclic imines has not been only limited to the use of SynPhos-based system. Thus, for instance, a range of quinoline derivatives have been successfully hydrogenated with diphosphines and monophosphoroamidites. (*S,S*)-f-Binaphane ligand has been also efficiently used in the reduction of 1-substituted 3,4-dihydroisoquinolines (ees up to >99%) [259]. Similarly, a range of 1-aryl-substituted tetrahydroisoquinolines were obtained in ees of up to >99% and good TON (up to 4,000) using Josiphos-type binaphane ligand [260]. Higher TONs (up to 43,000) were achieved in the reduction of a range of quinolines using both Ir/(*R*)-Difluorphos [261] and Ir/(*R*)-P-Phos [262] catalytic systems (Fig. 40). A successful example of the use of monophosphoramidite ligand can be found in the use of (R^{ax},*S,S*)-Siphos-pe ligand (Fig. 40) in the hydrogenation of 1-alkyl-dihydroisoquinolines with ees of up to 96%. The usefulness of the latter reaction was demonstrated with the synthesis of the tetracyclic alkaloid (*S*)-xylopinine in 85% yield and 96% ee [263].

Besides quinoline derivatives, the range of 6-membered ring cyclic amines successfully hydrogenated has been extended. Thus, for instance, Ir-SegPhos catalyst was successfully used in the hydrogenation of 1,4-benzoxazines [264] and quinazolines [265] (ees up to 98%, Fig. 41). The utility of the method was demonstrated with the synthesis of the bioactive compounds Eg5 inhibitor and (−)-SDZ 267-489 in excellent enantioselectivities (>99% and 99% ee, respectively) [265]. Another example can be found in the use of cationic dinuclear iridium(III) chloride catalyst $\{[IrH((S)\text{-Difluorphos})]_2(\mu\text{-}Cl)_3\}Cl$ for the reduction of 2-alkyl and 2-aryl- substituted dihydroquinoxalines [266–268] and tosylamido-substituted pyrazines [269] (ees up to 95%, Fig. 41). Ir-catalyst modified with phosphine-phosphite ligand **L45** proved to be highly efficient in the asymmetric hydrogenation of benzoxazines, benzoxazinones, benzothiazones, and quinoxalinones (ees up to 99%; Fig. 41) [270, 271].

Most of the literature dealing with the hydrogenation of cyclic imines reports examples on 6-membered ring systems. Examples on the successful hydrogenation of 7-membered cyclic imines are rare. In particular, Ir/C3*-TunePhos catalysts

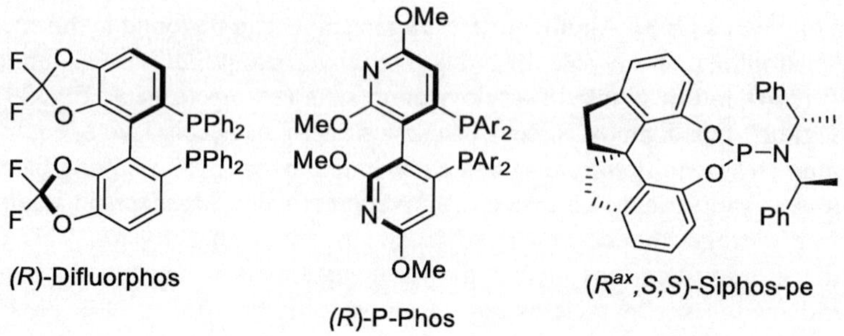

Fig. 40 (*R*)-Difluorphos, (*R*)-P-Phos, and (R^{ax},*S*,*S*)-Siphos-pe ligands successfully used in the AH of quinoline derivatives

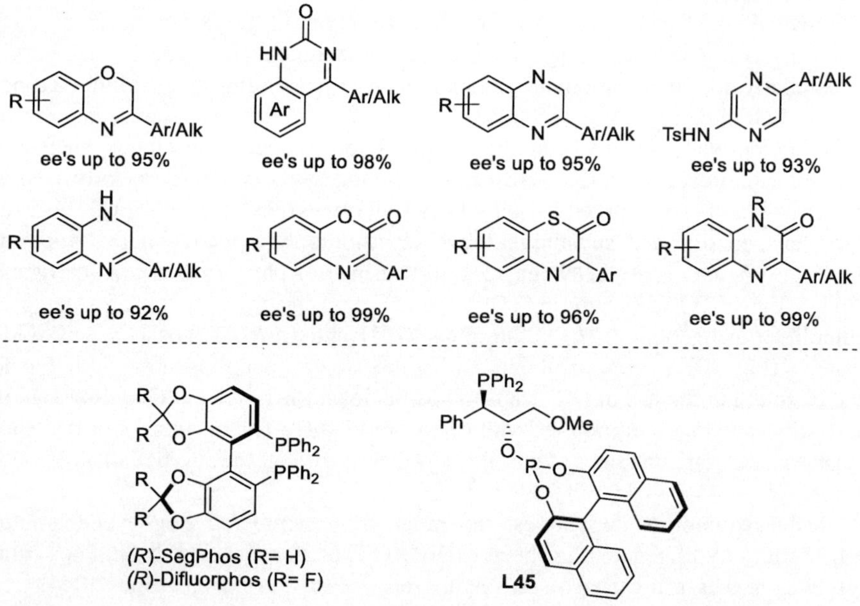

Fig. 41 Representative examples of 6-membered cyclic imines other than quinoline derivatives and iminium salts successfully hydrogenated using SegPhos, Difluorphos, and phosphine-phosphite ligand **L45**

proved to be highly efficient in the asymmetric hydrogenation of substituted dibenzo [b,f][1,4]oxazepines, benzodiazepinones, and benzodiazepines (ees up to 96%, Fig. 42a) [272, 273]. More recently, Zhou et al. reported the asymmetric hydrogenation of 6-substituted 5H-benzo[d]benzofuro[3,2-b]azepines using a BIPHEP-type ligand **L46** (ees up to 91%, Fig. 42b) [274]. 2,4-Diaryl-1,5-benzodiazepines and 2,4-diaryl-3H-benzo[b]azepines has been also successfully hydrogenated (ees up to 99% and drs up to >20:1) using aminophosphine-pyridine ligand **L47** (Fig. 42c) [275] and a dendritic PHOX derivative [276, 277]. Benzoxazinone derivatives have

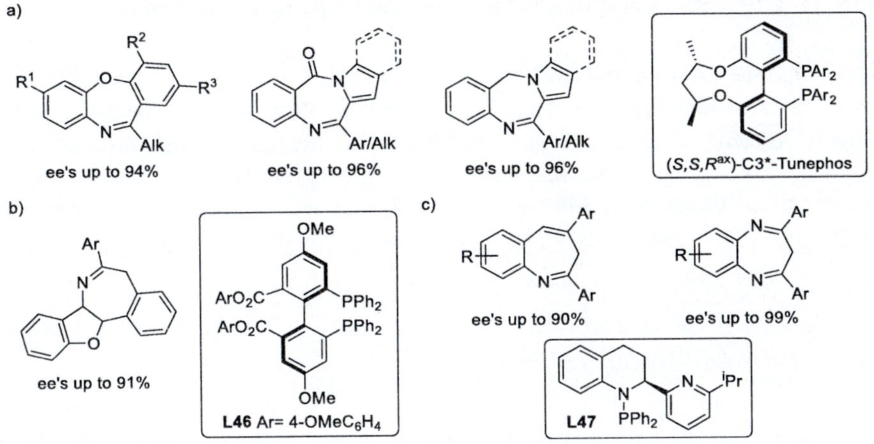

Fig. 42 Representative examples of 7-membered cyclic imines successfully hydrogenated using (**a**) C3*-TunePhos ligands, (**b**) BIPHEP-type ligand **L46**, and aminophosphine-pyridine ligand **L47**

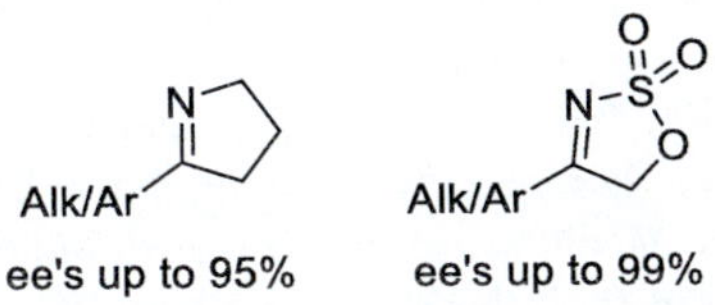

Fig. 43 Representative examples of 5-membered cyclic imines hydrogenated

also been recently reduced using ZhaoPhos ligand (Fig. 27) achieving excellent enantioselectivities (up to 99% ee) [278].

There are very few examples on the asymmetric hydrogenation of 5-membered ring cyclic imines. In this context, the use of iodine-bridged dimeric [{Ir(H)[(S,S)-(f)-binaphane]}$_2$(μ-I)$_3$]I complex catalyzed the asymmetric hydrogenation of a series of 2-aryl-1-pyrrolines in good ees (up to 86%) and high turnover numbers (TONs up to 5,000) [279]. More recently, the use of (*R*,*R*)-f-SpiroPhos ligand (Fig. 26) improved enantioselectivities to up to 98% (Fig. 43) [280]. In contrast to Ir/(*S*,*S*)-f-binaphane catalyst, the enantioselectivity was only affected when a 2-alkyl substituent was present on the imine instead of an aryl group (ees up to 77%). Moreover, this method was successfully applied to the synthesis of (+)-(6S,10bR)-McN-4612-Z, a potent inhibitor for the uptake of important central neurotransmitters norepinephrine, dopamine, and serotonin into nerve cells. Very recently a bifunctional bisphosphine-thiourea ZhaoPhos ligand (Fig. 27) was successfully applied in the Ir-catalyzed asymmetric hydrogenation of cyclic sulfamidate imines (ees up to 99%; Fig. 43) [281].

5 Ir-Catalyzed Asymmetric Hydrogenation of Ketones

Over the last decades, the Ir-catalyzed asymmetric hydrogenation of ketones has experienced a huge advance. Thus, a range of Ir-catalysts have demonstrated to be highly efficient, achieving excellent enantioselectivities and activities for a range of ketones. Such progress has made Ir-catalysts good alternatives to the most commonly used Ru-catalysts. Most of the key ligands are phosphine-based, although there are some important examples that do not contain a P-donor group.

5.1 P-Donor-Based Ligands

One of the key P-containing ligands was disclosed by Poli's group [282]. They developed a ferrocenyl-based phosphine-thioether ligand **L48** that provided high enantioselectivities (ees up to 99%) for a range of alkyl aryl ketones (Fig. 44), albeit activities (TOFs up to 250 h^{-1}) do not compare well with the current state of the art.

In 2010, Xie and Zhou et al. developed chiral spiro aminophosphine ligands **L49** (Fig. 44) [283, 284]. The use of Ir/**L49** catalyst provided high ees (up to 97%) and activities (TOFs up to 3.7×10^4 h^{-1}) in the hydrogenation of aryl alkyl ketones and exo-cyclic α,β-unsaturated ketones. Nevertheless, the catalyst deactivates under hydrogenation conditions. Mechanistic investigations indicated that the deactivation was due to the formation of $[IrH_2(\mathbf{L49})_2]^+$ complex [284].

To prevent the formation of such inactive species, the same group introduced in **L49** a pyridine group as a third coordinating position, leading to the tridentate P,N,N spiro pyridine-aminophosphine ligands SpiroPAP (Fig. 43) [285–287]. Ir/SpiroPAP (Ar = 3,5-^{t}Bu$_2$-C$_6$H$_3$ and R = 3-Me) system proved to be highly efficient in the AH of aryl alkyl ketones and aryl β- and δ-ketoesters (ees up to 99.8% and TONs up to 4.550.000; Fig. 45) [285–287]. The excellent performance of SpiroPAP ligands has been further extended to acyclic α,β-unsaturated acyclic ketones [288]; α-, γ-, and δ-keto acids [289, 290]; α-amino ketones [291]; and α,β-unsaturated-α-ethoxycarbonyl-β-substituted cyclic ketones [292] (Fig. 45). This catalytic system has been also used in the deracemization of α-substituted lactones [293, 294] (Fig. 45) via dynamic kinetic resolution (DKR) as well as in the kinetic resolution of aliphatic alcohols [295]. The synthetic versatility of Ir/SpiroPAP catalyst was demonstrated with the synthesis of drugs and natural products such as (−)-mesembrine, rivastigmine, and (−)-Hamerigan B [288, 296–300].

Xie's and Zhou's group developed a P,N,S variant (SpiroSAP; Fig. 44) with a 1,3-dithiane group instead of the pyridine group [301]. Ir/SpiroSAP catalyst proved to be highly efficient in the asymmetric hydrogenation of β-alkyl-β-ketoesters [301] as well as in the dynamic kinetic resolution (DKR) of β-ketolactams [302] (Fig. 46). Ir/SpiroSAP catalyst was also used in the formal total synthesis of (−)-cyanolide A and (−)-doluculine [303, 304]. Recently, a new SpiroPAP variant has been developed by replacing the pyridine group by an oxazoline moiety (SpiroOAP; Fig. 43)

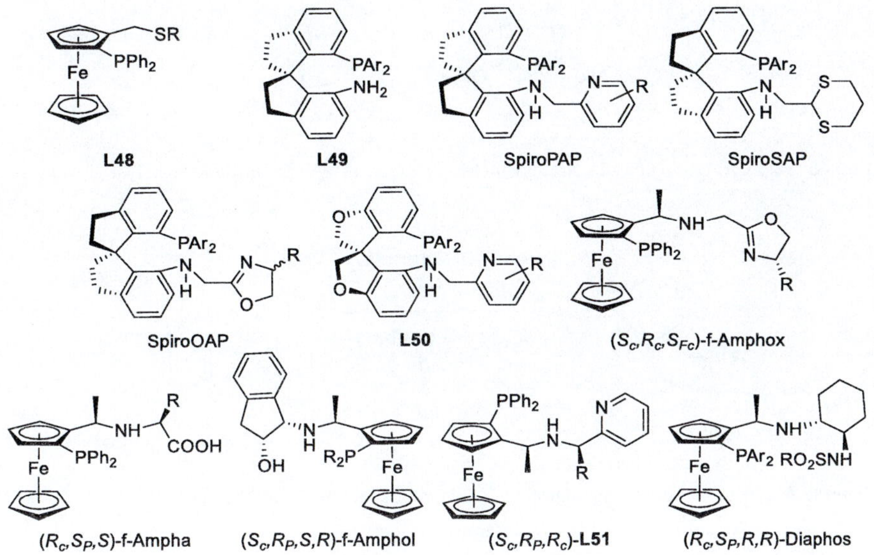

Fig. 44 Selected P-containing ligands developed for the Ir-catalyzed asymmetric hydrogenation of ketones

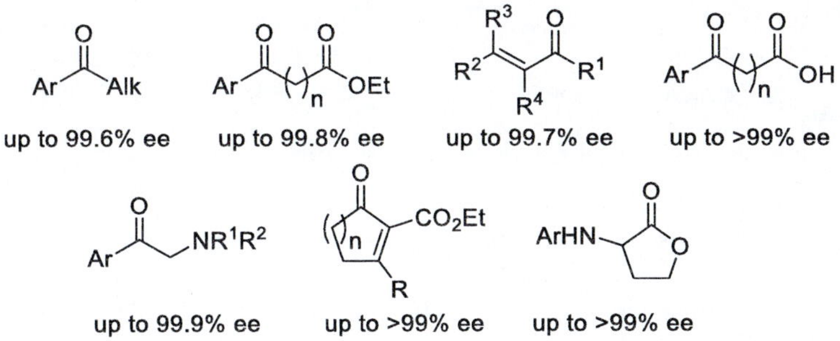

Fig. 45 Representative ketones successfully hydrogenated with Ir/SpiroPAP catalyst

[305]. SpiroOAP ligands proved to be efficient in the reduction of α-ketonamides (ees up to 98%).

Yin and Zhang et al. developed a variant of SpiroPAP ligands containing an oxa-spirocyclic ligand **L50** (Fig. 44) [306]. Ir/**L50** catalyst was successfully used in the asymmetric hydrogenation of Bringmann's lactones via DKR to yield enantiopure chiral biaryl diols in high ees (up to >99%; Scheme 2).

Dong and Zhang et al. developed novel tridentate ferrocene aminophosphoxazoline ligands (f-Amphox, Fig. 44). Ir/f-Amphox catalyst proved to be highly efficient in the reduction of simple aryl alkyl and dialkyl ketones (ees up to >99% and TONs up to 1,000,000; Fig. 47) [307]. f-Amphox-based catalytic systems were also successfully used in the hydrogenation of α-, β-, γ-, and δ-keto

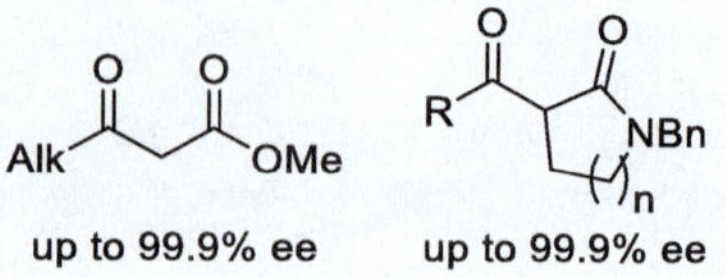

Fig. 46 Representative ketones successfully hydrogenated with Ir/SpiroSAP catalyst

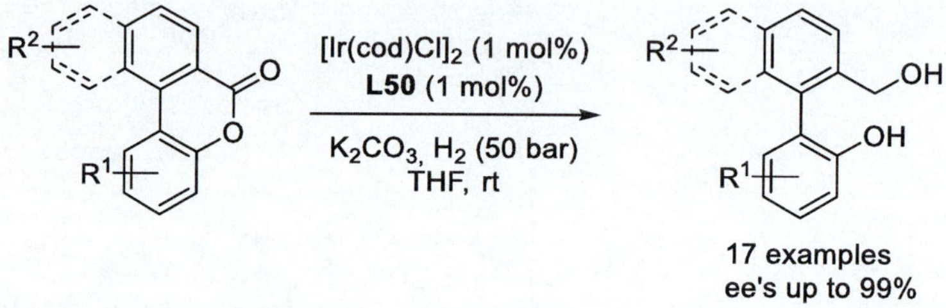

Scheme 2 Asymmetric hydrogenation of Bringmann's lactones via DKR with Ir/**L50**

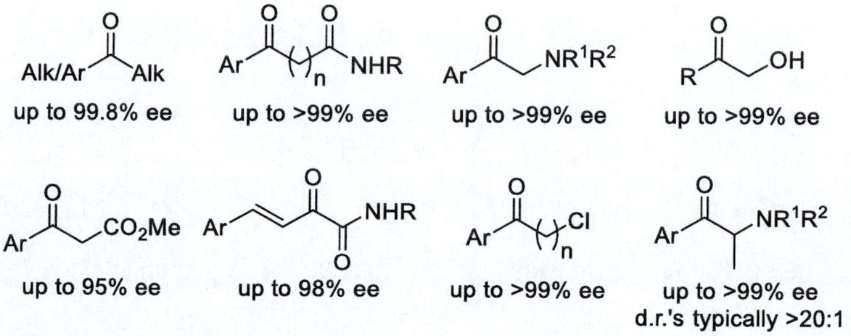

Fig. 47 Representative ketones successfully hydrogenated with the Ir/f-Amphox catalyst

amides [308, 309], α-amino ketones [310], α-hydroxy ketones [311], β-ketoesters [312], styrylglyoxylamides [313], and halohydrins [314] as well as in the deracemization of α-amino β-unfunctionalized ketones via DKR [315] (Fig. 47). Hou's group also developed a modification in which Uggi's amine was replaced by phenethylamine [316]. However, only moderate ees were achieved in the reduction of β-ketoesters.

Dong's and Zhang's groups developed a P,N,O variant (f-Ampha; Fig. 44) with a chiral carboxylic acid instead of the oxazoline moiety [317]. f-Ampha ligands exhibited excellent catalytic performance for a range of aryl alkyl ketones [317] and α-ketoesters [318] as well as in the desymmetrization of cyclic 1,3-diketones [319]. For these catalytic systems, the hydroxyl group of the carboxylic acid group is involved with the formation of O-H···substrate interaction with a new catalytic bifunctional mode. At the same time, a new ferrocene-based amino-phosphine-alcohol was developed (f-Amphol; Fig. 44) [320]. Ir/f-Amphol catalysts also showed excellent enantioselectivities for simple aryl alkyl ketones [320–322], albeit the turnover numbers are lower than for Ir/f-Amphox catalysts. Ir/f-Amphol ligands were also successfully used in the asymmetric hydrogenation of β-keto sulfones [323] and of α-substituted β-ketoesters via DKR [324]. For these catalytic systems, DFT studies indicated that the hydroxyl group of the f-Amphol ligands plays a key role in the reduction process.

Hou's group developed a sterically hindered ferrocenyl P,N,N-ligands in which the oxazoline in the f-Amphox ligands was replaced by a pyridinylmethyl group with an extra coordinating stereogenic center (ligands **L51**; Fig. 44) [325]. Ir/**L51** (R = 2-tolyl) catalyst demonstrated to be highly efficient in the asymmetric hydrogenation of α-alkyl-substituted β-aryl-β-ketoesters via DKR yielding the corresponding alcohols in high diastereo- and enantioselectivities (drs up to >95/5 and ees up to 99%).

Zhong's group developed tridentate ferrocene-based diamine-phosphine sulfonamide ligands (f-Diaphos; Fig. 44) [326]. The f-Diaphos ligands provided excellent reactivity and enantioselectivities in the Ir-catalyzed hydrogenation of diaryl and 2-pyridyl aryl ketones (ees up to >99%; TONs up to 19,600) [326, 327].

5.2 *Non-P-Donor-Based Ligands*

Ir-complexes bearing chiral phosphine-free ligands have also been successfully used in the asymmetric hydrogenation of ketones. Albeit such catalytic systems are able to induce high enantioselectivities, the turnover numbers are still lower than the most active P-based ligands such as SpiroPAP and f-Amphox.

In this context Ohkuma's group demonstrated that Cp*Ir(III) complex containing MsDPEN ligand **L52** (Fig. 48) efficiently hydrogenated α-hydroxy ketones [328]. A range of 1-aryl-1,2-ethanediols were therefore achieved in high ees (up to 99%) with TONs as high as 6,000. Later their high performance was extended to the use of aromatic (hetero)cyclic ketones (e.g., indanones, benzofuranones, chromanones, etc.) achieving an excellent enantiocontrol (ees up to >99%) [329]. At the same time, Ikariya's group developed a bifunctional triflylamide-tethered CpIr version of [Cp*Ir(OTf)**L52**], albeit it provided only good ees (up to 93%) in the reduction of benzophenone [330].

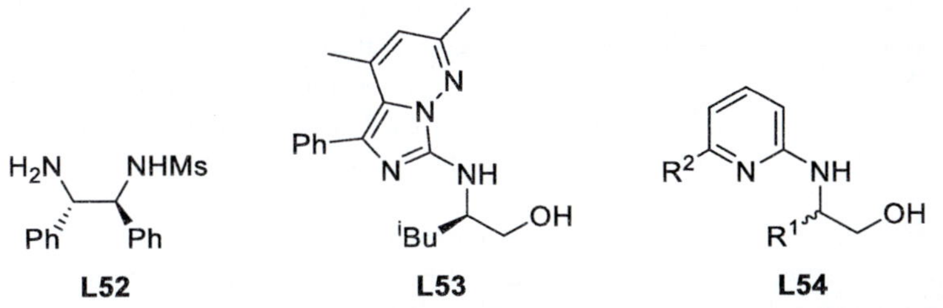

Fig. 48 Non-P-donor-based ligands for the Ir-catalyzed asymmetric hydrogenation of ketones

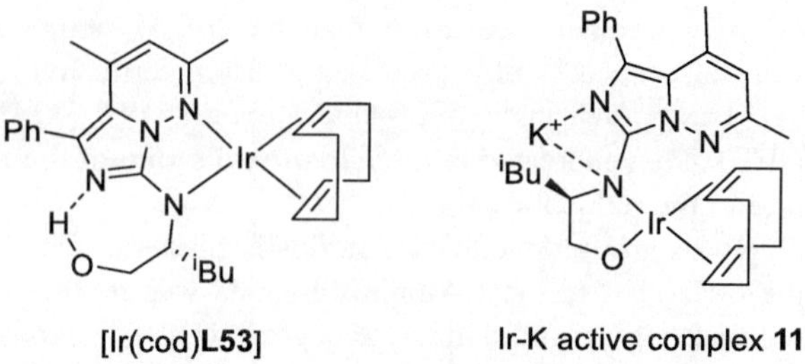

Fig. 49 Catalyst precursor [Ir(cod)**L53**] and its bimetallic active species **11**

Kempe's group introduced an imidazo[1,5-b]pyridazine-type ligand **L53** (Fig. 47) for the Ir-catalyzed hydrogenation of simple ketones [331]. A range of aryl alkyl and aryl aryl ketones were therefore efficiently hydrogenated providing excellent ees (up to >99%) and TONs (up to 200,000) using Ir/**L53** catalyst. It should be pointed out that the catalyst precursor [Ir(cod)**L53**] rapidly evolves under catalytic conditions to the formation of Ir-K catalyst complex **11** in which **L53** is coordinated thought the amino-alcohol (Fig. 49). In 2016, Kempe et al. developed the pyridylalkylamine ligands **L54** (Fig. 48) [332]. Nevertheless, the efficiency of Ir/**L54** catalysts proves to be somewhat lower than Ir/**L53** catalyst in the reduction of simple ketones (ees up to 96%).

6 Conclusions

Due to perfect atom economy and simplicity, the metal-catalyzed asymmetric hydrogenation continues to be one of the most sustainable and straightforward reactions for creating stereogenic centers in target molecules. In the asymmetric hydrogenation of unfunctionalized olefins or with poorly coordinative groups, Ir containing heterodonor P,X ligands continues to be the catalysts of choice. Examining the last progress in this field, the substrate scope has been largely extended, with the successful hydrogenation of challenging and relevant substrates such as polyene substrates, with the formation of multiple stereocenters, olefins that contain non-coordinating groups such as halides avoiding the dehalogenation side reaction, and allylsilanes. A lot of progress has also been achieved in the regio- and stereoselective monohydrogenation, as an efficient route for preparing natural products and complex organic molecules (e.g., juvabione). This has been achieved, thanks to the recent development of powerful catalytic systems that have also allowed to advance in the search of a single catalysts able to tolerate a large number of olefin types. However, some limitations have still to be solved such as the necessity to work with pure geometrical isomers (e.g., *E*- and *Z*-trisubstituted alkene produced opposite enantiomers of the hydrogenated products). Other demanding substrates are hindered tetrasubstituted olefins, whose hydrogenation can generate two vicinal stereocenters in a single reaction and thus give rise to products of higher stereochemical complexity. The first significant progress has been made in the past 5 years, but the scope of the substrate needs to be further improved, and this area of research is likely to expand in the future as well. The advances of Ir-catalysts in the last 10 years have not only been limited to unfunctionalized olefins, but the development of Ir-catalysts has also notably expanded the field of the asymmetric hydrogenation of unresolved functionalized substrates, providing in some cases complementarity to Rh- and Ru-catalysts and in other cases surpassing the widely used Rh- and Ru-catalysts.

Acknowledgments We gratefully acknowledge financial support from the Spanish Ministry of Economy and Competitiveness (CTQ2016-74878-P and PID2019-104904GB-I00), the European Regional Development Fund (AEI/FEDER, UE), the Catalan Government (2017SGR1472), the ICREA Foundation (ICREA Academia award to MD) and from "La Caixa" Foundation.

References

1. Blaser H-U, Federsel H-J (2010) Asymmetric catalysis in industrial scale: challenges, approaches and solutions.2nd edn. Wiley, Weinheim
2. Shang G, Li W, Zhang X (2000) Ojima I (ed) Catalytic asymmetric synthesis3rd edn. Wiley, Hoboken, p 343
3. Brown JM (1999) Jacobsen EN, Pfaltz A, Yamamoto H (eds) Comprehensive asymmetric catalysis, vol 1. Springer, Berlin, p 121
4. Noyori R (1994) Asymmetric catalysis in organic synthesis. Wiley, New York
5. Cornils B, Herrmann WA (2002) Applied homogeneous catalysis with organometallic compounds.2nd edn. Wiley-VCH, Weinheim
6. Genêt JP (2008) Andersson PG, Munslow IJ (eds) Modern reduction methods. Wiley-VCH, Weinheim, p 3
7. Tang W, Zhang X (2003) Chem Rev 103:3029
8. Kitamura M, Noyori R (2004) Murahashi S-I (ed) Ruthenium in organic synthesis. Wiley-VCH, Weinheim, p 3
9. Weiner B, Szymański W, Janssen DB, Minnaard AJ, Feringa BL (2010) Chem Soc Rev 39:1656
10. Cui X, Burgess K (2005) Chem Rev 105:3272
11. Roseblade S, Pfaltz A (2007) Acc Chem Res 40:1402
12. Woodmansee DH, Pfaltz A (2011) Chem Commun 47:7912
13. Zhu Y, Burgess K (2012) Acc Chem Res 45:1623
14. Verendel JJ, Pàmies O, Diéguez M, Andersson PG (2014) Chem Rev 114:2130
15. Margarita C, Andersson PG (2017) J Am Chem Soc 139:1346
16. Knowles WS, Sabacky MJ, Vineyard BD (1972) J Chem Soc Chem Commun 10
17. Knowles WS (2002) Angew Chem Int Ed 41:1998
18. Noyori R (2002) Angew Chem Int Ed 41:2008
19. Shultz CS, Krska SW (2007) Acc Chem Res 40:1320
20. Blaser HU (2002) Adv Synth Catal 344:17
21. Pharm DQ, Nogid A (2008) Clin Ther 30:813. (Rotigotine)
22. Osende JI, Shimbo D, Fuster V, Dubar M, Badimon JJ (2004) J Thromb Haemost 2:492. (Terutroban)
23. Ross SB, Thorberg SO, Jerning E, Mohell N, Stenfors C, Wallsten C, Milchert IG, Ojteg GA (1999) CNS Drug Rev 5:213. (Robalzotan)

24. Astier B, Lambás Señas L, Soulière F, Schmitt P, Urbain N, Rentero N, Bert L, Denoroy L, Renaud B, Lesourd M, Muñoz C, Chouvet G (2003) Eur J Pharmacol 459:17. (Alnespirone)
25. Salom E, Orgué S, Riera A, Verdaguer X (2016) Angew Chem Int Ed 55:7988
26. Magre M, Pàmies O, Diéguez M (2016) ACS Catal 6:5186
27. Biosca M, Magre M, Pàmies O, Diéguez M (2018) ACS Catal 8:10316
28. Renaud JL, Dupau P, Hay AE, Guingouain M, Dixneouf PH, Bruneau C (2003) Adv Synth Catal 345:230
29. Hoen R, van den Berg M, Bernsmann H, Minnaard AJ, de Vries JG, Feringa BL (2004) Org Lett 6:1433
30. Jiang XB, Lefort L, Goudriaan PE, de Vries AHM, van Leeuwen PWNM, Reek JNH (2006) Angew Chem Int Ed 45:1223
31. Sandee AJ, van der Burg AM, Reek JNH (2007) Chem Commun:864
32. Revés M, Ferrer C, León T, Doran S, Etayo P, Vidal-Ferran A, Riera A, Verdaguer X (2010) Angew Chem Int Ed 49:9452
33. Wu Z, Ayad T, Ratovelomanana-Vidal V (2011) Org Lett 13:3782
34. Pignataro L, Boghi M, Civera M, Carboni S, Piarulli U, Gennari C (2012) Chem Eur J 18:1383
35. Frank DJ, Franzke A, Pfaltz A (2013) Chem Eur J 19:2405
36. Bravo MJ, Ceder RM, Muller G, Rocamora M (2013) Organometallics 32:2632
37. Arribas I, Rubio M, Kleman P, Pizzano A (2013) J Org Chem 78:3997
38. Liu G, Liu X, Cai Z, Jiao G, Xu G, Tang W (2013) Angew Chem Int Ed 52:4235
39. Ohta T, Ikegami H, Miyake T, Takaya H (1995) J Organomet Chem 502:169
40. Schmid R, Broger EA, Cereghetti M, Crameri Y, Foricher J, Lalonde M, Müller RK, Scalone M, Schoettel G, Zutter U (1996) Pure Appl Chem 68:131
41. Forman GS, Ohkuma T, Hems WP, Noyori R (2000) Tetrahedron Lett 41:9471
42. Schnider P, Koch G, Prétôt R, Wang G, Bohnen FM, Krüger C, Pfaltz A (1997) Chem Eur J 3:887
43. Lightfoot A, Schnider P, Pfaltz A (1998) Angew Chem Int Ed 37:2897
44. Hilgraf R, Pfaltz A (1999) Synlett:1814
45. Blackmond DG, Lightfoot A, Pfaltz A, Rosner T, Schnider P, Zimmermann N (2000) Chirality 12:442
46. Bernardinelli GH, Kündig EP, Meier P, Pfaltz A, Radkowski K, Zimmermann N, Neuburger-Zehnder M (2001) Helv Chim Acta 84:3233
47. Blankenstein J, Pfaltz A (2001) Angew Chem Int Ed 40:4445
48. Menges F, Neuburger M, Pfaltz A (2002) Org Lett 4:4713
49. Menges F, Pfaltz A (2002) Adv Synth Catal 344:40
50. Smidt SP, Menges F, Pfaltz A (2004) Org Lett 6:2023
51. Hilgraf R, Pfaltz A (2005) Adv Synth Catal 347:61
52. Schönleber M, Hilgraf R, Pfaltz A (2008) Adv Synth Catal 350:2033
53. Drury WJ, Zimmermann N, Keenan M, Hayashi M, Kaiser S, Goddard R, Pfaltz A (2004) Angew Chem Int Ed 43:70
54. Bell S, Wüstenberg B, Kaiser S, Menges F, Netscher T, Pfaltz A (2006) Science 311:642
55. Woodmansee DH, Müller MA, Neuburger M, Pfaltz A (2010) Chem Sci 1:72
56. Woodmansee DH, Müller MA, Tröndlin L, Hörmann E, Pfaltz A (2012) Chem Eur J 18:13780
57. Kaiser S, Smidt SP, Pfaltz A (2006) Angew Chem Int Ed 45:5194
58. Ganič A, Pfaltz A (2012) Chem Eur J 18:6724
59. Schrems MG, Neumann E, Pfaltz A (2007) Angew Chem Int Ed 46:8274
60. Biosca M, Pàmies O, Diéguez M (2020) Cat Sci Technol 10:613
61. Pàmies O, Magre M, Diéguez M (2016) Chem Rec 16:1578
62. Brandt P, Hedberg C, Andersson PG (2003) Chem Eur J 9:339
63. Fan Y, Cui X, Burgess K, Hall MB (2004) J Am Chem Soc 126:16688
64. Cui X, Fan Y, Hall MB, Burgess K (2005) Chem Eur J 11:6859
65. Church TL, Rasmussen T, Andersson PG (2010) Organometallics 29:6769
66. Hopmann KH, Bayer A (2011) Organometallics 30:2483

67. Mazuela J, Norrby PO, Andersson PG, Pàmies O, Diéguez M (2011) J Am Chem Soc 133:13634
68. Gruber S, Pfaltz A (2014) Angew Chem Int Ed 53:1896
69. Schrems MG, Pfaltz A (2009) Chem Commun:6210
70. Hou DR, Reibenspies J, Colacot TJ, Burgess K (2001) Chem Eur J 7:5391
71. Liu D, Tang W, Zhang X (2004) Org Lett 6:513
72. Cozzi PG, Menges F, Kaiser S (2003) Synlett:833
73. Lu WJ, Chen YW, Hou XL (2010) Adv Synth Catal 352:103
74. Lu WJ, Chen YW, Hou XL (2008) Angew Chem Int Ed 47:10133
75. Li X, Li Q, Wu X, Gao Y, Xu D, Kong L (2007) Tetrahedron Asymmetry 18:629
76. Shang J, Han Z, Li Y, Wang Z, Ding K (2012) Chem Commun 48:5172
77. Wang X, Han Z, Wang Z, Ding K (2012) Angew Chem Int Ed 51:936
78. Wang Q, Zhang Z, Chen C, Yang H, Han Z, Dong XQ, Zhang X (2017) Org Chem Front 4:627
79. Meng K, Xia J, Wang Y, Zhang X, Yang G, Zhang W (2017) Org Chem Front 4:1601
80. Cozzi PG, Zimmermann N, Hilgraf R, Schäffner S, Pfaltz A (2001) Adv Synth Catal 343:450
81. Xu G, Gilbertson SR (2003) Tetrahedron Lett 44:953
82. Trifonova A, Diesen JS, Andersson PG (2006) Chem Eur J 12:2318
83. Chakka SK, Peters BK, Andersson PG, Maguire GEM, Kruger HG, Govender T (2010) Tetrahedron Asymmetry 21:2295
84. Cheruku P, Diesen J, Andersson PG (2008) J Am Chem Soc 130:5595
85. Cheruku P, Gohil S, Andersson PG (2007) Org Lett 9:1659
86. Källström K, Munslow IJ, Hedberg C, Andersson PG (2006) Adv Synth Catal 348:2575
87. Engman M, Diesen JS, Paptchikhine A, Andersson PG (2007) J Am Chem Soc 129:4536
88. Paptchikhine A, Cheruku P, Engman M, Andersson PG (2009) Chem Commun:5996
89. Verendel JJ, Li JQ, Quan X, Peters B, Zhou T, Gautun OR, Govender T, Andersson PG (2012) Chem Eur J 18:6507
90. Li JQ, Liu J, Krajangsri S, Chumnanvej N, Singh T, Andersson PG (2016) ACS Catal 6:8342
91. Zheng J, Jongcharoenkamol J, Peters BBC, Guhl J, Ponra S, Ahlquist MSG, Andersson PG (2019) Nat Catal 2:1093
92. McIntyre S, Hörmann E, Menges F, Smidt SP, Pfaltz A (2005) Adv Synth Catal 347:282
93. Baeza A, Pfaltz A (2009) Chem Eur J 15:2266
94. Müller MA, Pfaltz A (2014) Angew Chem Int Ed 53:8668
95. Maurer F, Huch V, Ullrich A, Kazmaier U (2012) J Org Chem 77:5139
96. Verevkin S, Preetz A, Börner A (2007) Angew Chem Int Ed 46:5971
97. Verevkin SP, Emelyanenko VN, Bayardon J, Schäffner B, Baumann W, Börner A (2011) Ind Eng Chem Res 51:126
98. van Leeuwen PWNM, Kamer PCJ, Claver C, Pàmies O, Diéguez M (2011) Chem Rev 111:2077
99. Diéguez M, Pàmies O (2010) Acc Chem Res 43:312
100. Diéguez M, Pàmies O (2012) Isr J Chem 52:572
101. Diéguez M, Mazuela J, Pàmies O, Verendel JJ, Andersson PG (2008) J Am Chem Soc 130:7208
102. Mazuela J, Pàmies O, Diéguez M (2013) Eur J Inorg Chem:2139
103. Rovner ES, Wein A (2002) J Eur Urol 41:6
104. Wefer J, Truss MC, Jonas U (2001) World J Urol 19:312
105. Hills CJ, Winter SA, Balfour JA (1998) Drugs 55:813
106. McRae AL, Brady KT (2001) Expert Opin Pharmacother 2:883
107. Mazuela J, Verendel JJ, Coll M, Schäffner B, Börner A, Andersson PG, Pàmies O, Diéguez M (2009) J Am Chem Soc 131:12344
108. Krajangsri S, Wu H, Liu J, Rabten W, Singh T, Andersson PG (2019) Chem Sci 10:3649
109. Diéguez M, Mazuela J, Pàmies O, Verendel JJ, Andersson PG (2008) Chem Commun:3888
110. Biosca M, Magre M, Coll M, Pàmies O, Diéguez M (2017) Adv Synth Catal 359:2801

111. Bunlaksananusorn T, Polborn K, Knochel P (2003) Angew Chem Int Ed 42:3941
112. Liu QB, Zhou YG (2007) Tetrahedron Lett 48:2101
113. Zalubovskis R, Hörmann E, Pfaltz A, Moberg C (2008) ARKIVOC 14:58
114. Wang A, Fraga RPA, Hörmann E, Pfaltz A (2011) Chem Asian J 6:599
115. Liu QB, Yu CB, Zhou YG (2006) Tetrahedron Lett 47:4733
116. Netscher T (1996) Chimia 50:563
117. Verendel JJ, Andersson PG (2007) Dalton Trans:5603
118. Meng X, Li X, Xu D (2009) Tetrahedron Asymmetry 20:1402
119. Chelucci G, Marchetti M, Malkov AV, Friscourt F, Swarbrick ME, Kočovský P (2011) Tetrahedron 67:5421
120. Li X, Kong L, Gao Y, Wang X (2007) Tetrahedron Lett 48:3915
121. Han Z, Wang Z, Zhang X, Ding K (2010) Tetrahedron Asymmetry 21:1529
122. Mazuela J, Pàmies O, Diéguez M (2013) Adv Synth Catal 355:2569
123. Margalef J, Lega M, Ruffo F, Pàmies O, Diéguez M (2012) Tetrahedron Asymmetry 23:945
124. Biosca M, Pàmies O, Diéguez M (2019) J Org Chem 84:8259
125. Tolstoy P, Engman M, Paptchikhine A, Bergquist J, Church TL, Leung AWM, Andersson PG (2009) J Am Chem Soc 131:8855
126. Kaukoranta P, Engman M, Hedberg C, Bergquist J, Andersson PG (2008) Adv Synth Catal 350:1168
127. Paptchikhine A, Itto K, Andersson PG (2011) Chem Commun 47:3989
128. Rabten W, Margarita C, Eriksson L, Andersson PG (2018) Chem Eur J 24:1681
129. Peters BK, Liu J, Margarita C, Rabten W, Kerdphon S, Orebom A, Morsch T, Andersson PG (2016) J Am Chem Soc 138:11930
130. Ponra S, Yang J, Kerdphon S, Andersson PG (2019) Angew Chem Int Ed 58:9282
131. Zhou T, Peters B, Maldonado MF, Govender T, Andersson PG (2012) J Am Chem Soc 134:13592
132. Peters BK, Zhou T, Rujirawanich J, Cadu A, Singh T, Rabten W, Kerdphon S, Andersson PG (2014) J Am Chem Soc 136:16557
133. Liu J, Krajangsri S, Yang J, Li JQ, Andersson PG (2018) Nat Catal 1:438
134. Liu J, Krajangsri S, Singh T, De Seriis G, Chumnanvej N, Wu H, Andersson PG (2017) J Am Chem Soc 139:14470
135. Zheng J, Margarita C, Krajangsri S, Andersson PG (2018) Org Lett 20:5676
136. Mazuela J, Pàmies O, Diéguez M (2013) ChemCatChem 5:2410
137. Powell MT, Hou DR, Perry MC, Cui X, Burgess K (2001) J Am Chem Soc 123:8878
138. Perry MC, Cui X, Powell MT, Hou DR, Reibenspies JH, Burgess K (2003) J Am Chem Soc 125:113
139. Bolm C, Focken T, Raabe G (2003) Tetrahedron Asymmetry 14:1733
140. Källström K, Andersson PG (2006) Tetrahedron Lett 47:7477
141. Nanchen S, Pfaltz A (2006) Chem Eur J 12:4550
142. Chen D, Banphavichit V, Reibenspies J, Burgess K (2007) Organometallics 26:855
143. Khumsubdee S, Fan Y, Burgess K (2013) J Org Chem 78:9969
144. Schumacher A, Bernasconi M, Pfaltz A (2013) Angew Chem Int Ed 52:7422
145. Rageot D, Woodmansee DH, Pugin B, Pfaltz A (2011) Angew Chem Int Ed 50:9598
146. Rageot D, Pfaltz A (2012) Helv Chim Acta 95:2176
147. Elías-Rodríguez P, Borràs C, Carmona AT, Faiges J, Robina I, Pàmies O, Diéguez M (2018) More recently another example of P,O ligand has been reported with successful application in the Ir-catalyzed asymmetric hydrogenation of unfunctionalized olefins. ChemCatChem 10:5414
148. Coll M, Pàmies O, Diéguez M (2011) Chem Commun 47:9215
149. Coll M, Pàmies O, Diéguez M (2013) Adv Synth Catal 355:143
150. Borràs C, Biosca M, Pàmies O, Diéguez M (2015) Organometallics 34:5321
151. Margalef J, Borràs C, Alegre S, Alberico E, Pàmies O, Diéguez M (2019) ChemCatChem 11:2142

152. Biosca M, Coll M, Lagarde F, Brémond E, Routaboul L, Manoury E, Pàmies O, Poli R, Diéguez M (2016) Tetrahedron 72:2623
153. Margalef J, Caldentey X, Karlsson EA, Coll M, Mazuela J, Pàmies O, Diéguez M, Pericàs MA (2014) Chem Eur J 20:12201
154. Kraft S, Ryan K, Kargbo RB (2017) J Am Chem Soc 139:11630
155. Troutman MV, Appella DH, Buchwald SL (1999) J Am Chem Soc 121:4916
156. Zhang Z, Wang J, Li J, Yang F, Liu G, Tang W, He W, Fu JJ, Shen YH, Li A, Zhang WD (2017) J Am Chem Soc 139:5558
157. Busacca CA, Qu B, Grĕt N, Fandrick KR, Saha AK, Marsini M, Reeves D, Haddad N, Eriksson M, Wu JP, Grinberg N, Lee H, Li Z, Lu B, Chen D, Hong Y, Ma S, Senanayake CH (2013) Adv Synth Catal 355:1455
158. Ponra S, Rabten W, Yang J, Wu H, Kerdphon S, Andersson PG (2018) J Am Chem Soc 140:13878
159. Biosca M, Salomó E, de la Cruz-Sánchez P, Riera A, Verdaguer X, Pàmies O, Diéguez M (2019) Org Lett 21:807
160. Bigler R, Mack KA, Shen J, Tosatti P, Han C, Bachmann S, Zhang H, Scalone M, Pfaltz A, Denmark SE, Hildbrand S, Gosselin F (2020) Angew Chem Int Ed 59:2844
161. Shen TY (1972) Angew Chem Int Ed Engl 11:460
162. Lednicer D, Mitscher LA (1977) The organic chemistry of drug synthesis. Wiley, New York
163. Lu Y, Nguyen TMD, Weltrowska G, Berezowska I, Lemieux C, Chung NN, Schiller PWJ (2001) Med Chem 44:3048
164. Churcher I, Ashton K, Butcher JW, Clarke EE, Harrison T, Lewis HD, Owens AP, Teall MR, Williams S, Wrigley JDJ (2003) Bioorg Med Chem Lett 13:179
165. Henke BR (2004) J Med Chem 47:4118
166. Kasuga J, Makishima M, Hashimoto Y, Miyachi H (2006) Bioorg Med Chem Lett 16:554
167. Ono N, The Nitro Group (2001) Organic synthesis. Wiley-VCH, New York
168. Ballini R, Petrini M (2004) Tetrahedron 60:1017
169. Scrivanti A, Bovo S, Ciappa A, Matteoli U (2006) Tetrahedron Lett 47:9261
170. Zhou J, Ogle JW, Fan Y, Banphavichit V, Zhu Y, Burgess K (2007) Chem Eur J 13:7162
171. Smidt SP, Pfaltz A, Martínez-Viviente E, Pregosin PS, Albinati A (2003) Organometallics 22:1000
172. Zhu SF, Xie JB, Zhang YZ, Li S, Zhou QL (2006) SIPHOX ligands were developed for Ir-catalyzed asymmetric imine hydrogenation. J Am Chem Soc 128:12886
173. Li S, Zhu SF, Zhang CM, Song S, Zhou QL (2008) J Am Chem Soc 130:8584
174. Li S, Zhu SF, Xie JH, Song S, Zhang CM, Zhou QL (2010) J Am Chem Soc 132:1172
175. Zhu SF, Zhou QL (2017) Acc Chem Res 50:988
176. Yang S, Che W, Wu HL, Zhu SF, Zhou QL (2017) Chem Sci 8:1977
177. Yang S, Zhu SF, Zhang CM, Song S, Yu YB, Li S, Zhou QL (2012) Tetrahedron 68:5172
178. Li ZY, Song S, Zhu SF, Guo N, Wang LX, Zhou QL (2014) Chin J Chem 32:783
179. Song S, Zhu SF, Pu LY, Zhou QL (2013) Angew Chem Int Ed 52:6072
180. Song S, Zhu SF, Li Y, Zhou QL (2013) Org Lett 15:3722
181. Song S, Zhu SF, Yang S, Li S, Zhou QL (2012) Angew Chem Int Ed 51:2708
182. Song S, Zhu SF, Yu YB, Zhou QL (2013) Angew Chem Int Ed 52:1556
183. Yang S, Zhu SF, Guo N, Song S, Zhou QL (2014) Org Biomol Chem 12:2049
184. Zhu SF, Yu YB, Li S, Wang LX, Zhou QL (2012) Angew Chem Int Ed 51:8872
185. Li ML, Yang S, Su XC, Wu HL, Yang LL, Zhu SF, Zhou QL (2017) J Am Chem Soc 139:541
186. Zhang Y, Han Z, Li F, Ding K, Zhang A (2010) Chem Commun 46:156
187. Xu L, Zhaobin H, Zheng W, Kuiling D (2014) Acta Chim Sin 72:849
188. Li S, Huang K, Cao B, Zhang J, Wu W, Zhang X (2012) Angew Chem Int Ed 51:8573
189. Li S, Huang K, Zhang J, Wu W, Zhang X (2013) Chem Eur J 19:10840
190. Zhao Q, Li S, Huang K, Wang R, Zhang X (2013) Org Lett 15:4014
191. Yan Q, Liu M, Kong D, Zi G, Hou G (2014) Chem Commun 50:12870
192. Liu M, Kong D, Li M, Zi G, Hou G (2015) Adv Synth Catal 357:3875

193. Yu YB, Cheng L, Li YP, Fu Y, Zhu SF, Zhou QL (2016) Chem Commun 52:4812
194. Li S, Xiao T, Li D, Zhang X (2015) Org Lett 17:3782
195. Xie JH, Zhu SF, Zhou QL (2012) Chem Soc Rev 41:4126
196. Maire P, Deblon S, Breher F, Geier J, Böhler C, Rüegger H, Schönberg H, Grützmacher H (2004) Chem Eur J 10:4198
197. Giacomina F, Meetsma A, Panella L, Lefort L, de Vries AHM, de Vries JG (2007) Angew Chem Int Ed 46:1497
198. Enthaler S, Erre G, Junge K, Schröder K, Addis D, Michalik D, Hapke M, Redkin D, Beller M (2008) Eur J Org Chem:3352
199. Erre G, Enthaler S, Junge K, Addis D, Beller M (2009) Adv Synth Catal 351:1437
200. Hou G, Li W, Ma M, Zhang X, Zhang X (2010) J Am Chem Soc 132:12844
201. Busscher GF, Lefort L, Cremers JGO, Mottinelli M, Wiertz RW, de Lange B, Okamura Y, Yusa Y, Matsumura K, Shimizu H, de Vries JG, de Vries AHM (2010) Tetrahedron Asymmetry 21:1709
202. Han Z, Guan YQ, Liu G, Wang R, Yin X, Zhao Q, Cong H, Dong XQ, Zhang X (2018) Org Lett 20:6349
203. Patureau FW, Worch C, Siegler MA, Spek AL, Bolm C, Reek JNH (2012) Adv Synth Catal 354:59
204. Margalef J, Pàmies O, Diéguez M (2017) Chem Eur J 23:813
205. Cabré A, Verdaguer X, Riera A (2019) Adv Synth Catal 361:4196
206. Cabré A, Romagnoli E, Martínez-Balart P, Verdaguer X, Riera A (2019) Org Lett 21:9709
207. Lazarevski G, Kobrehel G, Metelko B, Duddeck H (1996) J Antibiot 49:1066
208. Nicholas GM, Molinski TF (2000) Tetrahedron 56:2921
209. Davies HML, Ni A (2006) Chem Commun:3110
210. Harris RN, Stabler RS, Repke DB, Kress JM, Walker KA, Martin RS, Brothers JM, Ilnicka M, Lee SW, Mirzadegan T (2010) Bioorg Med Chem Lett 20:3436
211. Ramesh P, Suman D (2018) Synthesis 50:211
212. Fleury-Brégeot N, de la Fuente V, Castillón S (2010) ChemCatChem 2:1346
213. Xie JH, Zhu SF, Zhou QL (2011) Chem Rev 111:1713
214. James BR (1997) Catal Today 37:209
215. Blaser HU, Buser HP, Loers K, Hanreich R, Jalett HP, Jelsch E, Pugin B, Schneider HD, Spindler F, Wagmann A (1999) Chimia 53:275
216. Xiao D, Zhang X (2001) Angew Chem Int Ed 40:3425
217. Mršić N, Panella L, Ijpeij EJ, Minnaard AJ, Feringa BL, De Vries JG (2011) ChemCatChem 3:1139
218. Baeza A, Pfaltz A (2010) Chem Eur J 16:4003
219. Schramm Y, Barrios-Landeros F, Pfaltz A (2013) Chem Sci 4:2760
220. Salomó E, Gallen A, Sciortino G, Ujaque G, Grabulosa A, Lledós A, Riera A, Verdaguer X (2018) J Am Chem Soc 140:16967
221. Salomó E, Rojo P, Hernández-Lladó P, Riera A, Verdaguer X (2018) J Org Chem 83:4618
222. Han Z, Wang Z, Zhang X, Ding K (2009) Angew Chem Int Ed 48:5345
223. Welch WM, Kraska AR, Sarges R, Koe BK (1984) J Med Chem 27:1508
224. Qu B, Samankumara LP, Ma S, Fandrick KR, Desrosiers JN, Rodriguez S, Li Z, Haddad N, Han ZS, McKellop K, Pennino S, Grinberg N, Gonnella NC, Song JJ, Senanayake CH (2014) Angew Chem Int Ed 53:14428
225. Hou CJ, Wang YH, Zheng Z, Xu J, Hu XP (2012) Org Lett 14:3554
226. Li Q, Hou CJ, Liu XN, Huang DZ, Liu YJ, Yanga RF, Hu XP (2015) RSC Adv 5:13702
227. Bishop MJ, McNutt RW (1995) Bioorg Med Chem Lett 5:1311
228. Spencer CM, Foulds D, Peters DH (1993) Drugs 46:1055
229. Sakurai S, Ogawa N, Suzuki T, Kato K, Ohashi T, Yasuda S, Kato H, Ito Y (1996) Chem Pharm Bull 44:765
230. Tulshian D, Ho GD, Silverman L, Matasi JJ, McLeod RL, Hey JA, Chapman RW, Bercovici A, Cuss FM, WO 01/07050 A1 (Schering-Plough)
231. Jolidon S, Narquizian R, Pinard E. WO 2008/022938 A1 (Hoffman-LaRoche)
232. Ducray P, Cavaliero T, Lorhmann M, Bouvier J WO 2008/062006 A1 (Novartis)

233. Baker RK, Hale JJ, Maio S, Rupprecht KM. WO 2007/062193 A1 (Merck). Cetirizine HCl, a istamine H1-receptor inverse agonist is commercialized as Zyrtec/Reactine by Pfizer
234. Hou G, Tao R, Sun Y, Zhang X, Gosselin F (2010) J Am Chem Soc 132:2124
235. Kong D, Li M, Zi G, Hou G, He Y (2016) J Org Chem 81:6640
236. Hu XH, Hu XP (2019) Adv Synth Catal 361:5063
237. Mazuela J, Antonsson T, Knerr L, Marsden SP, Munday RH, Johansson MJ (2019) Adv Synth Catal 361:578
238. Wang WB, Lu SM, Yang PY, Han XW, Zhou YG (2003) J Am Chem Soc 125:10536
239. Yang PY, Zhou YG (2004) Tetrahedron Asymmetry 15:1145
240. Wang DW, Wang XB, Wang DS, Lu SM, Zhou YG, Li YX (2009) J Org Chem 74:2780. Later, it was shown that an electronically deficient ligand synthesized from (*S*)-MeO-BiPhep showed an increase of the activity and enantioselectivity: Zhang DY, Wang DS, Wang MC, Yu CB, Gao K, Zhou YG (2011) Synthesis 2796
241. Lu SM, Wang YQ, Han XW, Zhou YG (2006) Angew Chem Int Ed 45:2260
242. Shi L, Ye ZS, Cao LL, Guo RN, Hu Y, Zhou YG (2012) Angew Chem Int Ed 51:8286
243. Hu SB, Zhai XY, Shen HQ, Zhou YG (2018) Adv Synth Catal 360:1334
244. Guo RN, Gao K, Ye ZS, Shi L, Li Y, Zhou YG (2013) Pure Appl Chem 85:843
245. Ye ZS, Chen MW, Chen QA, Shi L, Duan Y, Zhou YG (2012) Angew Chem Int Ed 51:10181
246. Guo RN, Cai XF, Shi L, Ye ZS, Chen MW, Zhou YG (2013) Chem Commun 49:8537
247. Ji Y, Shi L, Chen MW, Feng GS, Zhou YG (2015) J Am Chem Soc 137:10496
248. Ye ZS, Guo RN, Cai XF, Chen MW, Shi L, Zhou YG (2013) Angew Chem Int Ed 52:3685
249. Chen MW, Ye ZS, Chen ZP, Wu B, Zhou YG (2015) Org Chem Front 2:586
250. Huang WX, Yu CB, Shi L, Zhou YG (2014) Org Lett 16:3324
251. Huang WX, Liu LJ, Wu B, Feng GS, Wang B, Zhou YG (2016) Org Lett 18:3082
252. Hu SB, Chen ZP, Song B, Wang J, Zhou YG (2017) Adv Synth Catal 359:2762
253. Huang WX, Yu CB, Ji Y, Liu LJ, Zhou YG (2016) ACS Catal 6:2368
254. Ji Y, Feng GS, Chen MW, Shi L, Duc H, Zhou YG (2017) Org Chem Front 4:1125
255. Chang M, Huang Y, Liu S, Chen Y, Krska SW, Davies IW, Zhang X (2014) Angew Chem Int Ed 53:12761
256. Chen MW, Ji Y, Wang J, Chen QA, Shi L, Zhou YG (2017) Org Lett 19:4988
257. Qu B, Mangunuru HPR, Wei X, Fandrick KR, Desrosiers JN, Sieber JD, Kurouski D, Haddad N, Samankumara LP, Lee H, Savoie J, Ma S, Grinberg N, Sarvestani M, Yee NK, Song JJ, Senanayake CH (2016) Org Lett 18:4920
258. Qu B, Mangunuru HPR, Tcyrulnikov S, Rivalti D, Zatolochnaya OV, Kurouski D, Radomkit S, Biswas S, Karyakarte S, Fandrick KR, Sieber JD, Rodriguez S, Desrosiers JN, Haddad N, McKellop K, Pennino S, Lee H, Yee NK, Song JJ, Kozlowski MC, Senanayake CH (2018) Org Lett 20:1333
259. Chang M, Li W, Zhang X (2011) Angew Chem Int Ed 50:10679
260. Nie H, Zhu Y, Hu X, Wei Z, Yao L, Zhou G, Wang P, Jiang R, Zhang S (2019) Org Lett 21:8641
261. Tang W, Sun Y, Xu L, Wang T, Fan O, Lamc KH, Chan ASC (2010) Org Biomol Chem 8:3464
262. Tang WJ, Tan J, Xu LJ, Lam KH, Fan QH, Chan ASC (2010) Adv Synth Catal 352:1055
263. Xie JH, Yan PC, Zhang QQ, Yuan KX, Zhou QL (2012) ACS Catal 2:561
264. Gao K, Yu CB, Wang DS, Zhou YG (2012) Adv Synth Catal 354:483
265. Feng GS, Zhao ZB, Shi L, Zhou YG (2019) Org Chem Front 6:2250
266. Cartigny D, Nagano T, Ayad T, Genêt JP, Ohshima T, Mashima K, Ratovelomanana-Vidal V (2010) Adv Synth Catal 352:1886
267. Cartigny D, Berhal F, Nagano T, Phansavath P, Ayad T, Genêt JP, Ohshima T, Mashima K, Ratovelomanana-Vidal V (2012) J Org Chem 77:4544
268. Nagano T, Iimuro A, Schwenk R, Ohshima T, Kita Y, Togni A, Mashima K (2012) Chem Eur J 18:11578
269. Higashida K, Nagae H, Mashima K (2016) Adv Synth Catal 358:3949

270. Núñez-Rico JL, Fernández-Pérez H, Benet-Buchholz J, Vidal-Ferran A (2010) Organometallics 29:6627
271. Núñez-Rico JL, Vidal-Ferran A (2013) Org Lett 15:2066
272. Gao K, Yu CB, Li W, Zhou YG, Zhang X (2011) Chem Commun 47:7845
273. Gao K, Wu B, Yu CB, Chen QA, Ye ZS, Zhou YG (2012) Org Lett 14:3890
274. Shen HQ, Gao X, Liu C, Hu SB, Zhou YG (2016) Org Lett 18:5920
275. Liu Y, Chen F, He YM, Lia C, Fan QH (2019) Org Biomol Chem 17:5099
276. Ma B, Ding Z, Liu J, He Y, Fan QH (2013) Chem Asian J 8:1101
277. Miao T, Ma B, Ding Z, Liu Y, He YM, Fan QH (2017) Asian J Org Chem 6:1219
278. Han Z, Liu C, Wang R, Dong XQ, Zhang X (2019) Chem Sci 10:4328
279. Chang M, Li W, Hou G, Zhang X (2010) Adv Synth Catal 352:3121
280. Zhang Y, Kong D, Wang R, Hou G (2017) Org Biomol Chem 15:3006
281. Liu Y, Huang Y, Yi Z, Liu G, Dong XQ, Zhang X (2019) Adv Synth Catal 361:1582
282. Roux EL, Malacea R, Manoury E, Poli R, Gonsalvi L, Peruzzini M (2007) Adv Synth Catal 349:309
283. Xie JB, Xie JH, Liu XY, Kong WL, Li S, Zhou QL (2010) J Am Chem Soc 132:4538
284. Xie JB, Xie JH, Liu XY, Zhang QQ, Zhou QL (2011) Chem Asian J 6:899
285. Xie JH, Liu XY, Xie JB, Wang LX, Zhou QL (2011) Angew Chem Int Ed 50:7329
286. Xie JH, Liu XY, Yang XH, Xie JB, Wang LX, Zhou QL (2012) Angew Chem Int Ed 51:201
287. Yang XH, Xie JH, Liu WP, Zhou QL (2013) Angew Chem Int Ed 52:7833
288. Zhang QQ, Xie JH, Yang XH, Xie JB, Zhou QL (2012) Org Lett 14:6158
289. Yan PC, Xie JH, Zhang XD, Chen K, Li YQ, Zhou QL, Che DQ (2014) Chem Commun 50:15897
290. Hua YY, Bin HY, Wei T, Cheng HA, Lin ZP, Fu XF, Li YQ, Xie JH, Yan PC, Zhou QL (2020) Org Lett 22:818
291. Yuan ML, Xie JH, Yang XH, Zhou QL (2014) Synthesis 46:2910
292. Liu YT, Chen JQ, Li LP, Shao XY, Xie JH, Zhou QL (2017) Org Lett 19:3231
293. Gu XS, Yu N, Yang XH, Zhu AT, Xie JH, Zhou QL (2019) Org Lett 21:4111
294. Yang XH, Yue HT, Yu N, Li YP, Xie JH, Zhou QL (2017) Chem Sci 8:1811
295. Yang XH, Wang K, Zhu SF, Xie JH, Zhou QL (2014) J Am Chem Soc 136:17426
296. Yan PC, Zhu GL, Xie JH, Zhang XD, Zhou QL, Li YQ, Shen WH, Che DQ (2013) Org Process Res Dev 17:307
297. Lin H, Xiao LJ, Zhou MJ, Yu HM, Xie JH, Zhou QL (2016) Org Lett 18:1434
298. Zhu GL, Zhang XD, Yang LJ, Xie JH, Che DP, Zhou QL, Yan PC, Li YQ (2016) Org Process Res Dev 20:81
299. Zuo XD, Guo SM, Yang R, Xie JH, Zhou QL (2017) Chem Sci 8:6202
300. Bin HY, Wang K, Yang D, Yang XH, Xie JH, Zhou QL (2019) Angew Chem Int Ed 58:1174
301. Bao DH, Wu HL, Liu CL, Xie JH, Zhou QL (2015) Angew Chem Int Ed 54:8791
302. Bao DH, Gu XS, Xie JH, Zhou QL (2017) Org Lett 19:118
303. Che W, Li YZ, Liu JC, Zhu SF, Xie JH, Zhou QL (2019) Org Lett 21:2369
304. Che W, Wen DC, Zhu SF, Zhou QL (2019) Helv Chim Acta 102:e1900023
305. Zhang FH, Wang C, Xie JH, Zhou QL (2019) Adv Synth Catal 361:2832
306. Chen GQ, Lin BJ, Huang JM, Zhao LY, Chen QS, Jia SP, Yin Q, Zhang X (2018) J Am Chem Soc 140:8064
307. Wu W, Liu S, Duan M, Tan X, Chen C, Xie Y, Lan Y, Dong XQ, Zhang X (2016) Org Lett 18:2938
308. Gu G, Yang T, Yu O, Qian H, Wang J, Wen J, Dang L, Zhang X (2017) Org Lett 19:5920
309. Hu Y, Yin X, Chen Z, Dong XQ, Zhang X (2018) Org Chem Front 5:2000
310. Hu Y, Wu W, Dong XQ, Zhang X (2017) Org Chem Front 4:1499
311. Wu W, Xie Y, Li P, Li X, Liu Y, Dong XQ, Zhang X (2017) Org Chem Front 4:555
312. Qin C, Chen XS, Hou CJ, Liu H, Liu YJ, Huang DZ, Hu XP (2018) Synth Comun 48:672
313. Wang S, Yu Y, Wen J, Zhang X (2018) Synlett 29:2203

314. Yin C, Wu W, Hu Y, Tan X, You C, Liu Y, Chen Z, Dong XQ, Zhang X (2018) Adv Synth Catal 360:2119
315. Wu W, You C, Yin C, Liu Y, Dong XQ, Zhang X (2017) Org Lett 19:2548
316. Wei DQ, Chen XS, Hou CJ, Hu XP (2019) Synth Commun. https://doi.org/10.1080/00397911.2018.1550203
317. Yu J, Long J, Yang Y, Wu W, Xue P, Chung LW, Dong XQ, Zhang X (2017) Org Lett 19:690
318. Gu G, Yang T, Lu J, Wen J, Dang J, Zhang X (2018) Org Chem Front 5:1209
319. Gong Q, Wen J, Zhang X (2019) Chem Sci 10:6350
320. Yu J, Duan M, Wu W, Qi X, Xue P, Lan Y, Dong XQ, Zhang X (2017) Chem Eur J 23:970
321. Yin C, Dong XQ, Zhang X (2018) Adv Synth Catal 360:4319
322. Liang Z, Yang T, Gu G, Dang L, Zhang X (2018) Chin J Chem 36:851
323. Tao L, Yin C, Dong XQ, Zhang X (2019) Org Biomol Chem 17:785
324. Gu G, Lu J, Yu O, Wen J, Yin Q, Zhang X (2018) Org Lett 20:1888
325. Hou CJ, Hu XP (2016) Org Lett 18:5592
326. Ling F, Nian S, Chen J, Luo W, Wang Z, Lv Y, Zhong W (2018) J Org Chem 83:10749
327. Nian S, Ling F, Chen J, Wang Z, Shen H, Yi X, Yang YF, She Y, Zhong W (2019) Org Lett 21:5392
328. Ohkuma T, Utsumi N, Watanabe M, Tsutsumi K, Arai N, Murata K (2007) Org Lett 9:2565
329. Utsumi N, Tsutsumi K, Watanabe M, Murata K, Arai N, Kurono N, Ohkuma T (2010) Heterocycles 80:141
330. Ito M, Endo Y, Tejima N, Ikariya T (2010) Organometallics 29:2397
331. Irrgang T, Friedrich D, Kempe R (2011) Angew Chem Int Ed 50:2183
332. Kumar P, Irrgang T, Kostakis GE, Kempe R (2016) RSC Adv 6:39335

Top Organomet Chem (2021) 69: 207–226
https://doi.org/10.1007/3418_2020_53

Published online: 30 July 2020

Iridium-Catalyzed Undirected Homogeneous C–H Borylation Reaction

Elena Fernández

Contents

Abstract The present chapter describes the intense efforts devoted to develop new concepts from 2014 up today in the iridium-catalyzed undirected homogeneous C–H borylation of heteroarenes, arenes, and alkanes. Selectivity issues are principally highlighted in this chapter since no directed groups are included in these approaches, but improved ligands are responsible of the new trends instead. In parallel, mechanistic insights on the C–B bond formation analyzed through density functional theory orientate the suggestion of alternative catalytic cycles, to understand the high level of selectivity on Csp^2-H and Csp^3-H borylation. This area of work keeps very active because of the inherent interest on the C–B bond formation allowing access to multifunctionalized products, in a straightforward manner.

Keywords Alkanes · Arenes · Bis(pinacolato)diboron · Catalytic cycle · C–H borylation · Heteroarenes · Iridium · Phenanthroline · Pinacolborane · Selectivity

The original version of this chapter was revised. A correction to this chapter can be found at https://doi.org/10.1007/3418_2020_73

E. Fernández (✉)
Dept. Química Física i Inorgànica, Faculty of Chemistry, Tarragona, Spain
e-mail: mariaelena.fernandez@urv.cat

1 Introduction

The present chapter collects the most recent advances on undirected homogeneous C–H borylation catalyzed by iridium complexes since 2014, due to the tremendous development of the field within the last 5 years. Previous works are not described here to minimize unnecessary overlap with other reviews [1–5]. The criterion to classify the present chapter is based on the type of substrates such as heteroarenes, arenes, and alkanes. Precursors of catalysts are based on [Ir(μ-Cl)(COD)]$_2$ (COD = cyclooctadiene), [Ir(μ-OMe)(COD)]$_2$, and [Ir(η^6-mes)(Bpin)$_3$] although preformed catalysts have also been studied. The ligands involved in the modification of the iridium complexes are principally 4,4′-di-tert-butyl bipyridine (dtbpy), 3,4,7,8- tetramethyl-1,10-phenanthroline (Me$_4$Phen), and 1,10-phenanthroline (Phen) among others (Fig. 1). Borylating reagents covered in this chapter include pinacolborane (HBpin) and bis(pinacolato)diboron (B$_2$pin$_2$) (Fig. 1). The relevance of this undirected homogeneous iridium-catalyzed C–H borylation reaction is based on the high selectivity showed by the recent developed catalysts without the assistance of directing groups.

The most accepted catalytic cycle for undirected homogeneous iridium-catalyzed C–H borylation is based on experimental studies by Hartwig and co-workers from 2005, based on Ir(dtbpy)(Bpin)$_3$(COE) (COE = cyclooctene) as a precatalyst indicating that an Ir(III) species [Ir(dtbpy)(Bpin)$_3$] is the active catalyst that mediates C–H cleavage via rate-limiting oxidative addition followed by reductive elimination step yielding the corresponding organoboron compound [6]. In this mechanism, regeneration of the active catalyst can occur with either HBpin or B$_2$pin$_2$ reagents. The mechanism involves Ir(III) and Ir(V) species (Scheme 1).

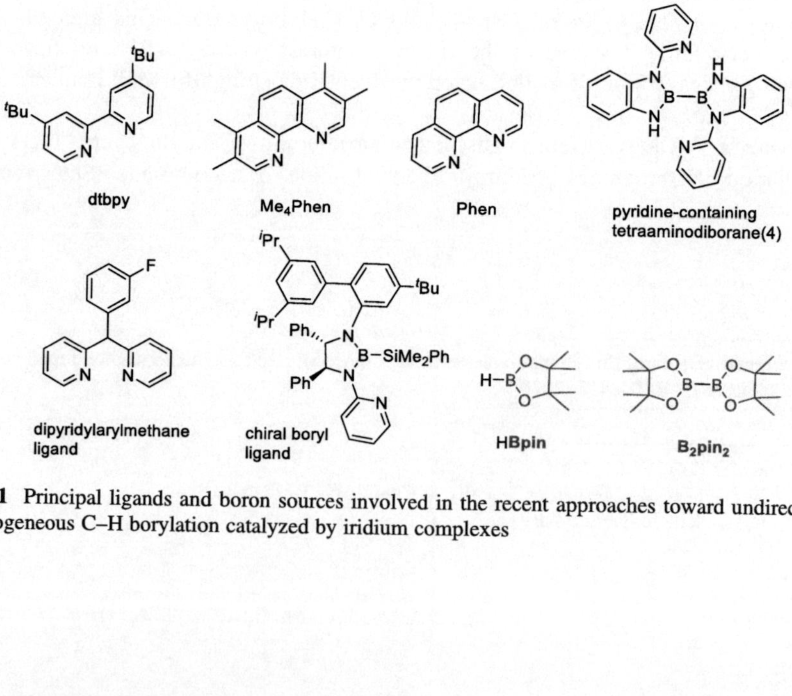

Fig. 1 Principal ligands and boron sources involved in the recent approaches toward undirected homogeneous C–H borylation catalyzed by iridium complexes

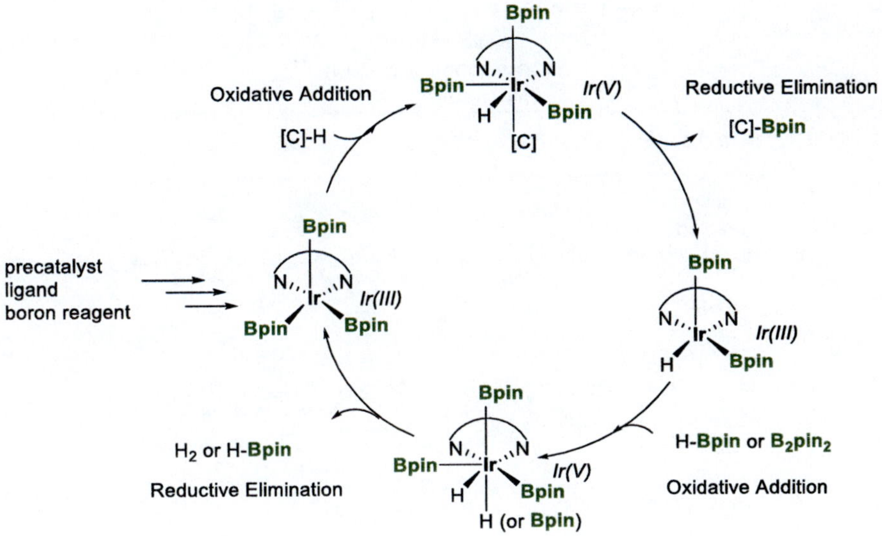

Scheme 1 Accepted catalytic cycle to understand the mechanism of the undirected homogeneous iridium-catalyzed C–H borylation

Alternative catalytic cycles have been suggested within the last years, both from experimental and theoretical studies, and will be discussed in due course along this chapter.

2 C–H Borylation of Heteroarenes

Early works on C–H borylation of aromatic heterocycles containing one heteroatom were conducted by Miyaura, Ishiyama, and co-workers in 2002 with bis(pinacolato) diboron (B_2pin_2) and $[Ir(\mu\text{-}Cl)(COD)]_2$ as precursor of catalyst [7–10]. The borylation of six-membered heterocycles including pyridine and quinoline selectively occurred at the 3-position, and the extension of the reaction to five-membered substrates such as thiophene, furan, pyrrole, and their benzo-fused derivatives exclusively produced 2-borylated products.

The use of $[Ir(\mu\text{-}OMe)(COD)]_2$ as precursor of catalyst was further extended to the borylation of heteroarenes containing multiple heteroatoms as well as high nitrogen content with pinacolborane (HBpin) [11, 12] or B_2pin_2 [13–15] as borylative reagents.

Iridium complex $[Ir(\mu\text{-}OMe)(COD)]_2$ has been modified with 4,4′-di-tert-butyl bipyridine (dtbpy) in mostly of the optimized reactions [16]; however, recently the use of 3,4,7,8-tetramethyl-1,10-phenanthroline (Me_4Phen) as ligand has provided higher yields presumably due to the greater electron-donating ability and backbone rigidity of Me_4phen compared to those of dtbpy [17]. Hartwig and co-workers have

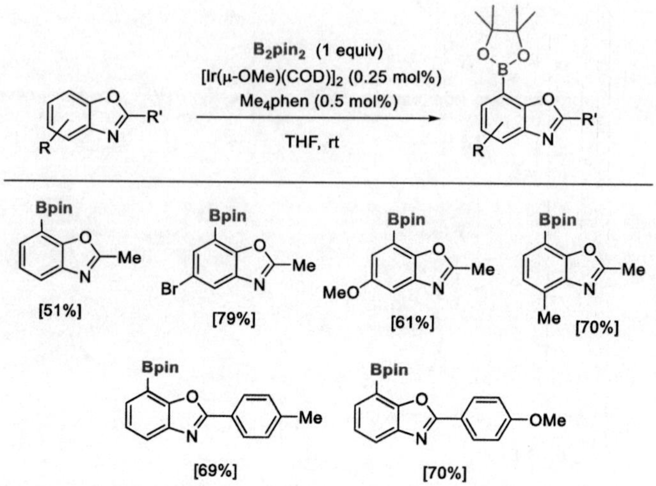

Scheme 2 Iridium-catalyzed C–H borylation of benzoxazoles

developed a detailed study about the effect of the electronic properties of phenanthroline ligands on the electron-donating ability of these ligands and on the rates for cleavage of C–H bonds by iridium–trisboryl complexes that catalyze the borylation of C–H bonds [18].

Hartwig and co-workers developed in 2014 the iridium-catalyzed C–H borylation of several heteroarenes containing multiple heteroatoms, such as benzoxazoles (Scheme 2), pyrimidines, and unprotected benzimidazoles (Scheme 3), pyrazoles, and azaindoles (Scheme 4), by the combination of an iridium (I) precursor [Ir(μ-OMe)(COD)]$_2$/Me$_4$Phen and B$_2$pin$_2$ [19].

In this work, the authors led to the development of powerful rules for predicting the regioselectivity of borylation of heteroarenes showing that borylation occurs distal to N–H bonds due to rapid N–H borylation, creating an unfavorable steric

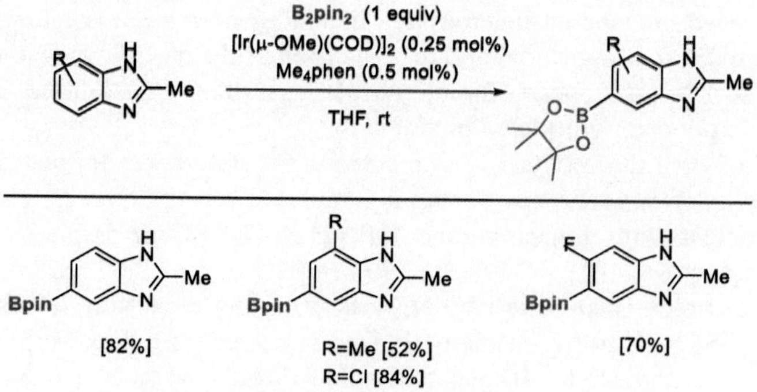

Scheme 3 Iridium-catalyzed C–H borylation of benzimidazoles

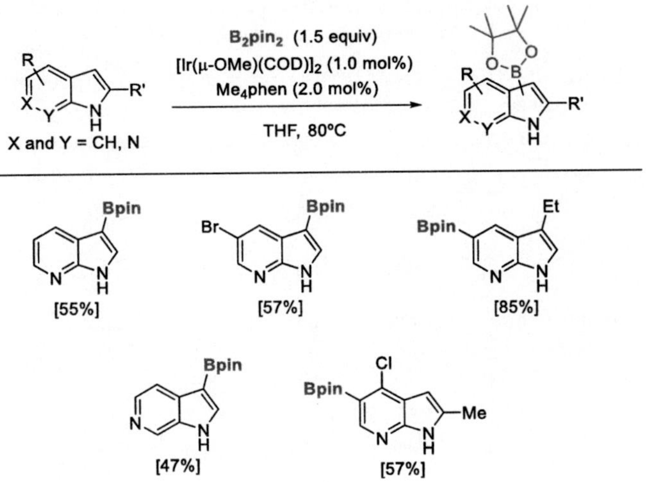

Scheme 4 Iridium-catalyzed C–H borylation of azaindoles

environment for borylation adjacent to these bonds. They also proved, via DFT calculations, that borylation or heteroarenes do not occur alpha or ortho to basic nitrogen moieties probably due to the combination of a higher-energy pathway for the borylation adjacent to basic nitrogen relative to the borylation at other sites and the instability of the products from borylation adjacent to basic nitrogen under the reaction conditions.

Simultaneously to the previous example, Colacot and co-workers developed a cheaper catalytic system based on Ir(I) modified with 1,10-phenantroline ligand (Phen). This work is remarkable because the authors studied the effect of preformed versus in situ formed iridium complexes for C–H borylation of heteroarenes, with a ligand that is 5 and 13 times cheaper than dtbpy and Me_4Phen, respectively [20]. They developed a robust and operationally simple process with relatively low Ir loading to borylate C–H of heterocycles (Scheme 5). One of the key points

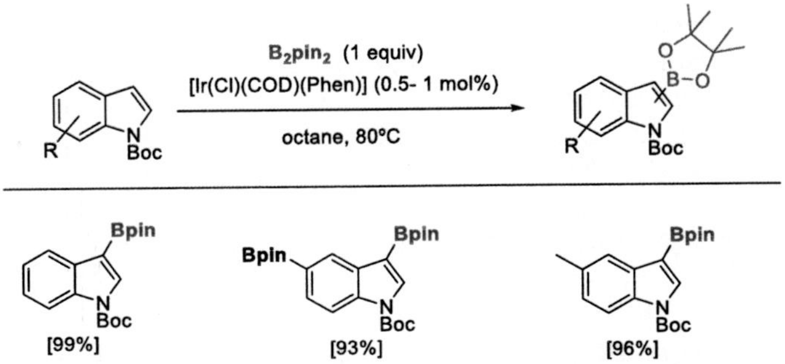

Scheme 5 Preformed iridium-catalyzed C–H borylation of N-containing heterocycles

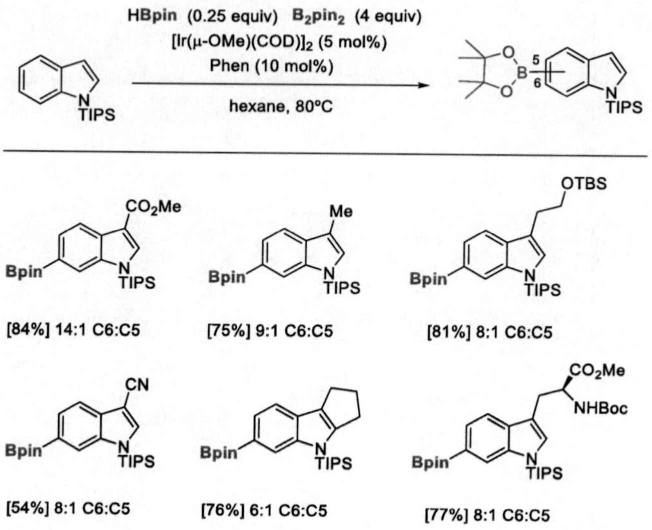

Scheme 6 Iridium-catalyzed C–H borylation of indoles with enhanced regioselectivity

of this work is that authors demonstrated that the neutral isolated precatalyst [Ir(Cl)(COD)(phen)] resulted more efficient for direct borylation of heteroarenes than the analogue cationic system $[Ir(COD)(phen)]^{+}Cl^{-}$. It has been justified by the fact that in situ protocol using $[Ir(\mu\text{-}Cl)(COD)]_2$ and 1,10-phenanthroline gave irreproducible results due to the competing formation of catalytically inactive cationic Ir species $[Ir(COD)(phen)]^{+}Cl^{-}$ in aliphatic solvents. This work demonstrates for the first time the relevance of neutral to cationic switch to understand the structure–activity relationship of the catalyst under the reaction conditions.

The benefits of 1,10-phenantroline ligand were also highlighted by Baran and co-workers thorough the iridium-catalyzed C–H borylation of indoles and in particular of tryptophan proving to be a general way to functionalize the C6 position of a N, C3-disubstituted substrates (Scheme 6) [21]. In order to achieve the desired regioselectivity, the catalyst loading was up to 5 mol% and the phen ligand 10 mol%. The boron source required a mixture of HBpin and B_2pin_2. The use of ligands dtbpy and Me_4Phen resulted in a very low conversion and regioselectivity on the desired product, under the same reaction conditions.

Alternatively, Li and co-workers have designed a double N,B-type of ligand based on dipyridinyl tetraaminodiborane(4) that proved to be efficient preligand in the iridium $[Ir(\mu\text{-}OMe)(COD)]_2$ catalyzed C–H borylation of various heteroarenes, including highly electron-rich ones and sterically hindered ones (Scheme 7) [22]. Considering the general catalytic cycle of Ir catalyzed C–H borylation shown in Scheme 1, a bipyridine-coordinated Ir(III) trisboryl complex has been accepted to be the real catalyst; however only one of the three boryl ligands is needed for the product formation in the catalytic cycle. In that context the remaining two boryl ligands can be selectively preinstalled by recombining the bipyridine ligand and two

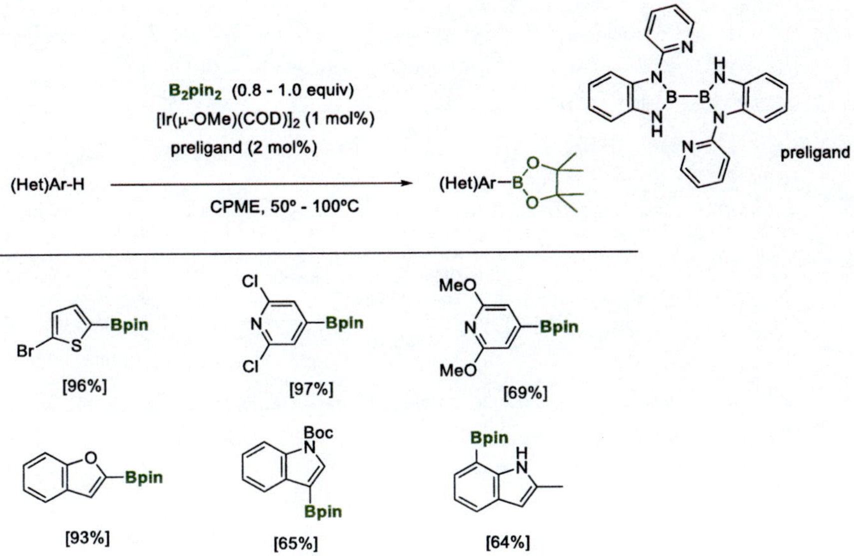

Scheme 7 Iridium-catalyzed C–H borylation of thiophene, benzofuran, pyridines, and indoles in the presence of N,B bidentate preligand

boryls into two N,B-type bidentate ligands, providing new opportunities in tuning the electronic and steric properties of the catalyst center.

In the absence of steric groups, iridium-catalyzed C–H borylation of N-protected indazoles occurs rapidly and selectively at C-3. Burton and co-workers found that N1-protected indazole could be transformed efficiently toward three boryl indazoles in the presence of $[Ir(\mu\text{-}OMe)(COD)]_2$/dtbpy and B_2pin_2 (Scheme 8a) [23]. Steel and co-workers extended the application of this protocol to a large number of N2-protected indazoles with exclusive borylation at C-3 even in the presence of bulky benzyl or THP-protecting groups (Scheme 8b) [24].

Ito, Ishiyama, and co-workers have developed a rational regiodivergent C–H borylation of multifunctionalized heteroarenes, such as furans, thiophenes, and pyrroles, by using two different iridium catalytic systems. The borylation proceeded regioselectively at the 3-position under the catalytic system $[Ir(\mu\text{-}OMe)(COD)]_2$/dtbpy and B_2pin_2, whereas $[Ir(\mu\text{-}OMe)(COD)]_2/2AsPh_3$ afforded the 4-borylated product (Scheme 9) [25].

A combination of iridium catalyst and bulky aluminum-based Lewis acid catalysts has been designed to control the *para*-selectivity in C–H borylation of benzamides and pyridines (Scheme 10) [26]. The success of this strategy is based on the complexation of an heteroarene bearing a Lewis basic functionality with a Lewis acid (LA) resulting in charge transfer, making the heteroarene core more electron-deficient and thus more reactive. But also steric repulsion between a ligand on iridium and LA would block the *ortho*- and *meta*-positions and force the C–H borylation to proceed selectively at the *para*-position.

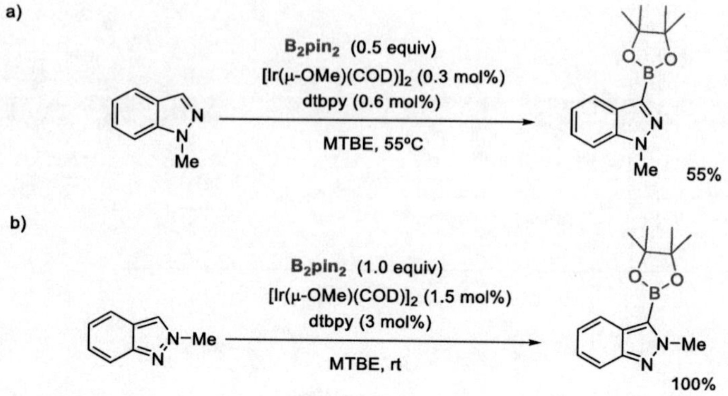

Scheme 8 Iridium-catalyzed C–H borylation of (**a**) N1-protected indazoles and (**b**) N2-protected indazoles

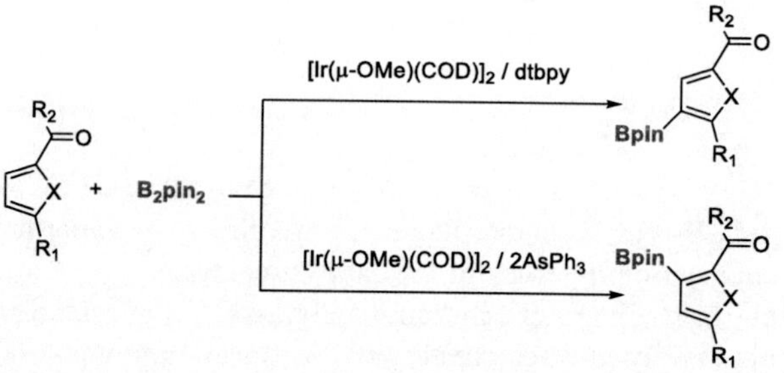

Scheme 9 Iridium-catalyzed regiodivergent C–H borylation of heteroarenes

3 C–H Borylation of Arenes

The advantage of C–H borylation of arenes is highlighted by its ability to produce highly versatile aryl organoboronate ester intermediates from abundant substrates without the need for reactive groups, such as halides or sulfonates.

The origins of regioselectivity in iridium-catalyzed C–H borylation reaction of substituted benzene rings have been studied by Merlic, Houk, and co-workers [27]. Distortion/interaction analyses show that regioselectivity is determined primarily by the interaction energy between the iridium catalyst and the substrate in the oxidative addition transition state. As a result, Ir–C bond energies in the aryl iridium hydride intermediates or in the C–H oxidative addition transition states both correlate well with the activation energy and can be used to determine the regioselectivity of the C–H activation reaction. Musaev, Itami, and co-workers used DFT calculations to elucidate the mechanism and source of regioselectivity for Ir-catalyzed *para*-selective C–H borylation with the bulky Xyl-MeO-BIPHEP((6,6′-Dimethoxybiphenyl-2,2′-diyl)bis[bis(3,5-dimethylphenyl)phosphine]) diphosphine ligand. They found that the bulky diphosphine–Ir complex forms a flexible reaction pocket that roughly

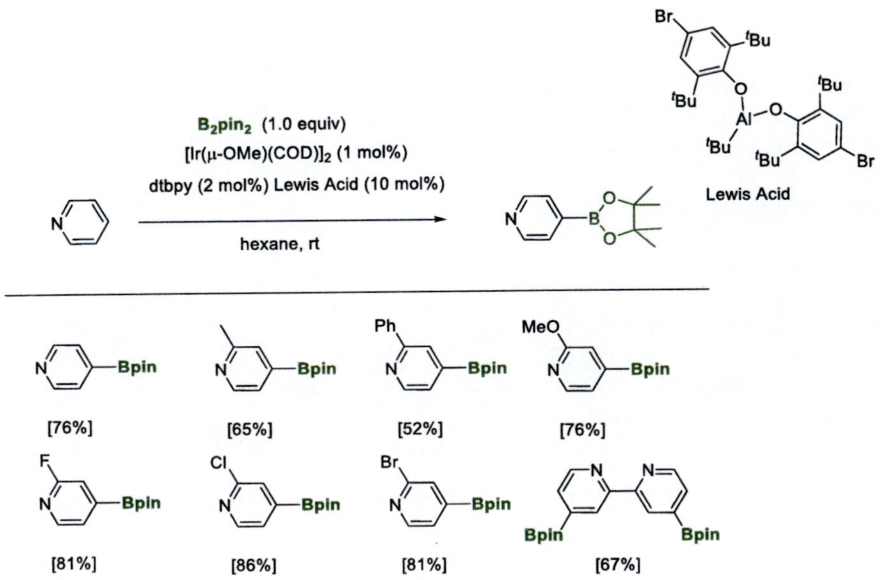

Scheme 10 Iridium-catalyzed selective *para*-C–H borylation of pyridines with the aid of a bulky Lewis acid

mimics the role of an enzyme active site by modulating access of the substrate to the Ir center. It is shown that the regioselectivity of the reaction arises from a balance of the attractive and repulsive interactions between the substrate and ligand and their corresponding entropic penalties across the high-energy C–H activation and C–B bond formation transition states in the reaction pocket [28]. Lan, Bai, and co-workers have proposed a novel iridium(I)/iridium(III)-based catalytic cycle for this transformation induced by these steric effects including (1) the oxidative addition of the C–H bond of the substrate to an active iridium(I) boryl complex; (2) the reductive elimination of a C–B bond; (3) the oxidative addition of B_2pin_2 to an iridium(I) hydride complex; and (4) the reductive elimination of a B–H bond (Scheme 11) [29]. The high *para*-selectivity of this reaction was also explained using structural analysis and a 2D contour model, which revealed that the strong steric repulsion between the diphosphine ligand and the *meta*-substituents resulted in a higher-energy barrier for *meta*-C–H activation.

Maseras and Jover have studied with DFT calculations the Ir-catalyzed C–H borylation of methyl benzoate in order to understand the experimentally observed ligand-induced regioselectivity and activity when different [Ir(ligand)(Bpin)$_3$] catalysts are employed [30]. They observed that while bidentate ligands such dtbpy completely inhibit *ortho*-borylation, the use of selected triphenylphosphine derivatives enables the reaction on that position, avoiding the *meta*- and *para*-regioisomers. The analysis of the catalytic cycles for the borylation reactions with dtbpy, PPh_3, $P(p\text{-}CF_3C_6H_4)_3$, and $P(m,m\text{-}(CF_3)_2C_6H_3)_3$ can also be rationalized in terms of catalyst stability. The stability of the iridium(V) intermediates toward ligand

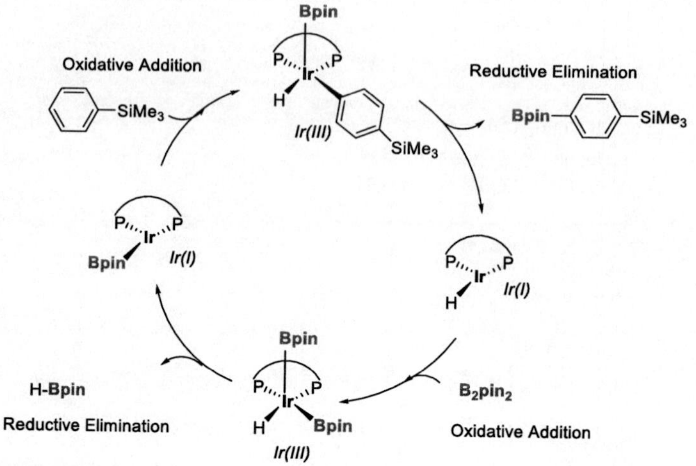

Scheme 11 Plausible iridium(I)/iridium(III) catalytic cycle for the C–H selective *para* borylation of trimethyl(phenyl)silane

dissociation plays a crucial role in the reaction, since PPh_3 and $P(p\text{-}CF_3C_6H_4)_3$ show higher ligand ability.

With this trend in mind, Kanai, Kuninobu, and co-workers have recently found that the introduction of an electron withdrawing substituent on the bipyridine-type ligand and a methylthiomethyl group on the hydroxy and amino groups of phenol and aniline substrates dramatically altered the regioselectivity, affording exclusively *ortho*-borylated products, using B_2pin_2 and HBpin as the boron source (Scheme 12) [31].

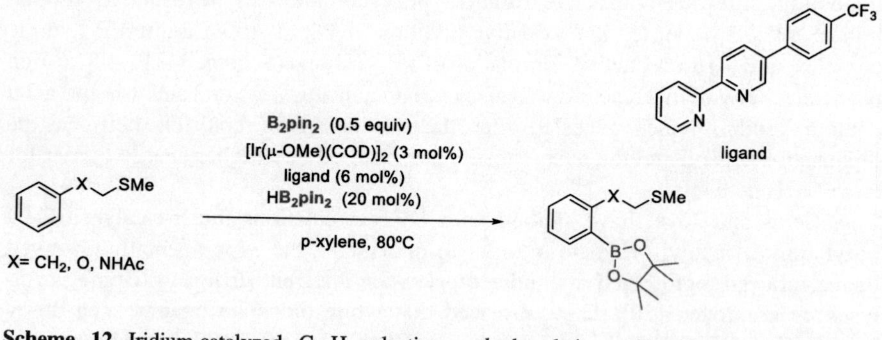

Scheme 12 Iridium-catalyzed C–H selective *ortho*-borylation with electron withdrawing bipyridine-type ligand

4 C–H Borylation of Alkanes

New methodologies involving iridium-catalyzed C–H activation of alkanes have been mostly addressed under directed iridium-catalyzed protocols to terminal C (sp^3)–H bonds and in less extension to secondary or tertiary $C(sp^3)$–H bonds, particularly [32, 33].

Undirected borylation of methylarenes has been efficiently developed by Hartwig and co-workers using an iridium complex modified with an electron-deficient phenanthroline as ligand and silylborane as reagent. This catalytic system allows the selective formation of benzylic boronate esters over the corresponding aryl boronate esters (Scheme 13) [34]. The authors isolated an iridium diboryl monosilyl complex with the phenanthroline coordinated, suggesting that it might be the resting state. That species is more electron-deficient than the analogue trisboryl complex previously accepted on mechanistic studies for iridium-catalyzed C–H borylation, and consequently, the reduced electron density at the metal center reduces the rate of the aryl C–H borylation versus the rate of the benzylic C–H borylation (Scheme 14).

The same authors have identified a notably influence on the regioselectivity of the iridium-catalyzed C–H borylation of alkylamines and alkyl ethers with a preference at the beta position to oxygen or nitrogen with respect to other aliphatic C–H bonds (Scheme 15) [35]. Experimental studies and computational results show that C–H bond cleavage becomes the rate-determining step where the substate participates in a

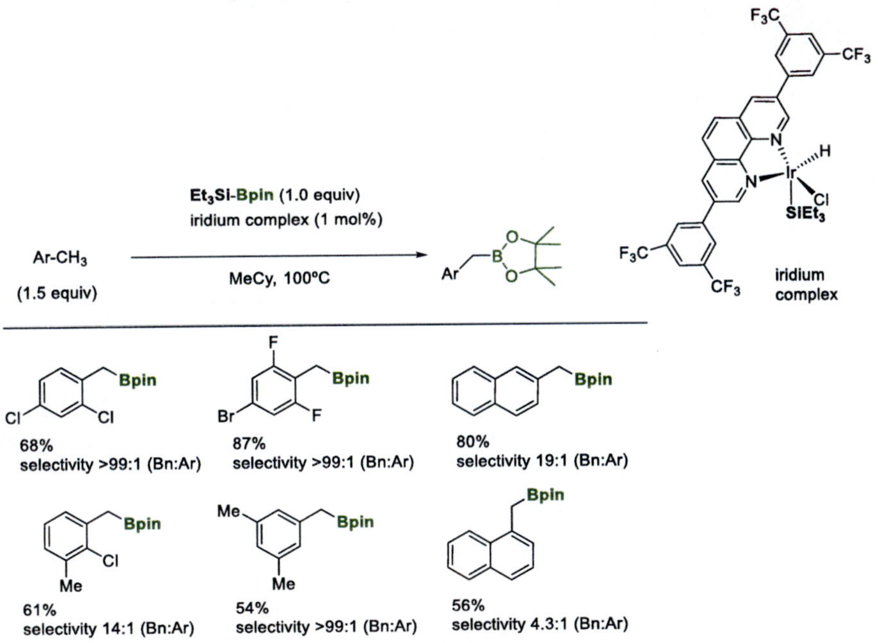

Scheme 13 Iridium-catalyzed C–H borylation of methylarenes

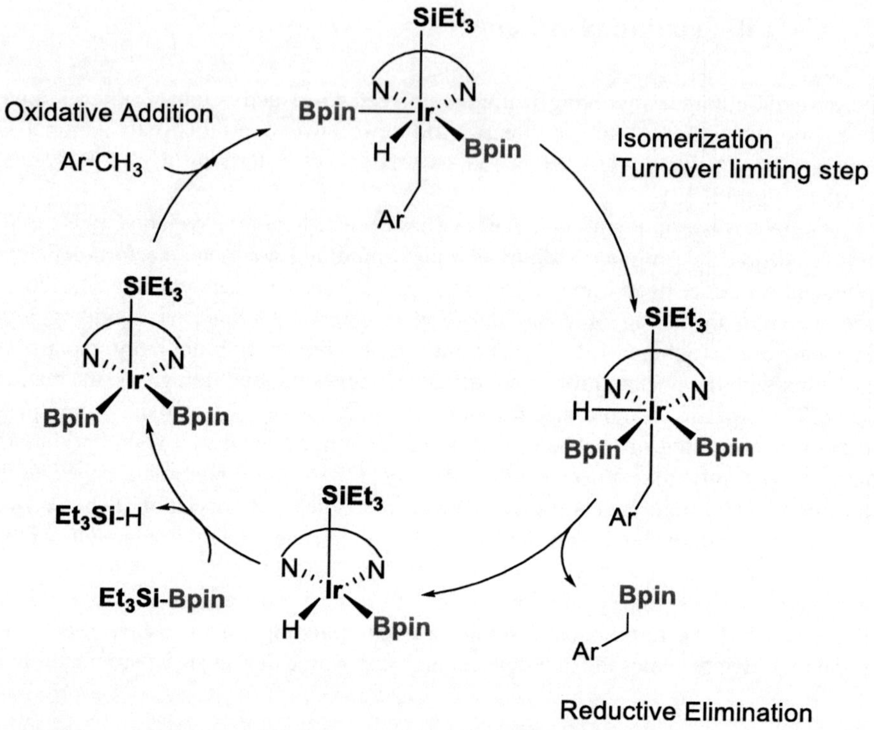

Scheme 14 Suggested iridium-catalyzed C–H borylation of methylarenes

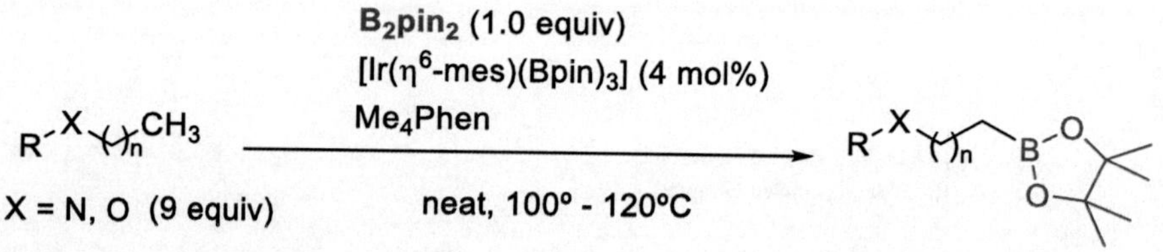

Scheme 15 Iridium-catalyzed regioselective C–H borylation of primary C–H bonds beta to nitrogen and oxygen

set of weak interactions leading to a significant stabilization of the transition state responsible for the major product formation. However, the iridium-catalyzed alkane C–H borylation suffered from poor atom economy, resulting from both the inclusion of only one equiv. of boron from the diboron reagent and a requirement for neat substrate.

A rate acceleration on the iridium-catalyzed alkane C–H borylation, particularly on C(sp^3)–H bonds located on the methyl groups of isopropyl groups, has been promoted by the addition of a catalytic amount of t-BuOK (Scheme 16) [36]. Although no mechanistic explanation was given for this observation, authors

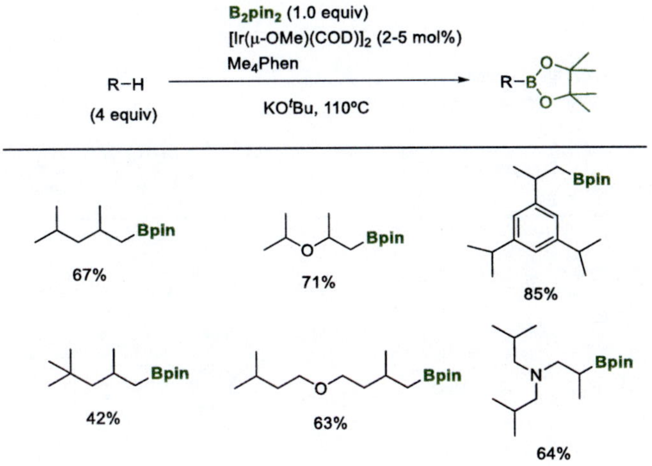

Scheme 16 Iridium-catalyzed C–H borylation of C(sp^3)–H bonds in the presence of a catalytic amount of t-BuOK

noted that the effect was highly sensitive to the amount of t-BuOK. This acceleration was not observed with 12 or 8 mol% catalyst loading (Ir: t-BuOK = 1: 1.5–1), whereas efficient borylation took place in the presence of less than 6 mol% catalyst loading (Ir: t-BuOK = 1: 0.75–0.125). It is also worthy to mention that Me_4phen resulted an essential ligand for the t-BuOK-accelerated C(sp^3)–H borylation since other phenanthroline ligands were less active.

Hartwig and co-workers have explored the influence of phenanthroline-type ligands in the iridium-catalyzed C–H borylation of alkanes [18]. They found that the donor ability of phenanthroline-type ligands correlated positively with the rate of C–H borylation catalyzed by the complexes containing those ligands. However, ligands possessing similar donor properties might suffer different interactions between the phenanthroline ligand and the boryl ligands attached to Ir in the transition state for C–H oxidative addition, accounting for significant differences in the activity of the catalysts. Remarkably, the effect of these interactions on the borylation of secondary alkyl C–H bonds is larger than it is on the borylation of primary alkyl C–H bonds.

More recently Schley and co-workers have developed an appropriately substituted dipyridylarylmethane ligand that efficiently modifies iridium complex providing a highly active alkane borylation catalysts [37]. This system works until complete consumption of the diboron reagent, producing two molar equivalents of product at low catalyst loadings (Scheme 17). The superior efficacy of this system also enables borylation of unactivated alkanes in hydrocarbon solvent as well as borylation of substrates containing polar functionalities, which are unreactive toward C–H borylation under neat conditions.

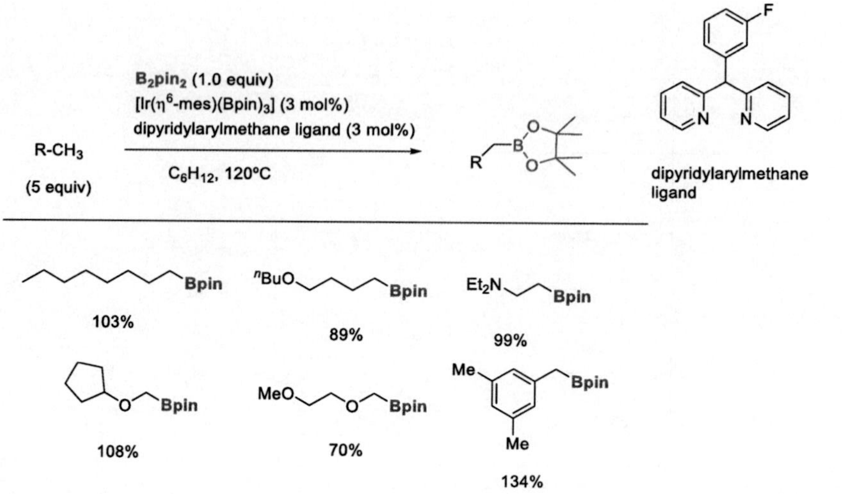

Scheme 17 Iridium-catalyzed C−H borylation of alkanes enabled by 2,2′-dipyridylarylmethane ligand

Enantioselective undirected C−H borylation reactions are much less developed than enantioselective directed C−H borylations, but not for lack of effort [38]. Bridging the gap between directed and completely undirected enantioselective alkane borylation, the exploitation of noncovalent interactions in the narrow chiral pocket around the iridium complex has been demonstrated to be useful. Sawamura and co-workers have developed a highly enantioselective borylation of unactivated methylene C(sp^3)−H bonds in 2-alkylpyridines and 2-alkyl-1,3-azole derivatives using an iridium-BINOL-based chiral monophosphite catalyst system [39]. The differentiation of the enantiotopic methylene C−H bonds is accomplished through an assembly of multiple noncovalent interactions.

Recently, Xu and co-workers have developed for the first time the use of a chiral bidentate boryl ligand to be efficiently employed in the Ir-catalyzed enantioselective C(sp^3)−H borylation of cyclopropanecarboxamides [40]. The reaction of this ligand with [Ir(μ-Cl)(COD)]$_2$ and B$_2$pin$_2$ tolerates a wide range of functional groups, providing a series of cyclopropylboronates in good yields with good to excellent enantioselectivities (Scheme 18).

Eventually, Mindiola and co-workers have suggested a rational design of a well-defined homogeneous and monomeric catalyst, [IrCl(dmpe)(COD)], (dmpe = 1,2-Bis(dimethylphosphino)ethane) that is readily available from commercial precursors, becoming very efficient in the borylation of methane with B$_2$pin$_2$ [41]. A new catalytic cycle has been suggested in basis to the lower energetic barrier of the oxidative addition to afford a seven-coordinate iridium(V) intermediate (Scheme 19). It has been suggested that pinacolborane, a side-product from methane borylation with bis(pinacolato)diboron, might inhibit the catalytic activity.

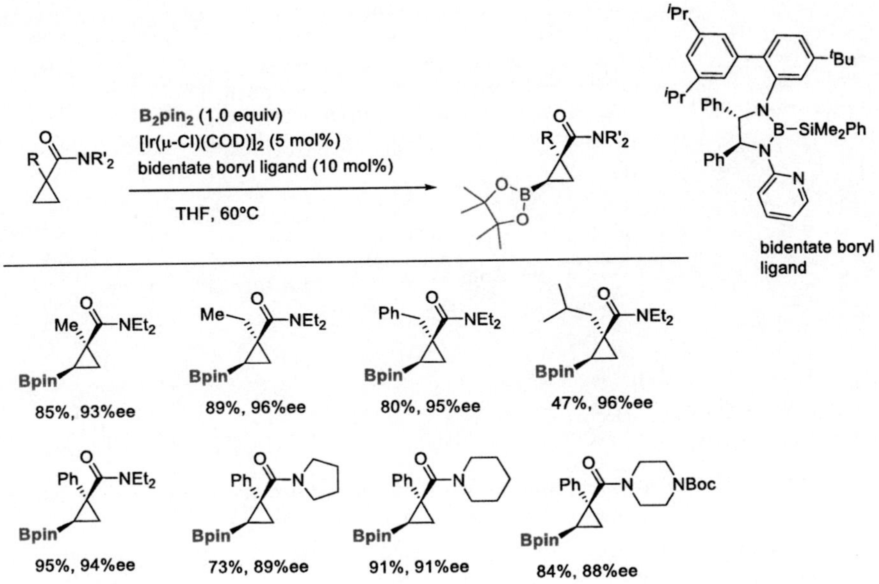

Scheme 18 Iridium-catalyzed enantioselective C–H borylation of cyclopropanecarboxamides

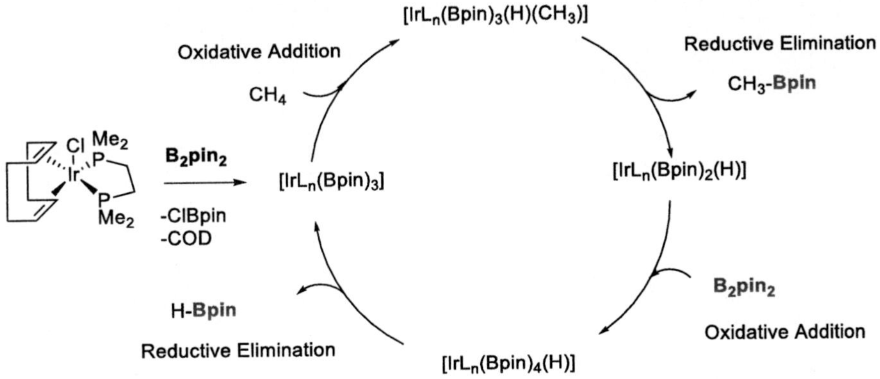

Scheme 19 C–H borylation of methane with [IrCl(dmpe)(COD)]

Sakaki and co-workers have recently studied the iridium-catalyzed sp^3 C–H borylation of THF using DFT [42]. They suggested that the active species is an iridium(III) trisboryl complex, and the catalytic cycle consists of the C–H activation through oxidative addition of the C–H bond to the iridium(III) atom and reductive elimination between boryl and oxolanyl groups, which is the same as that proposed for sp^2 C–H borylation of benzene. The reductive elimination of B–C bond is rate-determining step, and the β-regioselectivity is determined at this reductive

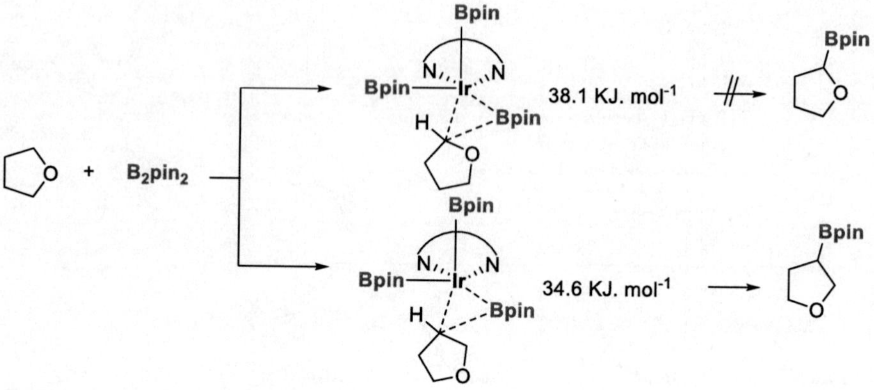

Scheme 20 Iridium beta-selective C–H borylation of THF

elimination. The reductive elimination in the β-regioselective borylation occurs through a TS lower in energy than that in the α-regioselective one (Scheme 20).

Himo and co-workers investigated the iridium-catalyzed C(sp^3)–H borylation of methylchlorosilanes through density functional theory [43]. The calculations established that the resting state of the catalyst might be a seven-coordinate Ir (V) species that has to be converted into an Ir(III)tris(boryl) complex in order to effect the oxidative addition of the C–H bond. This is then followed by a C–B reductive elimination to yield the borylated product, and the catalytic cycle is finally closed by the regeneration of the active catalyst over two steps. The calculations reproduce quite well the experimental trends in the reactivities of substrates with different substituents. It is demonstrated that the reactivity can be correlated to the Ir–C bond dissociation energies of the corresponding Ir(V) hydride intermediates. The effect of the chlorosilyl group is identified to originate from the α-carbanion-stabilizing effect of the silicon, which is further reinforced by the presence of an electron-withdrawing chlorine substituent (Scheme 21). Furthermore, the source of selectivity for the borylation of primary over secondary C(sp^3)–H can be explained by steric properties, in particular by repulsion between the alkyl group and the Ir/ligand moiety.

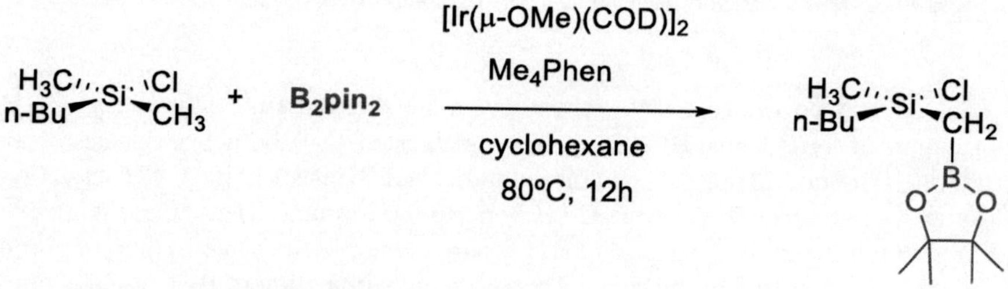

Scheme 21 Iridium-catalyzed C(sp^3)–H borylation of methylchlorosilanes

As concluding remarks it has to be said that iridium-catalyzed C–H borylation keeps being a source of inspiration for chemists around the world that search efficient and suitable methods to activate C(sp^2)–H and C(sp^3)–H bonds and convert them into versatile C–B bonds. [Ir(μ-OMe)(COD)]$_2$ can activate B_2pin_2 as well as HBpin, in the presence of 4,4′-di-tert-butyl bipyridine (dtbpy). But new challenges have been acquired in recent years by using alternative ligands such as 3,4,7,8-tetramethyl-1,10-phenanthroline (Me_4Phen) and 1,10-phenanthroline (phen), among others. The selective formation of the new C–B bond can be rationalized by a conjunction of steric and electronic issues. Essentially, the lack of directing groups increases the interest of the new catalytic systems able to afford high levels of chemo-, regio-, and stereoselectivity. The preference of C(sp^3)–H borylation versus C(sp^2)–H borylation opens an unprecedented selective functionalization of C–H bonds. In mostly of the new approaches, theoretical calculations have deeply studied all the mechanistic pathways to understand the rate-limiting steps and propose alternative catalytic cycles.

The future of the field might concern the development of improved catalytic systems to generalize even more the reaction trends, but the exploration of new reaction media to improve the sustainability of iridium C–H borylation remains crucial. Mechanochemistry has recently been applied for the first time to iridium(I)-catalyzed C–H borylation by the group of Ito and co-workers [44]. By using either none or just a catalytic amount of a liquid, the mechanochemical borylation can be performed in an efficient way. Therefore, the future might be orientated to determine a careful choice of the reaction milling, jar, ball, and additives that can enable the development of a solvent-free iridium(I)-catalyzed C–H borylation in air. Alternatively, the immobilization of iridium complexes, functionalized with bpy ligands, into oligomers has been shown by Madrahimov and co-workers to efficiently catalyze the C–H borylation of arenes with bis(pinacolato)diboron under mild conditions to produce a variety of arylboronate compounds [45]. The activity of this heterogenized catalyst is similar to that of an original non-recyclable catalyst which allows it to be used under milder conditions than other reported recyclable catalysts. The challenging oligomer that supports the Ir catalyst can be successfully recovered through biphasic extraction and reused for several cycles without a loss of activity.

Globally, the idea to borylate the most abundant yet inert bonds represents an exciting area of work. But if that strategy allows natural products to be created in a selective way, all the efforts are justified. Liu and Yuan have collected the most representative synthetic examples of high degree of molecular complexity from relatively simple building blocks via C–H borylation with iridium complexes [46]. The most significant approaches from 2014 are Baran's total synthesis of verruculogen and fumitremorgin A [21], Jia's total synthesis of (−)-goniomitine [47], Sperry's total synthesis of scalaridine A [48], or Shibata's total synthesis of cis-clavicipitic acid [49].

References

1. Mkhalid IAI, Barnard JH, Marder TB, Murphy JM, Hartwig JF (2010) Chem Rev 110:890
2. Hartwig JF (2011) Chem Soc Rev 40:1992
3. Hartwig JF (2011) Acc Chem Res 45:864
4. Hartwig JF, Larsen MA (2016) ACS Cent Sci 2:281
5. Xu L, Wang G, Zhang S, Wang H, Wang L, Liu L, Jiao J, Li P (2017) Tetrahedron 73:7123
6. Boller TM, Murphy JM, Hapke M, Ishiyama T, Miyaura N, Hartwig JF (2005) J Am Chem Soc 127:14263
7. Takagi J, Sato K, Hartwig JF, Ishiyama T, Miyaura N (2002) Tetrahedron Lett 43:5649
8. Ishiyama T, Takagi J, Yonekawa Y, Hartwig JF, Miyaura N (2003) Adv Synth Catal 345:1103
9. Ishiyama T, Nobuta Y, Hartwig JF, Miyaura N (2003) Chem Commun:2924
10. Ishiyama T, Miyaura N (2006) Pure Appl Chem 78:1369
11. Kallepalli VA, Shi F, Paul S, Onyeozili EN, Maleczka RE, Smith MR (2009) J Org Chem 74:9199
12. Preshlock SM, Plattner DL, Maligres PE, Krska SW, Maleczka RE, Smith MR (2013) Angew Chem Int Ed 52:12915
13. Klecka M, Pohl R, Klepetarova B, Hocek M (2009) Org Biomol Chem 7:866
14. Mkhalid IAI, Coventry DN, Albesa-Jove D, Batsanov AS, Howard JAK, Perutz RN, Marder TB (2006) Angew Chem Int Ed 45:489
15. Harrisson P, Morris J, Marder TB, Steel PG (2009) Org Lett 11:3586
16. Klecka M, Postova Slavetínska L, Hocek M (2015) Eur J Org Chem:7943
17. Oeschger RJ, Larsen MA, Bismuto A, Hartwig JF (2019) J Am Chem Soc 141:16479
18. Larsen MA, Oeschger RJ, Hartwig JF (2020) ACS Catal 10:3415
19. Larsen MA, Hartwig JF (2014) J Am Chem Soc 136:4287
20. Seechurn CCCJ, Sivakumar V, Satoskar D, Colacot TJ (2014) Organometallics 33:3514
21. Feng Y, Holte D, Zoller J, Umemiya S, Simke LR, Baran PS (2015) J Am Chem Soc 137:10160
22. Wang G, Xu L, Li P (2015) J Am Chem Soc 137:8058
23. Egan BA, Burton PM (2014) RSC Adv 4:27726
24. Sadler SA, Hones AC, Roberts B, Blakemore D, Marder TB, Steel PG (2015) J Org Chem 80:5308
25. Sakaki I, Taguchi J, Hiraki S, Ito H, Ishiyama T (2015) Chem Eur J 21:9236
26. Yang L, Semba K, Nakao Y (2017) Angew Chem Int Ed 18:4853
27. Green AG, Liu P, Merlic CA, Houk KN (2014) J Am Chem Soc 136:4575
28. Haines BE, Saito Y, Segawa Y, Itami K, Musaev DG (2016) ACS Catal 6:7536
29. Zhu L, Qi X, Li Y, Duan M, Zou L, Bai R, Lan Y (2017) Organometallics 36:2107
30. Jover J, Maseras F (2016) Organometallics 35:3221
31. Li HL, Kanai M, Kuninobu Y (2017) Org Lett 19:5944
32. Kawamorita S, Murakami R, Iwai T, Sawamura M (2013) J Am Chem Soc 135:2947
33. Mita T, Ikeda Y, Michigami K, Sato Y (2013) Chem Commun 49:5601
34. Larsen MA, Wilson CV, Hartwig JF (2015) J Am Chem Soc 137:8633
35. Li Q, Liskey CW, Hartwig JF (2014) J Am Chem Soc 136:8755
36. Ohmura T, Torigoe T, Suginome M (2014) Chem Commun 50:6333
37. Jones MR, Fast CD, Schley ND (2020) J Am Chem Soc 142:6488
38. Reyes RL, Harada T, Taniguchi T, Monde K, Iwai T, Sawamura M (2017) Chem Lett 46:1747
39. Reyes RL, Iwai T, Maeda S, Sawamura M (2019) J Am Chem Soc 141:6817
40. Shi Y, Gao Q, Xu S (2019) J Am Chem Soc 141:10599
41. Ahn S, Sorche D, Berritt S, Gau MR, Mindiola DJ, Baik MH (2018) ACS Catal 8:10021
42. Zhong RL, Sakaki S (2019) J Am Chem Soc 141:9854
43. Huang G, Kalek M, Liao RZ, Himo F (2015) Chem Sci 6:1735
44. Pang Y, Ishiyama T, Kubota K, Ito H (2019) Chem Eur J 25:4654
45. Mamlouk H, Suriboot J, Manyam PK, Alyazidi A, Bergbreiter DE, Madrahimov ST (2018) Catal Sci Technol 8:124

46. Yuan CH, Liu B (2017) Org Chem Front 5:106
47. Zhou S, Jia Y (2014) Org Lett 16:3416
48. Kim H, Sperry J (2015) Tetrahedron Lett 56:5914
49. Tahara YK, Ito M, Kanyiva KS, Shibata T (2015) Chem Eur J 21:11340

Top Organomet Chem (2021) 69: 227–270
https://doi.org/10.1007/3418_2020_55

Published online: 8 August 2020

Iridium-Catalyzed Silylation

Manuel Iglesias and Luis A. Oro

Contents

Abstract In this chapter, homogeneous iridium-catalyzed silylation reactions are reviewed, focusing primarily on their synthetic utility. Additionally, relevant catalytic cycles are commented, paying especial attention to those that are more representative of each type of process. The chapter is divided into two main types of reactions, namely, hydrosilylation and C–H bond silylation. The former deals with the hydrosilylation of polar unsaturated bonds (ketones and imines) and non-polar unsaturated bonds (alkenes and alkynes). The latter covers the directed and non-directed C–H bond silylation of alkenes, alkynes, arenes, and alkanes – mainly comprising dehydrogenative silylation reactions, which may occur in the presence or absence of a hydrogen acceptor.

The original version of this chapter was revised. A correction to this chapter can be found at https://doi.org/10.1007/978-3-030-69083-0_74

M. Iglesias (✉) and L. A. Oro (✉)
Department of Inorganic Chemistry, ISQCH, University of Zaragoza-CSIC, Zaragoza, Spain
e-mail: miglesia@unizar.es; oro@unizar.es

Keywords Catalysis · C–H bond functionalization · Dehydrogenative silylation · Hydrosilylation · Iridium

1 Introduction

In the past few decades, the synthesis of organosilanes by catalytic methods has undergone an outstanding development propelled by the potential use of these molecules as materials and convenient building blocks in organic synthesis. The preparation of compounds that contain C–Si bonds by means of catalytic processes is a sustainable alternative to conventional stoichiometric reactions. In addition, the development of new catalysts and synthetic methodologies might open the door to organosilanes hitherto unattainable by current procedures.

Organosilicon compounds are interesting intermediates for the preparation of more intricate molecular architectures because of their stability, low toxicity, availability, and versatile reactivity [1–7]. C–Si bonds can be converted into a great variety of new bonds that include C–C bonds (Hiyama-Denmark coupling [8]), C–O bonds (Tamao-Fleming oxidation [9]), C–N bonds (amination [10]), or C–halogen bonds (halogenation [11, 12]). Moreover, organosilicon compounds have found application, for example, as OLEDs (organic light-emitting diodes) or as conducting polymers [13–20]. On the other hand, organic molecules that contain Si–heteroatom bonds can be prepared by hydrosilane-mediated reductions. The most synthetically relevant catalytic process for the formation of these molecules is the hydrosilylation of ketones or amines. The dehydrogenative silylation of alcohols or amines and the reduction of ethers, esters, or amides with hydrosilanes are also interesting reactions that give rise to, for example, N-, O-, and S-silylated compounds. A related hydrosilane-mediated reduction is the dehalogenation of organohalides using hydrosilanes as reducing agent; however, in this case, it is the hydrogen atom of the hydrosilane – not the silicon – the one that is transferred to the substrate.

Homogeneous iridium catalysts have shown excellent activities in the synthesis of fine chemicals as well as in large-scale industrial process [21]. Regarding the former, organometallic iridium complexes have found remarkable success as catalysts in the hydrogenation of unsaturated bonds. Outstanding results have been reported on the asymmetric hydrogenation of imines and alkenes, with the catalysts developed by Pfaltz and co-workers being the most prominent examples [22–24]. Moreover, Crabtree's catalyst, i.e., $[Ir(cod)(PCy_3)(py)]PF_6$ (cod = 1,5-cyclooctadiene), is among the most active catalysts for the challenging hydrogenation of encumbered olefins [25, 26]. Large-scale industrial processes also have prominent examples such as the Cativa™ process for the manufacture of acetic acid, by which $[Ir(CO)_2I_2]^-$ catalyzes the carbonylation of methanol [27], or the synthesis of (*S*)-metolachlor, which requires the asymmetric hydrogenation of a C=N bond – a crucial reaction catalyzed by an Ir-xyliphos complex [28–31].

In this chapter, we intend to compile the most relevant iridium-catalyzed homogeneous processes for the silylation of organic molecules hitherto reported. Undoubtedly, the sheer number of hydrosilane-mediated reductions so far disclosed makes it impossible to deal with all of them in this revision. For this reason, we will

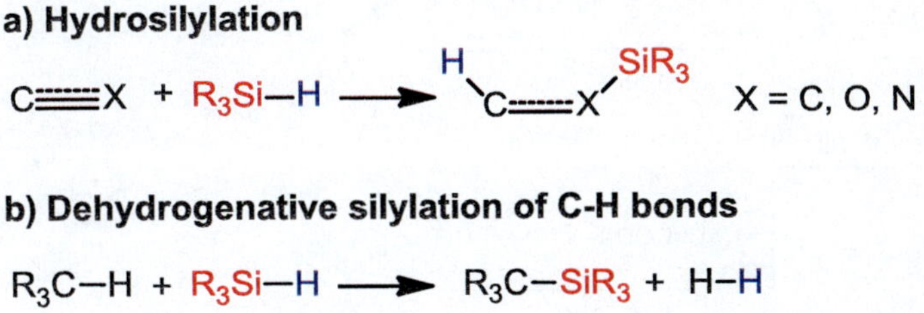

Scheme 1 Generic depiction of hydrosilylation and dehydrogenative silylation reactions

focus on the most relevant silylation processes from a synthetic viewpoint, which, in our opinion, are the hydrosilylation of unsaturated bonds (e.g., alkenes, alkynes, ketones, or imines; Scheme 1a) and the silylation of C–H bonds (e.g., alkanes or arenes; Scheme 1b).

We also aim to transmit to the reader an understanding of the activity-structure patterns observed for the catalytic systems discussed herein. For this purpose, we will discuss some of the most relevant reaction mechanisms, delivering an analysis of the role played by the metal center and the ligand system.

2 Hydrosilylation Reactions

The hydrosilylation reaction is formally the addition of a Si–H bond across a multiple bond. The electronegativity of the atoms that form the multiple bond is crucial to determine the reactivity, the selectivity, and even the mechanism of the reaction. In this regard, two main types of unsaturated substrates susceptible of undergoing hydrosilylation may be defined according to the polarity of the multiple bond: polar (carbon-heteroatom multiple bond, e.g., ketones or imines) and non-polar (carbon-carbon multiple bond, e.g., alkenes or alkynes). It is noteworthy that, in the case of the former, O–Si and N–Si bonds are formed, while C–Si bonds are generated in the case of the latter.

2.1 Hydrosilylation of Ketones

The hydrosilylation of ketones gives silyl ethers, which can be converted straightforwardly into alcohols. It is worth highlighting here that the substrate is often prochiral in these reactions and, consequently, the use of a suitable chiral catalyst may allow the enantioselective synthesis of secondary alcohols.

The first reports on Ir catalysts for the hydrosilylation of ketones appeared at the end of the twentieth century [32–36], slightly later than their successful Rh analogues [37]. The hydrosilylation of ketones can be achieved by using [Ir(μ-Cl)(COD)]$_2$ as catalyst, although in low yields. The addition of 1 equivalent of PPh_3 improves the catalytic activity, probably due to an increased stability of the catalyst,

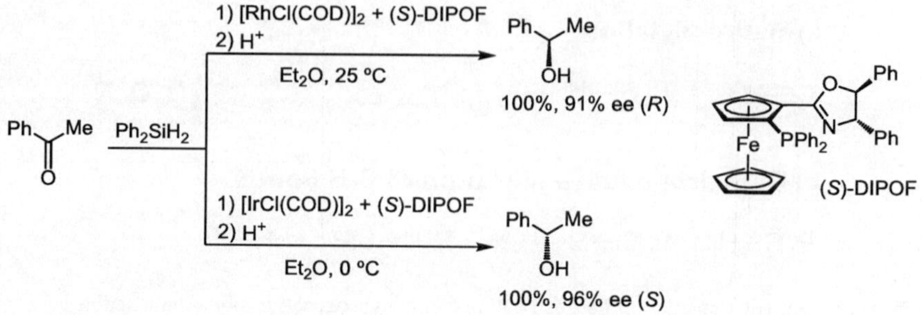

Scheme 2 Asymmetric hydrogenation of ketones with (S)-DIPOF and [Ir(μ-Cl)(COD)]$_2$ or [Rh(μ-Cl)(COD)]$_2$

which prevents its decomposition to iridium black. However, addition of another equivalent of PPh_3 leads to catalyst deactivation [36, 38]. This is due to the fact that the Ir(III) species obtained upon oxidative addition of the hydrosilane is overstable and precludes the catalytic cycle, which is in sharp contrast with the behavior of related rhodium catalysts. However, it must be noted that other Ir(III) complexes bearing phosphines have shown catalytic activity in the hydrosilylation of ketones [39].

A crucial breakthrough in the development of Ir hydrosilylation catalysts came with the use of the chiral ligand (*S*)-DIPOF ((*S*,*S*,*S*)-[2-(4,5-diphenyl-4,5-dihydro-1,3-oxazol-2-yl)ferrocenyl]diphenylphosphine), which had been used previously with excellent results in the Rh-catalyzed asymmetric hydrosilylation of ketones [32]. Remarkably, a previous report on the use of a chiral phosphine ((*S*)-amphos) with [Ir(μ-Cl)(COD)]$_2$ afforded the alcohol with opposite absolute configuration when compared to [Rh(μ-Cl)(COD)]$_2$, under the same reaction conditions. This was the first example of a transition-metal-controlled asymmetric hydrosilylation, although it must be mentioned that a low enantiomeric excess (ee) was obtained [35]. Shortly after, the use of (*S*)-DIPOF by the group of Uemura brought about a significant increase of the ee's (Scheme 2) [32, 35]. This methodology was proved effective for a variety of alkyl aromatic ketones, and, what is more, the use of (*R*)-DIPOF yielded the alcohols with inverse absolute configuration for both, Rh and Ir, affording the (*S*)- and (*R*)-alcohols, respectively. It is noteworthy that no mechanistic investigation that explains the inversion of absolute configuration observed upon changing from Rh to Ir has been reported so far.

Bis(oxazolines) were also explored as chiral ligands in the Ir-catalyzed hydrosilylation of ketones (Fig. 1), although less successfully than DIPOF [40]. The hydrosilylation of acetophenone was performed in situ employing [Ir(μ-Cl)(COD)]$_2$ and ligands **1** and **2** (other ligands were tested unsuccessfully) with a catalyst/substrate ratio of 1/100 in the presence of excess Ph_2SiH_2. Good yields (ca. 90%) were reported for both ligands, but asymmetric induction was only achieved with **1** (50% ee). The use of [Rh(μ-Cl)(COD)]$_2$ as metal precursor under analogous conditions gave rise to significantly lower conversions (below 50%) and selectivities (33% and 0% ee with **1** and **2**, respectively).

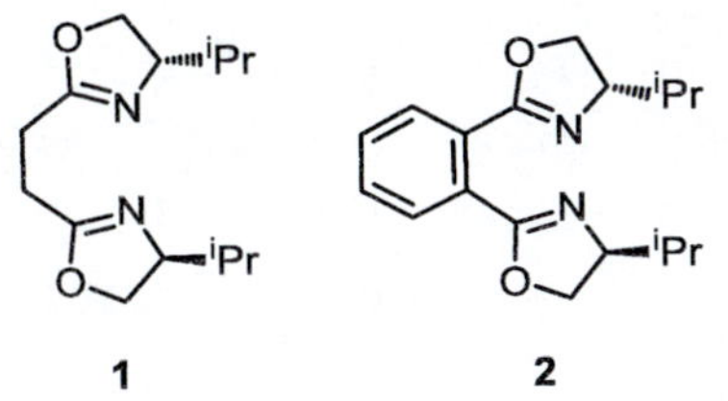

Fig. 1 Depiction of chiral (*S*,*S*)-bis(oxazolines) **1** and **2**

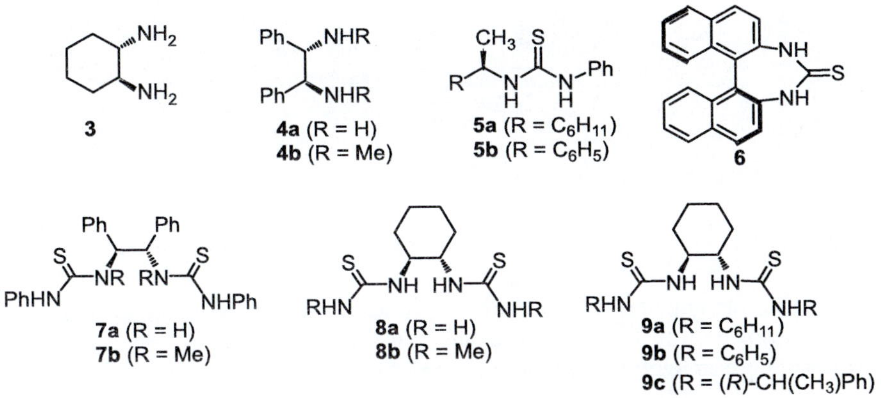

Fig. 2 Depiction of the diamine, thiourea, and dithiourea ligands tested by Lemaire et al.

A variety of chiral diamine, thiourea, and dithiourea ligands were studied in the hydrosilylation of ketones with Ph_2SiH_2 (Fig. 2) [41, 42]. Diamine ligands **3** and **4** bring about excellent conversions for the hydrosilylation of acetophenone, but enantiomeric excesses lower than 13% in all cases. Similar results were observed for the thioureas **5** and **6**, with the ee's reaching 25% with a 6/1 **6**/Ir ratio.

The best selectivities were achieved with dithiourea ligands, with ee's close to 60% for **9a** under conditions analogous to those used for diamine and thiourea ligands. In this work [41, 42], the authors also describe an elegant experiment where they evaluate the influence of the **9a**/Ir ratio on the conversion and selectivity of the reaction. On varying this ratio from 2 to 10, the authors observe a progressive increase of the selectivity and a decrease of the conversion, obtaining ee's over 70% with a tenfold excess of ligand. However, under these conditions, a low conversion (30%) is obtained.

The work on phosphine-oxazoline (PHOX) ligands was further expanded in 2007 by Molberg's group [43]. A range of PHOX ligands with different substitution at the oxazoline were evaluated (Fig. 3). Quantitative or almost quantitative yields were observed using ligands **10–12** with $[M(\mu\text{-Cl})(COD)]_2$ (M = Ir or Rh) and 1.2 equivalents of diphenyl silane. The best enantiomeric excesses were obtained with ligand **11a** and $[Rh(\mu\text{-Cl})(COD)]_2$ as metal precursor at 0 °C, reaching a 95%. The use of $[Ir(\mu\text{-Cl})(COD)]_2$ resulted in significantly lower ee's and the formation of alcohols with opposite absolute configuration; however, the best ee's were still

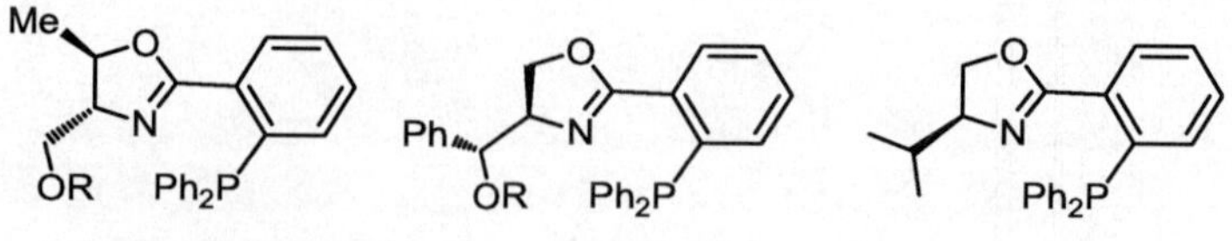

10a (R = H)
10b (R = Me)

11a (R = H)
11b (R = Me)

12

Fig. 3 Depiction of phosphine-oxazoline ligands **10–12**

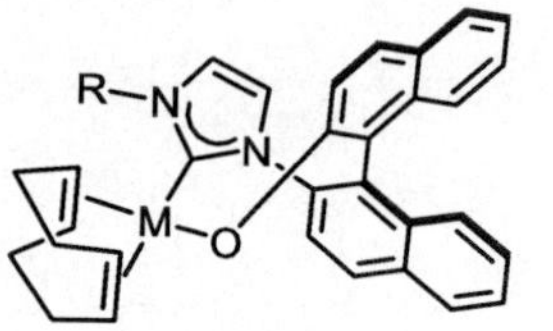

13a (M= Rh; R = ^{i}Pr)
13b (M= Rh; R = Bn)
14a (M= Ir; R = ^{i}Pr)
14b (M= Ir; R = Bn)

Fig. 4 Rh and Ir complexes featuring an NHC-naphthoxy ligand

obtained with **11a**, reaching almost 80%. The authors attributed the good performance of ligand **11a** to the presence of an OH moiety that can coordinate to the metal center. The addition of $AgBF_4$ improved significantly the catalytic activity, plausibly due to abstraction of the chloride ligand and formation of a cationic species that allows a more straightforward coordination of the OH.

The functionalization of the (*S*)-binaphthyl scaffold with NHC (*N*-heterocyclic carbenes) ligands led to the preparation of NHC-naphthoxy complexes of Rh and Ir (**13** and **14**, respectively; Fig. 4). These complexes show excellent activities in the hydrosilylation of acetophenone with diphenylsilane, allowing conversions over 98%. The reaction times required to achieve these conversions were significantly lower for the Rh complexes. Moreover, considerably high amounts of a by-product that arises from the dehydrogenative silylation of the enol of acetophenone was observed for the iridium catalysts (15% for Ir and 3% for Rh). However, the asymmetric induction was much higher in the case of the Ir complexes (12 and 13% ee's for **13a** and **13b**, respectively; 50 and 60% ee's for **14a** and **14b**, respectively). Analogous to previous examples, the absolute configuration of the product obtained with Ir catalysts was inverted compared to their Rh counterparts [44].

A tetradentate biaryl bis-NHC system (**15**) based on the (*S*)-2,2′-diamino-6,6′-dimethyl-1,1′-biphenyl scaffold was also studied in the hydrosilylation of acetophenone derivatives (Fig. 5) [45]. The reaction was performed in CH_2Cl_2 at 20 °C with Ph_2SiH_2 as silicon source and a 1 mol% loading of **15**, giving rise to excellent yields and enantiomeric excesses that range from 54 to 73% of the (*S*)-alcohols after hydrolysis.

The use of iridium complexes based on monodentate NHC ligands with a chiral center located at one of the wingtip groups (Fig. 6) brings about interesting results in the hydrosilylation of a variety of ketones [46–49]. The use of $(EtO)_2MeSiH$ in the presence of catalytic amounts of $AgBF_4$, or the generation of the catalyst in situ from $[Ir(COD)_2]BF_4$, affords excellent yields (up to 99%) and enantiomeric excesses

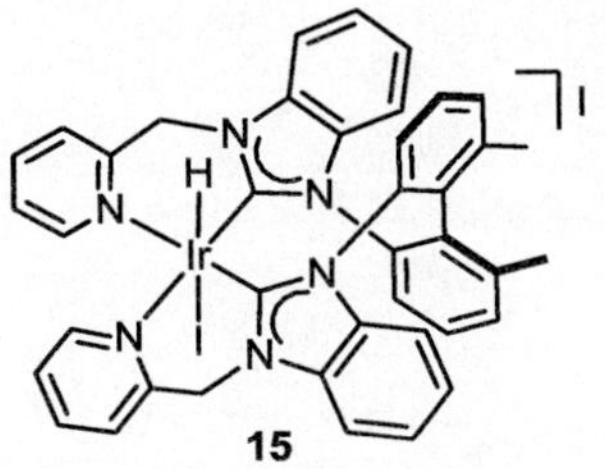

Fig. 5 Depiction of biaryl bis-NHC complex **15**

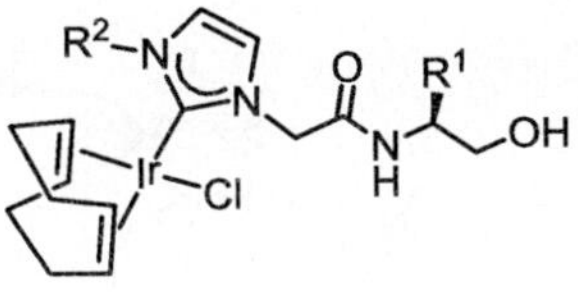

16a (R^1 = Bn; R^2 = Bn)
16b (R^1 = ^{i}Bu; R^2 = Bn)
16c (R^1 = Me; R^2 = Bn)
16d (R^1 = Et; R^2 = Bn)
16e (R^1 = ^{t}Bu; R^2 = Bn)
16f (R^1 = ^{s}Bu; R^2 = Bn)
16g (R^1 = ^{i}Pr; R^2 = Bn)

Fig. 6 Depiction of monodentate Ir-NHC catalysts **16a-g**

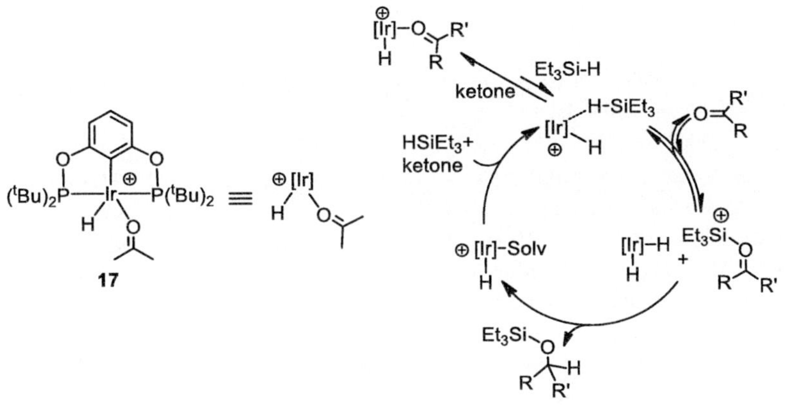

Scheme 3 Ionic mechanism postulated by Brookhart et al. for the hydrosilylation of ketones using catalyst **17**

(up to 96%) at room temperature, with **16b** being the best performing catalyst. The need for the silver additive to achieve asymmetric induction was attributed to the fact that chloride abstraction permits the coordination of the wingtip group that contains the chiral center. Other noteworthy Ir catalysts based on non-chiral NHC ligands that have shown interesting activities in the hydrosilylation of ketones have also been reported [50, 51].

In 2010, Brookhart et al. reported an ionic mechanism for the hydrosilylation of ketones employing the POCOP-Ir catalyst **17** (Scheme 3) [52]. This catalyst proved very active for a wide variety of ketones using a monohydrosilane (Et_3SiH), this being the first example where a dihydrosilane is not required to achieve good activities with an iridium catalyst in the hydrosilylation of ketones. The reaction

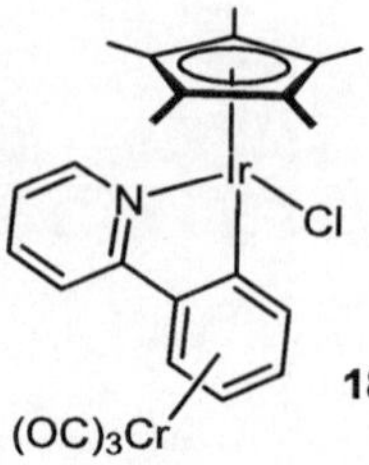

18

Fig. 7 Depiction of Ir(III)-metallacycle complex **18**

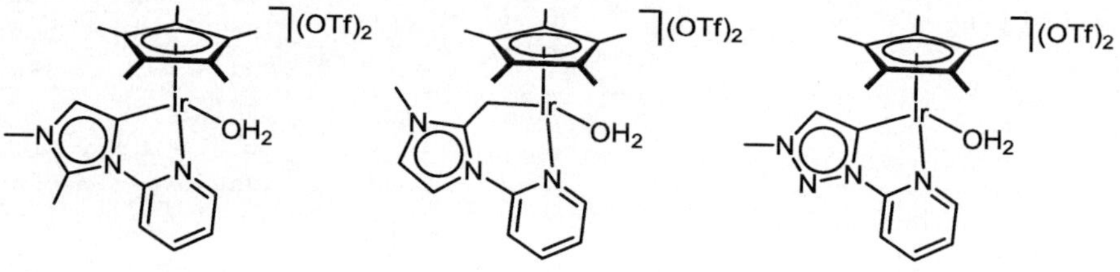

19 20 21

Fig. 8 Depiction of catalysts **19**, **20**, and **21** featuring an imidazolylidene, an *N*-heterocyclic olefin, and a triazolylidene ligand, respectively

mechanism entails essentially the heterolytic splitting of the Si–H bond via an end-on hydrosilane species. The electrophilic iridium center abstracts a hydride, while the ketone stabilizes a R_3Si^+ cation by interaction with the Lewis basic oxygen atom. Finally, the oxocarbenium ion thus formed abstracts one of the hydrides from the dihydride complex to yield the hydrosilylated product. This mechanism was later studied in more detail by Oestreich and co-workers [53]. The discussion of this mechanism, however, exceeds the scope of this chapter, and, for further insights into ionic mechanisms, we refer the reader to recent reviews [54, 55].

The iridium(III) metallacycle **18**, in the presence of sodium tetrakis [(3,5-trifluoromethyl)phenyl]borate ($NaBArF_{24}$), allows the hydrosilylation of a wide variety of ketones and aldehydes in excellent yields under mild conditions (Fig. 7). The authors proposed an ionic mechanism similar to that described above, where the active species is generated by abstraction of the chloride ligand in **18** by $NaBArF_{24}$ [56]. This catalyst is also able to efficiently catalyze the reduction of imines, esters, carboxylic acids, and amides with hydrosilanes.

In 2014, Albrecht et al. reported a comparative study on the activity of complexes **19**–**21** in the hydrosilylation of ketones with Et_3SiH (Fig. 8) [57]. The three catalysts, especially the one featuring the imidazolylidene **19**, were active under low catalyst loadings (<0.1 mol%). In a recent publication [58], Albrecht and co-workers showed that when Ph_2SiH_2 was employed as reducing agent instead of Et_3SiH, catalysts **19** and **20** still afforded selectively the expected silyl ether; however, catalyst **19**, which contains a triazolylidene ligand, underwent the reductive condensation reaction to render dialkyl ethers in excellent yields (Scheme 4a). The etherification of alcohols was also achieved in excellent yields under similar

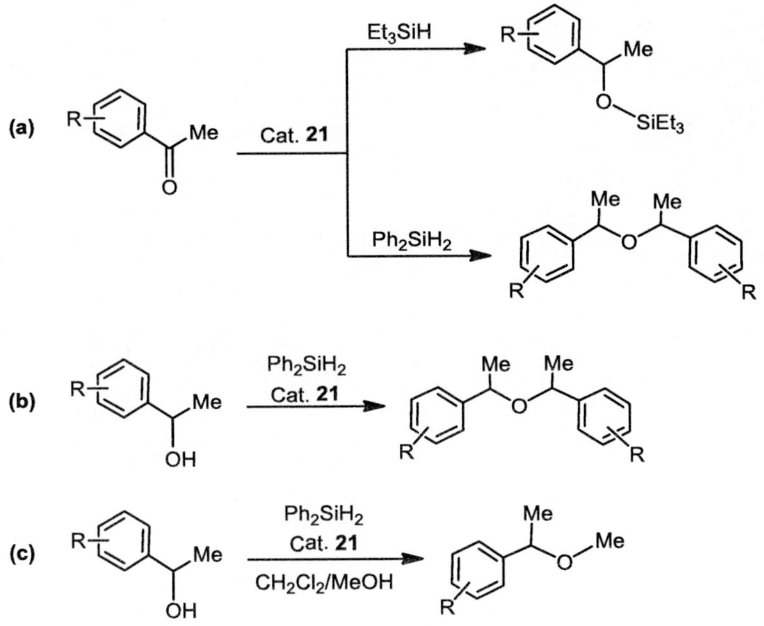

Scheme 4 Summary of etherification reactions catalyzed by complex **21**

conditions (Scheme 4b). This methodology was expanded to the preparation of unsymmetrical methyl ethers by performing the reaction in the presence of methanol, although only low yields were attained (Scheme 4c).

2.2 *Hydrosilylation of Imines*

The hydrosilylation of imines is a facile method for the preparation of amines, which are ubiquitously present in biologically active compounds. Analogous to the hydrosilylation of ketones, if the starting imines are prochiral, enantiomerically pure amines may be synthesized with appropriate catalysts.

Chiral PHOX ligands have been successfully employed for the hydrosilylation of cyclic imines with Ph_2SiH_2, using $[Ir(\mu\text{-}Cl)(COD)]_2$ as metal precursor (Scheme 5) [59]. Excellent yields were obtained with (S)-DIPOF, **14**, **22a**, and, **22b**; however, only traces were observed with **22c**. These ligands also bring about high enantiomeric excesses with iridium (ca. 90% in some cases). Conversely, noticeably lower ee's were obtained with Rh, and, with this system, the same absolute configuration was achieved for Ir and Rh.

Messerle et al. reported on the activity of complexes **23** and **24**, which efficiently catalyze the hydroamination of 4-pentyn-1-amine to 2-methyl-1-pyrroline, followed

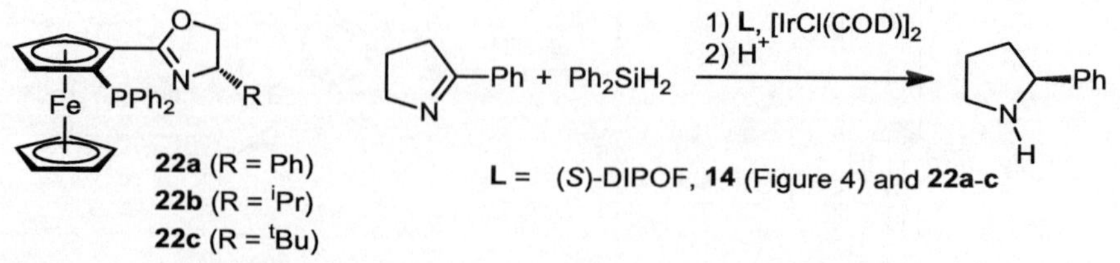

Scheme 5 Enantioselective hydrosilylation of 2-phenyl-1-pyrroline using chiral PHOX ligands and $[Ir(\mu\text{-}Cl)(COD)]_2$

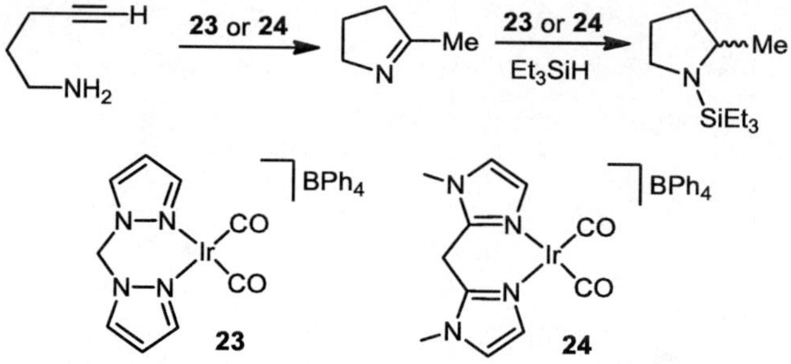

Scheme 6 One-pot hydroamination-hydrosilylation of 4-pentyn-1-amine catalyzed by **23** and **24**

Fig. 9 Imine substrates hydrosilylated by catalyst **23**

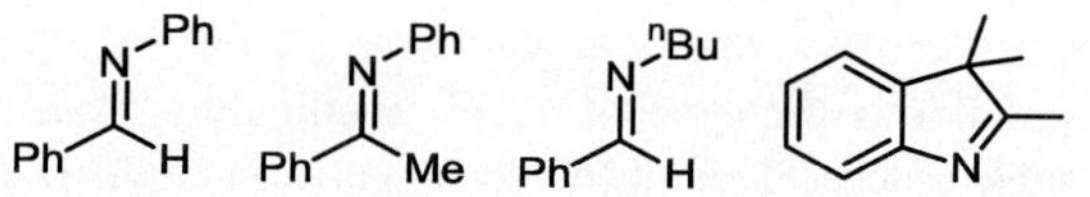

by hydrosilylation in the presence of Et_3SiH in a one-pot, two-step reaction (Scheme 6) [60].

Complex **23**, which showed better activities than **24**, was further tested as hydrosilylation catalyst for *N*-phenyl imines, an *N*-*n*-butyl imine, and an indole with Et_3SiH (Fig. 9). The *N*-phenyl imines were converted quantitatively, and good yields were achieved for the *N*-aliphatic aldimine and the indole. In this work, the authors also propose a catalytic cycle that entails initially the oxidative addition of the Si–H bond, followed by coordination of the imine, and subsequent migratory insertion into the Ir–H bond. Finally, reductive elimination renders the *N*-silylamine and restarts the catalytic cycle.

The iridacycle complex **25**, which is a variation of **18**, is capable of performing the hydrosilylation of a variety of imines prepared by hydroamination according to a tandem methodology (Scheme 7) [61]. It is noteworthy that this reaction can be carried out under mild conditions (25 °C) in the presence of a small excess of monohydrosilane Et_3SiH to afford the corresponding amines quantitatively.

The catalytic activity of this type of complexes was further evaluated in the hydrosilylation of imines [62]. A comparative study of the activity **18** and its related complex **26** (Fig. 10) showed that the latter was a more efficient catalyst in the

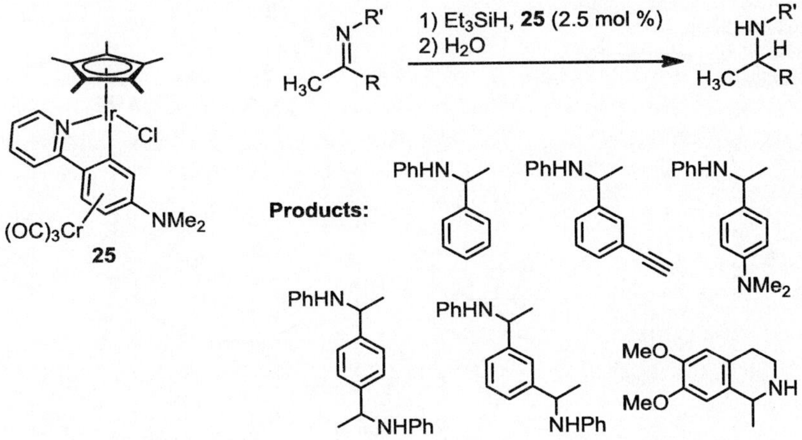

Scheme 7 Depiction of catalyst **25** and substrate scope for the hydrosilylation/hydrolysis of imines catalyzed by **25**

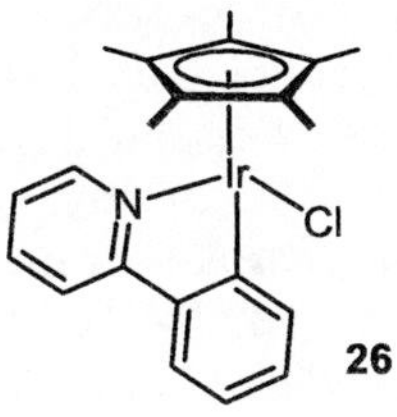

Fig. 10 Depiction of iridacycle **26**

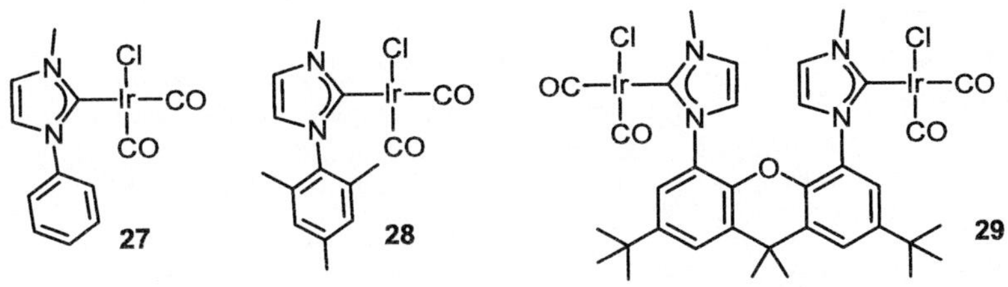

Fig. 11 Depiction of NHC-Ir catalysts **27–29**

hydrosilylation of aldimines and ketimines in the presence of $NaBArF_{24}$. A comprehensive study of the activity **26** proved that it performed efficiently for a very broad range of acyclic imines, employing Et_3SiH as reducing agent under mild conditions. More recently, related Cp*Ir(III)Cl complexes featuring cyclometallated phosphine ligands have been described, although modest yields were obtained even in the presence of $NaBArF_{24}$ [63].

Further work on Ir-catalyzed hydroamination/hydrosilylation tandem reactions has been reported by Messerle's group using NHC-containing complexes **27**, **28**, and **29** as catalysts and $NaBArF_{24}$ as additive (Fig. 11) [64]. The use of Ph_2SiH_2 as

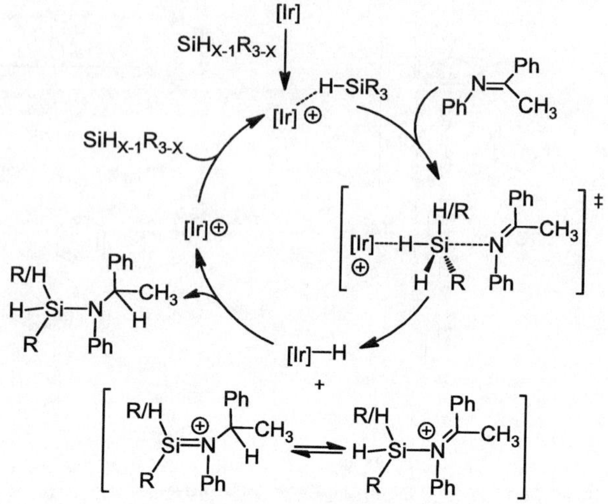

Scheme 8 Catalytic cycle postulated for the hydrosilylation of enolizable imines by **26**

reducing agent allowed excellent activities and selectivities for this tandem transformation with catalysts **27–29**. Regarding the relative activity of these catalysts, the bimetallic nature of complex **29** offers no advantage over their monometallic counterparts **27** and **28**. In fact, it is the electron density at the Ir center that seems to govern the activity trend, with electron-poor metal centers being more active.

Lately, a thorough mechanistic study by Freixa et al. shed light on the catalytic cycle that operates on the hydrosilylation of enolizable imines catalyzed by **26** and related iridacycle complexes (Scheme 8) [62–65]. The postulated ionic mechanism is similar to those previously reported for the hydrosilylation of ketones (vide supra) [66]. It is noteworthy that strong evidence of a [Ir–H] → SiR_3 catalytic intermediate has been reported by Djuckic et al. for this type of iridacycles [67].

A similar ionic mechanism has been proposed for the hydrosilylation of nitriles by cationic iridacyles **30–32**, this being the first example of this reaction catalyzed by an iridium complex [68]. The best results for the double hydrosilylation reaction were observed for **31** with Et_3SiH at 80 °C (Scheme 9). The monohydrosilylated product can be obtained selectively only for benzonitrile on applying milder reaction conditions (40 °C and short reaction times). Optimization of the reaction conditions allowed the preparation of a large array of di-silylated amines from aromatic nitriles in quantitative or nearly quantitative yields in less than 24 h, with a catalyst loading of 0.5 mol% of [**31**][$BArF_{24}$]. The use of other counterions such as $[OTf]^-$ or $[BPh_4]^-$ led to drastically lower yields.

The 1,2-hydrosilylation of *N*-heteroaromatic compounds is closely related to the hydrosilylation of imines. In fact, an example has been included in this section (Fig. 7), namely, the hydrosilylation of indole. The 1,2-hydrosilylation with Et_2SiH_2 of a comprehensive selection of *N*-heteroaromatic compounds, catalyzed by [IrCl $(COE)_2]_2$ (COE = cyclooctene), has been recently published by Park and Chang [69]. The authors present in this work the first examples of 1,2-hydrosilylation of

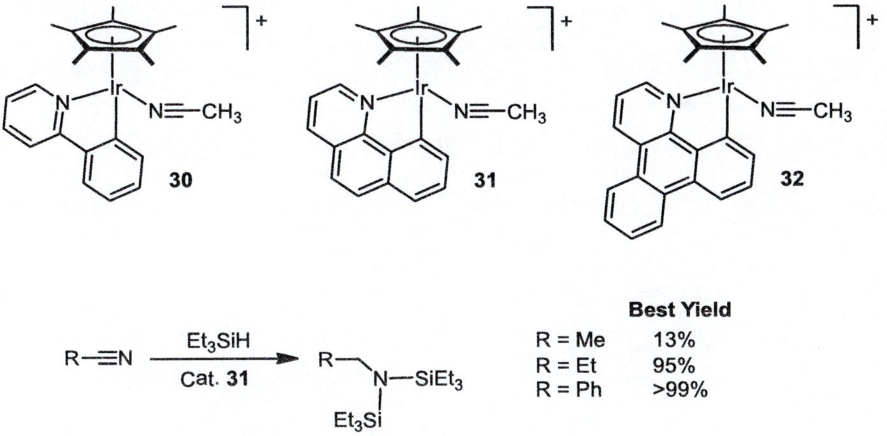

Scheme 9 Hydrosilylation on nitriles with iridacycles **30–32**

pyridines and other *N*-heterocycles catalyzed by transition-metal complexes. This methodology is very selective toward 1,2-hydrosilylation, avoiding side reactions such as dehalogenation or 1,4-hydrosilylation. In this regard, it is worth mentioning that metal-catalyzed examples of the latter have been explored by several groups employing transition metals different from iridium. [Selected examples of transition-metal-catalyzed 1,4-hydrosilylation reactions: [70–74].]

The proposed catalytic cycle (Scheme 10) requires an initiation step that involves the reaction of $[Ir(\mu\text{-}Cl)(COE)_2]_2$ with Et_2SiH_2 to give an Ir(V)-Ir(V) dissymmetric dinuclear polyhydride complex. Subsequently, the active species is generated by substitution of the COE ligand by the substrate (pyridine in this case) at one of the Ir centers, where all the steps of the catalytic cycle occur. The next step involves the migratory insertion of the C=N bond into one of the Ir–H bonds to afford the reduced pyridine (bound to the Ir center). This intermediate renders the hydrosilylated product by reductive elimination, with simultaneous formation of an Ir(V)-Ir(III) species. Finally, reaction of the resulting Ir(V)-Ir(III) species with Et_2SiH_2 regenerates the active species.

2.3 *Hydrosilylation of Alkenes*

The catalytic hydrosilylation of alkenes is one of the most important synthetic tools for the preparation organosilicon compounds, such as silicones, elastomers, and silicone-based coatings, in the chemical industry. The application of late transition metal catalysts in the hydrosilylation of alkenes from an industrial viewpoint has been thoroughly reviewed, and, therefore, this subject will not be dealt with in detail this section [75].

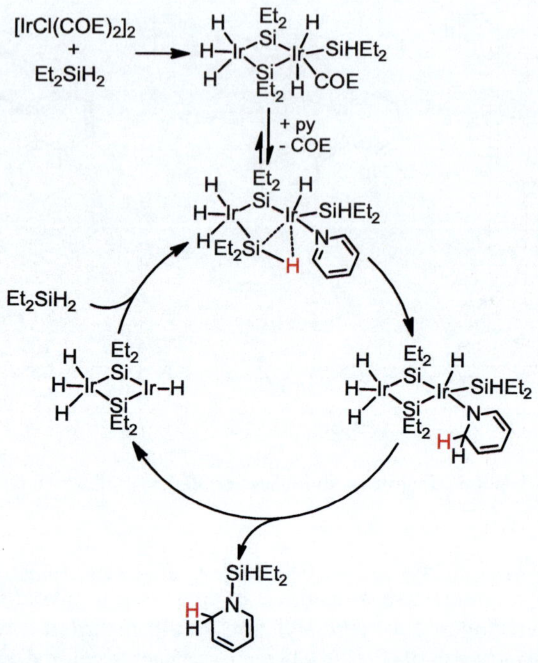

Scheme 10 Catalytic cycle postulated for the 1,2-hydrosilylation of pyridine

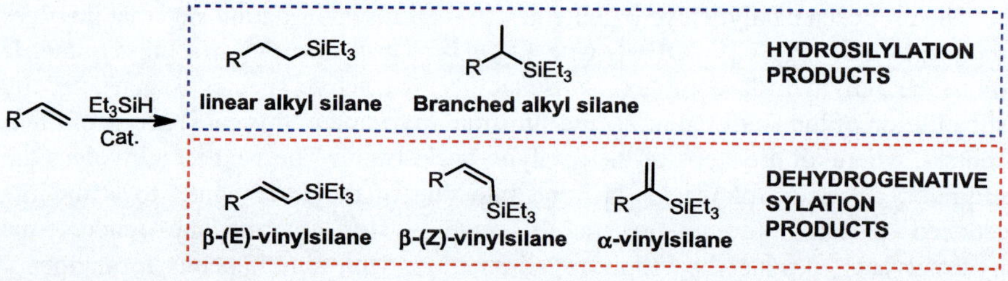

Scheme 11 Possible silicon-containing products obtained from the hydrosilylation of alkenes

The hydrosilylation of alkenes may afford only two different products, linear (anti-Markovnikov addition) and branched alkyl silanes (Markovnikov addition). However, under hydrosilylation conditions, it is frequent to observe the formation of other silicon-containing products, such as vinylsilanes resulting from dehydrogenative silylation (Scheme 11).

Pioneering work on the use of iridium complexes for the hydrosilylation of alkenes comprises two main works, the hydrosilylation of 2,3-dimethylbuta-1,3-diene with Et_3SiH by $[IrCl(COE)_2]_2$ (COE = cyclooctene) [35] and the hydrosilylation of 1-hexene catalyzed by $[Ir(\mu\text{-}X)(COD)]_2$ (X = Cl or OMe) and pnictogen-based monodentate ligands (NPh_3, PPh_3, $AsPh_3$, and $SbPh_3$) in 1:1 or 1:2 molar ratios. In the case of the latter, the reactions of 1-hexene and Et_3SiH with the different catalysts were performed at 60 °C in CH_2Cl_2, affording the

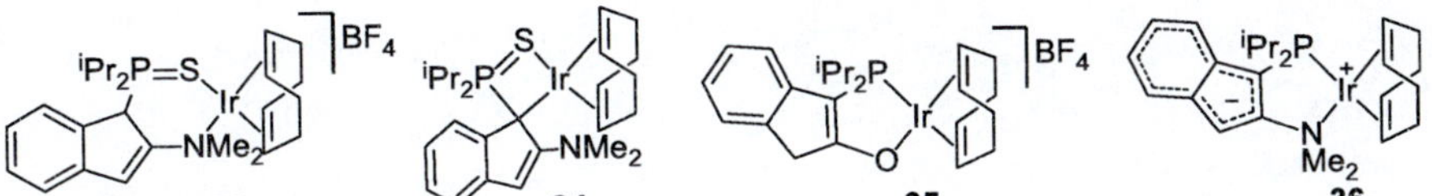

Fig. 12 Depiction of Indole base catalysts **33–36**

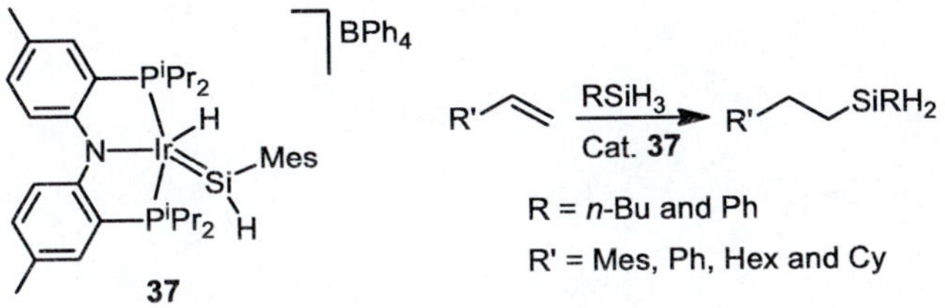

Scheme 12 Anti-Markovnikov hydrosilylation of alkenes by **37**

hydrosilylation product,1-hexyl(triethyl)silane, in low yields (3–17%). Remarkably, mixtures of the three possible hexenyl(triethyl)silanes, presumably resulting from dehydrogenative silylation, were obtained as the major product (12–85% yield) [76].

After these works, several catalysts for the hydrosilylation of ethylene were described, such as the heterobimetallic complex [Ir(CO)$_2$(μ-CH$_2$)$_2$TaCp$_2$] described by Bergman's group [77, 78] or Shapley's indenyl complex [Ir(η^5-C$_9$H$_7$)(COE)(CO)] [79].

An interesting application of the hydrosilylation reaction was described by Marciniec and co-workers, who developed a catalytic system that uses the complex [Ir(OTMS)(COD)]$_2$ in the presence of a phosphine ligand to prepare silicon-containing polymers by reaction of a bis-silylarene with a diolefin. This catalyst also proved efficient for the hydrosilylation of silylalkenes with heptamethyltrisiloxane [80–83].

Stradiotto et al. developed a series of bidentate Ir and Rh catalysts for the hydrosilylation of alkenes based on an indole scaffold (Fig. 12) [84–86]. Catalysts **33**, **34**, and **35** were able to hydrosilylate styrene, but important amounts of β-(*E*)-vinylsilanes (resulting from dehydrogenative silylation) and relatively low yields in some cases were observed. In sharp contrast, zwitterionic complex **36** leads to a quantitative yield and a 99% selectivity toward the linear hydrosilylation product, which was obtained reacting styrene with triethylsilane at 60 °C in 1,2-dichloroethane using styrene/silane ratio of 5/1 and a 5.0 mol% loading of **36**.

Silylene complex **37** is able to catalyze the hydrosilylation of alkenes exclusively by reaction with primary silanes, this being one of the few examples of silylane-containing catalysts for this reaction [87]. The exclusive anti-Markovnikov selectivity of the process – up to 77% yields of linear silane were obtained with **37** – and the fact that only primary silanes were able to induce this reaction suggested a catalytic cycle that would proceed according to a Glaser-Tilley-type mechanism, analogous to related Ru-silylene catalysts [88–94] (Scheme 12).

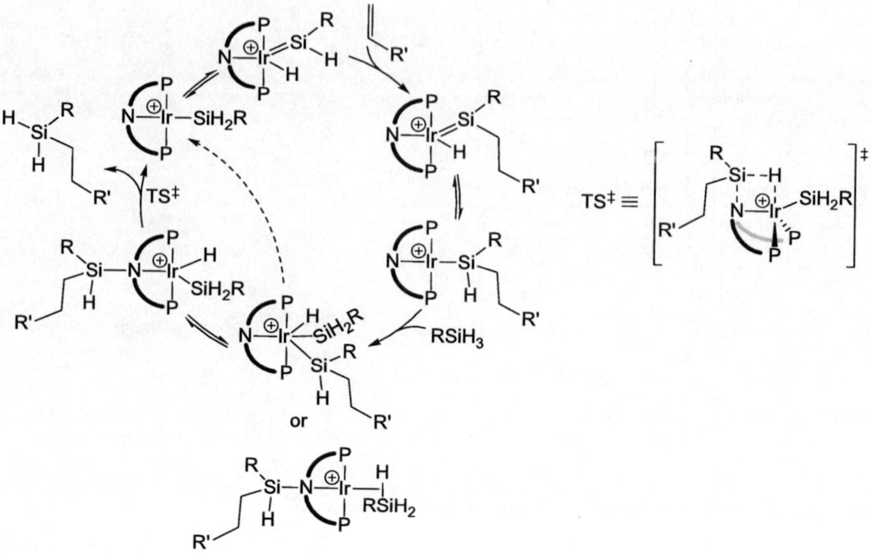

Scheme 13 Glaser-Tilley-type catalytic cycle proposed for the hydrosilylation of alkenes with **37**

Further work on this system allowed to improve the catalytic activity of **37** by modifying the PNP-Ir scaffold, which brought about yields of the linear product up to 84% [95]. Stoichiometric experiments on these complexes showed the formation of Ir species that feature Si–N bonds and are competent catalyst for this reaction. The authors postulated a Glaser-Tilley-type catalytic cycle where these Si-N species acted as intermediates (Scheme 13). The first step involves the direct Si–H bond addition at the silylene across the C=C double bond, this step being the cornerstone of this type of mechanism. The resulting Ir-silylene-hydride is in equilibrium with an Ir-silyl intermediate that reacts either with the primary silane to give an end-on complex or via oxidative addition of one of its Si–H bonds. This reaction may afford two different species, depending on whether the Si(H)(R)(CH_2CH_2R') moiety is bound to the Ir center or to the nitrogen of the PNP ligand. Either way, these species are in equilibrium with a complex that features a Si(H)(R)(CH_2CH_2R') moiety *N*-bound to a PNP-Ir(H)(SiH_2R) complex. The hydrosilylated alkene is formed from the latter via a transition state where simultaneous splitting of the Si–N and Ir–H bonds and formation of the Si–H bond take place. The formation of the linear alkylsilane yields, concomitantly, a PNPIr(SiH_2R) complex that is in equilibrium with the silylane complex, thus restarting the catalytic cycle.

The selective anti-Markovnikov hydrosilylation of a variety of functionalized aliphatic alkenes with ethynylsilanes was achieved using $[Ir(\mu\text{-}Cl)(COD)]_2$ (Scheme 14) [96]. The reaction was performed at room temperature under solventless conditions, according to a modified literature procedure [97], the best yields being obtained in the presence of a large excess of 1,5-cyclooctadiene (4 equivalents

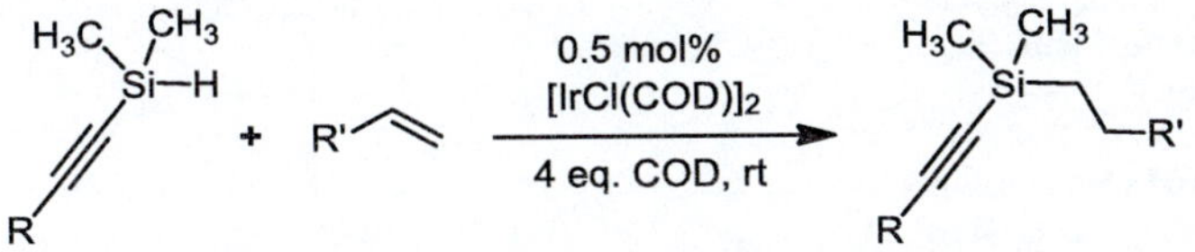

Scheme 14 Anti-Markovnikov hydrosilylation of alkenes with ethynylsilanes

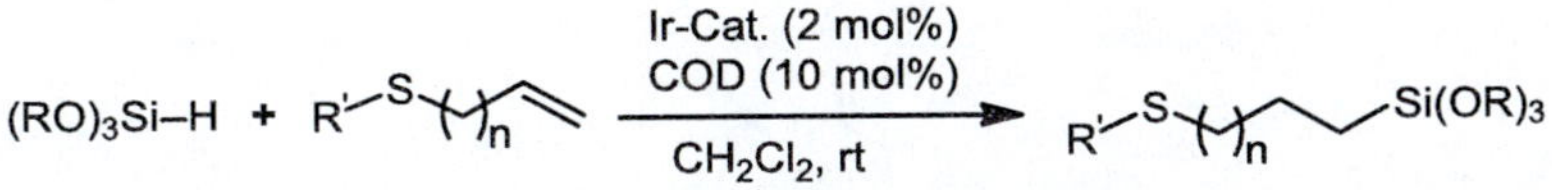

Scheme 15 Anti-Markovnikov hydrosilylation of sulfur-containing olefins with trialkoxysilanes

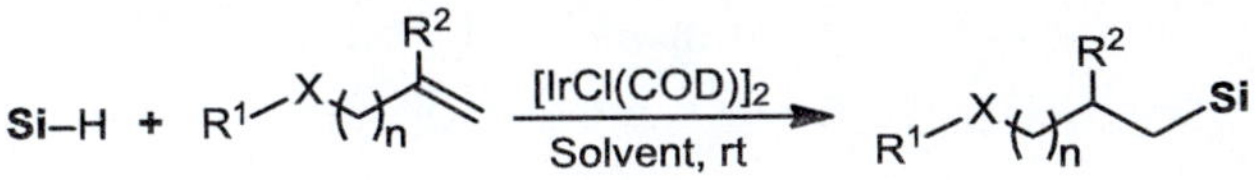

Scheme 16 Anti-Markovnikov hydrosilylation of therapeutically relevant olefins with various hydrosilanes

relative to the substrates). The amount of COD employed can be reduced to 0.5 equivalents by using a solvent such as 1,2-dichloroethane.

A similar methodology was employed for the hydrosilylation of a broad scope of sulfur-containing olefins with trialkoxysilanes. As catalyst, the authors successfully evaluated $[Ir(\mu\text{-}X)(COD)]_2$ (X = Cl or SAr) and $[IrCl_2(allyl)(COD)]$, with the three showing excellent selectivities toward the linear alkylsilanes (Scheme 15) [98].

Further work on the use of trialkoxysilanes as reducing agent entails the hydrosilylation of unactivated alkenes using $[Ir(\mu\text{-}Cl)(COD)]_2$ as catalyst, allowing excellent anti-Markovnikov selectivities (Scheme 16) [99]. This methodology is applicable to a wealth of alkenes, including therapeutically relevant substrates.

Similar mechanisms were proposed for the last two methodologies (Scheme 17). The first step requires the activation of the precatalyst, i.e., the abstraction of the chloride ligands from the iridium center by the hydrosilane and concomitant formation of an Ir(I)-H species. Subsequently, coordination of the olefin followed by migratory insertion into the Ir–H bond takes place. The last two steps of the cycle are (1) the oxidative addition of the Si–H bond of a molecule of silane to give an Ir (III) intermediate and (2) reductive elimination to afford the silylated product with concomitant regeneration of the Ir(I) hydride species.

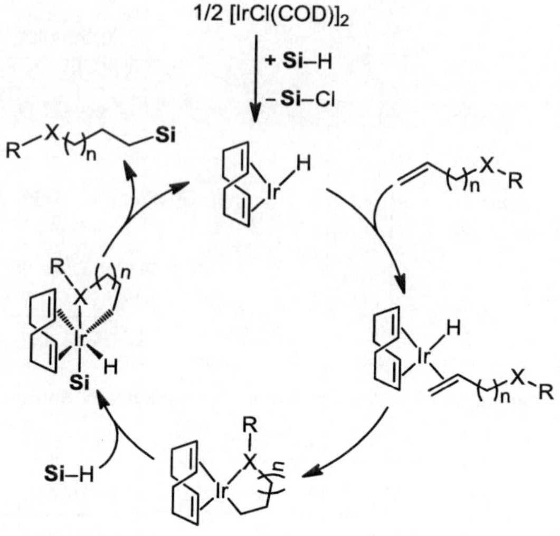

Scheme 17 Catalytic cycle postulated for the anti-Markovnikov hydrosilylation of functionalized olefins using $[Ir(\mu\text{-}Cl)(COD)]_2$ as catalyst

2.4 *Hydrosilylation of Alkynes*

The hydrosilylation of terminal alkynes is an atom-efficient and usually high-yield methodology for the preparation of vinylsilanes. The main disadvantage is the lack of selectivity of the process and the difficult separation of the product mixture. Therefore, the development of highly selective catalysts is essential to use this methodology for synthetic purposes.

Three hydrosilylation products may be obtained, namely, the β-(*E*)-, β-(*Z*)-, and α-vinylsilane. The β-(*E*)-vinylsilane (*syn* addition product) is the most thermodynamically stable product, and usually the major isomer, the β-(*Z*)-vinylsilane (*anti*-addition product), is somewhat less frequent, and the α-vinylsilane is rarely obtained selectively. Alongside vinylsilanes, dehydrogenative silylation and hydrogenation products (silylalkynes and alkenes, respectively) are often obtained (Scheme 18) – the latter is formed as a by-product of the dehydrogenative silylation reaction, since the alkyne plays the role of substrate and hydrogen acceptor. Although less frequent, cyclotrimerization or dimerization of alkynes is also observed.

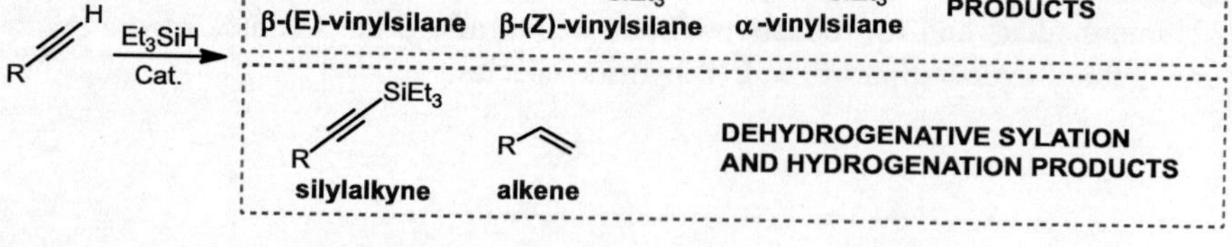

Scheme 18 Possible products obtained from the hydrosilylation of alkynes

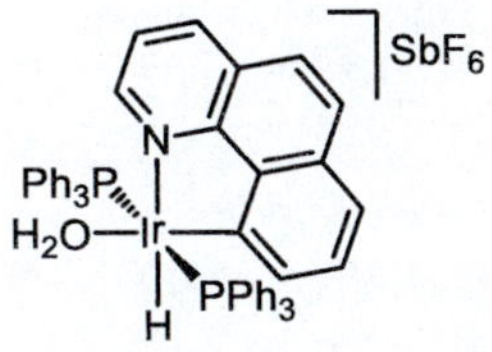

Fig. 13 Depiction of Ir catalyst **38**

The formation of β-(*E*)-vinylsilanes – obtained selectively by Pt-catalyzed processes – was explained invoking the Chalk-Harrod mechanism, which involves the oxidative addition of the silane's Si–H bond, followed by alkyne coordination and subsequent 1,2-migratory insertion into the Ir-H. Finally, reductive elimination generates the β-(*E*)-vinylsilane [100]. However, the discovery of rhodium hydrosilylation catalysts unveiled an intricate mechanistic riddle, as formation of β-(Z)-vinylsilanes was observed in significant amounts. The so-called modified Chalk-Harrod mechanism was then proposed [101–107]. In contrast with the Chalk-Harrod mechanism, a 2,1-migratory insertion into the Ir–Si bond takes place in this case. Subsequently, the alkenyl complex thus formed isomerizes to minimize steric repulsion between the silyl moiety and the metal-ligand system, resulting in the formation of a formal *anti*-addition intermediate [108, 109]. Finally, reductive elimination delivers the β-(Z)-vinylsilane.

Iridium catalysts for the hydrosilylation of terminal alkynes often give important amounts of β-(Z)-vinylsilane, even allowing excellent selectivities in some cases. At the beginning of the 1990s, Crabtree and co-workers reported the excellent selectivity of Ir-triso (triso = $[C\{Ph_2P{=}O\}_3]^-$) complexes toward the β-(Z)-vinylsilane [108, 110]. Complex [Ir(triso)$(COE)_2$] in particular is able to catalyze the hydrosilylation of phenylacetylene with Et_3SiH chemoselectively – internal ones are not converted – and with *Z*/*E* ratios up to 190/1 with no detectable α-isomer. This reaction allows high yields, but the use of other substrates, such as *t*-BuCCH, $C_5H_{11}CCH$, or $HCCCO_2Me$, leads to significantly lower yields. The use of Ph_3SiH, on the other hand, affords exclusively the β-(Z)-vinylsilane.

Complex **38** (Fig. 13) was studied later, proving efficient for a wider substrate scope [111]. Different terminal alkynes and hydrosilanes were evaluated, showing excellent selectivities toward the β-(Z)-vinylsilane at room temperature. On increasing the reaction temperature to 65 °C, a drastic increase of dehydrogenative silylation and β-(*E*)-isomer was observed – also the use of bulky substituents at the alkyne gives rise to a loss of selectivity.

Complexes of the type $[Ir(H)_2(SiEt_3)(TFB)(PR_3)]$ (TFB = tetrafluorobenzobarralene) (**39a-c**; Fig. 14) catalyze the hydrosilylation and dehydrogenative silylation of phenylacetylene [112]. In contrast with the [Ir(triso)$(COE)_2$] catalyst described above, an increase of the temperature from 20 to 60 °C has little effect on the hydrosilylation/dehydrogenative silylation product ratio, but it does have an impact on the *syn*/*anti* ratio, especially in the case of **39a**. Higher temperatures bring

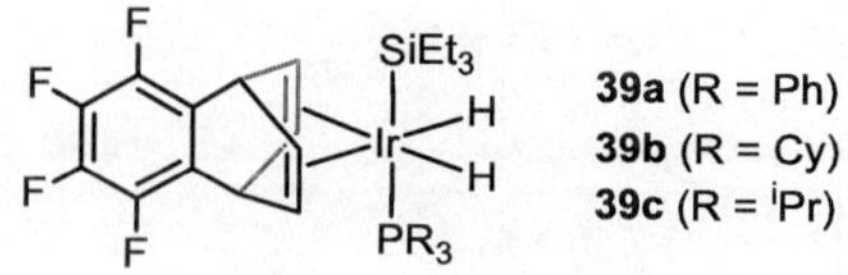

Fig. 14 Depiction of Ir catalyst **39a-c**

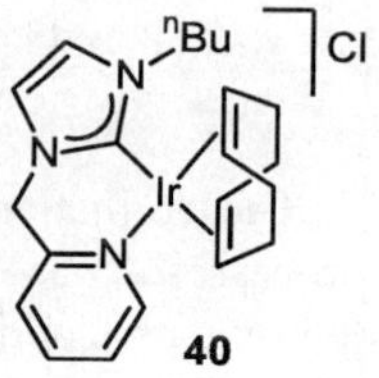

Fig. 15 Depiction of Ir (NHC-py) catalysts **40**

about a decrease of the β-(Z)-isomer, with **39b** giving the best *Z*/*E* ratio (3.5/1) at 20 °C.

Complexes [Ir(L$_2$)(^{i}Pr$_2$PCH$_2$CH$_2$X)][BF$_4$] (L$_2$ = COD or TFB; X = NMe$_2$ or OMe), [Ir(C$_2$Ph)(CO)$_2$(PCy$_3$)], and [Ir(C$_2$Ph)(TFB)(PCy$_3$)] are efficient catalysts for the hydrosilylation of phenylacetylene with Et$_3$SiH, but yield mainly mixtures of β-(Z)- and β-(*E*)-vinylsilanes and dehydrogenative silylation products [113, 114].

Ir-carbonyl complexes [Ir(dppe)(CO)Br] (dppe = 1,2-Bis(diphenylphosphino) ethane) and [Ir(PPh$_2$Me)$_2$(CO)Cl] were evaluated as hydrosilylation catalysts for 1-hexyne, phenylacetylene, and 1-phenyl-1-propyne. Reaction times around 24 h are required to achieve full conversion at 65 °C with Et$_3$SiH for terminal alkynes. A yield of 52% was achieved for the hydrosilylation of internal alkyne 1-phenyl-1-propyne with [Ir(dppe)(CO)Br] after prolonged reaction times. Only β-(*Z*)- and β-(*E*)-vinylsilanes were obtained, with the former being the major product (α-(*Z*) and α-(*E*) isomers were not observed). It is worth mentioning that [Ir(PPh$_2$Me)$_2$(CO) Cl] allows the formation of α-vinylsilane in relatively high selectivity when phenylacetylene is used as substrate. The use of internal alkynes as substrate, namely, 1-phenyl-1-propyne, resulted in yields below 3% for the vinylsilanes after 36 h [115].

A series of Ir-COD complexes containing chelating NHC ligands were evaluated in the hydrosilylation of phenylacetylene with dimethylphenylsilane by Peris et al. However, low yields and selectivities were reported, except in the case of complex **40** (Fig. 15), which allowed quantitative conversion with a 0.1 catalyst loading at 60 °C, over a 48 h period, leading to a β-(*E*)/β-(*Z*)/α ratio of 48/34/18 [116–119].

Neutral Ir(I) complexes of the type [IrCl(COD)(NHC)] catalyze the hydrosilylation of phenylacetylene with Me$_2$PhSiH at 60 °C, affording in all cases quantitative yields after 2 h with good β-(*Z*) selectivity (65–90%) [120]. The best selectivities were obtained with catalysts **41b** and **41c** (Fig. 16).

The use of Ir(III) precatalyst [Ir(μ-Cl)(Cl)(Cp*)]$_2$ permits the preparation of β-(*Z*)-vinylsilanes in excellent yields (up to 92% isolated yield) for terminal alkynes featuring Ph, Cy, and ^{n}Bu substituents with HSiEt$_3$ and HSiPh$_3$ [121].

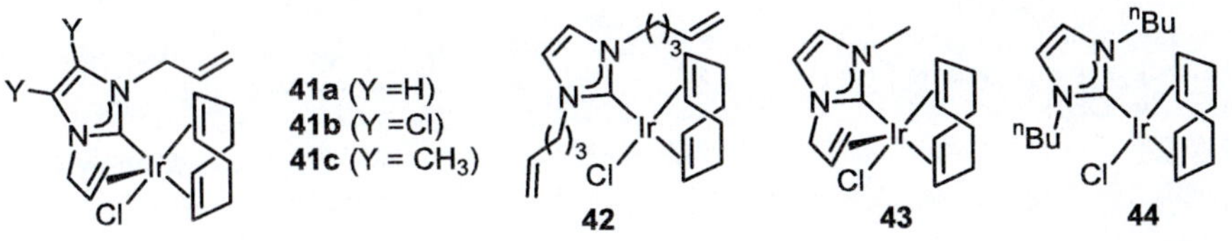

Fig. 16 Depiction of [IrCl(COD)(NHC)] catalysts **41–44**

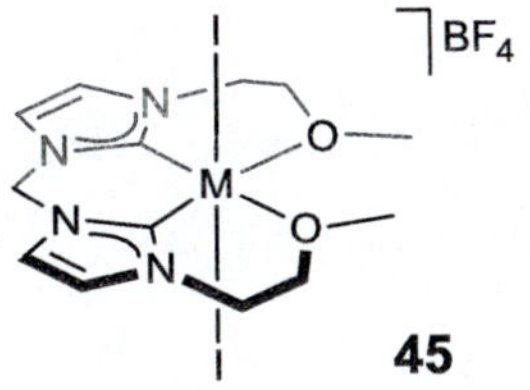

Fig. 17 Depiction of $[Ir(L\text{-}C,C,O,O)I_2]BF_4$ catalysts **45**

More recently, another Ir(III) complex $[Ir(L\text{-}C,C,O,O)I_2]BF_4$ (**45**, Fig. 17), this time featuring a bis-NHC ligand (L = methylenebis[(N-2-methoxyethyl)imidazole-2-ylidene]), has shown excellent yields and β-(Z) selectivities for a variety of alkynes and hydrosilanes at 50 °C in 5 h, using acetone as solvent. The best yields and selectivities were obtained for aromatic alkynes, with quantitative conversions and exclusive formation of the β-(Z)-vinylsilane in some cases [122, 123]. Mechanistic studies, based on DFT calculations and experimental data, suggest an ionic mechanism similar to that described above for the hydrosilylation of ketones. In this case, the first step entails the heterolytic splitting of the Si–H bond. This occurs via initial end-on coordination of the hydrosilane to the Ir center by the H atom and simultaneous interaction the acetone's oxygen with the silicon atom. This yields an oxocarbenium ion and a metal hydride species. The former transfers the R_3Si^+ cation from the hydrosilane to the alkyne to give the internal silylcarbocation ($[R_3Si\text{-}CH{=}C^+\text{-}R]$). Finally, transfer of the hydride ligand to the carbocation affords the hydrosilylated product. The least sterically hindered transition state for the formation of the vinylsilane involves a *cis* conformation for the carbocation when this approaches the hydride, which leads to formation of the β-(Z)-isomer [123].

Ir(III) complex **18** was tested successfully as catalyst for a range of silanes and alkynes (terminal and internal) [124]. Yields that range from 22 to 100% were obtained after 2 h at 40 °C with a catalyst loading of 2.5 mol%. Remarkably, 100% selectivities either for the β-(*E*) or β-(*Z*) were obtained depending on the type of alkyne. Longer reaction times (20 h) and higher temperatures were required for the hydrosilylation on internal alkynes; nonetheless, low yields were obtained in general. Another interesting example of Ir(III) catalysts that allow excellent selectivities toward β-(Z)-vinylsilanes are those based on the general formula [Ir(Cl)(Cp*)(L_2)], where L_2 is a C,N-bidentate pyridylideneamide ligand [125].

The hydrosilylation of internal thioalkynes was achieved in a regio- and stereoselective manner by using $[Ir(\mu\text{-}Cl)(COD)]_2$ (2 mol%) as catalyst and $(EtO)_3SiH$ as reducing agent (Scheme 19) [126]. To highlight the importance of this iridium-based

Scheme 19 Selective hydrosilylation of internal thioalkynes catalyzed by [Ir(μ-Cl)(COD)]$_2$

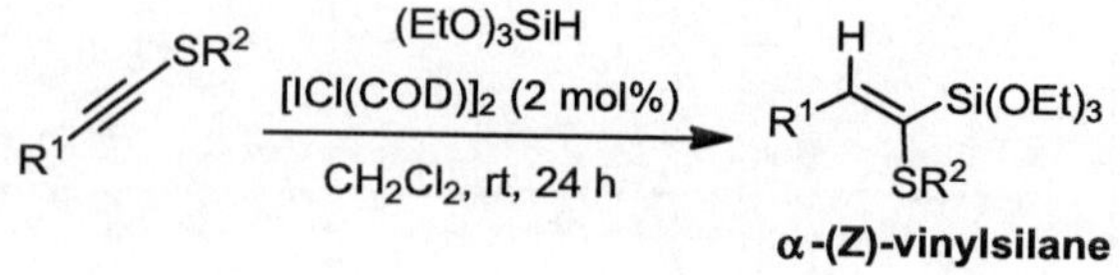

Scheme 20 Catalytic cycle proposed for the selective formation of the α-(Z)-vinylsilane

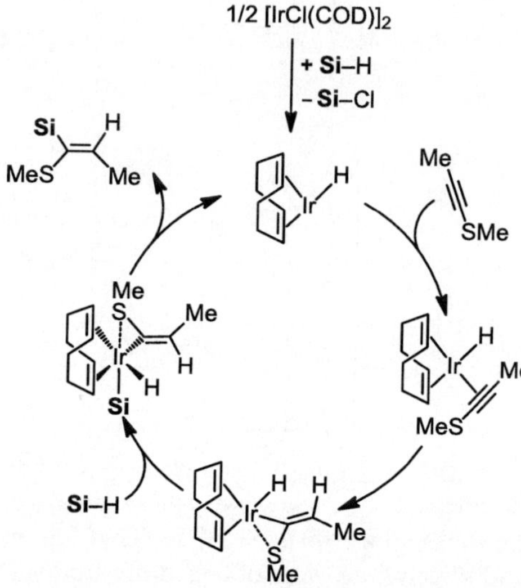

catalyst, it is worth mentioning that Pt, Co, and Ru catalysts showed no activity for this reaction. The hydrosilylation of unsymmetrical internal alkynes may afford four different vinylsilanes, the β-(*E*), β-(*Z*), α-(*E*), and α-(*Z*). Remarkably, this methodology allows the selective preparation of the α-(*Z*) isomer, i.e., the reaction is α regio- and *syn* stereo-selective. A broad variety of thioalkynes with several synthetically relevant hydrosilanes were tested, leading to yields up to 89% and selectivities β/α and *Z*/*E* up to >30/1.

A mechanism that explains the selectivity of the reaction was proposed and substantiated by DFT studies (Scheme 20). Chalk-Harrod-type mechanisms were discarded based on the high energy barrier of the processes. The postulated mechanism consists of the formation of an unsaturated Ir-hydride species by reaction with two equivalents of the silane. This species is able to coordinate the alkyne, which evolves to give the more stable *S*-chelated intermediate. Finally, oxidative addition of the Si–H bond leads to formation of an Ir(III) intermediate, which upon reductive elimination affords the vinylsilane and regenerates the Ir-H active species.

Complex [IrH(κ^3-*O*,*N*,*O*′-hqca(COE)] (**46**) (COE = cyclooctene; hqca = 8-oxidoquinoline-2-carboxylate), an iridium(III) hydride complex featuring an ONO-pincer ligand, proved an efficient catalyst for the hydrosilylation of a variety of aromatic and aliphatic terminal alkynes (Fig. 18). The reaction was selective toward the β-(*Z*)-vinylsilane in general, but bulky substituents, such as *t*-butyl, led to preferential formation of the β-(*E*) isomer. It is noteworthy that, after consumption of

Fig. 18 Depiction of ONO-Ir(III) catalyst **46**

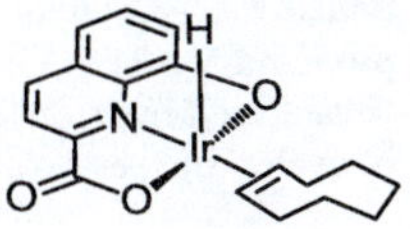

46

the alkynes, β-(*Z*) to β-(*E*) isomerization was observed. Besides, in some cases, important amounts of dehydrogenative silylation products were obtained. The formation of these products was explained by means of a reaction mechanism substantiated by DFT calculations. The intermediate that determines whether a hydrosilylation or dehydrogenative silylation reaction occurs is a metallacyclopropene species, which may evolve upon coordination of a silane molecule to give an alkenyl complex and eventually the corresponding vinylsilane. Alternatively, the metallacyclopropene may undergo β-hydride elimination to afford the silylalkyne [127].

The selective Markovnikov hydrosilylation of terminal alkynes catalyzed by iridium complexes has been recently reported [128]. This methodology involves the use of $(TMSO)_3SiH$ (TMS = tris(trimethylsilyl)silane) as silicon source, which allows α/β selectivities over 20/1 using $[Ir(\mu\text{-}Cl)(cod)]_2$ as catalyst. The use of this silane inverts the regioselectivity of the process when compared to other silanes, e.g., Et_3SiH, which renders α/β selectivities over 1/20. It is noteworthy that the use of other Ir precatalysts, namely, $[Ir(\mu\text{-}OMe)(cod)]_2$, $[Ir(\mu\text{-}Cl)(coe)_2]_2$, and $[Cp^*IrCl_2]_2$, led to yields lower than 5%. This methodology can be applied to a broad range of terminal alkynes, is compatible with a variety of functional groups, and is applicable to late-stage functionalization of a number of bio-relevant organic compounds.

2.5 *Hydrosilylation of Cyclopropanes*

The ring-opening of cyclopropanes via hydrosilylation, catalyzed by $[Ir(\mu\text{-}Cl)(COD)]_2$, has been recently reported [129]. This reaction requires the presence of an *N*-directing group, which also controls the regiochemistry of the Si-H addition across the C–C bond (Scheme 21).

The proposed mechanism involves the formation of an Ir-hydride that acts as the active species. Coordination of the pyridyl directing group situates the cyclopropane

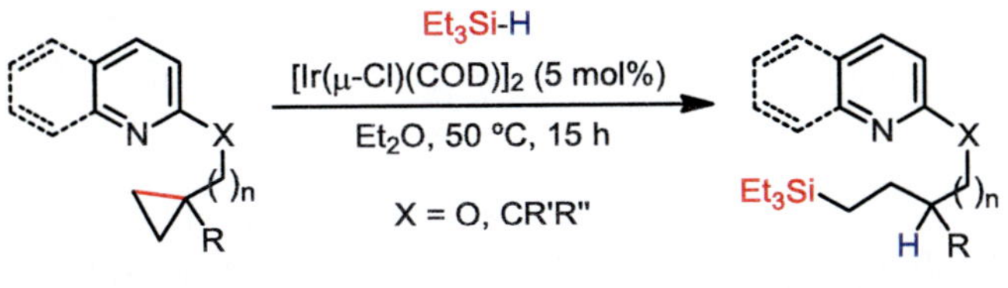

Scheme 21 Ring-opening hydrosilylation of cyclopropanes

Scheme 22 Catalytic cycle proposed for the Ir-catalyzed hydrosilylation of cyclopropanes

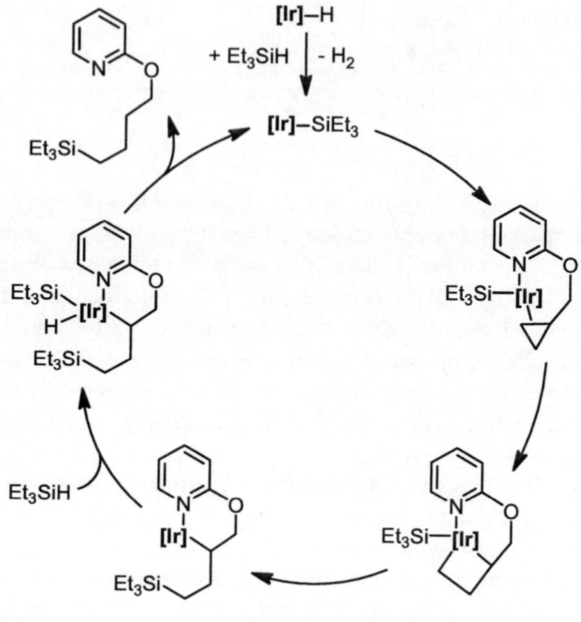

in the proximity of the Ir center. Oxidative addition of the C–C bond yields an iridacyclobutane intermediate, which, upon C-Si reductive elimination, affords an Ir-alkyl species. Finally, oxidative addition of Et_3SiH, followed by C-H reductive elimination, furnishes the linear alkylsilane and regenerates the active species (Scheme 22).

3 Silylation of C–H Bonds

The functionalization of C–H bonds with hydrosilanes by dehydrogenative silylation involves the formation of H_2 as co-product, which often calls for the use of an open system or a sacrificial hydrogen acceptor to favor the reaction thermodynamically. Alternatively, the use of disilanes as reducing agents circumvents the need for a driving force to compensate the unfavorable thermodynamics of the process; however, the availability of disilanes is low compared to that of hydrosilanes [130].

3.1 Silylation of Alkynes and Alkenes

The first examples of dehydrogenative silylation were observed as by-products of hydrosilylation reactions of alkenes and alkynes (vide supra) [76, 131–134], thus obtaining the silylated product together with equimolar amounts of the reduced alkyne or alkene, which act as hydrogen acceptors. One of the first examples of a selective catalytic systems was the reaction of ethylene or 1-hexene with Et_3SiH using $[IrH_2(SiEt_3)(COD)L]$ (L = PPh_3 or $AsPh_3$) as catalyst. The dehydrogenative silylation products were obtained in relatively good selectivities. Only 20% and 8% hydrosilylation products were obtained for ethylene and 1-hexene, respectively, with L = $AsPh_3$ [131]. When $[Ir(\mu\text{-}OMe)(COD)]_2$ + 2 L systems were used as catalysts, $[IrH_2(SiR_3)(COD)L]$ species were formed by the oxidative addition of Et_3SiH to the [Ir(OMe)(COD)L] intermediates [132].

Excellent selectivities toward the β-(Z)-vinylsilane were obtained by dehydrogenative silylation of a broad range of terminal alkenes using norbornene derivatives as hydrogen acceptors. The catalyst was generated "in situ" from $[Ir(\mu\text{-}OMe)(COD)]_2$ and dtbpy (4,4-di-tert-butyl-2,2-bipyridine). The best results were obtained at 40 °C, in tetrahydrofurane, employing 3 equivalents of 2-norbornene and Et_3SiH. Under these conditions, this methodology gives rise to *Z/E* ratios up to 10/1 and almost quantitative yields in 2 h. Moreover, the presence of a variety of functional groups – namely, ketones, ketals, amides, esters, alcohols, halides, epoxides, and silanes – was evaluated, showing no interference [135].

Hartwig et al. expanded this methodology to the use of more synthetically interesting silanes, namely, $(TMSO)_2MeSiH$ [136]. In this study, several 1,10-phenanthroline (phen) derivatives were tested as ligands with $[Ir(\mu\text{-}OMe)(COD)]_2$, $[Ir(\mu\text{-}OH)(COE)_2]_2$, or $[Ir(\mu\text{-}Cl)(COE)_2]_2$, showing that the *Z/E* diastereoselectivity may be controlled by tuning the substituents at the phenanthroline scaffold. The mechanistic studies presented in this work suggest the formation of the silyl-hydrido-Ir(III) active species upon oxidative addition of the Si–H bond. Subsequently, alkene coordination takes place, followed by a *syn* insertion of the alkene into the Ir–Si bond. Rotation around the single C–C bond of the silylalkyl intermediate and subsequent β-hydrogen elimination yields the β-(Z)-vinylsilane. Finally, hydrogenation of NBE (norbornene) to NBA (norbornane) and the ensuing activation of the silane regenerate the active species (Scheme 23).

Ir(III) complex **47** is able to catalyze a rare case of tandem isomerization-dehydrogenative silylation of alkenes; conversely, the related Rh **48** complex catalyzes very selectively the tandem isomerization/hydrosilylation of internal alkenes to afford linear alkylsilanes (Fig. 19) [137]. For example, 1-octene, *trans*-2-octene, *trans*-3-octene, and *trans*-4-octene can be converted by reaction with Et_3SiH, into a *Z*/*E* mixture of the allylic silane, i.e., triethyl(oct-2-en-1-yl)silane, in conversions that range from 67 to 73%.

An interesting application of the dehydrogenative silylation of terminal alkenes is the functionalization of monounsaturated fatty acids from vegetable oils, which may be employed as renewable feedstock. In this regard, the system $[Ir(\mu\text{-}OMe)(COD)]_2$/bipy acts as an efficient catalyst for the dehydrogenative silylation of methyl oleate with Et_3SiH to give mixtures of β-(*Z*)- and β-(*E*)-vinylsilanes in the presence of NBE [138].

High selectivities toward the dehydrogenative silylation of several terminal alkynes were achieved using a mixture $Ir_4(CO)_{12}/PPh_3$ as catalyst at 100 °C, with hydrosilanes. The use of 2.3 equivalents of the terminal alkyne, which acts as

Scheme 23 Mechanism proposed by Hartwig et al. for the dehydrogenative silylation of alkenes

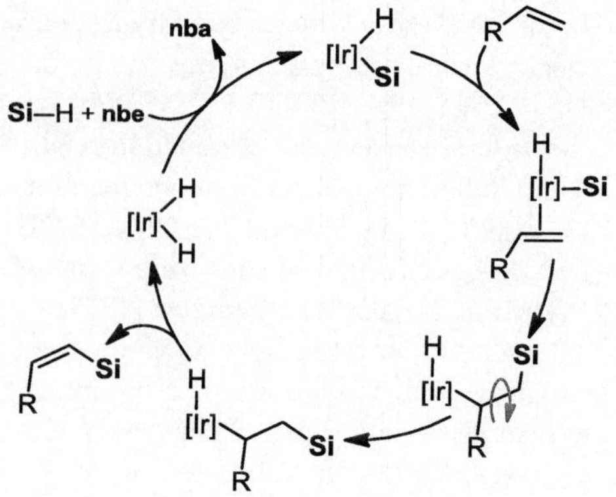

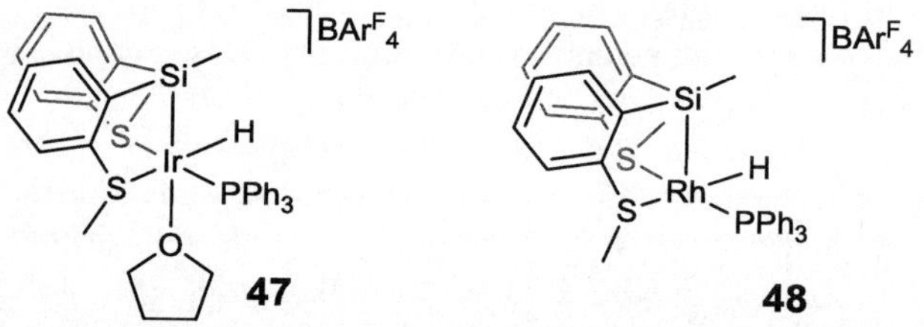

47 **48**

Fig. 19 Depiction of complexes **47** and its related Rh complex **48**

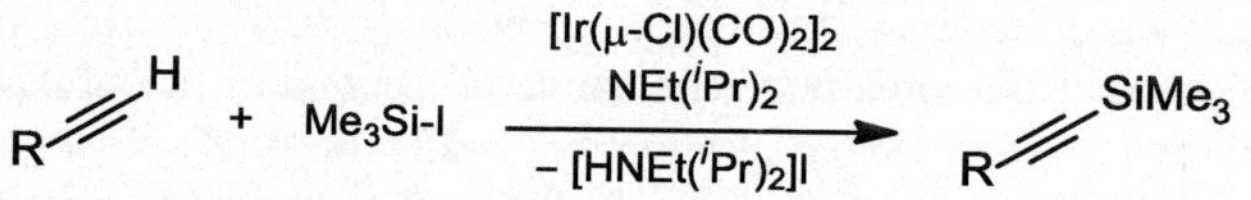

Scheme 24 Silylation of terminal alkynes with iodosilanes

substrate and hydrogen acceptor, affords the corresponding silylalkynes in yields up to 96% [139].

An alternative methodology for the preparation of silyl alkynes from terminal alkynes has been reported by the group of Marciniec, which consists of the use of iodosilanes instead of hydrosilanes in the presence of iridium catalysts and a base (Scheme 24) [140–143]. The reaction renders the silylated alkynes in nearly quantitative yields for a vast variety of alkynes using $[Ir(\mu\text{-}Cl)(CO)_2]_2$ as catalyst and NEt(^{i}Pr)$_2$ as base. Remarkably, this methodology also proved efficient using chlorosilanes as silicon source [143].

Based on stoichiometric experiments, the authors proposed a catalytic cycle that involves the formation in situ of the active species, namely, *trans*-[Ir(I)(CO){NEt(^{i}Pr)$_2$}$_2$], which undergoes oxidative addition of the alkyne's C–H bond to afford an Ir(III) intermediate. The base (NEt(^{i}Pr)$_2$) deprotonates the complex to give [HNEt(^{i}Pr)$_2$] and the Ir(I) intermediate *cis*-[Ir(R-C ≡ C)(CO){NEt(^{i}Pr)$_2$}$_2$]. Subsequent oxidative addition of Me_3Si-I leads to the formation of an Ir(III) species that, via

Scheme 25 Catalytic cycle proposed by Marciniec et al. for the coupling of terminal alkynes with Me_3Si-I

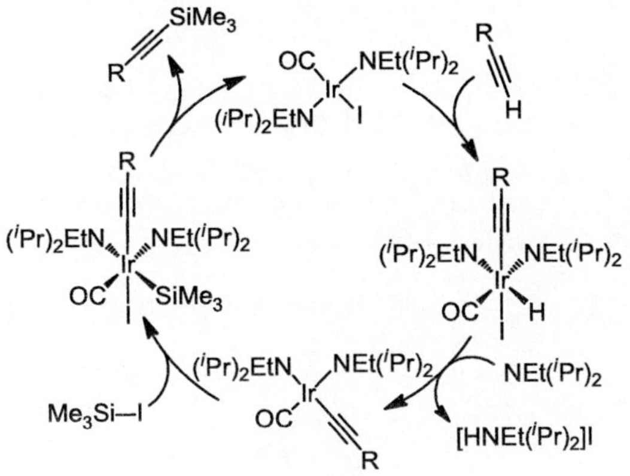

reductive elimination, yields the silylalkyne with concomitant regeneration of the active species (Scheme 25).

3.2 *Silylation of Arenes and Heteroarenes*

Pioneering work on the dehydrogenative silylation of arenes was disclosed by Curtis in 1982. In this work, the authors describe the use of Vaska's complex ([Ir(Cl)(CO)$(PPh_3)_2$]) as catalyst for the dehydrogenative coupling of benzene and hydrosilanes to yield phenylsilanes [144]. The reactivity studies suggest a mechanism whose first step consists of the abstraction of the chloride ligand from Vaska's complex by the silane with concomitant formation of the Ir^I-H active species. Subsequently, oxidative addition of one of the C–H bonds of benzene followed by reductive elimination of H_2 affords the Ir^I-Ph intermediate, which undergoes oxidative addition of the hydrosilane's Si–H bond. The resulting Ir^{III} intermediate yields the silylated product by reductive elimination and restarts the catalytic cycle (Scheme 26).

It was not until 2003 that the next Ir-catalyzed silylation of C–H bonds was described [145]. A variety of functionalized arenes (in large excess) were silylated using disilane $(^tBuF_2Si)_2$ by means of an iridium catalyst generated in situ from [Ir(μ-OMe)(COD)]$_2$ (1.5 mol%) and dtbp (3 mol%). This methodology allows the selective preparation of the least sterically hindered (*meta* or *para*) arylhalosilanes in high yields. Remarkably, these silylated compounds can be easily transformed via C-C coupling reactions. This work was later expanded to five-membered heteroarenes using the same methodology, but exploring phenanthroline-derived ligands, with the best results being obtained for the *ortho*-substituted tbphen ligand (2-*tert*-butyl-1,10-phenanthroline) and the bipyridine ligand dtbpy. The use of ^{t}Bu-*ortho*-disubstituted phenanthrolines precluded the catalytic activity for certain substrates, plausibly due to the high steric hindrance about the iridium center. The less encumbered tbphen (^{t}Bu-*ortho*-monosubstituted) ligand allows good activities while

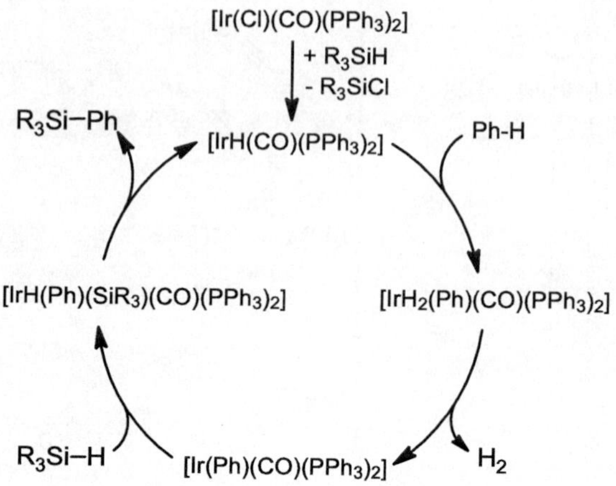

Scheme 26 Catalytic cycle proposed by Curtis et al. for the dehydrogenative silylation of benzene catalyzed by Vaska's complex

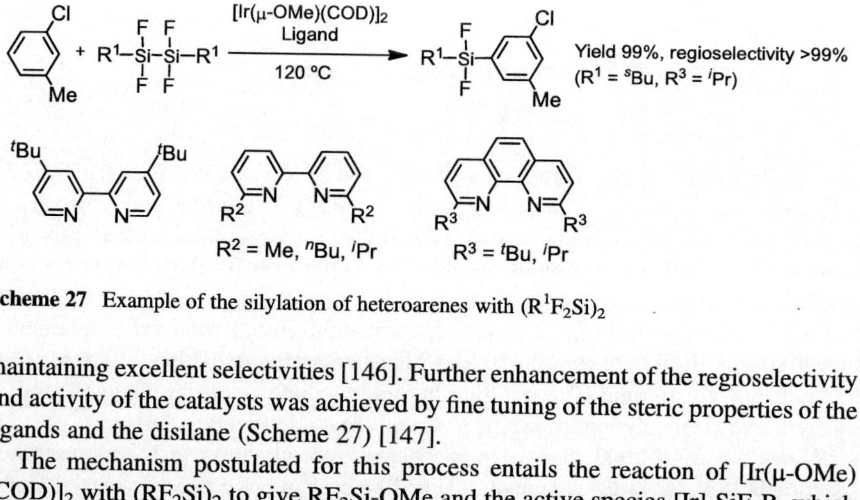

Scheme 27 Example of the silylation of heteroarenes with $(R^1F_2Si)_2$

maintaining excellent selectivities [146]. Further enhancement of the regioselectivity and activity of the catalysts was achieved by fine tuning of the steric properties of the ligands and the disilane (Scheme 27) [147].

The mechanism postulated for this process entails the reaction of [Ir(μ-OMe)(COD)]$_2$ with $(RF_2Si)_2$ to give RF_2Si-OMe and the active species [Ir]-SiF_2R, which may explain the need for this metal precursor. Subsequently, oxidative addition of the arene's C–H bond affords the Ir(III) intermediate [Ir]-(SiF_2R)(Ar)(H), which upon reductive elimination yields the silylated product and an Ir(I)-hydride intermediate. Alternatively, σ-bond metathesis would directly afford [Ir]-H and Ar-SiF_2R. [Ir]-H reacts now with a new molecule of disilane, via σ-bond metathesis or via an oxidative addition/reductive elimination process, to generate the active species and RF_2Si-H [147].

The silylation of indoles with $(^tBuF_2Si)_2$ was efficiently carried out with [Ir(μ-OMe)(COD)]$_2$ and dtbpy (1.5 and 3 mol%, respectively) prior protection of the NH moiety [146]. No protection of the nitrogen was required by using [Ir(μ-OMe)(COD)]$_2$ and dtbpy (5 and 10 mol%, respectively) with an affordable hydrosilane, Et_3SiH, in the presence of excess NBE, reaching yields up to 87%. This

Fig. 20 Depiction of [Bis (8-quinolyl)methylsilyl] iridium(III) complex **49**

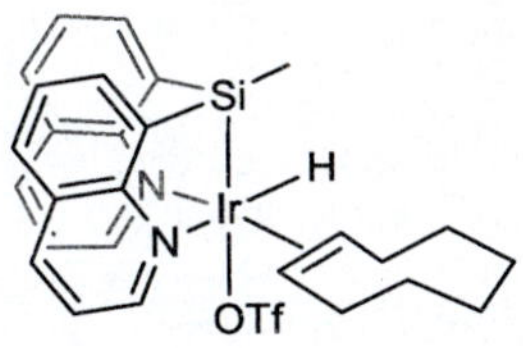

49

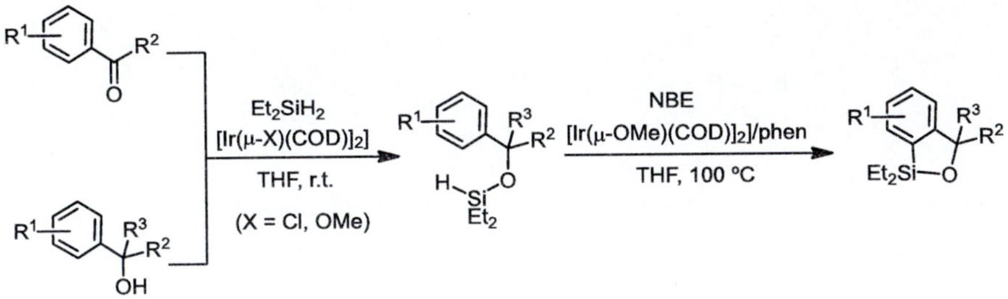

Scheme 28 One-pot *ortho*-silylation of benzyl alcohols, aryl ketones, and benzaldehyde derivatives

methodology also proved efficient for the selective *ortho*-silylation of a variety of thiophene and furan derivatives [148].

The dehydrogenative silylation of various deuterated aromatic solvents, namely, C_6D_6, C_6D_5Cl, $C_6D_5CF_3$, and $C_6D_5CH_3$, was reported in 2007 by Tilley et al. using complex **49** (Fig. 20) as catalyst and Ph_3SiH as silicon source. TBE and NBE were used as sacrificial hydrogen acceptors, with the best results being obtained for NBE, because TBE may be silylated. A 10 mol% catalyst loading at 120 °C for 48 h afforded $Ph_3SiC_6D_5$ in 100% yield from C_6D_6; $Ph_3SiC_6H_4CF_3$ in 81% yield from $C_6H_5CF_3$ with a 15% of Ph_3SiF (dehalogenation side-product); $Ph_3SiC_6H_4Cl$ in 45% from C_6H_5Cl; and $Ph_3SiC_6H_4CH_3$ in 71% yield from $C_6H_5CH_3$ [149].

In the last decade, the number of Ir-based C-H silylation methodologies has experienced an outstanding growth ([130]; for recent reviews on C-H silylation, see [150, 151]). A commonly used strategy comprises the use of directing groups to facilitate the selective activation of *ortho*-C–H bonds by formation of a cyclometallated intermediate. An alternative strategy makes use of tethered hydrosilanes, which can be introduced in the substrate by hydrosilylation or dehydrogenative coupling (Scheme 28). Hartwig et al. developed an Ir-catalyzed methodology for the *ortho*-silylation of benzyl alcohols, aryl ketones, and benzaldehyde derivatives. Benzyl alcohols can be transformed into tethered hydrosilyl ethers by dehydrogenative coupling with Et_2SiH_2 catalyzed by $[Ir(\mu\text{-}Cl)(COD)]_2$, while carbonyl compounds are reduced to the corresponding hydrosilyl ethers using $[Ir(\mu\text{-}X)(COD)]_2$ (X = OMe, Cl) as catalyst. Subsequently, the *ortho*-selective dehydrogenative silylation to give the corresponding benzoxasilole derivative was catalyzed by $[Ir(\mu\text{-}OMe)(COD)]_2$/phen (phen = 1,10-phenanthroline) in the

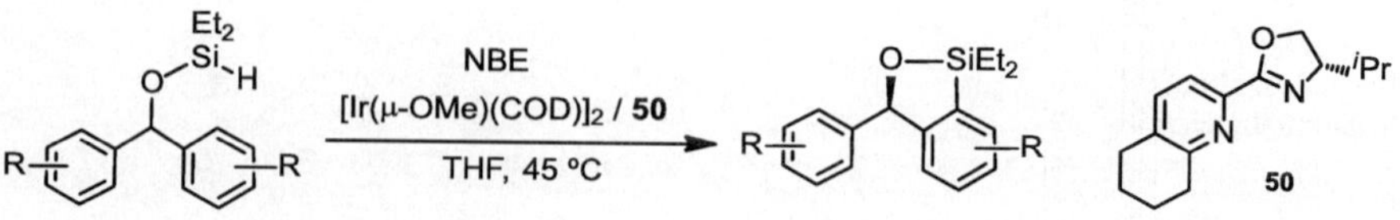

Scheme 29 Silylation of symmetric diarylmethoxy diethylsilanes with chiral pyridinyloxazoline ligand **50**

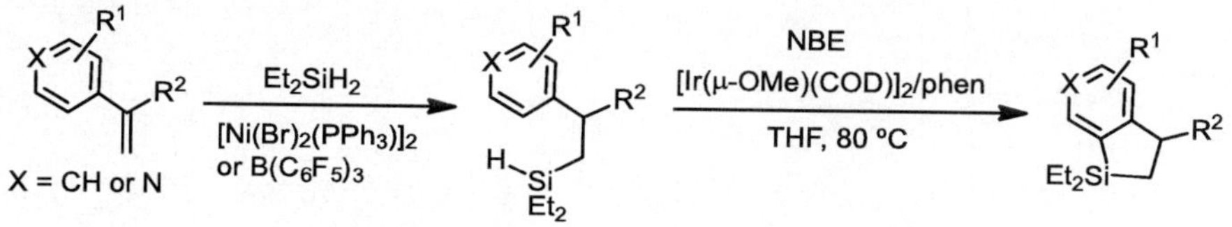

Scheme 30 One-pot *ortho*-silylation of styrenes to give dihydrobenzosiloles

presence of NBE (1.2 equivalents). This reaction can be performed in the absence of a hydrogen acceptor using $[Ir(\mu\text{-}OMe)(COD)]_2$/dtbpy as catalyst; however, somewhat lower yields were obtained. It is noteworthy that, in this work, the benzoxasilole derivatives were prepared by a one-pot procedure and functionalized successively by Hiyama coupling and Tamao-Fleming oxidation [152].

An asymmetric version of this methodology has also been developed employing a chiral pyridinyloxazoline ligand (**50**) instead of 1,10-phenanthroline, reducing the temperature to 45 °C. The system proved efficient for the silylation of symmetric diarylmethoxy diethylsilanes achieving enantiomeric excesses up to 93% and yields up to 93% (Scheme 29). Moreover, this catalytic system was able to resolve racemic mixtures of unsymmetric diarylmethoxy diethylsilanes at 35 °C, thus affording enantioenriched benzoxasiloles by kinetic resolution in enantiomeric excesses up to 99% and yields up to 42% [153].

Following a similar methodology, dihydrobenzosiloles were prepared from styrene derivatives or vinylpyridines in a one-pot procedure. The first step requires the selective formation of the β-alkyl(hydro)silane by hydrosilylation of the styrene derivative with diphenylsilane catalyzed by $[Ni(Br)_2(PPh_3)_2]$ or $B(C_6F_5)_3$ (for α-substituted styrenes). Subsequently, the dehydrogenative *ortho*-silylation was achieved in the presence of 1.2 equivalents of NBE catalyzed by $[Ir(\mu\text{-}OMe)(COD)]_2$/dtbpy (Scheme 30). This procedure shows excellent tolerance for a variety of functional groups, and the dihydrobenzosiloles thus formed were further functionalized to generate benzosiloles, furanes, 2-hydroxyphenethyl alcohols, or halides [154, 155].

The dehydrogenative *ortho*-silylation of a variety of arenes with excess Et_3SiH and NBE at 115 °C was achieved using an Ir(NHC) catalyst. This selectivity is due to the presence of *N*-directing functional groups, such as pyridines, pyrazoles, or imines, which promote the activation of the *ortho*-C–H bond upon Ir-N coordination and subsequent cyclometalation. The perimidine-based NHC (**51**) features xylyl

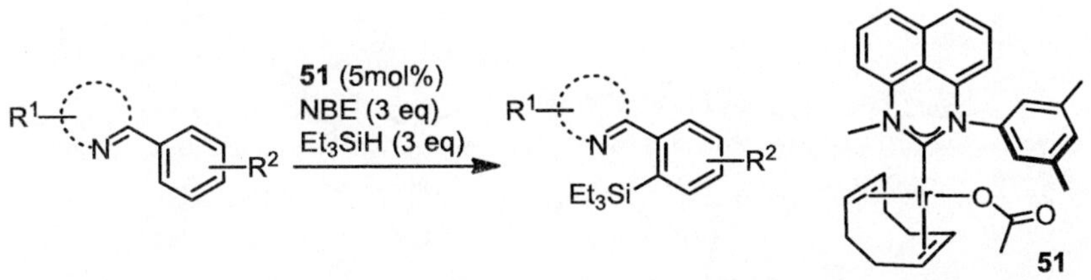

Scheme 31 Directed dehydrogenative silylation of N-functionalized arenes catalyzed by **51**

wingtip groups that can undergo cyclometalation reversibly, which is crucial for the activity of the catalyst (Scheme 31) [156].

The proposed mechanism requires the activation of **51** by reaction with 2-phenylpyridine to give a doubly cyclometalated complex that reacts with Et_3SiH to afford the Ir-hydride active species. Subsequently, a new molecule of Et_3SiH oxidatively adds, which prompts the decyclometallation of the wingtip group. Reaction with NBE yields a silyl species with a cyclometalated xylyl moiety and NBA. Then, Et_3SiH coordinates to the vacant coordination site, and, by a σ-CAM pathway (TS in Scheme 32), the silylated arene is formed. Finally, the catalytic cycle is restarted upon reaction with 2-phenylpyridine. Therefore, the xylyl moiety of the NHC behaves as an anionic hemilabile substituent throughout the catalytic cycle, thus promoting the dehydrogenative silylation reaction.

Another interesting system for the dehydrogenative silylation of arenes, directed by *N*-containing functional groups, is the use of Ir-complex [Ir(acac)(COD)] (acac = acetylacetonate) as catalyst in the presence of NBE using Ph_2FSiH as reducing agent. This permits the gram-scale *ortho*-selective C–H silylation of several types of substrates in excellent yields. The fluorinated silicon and nitrogen atoms of the substrates bind by means of Lewis acid-Lewis base interaction, which gives rise to fluorescent compounds with high quantum yields [157].

The acceptorless dehydrogenative silylation of neat arenes was achieved at 120 °C with a 2/1 mixture of $[Ir(\mu\text{-}OMe)(COD)]_2$ and 2,9-dimethyl-1,10-phenanthroline (dmphen) as catalyst precursor (Scheme 33). The process was selective toward the least sterically hindered *meta* or *para* positions, e.g., 1,3- and 1,2-chloroarenes undergo exclusively *meta*-silylation. A protected triethoxysilane, namely, 1-hydrosilatrane, was used as reducing agent, thus allowing further functionalization of the silylated products via Hiyama cross-coupling reaction [158].

The use of synthetically relevant hydrosilane $HSiMe(OSiMe_3)_2$ was successfully explored in the non-directed dehydrogenative silylation of arenes and heteroarenes using $[Ir(\mu\text{-}OMe)(COD)]_2$ and 2,4,7-trimethylphenanthroline as catalyst precursor in the presence of 1 equivalent of a hydrogen acceptor [159]. This methodology allows the use of the aromatic compound as the limiting reagent, proceeds with high regioselectivity, and shows an excellent functional groups tolerance, improving that of the related Rh-catalyzed system [10]. Moreover, a broad variety of substrates, including pharmaceutically relevant compounds, were silylated efficiently. The resulting silylated compounds proved easy to functionalize by Hiyama cross-

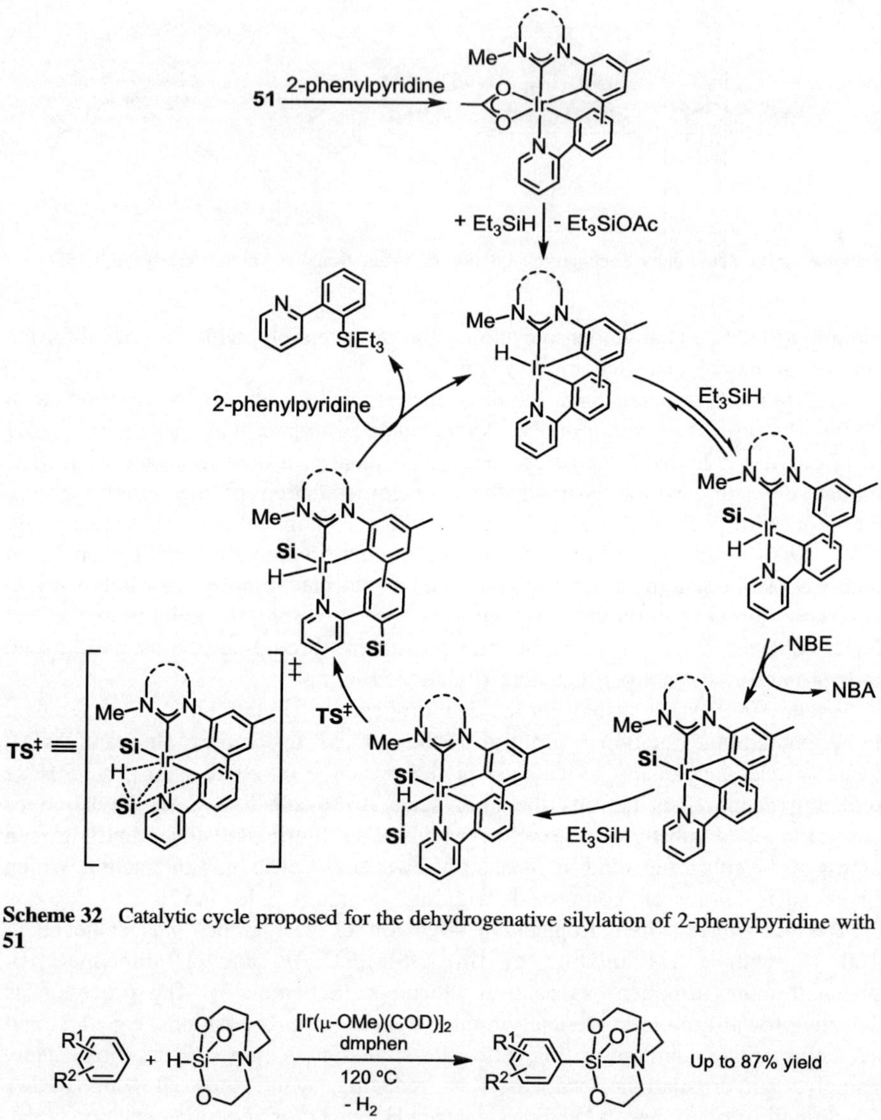

Scheme 32 Catalytic cycle proposed for the dehydrogenative silylation of 2-phenylpyridine with **51**

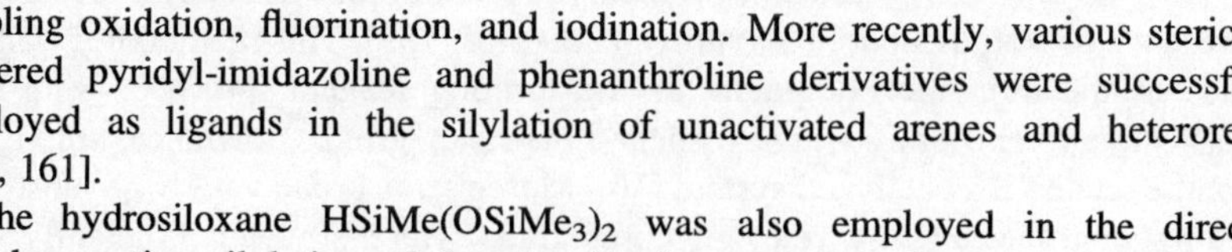

Scheme 33 Acceptorless silylation of arenes with 1-hydrosilatrane

coupling oxidation, fluorination, and iodination. More recently, various sterically hindered pyridyl-imidazoline and phenanthroline derivatives were successfully employed as ligands in the silylation of unactivated arenes and heterorenes [160, 161].

The hydrosiloxane $HSiMe(OSiMe_3)_2$ was also employed in the directed dehydrogenative silylation of functionalized arenes and alkanes, using [Ir(μ-Cl)

Scheme 34 Ir-catalyzed intramolecular silylation of arenes featuring tethered hydrosiloxanes

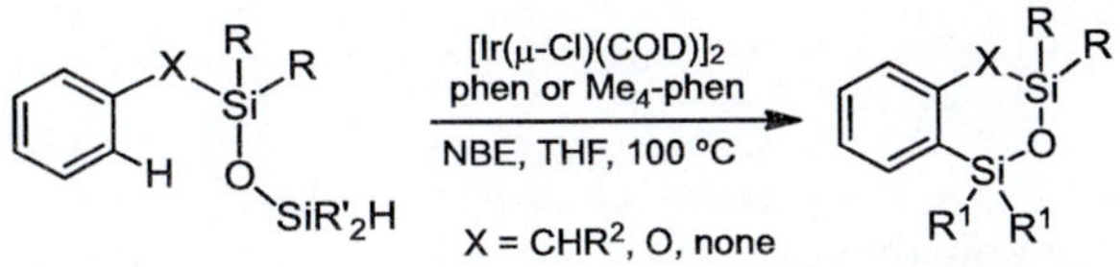

(COD)]$_2$ (2 mol%) as catalyst precursor, 20 mol% DIPEA (diisopropylethylamine) as additive, and 2 equivalents of TBE as hydrogen acceptor. Several nitrogen-based directing groups (imines, hydrazone, oxazoline, pyridine, *N,N*-dimethylamine, and oxime ether) proved efficient to give the corresponding monosubstituted organosilane in hexane at 80 °C [162].

The use of [Ir(μ-OMe)(COD)]$_2$ (5 mol%) as catalyst precursor with phenanthroline-derived ligands also proved efficient for the silylation of polycyclic aromatic molecules (2 equivalents) with Et_3SiH and 3,3-dimethyl-1-butene as hydrogen acceptor at 100 °C in dioxane. The regioselectivity of the process was sterically and electronically controlled, leading to a variety of selectively mono- and di-silylated polyaromatic compounds based on naphthalene, phenanthrene, anthracene, pyrene, and azulene scaffolds, with a divergent chemo- and regioselectivity when compared to that expected for electrophilic functionalization [163, 164].

Hydrosiloxane-tethered arenes, prepared from aryl and benzyl silanols, were converted by intramolecular dehydrogenative *ortho*-silylation into 5- or 6-membered cyclic siloxanes, respectively, fused with the arene (Scheme 34). The reaction shows a wide scope, including bio-relevant compounds that can be additionally functionalized by, for example, selective and simultaneous iodination/chlorination of the two newly formed C–Si bonds. This methodology employs a mixture of [Ir(μ-Cl)(COD)]$_2$ and phen or Me$_4$phen (3,4,7,8-tetramethyl-1,10-phenanthroline) as catalyst, NBE as hydrogen acceptor, and THF as solvent at 100 °C [165].

The NHC-Ir(III) complex **52**, [Ir(H)$_2$(IPr)(py)$_3$][BF$_4$] (IPr = 1,3-bis-(2,6-diisopropylphenyl)imidazol-2-ylidene), catalyzes the directed and non-directed dehydrogenative silylation of a broad range of arenes and heteroarenes, which were used in all cases as limiting reagent [166]. This methodology is amenable to a variety of hydrosilanes, namely, Et_3SiH, Ph_2MeSiH, $PhMe_2SiH$, Ph_3SiH, and, for the first time, $(EtO)_3SiH$. The reaction proceeds without the need for a hydrogen acceptor; however, the reaction rates are significantly higher in the presence of NBE. Moreover, the use of bisarylated bis(silanes) led to the selective formation of bis(hydrosilane)s by reaction with arenes and heteroarenes, which contrasts with previous examples where no monoarylation was described [167]. The proposed catalytic cycle requires as first step the dehydrogenation of **52** (thermic or via hydrogen acceptor) to afford an Ir(I) species, which is able to coordinate the substrate (2-phenylpyridine) by substitution of a pyridine ligand. Subsequently, cyclometalation of 2-phenylpyridine takes place to give an Ir(III) intermediate, with concomitant pyridine decoordination. The vacant coordination site left by the pyridine ligand is occupied by the hydrosilane, which coordinates end-on. The silylation of the *ortho*-C–H bond takes place by a σ-CAM pathway

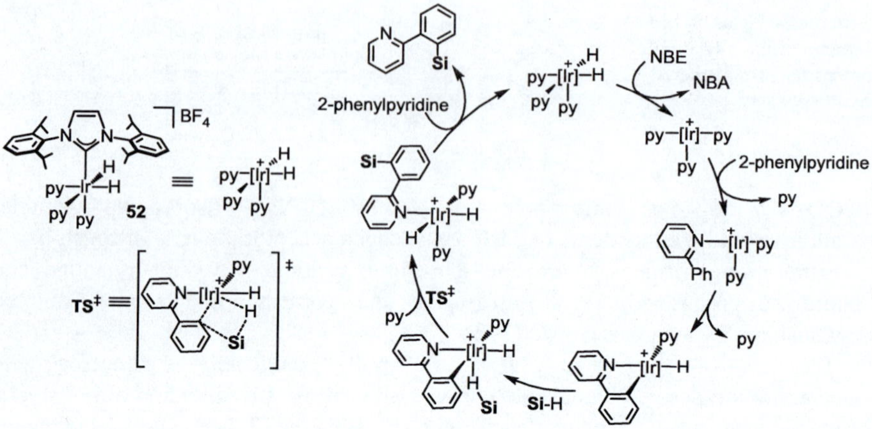

Scheme 35 Mechanism proposed for the dehydrogenative silylation of 2-phenylpyridine catalyzed by **52**

(transition state; Scheme 35). Finally, substitution of the silylated substrate by pyridine regenerates **52**.

A one-pot methodology has been developed for the sequential silylation-borylation of heteroarenes using [Ir(μ-OMe)(COD)]$_2$ as catalyst and Et_3SiH, in the presence of TBE as hydrogen acceptor. Several substrates were studied, among them substituted quinolines, which were selectively silylated in the 2-position, presumably due to Ir-N coordination. Subsequent borylation was less selective, plausibly because the remaining C–H bonds are similar from an electronic and steric viewpoint. Thiophenes and furanes substituted at the 2-position were selectively silylated in the presence of dtbpy (4,4′-di-tert-butyl-2,2′-bipyridyl) at the 5-position and borylated at the 3-position. Indene was silylated selectively at the C-H adjacent to the heteroatom, while borylation occurred at the 7-position [168]. Further development of this methodology led to the sequential silylation-borylation of a variety of heteroaromatic substrates featuring functional groups that permit a regioselective silylation step [169].

The [Ir(μ-OMe)(COD)]$_2$/dtbpy system also showed good activity for the *ortho*-C–H silylation of the phenyl group of 6-phenyl-7-deazapurines and 6-phenyl-9-deazapurines with Et_3SiH in the presence of NBE. This selectivity has been ascribed to the directing effect of the N1 nitrogen, being silylation at the deazapurine scaffold a minor product (Scheme 36) [170].

The use of [Ir(μ-OMe)(COD)]$_2$/Me$_4$phen in the presence of NBE allows the silylation and bis-silylation of heteroarenes to yield organo-{2-(hydroxymethyl) phenyl}-dimethylsilanes (organo-HOMSi) that were used as efficient cross-coupling reagents (Scheme 37) [171]. The protection of the hydrosilane's OH group (HOMSi-H) was required, being tetrahydropyran (THP) the preferred protecting group.

The intramolecular dehydrogenative C–H silylation of 2′,6′-diaryl-2-(hydrosilyl) biphenyls catalyzed by [Ir(μ-Cl)(COD)]$_2$ (10 mol%) and DPPBen (1,2-bis

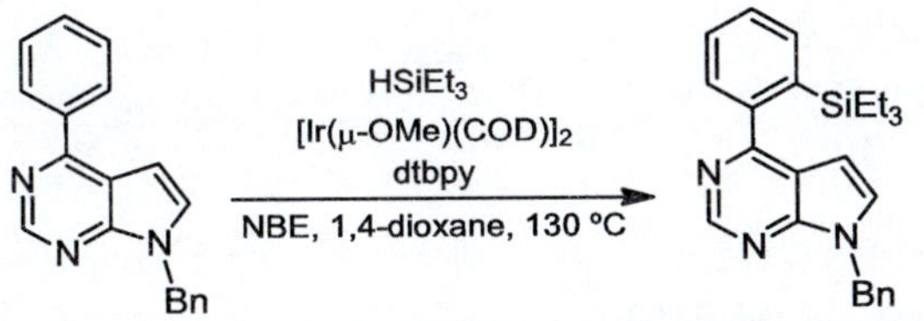

Scheme 36 Dehydrogenative silylation of 6-phenyl-7-deazapurine

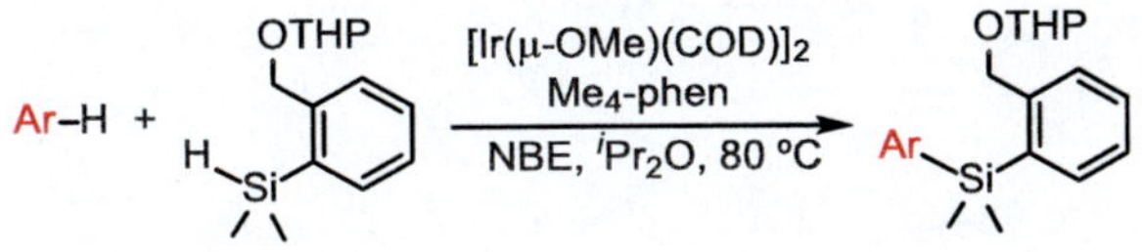

Scheme 37 Dehydrogenative silylation of heteroarenes with THP-protected HOMSi-H

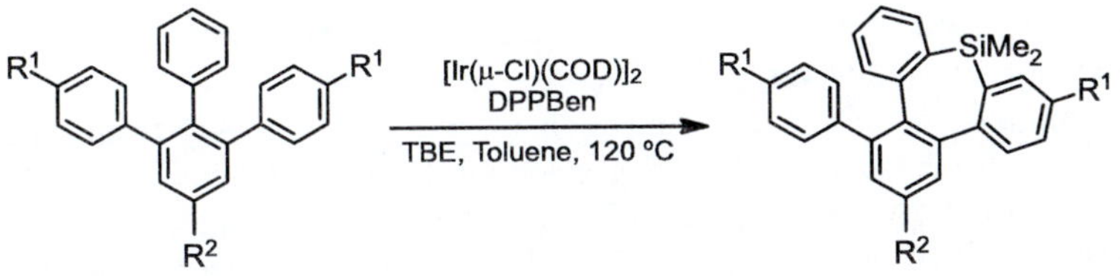

Scheme 38 Catalytic preparation of tribenzosilepins by dehydrogenative silylation

(diphenylphosphino)-benzene) (20 mol%) permitted, for the first time, the synthesis of a several tribenzosilepins by a catalytic process (Scheme 38) [172].

3.3 *Silylation of Alkanes*

The selective silylation of primary C–H bonds was achieved by intramolecular dehydrogenative silylation catalyzed by [Ir(μ-OMe)(COD)]$_2$ and phenanthroline-based ligands – the best results being obtained with Me$_4$phen. The tethered hydrosilane was introduced in the molecule via hydrosilylation of ketones or dehydrogenative coupling with alcohols using dihydrosilanes (Scheme 39). Subsequently, intermolecular dehydrogenative silylation of the primary γ-C–H bond yields the C(sp^3)-Si functionalized product. The oxasilolane derivatives thus obtained were transformed into diols by Tamao-Fleming oxidation. This strategy for the directed γ-functionalization is compatible with a broad scope of alcohols and ketones with different auxiliary groups and proved efficient for the functionalization of intricate natural products [173–175].

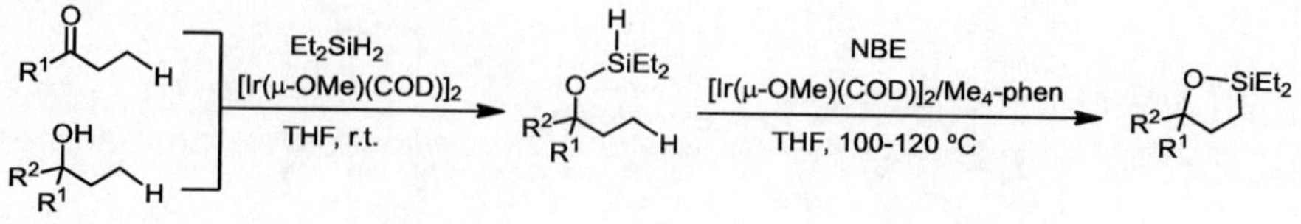

Scheme 39 Hydroxyl- or carbonyl-directed γ-silylation of primary C–H bonds

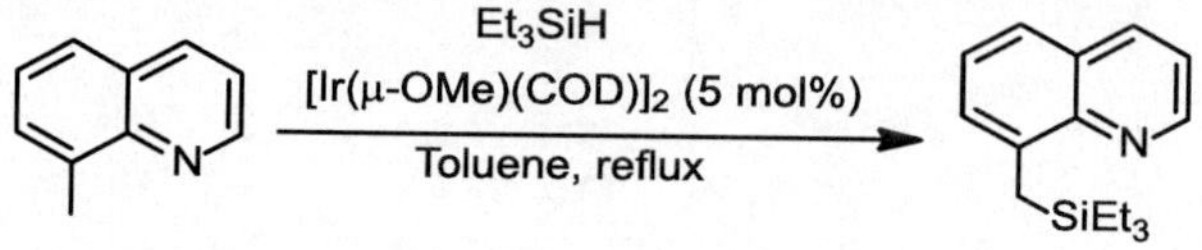

Scheme 40 Directed dehydrogenative silylation of 8-methylquinoline

A related methodology was employed for the synthesis of L-sugars from their parent deoxy derivatives. For example, L-glycopyranosyl donors were prepared via selective silylation of the methyl group's C-H in γ-position to the OH at the C4 of their parent 6-deoxy-l-hexoses [9, 176].

The mechanism of γ-functionalization of primary C(sp^3)–H bonds has been studied by DFT calculations resulting in two different postulations, an Ir(I)-Ir(III) and an Ir(III)-Ir(V) catalytic cycle [177, 178]. The latter was found to be more likely due to the fact that the formation of [Ir{Si(OR)Et_2}(Me_4phen)] is more favorable than that of [Ir(H)(Me_4phen)(norbornene)] – the Ir-hydride complex being the active species in the Ir(I)-Ir(III) catalytic cycle. The Ir(III)-Ir(V) catalytic cycle comprises initially the formation of the active species by reaction of [Ir{Si(OR)Et_2}(Me_4phen)] with 1 equivalent of HSi(OR)Et_2 to give [Ir(H){Si(OR)Et_2}$_2$(Me_4phen)]. Then, NBE coordination and migratory insertion into the Ir–H bond renders [Ir(norbornyl){Si(OR)Et_2}$_2$(Me_4phen)], which reacts with a new molecule of hydrosilane by oxidative addition. Reductive elimination gives [Ir{Si(OR)Et_2}$_3$(Me_4phen)] with concomitant formation of NBA. Subsequently, intramolecular C–H oxidative addition of the γ-C–H bond followed by reductive elimination affords the oxasilolane derivatives and regenerates the active species.

The directed intermolecular dehydrogenative silylation of the aliphatic C–H bond of 8-methylquinoline (Scheme 40) and related fused ring systems was achieved using [Ir(μ-Cl)(COD)]$_2$ as catalyst, with Et_3SiH in toluene under reflux. The resulting organosilanes were converted into esters via carboxylation with CO_2 gas in the presence methyl iodide [179]. A similar methodology was employed for the *N*-directed silylation of C(sp^3)–H bonds with HSiMe($OSiMe_3$)$_2$ and TBE as hydrogen acceptor [162].

The undirected intermolecular dehydrogenative silylation of aliphatic C–H bonds adjacent to the nitrogen of 2-dimethylaminopyridine was achieved using [Ir(μ-Cl)(COD)]$_2$ as catalyst with Et_3SiH. The monosilylated compound was obtained in yields up to 76% without the need for a hydrogen acceptor (Scheme 41). The bis-silylated product was also observed in small amounts together with other by-products. Significantly, the silylated products were converted into α-amino

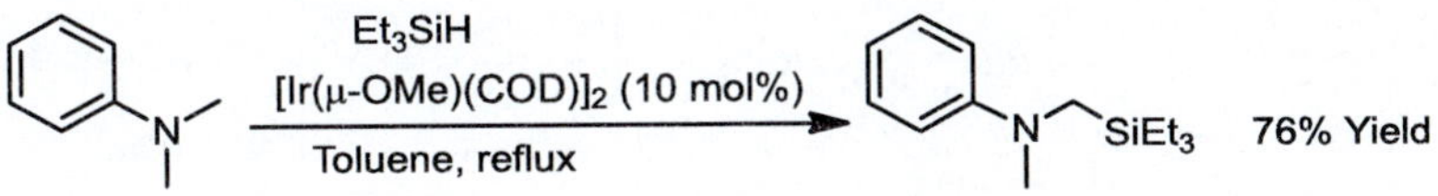

Scheme 41 Intermolecular dehydrogenative silylation of N–C(sp^3)–H bonds

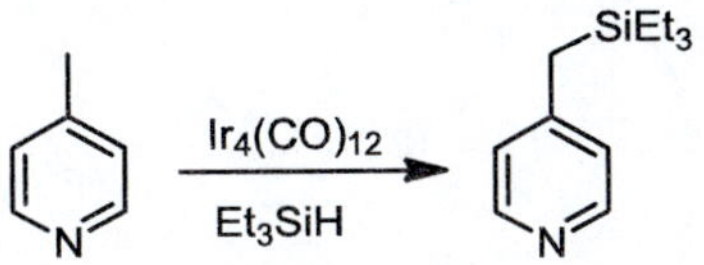

Scheme 42 Benzylic C-H silylation of 4-methylpyridine catalyzed by $Ir_4(CO)_{12}$

acids under CO_2 (1 atm). The silylation of 2-alkylpyridines (R = Et, ^{i}Pr) was also accomplished, although yields below 34% were obtained [180].

The dehydrogenative silylation of C(sp^3)–H bonds at the benzylic position of *ortho-* and *para*-substituted pyridine derivatives was achieved employing $Ir_4(CO)_{12}$ as catalyst, NBE as hydrogen acceptor, and Et_3SiH as silicon source (Scheme 42) [181, 182]. Substrates that present *ortho-* and *para*-benzylic C–H bonds, such as 2,4-dimethylpyridine, undergo selective silylation at the *para*-position. In the case of *ortho*-alkyl pyridines, the use of 3,5-dimethylpyridine as an additive improves significantly the activity of the catalytic system. This behavior has been attributed to the ability of 3,5-dimethylpyridine to form a silylpyridinium intermediate that may transfer the silyl moiety to the benzylic position (vide infra) [182].

The proposed mechanism requires the coordination (end-on or side-on) of the hydrosilane, followed heterolytic splitting of the Si–H bond aided by the Lewis acid nature of the Ir center and the Lewis basic pyridine. Thus, an *N*-silylpyridinium and an Ir-H species are formed. Abstraction of the benzylic hydrogen from the *N*-silylpyridinium by the Ir-hydride affords a dihydride intermediate and a *N*-silylenamine. The *N*-silylenamine is silylated by the *N*-silylpyridinium or the hydrosilane-Ir species. Finally, dehydrogenation of the dihydride species restarts the catalytic cycle (Scheme 43). This mechanism explains why, in the case of 2,4-dimethylpyridine, the silylation is selective at the *para*-benzylic position – the silyl group at the *N*-silylpyridinium hinders the H-abstraction at the *ortho*-position.

Complex **53** combined with $NaBAr^F{}_4$ catalyzes the silylation of 2-alkyl-1,3-azoles at the α-C(sp^3)–H bond of the 2-alkyl group. As mentioned above in the previous example, the use of 3,5-dimethylpyridine as additive is crucial. It is noteworthy that the reaction proceeds in the absence of a hydrogen acceptor; however, this methodology is more efficient when cyclopentene or cyclohexene is employed (Scheme 44). The catalytic cycle was proposed to occur according to an ionic mechanism analogous to that depicted in Scheme 43 [183].

The dimerization of benzylmethylsilanes via dehydrogenative silylation was achieved by a successive functionalization of the methyl's C(sp^3)-H and the *ortho*'s C(sp^2)–H bonds. The reaction proceeds efficiently in the absence of a hydrogen acceptor using [Ir(μ-OMe)(COD)]$_2$ and 5,6-Me_2phen to generate the catalyst in situ (Scheme 45) [184].

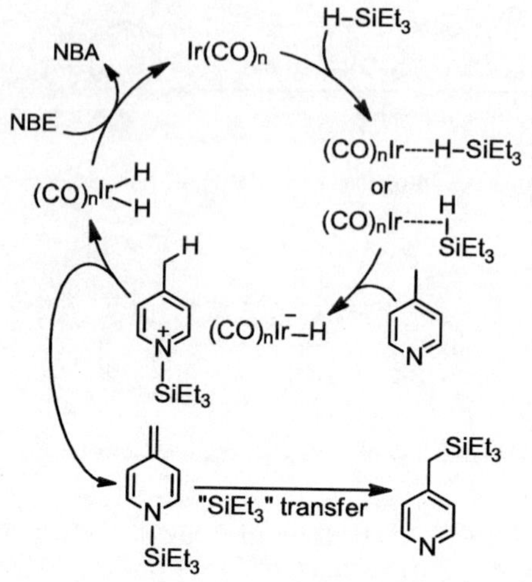

Scheme 43 Catalytic cycle proposed for the benzylic silylation of 4-methylpyridine

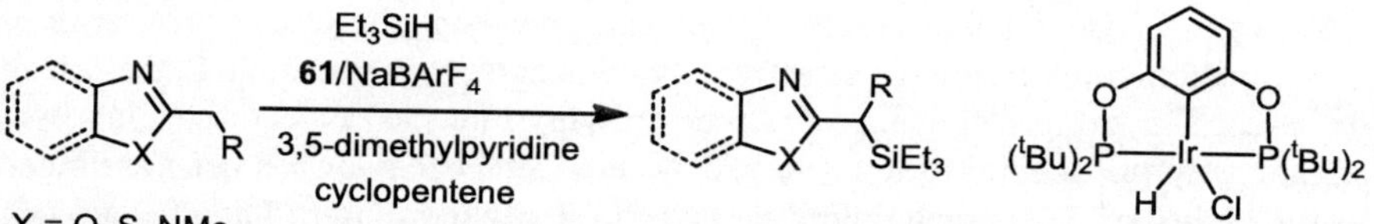

Scheme 44 Silylation of 2-alkyl-1,3-azoles catalyzed by **53**/NaBAr$^F{}_4$

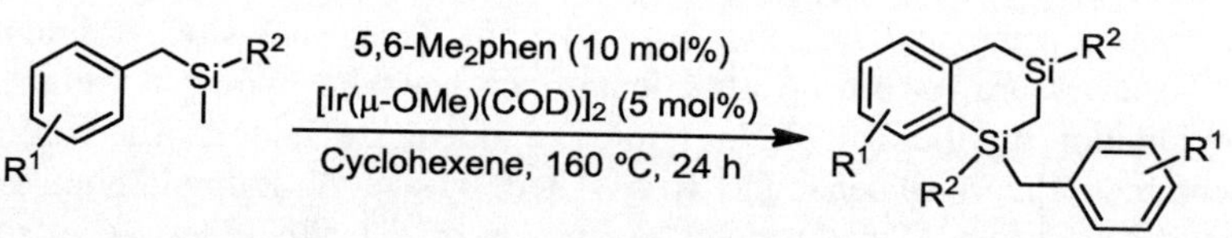

Scheme 45 Dimerization of benzylmethylsilanes by dehydrogenative silylation

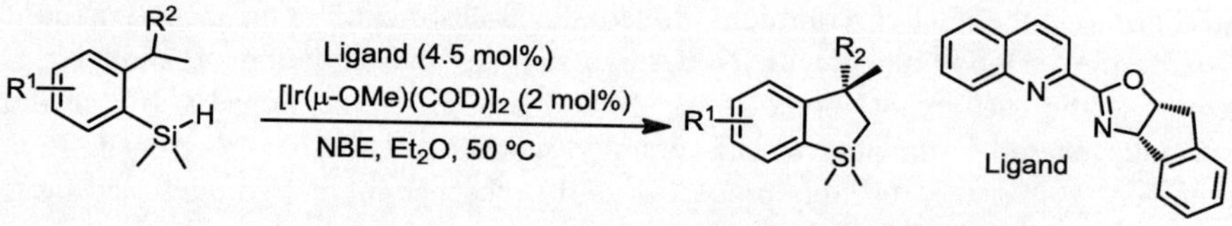

Scheme 46 Preparation of dihydrobenzosiloles by intramolecular C(sp^3) – H silylation

Dihydrobenzosiloles were prepared in high yields and excellent enantioselectivities by the dehydrogenative silylation of primary C(sp^3)–H bonds, using $[Ir(\mu\text{-}OMe)(COD)]_2$ and adequate *N,N*-chiral ligands in the presence of NBE (Scheme 46). The resulting silylated products were transformed into halocarbon and hydroxylated compounds. It is noteworthy that this methodology proved efficient for

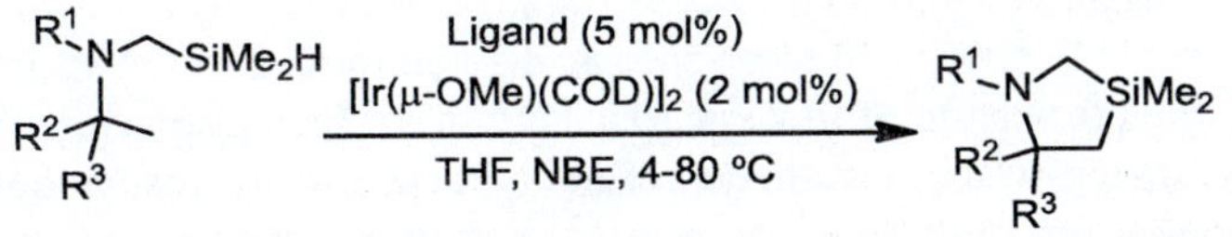

Scheme 47 *β*-Functionalization of aliphatic amines

the functionalization of natural products [185]. The mechanism of this reaction has been recently studied by DFT calculations [186], suggesting an Ir(III)-Ir(V) catalytic cycle. The active species, a pentacoordinate $[Ir(H)_2(SiMe_2Ar)(N,N\text{-ligand})]$ complex, undergoes dehydrogenation by NBE and concerted Si-H activation to give $[Ir(H)(SiMe_2Ar)_2(N,N\text{-ligand})]$. Subsequently, intramolecular $C(sp^3)$–H bond activation affords the corresponding cyclometalated Ir(V) species, which via reductive elimination affords the organosilane and regenerates de active species.

The *β*-functionalization of unactivated $C(sp^3)$–H bonds of aliphatic amines was accomplished via intramolecular dehydrogenative silylation employing a cleavable directing group, which is removed upon Tamao-Fleming oxidation with concomitant formation of 1,2-amino alcohols. The directing group was introduced by reaction of the amine with chloromethyl(dimethyl)silane. The resulting methylene-linked hydrosilane undergoes intramolecular dehydrogenative silylation catalyzed by $[Ir(\mu\text{-OMe})(COD)]_2$ (2 mol%) to form silapyrrolidines enantioselectively in the presence of chiral pyridyl-imidazoline ligands (Scheme 47).

4 Concluding Remarks

This chapter summarizes the most relevant advances in iridium-catalyzed hydrosilylation and C–H bond silylation hitherto reported. The hydrosilylation of unsaturated polar bonds permits the synthesis of a wide range of alcohols and amines, while the hydrosilylation of carbon–carbon multiple bonds and the silylation of C–H bonds furnish alkyl, aryl, alkenyl, and alkynyl silanes. Regarding the hydrosilylation reactions, Ir catalysts have shown interesting performances, even surpassing their extensively studied Rh counterparts in some cases. Interestingly, an inversion of the absolute configuration of silyl ethers – compared to their related Rh catalysts – has been often observed in the hydrosilylation of ketones. The Ir-catalyzed hydrosilylation of alkenes and alkynes usually occurs according to an anti-Markovnikov addition. In the case of the latter, selective formation of β-(Z)-vinylsilanes has been achieved. Concerning the silylation of C–H bonds, iridium complexes proved exceptionally efficient catalysts, especially those generated in situ from $[Ir(\mu\text{-OMe})(COD)]_2$ and *N,N*-chelating ligands. In this regard, it is worth highlighting that prominent examples of selective $C(sp^3)$–H bond silylation have been reported, even allowing the synthesis of bio-relevant molecules. A variety of reaction mechanisms have been suggested to operate in the reactions commented above, for instance, Glaser-Tilley-type, ionic, and classical inner-sphere mechanisms, where Ir(I), Ir(III) and Ir(V) species may be formed. σ-CAM pathways that circumvent the formation of Ir(V) intermediates have been also invoked.

Finally, owing to their synthetic versatility, selective hydrosilylation and C–H bond silylation reactions show great promise for the functionalization of organic molecules – where Ir catalysts may play a central role. However, intense research on reactivity trends, substrate scope, and mechanistic understanding are still required to improve the applicability of these methodologies.

Acknowledgements This work was supported by the "Ramón y Cajal" program (RYC2016-20864) (FSE/AgenciaEstatal de Investigación) (M. I.) and the Spanish Ministry of Science, Innovation and Universities (RTI2018-099136-A-I00).

References

1. Marciniec B, Pietraszuk C, Kownacki I, Zaidlewicz M (2005) Katrizky AR, Taylor JK (eds) Comprehensive organic functional group transformations. Elsevier, Oxford, p 941
2. Nakao Y, Hiyama T (2011) Chem Soc Rev 40:4893
3. Ojima I (1989) Patai S, Rappoport Z (eds) The chemistry of organic silicon compounds. Wiley, New York, p 1479
4. Ojima I, Li Z, Zhu J (1998) Rappoport Z, Apeloig Y (eds) The chemistry of organic silicon compounds. Wiley, New York, p 1687
5. Gribble GW, Li JJ (2006) Gribble GW, Li JJ (eds) Palladium in heterocyclic chemistry: a guide for the synthetic chemist. Elsevier, Oxford, p 12
6. Fleming I, Dunogues J, Smithers R (1989) Kende AS (ed) Organic reactions, vol 2. Wiley, New York, p 57
7. Luh TY, Liu ST (2003) Rappoport YAZ (ed) The chemistry of organic silicon compounds, vol 2. Wiley, Chichester, p 1793
8. Lam PYS, Deudon S, Averill KM, Li R, He MY, DeShong P, Clark CG (2000) J Am Chem Soc 122:7600
9. Frihed TG, Heuckendorff M, Pedersen CM, Bols M (2012) Angew Chem Int Ed 51:12285
10. Cheng C, Hartwig JF (2014) Science 343:853
11. Ihara H, Suginome M (2009) J Am Chem Soc 131:7502
12. Tang P, Ritter T (2011) Tetrahedron 67:4449
13. Ponomarenko SA, Kirchmeyer S (2010) Muzafarov AM (ed) Silicon polymers. Springer-Verlag, Heidelberg, p 36
14. Holmes RJ, D'Andrade BW, Forrest SR, Ren X, Li J, Thompson ME (2003) Appl Phys Lett 83:3818
15. Ren X, Li J, Holmes RJ, Djurovich PI, Forrest SR, Thompson ME (2004) Chem Mater 16:4743
16. Bin JK, Cho NS, Hong JI (2012) Adv Mater 24:2911
17. Lin JJ, Liao WS, Huang HJ, Wu FI, Cheng CH (2008) Adv Funct Mater 18:485
18. Tsai MH, Lin HW, Su HC, Ke TH, Wu CC, Fang FC, Liao YL, Wong KT, Wu CI (2006) Adv Mater 18:1216
19. Leung MK, Yang WH, Chuang CN, Lee JH, Lin CH, Wei MK, Liu YH (2012) Org Lett 14:4986
20. Han WS, Son HJ, Wee KR, Min KT, Kwon S, Suh IH, Choi SH, Jung DH, Kang SO (2009) J Phys Chem C 113:19686

21. Oro LA, Claver C (2009) Iridium complexes in organic synthesis. Wiley-VCH Verlag GmbH & Co. KGaA, Weinheim
22. Lightfoot A, Schnider P, Pfaltz A (1998) Angew Chem Int Ed 37:2897
23. Pfaltz A (2007) Acc Chem Res 40:1402
24. Bell S, Wüstenberg B, Kaiser S, Menges F, Netscher T, Pfaltz A (2006) Science 311:642
25. Crabtree RH, Felkin H, Fillebeen-Khan T, Morris GE (1979) J Organomet Chem 168:183
26. Crabtree RH (1979) Acc Chem Res 12:331
27. Jones JH (2000) Platin Met Rev 44:94
28. Blaser HU, Buser HP, Coers K, Hanreich R, Jalett HP, Jeslch E, Pugin B, Schneider HD, Splinder F, Wegmann A (1999) Chimia 53:275
29. Blaser HU (2002) Adv Synth Catal 344:17
30. Hofer R (2005) Chimia 59:10
31. Dorta R, Broggini D, Stoop R, Rüegger H, Spindler F, Togni A (2004) Chem Eur J 10:267
32. Nishibayashi Y, Segawa K, Ohe K, Uemura S (1995) Organometallics 14:5486
33. Faller JW, Chase KJ (1994) Organometallics 13:989
34. Kinting A, Kreuzfeld HJ (1989) Organomet Chem 370:343
35. Apple DC, Brady KA, Chance JM, Heard NE, Nile TA (1985) J Molec Catal 29:55
36. Nishibayashi Y, Segawa K, Takada H, Ohe K, Uemura S (1996) Chem Commun:847
37. Nishiyama H (1999) Jacobsen E, Pfaltz A, Yamamoto H (eds) Comprehensive asymmetric catalysis I-III. Springer-Verlag, Berlin, p 267
38. Carpentier JF, Bette V (2002) Curr Org Chem 6:913
39. Klei SR, Tilley TD, Bergman RG (2002) Organometallics 21:4648
40. Gómez M, Jansat S, Muller G, Bonnet MC, Breuzard JAJ, Lemaire M (2002) J Organomet Chem 659:186
41. Karamé I, Tommasino ML, Lemaire M (2003) J Mol Catal A 196:137
42. Santacruz E, Huelgas G, Angulo SK, Mastranzo VM, Hernández-Ortega S, Aviña JA, Juaristi E, Anaya de Parrodi C, Walsh PJ (2009) Tetrahedron Asymmetry 20:2788
43. Frölander A, Moberg C (2007) Org Lett 9:1371
44. Chianese AR, Crabtree RH (2005) Organometallics 24:4432
45. Chen L, Liu Y, Hou G, Song H, Zi G (2013) Inorg Chem Commun 29:141
46. Kawabata S, Tokura H, Chiyojima H, Okamoto M, Sakaguchi S (2012) Adv Synth Catal 354:807
47. Shinohara K, Kawabata S, Nakamura H, Manabe Y, Sakaguchi S (2014) Eur J Org Chem:5532
48. Manabe Y, Shinohara K, Nakamura H, Teramoto H, Sakaguchi S (2016) J Mol Cat A Chem 421:138
49. Teramoto H, Sakaguchi S (2018) J Organomet Chem 875:52
50. Tapu D, Buckner OJ, Boudreaux CM, Norvell B, Vasiliu M, Dixon DA, McMillen CD (2016) J Organomet Chem 823:40
51. Chianese AR, Mo A, Datta D (2009) Organometallics 28:465
52. Park S, Brookhart M (2010) Organometallics 29:6057
53. Metsänen TT, Hrobárik P, Klare HFT, Kaupp M, Oestreich M (2014) J Am Chem Soc 136:6912
54. Iglesias M, Fernández-Alvarez FJ, Oro LA (2014) ChemCatChem 6:2486
55. Iglesias M, Fernández-Alvarez FJ, Oro LA (2019) Coord Chem Rev 386:240
56. Corre Y, Rysak V, Trivelli X, Agbossou-Niedercorn F, Michon C (2017) Eur J Org Chem:4820
57. Petronilho A, Woods JA, Mueller-Bunz H, Bernhard S, Albrecht M (2014) Chem Eur J 20:15775
58. Petronilho A, Vivancos A, Albrecht M (2017) Cat Sci Technol 7:5766
59. Takei I, Nishibayashi Y, Arikawa Y, Uemura S, Hidai M (1999) Organometallics 18:2271
60. Field LD, Messerle BA, Wren SL (2003) Organometallics 22:4393

61. Iali W, La Paglia F, Le Goff XF, Sredojević D, Pfeffer M, Djukic JP (2012) Chem Commun 48:10310
62. Corre Y, Iali W, Hamdaoui M, Trivelli X, Djukic JP, Agbossou-Niedercornabe F, Michon C (2015) Cat Sci Technol 5:1452
63. Li T, Wang Z, Liu K, Xing S, Zhu B (2018) Eur J Inorg Chem:3195
64. Gatus MRD, Pernik I, Tompsett JA, Binding SC, Peterson MB, Messerle BA (2019) Dalton Trans 48:4333
65. Pérez-Miqueo J, San Nacianceno V, Urquiola FB, Freixa Z (2018) Cat Sci Technol 8:6316
66. Blackwell JM, Sonmor ER, Scoccitti T, Piers WE (2000) Org Lett 2:3921
67. Hamdaoui M, Ney M, Sarda V, Karmazin L, Bailly C, Sieffert N, Dohm S, Hansen A, Grimme S, Djukic JP (2016) Organometallics 35:2207
68. Hamdaoui M, Desrousseaux C, Habbita H, Djukic JP (2017) Organometallics 36:4864
69. Jeong J, Park S, Chang S (2016) Chem Sci 7:5362
70. Gutsulyak DV, van der Est A, Nikonov GI (2011) Angew Chem Int Ed 50:1384. (2011) Angew Chem 123:1420
71. Shaw AP, Ryland BL, Franlkin ML, Norton JR, Chen JYC, Hall ML (2008) J Org Chem 73:9668
72. Guan H, Saddoughi SA, Shaw AP, Norton JR (2005) Organometallics 24:6358
73. Königs CDF, Klare HFT, Oestreich M (2013) Angew Chem Int Ed 52:10076. (2013) Angew Chem 125:10260
74. Wgbbolt S, Oestreich M (2015) Angew Chem Int Ed 54:15876; (2015) Angew Chem 127:16103. For a review see: Park S, Chang S (2017) Angew Chem Int Ed 56:7720
75. Troegel D, Stohrer J (2011) Coord Chem Rev 255:1440
76. Oro LA, Fernández MJ, Esteruelas MA, Jimenez MS (1986) J Mol Catal 37:151
77. Hostetler MJ, Bergman RG (1990) J Am Chem Soc 112:8621
78. Hostetler MJ, Butts MD, Bergman RG (1993) Organometallics 12:65
79. Szajek LP, Shapley JR (1994) Organometallics 13:1395
80. Kownacki I, Kubicki M, Marciniec B (2002) Inorg Chim Acta 334:301
81. Marciniec B, Kownacki I, Kubicki M (2002) Organometallics 21:3263
82. Pawluc P, Marciniec B, Kownacki I, Maciejewski H (2005) Appl Organometal Chem 19:49
83. Kownacki I, Marciniec B, Macina A, Rubinsztajn S, Lamb D (2007) Appl Catal A Gen 317:53
84. Wechsler D, Myers A, McDonald R, Ferguson MJ, Stradiotto M (2006) Inorg Chem 45:4562
85. Cipot J, McDonald R, Ferguson MJ, Schatte G, Stradiotto M (2007) Organometallics 26:594
86. Hesp KD, Wechsler D, Cipot J, Myers A, McDonald R, Ferguson MJ, Schatte G, Stradiotto M (2007) Organometallics 26:5430
87. Calimano E, Tilley TD (2008) J Am Chem Soc 130:9226
88. Hayes PG, Beddie C, Hall MB, Waterman R, Tilley TD (2006) J Am Chem Soc 128:428
89. Tuttle T, Wang D, Thiel W, Köhler J, Hofmann M, Weis J (2007) J Organomet Chem 692:2282
90. Beddie C, Hall MB (2004) J Am Chem Soc 126:13564
91. Zhang XH, Chung LW, Lin Z, Wu YD (2008) J Org Chem 73:820
92. Glaser PB, Tilley TD (2003) J Am Chem Soc 125:13640
93. Fasulo ME, Glaser PB, Tilley TD (2011) Organometallics 30:5524
94. Rankin MA, MacLean DF, Schatte G, McDonald R, Stradiotto M (2007) J Am Chem Soc 129:15855
95. Calimano E, Tilley TD (2009) J Am Chem Soc 131:11161
96. Muchnij JA, Kwaramba FB, Rahaim RJ (2014) Org Lett 16:1330
97. Aoki K, Shimada T, Hayashi T (2004) Tetrahedron Asymmetry 15:1771
98. Srinivas V, Nakajima Y, Sato K, Shimada S (2018) Org Lett 20:12
99. Xie X, Zhang X, Yang H, Ji X, Li J, Ding S (2019) J Org Chem 84:1085
100. Chalk AJ, Harrod JF (1965) J Am Chem Soc 87:16
101. Harrod JF, Chalk AJ (1977) Wender I, Pino P (eds) Organic synthesis via metal carbonyls, vol 2. Wiley, New York, p 673

102. Schroeder MA, Wrighton MS (1977) J Organomet Chem 128:345
103. Reichel CL, Wrighton MS (1980) Inorg Chem 19:3858
104. Randolph CL, Wrighton MS (1986) J Am Chem Soc 108:3366
105. Millan A, Towns E, Maitlis PM (1981) J Chem Soc Chem Commun:673
106. Milan A, Fernandez MJ, Bentz P, Maitlis PM (1984) J Mol Catal 26:89
107. Ojima I, Yatabe M, Fuchikami T (1984) J Organomet Chem 260:335
108. Tanke RS, Crabtree RH (1990) J Am Chem Soc 112:7984
109. Ojima I, Clos N, Donovan RJ, Ingallina P (1990) Organometallics 9:3127
110. Tanke RS, Crabtree RH (1990) J Chem Soc Chem Commun:1056
111. Jun CH, Crabtree RH (1993) J Organomet Chem 447:177
112. Esteruelas MA, Nurnberg O, Olivan M, Oro LA, Werner H (1993) Organometallics 12:3264
113. Esteruelas MA, Olivan M, Oro LA, Tolosa JI (1995) J Organomet Chem 487:143
114. Esteruelas MA, Olivan M, Oro LA (1996) Organometallics 15:814
115. Field LD, Ward AJ (2003) J Organomet Chem 681:91
116. Mas-Marzá E, Poyatos M, Sanaú M, Peris E (2004) Inorg Chem 43:2213
117. Mas-Marzá E, Sanaú M, Peris E (2005) Inorg Chem 44:9961
118. Viciano M, Mas-Marzá E, Sanaú M, Peris E (2006) Organometallics 25:3063
119. Vicent C, Viciano M, Mas-Marzá E, Sanaú M, Peris E (2006) Organometallics 25:3713
120. Zanardi A, Peris E, Mata JA (2008) New J Chem 32:120
121. Sridevi VS, Fan WY, Leong WK (2007) Organometallics 26:1157
122. Iglesias M, Pérez-Nicolás M, Sanz Miguel PJ, Polo V, Fernández-Alvarez FJ, Pérez-Torrente JJ, Oro LA (2012) Chem Commun 48:9480
123. Iglesias M, Sanz Miguel PJ, Polo V, Fernández-Alvarez FJ, Pérez-Torrente JJ, Oro LA (2013) Chem Eur J 19:17559
124. Corre Y, Werlé C, Brelot-Karmazin L, Djukic JP, Agbossou-Niedercorn F, Michon C (2016) J Molec Catal A Chem 423:256
125. Navarro M, Smith CA, Albrecht M (2017) Inorg Chem 56:11688
126. Song LJ, Ding S, Wang Y, Zhang X, Wu YD, Sun J (2016) J Org Chem 81:6157
127. Pérez-Torrente JJ, Nguyen DH, Jiménez MV, Modrego FJ, Puerta-Oteo R, Gómez-Bautista D, Iglesias M, Oro LA (2016) Organometallics 35:2410
128. Xie X, Zhang X, Gao W, Meng C, Wang X, Ding S (2019) Commun Chem 2:101
129. Murai M, Nishiyama A, Nishinaka N, Morita H, Takai K (2017) Chem Commun 53:9281
130. Cheng C, Hartwig JF (2015) Chem Rev 115:8946
131. Fernández MJ, Esteruelas MA, Jiménez MS, Oro LA (1986) Organometallics 5:1519
132. Fernández MJ, Esteruelas MA, Oro LA, Apreda MC, Foces-Foces C, Cano FH (1987) Organometallics 6:1751
133. Tanke RS, Crabtree RH (1991) Organometallics 10:415
134. Fernández MJ, Oro LA, Manzano BR (1988) J Mol Catal 45:7
135. Lu B, Falck JR (2010) J Org Chem 75:1701
136. Cheng C, Simmons EM, Hartwig JF (2013) Angew Chem Int Ed 52:8984
137. Azpeitia S, Rodriguez-Dieguez A, Garralda MA, Huertos MA (2018) ChemCatChem 10:2210
138. Huber T, Firlbeck D, Riepl HM (2013) J Organomet Chem 744:44
139. Shimizu R, Fuchikami T (2000) Tet Lett 41:907
140. Kownacki I, Marciniec B, Dudziec B, Kubicki M (2011) Organometallics 30:2539
141. Kownacki I, Orwat B, Marciniec B, Kownacka A (2014) Tet Lett 55:548
142. Rzonsowska M, Dudziec B, Kownacki I, Marciniec (2015) J Organomet Chem 775:20
143. Kownacki I, Orwat B, Marciniec B (2014) Organometallics 33:3051
144. Gustavson WA, Epstein PS, Curtis MD (1982) Organometallics 1:884
145. Ishiyama T, Sato K, Nishio Y, Miyaura N (2003) Angew Chem Int Ed 42:5346
146. Ishiyama T, Sato K, Nishio Y, Saiki T, Miyaura N (2005) Chem Commun:5065
147. Saiki T, Nishio Y, Ishiyama T, Miyaura N (2006) Organometallics 25:6068
148. Lu B, Falck JR (2008) Angew Chem Int Ed 47:7508
149. Sangtrirutnugul P, Tilley TD (2007) Organometallics 26:5557

150. Xu Z, Huang WS, Zhanga J, Xu LW (2015) Synthesis 47:3645
151. Sharma R, Kumar R, Kumar I, Singh B, Sharma U (2015) Synthesis 47:2347
152. Simmons EM, Hartwig JF (2010) J Am Chem Soc 132:17092
153. Su B, Zhou TG, Li XW, Shao XR, Xu PL, Wu WL, Hartwig JF, Shi ZJ (2017) Angew Chem Int Ed 56:1092
154. Kuznetsov A, Gevorgyan V (2012) Org Lett 14:914
155. Kuznetsov A, Onishi Y, Inamoto Y, Gevorgyan V (2013) Org Lett 15:2498
156. Choi G, Tsurugi H, Mashima K (2013) J Am Chem Soc 135:13149
157. Wakaki T, Kanai M, Kuninobu Y (2015) Org Lett 17:1758
158. Ishiyama T, Saiki T, Kishida E, Sasaki I, Ito H, Miyaura N (2013) Org Biomol Chem 11:8162
159. Cheng C, Hartwig JF (2015) J Am Chem Soc 137:592
160. Karmel C, Chen Z, Hartwig JF (2019) J Am Chem Soc 141:7063
161. Karmel C, Rubel CZ, Kharitonova EV, Hartwig JF (2020) Angew Chem Int Ed 59:6074
162. Wang H, Wang G, Li P (2017) Org Chem Front 4:1943
163. Murai M, Takami K, Takai K (2015) Chem Eur J 21:4566
164. Murai M, Takami K, Takeshima H, Takai K (2015) Org Lett 17:1798
165. Lin Y, Jiang KZ, Cao J, Zheng ZJ, Xu Z, Cui YM, Xu LW (2017) Adv Synth Catal 359:2247
166. Rubio-Pérez L, Iglesias M, Munárriz J, Polo V, Passarelli V, Pérez-Torrente JJ, Oro LA (2017) Chem Sci 8:4811
167. Toutov A, Liu WB, Betz KN, Fedorov A, Stoltz BM, Grubbs RH (2015) Nature 518:80
168. Murai M, Nishinaka N, Takai K (2018) Angew Chem Int Ed 57:5843
169. Murai M, Nishinaka N, Enoki T, Takai K (2020) Org Lett 22:316
170. Sabat N, Slavětínská LP, Hocek M (2015) Tet Lett 56:6860
171. Minami Y, Komiyama T, Hiyama T (2015) Chem Lett 44:1065
172. Shibata T, Uno N, Sasaki T, Takano H, Sato T, Kanyiva KS (2018) J Org Chem 83:3426
173. Simmons EM, Hartwig JF (2012) Nature 483:70
174. Li B, Driess M, Hartwig JF (2014) J Am Chem Soc 136:6586
175. Hartwig JF, Romero EA (2019) Tetrahedron 75:4059
176. Frihed TG, Pedersen CM, Bols M (2014) Angew Chem Int Ed 53:1388
177. Parija A, Sunoj RB (2013) Org Lett 15:4066
178. Li Z, Xia M, Boyd RJ (2016) Can J Chem 94:1028
179. Mita T, Michigami K, Sato Y (2012) Org Lett 14:3462
180. Mita T, Michigami K, Sato Y (2013) Chem Asian J 8:2970
181. Fukumoto Y, Hirano M, Chatani N (2017) ACS Catal 7:3152
182. Fukumoto Y, Hirano M, Matsubara N, Chatani N (2017) J Org Chem 82:13649
183. Hirano M, Fukumoto Y, Matsubara N, Chatani N (2018) Chem Lett 47:385
184. Murai M, Takeuchi Y, Takai K (2017) Chem Lett 46:1044
185. Su B, Hartwig JF (2017) J Am Chem Soc 139:12137
186. Zhang M, Liang J, Huang G (2019) J Org Chem 84:2372

Top Organomet Chem (2021) 69: 271–302
https://doi.org/10.1007/3418_2020_58

Published online: 8 August 2020

Iridium Catalysts for Hydrogen Isotope Exchange

Marc Reid

Contents

Abstract A history and summary of iridium-catalyzed hydrogen isotope exchange (HIE) is described. Owing to the wide range of applications served by installation of heavy and radioactive hydrogen isotopes, a wealth of synthetic labeling strategies have been forthcoming. Principle among all HIE methods are those developed using homogeneous iridium catalysts. This chapter covers major developments in (primarily homogeneous) iridium-centered catalysts for HIE. Connections to the broader fields of hydrogenation and C–H functionalization are also considered.

The original version of this chapter was revised. A correction to this chapter can be found at https://doi.org/10.1007/3418_2020_73

M. Reid (✉)
School of Chemistry, University of Bristol, Bristol, UK
e-mail: marc.reid@bristol.ac.uk

Keywords C–H functionalization · Deuterium · Iridium · Isotope exchange · Tritium

1 Introduction

1.1 Isotopes and Isotopic Labeling

Isotopes of a particular element have an identical number of protons in their respective nuclei but possess an unequal number of neutrons. Namely, they share the same atomic number but have different mass numbers, as exemplified for hydrogen (Scheme 1) [1]. The stability of an isotope is governed by the ratio of neutrons to protons within the nucleus, thus giving rise to two possible circumstances. Firstly, a *heavy isotope* of an element, such as ^{2}H or ^{13}C, has a stable nucleus and tends to be found in nature, albeit at lower abundances than their more common counterparts, ^{1}H and ^{12}C, respectively. In the alternative case, *radioisotopes*, such as ^{3}H or ^{14}C, have an unstable neutron/proton ratio and decay, via emission of radiation or particles, to form other elements, or different isotopes of the parent element.

The synthesis and supply of isotopically labeled molecules has a sustained importance in the study of metabolic processes, among myriad other processes [2]. It is therefore unsurprising that there is a large and growing body of research dedicated to the synthesis of isotopically labeled compounds. The labeling of molecules with ^{13}C or ^{14}C is most readily achieved through the use of commercially available, isotopically enriched starting materials. While such a technique ensures a regiospecific label will be present in the desired target molecule, it ultimately comes at the price of unwanted additional steps in the synthesis [3].

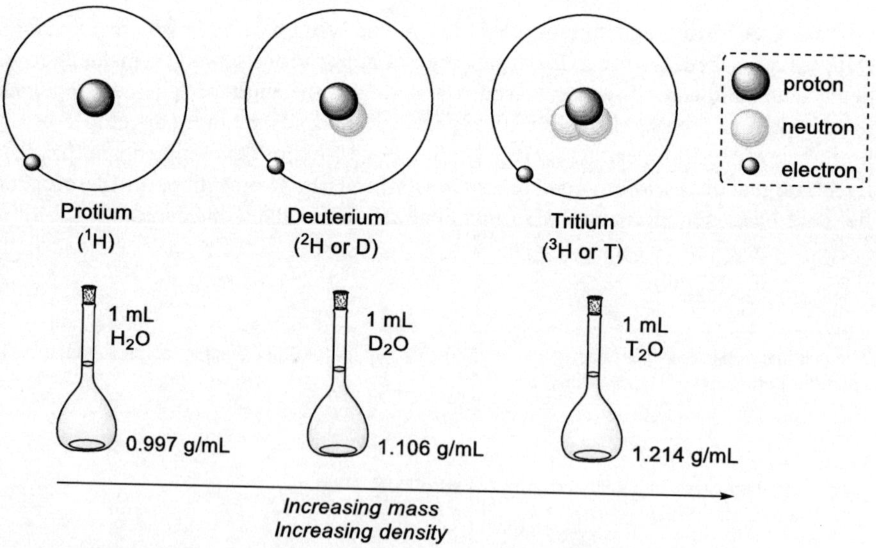

Scheme 1 Simplified Bohr representations of the isotopes of hydrogen

Research into deuterium (^{2}H or D) and tritium (^{3}H or T) labeling is more substantial than that for other isotopes and has been developed on a number of fronts over the past 60 years [2–22]. Further to this, key developments in synthetic strategies and analytical techniques over the past three decades are gradually making tritium labeling the preferred technique in many absorption, distribution, metabolism, excretion, and toxicology (ADMET) studies [10]. In one particularly active branch of such research, *hydrogen isotope exchange* (HIE) is commonly employed to deliver deuterium or radioactive tritium to pharmaceutical drug candidates in one synthetic step.

1.2 Applications of Hydrogen Isotope Exchange

The importance of hydrogen isotope exchange (HIE), for iridium catalysts and beyond, is manifest in the wealth of reviews published in the area [2–36]. As well as circumventing the requirement for isotopically enriched starting materials in synthesizing tritiated drug candidates [3, 10], HIE can also provide analogous deuterated compounds for use as internal standards for mass spectrometry [29, 37], for kinetic isotope studies, [21, 38, 39], and for the alteration of reaction pathways in total syntheses [40]. Additionally, HIE is applied within almost every sub-discipline in life science, in nuclear science, and beyond [2]. The ability for precise measurement of isotope ratios promotes a dynamic view on biosynthetic pathways, protein turnover, and systems-wide metabolic networks and, thus, has paved the way for a number of scientific breakthroughs in biomedical research. In assessing a drug candidate's metabolic fate, the chemist must first have a flexible technique with which to study it. Consequently, *isotopic labeling* is the gold standard method by which early stage drug discovery processes are optimized. The numerous application areas for HIE are summarized in Scheme 2.

1.3 Synthetic Methods in HIE

With a broad range of existing HIE applications, there exists a wide range of synthetic methods to achieve hydrogen isotope incorporation in an ever-expanding array of substrates. While the full gamut of chemistries developed for HIE is beyond the primary focus of this chapter, it is worth covering these in brief. Firstly, the source of deuterium and tritium has varied extensively from method to method; however, some patterns exist. For deuteration, many methods have applied D_2 gas,

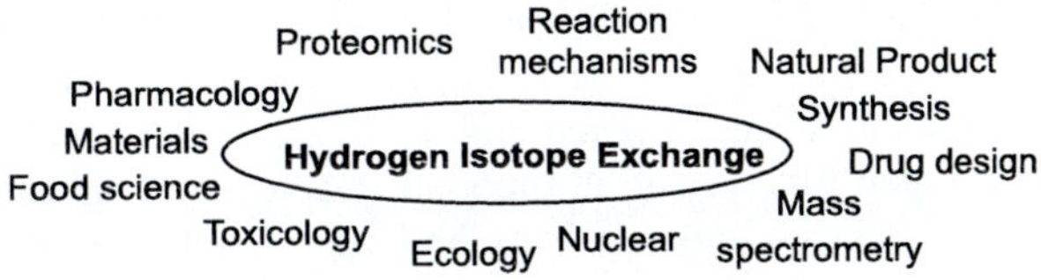

Scheme 2 Application areas served by hydrogen isotope exchange (HIE) technology

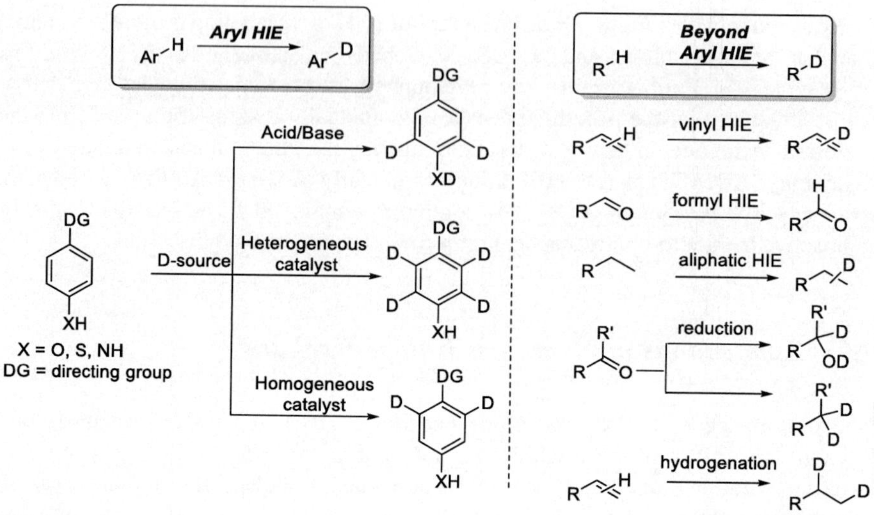

Scheme 3 Common synthetic transformations toward the installation of hydrogen isotopes in organic substrates

D_2O (heavy water), DCl, benzene-d$_6$, DMSO-d$_6$, and numerous deuterated alcohols. Of these isotope sources, and of direct relevance to the focus of this chapter, D_2 gas has been the preferred isotope source as it directly maps onto the preferred use of tritium gas for radiolabeling protocols [5, 7, 10, 17].

Hand in hand with the range of hydrogen isotope sources is a wide range of metal-mediated and other mediated processes for HIE (Scheme 3). Classically, these include acid- and base-mediated reactions, as well as modern variations using frustrated Lewis pairs (FLPs). Aryl labeling is most common, but many common organic transformations have been pivoted into labeling protocols. Nonetheless, metal-catalyzed HIE is dominant in HIE, covering heterogeneous and homogeneous catalysis. Such methods have been more fully reviewed elsewhere [11, 12].

2 *Ortho*-Directed Iridium-Catalyzed HIE

Among all transition metals employed in homogeneous HIE methods, iridium is arguably the most widely studied [2, 3, 5, 6, 11–13, 15, 17, 18, 20, 22, 26, 35, 36, 41, 42], which is, in part, due to the vast and ever-expanding literature precedent in related hydrogenation reactions [31, 43–68]. Iridium was also present in some of the earliest metal-centered catalysts applied to HIE chemistries [69]. In support of this analysis of iridium's popularity in HIE, Oro and co-workers estimated that iridium accounted for 33% of all reported HIE methods, greater than for any other metal [11]. While iridium catalysts have also been applied in heterogeneous catalysis

[70, 71], the focus of this chapter is on the far more expansive homogeneous iridium catalysis developments.

2.1 Early Developments in Ortho-Directed HIE

There is a clear dominance of *ortho*-HIE in the homogeneous iridium catalyst literature. In 1992, Heys demonstrated the successful *ortho*-directed deuteration of several substituted aromatic compounds using the 18-electron Ir(III) bis-phosphine dihydride complex **1** under very mild conditions (**2** → **3**, Scheme 4) [72]. Crucially for the time, Heys' investigations marked a significant advancement from Lockley's *ortho*-labeling work (with rhodium and ruthenium catalysts) [73–75]: D_2 gas replaced D_2O as the deuterium source (an advantage when considering the use of tritium), reactions operated efficiently at room temperature, and, perhaps most importantly, catalyst loadings were significantly reduced from 50 mol% to 2 mol %. Interestingly, it was noted that labeling was significantly affected by steric or electronic aspects of the substituents present on arene substrates. For example, *meta*-substituted ethyl benzoates, such as **5**, showed a consistent preference for labeling at C-2 over the less hindered C-6 position, presumably due to additional coordination assistance from *meta*-substituent lone pairs [76]. Steric hindrance from *ortho*-substitution reduced labeling efficiency (**4** vs. **9**); however, bulky α-substituted ketones such as **6** were not so adversely affected. Further to this, where substrates possessed more than one carbonyl directing group, the labeling site(s) changed according to which substituent could coordinate to the catalyst to the greatest extent (e.g., **4** vs. **10**).

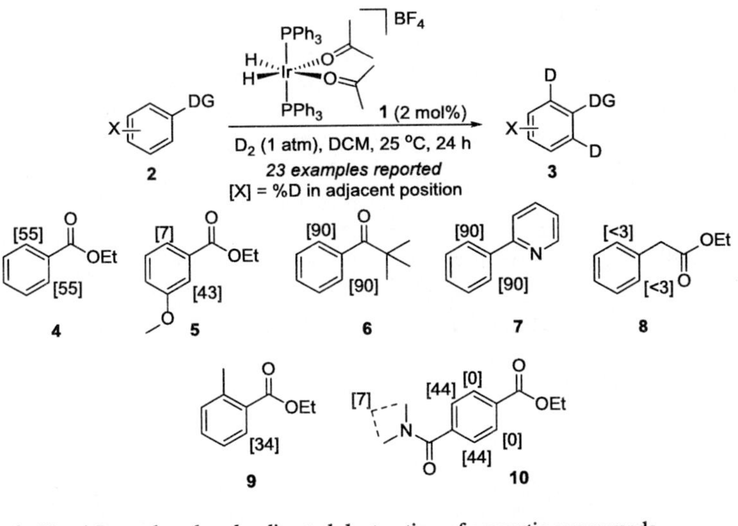

Scheme 4 Heys' Ir-catalyzed *ortho*-directed deuteration of aromatic compounds

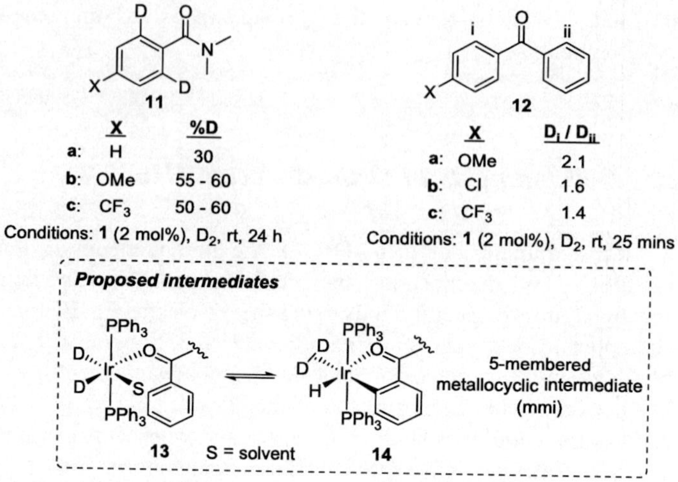

Scheme 5 Mechanistic investigations into Heys' Ir-catalyzed HIE protocol

The mild labeling conditions pioneered by Heys and co-workers, coupled with intriguing substrate-dependent regioselectivity, captured the combined interest of the industrial and academic HIE communities, resulting in a large number of subsequent studies aimed at understanding the catalytic properties of **1** and related Ir-based HIE catalysts. Firstly, Heys followed up his initial study with a more in-depth assessment of the aryl substituent effects in the labeling efficiency of ethyl benzoates and *N,N*-dimethylbenzamide substrates (Scheme 5) [72]. In a rather unexpected outcome, *para*-substitution improved the rate of labeling in both substrate types, *irrespective of substituent electronics* (e.g., **11a** vs. **11b** and **11c**). In an attempt to explain this effect, Heys monitored the rate of labeling in both rings of several monosubstituted benzophenones [72]. The substituted ring was labeled faster in every instance (**12a–12c**). As both rings are connected to the same carbonyl functionality, it appeared that the rate-limiting step of the overall reaction could *not* be ascribed to the initial coordination of the substrate, nor hydride fluxionality [23]. Instead, Heys suspected that some aspect of the C–H bond cleavage was rate-limiting, proposing key intermediates **13** and **14** based on available literature. At this time, the formal oxidation state of iridium intermediates involved in the C–H bond cleavage (Ir^{I} / Ir^{III} or Ir^{III}/Ir^{V}) was not clear.

Inspired by Heys, Hesk and co-workers probed the efficacy of the commercially available Crabtree hydrogenation catalyst, **15** [43], in labeling acetanilide derivatives, the first such substrates to be effectively labeled via a 6- rather than a 5-membered metallocyclic intermediate (mmi) [77]. Consistent with Heys' work, Hesk reported that deuteration was directed *ortho* to the coordinating functionality. Moreover, no clear relationship emerged regarding the electronics of *para*-substituents and labeling efficacy. Ketones **17** and **18** were also compatible with this mild labeling method; however, weakly coordinating benzenesulfonamide **19** and benzoic acid **20** were not (Scheme 6).

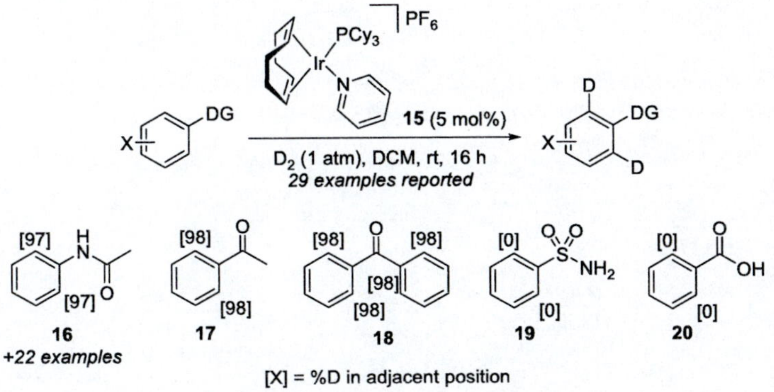

Scheme 6 Hesk's application of Crabtree's catalyst in *ortho*-HIE

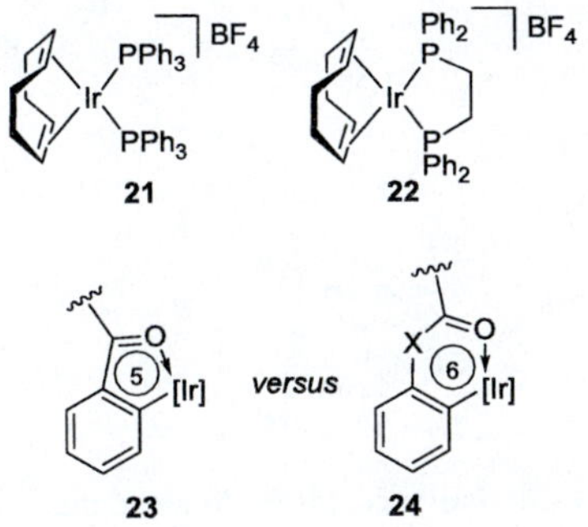

Scheme 7 Mono- vs. bidentate Ir–phosphine catalysts to study *ortho*-deuteration via 5- and 6-mmis

Since Heys' and Hesk's respective discoveries of iridium catalysts for *ortho*-directed HIE, complexes **1** and **15** (and derivatives thereof, vide infra) have remained topics of high interest in HIE research [26, 78–81]. In a further key development by Heys, **21**, a precatalytic Ir(I) variant of Ir(III) catalyst **1** was compared to related bidentate pre-catalyst, **22** (Scheme 7) [82]. By the mid-1990s, it had already been hypothesized by several researchers that both 5- and 6-mmis could be formed during the C–H bond cleavage step in the *ortho*-deuteration process (**23** vs. **24**), depending on the substrate being studied; this was to be the platform on which to compare iridium catalysts **21** and **22**.

Labeling a range of substrates enabled a comparison of the mono- and bidentate phosphine complexes to be made, highlighting a preference for monodentate **21** to react through a 5-mmi only, whereas bidentate **22** could react through both a 5- and a 6-mmi. This result was exemplified in the labeling of ethyl 1-naphthoate, **25** (Scheme 8, top). Of the two available labeling sites, the monodentate complex, **21**, labeled solely at C-2. Conversely, bidentate complex **22** demonstrated the capability to direct labeling at both C-2 and C-8. When Crabtree's catalyst, **15**, was exposed to

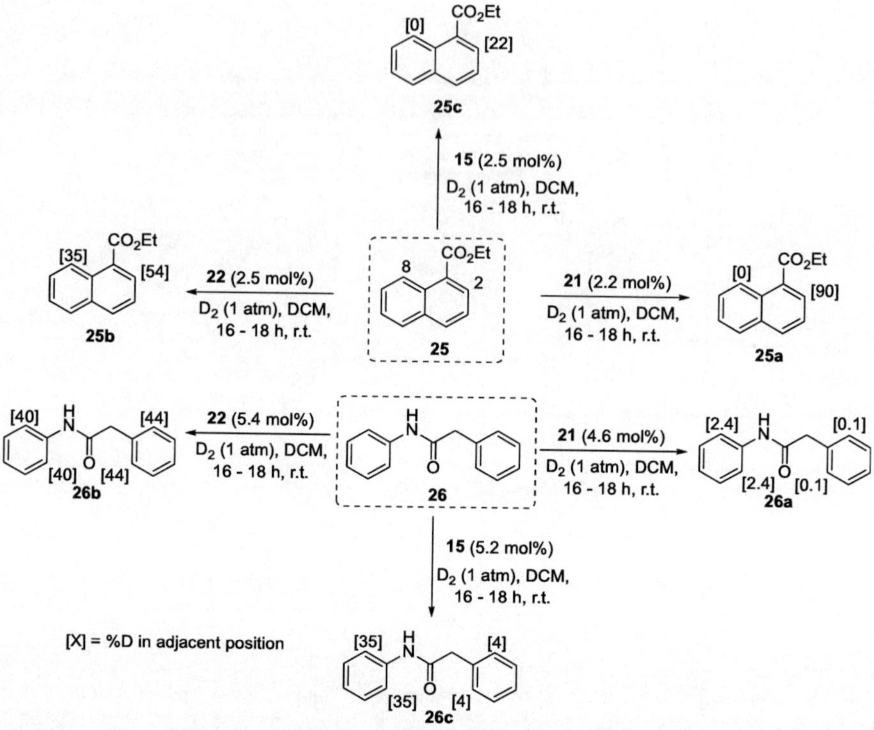

Scheme 8 Heys' vs. Hesk's *ortho*-HIE methods for 5- and 6-mmi substrates [77, 82]

similar reaction conditions, the regioselectivity in labeling was similar to monodentate complex **21**, albeit with reduced labeling efficiency. Labeling through a 6-mmi only was also investigated. Perhaps the most remarkable findings from this study were those concerning the labeling of *N*-phenyl phenylacetamide, **26** (Scheme 8, bottom). Interestingly, the less active monodentate complex, **21**, showed selectivity for the aromatic ring adjacent to the nitrogen, **26a**, an effect emulated more efficiently by Crabtree's catalyst in **26c**. However, the bidentate catalyst **22** was able to label both rings of **26** almost indiscriminately (see **26b**). This served to show that there was potential to distinguish not only between a 5- and 6-mmi, but also between different *types* of 6-mmi, depending on the ancillary ligands employed.

On accumulation of these data, Heys proposed a catalytic cycle by which these iridium complexes may be affecting the observed regioselective hydrogen isotope exchange (Scheme 9) [82]. Upon treatment of the Ir(I) pre-catalyst, **27**, with deuterium gas, hydrogenolysis of cyclooctadiene (COD) as d_4-cyclooctane generates the active Ir(III) catalyst, **28**, where ligands (L) are assumed to be arranged *trans* to one another when monodentate. Coordination of substrate displaces a solvent molecule (S) and is thus accepted into the coordination sphere of the iridium catalyst to give **29**. A second solvent molecule can then be displaced, allowing iridium to cleave the

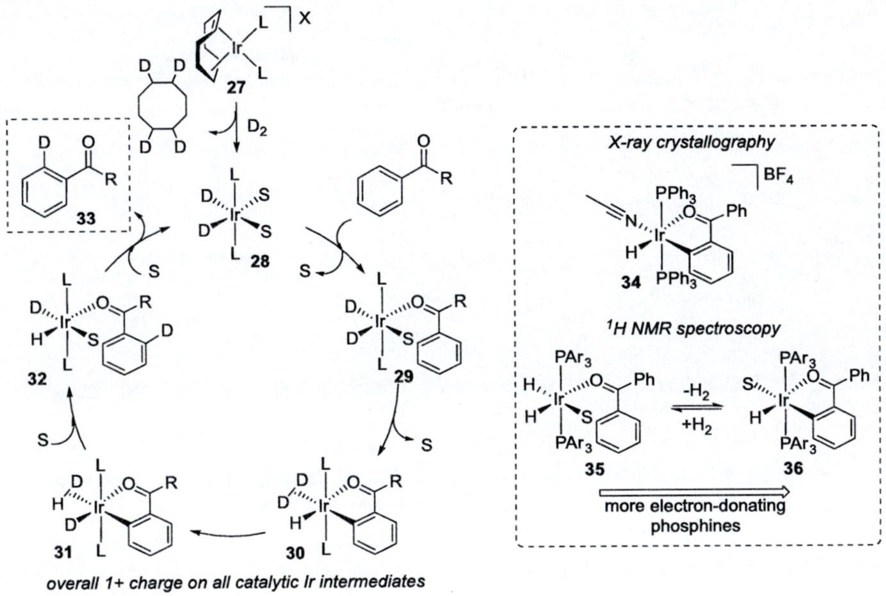

Scheme 9 Heys' mechanistic analysis for homogeneous Ir-catalyzed *ortho*-HIE

nearby *ortho* C–H bond of the aryl ring to yield **30**. Transformation of species **31** to **32** is driven by a process known as *hydride fluxionality* and is central to the isotope exchange process [23]. The overall effect brings a deuteride and the activated aryl carbon into a *cis* arrangement. Subsequent C–D elimination furnishes **32**, with a deuterium atom now installed *ortho* to the directing group. Finally, the release of deuterated substrate, **33**, regenerates the resting catalytic intermediate, **28**. This mechanism invokes an all-Ir(III) catalytic cycle with C–H activation as the rate-limiting step, supporting evidence for which would take another decade to accumulate. Said evidence involved isolation and crystallographic characterization of **34** (an acetonitrile-solvated analogue of **30**) and spectroscopic studies on the evolving nature of iridium hydride equilibria as a function of ancillary ligand electronics (Scheme 9, inset) [20].

In an extension of the theory of *ortho*-directed HIE, Heys postulated that the preference for the monodentate phosphine complex, **21**, to react only via a 5-mmi, **38**, as opposed to a 6-mmi, **37**, was based on steric effects (Scheme 10) [82]. By contrast, the bidentate complex, **22**, is forced to arrange the phosphines *cis* to one another. For substrates such as **25**, this opens up a second face on the iridium complex, offering greater spatial freedom for the formation of the less planar 6-mmi, **40**, as well as the 5-mmi, **39**. By the same thought, the monodentate Crabtree's catalyst, **15**, *can* facilitate labeling through a 6-mmi as the pyridine and tricyclohexylphosphine ligands present less steric bulk than the two triphenylphosphine ligands of complex **21** and may thus exist in *cis* or *trans* form. Herbert later capitalized on this rationale to further improve bidentate catalyst **22** in

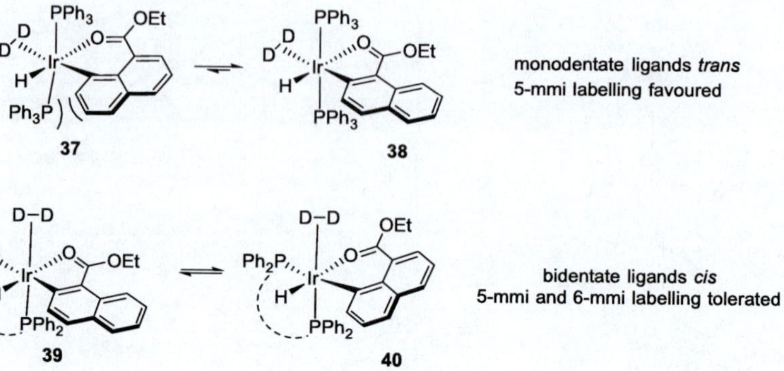

Scheme 10 Rationale for 5- vs. 6-mmi labeling selectivity with mono-/bidentate phosphine catalysts

the labeling of 6-mmi substrates, changing the diphenylphosphinoethane (dppe) ligand for the sterically less encumbered arsine analogue [81].

2.2 *Contemporary Methods in Ortho-Directed HIE*

Further synthetic developments by Herbert [28, 78, 83] and later Salter [26] showed that bis-phosphine catalysts like **22** may be generated in situ from the appropriate free phosphine and commercial iridium dimer, $[Ir(COD)Cl]_2$, with comparable activity to the isolated complexes. The same authors are also separately responsible for detailed studies into alteration of the phosphine structure [26, 78, 81]. However, both parties have remarked that strong correlations between ligand properties (such as sterics or electronics) and catalyst activity are difficult to detect. The number of such ligands applied to iridium-catalyzed HIE is now extensive and includes more elaborate catalyst system like **41** (Scheme 11).

Parallel with studies into bis-phosphine systems, Crabtree's catalyst **15** has also been the subject of intense study in deuteration and tritiation, since Hesk's discovery [76, 84–89]. In one of the largest of any such study, Herbert explored an expansive substrate scope, including ketones, amides, anilides, and various heterocycles [83]. Despite the impressive array of examples reported, this study employed *at least* stoichiometric quantities of **15** and a dual D_2/D_2O isotope source, making comparisons to related *ortho*-labeling methods difficult.

In a notable crossover between bis-phosphine catalysts and Crabtree's catalyst, Hickey and co-workers developed a polymer-supported variant of Heys' bis-phosphine catalyst, **42**, which showed comparable *ortho*-HIE activity to **15** and **22**, but with the practical benefit of simple catalyst filtration at the end of the reaction (**43** vs. **44** vs. **45**; Scheme 12) [71]. Solid-supported iridium catalysts for HIE have now been adapted to flow systems [80].

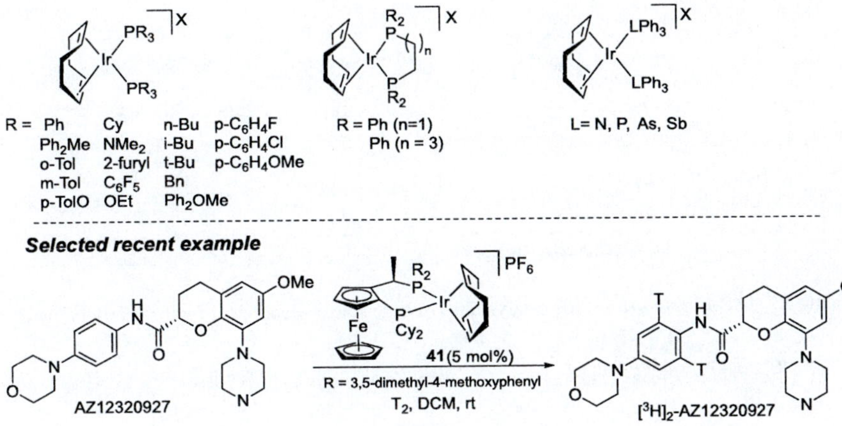

Scheme 11 An overview of bis-phosphine-ligated iridium catalysts applied in HIE

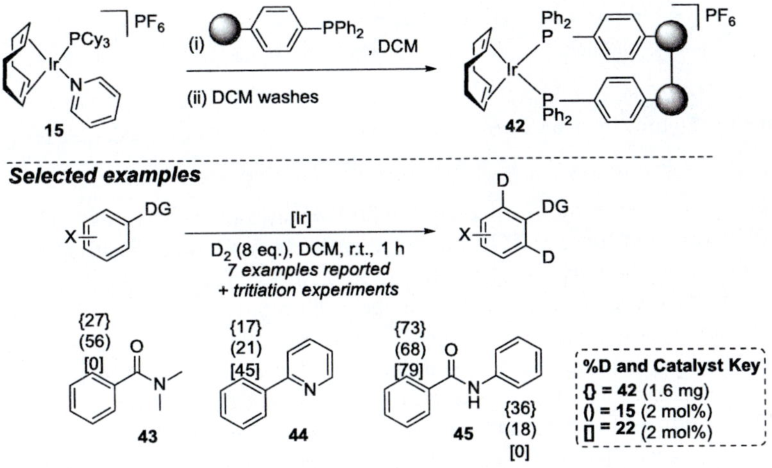

Scheme 12 Polymer-supported iridium catalyst in *ortho*-HIE

Exploring an altogether different ligand architecture, Lockley reported the application of hexafluoroacetylacetonate (hfacac)-ligated Ir(I) complex, **46**, in *ortho*-HIE (Scheme 13) [8, 10, 22, 90–92]. This catalyst has been successfully applied in the labeling of benzylic amines, benzoic acids, and primary sulfonamides, where few other Ir-based HIE catalysts have succeeded. The catalyst is one of the few iridium HIE catalysts operational in highly polar solvents such as DMF (desirable for poorly soluble drug molecules) and displays different labeling regioselectivities depending on the choice of isotope source (D_2 or D_2O; see **47** → **48** vs. **47** → **49**).

In the early 2000s, increasing interest in Crabtree's catalyst, **15**, in HIE was paralleled with investigations by other researchers to improve efficiency and

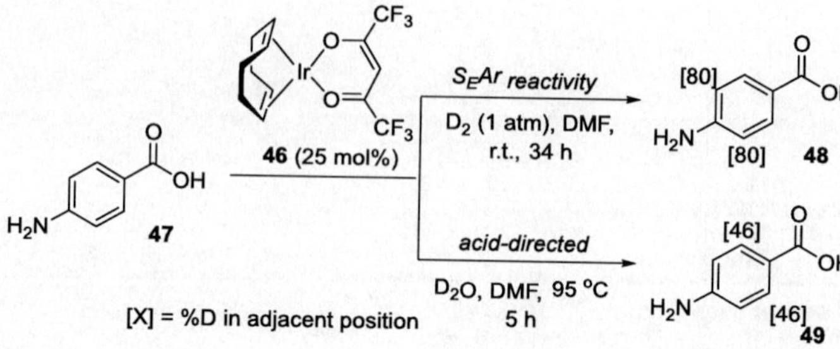

Scheme 13 Ir(I)-hfacac *ortho*-HIE catalyst and isotope source-dependent regioselectivity switch

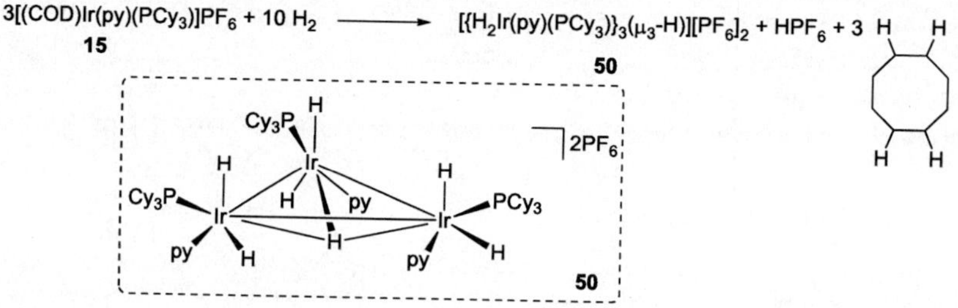

Scheme 14 Trimeric iridium cluster formed from thermal deactivation of Crabtree's catalyst

chemoselectivity in iridium-catalyzed olefin hydrogenation reactions [61]. Despite its widely reported success, **15** is known to suffer from thermal deactivation via the formation of inactive, hydride-bridged, iridium clusters (**50**, Scheme 14) [54]. Similar effects have been documented for other iridium-based complexes [66, 93].

Separate investigations by Nolan [62] and Buriak [94] toward improved thermal stability and predictable chemoselectivity of Crabtree-like hydrogenation catalysts resulted in a plethora of highly promising electron-rich, N-heterocyclic carbene (NHC)-ligated iridium catalysts (Scheme 15). Such species were first applied and published in *ortho*-HIE processes by Powell and co-workers [95]. In Powell's study, complexes **51a** and **52a–52c** were employed under stoichiometric (industrial "tritiation-like") conditions, with the most active variant, **52c**, shown to be superior to Crabtree's catalyst across the entire substrate range.

In a more interesting variant of this work, Kerr and co-workers studied the *catalytic* activity of complexes **51b–51f**, showing most active complex, **51e**, to be highly active over an appreciable substrate scope (5 mol% [Ir], 16 h, rt) and displaying a higher turnover frequency (TOF) than Heys' bis-phosphine catalyst, **22**. Interestingly, the smaller complexes in the series studied by Kerr (**51b** and **51c**) were completely inactive as HIE catalysts [86]. Similar investigations by the same group led to the discovery that small NHC/phosphine complexes such as **52c** were inactive as HIE catalysts, but larger variants **52d** and **52e** were active across a limited substrate scope [96].

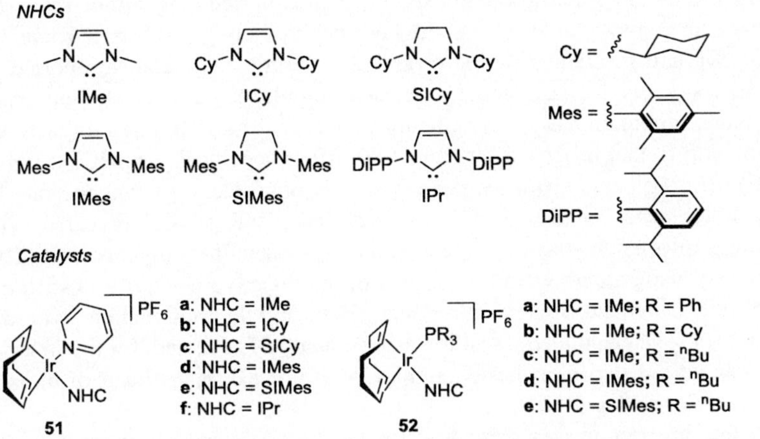

Scheme 15 NHC-ligated iridium catalysts for hydrogenation later explored by HIE chemists

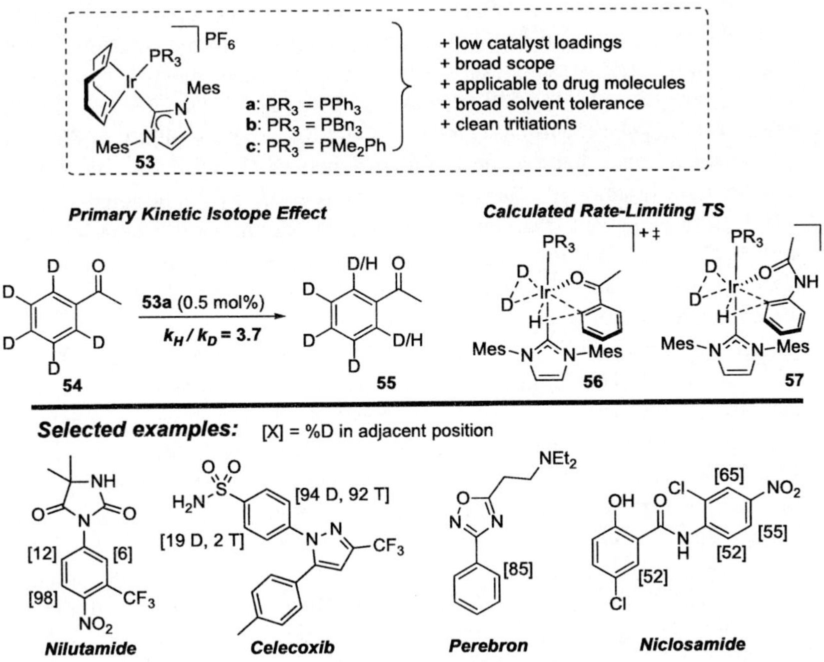

Scheme 16 Highly active NHC/phosphine *ortho*-HIE iridium catalysts

The exploration of NHC-ligated iridium HIE catalysts had revealed promising (proof-of-concept) developments beyond the popular and established works of Hesk and Heys. Kerr and co-workers later developed a synthesis of previously unattainable complexes **53a–53c**, bearing large phosphine *and* large NHC ligands

(Scheme 16) [97]. These complexes have proven seminal within the *ortho*-HIE domain and have among the highest activity [98], substrate/solvent scope [99, 100], and tritiation reaction cleanliness of any such catalyst reported to date. Additionally, *ortho*-HIE process with these complexes has been studied experimentally and computationally, strengthening the case for a Ir(III)-based reaction mechanism akin to that proposed by Heys [98]. More specifically, kinetic isotope effect (KIE) measurements [101] revealed that C–H bond cleavage was the rate-limiting step of the reaction (**54** → **55**), and detailed NMR studies revealed (via $^{2}J_{P\text{–}C}$ coupling information) the *trans*-geometry of the ancillary ligands [98]. The same study was also able to reveal the origins of the selective reactivity of such catalysts for 5- over 6-mmi substrates, citing dual kinetic and thermodynamic favorability for the 5-mmi. The calculated transition states **56** and **57** revealed, for the first time, the sigma-bond-assisted metathesis (sigma-CAM) process at the heart of the all-Ir(III) C–H activation step [98, 102].

While developing a rare method for labeling primary sulfonamides, Kerr and co-workers considered directing group chemoselectivity in detail [102]. It was observed that the sulfonamide vs. pyrazole selectivity in *celecoxib* **58** varied dramatically with catalyst choice (Scheme 17). Whereas encumbered and most-often used NHC/phosphine catalysts facilitated labeling adjacent to the pyrazole moiety, giving **58b**, neutral NHC/Cl catalysts, such as **59**, facilitated selective sulfonamide labeling, delivering **58a**, for the first time. Accompanying DFT studies revealed that the substrate binding event was likely to be product-determining (**60a** vs. **60b**), even though C–H activation remained rate-limiting (Scheme 17). A similar rationale was presented for multifunctional molecules containing esters as the targeted directing group [103]. Following this, Derdau and co-workers significantly expanded on the HIE studies of competing directing groups, showing once again that calculated binding energies could serve as a semiquantitative and predictive tool for rationalizing directing group chemoselectivity in HIE [42].

Building on Kerr's work, Ir(III)-catalyzed *ortho*-HIE has continued to flourish [3, 11, 13, 15, 35, 41, 42]. From the same group, and others applying the developments therefrom, the application of bulky NHC–phosphine systems in HIE has steadily advanced in terms of the applicable substrate and solvent scope [42, 102, 104–108]. With regard to solvent scope, Kerr and Tamm have reported complementary strategies toward modifying the solubility profile of existing iridium HIE catalysts. On the one hand, Kerr explored the use of the bulky tetrakis[3,5-bis (trifluoromethyl)phenyl]borate (BArF) counterion in place of the standard hexafluorophosphate (PF_6) [104, 109], and on the other, Tamm integrated a related borate anion into the backbone of an anionic carbene ligand (Scheme 18) [110]. The wide range of solvents made applicable in extending the Kerr catalyst series through **61a–61d** evidenced new opportunities to tune HIE regioselectivity through simple solvent switching [104]. From Tamm's most recent developments, catalysts **62a**, **62b**, and **62e** have been identified as competent HIE catalysts in hexane and cyclohexane for the first time [110].

A growing community of researchers have, in more recent times, contributed a wider range of elaborated ligand spheres around tractable iridium(I) pre-catalysts. In

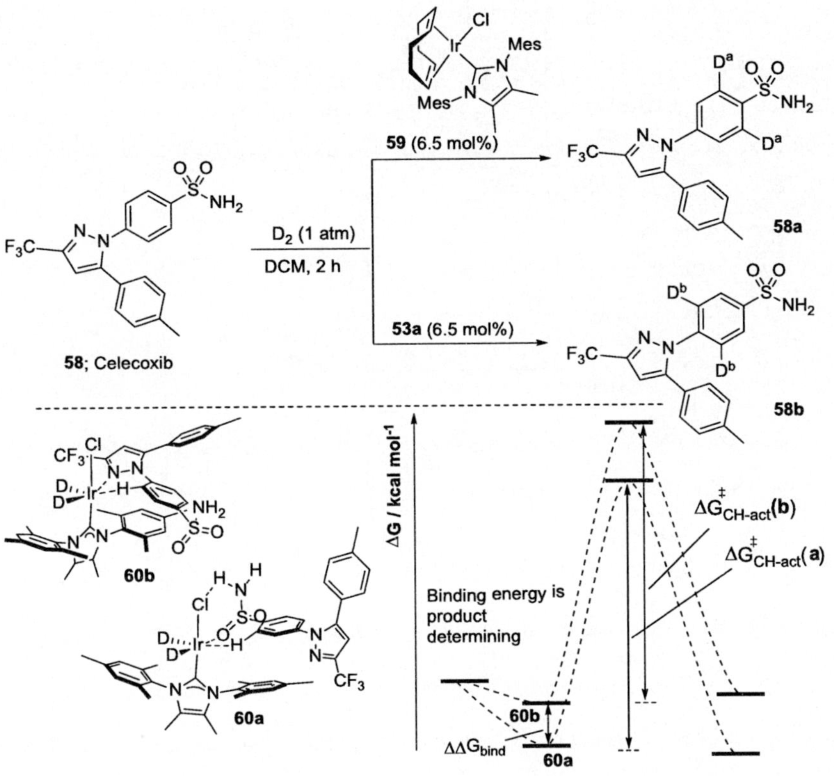

Scheme 17 DFT-calculated rationale for directing group selectivity using catalyst **59** and sulfonamide drug **58**

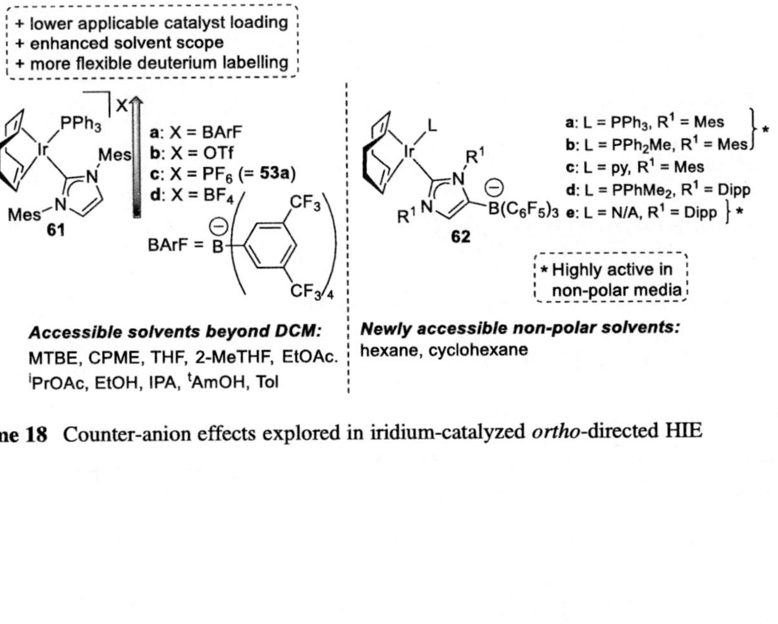

Scheme 18 Counter-anion effects explored in iridium-catalyzed *ortho*-directed HIE

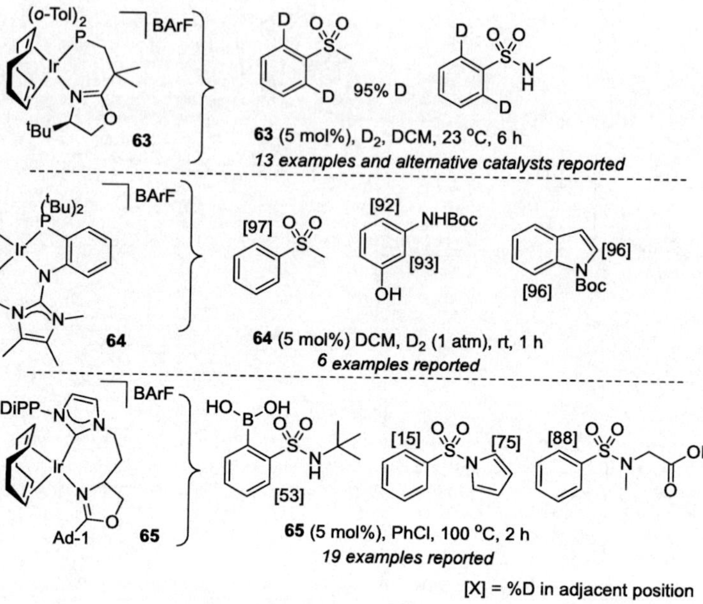

Scheme 19 Modern chelated iridium catalysts expanding the range of accessible ortho-directing groups in HIE

turn, more iridium HIE catalysts have enabled applications using more challenging directing groups. A recent contribution from Pfaltz and Muri showed the application of *P,N*-derived bidentate ligands [111]. Most notably, these latest iridium-based HIE catalysts have been developed to be able to label *ortho* to secondary benzenesulfonamides for the first time, albeit using high temperatures and synthetically intricate ligands [111]. Along similar lines, Tamm and Derdau have reported complementary P,N- and C,N-ligated iridium catalysts able to further expand the range of accessible directing groups applicable in *ortho*-directed HIE processes (**63**–**65**, Scheme 19) [110, 112, 113].

3 Beyond *Ortho*-Directed HIE

Far from the humble beginnings of homogeneous iridium-catalyzed HIE [69], labeling of organic molecules has continued to advance along complementary lines to *ortho*-directed HIE. While some instances have been discovered as unintended by-products of desired *ortho*-labeling, [111] or to assess non-innocent ancillary ligand behaviors, [114–125] contributions have been made to labeling global aromatic, sp^3, vinyl, formyl, and heteroatom positions in a strategic manner (cf. Scheme 3). In the application domain, such developments have given industrialists a more diverse palette of methods with which to incorporate hydrogen isotopes into an increasingly elaborate array of drug candidates.

3.1 Directed sp^3 C–H HIE Methods

Somewhat inspired by the deep understanding of iridium catalysts and compatible directing groups for *ortho*-directed HIE protocols, significant contributions have emerged toward labeling sp^3 centers rather than aromatic sp^2 centers [15, 111, 126, 127].

Using Kerr's commercially available catalyst **53a** and **61a**, Derdau and Kerr have developed expansions of the original *ortho*-labeling methodologies, showing that the same catalyst systems can effectively label sp^3 C–H positions in complex amides and a range of drug molecules (Scheme 20).

In a new paradigm for the field, MacMillan and co-workers developed a photoredox- and hydrogen atom transfer (HAT)-catalyzed method, employing an iridium(III) photocatalyst Ir(F-Meppy)$_2$ (dtbbpy)PF_6 [F-Meppy, 2-(4-fluorophenyl)-5-(methyl)pyridine; dtbbpy, 4,4′-di-tert-butyl-2,2′-bipyridine], **66** [128]. In combination with labeled water (D_2O or T_2O) as the isotope source, and a suitable

Scheme 20 Directed sp^3 HIE using the commercially-available iridium catalysts

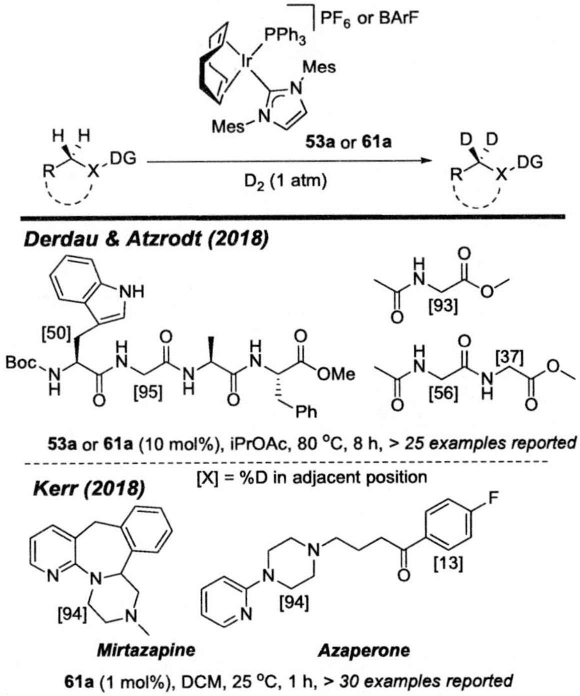

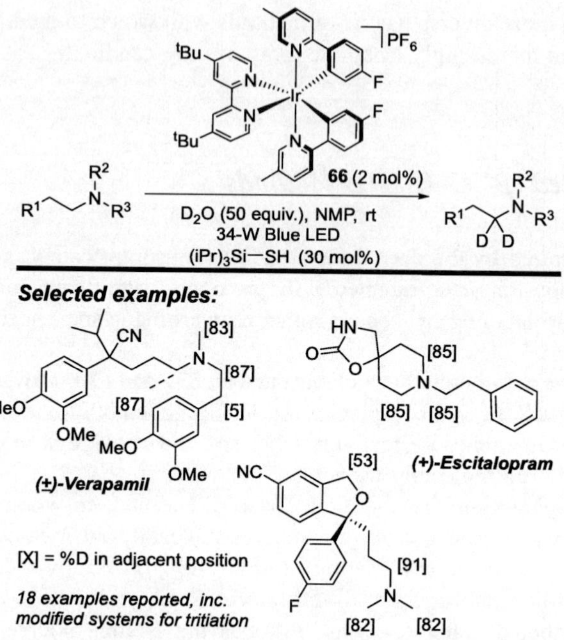

Scheme 21 Selected examples from MacMillan's photoredox and hydrogen atom transfer (HAT)-mediated alpha-selective sp^3 HIE process for drug-like amines

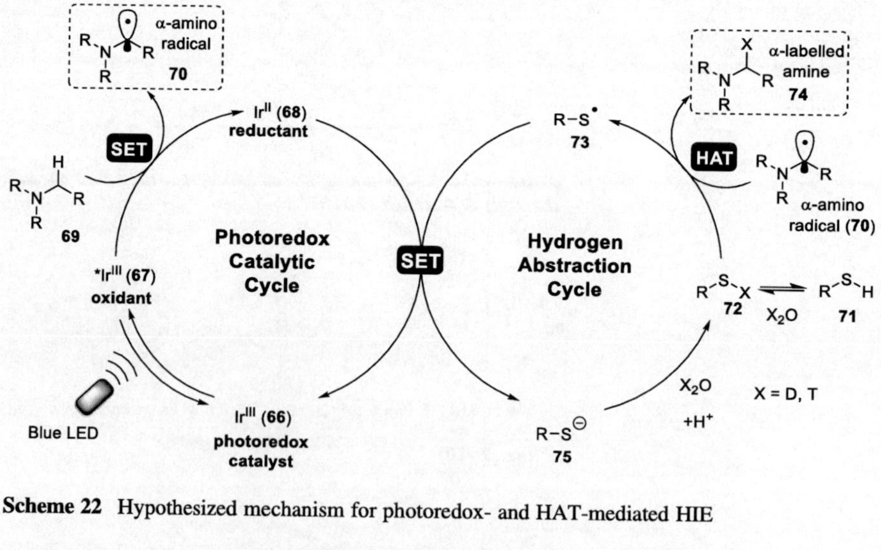

Scheme 22 Hypothesized mechanism for photoredox- and HAT-mediated HIE

hydrogen atom donor, this method selectively delivered isotope incorporation to the sp3 a-amino sites in 18 drug molecules (Scheme 21).

The reaction is proposed to operate via coupled photoredox and hydrogen atom transfer (HAT) cycles (Scheme 22). The photoredox catalyst **66** is excited by the blue light-emitting diode (LED) to generate a long-lived excited state triplet **67**, a strong single electron oxidant. The catalyst then generates an alpha-amino radical **70** from **69**, and the reduced Ir(II) catalyst **68**, which is now a strong reductant. Isotopic scrambling between the labeled water source and added thiol delivers the on-cycle labeled thiol **72** from **71**, judiciously chosen due to the favorably weak S–H bond. Labeled thiol **72** (polarity matched with the nucleophilic amino radical **69**) undergoes a HAT process to generate the alpha-labeled amine product **74** and thiol radical **73**. Thereafter, the photoredox and HAT catalytic cycles converge to generate the thiolate anion **75** and regenerate the photoredox catalyst **66**. Through adjustments in the choice of photocatalyst and thiol source, this method was applicable to both deuteration and tritiation processes.

3.2 Non-ortho-HIE on Aromatic Substrates

A range of cyclopentadienyl (Cp, and derivatives thereof)-ligated iridium complexes have been shown to be active in HIE (**76–84**, Scheme 23). Principally, several nondirected and global aromatic C–H deuteration strategies have been reported and improved over several iterations of catalyst design [114, 129–136]. In 2001, Bergman and co-workers showed that complexes of the type [(Cp*)Ir(PR$_3$)(H)(DCM)], such as **76**, and, later, [(Cp*)Ir(PMe$_3$)(H)$_3$]OTf, were active in HIE across a range of aromatic and aliphatic substrates [114, 129–131]. In further iterations, Peris [132] and Ison [134, 135] reported a range of NHC-ligated complexes based on the Cp-I core. In more practically facing contributions, Thieuleux and collaborators divulged solid-supported variants of [(Cp*)Ir(NHC)] cores, **82–84** [133, 136]. Across

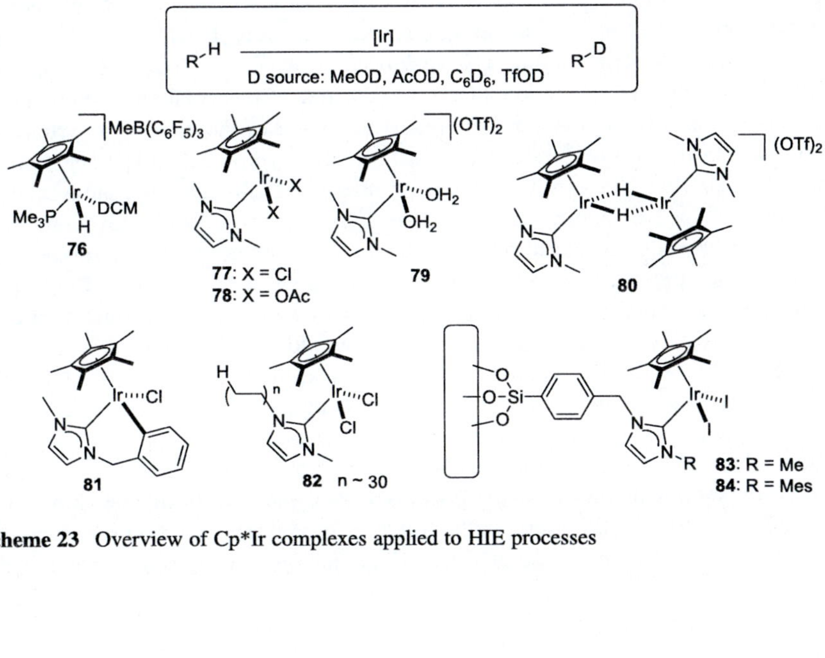

Scheme 23 Overview of Cp*Ir complexes applied to HIE processes

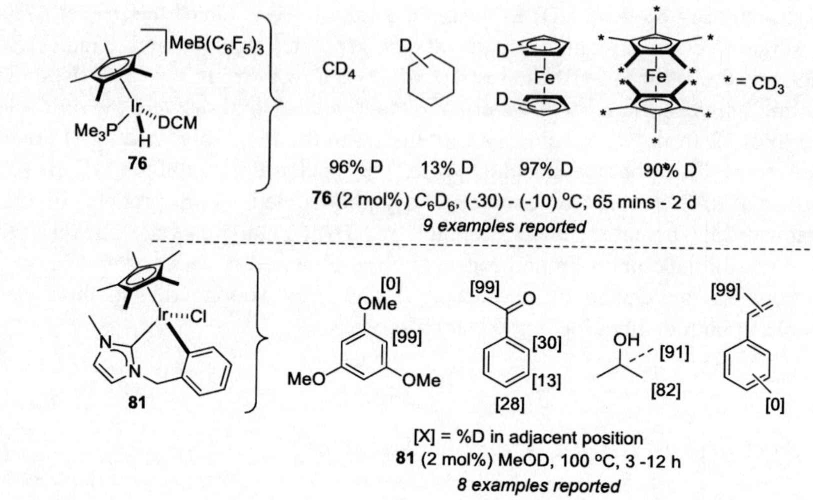

Scheme 24 Exemplar HIE processes enabled by Cp*Ir complexes

this series of publications, mechanisms of HIE were hypothesized to vary with deuterium source, solvent, and ancillary ligand combination (see Scheme 24 or exemplar transformations).

3.3 *Vinyl HIE Processes*

Expanding sp^2 labeling protocols beyond simple aromatic systems, a number of recent reports have shown the possibility of selectively labeling vinyl groups. Because many modern iridium HIE catalysts of the type $[(COD)Ir(L^1)(L^2)]X$ evolved from the hydrogenation literature [54, 62, 67], the labeling community has been aware of (and exploited) the reductive power of these catalyst systems to install isotopes across unsaturated moieties [10]. However, the dual HIE and hydrogenation reactivity of these iridium systems presents a challenge if the same catalyst is targeted for an HIE application, and *not* a hydrogenation. While designing HIE methods for labeling α,β-unsaturated substrates, Kerr and co-workers hypothesized that the competing reactivity could be rationalized by a equilibrating C–C bond rotation **85** to **86** upon substrate coordination (Scheme 25). For larger ligand spheres such as in catalyst **53a**, intermediate **86** would be favored, driving HIE (**86** → **87** → **88** → **89**). For smaller ligand systems, as has been observed in attempts to use Crabtree's catalyst for similar transformations [137], intermediate **85** is favored, driving hydrogenation over HIE (**85** → **90** → **91** → **92**).

Beyond re-optimizing HIE the use of catalysts in which competing hydrogenation is an issue, several methods for the chemoselective labeling of alkenes have also appeared in the iridium literature. In 2008, Hartwig reported a method where pincer complex **93** was shown to label vinyl C–H positions with selectivity largely

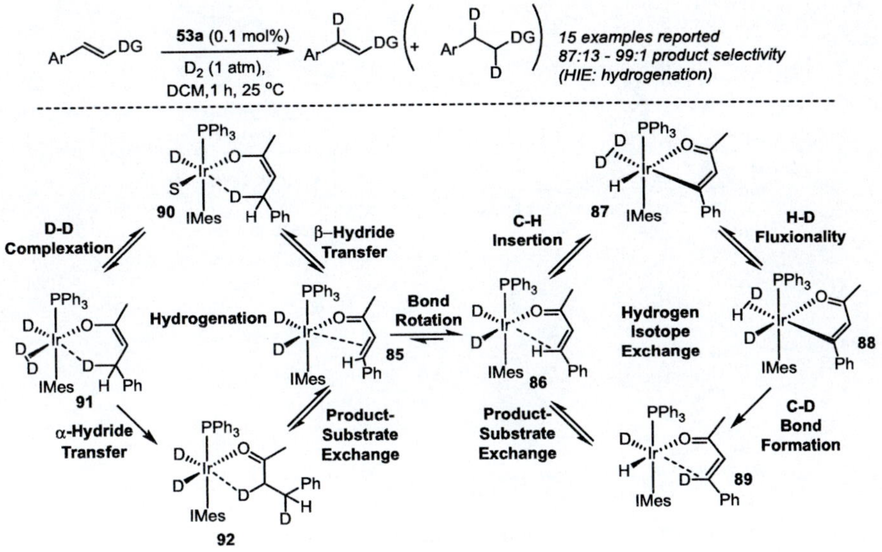

Scheme 25 Hypothesized competing HIE and hydrogenation pathways [99]

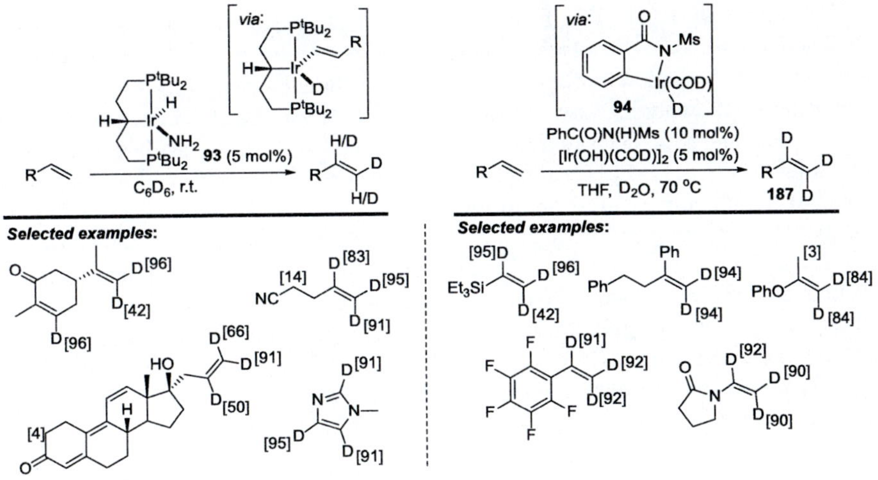

Scheme 26 Iridium-catalyzed vinyl HIE

dependent on the specific steric environment of the substrate, albeit under air and moisture sensitive conditions (Scheme 26, left) [138]. Notably, this method was applied to a series of both simple and complex organic molecules and included global labeling of aromatic and heteroaromatic substrates. A more practical variant of this method was divulged by Nishimura and co-workers [139]. Using an in situ-derived Ir(III) monohydride, **94**, and D_2O as the isotope source, an attractive range of

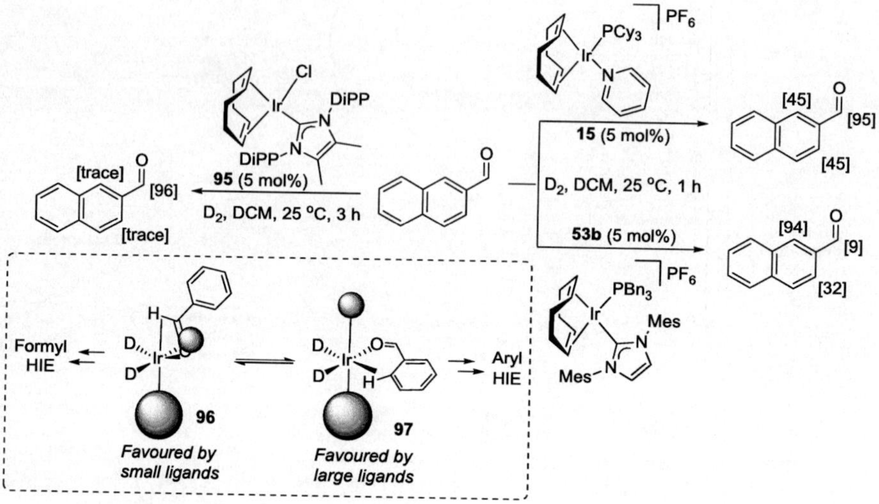

Scheme 27 Toward formyl-selective iridium-catalyzed HIE processes

mono-substituted alkenes could selectively deuterated at the vinyl or methylidene positions (Scheme 26, right).

In relation to vinyl HIE, formyl-selective methods of labeling benzaldehyde derivatives has been of notable interest, due, in part, to the synthetic handle of derivatization presented through the carbonyl functional group [140–145]. In 2010, Chapelle and co-workers showed that Crabtree's catalyst was able to deliver formyl-labeled benzaldehyde derivatives, albeit with variable selectivity against competing aryl ring labeling [142]. Kerr and co-workers used this work as inspiration to compare Crabtree's catalyst in formyl labeling vs. other competent *ortho*-HIE catalysts. Comparing catalysts **15** vs. **53b** vs. **95**, it was shown that the NHC/Cl system delivered superior formyl selectivity than either of the cationic iridium centers bearing larger ligand spheres (Scheme 27). The group accounted for these observations using a detailed mechanistic model centered around *cis-trans* isomerization of the activated Ir(III) catalyst. While intermediate **97** bearing *trans* ancillary ligands favors the approach trajectory of the aldehyde substrate that leads to aryl HIE, isomer **96** of the same catalyst enables the aldehyde to approach along a trajectory leading to formyl HIE [146].

3.4 Beyond C–H Labeling

Some of the most recent developments in isotopic labeling employing iridium catalysis have been applied to X–H moieties. While comparatively rare when compared to C–H HIE methods, heteroatom labeling can be insightful en route to establishing new carbon–heteroatom bonding-forming processes. Specifically,

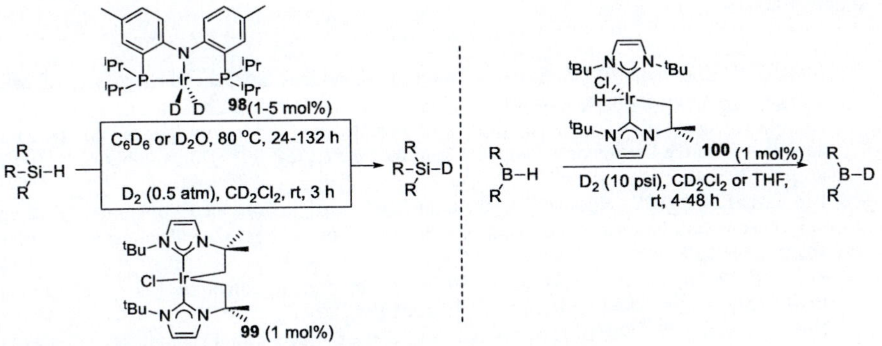

Scheme 28 Iridium-catalyzed HIE for Si–H and B–H bonds

Nolan and Grubbs have independently reported on silane labeling [147, 148]. Grubbs studied catalyst **98**, while Nolan investigated **99** and **100** in Si–H and B–H labeling, respectively (Scheme 28) [149].

4 Concluding Remarks

Notwithstanding earlier pioneering developments in the field [69, 73–75, 150–152], iridium-catalyzed HIE has undergone explosive growth since Heys' use of bis-phosphine systems in the early 1990s [153]. The main thrust of developments in the field have been in *ortho*-directed HIE domain. Such is the maturity and underlying mechanistic understanding of the *ortho*-labeling subfield, that it is now influencing catalyst design strategies in the broader C–H functionalization field. Considered alteration of the iridium ligand sphere – for both Ir(I) and Ir(III) systems – has now expanded the field of HIE well beyond its *ortho*-labeling comfort zone. Iridium-catalyzed methods to install heavy and radioactive hydrogen isotopes now span global aromatic labeling, sp^3 labeling, vinyl labeling, heteroatom labeling, and combinations thereof.

Iridium-catalyzed HIE is evolving at a time when computationally supported catalyst design is reaching unprecedented levels of sophistication [154–158]. It is expected, therefore, that forthcoming developments in iridium-catalyzed HIE will be enabled by deeper exploration of predictive methods of understanding substrate–catalyst compatibility.

Acknowledgments Dr. Marc Reid thanks the Leverhulme Trust for Early Career Fellowship funding (ECF-2016-264) and GlaxoSmithKline for their generous support. Thanks also go to Professor William J. Kerr and Dr. Richard Mudd for useful discussions relating to the preparation of this chapter.

References

1. Ebbing DD, Gammon SD (2005) Atoms, molecules, and ions. In: General chemistry, 8th edn. Houghton Mifflin Company, Boston
2. Atzrodt J, Derdau V, Kerr WJ, Reid M (2018) Deuterium- and tritium-labelled compounds: applications in the life sciences. Angew Chem Int Ed 57:1758–1784. https://doi.org/10.1002/anie.201704146
3. Isin EM, Elmore CS, Nilsson GN et al (2012) Use of radiolabeled compounds in drug metabolism and pharmacokinetic studies. Chem Res Toxicol 25:532–542. https://doi.org/10.1021/tx2005212
4. Filer CN (2010) Direct metal-catalyzed tritiation of organic compounds. J Label Compd Radiopharm 53:739–744. https://doi.org/10.1002/jlcr.1801
5. Nilsson GN, Kerr WJ (2010) The development and use of novel iridium complexes as catalysts for ortho-directed hydrogen isotope exchange reactions. J Label Compd Radiopharm 53:662–667. https://doi.org/10.1002/jlcr.1817
6. Allen PH, Hickey MJ, Kingston LP, Wilkinson DJ (2010) Metal-catalysed isotopic exchange labelling: 30 years of experience in pharmaceutical R&D. J Label Compd Radiopharm 53:731–738. https://doi.org/10.1002/jlcr.1825
7. Hesk D, Lavey CF, McNamara P (2010) Tritium labelling of pharmaceuticals by metal-catalysed exchange methods. J Label Compd Radiopharm 53:722–730. https://doi.org/10.1002/jlcr.1800
8. Lockley WJS, Heys JR (2010) Metal-catalysed hydrogen isotope exchange labelling: a brief overview. J Label Compd Radiopharm 53:635–644. https://doi.org/10.1002/jlcr.1851
9. Lockley WJS, Hesk D (2010) Rhodium- and ruthenium-catalysed hydrogen isotope exchange. J Label Compd Radiopharm 53:704–715. https://doi.org/10.1002/jlcr.1815
10. Lockley WJS, McEwen A, Cooke R (2012) Tritium: a coming of age for drug discovery and development ADME studies. J Label Compd Radiopharm 55:235–257. https://doi.org/10.1002/jlcr.2928
11. Di Giuseppe A, Castarlenas R, Oro LA (2015) Mechanistic considerations on catalytic H/D exchange mediated by organometallic transition metal complexes. C R Chim 18:713–741. https://doi.org/10.1016/j.crci.2015.02.006
12. Atzrodt J, Derdau V, Kerr WJ, Reid M (2018) C−H functionalisation for hydrogen isotope exchange. Angew Chem Int Ed 57:3022–3047. https://doi.org/10.1002/anie.201708903
13. Hesk D (2020) Highlights of C(sp^2)–H hydrogen isotope exchange reactions. J Label Compd Radiopharm 63:247–265. https://doi.org/10.1002/jlcr.3801
14. Junk T, Catallo WJ (1997) Hydrogen isotope exchange reactions involving C–H (D, T) bonds. Chem Soc Rev 26:401–406. https://doi.org/10.1039/CS9972600401
15. Valero M, Derdau V (2020) Highlights of aliphatic C(sp^3)–H hydrogen isotope exchange reactions. J Label Compd Radiopharm 63:266–280. https://doi.org/10.1002/jlcr.3783
16. Saljoughian M (2002) Synthetic tritium labeling: reagents and methodologies. Synthesis (Stuttg) 2002:1781–1801. https://doi.org/10.1055/s-2002-33907
17. Atzrodt J, Derdau V, Fey T, Zimmermann J (2007) The renaissance of H/D exchange. Angew Chem Int Ed 46:7744–7765. https://doi.org/10.1002/anie.200700039
18. Lockley WJS (2007) 30 Years with ortho-directed hydrogen isotope exchange labelling. J Label Compd Radiopharm 50:779–788. https://doi.org/10.1002/jlcr.1421
19. Hesk D, McNamara P (2007) Synthesis of isotopically labelled compounds at Schering-Plough, an historical perspective. J Label Compd Radiopharm 50:875–887. https://doi.org/10.1002/jlcr.1424
20. Heys JR (2007) Organoiridium complexes for hydrogen isotope exchange labeling. J Label Compd Radiopharm 50:770–778. https://doi.org/10.1002/jlcr.1428
21. Parkin G (2007) Applications of deuterium isotope effects for probing aspects of reactions involving oxidative addition and reductive elimination of H–H and C–H bonds. J Label Compd Radiopharm 50:1088–1114. https://doi.org/10.1002/jlcr.1435

22. Lockley WJS (2010) Hydrogen isotope labelling using iridium(I) dionates. J Label Compd Radiopharm 53:668–673. https://doi.org/10.1002/jlcr.1806
23. Gusev DG, Berke H (1996) Hydride fluxionality in Transition metal complexes: an approach to the understanding of mechanistic features and structural diversities. Chem Ber 129:1143–1155. https://doi.org/10.1002/cber.19961291002
24. Elander N, Jones JR, Lu S-Y, Stone-Elander S (2000) Microwave-enhanced radiochemistry. Chem Soc Rev 29:239–249. https://doi.org/10.1039/a901713e
25. Takeuchi R, Kezuka S (2006) Iridium-catalyzed formation of carbon-carbon and carbon-heteroatom bonds. Synthesis (Stuttg) 2006:3349–3366. https://doi.org/10.1055/s-2006-950284
26. Salter R (2010) The development and use of iridium(I) phosphine systems for ortho-directed hydrogen-isotope exchange. J Label Compd Radiopharm 53:645–657. https://doi.org/10.1002/jlcr.1814
27. Milstein D (2010) Discovery of environmentally benign catalytic reactions of alcohols catalyzed by pyridine-based pincer Ru complexes, based on metal–ligand cooperation. Top Catal 53:915–923. https://doi.org/10.1007/s11244-010-9523-7
28. Herbert JM (2010) Deuterium exchange promoted by iridium complexes formed in situ. J Label Compd Radiopharm 53:658–661. https://doi.org/10.1002/jlcr.1790
29. Atzrodt J, Derdau V (2010) Pd- and Pt-catalyzed H/D exchange methods and their application for internal MS standard preparation from a Sanofi-Aventis perspective. J Label Compd Radiopharm 53:674–685. https://doi.org/10.1002/jlcr.1818
30. Wiegerinck P (2011) 17th workshop of the international isotope society – central European division. The synthesis and applications of isotopes and isotopically labelled compounds. J Label Compd Radiopharm 54:278–288. https://doi.org/10.1002/jlcr.1859
31. Choi J, MacArthur AHR, Brookhart M, Goldman AS (2011) Dehydrogenation and related reactions catalyzed by iridium pincer complexes. Chem Rev 111:1761–1779. https://doi.org/10.1021/cr1003503
32. Gregson TJ, Herbert JM, Row EC (2011) Synthetic approaches to regiospecifically mono- and dilabelled arenes. J Label Compd Radiopharm 54:1–32. https://doi.org/10.1002/jlcr.1783
33. Suzuki T (2011) Organic synthesis involving iridium-catalyzed oxidation. Chem Rev 111:1825–1845. https://doi.org/10.1021/cr100378r
34. Aboagye EO, Aigbirhio FI, Allen P et al (2014) Abstracts of the 22nd international isotope society (UK Group) symposium: synthesis and applications of labelled compounds 2013. J Label Compd Radiopharm 57:178–190. https://doi.org/10.1002/jlcr.3173
35. Iglesias M, Oro LA (2018) A leap forward in iridium–NHC catalysis: new horizons and mechanistic insights. Chem Soc Rev 47:2772–2808. https://doi.org/10.1039/C7CS00743D
36. Shevchenko VP, Nagaev IY, Myasoedov NF (2019) Effect of processes occurring in the presence of metal catalysts on the main characteristics of the hydrogen isotope labeled organic compounds obtained. Radiochemistry 61:257–292. https://doi.org/10.1134/S1066362219030019
37. Lin D, Chang W (2000) Chemical derivatization and the selection of deuterated internal standard for quantitative determination – methamphetamine example. J Anal Toxicol 24:275–280. https://doi.org/10.1093/jat/24.4.275
38. Heinekey DM (2007) Transition metal dihydrogen complexes: isotope effects on reactivity and structure. J Label Compd Radiopharm 50:1063–1071. https://doi.org/10.1002/jlcr.1385
39. Lloyd-Jones GC, Muñoz MP (2007) Isotopic labelling in the study of organic and organometallic mechanism and structure: an account. J Label Compd Radiopharm 50:1072–1087. https://doi.org/10.1002/jlcr.1382
40. Quasdorf KW, Huters AD, Lodewyk MW et al (2012) Total synthesis of oxidized welwitindolinones and (−)- N -methylwelwitindolinone C isonitrile. J Am Chem Soc 134:1396–1399. https://doi.org/10.1021/ja210837b

41. Kerr WJ, Knox GJ, Paterson LC (2020) Recent advances in iridium(I) catalysis towards directed hydrogen isotope exchange. J Label Compd Radiopharm 63:281–295. https://doi.org/10.1002/jlcr.3812
42. Valero M, Kruissink T, Blass J et al (2020) C−H functionalization – prediction of selectivity in iridium(I)-catalyzed hydrogen isotope exchange competition reactions. Angew Chem Int Ed 59:5626–5631. https://doi.org/10.1002/anie.201914220
43. Crabtree RH, Felkin H, Morris GE (1977) Cationic iridium diolefin complexes as alkene hydrogenation catalysts and the isolation of some related hydrido complexes. J Organomet Chem 141:205–215. https://doi.org/10.1016/S0022-328X(00)92273-3
44. Zhu Y, Fan Y, Burgess K (2010) Carbene-metal hydrides can be much less acidic than phosphine-metal hydrides: significance in hydrogenations. J Am Chem Soc 132:6249–6253. https://doi.org/10.1021/ja101233g
45. Dobereiner G, Nova A (2011) Iridium-catalyzed hydrogenation of N-heterocyclic compounds under mild conditions by an outer-sphere pathway. J Am Chem Soc 133(19):7547–7562. https://doi.org/10.1021/ja2014983
46. Bennie LS, Fraser CJ, Irvine S et al (2011) Highly active iridium(I) complexes for the selective hydrogenation of carbon-carbon multiple bonds. Chem Commun 47:11653–11655. https://doi.org/10.1039/c1cc14367k
47. Semeniuchenko V, Exner TE, Khilya V, Groth U (2011) Homogeneous hydrogenation of electron-deficient alkenes by iridium complexes. Appl Organomet Chem 25:804–809. https://doi.org/10.1002/aoc.1841
48. Hopmann KH, Bayer A (2011) On the mechanism of iridium-catalyzed asymmetric hydrogenation of imines and alkenes: a theoretical study. Organometallics 30:2483–2497. https://doi.org/10.1021/om1009507
49. Woodmansee DH, Pfaltz A (2011) Asymmetric hydrogenation of alkenes lacking coordinating groups. Chem Commun (Camb) 47:7912–7916. https://doi.org/10.1039/c1cc11430a
50. Maurer F, Huch V, Ullrich A, Kazmaier U (2012) Development of catalysts for the stereoselective hydrogenation of α,β-unsaturated ketones. J Org Chem 77:5139–5143. https://doi.org/10.1021/jo300246c
51. Bert K, Noël T, Kimpe W et al (2012) Chiral imidate-ferrocenylphosphanes: synthesis and application as P,N-ligands in iridium(I)-catalyzed hydrogenation of unfunctionalized and poorly functionalized olefins. Org Biomol Chem 10:8539–8550. https://doi.org/10.1039/c2ob25871d
52. Song S, Zhu S-F, Yang S et al (2012) Enantioselective iridium-catalyzed hydrogenation of β,γ-unsaturated carboxylic acids: an efficient approach to chiral 4-alkyl-4-aryl butanoic acids. Angew Chem Int Ed 51:2708–2711. https://doi.org/10.1002/anie.201107802
53. Song S, Zhu S, Li Y, Zhou Q-L (2013) Iridium-catalyzed enantioselective hydrogenation of α,β-unsaturated carboxylic acids with tetrasubstituted olefins. Org Lett 15:3722–3725. https://doi.org/10.1021/ol401593a
54. Crabtree RH (1979) Iridium compounds in catalysis. Acc Chem Res 12:331
55. Song S, Zhu S-F, Yu Y-B, Zhou Q-L (2013) Carboxy-directed asymmetric hydrogenation of 1,1-diarylethenes and 1,1-dialkylethenes. Angew Chem Int Ed 125:1596–1599. https://doi.org/10.1002/ange.201208606
56. Kolychev EL, Kronig S, Brandhorst K et al (2013) Iridium(I) complexes with anionic N-heterocyclic carbene ligands as catalysts for the hydrogenation of alkenes in nonpolar media. J Am Chem Soc 135:12448–12459. https://doi.org/10.1021/ja406529c
57. Gruber S, Neuburger M, Pfaltz A (2013) Characterization and reactivity studies of dinuclear iridium hydride complexes prepared from iridium catalysts with N, P and C, N ligands under hydrogenation. Organometallics 32:4702–4711. https://doi.org/10.1021/om4007467
58. Horváth H, Kathó Á, Udvardy A et al (2014) New water-soluble iridium(I)–N-heterocyclic carbene–tertiary phosphine mixed-ligand complexes as catalysts of hydrogenation and redox isomerization. Organometallics 33:6330–6340. https://doi.org/10.1021/om5006148

59. Liu Y, Gridnev ID, Zhang W (2014) Mechanism of the asymmetric hydrogenation of exocyclic α,β-unsaturated carbonyl compounds with an iridium/BiphPhox catalyst: NMR and DFT studies. Angew Chem Int Ed 53:1901–1905. https://doi.org/10.1002/anie.201309677
60. Polo V, Al-Saadi AA, Oro LA (2014) Theoretical studies on the mechanism of iridium-catalyzed alkene hydrogenation by the cationic complex $[IrH_2(NCMe)_3(PiPr_3)]^+$. Organometallics 33:5156–5163. https://doi.org/10.1021/om500361e
61. Verendel JJ, Pàmies O, Diéguez M, Andersson PG (2014) Asymmetric hydrogenation of olefins using chiral Crabtree-type catalysts: scope and limitations. Chem Rev 114:2130–2169. https://doi.org/10.1021/cr400037u
62. Lee HM, Jiang T, Stevens ED, Nolan SP (2001) A cationic iridium complex bearing an imidazol-2-ylidene ligand as alkene hydrogenation catalyst. Organometallics 20:1255–1258
63. Vázquez-Serrano LD, Owens BT, Buriak JM (2002) Catalytic olefin hydrogenation using N-heterocyclic carbene–phosphine complexes of iridium. Chem Commun:2518
64. Perry M, Burgess K (2003) Chiral N-heterocyclic carbene-transition metal complexes in asymmetric catalysis. Tetrahedron Asymm 14:951–961
65. Zhu S-F, Xie J-B, Zhang Y-Z et al (2006) Well-defined chiral spiro iridium/phosphine-oxazoline cationic complexes for highly enantioselective hydrogenation of imines at ambient pressure. J Am Chem Soc 128:12886–12891. https://doi.org/10.1021/ja063444p
66. Smidt SP, Pfaltz A, Martínez-Viviente E et al (2003) X-ray and NOE studies on trinuclear iridium hydride phosphino oxazoline (PHOX) complexes. Organometallics 22:1000–1009. https://doi.org/10.1021/om020805a
67. Vazquez-Serrano LD, Owens BT (2006) The search for new hydrogenation catalyst motifs based on N-heterocyclic carbene ligands. Inorg Chim Acta 359:2786
68. Li S, Zhu S-F, Zhang C-M et al (2008) Iridium-catalyzed enantioselective hydrogenation of alpha, beta-unsaturated carboxylic acids. J Am Chem Soc 130:8584–8585. https://doi.org/10.1021/ja802399v
69. Garnett JL, Long MA, McLaren AB, Peterson KB (1973) Iridium(III) salts as homogeneous metal catalysts for hydrogen isotope exchange in organic compounds: a comparison with heterogeneous iridium for the deuteriation of alkylbenzenes. J Chem Soc Chem Commun 749:749. https://doi.org/10.1039/c39730000749
70. Vliegen M, Haspeslagh P, Verluyten W (2012) Alternative efficient tritium labeling of repaglinide. J Label Compd Radiopharm 55:155–157. https://doi.org/10.1002/jlcr.2913
71. Hickey MJ, Kingston LP, Lockley WJS et al (2007) Tritium-labelling via an iridium-based solid-phase catalyst. J Label Compd Radiopharm 50:286–289. https://doi.org/10.1002/jlcr.1233
72. Heys JR, Shu AYL, Senderoff SG, Phillips NM (1993) Deuterium exchange labelling of substituted aromatics using $[IrH_2(Me_2CO)_2(PPh_3)_2]BF_4$. J Label Compd Radiopharm 33:431–438. https://doi.org/10.1002/jlcr.2580330509
73. Lockley WJS (1985) Regioselective labelling of anilides with deuterium. J Label Compd Radiopharm 22:623–630. https://doi.org/10.1002/jlcr.2580220612
74. Lockley WJS (1984) Regioselective deuterium labelling of aromatic acids, amides and amines using group VIII metal catalysts. J Label Compd Radiopharm 21:45–57. https://doi.org/10.1002/jlcr.2580210105
75. Hesk D, Jones JR, Lockley WJS (1990) Regiospecific tritium labeling of aromatic acids, amides, amines and heterocyclics using homogeneous rhodium trichloride and ruthenium acetylacetonate catalysts. J Label Compd Radiopharm 28:1427–1436. https://doi.org/10.1002/jlcr.2580281211
76. Heys JR, Elmore CS (2009) Meta -substituent effects on organoiridium-catalyzed ortho -hydrogen isotope exchange. J Label Compd Radiopharm 52:189–200. https://doi.org/10.1002/jlcr.1588

77. Hesk D, Das PR, Evans B (1995) Deuteration of acetanilides and other substituted aromatics using [Ir(COD)(Cy_3P)(Py)]PF_6 as catalyst. J Label Compd Radiopharm 36:497–502. https://doi.org/10.1039/C39920000680
78. Ellames GJ, Gibson JS, Herbert JM et al (2004) Ligand effects upon deuterium exchange in arenes mediated by[Ir(PR_3)$_2$(cod)]$^+$$BF_4^-$. J Label Compd Radiopharm 47:1–10. https://doi.org/10.1002/jlcr.790
79. Herbert JM (2005) The mediation of aryl ketone deuteration by [Ir(PPh_3)$_3$(cod)]BF_4. J Label Compd Radiopharm 48:317–322. https://doi.org/10.1002/jlcr.925
80. Habraken E, Haspeslagh P, Vliegen M, Noël T (2015) Iridium(I)-catalyzed ortho-directed hydrogen isotope exchange in continuous-flow reactors. J Flow Chem 5:2–5. https://doi.org/10.1556/JFC-D-14-00033
81. Herbert JM, Kohler AD, McNeill AH (2005) An improved bidentate complex of iridium as a catalyst for hydrogen isotope exchange. J Label Compd Radiopharm 48:285–294. https://doi.org/10.1002/jlcr.921
82. Shu AYL, Chen W, Heys JR (1996) Organoiridium catalyzed hydrogen isotope exchange: ligand effects on catalyst activity and regioselectivity. J Organomet Chem 524:87–93. https://doi.org/10.1016/S0022-328X(96)06413-3
83. Ellames GJ, Gibson JS, Herbert JM et al (2001) Deuterium exchange mediated by an iridium–phosphine complex formed in situ. Tetrahedron Lett 42:6413–6416. https://doi.org/10.1016/S0040-4039(01)01266-7
84. Simonsson R, Stenhagen G, Ericsson C, Elmore CS (2013) Synthesis of ximelagatran, melagatran, hydroxymelagatran, and ethylmelagatran in H-3 labeled form. J Label Compd Radiopharm 56:334–337. https://doi.org/10.1002/jlcr.3028
85. Shu AYL, Heys JR (2000) Direct, efficient and selective tritiations of paclitaxel and photoaffinity taxoids. Tetrahedron Lett 41:9015–9019. https://doi.org/10.1016/S0040-4039(00)01652-X
86. Cross P, Herbert J, Kerr W et al (2015) Isotopic labelling of functionalised arenes catalysed by iridium(I) species of the [(cod)Ir(NHC)(py)]PF_6 complex class. Synlett 27:111–115. https://doi.org/10.1055/s-0035-1560518
87. Bushby N, Killick DA (2007) Hydrogen isotope exchange at alkyl positions using Crabtree's catalyst and its application to the tritiation of methapyrilene. J Label Compd Radiopharm 50:519–520. https://doi.org/10.1002/jlcr.1236
88. Ellames GJ, Gibson JS, Herbert JM, McNeill AH (2001) The scope and limitations of deuteration mediated by Crabtree's catalyst. Tetrahedron 57:9487–9497. https://doi.org/10.1016/S0040-4020(01)00945-0
89. Valsborg JS, Sørensen L, Foged C (2001) Organoiridium catalysed hydrogen isotope exchange of benzamide derivatives. J Label Compd Radiopharm 44:209–214. https://doi.org/10.1002/jlcr.446
90. Kingston LP, Lockley WJS, Mather AN et al (2000) Parallel chemistry investigations of ortho-directed hydrogen isotope exchange between substituted aromatics and isotopic water: novel catalysis by cyclooctadienyliridium(I)pentan-1,3-dionates. Tetrahedron Lett 41:2705–2708. https://doi.org/10.1016/S0040-4039(00)00244-6
91. McAuley B, Hickey MJ, Kingston LP et al (2003) Convenient and efficient deuteration of functionalized aromatics with deuterium oxide: catalysis by cycloocta-1,5-dienyliridium (I) 1,3-dionates. J Label Compd Radiopharm 46:1191–1204. https://doi.org/10.1002/jlcr.780
92. Jones JR, Lockley WJS, Lu S-Y, Thompson SP (2001) Microwave-enhanced aromatic dehalogenation studies: a rapid deuterium-labelling procedure. Tetrahedron Lett 42:331–332. https://doi.org/10.1016/S0040-4039(00)01941-9
93. Gruber S, Neuburger M, Pfaltz A (2013) Characterization and reactivity studies of dinuclear iridium hydride complexes prepared from iridium catalysts with N,P and C,N ligands under hydrogenation conditions. Organometallics 32:4702–4711. https://doi.org/10.1021/om4007467

94. Vazquez-Serrano LD, Owens BT, Buriak JM (2006) The search for new hydrogenation catalyst motifs based on N-heterocyclic carbene ligands. Inorg Chim Acta 359:2786–2797. https://doi.org/10.1016/j.ica.2005.10.049
95. Powell ME, Elmore CS, Dorff PN, Heys JR (2007) Investigation of isotopic exchange reactions using N-heterocyclic iridium (I) complexes. J Label Compd Radiopharm 50:523–525. https://doi.org/10.1002/jlcr.1239
96. Cross PWC, Ellames GJ, Gibson JS et al (2003) Conditions for deuterium exchange mediated by iridium complexes formed in situ. Tetrahedron 59:3349–3358. https://doi.org/10.1016/S0040-4020(03)00422-8
97. Brown JA, Irvine S, Kennedy AR et al (2008) Highly active iridium(i) complexes for catalytic hydrogen isotope exchange. Chem Commun 1115. https://doi.org/10.1039/b715938b
98. Brown JA, Cochrane AR, Irvine S et al (2014) The synthesis of highly active iridium (I) complexes and their application in catalytic hydrogen isotope exchange. Adv Synth Catal 356:3551–3562. https://doi.org/10.1002/adsc.201400730
99. Kerr WJ, Mudd RJ, Paterson LC, Brown JA (2014) Iridium(I)-catalyzed regioselective C–H activation and hydrogen-isotope exchange of non-aromatic unsaturated functionality. Chem A Eur J 20:14604–14607. https://doi.org/10.1002/chem.201405114
100. Cochrane ARX, Idziak C, Kerr WJ et al (2014) Practically convenient and industrially-aligned methods for iridium-catalysed hydrogen isotope exchange processes. Org Biomol Chem 12:3598–3603. https://doi.org/10.1039/C4OB00465E
101. Simmons EM, Hartwig JF (2012) On the interpretation of deuterium kinetic isotope effects in C-H bond functionalizations by transition-metal complexes. Angew Chem Int Ed 51:3066–3072. https://doi.org/10.1002/anie.201107334
102. Kerr WJ, Reid M, Tuttle T (2015) Iridium-catalyzed C–H activation and deuteration of primary sulfonamides: an experimental and computational study. ACS Catal 5:402–410. https://doi.org/10.1021/cs5015755
103. Devlin J, Kerr W, Lindsay D et al (2015) Iridium-catalysed ortho-directed deuterium labelling of aromatic esters – an experimental and theoretical study on directing group Chemoselectivity. Molecules 20:11676–11698. https://doi.org/10.3390/molecules200711676
104. Kennedy AR, Kerr WJ, Moir R, Reid M (2014) Anion effects to deliver enhanced iridium catalysts for hydrogen isotope exchange processes. Org Biomol Chem 12:7927–7931. https://doi.org/10.1039/C4OB01570C
105. Queen AE, Hesk D, Lindsay DM et al (2020) Synthesis of [^{3}H] and [^{14}C]genipin. J Label Compd Radiopharm 63:196–202. https://doi.org/10.1002/jlcr.3832
106. Kerr WJ, Lindsay DM, Reid M et al (2016) Iridium-catalysed ortho-H/D and -H/T exchange under basic conditions: C–H activation of unprotected tetrazoles. Chem Commun 52:6669–6672. https://doi.org/10.1039/C6CC02137A
107. Atzrodt J, Derdau V, Kerr WJ et al (2015) Expanded applicability of iridium(I) NHC/phosphine catalysts in hydrogen isotope exchange processes with pharmaceutically-relevant heterocycles. Tetrahedron 71:1924–1929. https://doi.org/10.1016/j.tet.2015.02.029
108. Kerr WJ, Lindsay DM, Owens PK et al (2017) Site-selective deuteration of N -heterocycles via iridium-catalyzed hydrogen isotope exchange. ACS Catal 7:7182–7186. https://doi.org/10.1021/acscatal.7b02682
109. Kerr WJ, Mudd RJ, Owens PK et al (2016) Hydrogen isotope exchange with highly active iridium(I) NHC/phosphine complexes: a comparative counterion study. J Label Compd Radiopharm 59:601–603. https://doi.org/10.1002/jlcr.3427
110. Koneczny M, Phong Ho L, Nasr A et al (2020) Iridium(I) complexes with anionic N-heterocyclic carbene ligands as catalysts for H/D exchange in nonpolar media. Adv Synth Catal. https://doi.org/10.1002/adsc.202000438
111. Parmentier M, Hartung T, Pfaltz A, Muri D (2014) Iridium-catalyzed H/D exchange: ligand complexes with improved efficiency and scope. Chem A Eur J 20:11496–11504. https://doi.org/10.1002/chem.201402078

112. Valero M, Burhop A, Jess K et al (2018) Evaluation of a P,N-ligated iridium(I) catalyst in hydrogen isotope exchange reactions of aryl and heteroaryl compounds. J Label Compd Radiopharm 61:380–385. https://doi.org/10.1002/jlcr.3595
113. Jess K, Derdau V, Weck R et al (2017) Hydrogen isotope exchange with iridium(I) complexes supported by phosphine-imidazolin-2-imine P,N ligands. Adv Synth Catal 359:629–638. https://doi.org/10.1002/adsc.201601291
114. Klei SR, Tilley TD, Bergman RG (2002) Iridium(III) and rhodium(III) complexes bearing chelating cyclopentadienyl−phosphine ligands as C−H activation catalysts for the deuteration of hydrocarbons in D 2 O. Organometallics 21:4905–4911. https://doi.org/10.1021/om020375o
115. Kloek SM, Goldberg KI (2007) Competitive C−H bond activation and β-hydride elimination at platinum(II). J Am Chem Soc 129:3460–3461. https://doi.org/10.1021/ja0669629
116. Polukeev AV, Marcos R, Ahlquist MSG, Wendt OF (2016) Iridium hydride complexes with cyclohexyl-based pincer ligands: fluxionality and deuterium exchange. Organometallics 35:2600–2608. https://doi.org/10.1021/acs.organomet.6b00324
117. Bhattacharjee R, Nijamudheen A, Karmakar S, Datta A (2016) Strain control: reversible H 2 activation and H 2 /D 2 exchange in Pt complexes. Inorg Chem 55:3023–3029. https://doi.org/10.1021/acs.inorgchem.5b02904
118. Butschke B, Schlangen M, Schröder D, Schwarz H (2008) Competitive intramolecular aryl- and alkyl-C–H bond activation and ligand evaporation from gaseous bisimino complexes [Pt(L)(CH_3)((CH_3)$_2$S)]$^+$ (L=C_6H_5N=C(CH_3)–C(CH_3)=NC_6H_5). Helv Chim Acta 91:1902–1915. https://doi.org/10.1002/hlca.200890204
119. Crosby SH, Clarkson GJ, Rourke JP (2009) A delicate balance between sp^2 and sp^3 C−H bond activation: a Pt(II) complex with a dual agostic interaction. J Am Chem Soc 131:14142–14143. https://doi.org/10.1021/ja905046n
120. Butschke B, Ghassemi Tabrizi S, Schwarz H (2010) Ion-molecule reactions of "Rollover" cyclometalated [Pt(bipy−H)]$^+$ (bipy=2,2′-bipyridine) with dimethyl ether in comparison with dimethyl sulfide: an experimental/computational study. Chem A Eur J 16:3962–3969. https://doi.org/10.1002/chem.200902742
121. Campos J, Espada MF, López-Serrano J, Carmona E (2013) Cyclometalated iridium complexes of Bis(aryl) phosphine ligands: catalytic C–H/C–D exchanges and C–C coupling reactions. Inorg Chem 52:6694–6704. https://doi.org/10.1021/ic400759r
122. Grellier M, Mason SA, Albinati A et al (2013) Probing highly selective H/D exchange processes with a ruthenium complex through neutron diffraction and multinuclear NMR studies. Inorg Chem 52:7329–7337. https://doi.org/10.1021/ic302307m
123. Goforth SK, Walroth RC, McElwee-White L (2013) Evaluation of multisite polypyridyl ligands as platforms for the synthesis of Rh/Zn, Rh/Pd, and Rh/Pt heterometallic complexes. Inorg Chem 52:5692–5701. https://doi.org/10.1021/ic301810y
124. Benedetti M, Barone CR, Girelli CR et al (2014) H/D exchange at sp3 carbons in the coordination sphere of platinum(III). Dalton Trans 43:3669. https://doi.org/10.1039/c3dt53216j
125. Rivada-Wheelaghan O, Roselló-Merino M, Ortuño MA et al (2014) Reactivity of coordinatively unsaturated bis(N-heterocyclic carbene) Pt(II) complexes toward H2. Crystal structure of a 14-electron Pt(II) hydride complex. Inorg Chem 53:4257–4268. https://doi.org/10.1021/ic500705t
126. Kerr WJ, Mudd RJ, Reid M et al (2018) Iridium-catalyzed Csp^3–H activation for mild and selective hydrogen isotope exchange. ACS Catal 8:10895–10900. https://doi.org/10.1021/acscatal.8b03565
127. Valero M, Weck R, Güssregen S et al (2018) Highly selective directed iridium-catalyzed hydrogen isotope exchange reactions of aliphatic amides. Angew Chem Int Ed 57:8159–8163. https://doi.org/10.1002/anie.201804010

128. Loh YY, Nagao K, Hoover AJ et al (2017) Photoredox-catalyzed deuteration and tritiation of pharmaceutical compounds. Science 358:1182–1187. https://doi.org/10.1126/science.aap9674
129. Golden JT, Andersen RA, Bergman RG (2001) Exceptionally low-temperature carbon−hydrogen/carbon−deuterium exchange reactions of organic and organometallic compounds catalyzed by the Cp*(PMe_3)IrH($ClCH_2Cl$)$^+$ cation. J Am Chem Soc 123:5837–5838. https://doi.org/10.1021/ja0155480
130. Klei SR, Golden JT, Burger P, Bergman RG (2002) Cationic Ir(III) alkyl and hydride complexes: stoichiometric and catalytic C–H activation by Cp*(PMe_3)Ir(R)(X) in homogeneous solution. J Mol Catal A Chem 189:79–94. https://doi.org/10.1016/S1381-1169(02)00192-9
131. Skaddan MB, Yung CM, Bergman RG (2004) Stoichiometric and catalytic deuterium and tritium labeling of "Unactivated" organic substrates with cationic Ir(III) complexes. Org Lett 6:11–13. https://doi.org/10.1021/ol0359923
132. Corberán R, Sanaú M, Peris E (2006) Highly stable Cp*−Ir(III) complexes with N -heterocyclic carbene ligands as C−H activation catalysts for the deuteration of organic molecules. J Am Chem Soc 128:3974–3979. https://doi.org/10.1021/ja058253l
133. Maishal TK, Alauzun J, Basset JM et al (2008) A tailored organometallic-inorganic hybrid mesostructured material: a route to a well-defined, active, and reusable heterogeneous iridium-NHC catalyst for H/D exchange. Angew Chem Int Ed 47:8654–8656. https://doi.org/10.1002/anie.200802956
134. Feng Y, Jiang B, Boyle PA, Ison EA (2010) Effect of ancillary ligands and solvents on H/D exchange reactions catalyzed by Cp*Ir complexes. Organometallics 29:2857–2867. https://doi.org/10.1021/om100018x
135. Lehman MC, Gary JB, Boyle PD et al (2013) Effect of solvent and ancillary ligands on the catalytic H/D exchange reactivity of Cp IrIII(L) complexes. ACS Catal 3:2304–2310. https://doi.org/10.1021/cs400420n
136. Romanenko I, Norsic S, Veyre L et al (2016) Active and recyclable polyethylene-supported iridium-(N- heterocyclic carbene) catalyst for hydrogen/deuterium exchange reactions. Adv Synth Catal 358:2317–2323. https://doi.org/10.1002/adsc.201600045
137. Marek A, Pedersen MHF, Vogensen SB et al (2016) The labeling of unsaturated γ-hydroxybutyric acid by heavy isotopes of hydrogen: iridium complex-mediated H/D exchange by C⍰H bond activation vs reduction by boro-deuterides/tritides. J Label Compd Radiopharm 59:476–483. https://doi.org/10.1002/jlcr.3432
138. Zhou J, Hartwig JF (2008) Iridium-catalyzed H/D exchange at vinyl groups without olefin isomerization. Angew Chem Int Ed 47:5783–5787. https://doi.org/10.1002/anie.200801992
139. Hatano M, Nishimura T, Yorimitsu H (2016) Selective H/D exchange at vinyl and methylidene groups with D 2 O catalyzed by an iridium complex. Org Lett 18:3674–3677. https://doi.org/10.1021/acs.orglett.6b01721
140. Spletstoser JT, White JM, Georg GI (2004) One-step facile synthesis of deuterium labeled aldehydes from tertiary amides using $Cp_2Zr(D)Cl$. Tetrahedron Lett 45:2787–2789. https://doi.org/10.1016/j.tetlet.2004.02.030
141. Ariza X, Asins G, Garcia J et al (2010) Preparation of α-labeled aldehydes by base-catalyzed exchange reactions. J Label Compd Radiopharm 53:556–558. https://doi.org/10.1002/jlcr.1759
142. Chappelle MR, Hawes CR (2010) The use of metal-catalysed hydrogen isotope exchange in the contract supply of tritiated compounds. J Label Compd Radiopharm 53:745–751. https://doi.org/10.1002/jlcr.1821
143. Barnett DW, Refaei MS, Curley RW (2013) Chirally deuterated benzyl chlorides from benzyl alcohols via hexachloroacetone/polymer-supported triphenylphosphine: synthesis of protected (2 S , 3 S)-[3-^{2}H, ^{15}N]-tyrosine. J Label Compd Radiopharm 56:6–11. https://doi.org/10.1002/jlcr.3004

144. Korsager S, Taaning RH, Lindhardt AT, Skrydstrup T (2013) Reductive carbonylation of aryl halides employing a two-chamber reactor: a protocol for the synthesis of aryl aldehydes including 13 C- and D-isotope labeling. J Org Chem 78:6112–6120. https://doi.org/10.1021/jo400741t
145. Boga SB, Alhassan AB, Hesk D (2014) Efficient synthesis of ^{2}H & ^{13}C labeled benzaldehydes via regio-selective formylation. Tetrahedron Lett 55:4442–4444. https://doi.org/10.1016/j.tetlet.2014.06.053
146. Kerr WJ, Reid M, Tuttle T (2017) Iridium-catalyzed formyl-selective deuteration of aldehydes. Angew Chem Int Ed 56:7808–7812. https://doi.org/10.1002/anie.201702997
147. Fortman GC, Jacobsen H, Cavallo L, Nolan SP (2011) Catalytic deuteration of silanes mediated by N-heterocyclic carbene-Ir(III) complexes. Chem Commun 47:9723. https://doi.org/10.1039/c1cc13492b
148. Iluc VM, Fedorov A, Grubbs RH (2012) H/D exchange processes catalyzed by an iridium-pincer complex. Organometallics 31:39–41. https://doi.org/10.1021/om201049p
149. Nelson DJ, Egbert JD, Nolan SP (2013) Deuteration of boranes: catalysed versus non-catalysed processes. Dalton Trans 42:4105. https://doi.org/10.1039/c3dt33045a
150. Garnett JL, Hodges RJ (1967) Homogeneous metal-catalyzed exchange of aromatic compounds. Isotopic hydrogen labeling procedure. J Am Chem Soc 89:4546–4547. https://doi.org/10.1021/ja00993a067
151. Lockley WJS (1982) Regioselective deuteration of aromatic and α,β-unsaturated carboxylic acids via rhodium(III) chloride catalysed exchange with deuterium oxide. Tetrahedron Lett 23:3819–3822. https://doi.org/10.1016/S0040-4039(00)87716-3
152. Anderson GK, Saum SE, Cross RJ, Morris SA (1983) Homogeneous catalysts of hydrogen-deuterium exchange reactions involving cyclopentadienyl complexes of palladium and platinum. Organometallics 2:780–782. https://doi.org/10.1021/om00078a021
153. Heys R (1992) Investigation of $[IrH_2(Me_2CO)_2(PPh_3)_2]BF_4$ as a catalyst of hydrogen isotope exchange of substrates in solution. J Chem Soc Chem Commun 680. https://doi.org/10.1039/c39920000680
154. Maldonado AG, Rothenberg G (2010) Predictive modeling in homogeneous catalysis: a tutorial. Chem Soc Rev 39:1891–1902. https://doi.org/10.1039/b921393g
155. Houk KN, Cheong PH-Y (2008) Computational prediction of small-molecule catalysts. Nature 455:309–313. https://doi.org/10.1038/nature07368
156. Peng Q, Duarte F, Paton RS (2016) Computing organic stereoselectivity – from concepts to quantitative calculations and predictions. Chem Soc Rev 45:6093–6107. https://doi.org/10.1039/c6cs00573j
157. Sperger T, Sanhueza IA, Schoenebeck F (2016) Computation and experiment: a powerful combination to understand and predict reactivities. Acc Chem Res 49:1311–1319. https://doi.org/10.1021/acs.accounts.6b00068
158. Balcells D, Clot E, Eisenstein O et al (2016) Deciphering selectivity in organic reactions: a multifaceted problem. Acc Chem Res 49:1070–1078. https://doi.org/10.1021/acs.accounts.6b00099

Top Organomet Chem (2021) 69: 303–324
https://doi.org/10.1007/3418_2020_52

Published online: 7 August 2020

Iridium-Catalyzed Homogeneous Hydrogenation and Hydrosilylation of Carbon Dioxide

Francisco J. Fernández-Alvarez and Luis A. Oro

Contents

Abstract The knowledge of the potential of transition metal-based complexes as catalysts for the reduction of CO_2 has grown significantly over the last few decades. This chapter focuses on the progress made during recent years in the field of homogeneous iridium-catalyzed reduction of CO_2 by using hydrogen and/or silicon hydrides as reducing agents, comparing them with homogeneous catalysts based on other transition metals.

The reported studies on iridium-catalyzed CO_2 reduction processes show that an important point to keep in mind when designing a catalyst is the nature of the reducing agent (hydrogen, hydrosilanes, and/or hydrosiloxanes). Thus, iridium(III)

The original version of this chapter was revised. A correction to this chapter can be found at https://doi.org/10.1007/978-3-030-69083-0_74

F. J. Fernández-Alvarez (✉) and L. A. Oro (✉)
Departamento de Química Inorgánica, Facultad de Ciencias, Instituto de Síntesis Química y Catálisis Homogénea (ISQCH), Universidad de Zaragoza – CSIC, Zaragoza, Spain
e-mail: paco@unizar.es; oro@unizar.es

half-sandwich complexes with 4,4′-dihydroxy-bipyridine (DHBP) or 4,7-dihydroxy-1,10-phenanthroline (DHPT) ligands, and iridium(III)-PNP pincer complexes have proven to be excellent catalysts for the hydrogenation of CO_2 to formic acid. However, Ir(III)-NSiN^{Me} (NSiN = *fac*-bis-(4-methylpyridine-2-yloxy)methylsilyl) and Ir(III)-NSi^{Me} (NSi^{Me} = 4-methylpyridine-2-yloxydimethylsilyl) species are not stable under hydrogen atmosphere but are effective catalysts for the reduction of CO_2 with hydrosiloxanes to silylformate under solvent-free conditions and moderate CO_2 pressures and temperatures. Moreover, while using iridium(III)-DHBP half-sandwich complexes, high CO_2 and H_2 pressures are required to achieve the catalytic CO_2 hydrogenation to methanol; Ir-NSi^{Me} species catalyze the reduction of CO_2 to methoxysilane with hydrosiloxanes under low CO_2 pressure.

Keywords CO_2 hydrogenation · CO_2 hydrosilylation · CO_2 reduction · Homogenous catalysis · Iridium

1 Introduction

Carbon dioxide is an abundant, easily available, cheap, and low toxic chemical. On the other hand, during the last decades, the concentration of CO_2 in the earth's atmosphere has reached historical values, which is generally considered one of the main reasons for the global warming. Therefore, both for economic and environmental reasons the development of sustainable processes that allow the transformation of CO_2 on an industrial scale into valuable chemicals could be considered one of the most important tasks for the sustainability of the modern chemical industry [1–5]. In this context, to achieve the goal of using CO_2 as raw material of the chemical industry there are several difficulties to face, among which its great thermodynamic stability stands out.

Catalysis has proven to be essential to overcome the challenge of CO_2 stability. Thus, in recent decades, great advances have been made in the field of catalytic CO_2 transformation into value added chemicals [6–16]. Particularly, catalytic hydrogenation [6, 9, 10, 17–21] and/or hydrosilylation [22–25] of CO_2 have proven to be efficient methodologies for its reduction to formate, formaldehyde, methanol, or methane level (Scheme 1). In this regard, it is remarkable that several homogeneous

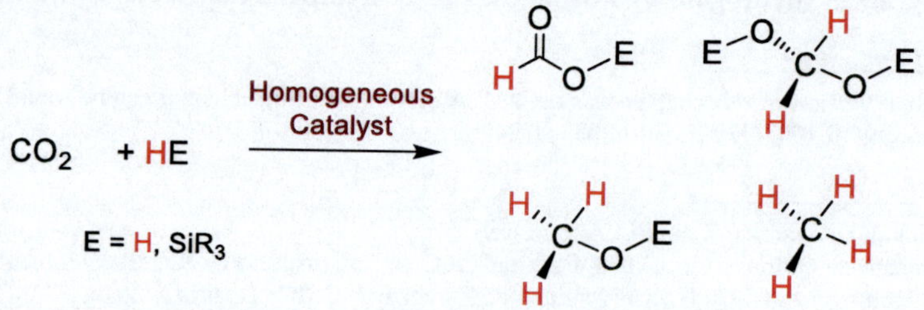

Scheme 1 Possible products from the catalytic reduction of CO_2 with hydrogen and/or silicon hydrides

catalytic systems based on iridium complexes have shown high catalytic performance as CO_2 reduction catalysts [18, 26–28]. This chapter will focus on the progress made during recent years in the field of iridium-catalyzed reduction of CO_2 by using hydrogen and/or hydrosilanes as reducing agents.

2 Recent Advances on Iridium-Catalyzed CO_2 Hydrogenation

During last decades, several examples of homogeneous catalysts effective for the hydrogenation of CO_2 have been reported, most of them are based on ruthenium (II) complexes but some examples of highly active iridium(III) catalysts have also been described. Among them are iridium(III) half-sandwich complexes with 4,4′-dihydroxy-bipyridine (DHBP) or 4,7-dihydroxy-1,10-phenanthroline (DHPT) ligands, which are excellent catalysts for the hydrogenation of CO_2 to formic acid and also have been used as catalysts for the direct hydrogenation of CO_2 to methanol. Moreover, iridium(III)-PNP pincer complexes have also been used as effective catalysts for the hydrogenation of CO_2 to formic acid. Conversely, the potential of iridium complexes as catalysts for the hydrogenation of CO_2 to formaldehyde, methyl carbonate, and/or methyl formate remains a challenge.

2.1 *Iridium-Catalyzed Formic Acid or Formate Preparation from CO_2 and H_2*

Catalytic hydrogenation of CO_2 to formic acid (FA) has been a research subject of great interest over the last decades [6, 9, 10, 18, 20, 21, 28]. The hydrogenation of CO_2 is endergonic in the gas phase ($\Delta G°_{298} = 32.9$ kJ mol^{-1}), however, in water solution and in presence of a base (NH_3), this reaction becomes thermodynamically favored ($\Delta G°_{298} = -35.4$ kJ mol^{-1}) [29].

The first studies of the potential of transition metal complexes as homogenous catalysts for the hydrogenation of CO_2 to FA were reported by Inoue et al. in 1976 [30]. These studies revealed that using NEt_3 water solutions under 50 atm of mixtures of CO_2 and H_2 (1:1) at r.t. the complex $[IrH_3(PPh_3)_3]$ catalyzes this transformation, however, its catalytic activity is low. Under the same conditions species $[RuH_2(PPh_3)_4]$ was found to be the most active of the studied catalyst precursors [30]. Some years later, Leitner et al. reported very efficient rhodium phosphane water soluble catalysts, which were able to promote the formation of FA in relatively high yields [31, 32]. After that, Noyori et al. described that the effectivity of ruthenium phosphane complexes as CO_2 hydrogenation catalysts improves when using supercritical carbon dioxide [33, 34]. Few years after that, Joó, Laurenczy et al. reported that the performance of catalytic systems based on

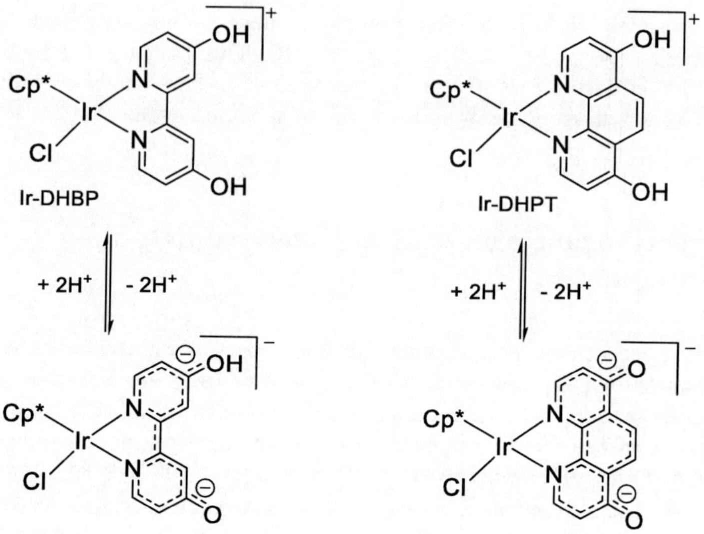

Scheme 2 Examples of Ir-DHBP and Ir-DHPT CO_2 hydrogenation catalyst precursors

water soluble Ir, Rh, Ru, and Pd phosphane complexes as CO_2 hydrogenation catalyst is strongly pH dependent [35]. In this regard, Jessop et al. found that using the complex [RuCl(O_2CMe)(PMe_3)$_4$], which is soluble in supercritical CO_2, as catalysts for the hydrogenation of CO_2 to FA in presence of the appropriate amine and one alcohol that has an aqueous scale pK_a below that of the protonated amine, it was possible to achieve an initial turnover frequency (TOF) for FA production of 95,000 h^{-1} [36]. Since then till the development of the highly active Himeda's catalysts [37], based on half-sandwich bipyridine iridium complexes, most of the homogeneous catalysts effective for the hydrogenation of CO_2 to FA were based on Ru- and Rh-phosphane complexes.

Early examples of highly active iridium CO_2 hydrogenation catalysts were based on iridium half-sandwich complexes with 4,4′-dihydroxy-bipyridine (DHBP) or 4,7-dihydroxy-1,10-phenanthroline (DHPT) ligands (Scheme 2) [38]. These catalysts are highly efficient for the hydrogenation of carbonate, in situ generated from CO_2 in basic KOH aqueous solutions, to formate. The oxyanions generated from the hydroxy group along the catalytic process play a key role on both the catalytic activity and water solubility of these catalysts (Scheme 2).

Initial turnover frequencies (TOF) of 42,000 and 35,000 h^{-1} were obtained for the Ir-DHPB and Ir-DHPT (Scheme 2) catalyzed reactions, respectively. The best performance was achieved heating at 120°C aqueous KOH (1.0 M) solutions of the corresponding iridium catalysts under 6 MPa of CO_2/H_2 (1:1). Moreover, these iridium catalysts could be reused for four cycles maintaining high catalytic performance [38].

Himeda et al. have extended their studies to iridium half-sandwich complexes with *N,N*-bidentate ligands different from bipyridine such as picolinamide- [39, 40],

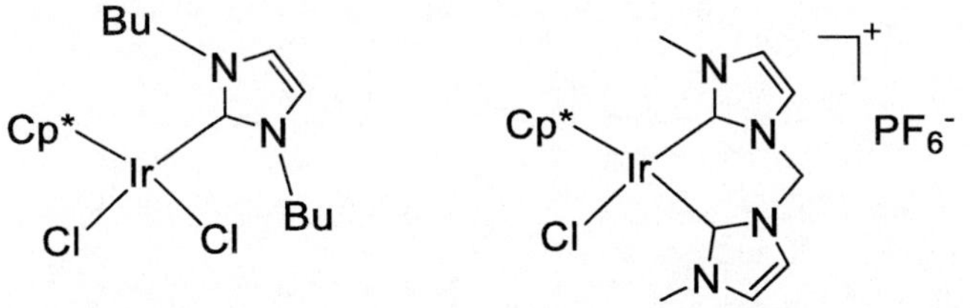

Fig. 1 CO_2 hydrogenation catalysts based on half-sandwich iridium(III) complexes with NHC ligands

azole- [41] and pyridyl-pyrazole derivatives [42]. Mechanistic studies have found that these catalysts promote the activation of CO_2 via an outer-sphere mechanism [41]. Interestingly, it has been found that using this type of iridium catalysts it is possible to achieve the pH-controlled reversible hydrogen storage [40, 42, 43].

Further support to the relevant role of oxyanions in these type of catalysts comes from the studies reported from Peris et al. [44], which showed that using half-sandwich iridium(III) complexes with strong donor NHC ligands (Fig. 1) or bipyridine derivatives without hydroxy substituents, as catalysts precursors for the hydrogenation of CO_2 lower activities (TOF = 1,600 h^{-1}) were observed.

Iridium-pincer complexes have also found to be active catalysts for the homogeneous hydrogenation of CO_2 to FA. The iridium(III)trihydride-PNP complex shown in Scheme 3 reached a TOF of 150,000 h^{-1} for the hydrogenation of CO_2 to FA in basic medium. The performance of this catalytic system is strongly influenced by the nature of the base, the temperature and the presence of THF in the reaction medium. Thus, the best results were obtained at 200°C, using 1.0 M KOH aqueous solution and adding 0.1 mL of THF [45]. Mechanistic studies showed that two reactions pathways are possible, one of them involving a deprotonative dearomatization of the pyridinic ring and other a hydroxy-assisted hydrogenolysis as the rate determining step, respectively. Moreover, an outer-sphere mechanism has been found for the CO_2 activation step (Scheme 3) [46].

Iridium-PNP catalysts showed the best performance in KOH aqueous solutions, however, under these conditions, the corresponding formate salt, not FA, is obtained as reaction product. Therefore, a neutralization step of the formate with a strong acid is required to obtain FA. Interestingly, when using amine derivatives as bases a simple distillation of the resulting ammonium formate allows separation of pure FA from the starting base. In this regard, Nozaki's group has studied the effect of both using triethanolamine aqueous solution as base and having different substituents at the pyridinic ring on the activity of Ir-PNP catalysts (Fig. 2). They have found that under these conditions the dichlorohydride derivative with a *p*-MeO substituent is the most active catalyst, indeed, using this species as catalyst precursor in a 1.0 M triethanolamine aqueous solution, in presence of THF and heating at 150°C, a TON for the conversion of CO_2 to FA of 160,000 (TOF = 12,000 h^{-1}) was obtained [47].

On the other hand, Hazari and coworkers have studied the activity of Ir-PN^{H}P (PN^{H}P = bis{(2-diisopropylphosphanyl)ethyl}amine) pincer species as CO_2

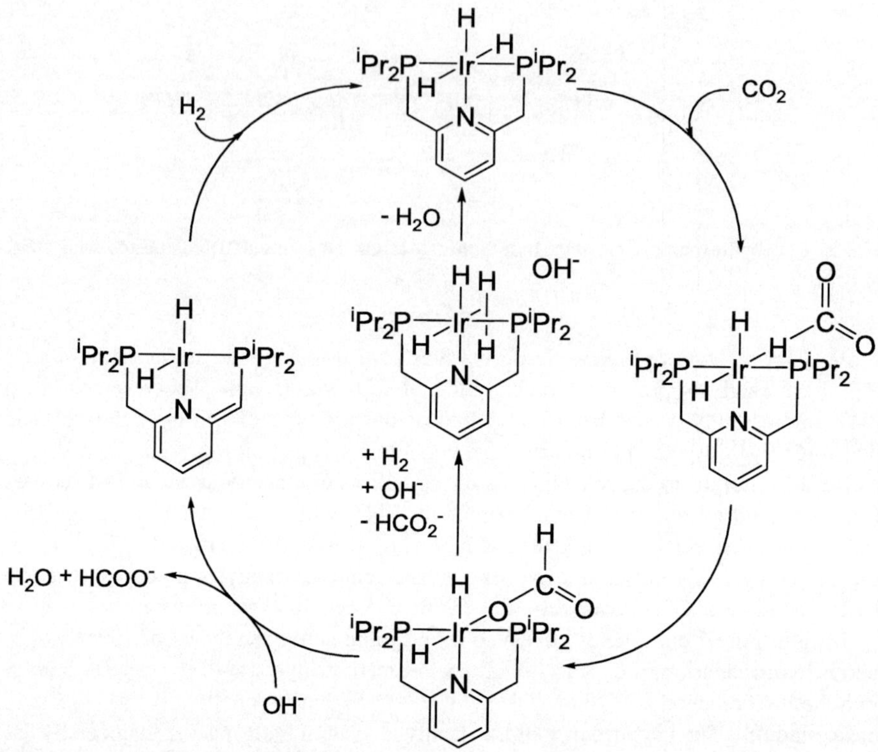

Scheme 3 Mechanism proposed for Ir(III)-PNP catalyzed CO_2 hydrogenation

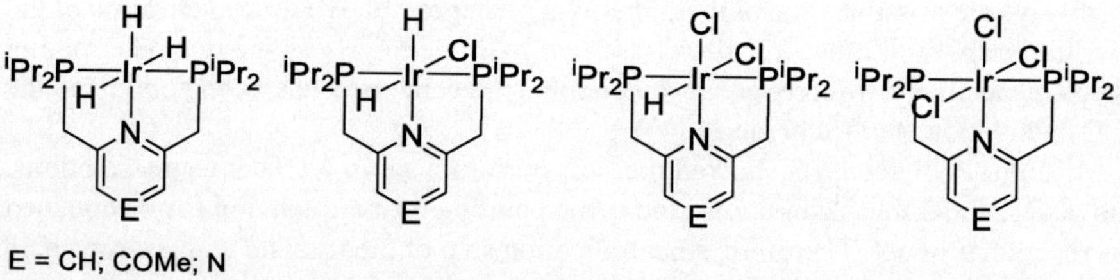

Fig. 2 Examples of Ir-PNP CO_2 hydrogenation catalysts. The species with E = COMe was found to be the most active catalyst

hydrogenation catalysts. They have shown that the insertion of CO_2 into one of the Ir-H bonds of the trihydride derivative [Ir(PN^HP)H_3] gives the corresponding [Ir (PN^HP)(HCO_2)H_2] species, which is stabilized by an intramolecular NH-OCO hydrogen bond (Scheme 4) [48]. This iridium-formate derivative catalyzes the hydrogenation of CO_2, in 1 M aqueous KOH solution at 185°C, with a TON and TOF values of 348,000 and 18,780 h^{-1}, respectively. DFT calculations show that the Ir-PN^HP-catalyzed CO_2 hydrogenation takes place through an outer-sphere

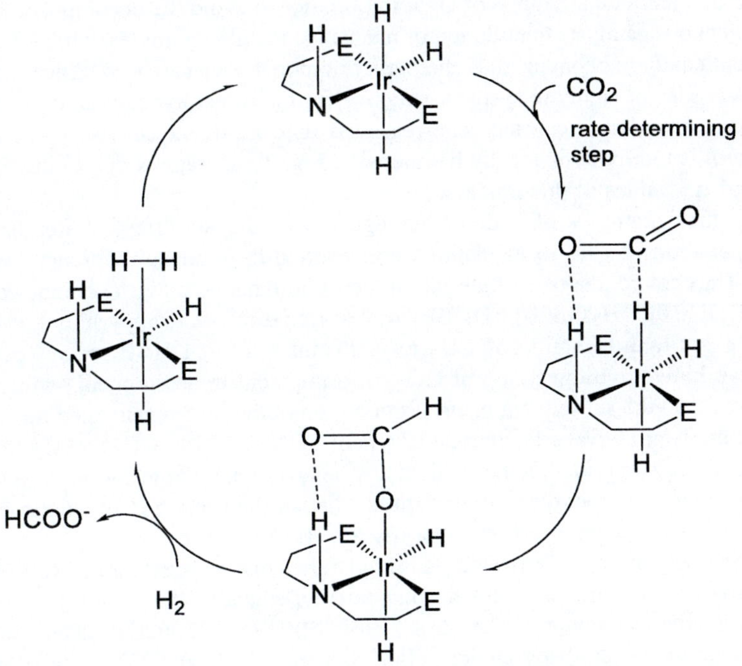

Scheme 4 Mechanism proposal for Ir-PN^{H}P catalyzed CO_2 hydrogenation

mechanism (Scheme 4). The rate determining step of the overall catalytic process corresponds to the NH-assisted CO_2 activation step [48].

2.2 *Iridium-Catalyzed Methanol Preparation from Direct Hydrogenation of CO_2*

Methanol is commonly produced on an industrial scale using fossil fuel-based syngas as the principal feedstock. The annual demand for methanol has grown steadily over the last decade, consequently the CO_2 emissions related to the industrial production of methanol have also grown [49, 50]. Therefore, the development of catalysts effective for the synthesis of methanol from renewable sources is attracting the interest of several research groups [50, 51]. In this regard, the production of methanol through carbon dioxide capture and recycling is one of the keys of the "Methanol Economy" concept [52]. The early example of a homogeneous catalyst effective for the direct hydrogenation of CO_2 to methanol was reported by Tominaga et al. in 1993 [53, 54]. They used [$Ru_3(CO)_{12}$] as catalyst precursor, KI as additive to prevent the formation of metallic nanoparticles, and *N*-methylpyrrolidone as solvent at 240°C under 80 bar of a 1:3 mixture of CO_2 and H_2. In this regard, it should be mentioned that it is of great importance to avoid the decomposition of the homogeneous catalysts to colloidal or nanosized metallic particles, which may have different catalytic behavior than the parent homogeneous catalysts. Since then only few examples of catalytic systems effective for the direct hydrogenation of CO_2 to methanol have been reported. The reason is that the direct conversion of CO_2 to methanol is thermodynamically hampered at high temperatures due to the negative ΔH and ΔS values of this process.

The first examples of iridium homogeneous catalysts effective for the direct hydrogenation of CO_2 to methanol were reported by Himeda, Laurenczy et al. in 2016. They found that the sulfate salt of the iridium half-sandwich cationic complex [IrCp*(DHBP)(OH_2)][SO_4] (DHBP = 4,4′-dihydroxy-2,2′-bypyridine) catalyzes the one pot hydrogenation of CO_2 to methanol. This Ir-DHBP species catalyzes the quantitative hydrogenation of CO_2 to formic acid in acidic media without any additives, and the subsequent disproportionation of the in situ generated formic acid to give methanol (96% selectivity; 47% yield; TON = 1,314), CO_2 and H_2O [55]. In this regard, it is important to be aware that whenever the hydrogenation of CO_2 takes place in basic solution, the question arises whether the actual reactive partner of the catalysts is carbonate, bicarbonate, or (hydrated) CO_2.

The activity of this iridium catalyst is higher than that reported for the Ru-(Triphos) (Triphos = 1,1,1-tris(diphenylphosphinomethyl)ethane) species (TON = 221) [56, 57], the ruthenium(II) species [Ru(PNP)(H)(H-BH_3)(CO)] (PNP = {Bis [2-(diphenylphosphino)ethyl]amine}) [58, 59], Co-(Triphos) (TON = 50) [60] and Mn-(PNP) (TON = 36) [61]. Being surpassed by that of the complex Fe-(κ^3-$^{H}T_{pm}$) ($^{H}T_{pm}$ = tris(pyrazolyl)methane; 44% yield; TON = 2,283) [62].

2.3 *Miscellaneous*

Examples of homogenous catalysts effective for the hydrogenation of CO_2 to other products, different of formic acid and/or methanol, are scarce. Indeed, to the best of our knowledge only few examples of ruthenium catalysts effective for the hydrogenation of CO_2 to dimethyl ether [63], formaldehyde [64, 65], or methyl formate [66] have been reported. Therefore, the potential of iridium complexes as catalysts for these types of processes remains unexplored.

3 Recent Advances on Iridium-Catalyzed CO_2 Hydrosilylation

The catalytic hydrogenation of CO_2 with H_2 requires high H_2 and CO_2 pressures and temperatures, as well as the addition of bases or other additives. Contrariwise, the catalytic reduction of CO_2 with hydrosilanes features several advantages such as

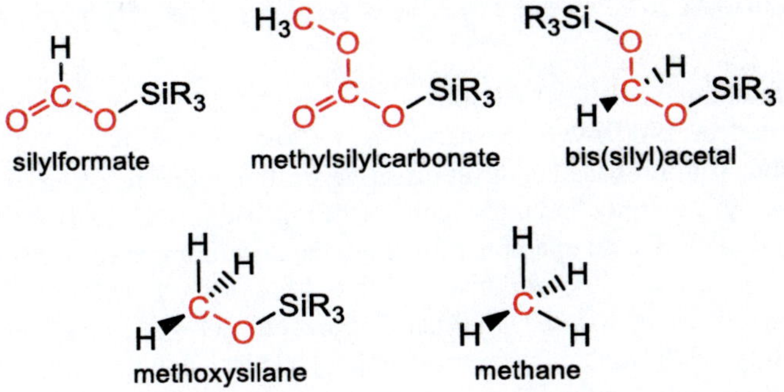

Scheme 5 Reported products from the catalytic reduction of CO_2 with silicon hydrides

being a thermodynamically favored process and the fact that silanes are easier and safer to handle and to store than molecular hydrogen [22, 23, 25, 26, 67]. However, the utilization of silicon hydrides as reductants for large-scale reduction of CO_2 faces some difficulties. One of them is the high price of hydrosilanes, which could be solved by using cheap hydrosiloxanes instead of hydrosilanes, another is the stoichiometric generation of siloxanes, which is unsustainable due to the challenge of Si-H regeneration from Si-O-Si bonds [24, 68]. Furthermore, differently to hydrogenation processes, the catalytic hydrosilylation cannot be performed in aqueous or alcoholic solutions since homogeneous hydrosilylation catalysts usually catalyzed the dehydrogenative hydrolysis and/or alcoholysis of silicon hydrides [69, 70].

The catalytic reaction of CO_2 with silicon hydrides allows its selective reduction to the corresponding silylformate, bis(silyl)acetal or methoxysilane, and to methane [22, 23, 25, 26] (Scheme 5). In addition, the formation of methyl carbonates from the iridium-catalyzed reduction of CO_2 with silicon hydrides has been recently reported (Scheme 5) [71].

The first examples of homogeneous catalytic reduction of CO_2 using hydrosilanes as reductants were reported in the 1980s [72–74]. However, it was during the year 2012 that the breakthrough of this chemistry took place. Since then until today, the number of catalytic systems effective for the reduction of CO_2 with hydrosilanes based on transition metal complexes as well as on metal-free catalysts or main elements derivatives that have proven to be effective in CO_2 hydrosilylation processes has considerably grow up [8, 22–25, 67, 75]. Among them, catalysts based on iridium complexes stand out not only for their activity but also for their versatility that allows selectivity control by choosing proper ligands and/or tuning the reaction conditions. Furthermore, some examples of iridium-based CO_2 hydrosilylation catalysts have proven to be effective under solvent-free conditions and using hydrosiloxanes as reductants.

3.1 *Iridium-Catalyzed CO_2 Hydrosilylation to Silylformate*

The iridium complex [Ir(CN)(CO)(dppe)] (dppe = 1,2-bis(diphenylphosphino)ethane), reported in 1989 by Eisenschmid and Eisenberg, is the first example of a homogeneous iridium-based catalyst effective for the hydrosilylation of CO_2. However, the catalytic activity and the selectivity of this iridium catalyst were low [74]. It was not until 2012 that an example of iridium catalyst, complex [Ir(CF_3SO_3)(NSiN)(SiR_3)($NCMe_3$)] (NSiN = *fac*-bis-(pyridine-2-yloxy)methylsilyl; SiR_3 = SiMe($OSiMe_3$)$_2$), efficient for the hydrosilylation of CO_2 to selectively give the corresponding silylformate was reported [76]. This catalytic system allows the solvent-free and gram-scale formation of silylformates under mild reaction conditions (3 bar, 298 K, TON = 97.5) but is slow (TON = 0.7 h^{-1}) [76]. Interestingly, using species [Ir(CF_3SO_3)(NSiN)(H)(coe)] (coe = *cis*-cyclooctene, Scheme 6), which is easier to prepare than the abovementioned Ir-NSiN-acetonitrile derivative, under the same reaction conditions (3 bar, 298 K) produces an increase of the reaction rate (TOF = 1.2 h^{-1}) [77, 78]. Further studies on the influence of reaction temperature [77] and CO_2 pressure [78] on the catalytic performance of this catalytic system showed that the activity is directly proportional to the temperature; however, increasing the temperature reduces the selectivity to silylformate [77]. On the other hand, it is more difficult to generalize the CO_2-pressure effect on the activity of the reaction. It is remarkable, that from the point of view of selectivity the CO_2- pressure has proven to be a parameter to consider. Indeed, for each temperature an enhancement of the CO_2-pressure results in increased the selectivity of the process [78]. Thus, using species [Ir(CF_3SO_3)(NSiN)(H)(coe)] as catalyst precursor the best reaction performance was achieved at 344 K and under 8 bar of CO_2 (99.9% conversion, 89.7% purity (GC-MS), TOF = 138 h^{-1}; TON = 87.5) (Scheme 6) [78].

The iridium(III) complex [Ir(H)(CF_3CO_2)($NSiN^{Me}$)(coe)] ($NSiN^{Me}$ = *fac*-bis-(4-methylpyridine-2-yloxy)methylsilyl), which contains a trifluoroacetate instead of a triflate ligand and a $NSiN^{Me}$ ligand with 4-methylated pyridinic rings (Fig. 3), has proven to be a highly effective CO_2 hydrosilylation catalyst [79]. Using this Ir-trifluoroacetate-$NSiN^{Me}$ species as catalyst precursor for the hydrosilylation of CO_2 to silylformate with HSiMe($OSiMe_3$)$_2$ the best results were achieved at 328 K

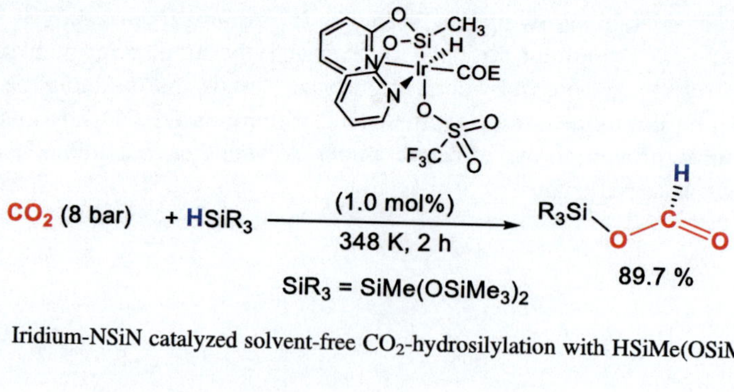

Scheme 6 Iridium-NSiN catalyzed solvent-free CO_2-hydrosilylation with HSiMe($OSiMe_3$)$_2$

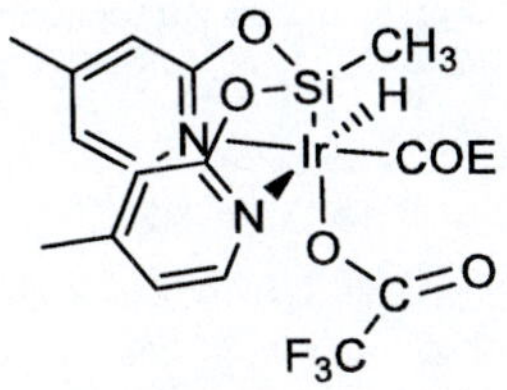

Fig. 3 Iridium(III) complex $[Ir(H)(CF_3CO_2)(NSiN^{Me})(coe)]$

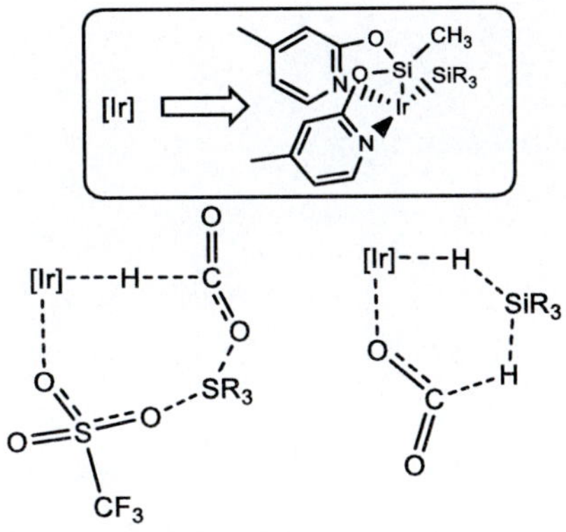

Fig. 4 Outer- and inner-sphere transition state (TS) found for Ir-triflate-$NSiN^{Me}$ and Ir-trifluoroacetate-$NSiN^{Me}$ catalysts precursors, respectively

and under 8 bar of CO_2 (100% conversion; 98.9% yield to SF by GC-MS; TOF = 99.3 h^{-1}), at temperatures above 328 K a decrease in catalytic selectivity and activity was observed [79].

Mechanistic studies based on theoretical calculations at DFT level showed that while Ir-trifluoroacetate-$NSiN^{Me}$ species catalyzes the CO_2 activation via an inner-sphere mechanism, an outer-sphere mechanism is favored for Ir-triflate-$NSiN^{Me}$ derivatives (Fig. 4) [80].

The presence of the Ir-silyl group of the $NSiN^R$ (R = H, Me) ligand *trans*-located to the trifluoroacetate (or triflate) ligand plays a key role on the catalytic activity of Ir-$NSiN^R$ catalysts. Based on this knowledge the catalyst precursor $[Ir(CF_3CO_2)(\kappa^2\text{-}NSi^{Me})_2]$ (NSi^{Me} = 4-methylpyridine-2-yloxydimethylsilyl), containing two Ir-Si bonds *trans*-located to the catalyst active positions was designed (Fig. 5) [81]. ^1H NMR studies on the activity of $[Ir(CF_3CO_2)(\kappa^2\text{-}NSi^{Me})_2]$ as CO_2 hydrosilylation catalyst using $HSiMe(OSiMe_3)_2$ show that at 298 K under 4 bar of CO_2 this catalyst is more active (TOF = 28.6 h^{-1}) [81] than the previously reported Ir-NSiN species, which at 298 K independently of the CO_2-pressure are low active with TOF values in the rage of 1.2–1.6 h^{-1} [78]. The higher activity of $[Ir(CF_3CO_2)(\kappa^2\text{-}NSi^{Me})_2]$ allows the selective formation methoxysilane from CO_2 and $HSiMe(OSiMe_3)_2$ as it is shown below [81].

Other iridium complex which have proven to be an active catalyst for the selective hydrosilylation of CO_2 (3 bar) to silylformates is the zwitterionic iridium (III) half-sandwich species $[IrClCp^*\{(MeIm)_2CHCOO\}]$ ((MeIm = 3-

Fig. 5 Iridium(III) complex $[Ir(CF_3CO_2)(\kappa^2\text{-}NSi^{Me})_2]$

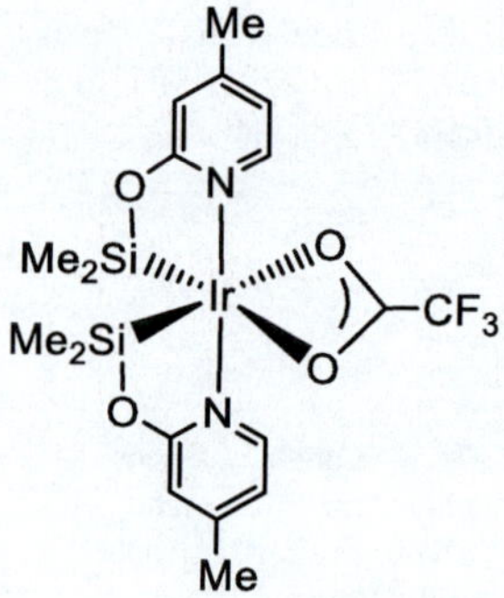

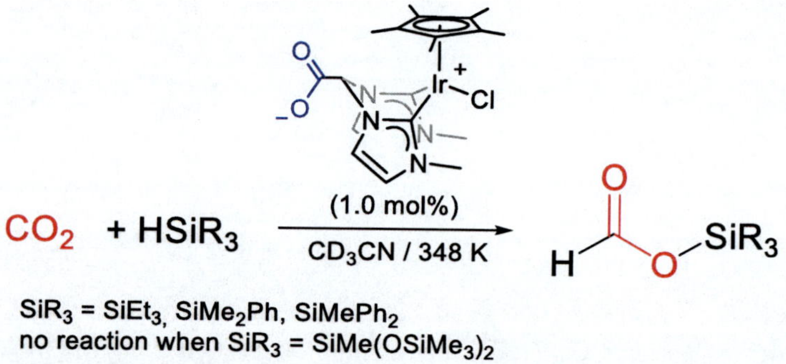

Scheme 7 CO_2 hydrosilylation catalyzed by the zwitterionic iridium species [Cp*IrCl{$(MeIm)_2CHCO_2$)}]

methylimidazol-2-yliden-1-yl; Cp* = pentamethylcyclopentadienyl) (Scheme 7) [82]. However, this catalytic system requires the use of acetonitrile as reaction solvent. It is relatively high active for the hydrosilylation of CO_2 with $HSiMe_2Ph$ (TOF = 51 h^{-1}), but under the same reaction conditions is not active when the hydrosiloxane $HSiMe(OSiMe_3)_2$ is used as reductant instead of $HSiMe_2Ph$ [82].

Other transition metal-based catalysts including Ru [83, 84], Co [85], Rh [86], Pd [87], Pt [88], Cu [89, 90], and Zn [91, 92] complexes effective for the selective hydrosilylation of CO_2 to the formate level have been reported. Among them, the catalytic system based on the Pd-PAlP complex shown in Scheme 8 has proven to be the most active catalyst for CO_2-hydrosilylation reported so far [87]. Indeed, using this Pd-PAlP catalyst in DMF as solvent in presence of Cs^tBuCO_2 (1.0 mol%) at 298 K, the selective reaction of CO_2 with $HSiMe_2Ph$ to give HCO_2SiMe_2Ph (92%, TOF = 19,300 h^{-1}) was achieved in 1 h (Scheme 8) [87].

Ir-NSiN and Ir-NSi^{Me} species are comparatively less active than some of the abovementioned catalysts; however, they have the advantage of being active under solvent-free conditions and are highly effective when using hydrosiloxanes, instead

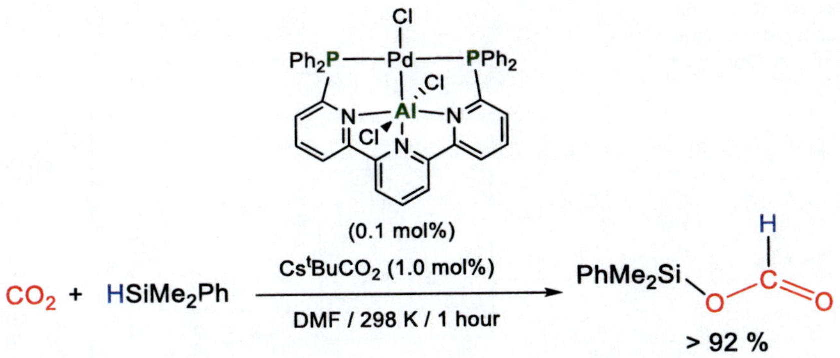

Scheme 8 Palladium-PAlP catalyzed CO_2-hydrosilylation with $HSiMe_2Ph$

of hydroorganosilanes, as reductants. Therefore, from the point of view of sustainability iridium species based on Ir-NSiN and Ir-NSi^{Me} species could be considered promising for future applications of the catalytic reduction of CO_2 with silicon hydrides.

3.2 *Iridium-Catalyzed Reduction of CO_2 to Methoxysilanes with Silicon-Hydrides*

Only few examples of homogeneous catalysts effective for the reduction of CO_2 to methanol level using silicon hydrides as reducing agents have been published to date. The first one was the abovementioned iridium complex [Ir(CN)(CO)(dppe)] (dppe = 1,2-bis(diphenylphosphino)ethane) [74]. This catalyst promotes the reduction of CO_2 with $HSiMe_3$ in C_6D_6 at 313 K to the corresponding methoxysilane, CH_3OSiMe_3. This reaction is slow, and 2 weeks are required to achieve the conversion of the starting hydrosilane into CH_3OSiMe_3. ^{13}C NMR studies of this process using $^{13}CO_2$ confirm that it entails in a stepwise progression with the initial formation of the corresponding silylformate HCO_2SiMe_3, which is further reduce to bis (silyl)acetal $CH_2(OSiMe_3)_2$, the later finally reacts with one equivalent of $HSiMe_3$ to give CH_3OSiMe_3 and $O(SiMe_3)_2$ (Scheme 9) [74].

The iridium(III) complex $[Ir(CF_3CO_2)(\kappa^2\text{-NSi}^{Me})_2]$ (Fig. 5) has proven to be an effective catalyst for the reduction of CO_2 with $HSiMe(OSiMe_3)_2$ to the methoxysilane $CH_3OSiMe(OSiMe_3)_2$ under mild reaction conditions. ^{1}H NMR studies of the reaction of CO_2 (1 bar) with $HSiMe(OSiMe_3)_2$ in C_6D_6 at 298 K evidenced the selective formation of the corresponding methoxysilane after 16 h (99.0%; TON = 33.6; TOF = 2.1 h^{-1}) [81]. Interestingly, increasing the CO_2 pressure to 4 bar the reaction stops in the corresponding silylformate, which under 4 bar is the major reaction product (93%; TON = 93; TOF = 2.9 h^{-1}) together with a 7% of $CH_3OSiMe(OSiMe_3)_2$ after 3.5 h. ^{1}H and ^{13}C NMR studies and theoretical

Scheme 9 Iridium-catalyzed reduction of CO_2 to the methoxysilane level with $HSiR_3$

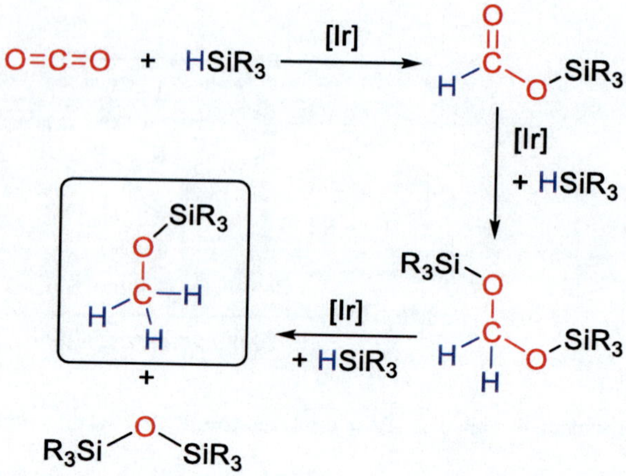

calculations at the DFT level on the Ir-NSi^{Me} catalyzed CO_2 reduction to methoxysilane with silicon hydrides, agree with an stepwise mechanism similar to that shown in Scheme 9.

The related complex $[Ir(\mu\text{-}CF_3SO_3)(\kappa^2\text{-}NSi^{Me})_2]_2$, which is a rare example of an iridium dinuclear species with triflate groups acting as bridges, catalyzed the reaction of CO_2 (3 bar) with $HSiMe(OSiMe_3)_2$ in C_6D_6 at 323 K to afford, after 3 h, a mixture of the corresponding silylformate (65.2%), methoxysilane (8.1%) and methylsilylcarbonate (26.7%) (Scheme 10) [71].

^{1}H and ^{13}C NMR studies of the reaction shown in Scheme 10 evidenced that at 323 K, once all the starting hydrosilane is consumed; the methylsilylcarbonate is slowly transformed into the corresponding methoxysilane. These outcomes prove that the formation of methoxysilanes during the catalytic reduction of CO_2 with silicon hydrides, which traditionally has been explained by the stepwise process shown in Scheme 9, could also be consequence of thermal decomposition of the corresponding methylsilylcarbonate (Scheme 11) [71].

Few examples of other homogeneous catalysts effective for the reduction of CO_2 to methanol level using silicon hydrides as reductants have been described, which include the anionic rhenium complex $[N(hexyl)_4][ReO_4]$ [93], the cationic zinc derivative $[Zn(Me)(IDipp)][C_6F_5)_3]$ (IDipp = 1,3-bis(2,6-diisopropylphenyl) imidazolin-2-ylidene) [94] and metal-free NHC-catalysts [95]. In this context, it is noteworthy that the activity of the Ir-trifluoroacetate-NSi^{Me} catalyst is similar to that reported for these Re-, Zn-, and NHC-based catalytic systems.

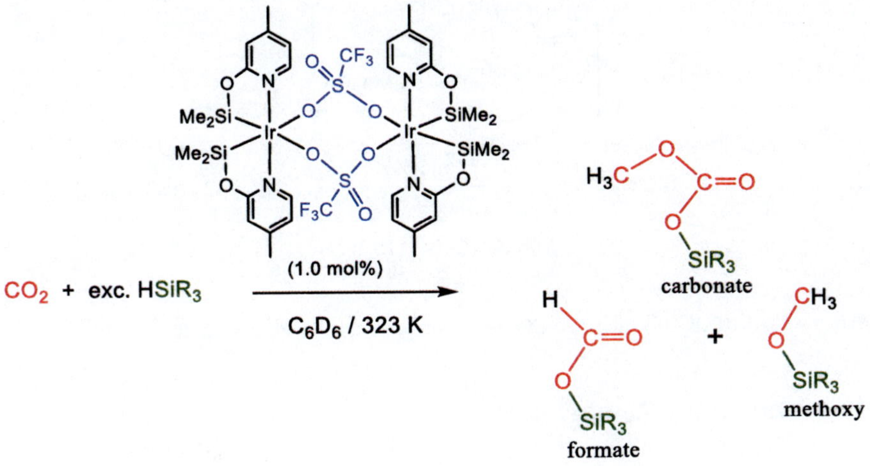

Scheme 10 Iridium-catalyzed reduction of CO_2 to the methoxysilane level with $HSiR_3$

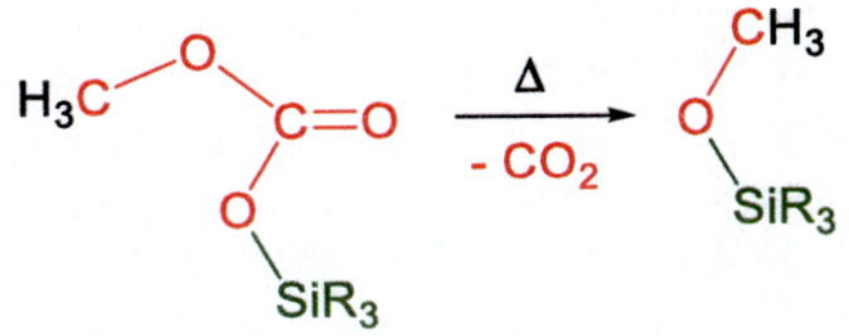

Scheme 11 Thermal decomposition of methylsilylcarbonates to give methoxysilanes and CO_2

3.3 *Iridium-Catalyzed Reduction of CO_2 to Methane with Silicon-Hydrides*

The catalytic reduction of CO_2 to methane using hydrosilanes as reducing agents remains a challenge. Examples of transition metal catalysts based on Zr [96, 97], Hf [97], Ir [98], Pd [99] and Pt [99] complexes as well as transition metal-free catalysts such as the frustrated Lewis pair $B(C_6F_5)_3$/TMP (TMP = 2,2,6,6-tetramethylpyperidine) [100] and other Lewis acids and ionic pairs [101–104]. Among them stands out the iridium(III) cationic species $[Ir(H)(\eta^1\text{-}HSiR_3)(POCOP)][B(C_6F_5)_4]$ (POCOP = 2,6-bis((di-tert-butylphosphanyl)oxy)benzen-1-yl) reported by Brookhart et al. in 2012 [98], which has proven to be effective for the reduction of CO_2 (1 bar, 296 K) to methane with different hydrosilanes ($HSiEt_3$, $HSiPh_3$, $HSiMe_2Et$, $HSiMe_2Ph$, and $HSiEt_2Me$) using C_6H_5Cl as solvent. This catalytic system works reasonably well with $HSiMe_2Ph$ at 296 K (TOF = 115 h^{-1}), moreover, increasing the temperature to 333 K produces a positive effect of the catalytic activity (TOF = 661 h^{-1}) (Scheme 12) [98].

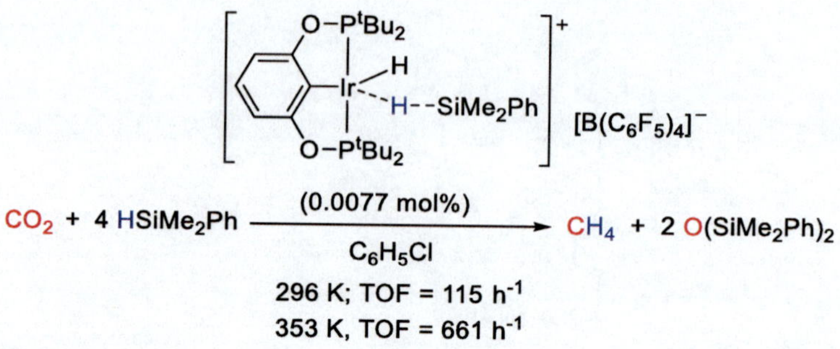

Scheme 12 Ir-(POCOP) catalyzed CO_2 reduction to methane with $HSiMe_2Ph$

4 Concluding Remarks

This chapter illustrates the progress made during recent years in the field of iridium-catalyzed reduction of CO_2 with hydrogen and/or silicon hydrides as reductants. It is difficult to draw general conclusions since not only the characteristics of the ligands but also the nature of the reducing agent (hydrogen, hydrosilanes, and/or hydrosiloxanes) strongly influences the reaction conditions and the mechanism. It has been observed that most of the iridium CO_2 hydrogenation and hydrosilylation catalysts are based on Ir(III) species. The selectivity is one of the challenges of homogeneous catalytic CO_2 reduction with hydrogen and silicon hydrides, this is because mixtures of different reduction products are frequently obtained. In this regard, it is worth mentioning that iridium(III) half-sandwich-DHBP species and iridium(III)-PNP pincer complexes have found to be highly efficient and selective CO_2 hydrogenation catalysts and that Ir(III)-NSiN and Ir(III)-NSi^{Me} species have proven to be highly selective CO_2 hydrosilylation catalysts. From the point of view of the mechanism, it is difficult to establish a general behavior trend. Thus, although most of the reported homogeneous Ir(III) catalysts follow an outer-sphere CO_2 activation mechanism, when using Ir-NSiN and Ir-NSi^{Me} trifluoroacetate derivatives as CO_2 hydrosilation catalysts, an inner-sphere CO_2 activation mechanism is preferred. Therefore, it could be concluded that iridium(III) complexes have great potential as homogeneous CO_2 reduction catalysts; however, there are still many mechanistic questions to answer and future applications to unveil.

References

1. Artz J, Müller TE, Thenert K, Kleinekorte J, Meys R, Sternberg A, Bardow A, Leitner W (2018) Sustainable conversion of carbon dioxide: an integrated review of catalysis and life cycle assessment. Chem Rev 118:434–504
2. Stahel WR (2016) Circular economy. Nature 531:435–438

3. Clark JH, Farmer TJ, Herrero-Davila L, Sherwood J (2016) Circular economy design considerations for research and process development in the chemical sciences. Green Chem 18:3914–3934
4. Grignard B, Gennen S, Jérôme C, Kleij AW, Detrembleur C (2019) Advances in the use of CO_2 as a renewable feedstock for the synthesis of polymers. Chem Soc Rev 48:4466–4514
5. Martens JA, Bogaerts A, De Kimpe N, Jacobs PA, Marin GB, Rabaey K, Saeys M, Verhelst S (2017) The chemical route to a carbon dioxide neutral world. ChemSusChem 10:1039–1055
6. Klankermayer J, Wesselbaum S, Beydoun K, Leitner W (2016) Selective catalytic synthesis using the combination of carbon dioxide and hydrogen: catalytic chess at the Interface of energy and chemistry. Angew Chem Int Ed 55:7296–7343
7. Liu Q, Wu L, Jackstell R, Beller M (2015) Using carbon dioxide as a building block in organic synthesis. Nat Commun 6:5933
8. Das Neves Gomes C, Jacquet O, Villiers C, Thuéry P, Ephritikhine M, Cantat T (2012) A diagonal approach to chemical recycling of carbon dioxide: organocatalytic transformation for the reductive functionalization of CO_2. Angew Chem Int Ed 51:187–190
9. Jessop PG, Joó F, Tai C-C (2004) Recent advances in the homogeneous hydrogenation of carbon dioxide. Coord Chem Rev 248:2425–2442
10. Wang W-H, Himeda Y, Muckerman JT, Manbeck GF, Fujita E (2015) CO_2 hydrogenation to formate and methanol as an alternative to photo- and electrochemical CO_2 reduction. Chem Rev 115:12936–12973
11. Cokoja M, Bruckmeier C, Rieger B, Herrmann WA, Kühn FE (2011) Transformation of carbon dioxide with homogeneous transition-metal catalysts: a molecular solution to a global challenge? Angew Chem Int Ed 50:8510–8537
12. Peters M, Köhler B, Kuckshinrichs W, Leitner W, Markewitz P, Müller TE (2011) Chemical technologies for exploiting and recycling carbon dioxide into the value chain. ChemSusChem 4:1216–1240
13. Aresta M, Dibenedetto A, Angelini A (2014) Catalysis for the valorization of exhaust carbon: from CO_2 to chemicals, materials, and fuels. Technological use of CO_2. Chem Rev 114:1709–1742
14. Goeppert A, Zhang H, Czaun M, May RB, Prakash GKS, Olah GA, Narayanan SR (2014) Easily regenerable solid adsorbents based on polyamines for carbon dioxide capture from the air. ChemSusChem 7:1386–1397
15. Sakakura T, Choi J-C, Yasuda H (2007) Transformation of carbon dioxide. Chem Rev 107:2365–2387
16. Centi G, Quadrelli EA, Perathoner S (2013) Catalysis for CO_2 conversion: a key technology for rapid introduction of renewable energy in the value chain of chemical industries. Energ Environ Sci 6:1711–1731
17. Ronda-Lloret M, Rothenberg G, Shiju NR (2019) A critical look at direct catalytic hydrogenation of carbon dioxide to olefins. ChemSusChem 12:3896–3914
18. Onishi N, Laurenzcy G, Beller M, Himeda Y (2018) Recent Progress for reversible homogeneous catalytic hydrogen storage in formic acid and in methanol. Coord Chem Rev 373:317–332
19. Jia J, Qian C, Dong Y, Li YF, Wang H, Ghoussoub M, Butler KT, Walsh A, Ozin GA (2017) Heterogeneous catalytic hydrogenation of CO_2 by metal oxides: defect engineering – perfecting imperfection. Chem Soc Rev 46:4631–4644
20. Prieto G (2017) Carbon dioxide hydrogenation into higher hydrocarbons and oxygenates: thermodynamic and kinetic bounds and Progress with heterogeneous and homogeneous catalysis. ChemSusChem 10:1056–1070
21. Wang W, Wang S, Ma X, Gong J (2011) Recent advances in catalytic hydrogenation of carbon dioxide. Chem Soc Rev 40:3703–3727
22. Fernández-Alvarez FJ, Oro LA (2018) Homogeneous catalytic reduction of CO_2 with silicon-hydrides, state of the art. ChemCatChem 10:4783–4796

23. Fernández-Alvarez FJ, Aitani AM, Oro LA (2014) Homogeneous catalytic reduction of CO_2 with Hydrosilanes. Cat Sci Technol 4:611–624
24. Chauvier C, Cantat T (2017) A viewpoint on chemical reductions of carbon-oxygen bonds in renewable feedstocks including CO_2 and biomass. ACS Catal 7:2107–2115
25. Chen J, McGraw M, Chen EY-X (2019) Diverse catalytic systems and mechanistic pathways for hydrosilylative reduction of CO_2. ChemSusChem 12:4543–4569
26. Fernández-Alvarez FJ, Iglesias M, Oro LA, Polo V (2013) CO_2 activation and catalysis driven by iridium complexes. ChemCatChem 5:3481–3494
27. Iglesias M, Oro LA (2018) A leap forward in iridium-NHC catalysis: new horizons and mechanistic insights. Chem Soc Rev 47:2772–2808
28. Gunasekar GH, Park K, Jung K-D, Yoon S (2016) Recent development in the catalytic hydrogenation of CO_2 to formic acid/formate using heterogeneous catalysts. Inorg Chem Front 3:882–895
29. Jessop PG, Ikariya T, Noyori R (1995) Homogeneous hydrogenation of carbon dioxide. Chem Rev 95:259–272
30. Inoue Y, Izumida H, Sasaki Y, Hashimoto H (1976) Catalytic fixation of carbon dioxide to formic acid by transition-metal complexes under mild conditions. Chem Lett 5:863–864
31. Graf E, Leitner W (1992) Direct formation of formic acid from carbon dioxide and dihydrogen using the $[\{Rh(cod)Cl\}_2]$-$Ph_2P(CH_2)_4PPh_2$ catalyst system. J Chem Soc Chem Commun:623–624
32. Gassner F, Leitner W (1993) Hydrogenation of carbon dioxide to formic acid using water-soluble rhodium catalysts. J Chem Soc Chem Commun:1465–1466
33. Jessop PG, Ikariya T, Noyori R (1994) Homogeneous catalytic hydrogenation of supercritical carbon dioxide. Nature 368:231–233
34. Jessop PG, Hsiao Y, Ikariya T, Noyori R (1996) Homogeneous catalysis in supercritical fluids: hydrogenation of supercritical carbon dioxide to formic acid, alkyl formates, and formamides. J Am Chem Soc 118:344–355
35. Joó F, Laurenczy G, Nádasdi L, Elek J (1999) Homogeneous hydrogenation of aqueous hydrogen carbonate to Formate under exceedingly mild conditions – a novel possibility of carbon dioxide activation. Chem Commun:971–972
36. Munshi P, Main AD, Linehan JC, Tai C-C, Jessop PG (2002) Hydrogenation of carbon dioxide catalyzed by ruthenium trimethylphosphine complexes: the accelerating effect of certain alcohols and amines. J Am Chem Soc 124:7963–7971
37. Himeda Y (2007) Conversion of CO_2 into formate by homogeneously catalyzed hydrogenation in water: tuning catalytic activity and water solubility through the acid-base equilibrium of the ligand. Eur J Inorg Chem:3927–3941
38. Himeda Y, Onozawa-Komatsuzaki N, Sugihara H, Kasuga K (2007) Simultaneous tuning of activity and water solubility of complex catalysts by acid-base equilibrium of ligands for conversion of carbon dioxide. Organometallics 26:702–712
39. Kanega R, Onishi N, Szalda DJ, Ertem MZ, Muckerman JT, Fujita E, Himeda Y (2017) CO_2 hydrogenation catalysts with deprotonated picolinamide ligands. ACS Catal 7:6426–6429
40. Kanega R, Onishi N, Wang L, Murata K, Muckerman JT, Fujita E, Himeda Y (2018) Picolinamide-based iridium catalysts for dehydrogenation of formic acid in water: effect of amide N substituent on activity and stability. Chem A Eur J 24:18389–18392
41. Suna Y, Himeda Y, Fujita E, Muckerman JT, Ertem MZ (2017) Iridium complexes with proton-responsive azole-type ligands as effective catalysts for CO_2 hydrogenation. ChemSusChem 10:4535–4543
42. Onishi N, Kanega R, Fujita E, Himeda Y (2019) Carbon dioxide hydrogenation and formic acid dehydrogenation catalyzed by iridium complexes bearing pyridyl-pyrazole ligands: effect of an electron-donating substituent on the pyrazole ring on the catalytic activity and durability. Adv Synth Catal 361:289–296

43. Hull JF, Himeda Y, Wang W-H, Hashiguchi B, Periana R, Szalda DJ, Muckerman JT, Fujita E (2012) Reversible hydrogen storage using CO_2 and a proton-switchable iridium catalyst in aqueous media under mild temperatures and pressures. Nat Chem 4:383–388
44. Sanz S, Benítez M, Peris E (2010) A new approach to the reduction of carbon dioxide: CO_2 reduction to formate by transfer hydrogenation in iPrOH. Organometallics 29:275–277
45. Tanaka R, Yamashita M, Nozaki K (2009) Catalytic hydrogenation of carbon dioxide using Ir (III)-pincer complexes. J Am Chem Soc 131:14168–14169
46. Tanaka R, Yamashita M, Chung LW, Morokuma K, Nozaki K (2011) Mechanistic studies on the reversible hydrogenation of carbon dioxide catalyzed by an Ir-PNP complex. Organometallics 30:6742–6750
47. Aoki W, Wattanavinin N, Kusumoto S, Nozaki K (2016) Development of highly active Ir-PNP catalysts for hydrogenation of carbon dioxide with organic bases. Bull Chem Soc Jpn 89:113–124
48. Schmeier TJ, Dobereiner GE, Crabtree RH, Hazari N (2011) Secondary coordination sphere interactions facilitate the insertion step in an iridium(III) CO_2 reduction catalysis. J Am Chem Soc 133:9274–9277
49. Pérez-Fortes M, Schöneberger JC, Boulamanti A, Tzimas E (2016) Methanol synthesis using captured CO_2 as raw material: techno-economic and environmental assessment. Appl Energy 161:718–732
50. Ali KA, Abdullah AZ, Mohamed AR (2015) Recent development in catalytic technologies for methanol synthesis from renewable sources: a critical review. Renew Sustain Energy Rev 44:508–518
51. Hertrich MF, Beller M (2018) Metal-catalysed hydrogenation of CO_2 into methanol. In: Dixneuf P, Soulé JF (eds) Organometallics for green catalysis. Topics in organometallic chemistry, vol 63. Springer, Cham
52. Olah GA (2005) Beyond oil and gas: the methanol economy. Angew Chem Int Ed 44:2636–2639
53. Tominaga K-I, Sasaki Y, Kawai M, Watanabe T, Saito M (1993) Ruthenium complex catalysed hydrogenation of carbon dioxide to carbon monoxide, methanol and methane. J Chem Soc Chem Commun:629–631
54. Tominaga K-I, Sasaki Y, Kawai M, Watanabe T, Saito M (1995) Homogeneous hydrogenation of carbon dioxide to methanol catalyzed by ruthenium cluster anions in the presence of halide anions. Bull Chem Soc Jpn 68:2837–2842
55. Sordakis K, Tsurusaki A, Iguchi M, Kawanami H, Himeda Y, Laurenczy G (2016) Carbon dioxide to methanol: the aqueous catalytic way at room temperature. Chem A Eur J 22:15605–15608
56. Wesselbaum S, vom Stein T, Klankermayer J, Leitner W (2012) Hydrogenation of carbon dioxide to methanol by using a homogeneous ruthenium–phosphine catalyst. Angew Chem Int Ed 51:7499–7502
57. Wesselbaum S, Moha V, Meuresch M, Brosinski S, Thenert KM, Kothe J, von Stein T, Englert U, Hölscher M, Klankermayer J, Leitner W (2015) Hydrogenation of carbon dioxide to methanol using a homogeneous ruthenium-triphos catalyst: from mechanistic investigations to multiphase catalysis. Chem Sci 6:693–704
58. Rezayee NM, Huff CA, Sanford MS (2015) Tandem amine and ruthenium-catalyzed hydrogenation of CO_2 to methanol. J Am Chem Soc 137:1028–1031
59. Kar S, Sen R, Goeppert A, Prakash GKS (2018) Integrative CO_2 capture and hydrogenation to methanol with reusable catalyst and amine: toward a carbon neutral methanol economy. J Am Chem Soc 140:1580–1583
60. Schneidewind J, Adam R, Baumann W, Jackstell R, Beller M (2017) Low-temperature hydrogenation of carbon dioxide to methanol with a homogeneous cobalt catalyst. Angew Chem Int Ed 56:1890–1893
61. Kar S, Goeppert A, Kothandaraman J, Prakash GKS (2017) Manganese-catalyzed sequential hydrogenation of CO_2 to methanol via formamide. ACS Catal 7:6347–6351

62. Ribeiro APC, Martins LMDRS, Pombeiro AJL (2017) Carbon dioxide-to-methanol single-pot conversion using a C-scorpionate iron(II) catalyst. Green Chem 19:4811–4815
63. Thenert K, Beydoun K, Wiesenthal J, Leitner W, Klankermayer J (2016) Ruthenium-catalyzed synthesis of dialkoxymethane ethers utilizing carbon dioxide and molecular hydrogen. Angew Chem Int Ed 55:12266–12269
64. Siebert M, Seibicke M, Siegle AF, Kräh S, Trapp O (2019) Selective ruthenium-catalyzed transformation of carbon dioxide: an alternative approach toward formaldehyde. J Am Chem Soc 141:334–341
65. Seibicke M, Siebert M, Siegle AF, Gutenthaler SM, Trapp O (2019) Application of hetero-triphos ligands in the selective ruthenium-catalyzed transformation of carbon dioxide to the formaldehyde oxidation state. Organometallics 38:1809–1814
66. Westhues N, Belleflamme M, Klankermayer J (2019) Base-free hydrogenation of carbon dioxide to methyl formate with a molecular ruthenium-phosphine catalyst. ChemCatChem 11:5269–5274
67. Zhang Y, Zhang T, Das S (2020) Catalytic transformation of CO_2 into C1 chemicals using hydrosilanes as reducing agents. Green Chem 22:1800–1820
68. Addis D, Das S, Junge K, Beller M (2011) Selective reduction of carboxylic acid derivatives by catalytic hydrosilylation. Angew Chem Int Ed 50:6004–6011
69. Garcés K, Fernández-Alvarez FJ, Polo V, Lalrempuia R, Pérez-Torrente JJ, Oro LA (2014) Iridium-catalyzed hydrogen production from hydrosilanes and water. ChemCatChem 6:1691–1697
70. Aliaga-Lavrijsen M, Iglesias M, Cebollada A, Garcés K, García N, Sanz Miguel PJ, Fernández-Alvarez FJ, Pérez-Torrente JJ, Oro LA (2015) Hydrolysis and methanolysis of silanes catalyzed by iridium(III) bis-N-heterocyclic carbene complexes: influence of the wingtip groups. Organometallics 34:2378–2385
71. Guzmán J, García-Orduña P, Lahoz FJ, Fernández-Alvarez FJ (2020) Unprecedent formation of methylsilylcarbonates from iridium-catalyzed reduction of CO_2 with hydrosilanes. RSC Adv 10:9582–9586
72. Koinuma H, Kawakami F, Kato H, Hirai H (1981) Hydrosilylation of carbon dioxide catalysed by ruthenium complexes. J Chem Soc Chem Commun:213–214
73. Süss-Fink G, Reiner J (1981) Anionische Rutheniumcluster als Katalysatoren bei der Hydrosilylierung von Kohlendioxid. J Organomet Chem 221:C36–C38
74. Eisenschmid TC, Eisenberg R (1989) The iridium complex catalyzed reduction of carbon dioxide to methoxide by alkylsilanes. Organometallics 8:1822–1824
75. Motokura K, Nakagawa C, Pramudita RA, Manaka Y (2019) Formate-catalyzed selective reduction of carbon dioxide to formate products using hydrosilanes. ACS Sustainable Chem Eng 7:11056–11061
76. Lalrempuia R, Iglesias M, Polo V, Sanz Miguel PJ, Fernández-Alvarez FJ, Pérez-Torrente JJ, Oro LA (2012) Effective fixation of CO_2 by iridium-catalyzed hydrosilylation. Angew Chem Int Ed 51:12824–12827
77. Jaseer EA, Akhtar MN, Osman M, Al-Shammari A, Oladipo HB, Garcés K, Fernández-Alvarez FJ, Al-Khattaf S, Oro LA (2015) Solvent-free iridium-catalyzed CO_2 hydrosilylation: experiments and kinetic modeling. Cat Sci Technol 5:274–279
78. Oladipo HB, Jaseer EA, Julián A, Fernández-Alvarez FJ, Al-Khattaf S, Oro LA (2015) Effect of the CO_2-pressure on the hydrosilylation of CO_2 catalyzed by [Ir (NSiN)] species. J CO_2 Utilization 12:21–26
79. Julián A, Jaseer EA, Garcés K, Fernández-Alvarez FJ, García-Orduña P, Lahoz FJ, Oro LA (2016) Tuning the activity and selectivity of iridium-NSiN catalyzed CO_2 hydrosilylation processes. Cat Sci Technol 6:4410–4417
80. Julián A, Guzmán J, Jaseer EA, Fernández-Alvarez FJ, Royo R, Polo V, García-Orduña P, Lahoz FJ, Oro LA (2017) Mechanistic insights on the reduction of CO_2 to silylformates catalyzed by Ir-NSiN species. Chem A Eur J 23:11898–11907

81. Guzmán J, García-Orduña P, Polo V, Lahoz FJ, Oro LA, Fernández-Alvarez FJ (2019) Ir-catalyzed selective reduction of CO_2 to the methoxy or formate level with HSiMe$(OSiMe_3)_2$. Cat Sci Technol 9:2858–2867
82. Ojeda-Amador AI, Munarriz J, Alamán-Valtierra P, Polo V, Puerta-Oteo R, Jiménez MV, Fernández-Alvarez FJ, Pérez-Torrente JJ (2019) Mechanistic insights on the functionalization of CO_2 with amines and hydrosilanes catalyzed by a zwitterionic iridium carboxylate-functionalized Bis-NHC catalyst. ChemCatChem 11:5524–5535
83. Jansen A, Görls H, Pitter S (2000) *trans*-$[Ru^{II}Cl(MeCN)_5][Ru^{III}Cl_4(MeCN)_2]$: a reactive intermediate in the homogeneous catalyzed hydrosilylation of carbon dioxide. Organometallics 19:135–138
84. Deglmann P, Ember E, Hofmann P, Pitter S, Walter O (2007) Experimental and theoretical investigations on the catalytic hydrosilylation of carbon dioxide with ruthenium nitrile complexes. Chem A Eur J 13:2864–2879
85. Scheuermann ML, Semproni SP, Pappas I, Chirik PJ (2014) Carbon dioxide hydrosilylation promoted by cobalt pincer complexes. Inorg Chem 53:9463–9465
86. Itagaki S, Yamaguchi K, Mizuno N (2013) Catalytic synthesis of silyl formates with 1 atm of CO_2 and their utilization for synthesis of formyl compounds and formic acid. J Mol Catal A Chem 366:347–352
87. Takaya J, Iwasawa N (2017) Synthesis, structure, and catalysis of palladium complexes bearing a group 13 metalloligand: remarkable effect of an aluminum-metalloligand in hydrosilylation of CO_2. J Am Chem Soc 139:6074–6077
88. Rios P, Díez J, López-Serrano J, Rodríguez A, Conejero S (2016) Cationic platinum(II) σ-SiH complexes in carbon dioxide hydrosilation. Chem A Eur J 22:16791–16795
89. Motokura K, Kashiwame D, Takahashi N, Miyaji A, Baba T (2013) Highly active and selective catalysis of copper diphosphine complexes for the transformation of carbon dioxide into silyl formate. Chem A Eur J 19:10030–10037
90. Zhang L, Cheng J, Hou Z (2013) Highly efficient catalytic hydrosilylation of carbon dioxide by an N-heterocyclic carbene copper catalyst. ChemCommun 49:4782–4784
91. Sattler W, Parkin G (2012) Zinc catalysts for on-demand hydrogen generation and carbon dioxide functionalization. J Am Chem Soc 134:17462–17465
92. Specklin D, Fliedel C, Gourlaouen C, Bruyere J-C, Avilés T, Boudon C, Ruhlmann L, Dagorne S (2017) N-heterocyclic carbene based tri-organyl-Zn–alkyl cations: synthesis, structures, and use in CO_2 functionalization. Chem A Eur J 23:5509–5519
93. Morris DS, Weetman C, Wennmacher JTC, Cokoja M, Drees M, Kühn FE, Love JB (2017) Reduction of carbon dioxide and organic carbonyls by hydrosilanes catalysed by the perrhenate anion. Cat Sci Technol 7:2838–2845
94. Specklin D, Hild F, Fliedel C, Gourlaouen C, Veiros LF, Dagorne S (2017) Accessing two-coordinate Zn^{II} Organocations by NHC coordination: synthesis, structure, and use as π-Lewis acids in alkene, alkyne, and CO_2 hydrosilylation. Chem A Eur J 23:15908–15912
95. Riduan SN, Zhang Y, Ying JY (2009) Conversion of carbon dioxide into methanol with silanes over N-heterocyclic carbene catalysts. Angew Chem Int Ed 48:3322–3325
96. Matsuo T, Kawaguchi H (2006) From carbon dioxide to methane: homogeneous reduction of carbon dioxide with hydrosilanes catalyzed by zirconium–borane complexes. J Am Chem Soc 128:12362–12363
97. Luconi L, Rossin A, Tuci G, Gafurov Z, Lyubov DM, Trifonov AA, Cicchi S, Ba H, Pham-Huu C, Yakhvarov D, Giambastiani G (2019) Benzoimidazole-pyridylamido zirconium and hafnium alkyl complexes as homogeneous catalysts for tandem carbon dioxide hydrosilylation to methane. ChemCatChem 11:495–510
98. Park S, Bézier D, Brookhart M (2012) An efficient iridium catalyst for reduction of carbon dioxide to methane with trialkylsilanes. J Am Chem Soc 134:11404–11407
99. Mitton SJ, Turculet L (2012) Mild reduction of carbon dioxide to methane with tertiary silanes catalyzed by platinum and palladium silyl pincer complexes. Chem A Eur J 18:15258–15262

100. Berkefeld A, Piers WE, Parvez M (2010) Tandem frustrated Lewis pair/tris (pentafluorophenyl)borane-catalyzed deoxygenative hydrosilylation of carbon dioxide. J Am Chem Soc 132:10660–10661
101. Rauch M, Parkin G (2017) Zinc and magnesium catalysts for the Hydrosilylation of carbon dioxide. J Am Chem Soc 139:18162–18165
102. Khandelwal M, Wehmschulte RJ (2012) Deoxygenative reduction of carbon dioxide to methane, toluene, and Diphenylmethane with $[Et_2Al]^+$ as catalyst. Angew Chem Int Ed 51:7323–7326
103. Berkefeld A, Piers WE, Parvez M, Castro L, Maron L, Eisenstein O (2013) Decamethylscandocinium-hydrido-(perfluorophenyl)borate: fixation and tandem tris (perfluorophenyl)borane catalysed deoxygenative hydrosilation of carbon dioxide. Chem Sci 4:2152–2162
104. Chen J, Falivene L, Caporaso L, Cavallo L, Chen EY-X (2016) Selective reduction of CO_2 to CH_4 by tandem hydrosilylation with mixed Al/B catalysts. J Am Chem Soc 138:5321–5333

Top Organomet Chem (2021) 69: 325–340
https://doi.org/10.1007/3418_2020_54

Published online: 4 August 2020

Electroreduction of Carbon Dioxide by Homogeneous Iridium Catalysts

Check for updates

Ryoichi Kanega

Contents

Abstract The electroreduction of carbon dioxide (CO_2) to chemical fuels provides not only a means to utilize CO_2 but also a solution to challenges relating to the storage and transport of renewable energy. For this purpose, a range of catalysts for the electroreduction of CO_2 have been studied, and recent progress in the context of tuning catalytic properties and understanding their mechanism of action has been remarkable. For example, molecular approaches allow fine-tuning of the catalyst behavior by the design of suitable ligands to suppress the overpotential for CO_2 conversion. This chapter focuses on homogeneous iridium catalysts for the electroreduction of CO_2, whereby the examples provided give mechanistic insight into the design of catalysts to efficiently and selectively produce electroreduced compounds from CO_2 using electricity.

The original version of this chapter was revised. A correction to this chapter can be found at https://doi.org/10.1007/3418_2020_73

R. Kanega (✉)
Research Institute of Energy Conservation, National Institute of Advanced Industrial Science and Technology, Tsukuba, Ibaraki, Japan
e-mail: r-kanega@aist.go.jp

Keywords Carbon dioxide · Carbon monoxide · Electroreduction · Formate · Hydride species · Oxalate

1 Introduction

To reduce the carbon dioxide (CO_2) emissions associated with the consumption of fossil fuels [1], it is necessary to utilize various technologies, such as renewable energy and energy-saving technologies. Recently, the price of renewable energy has drastically decreased, and so its usage is expected to increase [2]. However, the lack of established methods for the storage and transport of electricity derived from renewable energy sources remains a challenge that hinders its use. For example, the storage and transportation of electricity is essential if the electricity generated during the day is to be used at night, or in places without a power grid. In this context, chemical fuels produced electrochemically have the potential to address these issues compared to other methods, such as chemical and photochemical reactions [3, 4]. In particular, one attractive approach involves the use of electricity derived from renewable energy sources to produce chemical fuels from CO_2 and H_2O (an H^+ source) [5].

The redox potentials of typical chemical fuels obtained by the electroreduction of CO_2 are shown in Table 1 [6–8]. Generally, in the electroreduction of CO_2, proton-coupled multi-electron transfer is more favorable than multi-electron transfer, as thermodynamically more stable compounds are produced. The single-electron transfer of CO_2 to $CO_2^{\cdot-}$ requires a large redox potential of −1.90 V (vs the standard hydrogen electrode (SHE); all potentials are given with respect to this reference), a large reorganization energy between the linear molecule and bent radical anion [9]. On the other hand, proton-coupled multi-electron transfer is achieved using a potential lower than −1.0 V. However, the development of efficient catalysts is necessary to carry out the intended electrochemical transformations with low overpotentials and high current densities.

To date, significant efforts have been devoted to the exploration of potential homogeneous and heterogeneous catalysts. For example, heterogeneous catalysts based on various metals, such as copper, cobalt, and tin, have been reported [10];

Table 1 Selected redox potentials for the electroreduction of CO_2 and generation of hydrogen (vs SHE in aqueous solution at pH 7)

Product	n^a	Cathode reaction	E^0 [V]
CO	2	$CO_2 + 2H^+ + 2e^- \rightarrow CO + H_2O$	−0.53
HCO_2^-	2	$CO_2 + H^+ + 2e^- \rightarrow HCO_2^-$	−0.49
HCO_2H	2	$CO_2 + 2H^+ + 2e^- \rightarrow HCO_2H$	−0.61
CH_3OH	6	$CO_2 + 6H^+ + 6e^- \rightarrow CH_3OH + H_2O$	−0.38
CH_4	8	$CO_2 + 8H^+ + 8e^- \rightarrow CH_4 + 2H_2O$	−0.24
$C_2O_4^{2-}$	2	$2CO_2 + 2e^- \rightarrow C_2O_4^{2-}$	−1.00
$CO_2^{\cdot-}$	1	$CO_2 + e^- \rightarrow CO_2^{\cdot-}$	−1.90
H_2	2	$2H^+ + 2e^- \rightarrow H_2$	−0.41

[a] *n* Number of reaction electrons

however, in many cases, the conversion of CO_2 involves inefficient $CO_2^{\cdot-}$ formation. On the other hand, homogeneous catalysts, such as ruthenium-, rhodium-, and iridium-based complexes, often form metal hydrides as intermediates [11, 12]. This circumvents the $CO_2^{\cdot-}$ formation, and hence, more efficient electrochemical conversions are expected. For example, $[Rh(bpy_2(TFMS)_2)]^+$ (bpy, 2,2′-bipyridine; TFMS, trifluoromethanesulfonate anion) exhibited 80% Faradaic efficiency (FE) for HCO_2^- production in CH_3CN at −0.96 V. Moreover, electroreduction of CO_2 in aqueous solution was achieved with high FE at a very low overpotential using iridium catalysts. This chapter, therefore, describes the development of iridium-based homogeneous catalysts for the electroreduction of CO_2 to formate (HCO_2^-), carbon monoxide (CO), and oxalate ($C_2O_4^{2-}$).

2 Electroreduction of Carbon Dioxide to Formate

2.1 *Homogeneous Catalysts*

HCO_2^- (HCO_2H), a two-electron reduction product of CO_2, has recently attracted significant attention as a liquid organic hydrogen (H_2) carrier, in which H_2 is established as a new energy vector [13, 14]. Through HCO_2^- (HCO_2H) dehydrogenation (Eqs. 1 and 2), H_2 can be released with a lower energy consumption than those required for other H_2 storage media, such as methylcyclohexane or ammonia. Furthermore, in contrast to other chemical H_2 carriers, HCO_2H can produce high-pressure H_2 using only thermal reactions [15]. The potential for the CO_2 to HCO_2^- transformation is −0.49 V, which indicates a slightly higher energy consumption than H_2 generation (−0.41 V). If the electroreduction of CO_2 to HCO_2^- can be realized with a high FE (Eq. 3, where X is a product such as CO, HCO_2^-, $C_2O_4^{2-}$ or H_2, F is the Faraday constant, and n is the number of reaction electrons) and a low overpotential, it becomes an energy storage method comparable to water electrolysis. Additionally, C-H bond formation is one of the most essential elementary reactions in synthetic chemistry. The conversion of CO_2 to HCO_2^- (HCO_2H) involves C-H bond formation, which can be considered as an electrochemical method of C-H bond formation.

$$HCO_2^- + H_2O \rightarrow H_2 + CO_2 + OH^- \quad (1)$$

$$HCO_2H \rightarrow H_2 + CO_2 \quad (2)$$

$$\mathrm{FE(X)}\ [\%] = \mathrm{F}\left[\mathrm{C\ mol^{-1}}\right] \times n\,\mathrm{X\ [mol]/total\ Q\ [C]} \times 100 \quad (3)$$

In 1996, $[Ir_2(dimen)_4]^{2+}$ (dimen, 1,8-diisocyano-n-menthane) was studied using infrared spectro-electrochemistry, whereby HCO_2^- and bicarbonate were detected, although the products were not quantified [16]. In addition, the obtained results indicated that $[Ir_2(dimen)_4]^{2+}$ accepted 2 electrons to form $[Ir_2(dimen)_4]^0$, which then reacted with CO_2 and H_2O. In addition, the electroreduction of CO_2 to HCO_2^- was reported using $[Cp^*Ir(bpy)Cl]^+$ (Cp*, pentamethylcyclopentadienyl) in CH_3CN

Chart 1 Catalysts for the electroreduction of CO_2 to HCO_2^-

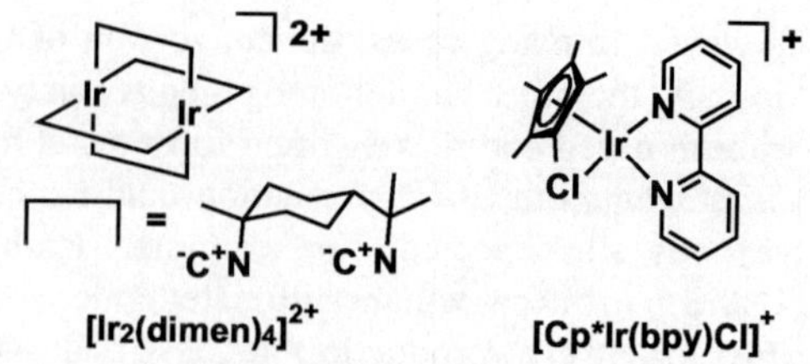

Table 2 Electroreduction of CO_2 to HCO_2^-

Catalyst	Solvent	E_{app} [a] [V]	FE(HCO_2^-) [%]	*j* [mA cm^{-2}]	Ref.
$[Ir_2(dimen)_4]^{2+}$	THF/H_2O	−1.62	–	–	[16]
$[Cp^*Ir(bpy)Cl]^+$	CH_3CN	−1.16	22	132	[17]

[a] E_{app} Applied potential

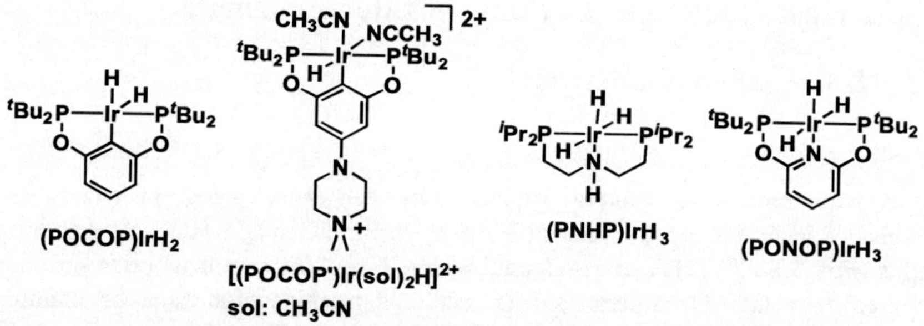

Chart 2 Catalysts bearing pincer ligands for the electroreduction of CO_2 to HCO_2^-

(Chart 1, Table 2) [17]. Electrolysis was performed at −1.16 V, and HCO_2^- was produced with an FE(HCO_2^-) of 22% (132 mA cm^{-2}). Although the formation of CO (FE(CO) < 1%) was confirmed, no other products were identified. Furthermore, based on the hydrogenation of CO_2, the electroreduction mechanism was speculated to involve reaction between the iridium-hydride intermediates and CO_2.

The reactivities and properties of a range of iridium-hydride species have been studied actively. Among them, iridium catalysts bearing a pincer ligand were found to exhibit a high activity and durability for the hydrogenation of CO_2 [18, 19]. In addition, through the use of related pincer ligands, electrochemically formed iridium-hydride species were investigated for the electroreduction of CO_2 (Chart 2, Table 3). For example, the potentiostatic electrolysis of (POCOP)IrH_2 was carried out at −1.45 V in a CH_3CN/5% H_2O solution to yield HCO_2^- in an FE(HCO_2^-) of 85% (1.07 mA cm^{-2}), whereby H_2 was detected as a by-product (FE(H_2) 15%) [20]. Moreover, through the introduction of 1,1-dimethyl-piperazinium to the pincer ligand, the FE(HCO_2^-) reached 93% in a water-based electrolyte ($NaHCO_3$/1% CH_3CN) [21]. Although CH_3CN was essential in the desorption of HCO_2^- from the iridium species and thus to promote catalyst turnover, $[(POCOP')Ir(sol)_2H]^{2+}$ (sol,

Table 3 Electroreduction of CO_2 to HCO_2^- using iridium complexes bearing pincer ligands

Catalyst	Solvent	E_{app} [a] [V]	FE(HCO_2^-) [%]	*j* [mA cm^{-2}]	Ref.
(POCOP)IrH_2	CH_3CN/5% H_2O	−1.45	85	1.07	[20]
[(POCOP′)IrH $(sol)_2]^{2+}$	$NaHCO_3$/1% CH_3CN	−1.41	93	0.60	[21]
(PNHP)IrH_3	CH_3CN/12% H_2O	−0.81	97	0.45	[22]
(PONOP)IrH_3	CH_3CN/5% H_2O	−1.15	97	2.10	[23]

[a] E_{app} Applied potential

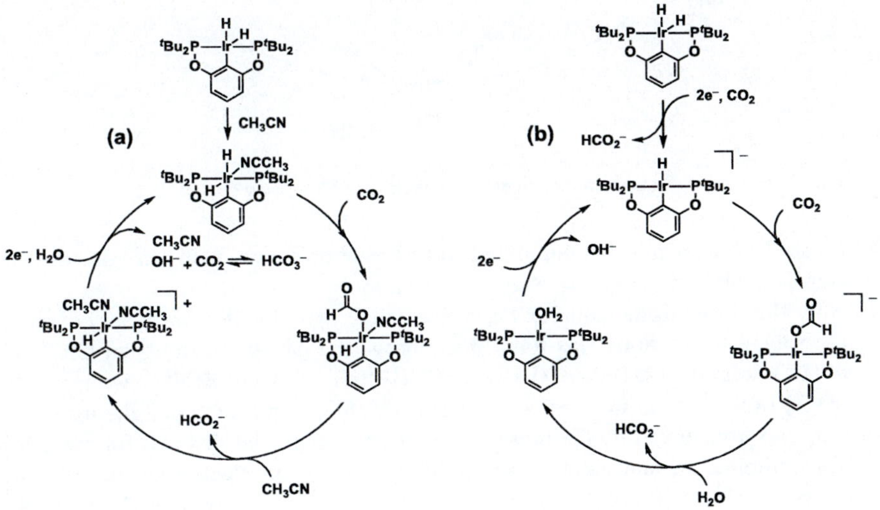

Scheme 1 Proposed mechanisms for the electroreduction of CO_2 with (**a**) (POCOP)IrH_2 by iridium(III) dihydride and (**b**) (POCOP)IrH_2 iridium(I) monohydride

CH_3CN) was also active, even in aqueous solution with a small amount of organic additive.

Subsequently, DFT calculations were employed to examine the reaction mechanism of the (POCOP)IrH_2 catalytic system [24]. As indicated in Scheme 1a, it was suggested that the reaction proceeds in three steps: (1) insertion of CO_2 into the iridium-hydride complex, (2) elimination of HCO_2^- from the resulting hydride-formate-iridium complex, and (3) regeneration of the active species. The reduction potential of the electrode reaction was calculated and was found to correspond with the experimental value. The solvents were also examined, and it was found that water was necessary for the transformation of CO_2 to HCO_2^-. Moreover, DFT calculations showed that HCO_2^- formation also involved iridium(I) monohydride as an active species formed in situ (Scheme 1b) [25]. It should be noted that the iridium(III) path can operate in parallel, but is associated with higher Gibbs free energies in the reaction between the iridium(III) dihydride species and CO_2 (i.e.,

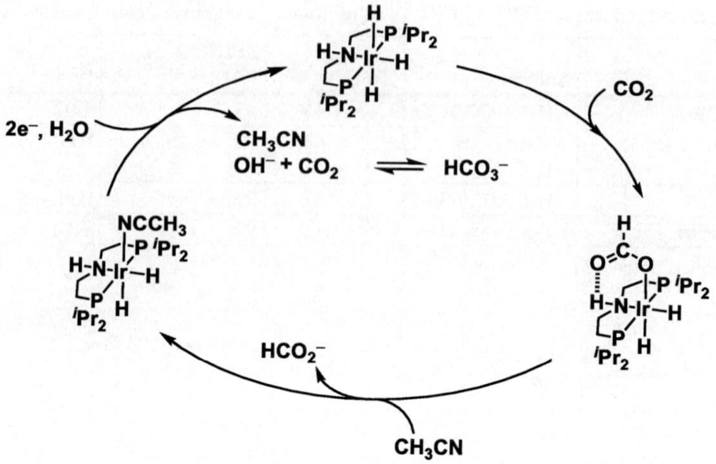

Scheme 2 Proposed mechanism for the electroreduction of CO_2 using (PNHP)IrH_3

16.6 and 19.0 kcal mol^{-1}), compared to the reaction between the in situ formed iridium(I) monohydride species (i.e., 12.3 kcal mol^{-1}).

An iridium trihydride complex bearing a PNHP ligand with a secondary coordination sphere interactions was found to improve the catalytic activity in CH_3CN/ 12% H_2O solution [22]. At −0.81 V, (PNHP)IrH_3 yielded HCO_2^- with an FE (HCO_2^-) of 97% (0.45 mA cm^{-2}). Mechanistic studies indicated that the insertion of CO_2 was accelerated by the presence of an NH group, and this was followed by HCO_2^- elimination and catalyst regeneration through an electroreduction/proton transfer mechanism (Scheme 2). Furthermore, the kinetic studies of (PNHP)IrH_3 appeared that the rate-determining step was HCO_2^- elimination, which can be enhanced by the addition of stabilizing agents such as water and Lewis acid.

Furthermore, an iridium trihydride complex bearing a PONOP pincer ligand was found to function not only in the electroreduction of CO_2 to HCO_2^- but also in the electrooxidation of HCO_2^- to CO_2 (Eq. 4) [23].

$$HCO_2^- \rightarrow CO_2 + H^+ + 2e^- \tag{4}$$

When the electroreduction of CO_2 in CH_3CN/5% H_2O was conducted at −1.15 V using (PONOP)IrH_3, the FE(HCO_2^-) reached 97% (2.1 mA cm^{-2}). In addition, HCO_2^- electrooxidation was carried out at 0.1 V in CH_3CN in the presence of $NBu_4HCO_2{\cdot}HCO_2H$, whereby CO_2 was detected as the sole product. Quantification of the evolved CO_2 indicated an FE(CO_2) 88%. Electrochemical and NMR spectroscopic studies suggested that the hydride species is critical for the bifunctional reactivity. A potential catalytic mechanism for the reaction involving (PONOP) IrH_3 is outlined in Scheme 3. More specifically, CO_2 was inserted into (PONOP) IrH_3 to yield a HCO_2^- complex (PONOP)IrH_2(HCO_2^-), which released HCO_2^- to give (PONOP)Ir(CH_3CN)H_2. Subsequent reduction yielded (PONOP)IrH_3. The

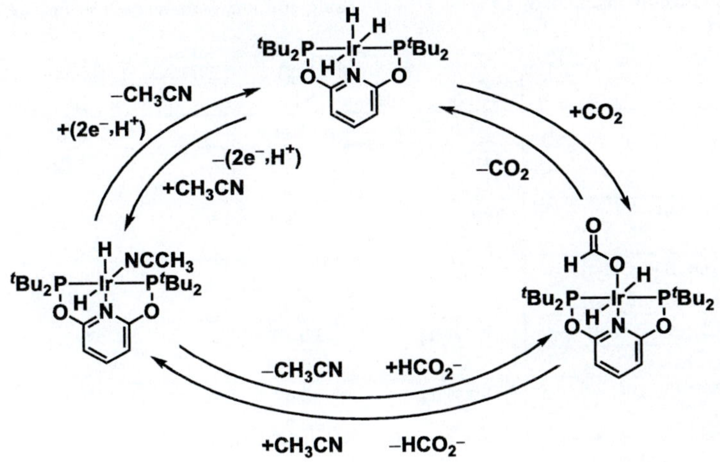

Scheme 3 Proposed mechanism for the electroreduction of CO_2 and the electrooxidation of HCO_2^- with (PONOP)IrH_3

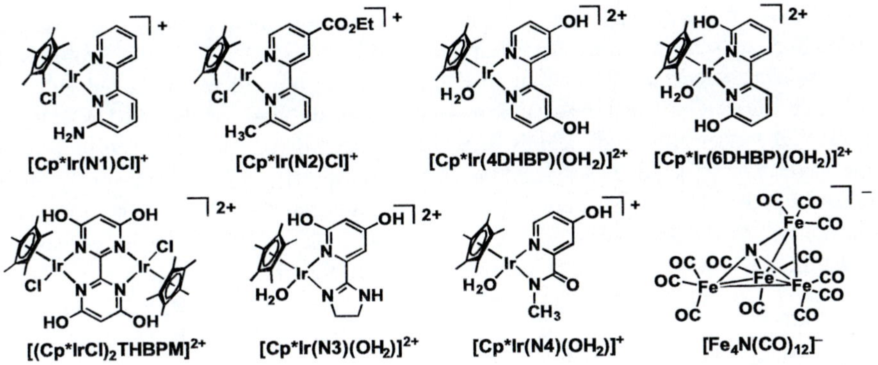

Chart 3 Catalysts bearing *N,N*-bidentate ligands and Fe carbonyl clusters for the electroreduction of CO_2 to HCO_2^-

HCO_2^- electrooxidation pathway is also illustrated in Scheme 3. More specifically, (PONOP)IrH_3 was oxidized electrochemically to form (PONOP)Ir(CH_3CN)H_2. In an equilibrium, (PONOP)Ir(CH_3CN)H_2 reacts rapidly with HCO_2^- to generate (PONOP)IrH_2(HCO_2^-), and this was followed by β-hydride elimination to generate (PONOP)IrH_3 and CO_2. The applicability of this homogeneous catalyst in both the electroreduction and electrooxidation steps is interesting due to such a system only having been reported once previously (i.e., $[Pt(depe)_2]^{2+}$; depe, 1,2-bis(diethylphosphino)ethane) [26].

In addition to pincer ligands, catalysts based on iridium-hydride species bearing *N, N*-bidentate ligands have also been developed (Chart 3). The influence of

Table 4 Electroreduction of CO_2 to HCO_2^- using iridium complexes bearing *N,N*-bidentate ligands

Catalyst	Solvent	E_{app}[a] [V]	$FE(HCO_2^-)$ [%]	*j* [mA cm^{-2}]	Ref.
$[Cp^*Ir(\mathbf{N1})cl]^+$	CH_3CN/5% H_2O	−1.60	6.6	0.18	[28]
	CH_3CN/50% CH_3OH	−1.60	22	1.08	
$[Cp^*Ir(\mathbf{N2})Cl]^+$	CH_3CN/50% CH_3OH	−1.60	13	0.84	
$[Cp^*Ir(4DHBP)(OH_2)]^{2+}$	$KHCO_3$ aq.	−0.49[b]	21	4.80	[29]
$[Cp^*Ir(6DHBP)(OH_2)]^{2+}$	$KHCO_3$ aq.	−0.49[b]	52	4.80	
$[(Cp^*IrCl)_2(THBPM)]^{2+}$	$KHCO_3$ aq.	−0.49[b]	62	5.00	
$[Cp^*Ir(\mathbf{N3})(OH_2)]^{2+}$	$KHCO_3$ aq.	−0.49[b]	89	5.20	
$[Cp^*Ir(\mathbf{N4})(OH_2)]^+$	$KHCO_3$ aq.	−0.49[b]	95	7.20	
$[Fe_4N(CO)_{12}]^-$	H_2O	−0.96[c]	96	3.80	[30]

[a] E_{app} Applied potential
[b] pH 8.3
[c] pH 7.0

substituent effects on the 2, 4, and 4′ positions of the bipyridine ligand was examined, as outlined in Table 4. More specifically, in CH_3CN/5% H_2O at −1.60 V, $[Cp^*Ir(\mathbf{N1})Cl]^+$ bearing an *ortho*-NH_2 group (N1, 6-amino-2,2′-bipyridine) produced HCO_2^- with an $FE(HCO_2^-)$ of 6.6% (0.18 mA cm^{-2}). In the case of the $[Cp^*Ir(\mathbf{N1})Cl]^+$ system, the electrolyte significantly affected the catalytic performance and improved the $FE(HCO_2^-)$ (i.e., to 21.9%) and the current density (i.e., to 1.08 mA cm^{-2}) when CH_3CN/50% CH_3OH was employed. Interestingly, in CH_3CN + 50% CH_3OH, formaldehyde (HCHO) was also detected (FE(HCHO) 20.2%), and the selectivity toward HCHO was increased in the case of $[Cp^*Ir(\mathbf{N2})Cl]^+$ bearing CH_3 and CO_2Et groups (N2, 4-ethoxycarbonyl-6′-methyl-2,2′-bipyridine; FE(HCHO) 32.2%). Furthermore, cyclic voltammetry studies gave information regarding the reaction mechanism (Scheme 4). Following an initial two-electron reduction of **a**, the resulting iridium(I) species (**b**) was protonated to form iridium(III) hydride (**c**). Subsequently, **c** underwent a bipyridine-centered one-electron reduction prior to transfer of the hydride from **d** to CO_2. The dissociation of HCO_2^- from **e** required a further one-electron reduction. An analogous mechanism was also proposed by Meyer and coworkers, which involves the use of a Ru-bipyridine complex in the electrocatalytic reduction of CO_2 [27].

Furthermore, $[Cp^*Ir(4DHBP)(OH_2)]^{2+}$ and $[Cp^*Ir(6DHBP)(OH_2)]^{2+}$ bearing *ortho*- and *para*-OH groups on the bpy ligand were reported to enable the electrochemical conversion of CO_2 to HCO_2^- under a low overpotential and without the requirement for organic solvents ($FE(HCO_2^-)$ 21.0 and 52.0%) [29]. In addition, the current density and FE_{HCO2-} of $[(Cp^*IrCl)_2(THBPM)]^{2+}$, whereby four OH groups were present on the ligand, were enhanced upon the introduction of OH groups. Furthermore, when one side of the bpy structure was replaced by an imidazoline ($[Cp^*Ir(\mathbf{N3})(OH_2)]^{2+}$) or amide species ($[Cp^*Ir(\mathbf{N4})(OH_2)]^+$), the catalytic activity

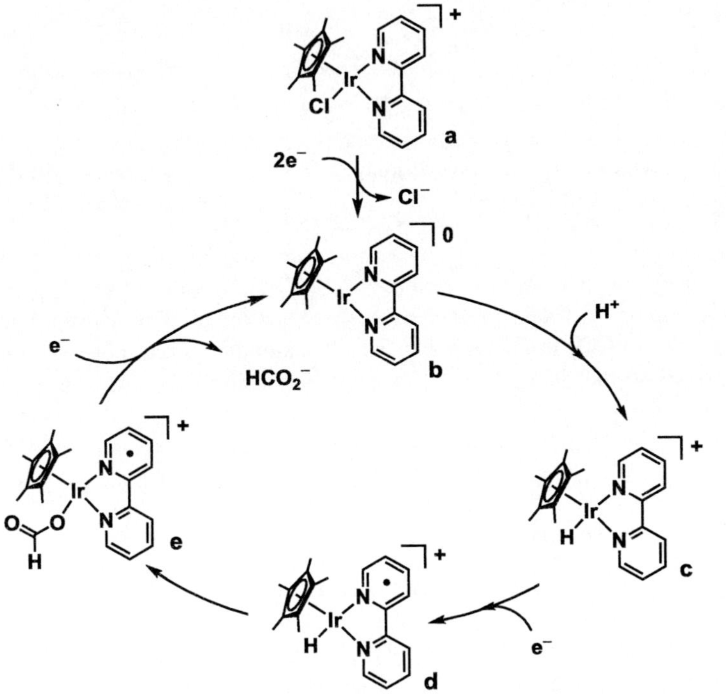

Scheme 4 Proposed mechanism for the electroreduction of CO_2, as proposed by Tzschucke and coworkers

was further enhanced. When the catalytic performances of [Cp*Ir(4DHBP)(OH_2)]$^{2+}$, [Cp*Ir(6DHBP)(OH_2)]$^{2+}$, [(Cp*IrCl)$_2$(THBPM)]$^{2+}$, [Cp*Ir(**N3**)(OH_2)]$^{2+}$, and [Cp*Ir(**N4**)(OH_2)]$^+$ were compared, the results were consistent with the order of activity for the hydrogenation of CO_2 [31], thereby indicating that common active species are involved in the electroreduction and hydrogenation of CO_2. Interestingly, combination of the iridium catalyst with different electrode materials indicated that electrodes exhibiting lower hydrogen overpotentials (i.e., a Pt black electrode) led to a remarkable decrease in the applied potential. As a result, the combination of [Cp*Ir (**N4**)(OH_2)]$^+$ and a Pt black working electrode gave a high current density (7.20 mA cm^{-2}) and a high FE(HCO_2^-) (95.0%). In addition, electrochemical measurements, electrokinetic analyses, and reaction mechanism studies revealed that the iridium-hydride species detected by ^{1}H NMR were generated via a two-electron reduction process. The above values represent the lowest overpotential and the highest current density among the previous reports using homogeneous catalysts for the electroreduction of CO_2 to HCO_2^-, and the overpotential was lower than that of the heterogeneous catalyst, thereby indicating that iridium-hydride species are key to the electroreduction of CO_2. A reaction mechanism was proposed in the case of [Cp*Ir(**N4**)(OH_2)]$^+$ as follows, whereby [Cp*Ir^I(**N4**)]$^-$ was initially

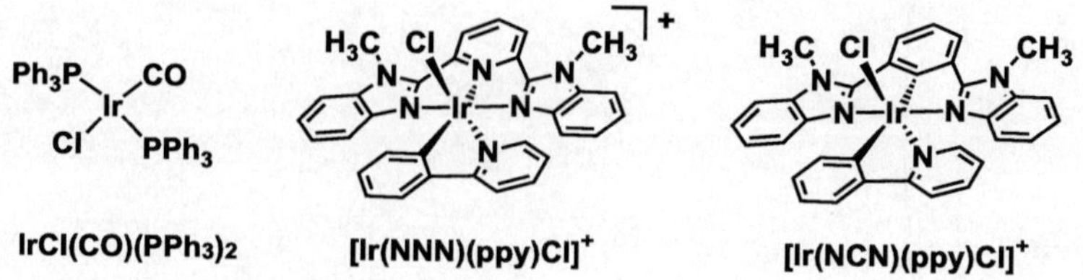

Chart 4 Catalysts for electroreduction of CO_2 to CO

generated by the two-electron reduction of $[Cp^{*}Ir^{III}(\mathbf{N4})(OH_2)]^+$ (Eq. 5). Subsequently, $[Cp^{*}Ir^{III}H(\mathbf{N4})]^+$ produced by the reaction of $[Cp^{*}Ir^{I}(\mathbf{N4})]^-$ with H_2O (Eq. 6) reduced CO_2 to HCO_2^- (Eq. 7). The cathodic current observed under CO_2 was derived so that $[Cp^{*}Ir^{III}(\mathbf{N4})]^+$ was regenerated by the reaction of $[Cp^{*}Ir^{III}H(\mathbf{N4})]^+$ with CO_2.

$$\left[Cp*Ir^{III}(\mathbf{N4})(OH_2)\right]^+ + 2e^- \rightarrow \left[Cp*Ir^{I}(\mathbf{N4})\right]^- + H_2O \tag{5}$$

$$\left[Cp*Ir^{I}(\mathbf{N4})\right]^- + H_2O \rightarrow \left[Cp*Ir^{III}H(\mathbf{N4})\right]^- + OH^- \tag{6}$$

$$\left[Cp*Ir^{III}H(\mathbf{N4})\right]^- + CO_2 + H_2O \rightarrow \left[Cp*Ir^{III}(\mathbf{N4})(OH_2)\right]^+ + HCO_2^- \tag{7}$$

Although the $[Fe_4N(CO)_{12}]^-$ catalyst has been reported to exhibit activity in aqueous solution (Chart 4) [30], the Ir catalyst (i.e., $[Cp^{*}Ir(\mathbf{N4})(OH_2)]^+$) is superior in terms of the overpotential and current density.

2.2 *Immobilized Catalysts*

Since electron transfer in electrochemical reactions is limited to the very vicinity of the electrode, the reaction is dominated by the diffusion of the catalyst to the electrode in the bulk electrolyte. Therefore, a method of immobilizing the catalyst to the electrode to eliminate the diffusion factor has been attempted [32]. More specifically, an iridium dihydride catalyst bearing a POCOP pincer ligand containing pyrene was immobilized on carbon nanotube-coated gas diffusion electrodes. Compared to $(POCOP)IrH_2$ (1.07 mA cm^{-2}, −1.45 V), (POCOP-pyrene)$Ir(CH_3CN)H_2$ exhibited and improved for the electroreduction by immobilization on the electrode in 0.5 M $LiClO_4$, 0.1 M $NaHCO_3$, and 1%v/vCH_3CN (3.60 mA cm^{-2}, −1.40 V). Furthermore, through optimization of the reaction system, the gas diffusion electrode interfaced both the gaseous and aqueous phases, significantly enhancing the current densities up to ~15.0 mA cm^{-2}, whereby a high FE_{HCO2-} was also maintained (Scheme 5). When the gas diffusion electrode interfaced both the gaseous and aqueous phases, (POCOP-pyrene)$Ir(CH_3CN)H_2$ was able to access sufficient CO_2 from the gas phase and readily release HCO_2^- into the aqueous phase, thereby

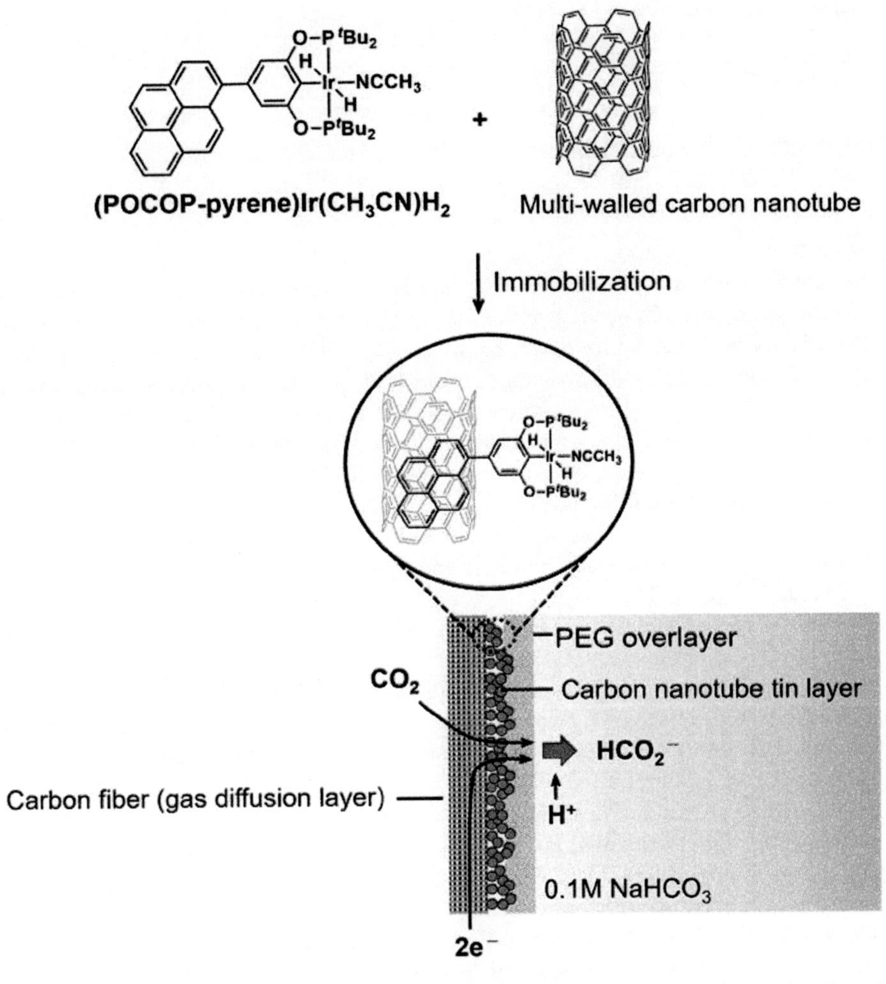

Scheme 5 Illustration of a carbon nanotube-coated gas diffusion electrode with a surface-bound (POCOP-pyrene)Ir(CH_3CN)H_2 catalyst for the electroreduction of CO_2 to HCO_2^-

relieving mass transport constraints. The proposed reaction mechanism was comparable to that presented in Scheme 1a.

3 Electroreduction of Carbon Dioxide to Carbon Monoxide

CO is another two-electron reduction product from CO_2 and is an important material in C1 chemistry. For example, a mixture of H_2 and CO, known as synthesis gas, can be used as a precursor for the production of methanol (CH_3OH) and hydrocarbons.

Table 5 Electroreduction of CO_2 to CO

Catalyst	Solvent	$E_{app}{}^{a}$ [V]	FE(CO) [%]	j [mA cm^{-2}]	Ref.
$IrCl(CO)(PPh_3)_2$	DMF	−1.31	58	0.16	[33]
	DMF/10% H_2O	−1.06	32	0.50	
$[Ir(NNN)(ppy)Cl]^+$	CH_3CN	−1.13	>99	–	[34]
[Ir(NCN)(ppy)Cl]	CH_3CN	−1.67	45 ± 5	–	

[a] E_{app} Applied potential

In general, since the reduction of CO_2 and H^+ on the cathode competes during the electroreduction of CO_2, the suppression of H_2 generation is necessary to improve the selectivity. On the other hand, from the viewpoint of synthesis gas production, H_2 generation is not a disadvantage; if the H_2/CO ratios can be controlled with a catalyst, not only CO but also synthesis gas can be produced directly. However, the reduction of CO_2 to CO requires a large amount of energy, and even in the presence of strong reducing agents, overcoming the O=CO bond enthalpy of 532 kJ mol^{-1} often presents kinetic difficulties.

The first electroreduction of CO_2 to CO using a homogeneous iridium catalyst (Chart 4, Table 5), reported by Pruchnik and coworkers [33], involved the reaction of $IrCl(CO)(PPh_3)_2$ in DMF or DMF + 10% H_2O. Using DMF, CO production at −1.31 V was observed with an FE(CO) of 58% (0.16 mA cm^{-2}). In addition, formic acid was found in the electrolyte. When 10% H_2O was added, although the current density increased to 0.5 mA cm^{-2}, the FE(CO) was 32%. In this case, along with CO and formic acid, a trace of H_2 was detected. They also proposed a mechanism for the reaction between the iridium-hydride intermediates and CO_2; however, spectroscopic data were not reported.

Fujita and coworkers described the electroreduction process and density functional theory (DFT) calculations using $[Ir(NNN)(ppy)Cl]^+$ and $[Ir(NCN)(ppy)Cl]^+$ (ppy, 2-phenylpyridine) [34]. When using $[Ir(NNN)(ppy)Cl]^+$, selective CO formation was achieved at −1.13 V (FE(CO) > 99%), while [Ir(NCN)(ppy)Cl] yielded CO with an FE(CO) of 45 ± 5% and HCO_2^- with an FE(HCO_2^-) of 5–10%. In both cases, H_2 production was not detected. The production of HCO_2^- is supported by the DFT calculations that show iridium-hydride intermediates exhibiting a high hydricity and where CO_2 insertion is thermodynamically favorable. Interestingly, the ligand exchange from Cl^- to CH_3CN, $[Ir(NNN)(ppy)(NCCH_3)]^{2+}$, functions as a photocatalyst and can convert CO_2 to CO.

However, the mechanism for CO production using iridium catalysts was not clear, despite reaction mechanisms being reported for other metal complexes. For example, using a Pd complex bearing a pincer ligand, cyclic voltammetry and potentiostatic electrolysis measurements revealed that Pd-CO_2^- species with η^1-C coordination were generated via an electrochemical one-electron reduction process, and Pd-COOH intermediates were produced by the reduction and protonation of these Pd-CO_2^- species. Furthermore, the intermediates were protonated to generate a Pd-CO species, and finally CO was eliminated to regenerate the active species [35].

4 Electroreduction of Carbon Dioxide to Oxalate

C-C bond formation is an important reaction in chemical synthesis that is commonly carried out through coupling reactions with organometallic reagents. In this context, the use of $C_2O_4^{2-}$ as a C2 compound is interesting from the viewpoint of electrochemical C-C bond formation. In addition, $C_2O_4^{2-}$ are useful chemicals; for example, dimethyl $C_2O_4^{2-}$ is a precursor for the production of ethylene glycol and methyl glycolate [36].

When the electroreduction of CO_2 was conducted in CH_3CN at −1.40 V, trinuclear iridium complex $[(Cp^*Ir)_3(\mu_3\text{-}S)_2]^{2+}$ gave $C_2O_4^{2-}$ (FE($C_2O_4^{2-}$) 64%) and $[(Cp^*Ir)_2(Ir(\eta^4\text{-}C_5(CH_3)_5)CH_2CN)(\mu_3\text{-}S)_2]^+$ ($[(Cp^*Ir)_3CH_2CN(\mu_3\text{-}S)_2]^+$) [37]. The crystal structure of $[(Cp^*Ir)_3CH_2CN(\mu_3\text{-}S)_2]^+$ was determined by X-ray diffraction, and it was found that a linear CH_2CN group was linked at the exo-position of a Cp* ligand, and the $C_5(CH_3)_5CH_2CN$ ligand was coordinated to an iridium atom in the η^4-mode. In addition, the cyclic voltammogram of $[(Cp^*Ir)_3CH_2CN(\mu_3\text{-}S)_2]^+$ recorded in CH_3CN in the presence of CO_2 exhibited a strong catalytic current due to the reduction of CO_2, although this was not observed for $[(Cp^*Ir)_3(\mu_3\text{-}S)_2]^{2+}$. As shown in Scheme 6, the reduced form of $[(Cp^*Ir)_3CH_2CN(\mu_3\text{-}S)_2]^+$ works as the active species in the reduction of CO_2. The direct attack of CO_2 on the Ir species of $[(Cp^*Ir)_3CH_2CN(\mu_3\text{-}S)_2]^+$ is blocked by the Cp*, (η^4-$C_5(CH_3)_5$)CH_2CN, and μ_3-S ligands, while there seems to be no serious

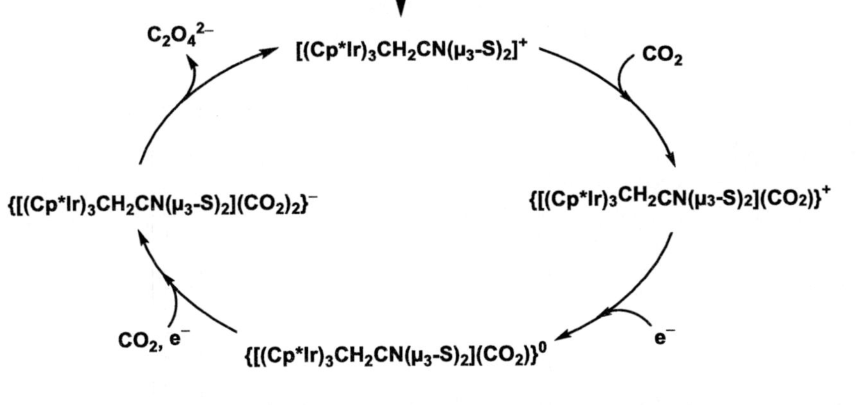

Scheme 6 Proposed mechanism for electroreduction of CO_2 to $C_2O_4^{2-}$ using $[(Cp^*Ir)_3(\mu_3\text{-}S)_2]^{2+}$

steric hindrance for an electrophilic attack of CO_2 to the μ_3-S ligand ({[(Cp*Ir)$_3$CH$_2$CN(μ_3-S)$_2$](CO_2)}$^+$). In addition, a two-electron reduction and subsequent reaction with CO_2 provided anionic species {[(Cp*Ir)$_3$CH$_2$CN(μ_3-S)$_2$] (CO_2)$_2$}$^-$. The coupling of two CO_2 molecules bonded on adjacent μ_3-S and Ir centers in {[(Cp*Ir)$_3$CH$_2$CN(μ_3-S)$_2$](CO_2)$_2$}$^-$ produced $C_2O_4^{2-}$. It should also be noted that the electrochemical conversion of CO_2 to $C_2O_4^{2-}$ has been achieved using Fe [38], Ni [39, 40], Cu [41], and Ru complexes [42], in addition to [(Cp*Ir)$_3$(μ_3-S)$_2$]$^{2+}$.

5 Conclusion

The development of iridium-based catalysts for the electrolytic reduction of CO_2 has been conducted for more than 30 years. Due to the increasing importance of CO_2 utilization technologies in recent years, selective CO_2 conversion has become possible, with dramatically improved catalyst performances being obtained through the development of new ligands. In addition, the overpotential and current density for the electroreduction of CO_2 have also received attention in terms of the energy consumption and reaction rate. In terms of the overpotential, it is important to understand the function of metal-hydride species, as discussed in this chapter. Furthermore, it has been found that the reaction system, and more specifically the electrode materials and mass diffusion, has a significant effect on the reaction efficiency. Hence, to achieve efficient reactions with high current densities and low overpotentials, the design of an appropriate catalyst and overall electrochemical system must be optimized.

In the context of future catalyst developments, compounds that require C-O bond cleavage and multiple C-H bond formation, such as CH_3OH, will be targeted. Indeed, the catalytic functions of single C-H bond formation and C-O bond cleavage, such as in the cases of HCO_2^- formation and CO formation, have been studied. In terms of the synthesis of CH_3OH, the integration of individual catalyst design principals will be necessary.

References

1. Rogelj J, den Elzen M, Höhne N, Fransen T, Fekete H, Winkler H, Schaeffer R, Sha F, Riahi K, Meinshausen M (2016) Nature 534:631–639
2. Ilas A, Ralon P, Rodriguez A, Taylor M (2018) International Renewable Energy Agency (IRENA). Abu Dhabi, UAE, pp 1–160
3. Gotz M, Lefebvre J, Mors F, Koch AM, Graf F, Bajohr S, Reimert R, Kolb T (2016) Renew Energy 85:1371–1390
4. Schemme S, Samsun RC, Peters R, Stolten D (2017) Fuel 205:198–221
5. Whipple DT, Kenis PJA (2010) J Phys Chem Lett 1:3451–3458
6. Benson EE, Kubiak CP, Sathrum AJ, Smieja JM (2009) Chem Soc Rev 38:89–99

7. Taheri A, Berben LA (2016) Chem Commun 52:1768–1777
8. Qiao J, Liu Y, Hong F, Zhang J (2014) Chem Soc Rev 43:631–675
9. Fujita E, Brunschwig BS (2001) Catalysis of electron transfer, heterogeneous systems, gas phase systems. In: Balzani V (ed) Electron transfer in chemistry, vol 1. Wiley, Weinheim, pp 88–126
10. Zhang L, Zhao Z-J, Gong J (2017) Angew Chem Int Ed 56:11326–11353
11. Francke R, Schille B, Roemelt M (2018) Chem Rev 118:4631–4701
12. Zhang S, Fan Q, Xia R, Meyer TJ (2020) Acc Chem Res 53:255–264
13. Wang W-H, Himeda Y, Muckerman JT, Manbeck GF, Fujita E (2015) Chem Rev 115:12936–12973
14. Boddien A, Gärtner F, Federsel C, Sponholz P, Mellmann D, Jackstell R, Junge H, Beller M (2011) Angew Chem Int Ed 50:6411–6414
15. Kawanami H, Iguchi M, Himeda Y (2020) Inorg Chem 59:4191–4199
16. Cheng SC, Blaine CA, Hill MG, Mann KR (1996) Inorg Chem 35:7704–7708
17. Caix C, ChardonNoblat S, Deronzier A (1997) J Electroanal Chem 434:163–170
18. Tanaka R, Yamashita M, Nozaki K (2009) J Am Chem Soc 131:14168–14169
19. Schmeier TJ, Dobereiner GE, Crabtree RH, Hazari N (2011) J Am Chem Soc 133:9274–9277
20. Kang P, Cheng C, Chen Z, Schauer CK, Meyer TJ, Brookhart M (2012) J Am Chem Soc 134:5500–5503
21. Kang P, Meyer TJ, Brookhart M (2013) Chem Sci 4:3497–3502
22. Ahn ST, Bielinski EA, Lane EM, Chen Y, Bernskoetter WH, Hazari N, Palmore GTR (2015) Chem Commun 51:5947–5950
23. Bi J, Hou P, Kang P (2019) ChemCatChem 11:2069–2072
24. Cao L, Sun C, Sun N, Meng L, Chen D (2013) Dalton Trans 42:5755–5763
25. Osadchuk I, Tamm T, Ahlquist MSG (2016) ACS Catal 6:3834–3839
26. Cunningham DW, Barlow JM, Velazquez RS, Yang JY (2020) Angew Chem Int Ed 59:4443–4447
27. Pugh JR, Bruce MRM, Sullivan BP, Meyer TJ (1991) Inorg Chem 30:86–91
28. Sypaseuth FD, Matlachowski C, Weber M, Schwalbe M, Tzschucke CC (2015) Chem Eur J 21:6564–6571
29. Kanega R, Onishi N, Wang L, Himeda Y (2018) ACS Catal 8:11296–11301
30. Taheri A, Thompson EJ, Fettinger JC, Berben LA (2015) ACS Catal 5:7140–7151
31. Kanega R, Onishi N, Szalda DJ, Ertem MZ, Muckerman JT, Fujita E, Himeda Y (2017) ACS Catal 7:6426–6429
32. Kang P, Zhang S, Meyer TJ, Brookhart M (2014) Angew Chem Int Ed 53:8709–8713
33. Szymaszek A, Pruchnik FP (1989) J Organomet Chem 376:133–140
34. Manbeck GF, Garg K, Shimoda T, Szalda DJ, Ertem MZ, Muckerman JT, Fujita E (2017) Faraday Discuss 198:301–317
35. Rakowski Dubois M, Dubois DL (2009) Acc Chem Res 42:1974–1982
36. Yue H, Zhao Y, Ma X, Gong J (2012) Chem Soc Rev 41:4218–4244
37. Tanaka K, Kushi Y, Tsuge K, Toyohara K, Nishioka T, Isobe K (1998) Inorg Chem 37:120–126
38. Pun S-N, Chung W-H, Lam K-M, Guo P, Chan P-H, Wong K-Y, Che C-M, Chen T-Y, Peng S-M (2002) J Chem Soc Dalton Trans:575–583
39. Rudolph M, Dautz S, Jäger E-G (2000) J Am Chem Soc 122:10821–10830
40. Udugala-Ganehenege MY, Dissanayake NM, Liu Y, Bond AM, Zhang J (2014) Transit Met Chem 39:819–830
41. Angamuthu R, Byers P, Lutz M, Spek AL, Bouwman E (2010) Science 327:313–315
42. Ali MM, Sato H, Mizukawa T, Tsuge K, Haga M, Tanaka K (1998) Chem Commun:249–250

Top Organomet Chem (2021) 69: 341–396
https://doi.org/10.1007/3418_2020_72

Published online: 12 January 2021

Homogenous Iridium Catalysts for Biomass Conversion

Sarah Kirchhecker, Brian Spiegelberg, and Johannes G. de Vries

Contents

S. Kirchhecker, B. Spiegelberg, and J. G. de Vries (✉)
Leibniz Institut für Katalyse e.V., Rostock, Germany
e-mail: johannes.devries@catalysis.de

Abstract The use of biomass as a sustainable feedstock for the production of chemicals has become more and more important in recent years. Homogeneous iridium catalysis offers great opportunities for the conversion of bio-derived platform molecules and even biomass components such as cellulose or lignin, due to the air, water, and acid stability of many iridium complexes. In this chapter, we review the application of iridium catalysts to the transformations of carbohydrate-derived compounds, fatty acids, and lignin.

Keywords Biomass conversion · Dehydrogenation · Fatty acids · Glycerol · HMF · Homogeneous catalysis · Hydrogenation · Hydrogen borrowing · Iridium catalysis · Levulinic acid · Lignin · Platform chemicals

1 Introduction

The diminishing supply of fossil resources, as well as the climate effects caused by the cumulative release of CO_2 from fossil energy carriers into the atmosphere, forces us to look for more sustainable alternatives. Thus, the conversion of biomass into useful chemicals and fuels has become a very important field of research. Whereas initially most work was focused on the conversion of easily accessible and chemically pure compounds such as sugars and fatty acids, nowadays also direct routes from lignocellulose to some platform chemicals have been established.

Lignocellulosic biomass is generally converted by acid hydrolysis (or in combination with enzymes) into its monomers (sugars from the cellulose and hemicellulose fractions), which are then further converted into a range of small platform molecules (Fig. 1) [1, 2]. Here, there is scope for the use of homogenous catalysis in the further transformation of these molecules into new and known compounds [3]. The great advantage of iridium catalysts in this regard is their high stability towards water, air, and acidic conditions, i.e. the exact conditions present in biomass processing. This opens up the possibility of performing several steps in one pot or at least without extensive and energy intensive purification of intermediates or drying of starting materials. Additionally, many biomass derived compounds are poorly soluble in solvents other than water.

In the following we will discuss the recent work on homogeneous iridium catalysis in the field of biomass conversion. As most current research focusses on the conversion of a few available platform chemicals, the chapter is divided according to the substrates. Most of these are derived from the cellulose/hemicellulose fraction, namely sugars, bio-derived alcohols, furanics (compounds derived from furfural or 5-hydroxymethylfurfural), and small organic acids. Another section deals with iridium-catalyzed transformations of fatty acids, which are derived, together with glycerol, from vegetable oils. Iridium-catalyzed conversion of lignin is discussed in Sect. 6.

Many of the transformations of biomass discussed here are hydrogenation or dehydrogenation reactions. In comparison with oil derived platform chemicals, biomass is highly functionalized with mostly oxygen containing groups. Hydrogenation/dehydrogenation strategies are therefore an efficient way of converting these compounds. The focus with dehydrogenation reactions can be either on the hydrogen generated or on the dehydrogenated product, while of course always both are formed. The generation of hydrogen gas from biomass as a renewable fuel will be discussed in Sect. 5, alongside dehydrogenative strategies for the synthesis of other compounds. Hydrogenation of CO_2 with H_2 gives formic acid, which has been widely researched and proposed as a reversible hydrogen carrier. While we mention the topic of formic acid dehydrogenation in the last section for the hydrogenation of other substrates, a full discussion of CO_2 hydrogenation was deemed out of scope for this chapter. For further information on this topic, we refer to this recent review [4].

2 Recent Trends in Iridium-Catalyzed Valorization of Bio-derived Alcohols

Alcohols, diols, and polyols are indispensable ingredients and reactive materials for polymers, cosmetics, and pharmaceuticals and their wide range of applications makes them one of the largest class of chemicals produced in the world [5]. The C2-C4 diols alone are produced on an annual scale of 18 million tons [6].

In the last two decades, the use of renewable resources for the benign production of industrially relevant alcohols (see Fig. 1) has received significant interest. In this regard, fermentative and catalytic processes were developed to obtain the bio-based versions of the existing fossil-based polyols as well as newly synthesized alcohols, usually not available from fossil feedstock material [6]. For a detailed overview on the production of bio-based alcohols, we refer to the recently published review articles [5, 7]. In this section the use of bio-based polyols in homogenously iridium-catalyzed reactions is reported.

2.1 Iridium-Catalyzed Reactions Using Ethylene Glycol

2.1.1 Iridium-Catalyzed Selective Cross-Coupling of Ethylene Glycol

Ethylene glycol (EG) is industrially produced from renewable resources on a multiton scale via the hydrogenolysis of sorbitol using a Ni/Ru catalyst [8]. This process leads to a mixture of EG and other bio-based polyols. Furthermore, the dehydration of bio-ethanol to ethylene and its further conversion to EG is an alternative bio-based route according to the conventional petro-based process [8]. Finally, the Dutch company Avantium has recently started to build a demonstration plant where EG will be produced directly from renewable sugars [9]. EG is

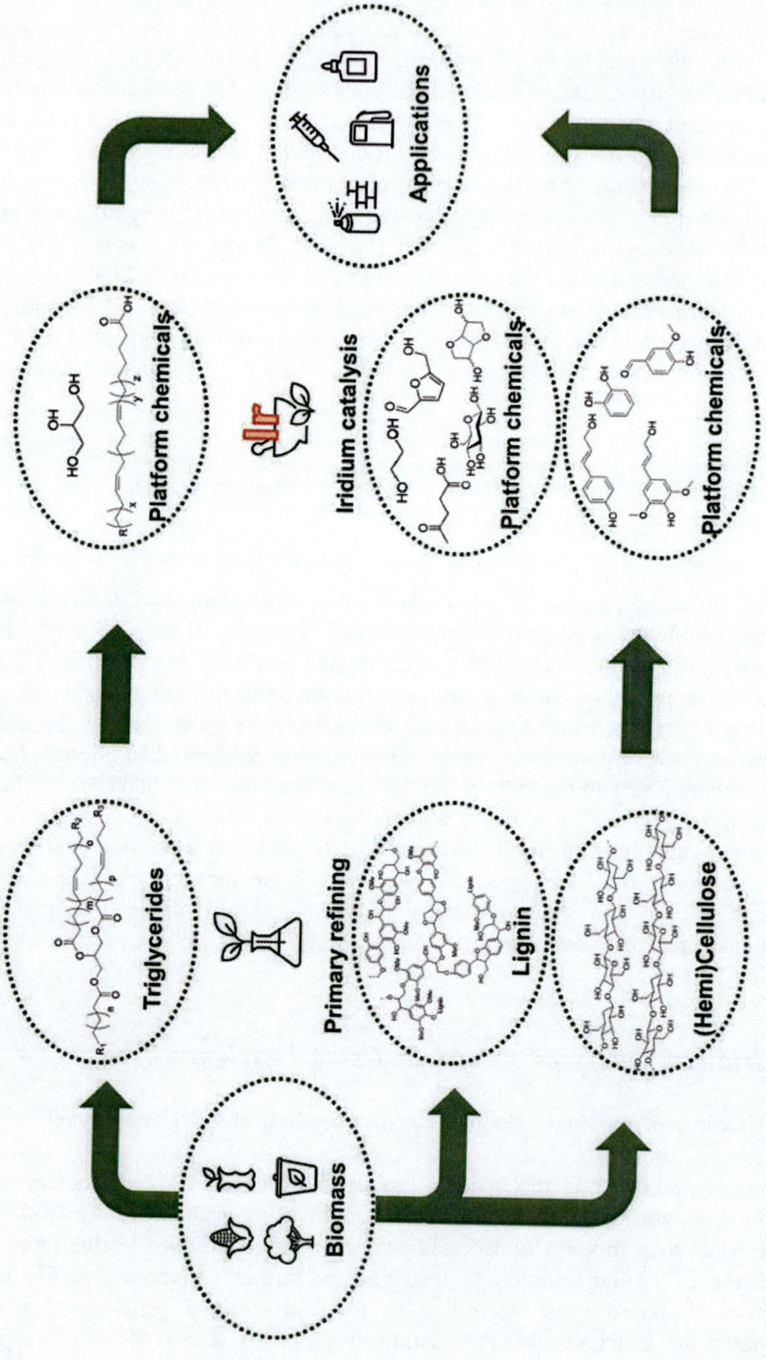

Fig. 1 Conversion of biomass to monomers and platform chemicals

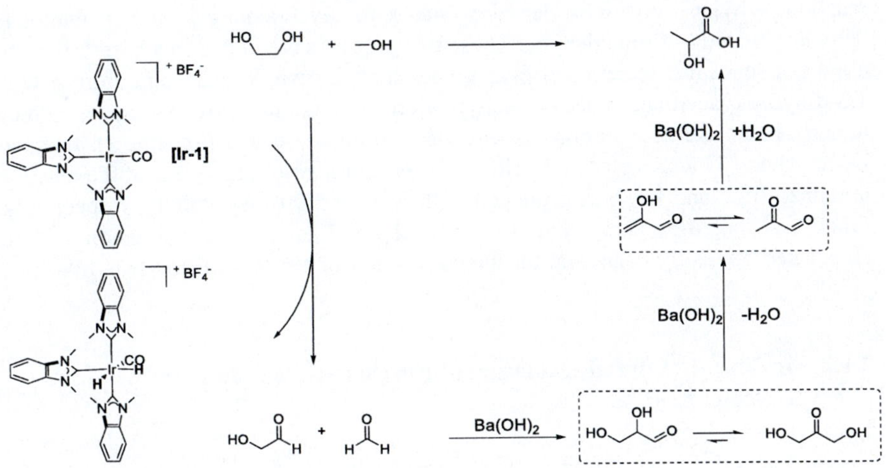

Scheme 1 Dehydrogenative cross-coupling of EG and MeOH in the presence of a novel iridium-tris-NHC complex

mostly known for its use as anti-freeze ingredient in liquid mixtures [5]. However, extremely relevant is its use as a building block for the synthesis of polyesters. Particularly, polyethylene terephthalate (PET) and, more recently, its green analogue polyethylene 2,5-furandicarboxylate (PEF) are the polyester materials in which EG finds its main use [6]. Additionally, other important derivates and platform chemicals can be obtained from EG. One of them is lactic acid (LA) which has attracted extensive attention over the last years due to its versatility and the useful properties of its derived polymer, polylactide [10]. Besides the existing conventional processes, based on the fermentation of carbohydrates, several new routes based on renewables have been developed for the production of LA. Tu and coworkers developed a new strategy to synthesize LA which uses bio-derived EG and MeOH as starting materials and a novel iridium complex **[Ir-1]** bearing three N-heterocyclic carbene ligands (NHC) as a catalyst (Scheme 1) [10].

The dehydrogenative cross-coupling between 5 mmol of EG and a large excess of MeOH (2 mL) to quantitatively form LA was performed at 140°C for 1 h in the presence of 500 ppm of **[Ir-1]** and 1.2 equiv. of $Ba(OH)_2 \cdot 8H_2O$. The use of the base $Ba(OH)_2 \cdot 8H_2O$ is crucial in this reaction since other alkali hydroxides such as calcium hydroxide led to low or no yield of the desired product. When the amount of base was lowered to 1.0 equiv., the yield was only slightly reduced from 100% to 95%. In contrast, by lowering the reaction temperature a significant loss in yield of LA was observed (58% of LA at 120°C for 1 h). Furthermore, the number of NHC ligands coordinated to the metal center is very important as the activity of the iridium (I) catalyst increases as follows: mono-NHC-Ir < bis-NHC-Ir < tris-NHC-Ir [10]. The authors postulated that the increasing number of NHC ligands enhances the steric bulkiness and the electron density which in turn facilitates the oxidation of Ir(I) to Ir(III) and thus improves catalytic performance. In the first step of the reaction, ethylene glycol is dehydrogenated to glycolaldehyde and methanol is dehydrogenated to formaldehyde. These two products are joined in a base-catalyzed aldol condensation reaction to give glyceraldehyde which is in equilibrium with 1,3-dihydroxyacetone. Further dehydration results in pyruvaldehyde, which undergoes an internal Cannizzaro reaction to form lactic acid (Scheme 1). Interestingly, when EG was replaced by either glycerol or sorbitol, the yield of the reaction remained high and LA was produced with 85% and 99% selectivity, respectively. Finally, strong σ-donating and weak π-acceptor ligands in combination with Ir (I) seemed generally beneficial for the conversion of bio-polyols towards LA.

2.1.2 Iridium-Catalyzed Oxidation of Ethylene Glycol Towards Glycolic Acid

Glycolic acid (GA) is a simple α-hydroxy acid which finds applications in the fields of fine chemicals and pharmaceutical chemistry [11]. Additionally, GA is of interest for the synthesis of polyglycolic acid (PGA), a material which could be a sustainable replacement for traditional plastics, due to its excellent biodegradability and biocompatibility. However, at present, GA is predominantly synthesized via the poorly sustainable process based on the hydrolysis of chloroacetic acid in alkaline solution [11]. In contrast, the synthesis of GA from biomass can be achieved through two different mechanisms both using EG as the starting material. In one method, GA can be synthesized from EG through an oxidation pathway which proceeds via the intermediates glycolaldehyde and glyoxal. However, the synthesis of GA through this process has been so far achieved only by using high temperatures and in the presence of an oxygen enriched atmosphere and metallic oxides [12]. The second strategy, alternative to the oxidation pathway for the conversion of EG into GA, is the use of a dehydrogenation reaction, which is superior to the oxidation pathway, because it avoids overoxidation. Tang and coworkers have used this strategy and developed an iridium-mediated cascade process in alkaline water which consists of two dehydrogenation steps followed by a Cannizzaro reaction (see Scheme 2) which transforms EG into GA [11].

Initially, two different metal catalysts based on Rh and Ir, respectively, were tested for the dehydrogenation of EG. The iridium complex [Cp*Ir(bpym)]Cl_2

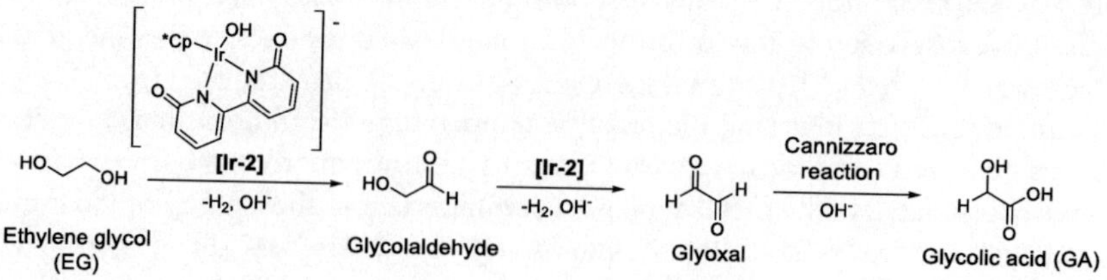

Scheme 2 Reaction pathway of the conversion of ethylene glycol into glycolic acid catalyzed by **[Ir-2]**

(Cp* = pentamethylcyclopentadienyl; bpym = bipyrimidine) led to yields which were two times higher than the ones obtained using the analogous rhodium complex [Cp*Rh(bpym)]Cl_2 [11]. For this reason, the authors further explored the reactivity of the iridium complex by replacing the ligands, and when the complex [Cp*Ir (bpyO)]OH^- (bpyO = α,α′-bipyridonate) **[Ir-2]** was used, an excellent conversion of 93% and a GA yield of 76% were obtained (Scheme 2). In order to maximize the concentration of the active dehydrogenation catalyst **[Ir-2]**, the pH of the solution was adjusted with a phosphate buffer and additional NaOH was added to compensate the acidifying effect of the produced GA. The best results were obtained (GA yield: 81.5%) when a stochiometric amount of NaOH was sequentially added to the reaction. Tang and coworkers were also able to reuse the **[Ir-2]** for several catalytic runs, although after the sixth run the conversion of EG and the yield of GA slightly dropped.

2.2 *Iridium-Catalyzed Hydrogen Transfer Initiated Dehydration of 1,3-Propanediol to Aldehydes*

Aldehydes like propionaldehyde are important intermediates for the production of fragrances, agrochemicals, plasticizers, or drugs [13]. 1,3-Propanediol (1,3-PDO) has been one of the first renewable platform chemicals which is produced by bacterial fermentation of sugars [6, 13]. Due to the relevance of 1,3-PDO, Marr and coworkers became interested in finding new chemocatalytic transformations of this bio-based alcohol into value-added chemicals. During their attempt to aminate 1,3-PDO with a Cp*Ir(III) N-heterocyclic carbene complex they discovered that, by working in ionic liquids, dehydration of 1,3-PDO occurred [14]. For that reason, the authors decided to use the same iridium catalyst (**[Ir-3]**, Fig. 2) in ionic liquids for the selective dehydration of 1,3-PDO (**1**) to propionaldehyde (**2**) [13]. This interesting conversion proceeds via a dehydrogenation to 3-hydroxypropionaldehyde, dehydration to acrolein, and hydrogenation to propionaldehyde. The authors refer to the sequence as a hydrogen transfer initiated dehydration (HTID). The reaction was initially tested in the presence of two different ionic liquids (Fig. 2), $EmmimNTf_2$ or $N_{1,8,8,8}NTf_2$, respectively, and at temperatures varying between 80°C and 150°C for 6 h. Next to the desired product **2**, side products **3**, **4**, and **5** were found in varying amounts (Scheme 3). The formation of **3** and **4** can be explained by the Aldol reaction of **2** followed by dehydration and consecutive hydrogenation. Hydrogenation of **2** leads to the formation of **5**. In order to prevent the occurrence of aldol condensation reactions, a reduced pressure (0.35 bar) was used to remove the highly volatile propionaldehyde (Schemes 3, **2**) which was collected in a cold trap.

The use of $EmmimNTf_2$ resulted in higher yields and selectivities towards **2** compared to $N_{1,8,8,8}NTf_2$. Furthermore, the use of ionic liquids led to better results for the dehydration with respect to neat conditions. Finally, it was shown that the base (K_2CO_3, KOH, Cs_2CO_3) itself had a minor effect on the product distribution

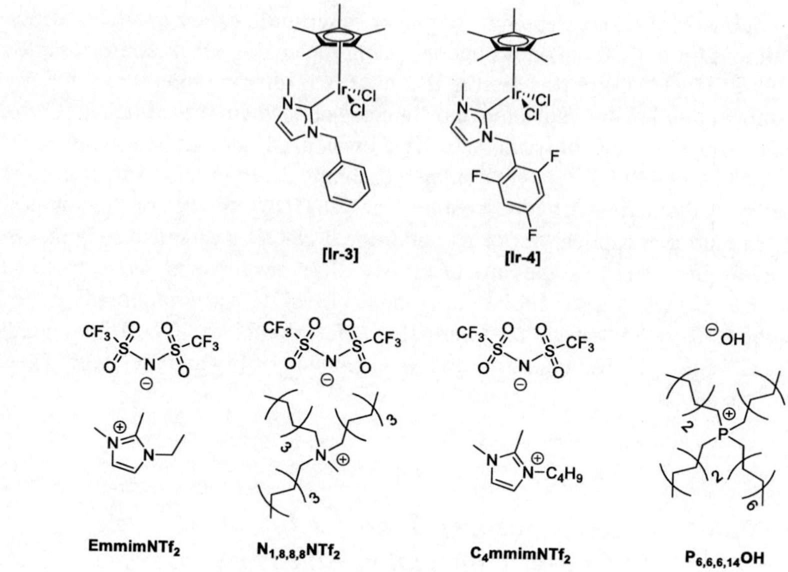

Fig. 2 Hydrogen borrowing catalysts and ionic liquids used as solvent for the HTID

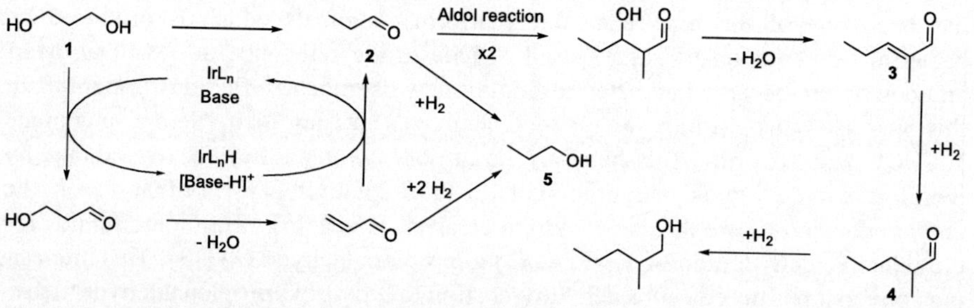

Scheme 3 Postulated reaction pathway of the iridium-catalyzed HTID of 1,3-PDO

although its concentration strongly affected the yield of the reaction. Specifically, it was noted that the higher the base concentration was, the lower the yield of **2**. The recyclability of **[Ir-3]** was proven in $EmmimNTf_2$, by reusing it for ten cycles in the presence of K_2CO_3 at 150°C and 0.35 bar [13]. The conversion of 1,3-PDO (75–99%) and the selectivity towards propionaldehyde (65–72%) remained high over all the runs. Both the used ionic liquid and the catalyst **[Ir-3]** exhibited excellent stability towards water, air, and temperature throughout the 10 runs albeit K_2CO_3 needed to be replenished.

Subsequent work of Marr and coworkers revealed that the use of the fluorinated analogue (**[Ir-4]**, Fig. 2) of their previous reported Ir(III) carbene catalyst **[Ir-3]** improved the conversion (quantitative) and the selectivity towards propanal (87%)

[15]. Intrigued by these promising results, they extended their investigations by using basic ionic liquids gel [16]. In this work, they used mixtures of [$P_{6,6,6,14}$][OH] and [C_4mmim][NTf_2] (Fig. 2) and added tetraethoxysilane which over time cured to form a gel. By using **[Ir-4]** in these gels, high conversions (80–99%) and good yields towards propanal (82–90%) were obtained. A 1:3 ratio of [C_4mmim][NTf_2]: [$P_{6,6,6,14}$][OH] was found to be the best mixture which allowed the optimal balance between basicity, stability, and activity with regard to the formation of propanal.

2.3 *Iridium-Catalyzed Direct Amination of Isohexides*

Isohexides like isosorbide and isomannide are produced from sorbitol and mannose, respectively (Fig. 3). They are of high interest for material science applications since their rigid skeleton can boost the glass transition temperature of the polymers in which they are incorporated [17]. Isohexides exist in three diastereomeric forms, but only the isomers isomannide and in particular isosorbide are used for research purposes due to their commercial availability (Fig. 3) [18]. The range of applications of these two diastereomers has been rapidly growing over the last years. Particularly, their corresponding amine derivatives have received a lot of attention in the field of polymer chemistry [19] and for asymmetric induction in organic synthesis [20, 21]. However, the reported synthetic route for isohexide-based amines which consists in the activation of the alcohol by transformation into better leaving groups followed by a substitution reaction with an amine moiety and a final deprotection step, has an overall low atom efficiency and generates large amounts of waste [17]. This problem was recognized by Popowycz and coworkers, who developed the diastereoselective amination of monobenzylated isohexides through a hydrogen borrowing synthetic pathway by using a combination of iridium complex **[Ir-5]** and diphenylphosphoric acid as catalyst (Scheme 4) [18]. The hydrogen borrowing methodology had already been applied for the diamination of isosorbide and isomannide by using homogenous ruthenium catalysts and excellent yields had been achieved [22, 23]. However, the product was obtained as a mixture of three diastereomers. Popowycz and coworkers achieved 60% conversion of the monobenzylated isosorbide and 50% yield of the corresponding amine [18] using diphenylphosphoric acid and **[Ir-6]**, a catalyst system that had been previously

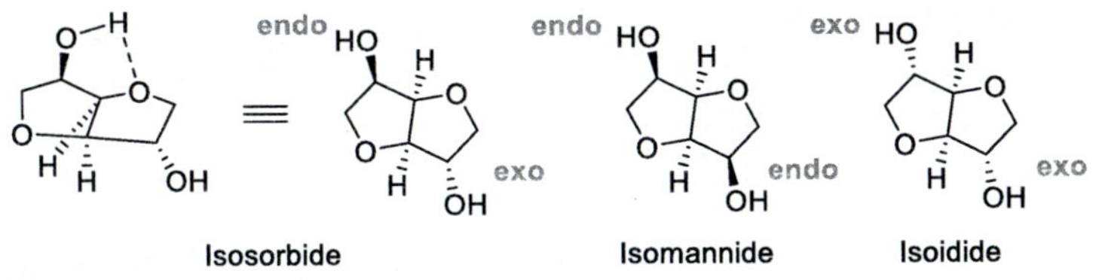

Fig. 3 Chemical structure of bio-based isohexides

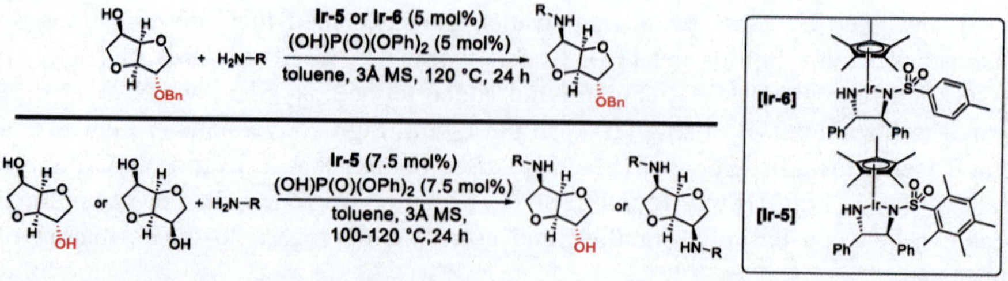

Scheme 4 Overview of direct amination of isohexides by using iridium catalysts coupled with diphenyl phosphoric acid

developed by Zhao and coworkers (Scheme 4) [24]. The diphenyl phosphoric acid facilitates the condensation of the used (4-methoxyphenyl)methanamine with the intermediate isosorbide ketone. Furthermore, by adding molecular sieves into the reaction mixture and by replacing **[Ir-6]** with **[Ir-5]** a further increase of the amount of amination product was observed (yield 67%). In the end, the best results were obtained with 5 mol% of **[Ir-5]** and 5 mol% of diphenylphosphoric acid in the presence of molecular sieves in toluene at 120°C for 24 h (68% yield; dr > 99:1) (Scheme 4, top). This amination protocol was then successfully applied by using various primary amines. The resulting monobenzylated isohexide amines were obtained in good yields (44–83%) and excellent diastereoselectivities.

DFT calculations revealed that it requires more energy to oxidize the alcohol in the exo-position than the alcohol in the endo-position due to the fact that the hydride that needs to be transferred is inside the envelope which is not easily reached by the bulky catalyst [25]. In agreement with this finding, Popowycz and coworkers experimentally confirmed the mono-amination of isosorbide into a chiral amino alcohol [18]. Intrigued by this first direct and regioselective amination of isosorbide, they further examined the scope of the selective transformation by using the hydrogen borrowing methodology [17]. Their optimization study led to slightly different conditions (7.5 mol% of **[Ir-5]** and diphenyl phosphoric acid) than their previous work (for comparison see Scheme 4). Subsequently, a range of different substituted amines were used for the regioselective amination of isosorbide. Depending on the amine, temperatures between 100°C and 120°C were used to achieve generally high conversions (65–99%) and good selectivities (46–95%) towards the desired chiral amino alcohols. Encouraged by these results, they also developed the first synthesis of the parent amino alcohol (Scheme 4, R = H) which they obtained in 54% yield and of the parent diamine (R = H) in 59% yield. They achieved this via the iridium-catalyzed hydrogen-borrowing reaction with benzylamine and isosorbide and isomannide, respectively, followed by the hydrogenation of the benzyl protecting groups using Pd on carbon [17]. Under the same conditions used for the synthesis of the amino alcohols of isosorbide, isomannide was also converted into a range of different diamines with similar good conversions (78–99%) and selectivities (66–91%).

2.4 Iridium-Catalyzed Conversion of Glycerol

2.4.1 Iridium-Catalyzed Oxidative Conversion of Glycerol

The by-product of the production of biodiesel from fats is glycerol. Thus, glycerol is readily available from biomass, and its use has received a lot of attention, not only with the aim of improving the entire value-chain of biomass transformation, but also as a consequence of the variety of reactions which can be done with this polyol. As mentioned before in Sect. 2.1.1, LA and its derivatives have numerous applications and the demand for polylactide materials is increasing steadily. Currently, lactic acid (LA) is produced via fermentation on a scale of 450,000 tons/year [26]. Nevertheless, the use of a low-value waste product like glycerol as a feedstock material could offer, when catalytically and economically feasible, another scalable and valuable alternative for the synthesis of lactic acid/lactate.

In recent years, several groups investigated the iridium-catalyzed acceptorless dehydrogenation of glycerol to LA. In the seminal work from Crabtree and coworkers the authors used iridium(I) bis NHC complexes for the dehydrogenation of glycerol to form either dihydroxyacetone or glyceraldehyde [27]. Afterwards, the latter products could undergo a dehydration followed by an intramolecular Cannizzaro reaction to afford LA (for comparison, see Scheme 1). They found that **[Ir-7]** (see Fig. 4) was the most active by resulting in high yields (91%) and selectivities (>95%) under solvent-free conditions. The same authors explored the use of sorbitol, mannitol, xylitol, and erythritol as starting materials which are readily available from biomass [28]. Unfortunately, the conversion of these polyols did not even come close to the conversions and selectivities which were obtained with glycerol.

Inspired by these results, Williams and coworkers developed the NHC-pyridine iridium(I) complex **[Ir-9]** which thus far is the most robust and active catalyst for the conversion of glycerol [29]. With this catalyst they were able to reach 4.5 million turnovers in 32 days with an average TOF of 6,000 h^{-1} at 145°C under neat

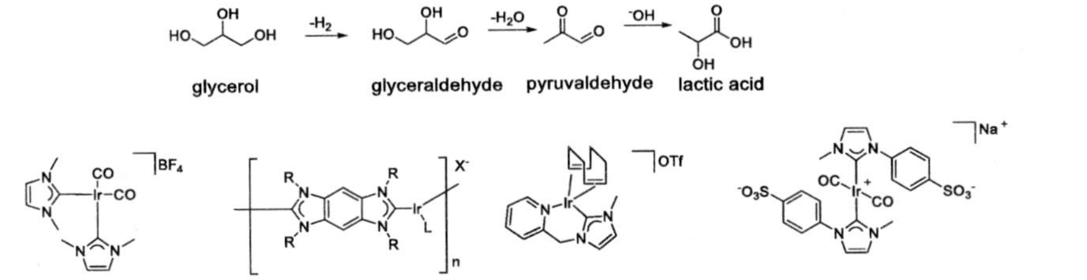

[Ir-7]	[Ir-8]	[Ir-9]	[Ir-10]
TOF$_{av}$: 337 h^{-1}, 115 °C	TOF$_{av}$: 3444 h^{-1}, 115 °C	TOF$_{av}$: 6000 h^{-1}, 145 °C	TOF$_{av}$: 3477 h^{-1}, 150 °C; 45592 h^{-1} (microwave) 150 °C
TON: 30100, 90 h	TON: 124000, 36 h	TON: 4.5 mil, 32 days	TON: 83448, 24 h; 293249 (microwave, 8 h)
Glycerol neat	Glycerol/water	Glycerol neat	Glycerol/water
91% yield, >95% sel.	92% yield, >99% sel.	>90% yield, 100% sel.	92% yield, 100% sel.

Fig. 4 Reported iridium catalysts for the acceptorless dehydrogenation of glycerol to lactic acid

conditions. Tu and coworkers developed a polymeric iridium catalyst (**[Ir-8]**, Fig. 4) which allowed the catalyst's recyclability [30]. A maximum turnover number of 124,000 was measured with this catalyst. The extraordinary performance of **[Ir-9]** was explained by the authors as a result of the high stability provided by the bidentate (pyridyl)carbene ligand [29]. In essence, the bidentate coordination mode reduces the probability of ligand scrambling processes which are the major cause of loss of catalytic activity [27]. Subsequently, the group of Voutchkova-Kostal reported other Ir(I)/Ir(III) complexes bearing NHC ligands with attached sulfonate groups [31]. The most active catalyst **[Ir-10]** reached a TOF of 45,592 h^{-1} under microwave irradiation at 150°C and of 3,477 h^{-1} when the reaction was conventionally heated at 150°C.

Jang and coworkers employed various biscarbene Ir(I) complexes (Fig. 5) for the transfer hydrogenation of glycerol with inorganic carbonates in order to obtain both formate and lactate (Scheme 5) [32]. Whereas lactate can be considered as intermediate to access LA and thus polylactic acid, formates are themselves a valuable class of chemicals. The best catalyst **[Ir-12]** reported by Jang and his group was more than two times as efficient than the biscarbene Ir(I) complex **[Ir-11]** of Crabtree [27, 32]. The reason for its superior performance is the high stability of the catalyst at 180°C, presumably due to the used biscarbene ligand. The latter is a chelating bidentate ligand which inhibits the formation of the inactive tris-carbene-coordinated iridium complexes [29]. Under optimized conditions Jang and coworkers explored the use of Na_2CO_3 and $KHCO_3$ as alternatives to K_2CO_3. In both cases, TONs with respect to both formate and lactate lower than the ones measured with K_2CO_3 were obtained. In the case of Na_2CO_3 the lower solubility compared to K_2CO_3 was considered responsible for the reduced catalytic performance. In the case of $KHCO_3$, the low pH value of 8.5 in comparison with the pH 11.9 of K_2CO_3 in aqueous glycerol was considered the main reason for the drop in TON. As a confirmation of the influence of the pH on catalytic activity, the addition of 2 mmol of KOH to a solution containing 4 mmol of $KHCO_3$ led to an increase in TON of more than twice.

Fig. 5 Iridium catalysts for the transfer hydrogenation of K_2CO_3 in glycerol. TON and TOF refer to the formation of formate

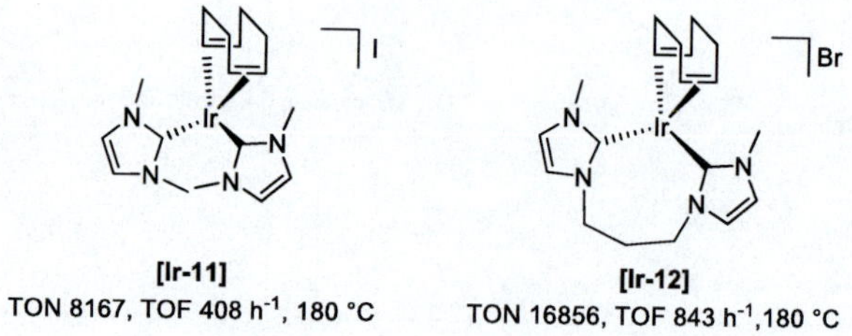

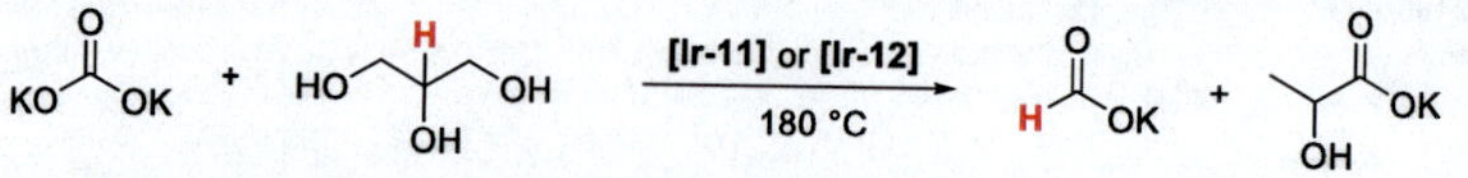

Scheme 5 Transfer hydrogenation of K_2CO_3 in glycerol

2.4.2 Iridium-Catalyzed Deoxygenation of Glycerol

Glycerol is like many other bio-derived chemicals, an oxygen-rich molecule and for this reason deoxygenation/dehydration reactions need to be performed to obtain molecules with low oxygen content. A glycerol derivative of potential interest is, for example, 1,3-propanediol (1,3-PD), a diol commonly used for the synthesis of polymers. The conversion of glycerol to 1,3-PD can be described as a tandem reaction consisting of an initial acid-catalyzed dehydration step followed by a metal catalyzed hydrogenation reaction. For the selective deoxygenation of glycerol mainly heterogenous catalysts are known, also because aqueous acidic conditions and high temperatures are very challenging conditions for many catalytic systems based on well-defined metal complexes. Nevertheless, the groups of Heinekey and Goldberg successfully used iridium pincer complexes (Fig. 6) for the hydrogenolysis of 1,2-propanediol (1,2-PD), a compound that can be made via hydrogenolysis of glycerol, in addition it can be seen as a model reaction for the selective deoxygenation of glycerol [33]. In the presence of 1.04 equivalents of triflic acid and 1.3 mol% **[Ir-13]**, 1,2-PD was effectively hydrogenolyzed using 7 bar of hydrogen to 1-propanol in 95% yield.

As solvent, an acidic aqueous/1,4-dioxane mixture was chosen and the best results were obtained at high water content and low acid concentrations. High acid concentrations or low amount of water led to the formation of several by-products. NMR studies performed after the completion of the reaction revealed that two different iridium species are formed: **[Ir-14]** and **[Ir-14]**-dihydride. Based on this finding, it was proposed that **[Ir-13]** catalyzes the decarbonylation of the intermediate aldehyde and it is presumably transformed in the two observed iridium species. In addition, the authors found that **[Ir-14]** is a catalytically active species that is more stable than the parent complex **[Ir-13]**, since it is still active after storage under air.

Encouraged by these results, the same groups extended their work to the deoxygenation of glycerol [34]. However, complex **[Ir-14]** was poorly soluble in acidic water/1,4-dioxane/glycerol mixtures, also when high temperatures were tested. Therefore, the authors decided to synthesize novel iridium pincer complexes. Complex **[Ir-15]** emerged as a suitable catalyst due to its high solubility in the glycerol containing aqueous dioxane mixture. Particularly, by increasing the amount of

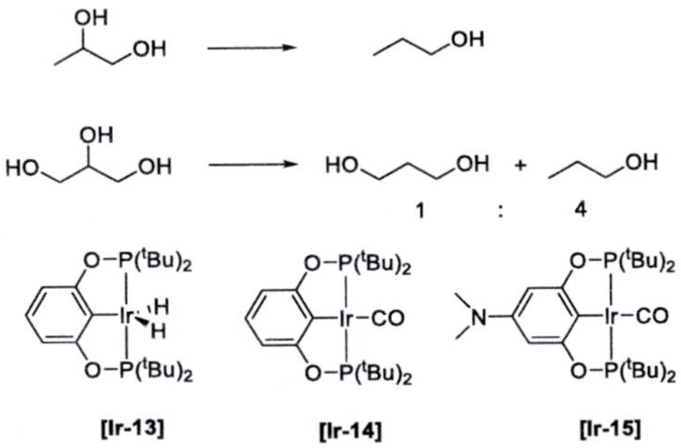

Fig. 6 Iridium catalysts for the deoxygenation of 1,2-propanediol and glycerol

sulfuric acid from 0.25% to 4% (w.r.t. glycerol), the conversion of glycerol increased from 8 to 38% over 24 h at 200°C. However, the increase of the acid concentration was also accompanied by a decrease in selectivity towards 1,3-PD in favor of 1-propanol (1-PO).

The iridium catalyst was also tested for the direct conversion of crude glycerol generated as a by-product of the biodiesel manufacturing process. Since this material is highly basic, the crude glycerol was acidified with 1 M sulfuric acid prior to reaction. The resulting biphasic system was allowed to settle, the layer of impurities was discarded, and the remaining mixture was used for the reaction. The recycled waste glycerol was deoxygenated in 20% yield after 24 h at 200°C under 80 bar H_2 by using 0.125 mol% of **[Ir-15]** with a high selectivity towards 1-propanol (1-PO: 1,3-PD = 9: 1). The observed selectivity was explained by the high acid concentration which remained after the purification of crude glycerol.

2.4.3 Iridium-Catalyzed Carbonylation of Glycerol to Monocarboxylic Acids

A very common deoxygenation approach is the so-called hydro-deoxygenation reaction, in which hydrogen is used as a co-reagent and water is formed as a by-product. In a similar way, CO can be used as a co-reagent instead of hydrogen in a reaction called carbodeoxygenation. Generally, carbonylation reactions are well-established in industry like the production of acetic acid from methanol. However, notwithstanding the progress which has been made in the last decades, the carbonylation of higher alcohols like glycerol using iridium catalysts remains elusive. The latter can be explained by the fact that iridium-based carbonylation catalysts rapidly deactivate due to the formation of $[IrI_4(CO)_2]^-$ in the presence of HI as a cocatalyst. As a consequence, a promoter is needed to reversibly bind the iodide and thus to regenerate the active iridium catalyst [35]. To this regard, the group of Britovsek found that by using $[Ru_3(CO)_{12}]$ as a promoter in the presence of 0.37 mol% of $[NBu_4][IrCl_2(CO)_2]$ **[Ir-16]** as a catalyst and 48 mol% of HI as a cocatalyst a very good conversion (96%) of triacetin (fully acetylated glycerol) after 5 h at 180°C was achieved (Scheme 6) [36].

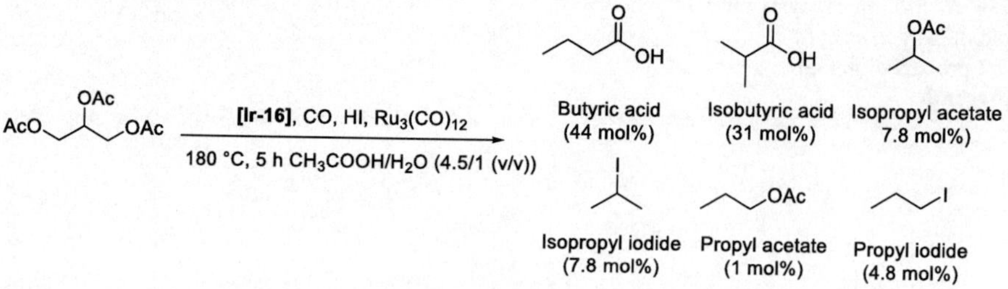

Scheme 6 Iridium-catalyzed carbonylation of triacetin

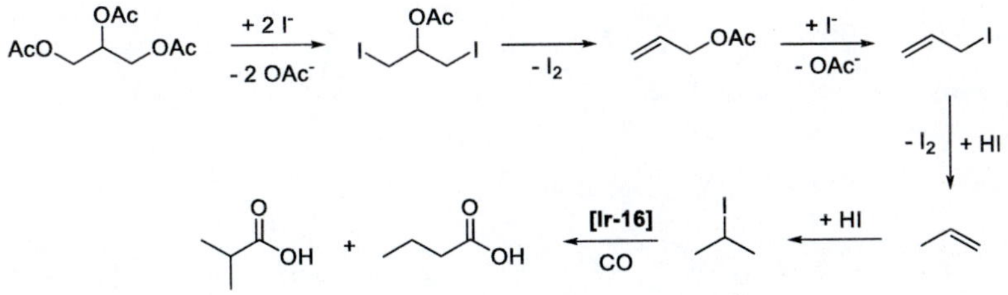

Scheme 7 Sequence of intermediates and products in the carbonylation reaction of triacetin

Triacetin was chosen for convenience as the starting material, because it is less viscous than glycerol and because the reaction is carried out in a water/acetic acid mixture which would anyway lead to the formation of a mixture of mono-, di-, and triacetoxyesters of glycerol. The main products of the reaction were butyric acid and isobutyric acid followed by isopropyl iodide and isopropyl acetate (Scheme 6). HI was found to be superior to methyl iodide as a cocatalyst. Additionally, the authors examined the effect of an increasing CO pressure which resulted in an increase of butyric acid while simultaneously the amount of isobutyric acid decreased.

Furthermore, the group was interested in understanding the selectivity to monocarboxylic acids rather than di- or tricarboxylic acids. Therefore, the authors performed their reaction by using DI in D_2O and CD_3COOD and they monitored the reaction by ^1H NMR spectroscopy. As a result, they discovered that triacetin undergoes double nucleophilic substitution by HI, which is followed by elimination of iodine and by a subsequent 2,1- shift of the acyl group to the terminal position to give the allyl acetate (Scheme 7). The latter is the key intermediate which subsequently leads via carbonylation of the in situ formed propene to the formation of the monocarboxylic acids butyric acid and isobutyric acid.

2.4.4 Iridium-Catalyzed Chemoselective Dehydrogenation of Glycerol

Glycerol is already widely used as a green solvent to replace other more common but at the same time more hazardous solvents in organic reactions [37]. However, the use of glycerol as a hydrogen donor solvent had remained unexplored until Crotti and coworkers developed a range of iridium complexes bearing a diene unit and a dinitrogen containing ligand, which can catalyze the dehydrogenation of glycerol.

The authors found that **[Ir-17]** was the most active catalyst for the selective dehydrogenation of glycerol towards dihydroxyacetone (Fig. 7) [38]. The selective formation of dihydroxyacetone over glyceraldehyde is important as glyceraldehyde tends to be easily decarbonylated and this could lead to the deactivation of the catalyst due to the formation of a stable carbonyl complex. As a model reaction, the dehydrogenation of glycerol to dihydroxyacetone was coupled with the hydrogenation of acetophenone to phenylethanol. The best results were found in the presence of 1 mol% of **[Ir-17]** and 2 mol% of K_2CO_3 at 100°C. The conversion of acetophenone proceeded albeit slow starting from 15% after 5 min and reaching

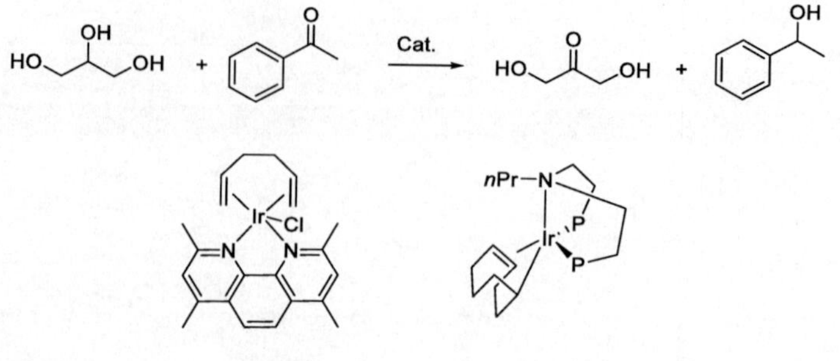

Fig. 7 Iridium catalysts used for the chemoselective dehydrogenation of glycerol to dihydroxyacetone reported by Farnetti and Crotti

34% after 180 min, while the selectivity of glycerol towards dihydroxyacetone dropped in total by 43% during the same period of time. Raising the temperature could further improve the conversion of acetophenone up to 47%, however again a low yield of dihydroxyacetone of only 12% was obtained. A possible explanation for this poor selectivity to dihydroxyacetone is its further base-catalyzed conversion to lactate, as seen before in several dehydrogenation reactions of glycerol [33]. In order to address this problem, Crotti and co. performed the same reaction in the absence of base and they used a new iridium catalyst bearing a pincer ligand **[Ir-18]** [39]. Different hydrogen acceptors, including acetophenone, were tested at 100°C for 1 h, by using 1 mol% of **[Ir-18]**. As a result, while with acetophenone only 3% yield of the corresponding phenylethanol was achieved, benzaldehyde was converted to benzyl alcohol in 37% yield to form dihydroxyacetone with a selectivity of 23%. Hence, the work of Crotti and Farnetti demonstrated for the first time that glycerol can be used as a solvent, but also as a hydrogen donor leading in this case to the dehydrogenation product dihydroxyacetone. Unfortunately, the latter easily undergoes side reactions at high temperatures. Nevertheless, many researchers have been inspired by the work of Crotti and Farnetti and the use of glycerol as H-donor has been recently reviewed by Crabtree [37]. The latter review article illustrates the progress made in the last decade in this field, and in particular it points out how iridium catalysts bearing carbene ligands emerged as highly stable and active complexes for the dehydrogenation of glycerol.

2.4.5 Iridium-Catalyzed Acetalization of Glycerol

Farnetti and coworkers used $[IrCp^{*}Cl_2]_2$ **[Ir-19]** as catalyst for the acetalization of glycerol with several ketones and aldehydes [40]. Due to the three hydroxyl groups in glycerol, the products of this reaction can exist in the dioxolane (5-membered ring) or dioxane (6-membered ring) structure (Scheme 8).

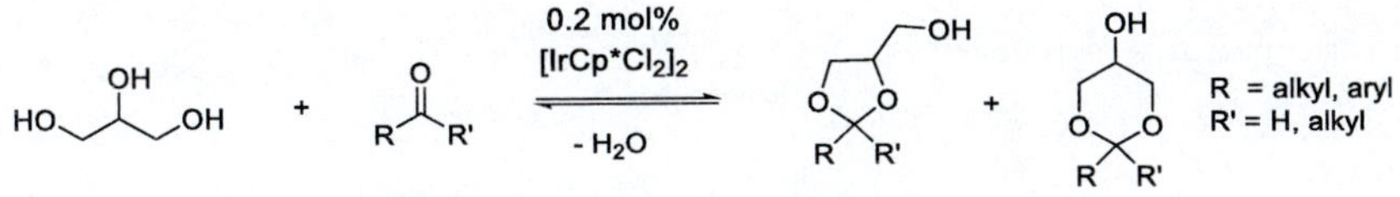

Scheme 8 Acetalization of glycerol catalyzed by iridium

Cyclic acetals and ketals of glycerol are used, for example, as diesel additives, and for the synthesis of surfactants and scents, flavors, and polymers [41]. The reaction can be catalyzed by acids, however where substrates are acid sensitive or contain other sensitive functional group, a more selective homogeneous catalyst may be the better choice. The authors tested other iridium precursors such as [Ir(COD)Cl]$_2$, which did not catalyze this reaction, and other iridium complexes, but [IrCp*Cl$_2$]$_2$ was by far the most active. With acetone the reaction showed high selectivity (98%) to the dioxolane product with 90% conversion in one hour. With other ketones conversions were a bit lower. The reaction was performed in excess glycerol and excess acetone, with the reaction in acetone giving slightly better conversions. However, using an excess of ketone or aldehyde may not always be possible. The substrate scope was fairly limited, however all ketones tested showed high selectivity towards the dioxolane product after 1 h reaction time. With the two aldehydes tested, the aliphatic aldehyde octanal gave higher conversions than benzaldehyde and also slightly higher selectivity towards the dioxolane. Compared to the ketones the selectivity was lower. This may be due to traces of acid present in the aldehyde which catalyze the reaction unselectively. It is known that the dioxolane is the kinetic product and that with longer reaction times a more equal mixture of the two products is formed [42]. This was also found in this case, when the reaction with benzaldehyde was allowed to proceed for 48 h, after which time the ratio of dioxolane: dioxane was 49:51. In most cases, the dioxane isomer (or one of the two E and Z isomers where applicable) can be isolated by crystallization [43], so that this protocol can be used to selectively synthesize the dioxolane product, and – via long reaction times for isomerization combined with repeated crystallization workup – also the dioxane product. The catalyst was also active for the transacetalization with methanol.

2.5 *Iridium-Catalyzed Cyclizations of Bio-Derived Alcohols*

Nitrogen-containing heterocycles are a key structural motif which can be found, for example, in drugs, dyes, and organic electronic materials [44]. Despite their great importance, the conventional synthetic methods of N-heterocycles still often require multiple steps, they sometimes use halogenation reagents, they result in the formation of salt wastes, and they occasionally need oxidants [44].

Bruneau and coworkers used an iridium-catalyzed acceptorless dehydrogenative condensation reaction to prepare julolidines, a class of N-heterocycles with applications in solar cells [45].

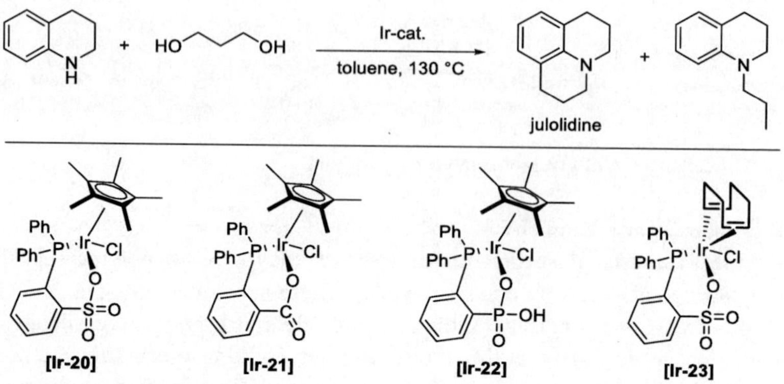

Scheme 9 Synthesis of julolidine catalyzed by different iridium catalysts

Their approach uses bio-derived 1,3-propanediol as a reagent, and it meets several criteria of the twelve principles of green chemistry: it avoids the use of halogenated reagents, it is a catalytic process, and it generates water as an ecofriendly by-product. For their study, they synthesized and tested a range of different well-defined iridium complexes. In particular complexes **[Ir-20-23]** derived from $[Ir(Cp^*)Cl_2]_2$ were found to be suitable catalysts for the N,C(sp^2)-dialkylation of tetrahydroquinoline with 1,3-propanediol (see Scheme 9). The reaction proceeds via dehydrogenation of 1,3-propanediol to 3-hydroxy-propanal, which reacts with tetrahydroquinoline to the iminium compound which can be converted to julolidine via three possible pathways. During their optimization, *N*-propyl-tetrahydroquinoline was identified as the main side product of the reaction and its formation was attributed to the acidity of the catalyst or the ligand in the case of **[Ir-22]** which favors a dehydration/reduction pathway [45]. Nevertheless, when the catalysts **[Ir-20]** and **[Ir-21]** were applied, good yields of 72% and 77% of the desired julolidine were observed. Interestingly, the ratio of amine to 1,3-propanediol had a significant impact on the yield which increased from 43% to 72% when the amine/1,3-PDO ratio was changed from 1/1.2 to 2/1. Catalyst **[Ir-21]** was found to be the best catalyst and no significant differences in terms of activity were found when the catalyst was generated in situ by simply mixing 1 mol% of $[Ir(Cp^*)Cl_2]_2$ and 2 mol% of DPPBA. Furthermore, the use of Butane-1,3-diol, another bio-based polyol, was found to be more challenging than 1,3-PDO, because the dehydration/reduction pathway led mainly to the formation of the undesired N-alkylated product (40%), and the julolidine yield was quite low (28%). Finally, since the variety of julolidine derivatives which can be obtained is limited due to the limited number of commercially available 1,3-diols, the authors described the possible post-functionalization of julolidines through hydrogen auto transfer reactions by using aldehydes as electrophiles [45]. The latter led to the β-alkylation of julolidines in a reaction mediated by the same iridium system **[Ir-21]** used for the dehydrogenative condensation step, leading to the possibility for a tandem reaction which combines the synthesis of julolidines with their subsequent β-alkylation.

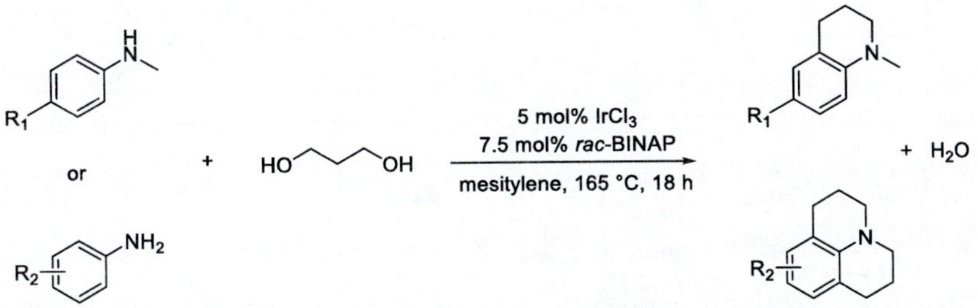

Scheme 10 Iridium-catalyzed direct cyclization of aromatic amines with diols

The group of Minakawa used anilines for the cyclization with diols [46]. In spite of the limited number of bio-based diols, the access to a wide range of different *N*-heterocycles is possible due to the large number of different substituted anilines available (Scheme 10).

Initially, the direct cyclization of 1,3-propanediol with *N*-methylaniline was studied. After finding the best conditions with 5 mol% of iridium(III) chloride and 7.5 mol% *rac*-BINAP as the ligand **[Ir-24]**, a range of different methylated anilines were tested. Uninfluenced by the electron-withdrawing or electron-donating nature of R_1 in the *para*-position, a good isolated yield was obtained (73–83%). Under the same reaction conditions 1,3-propanediol was then reacted with anilines bearing different substituents in the *para*- or *meta*-position to afford various julolidines in moderate yields (26–76%). In contrast, when EG or 1,4-butanediol were reacted with *N*-methylaniline, an indole and a diamine in 45% and 61% yields were found, respectively (Scheme 11) [46].

In a similar fashion, Wan et al. reported the iridium-catalyzed synthesis of quinoxalines using benzene-1,2-diamine and vicinal diols like 1,2-propanediol and 2,3-butanediol [47]. This transformation relies on a dehydrogenation-condensation strategy where first one alcohol group is dehydrogenated by complex **[Ir-25]** whereupon the resulting ketone reacts with one amine of the benzene-1,2-diamine to form an imine. Subsequently, the remaining alcohol is dehydrogenated and reacts in a similar manner with the remaining amine which results in a ring-closure and in the formation of the quinoxaline (Scheme 12).

Both 1,2-propanediol and 2,3-butanediol were successfully used with different substituted aromatic diamines to form the corresponding quinoxalines in generally good yields (56–73%). However, the reported synthetic protocol suffers from the use of a relatively high catalyst loading of 5 mol%, from the use of a large excess of base (3 equiv. Cs_2CO_3) and from the use of xylene as a non-environmental benign

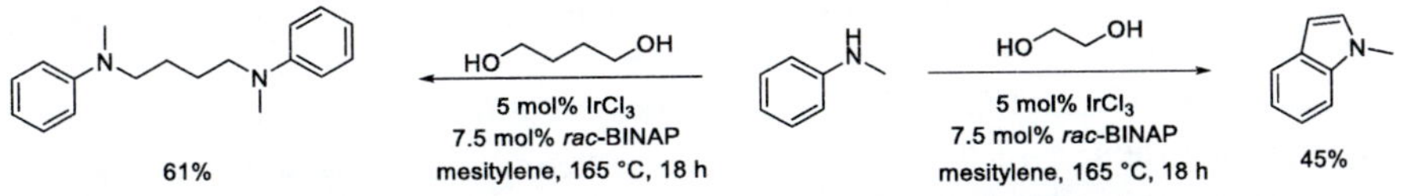

Scheme 11 Direct cyclization of different bio-based diols with *N*-methylaniline

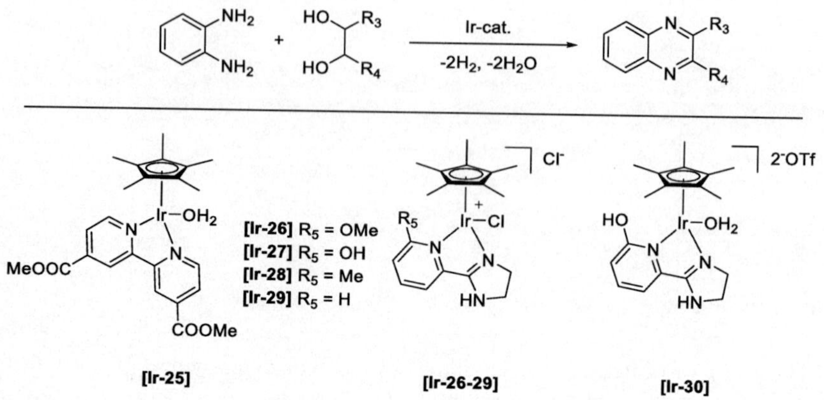

Scheme 12 Iridium-catalyzed synthesis of quinoxalines by using vicinal bio-derived alcohols

solvent. The latter problem was addressed by Kundu and coworkers who studied the water-soluble iridium catalysts **[Ir-26-30]** for the synthesis of quinoxalines under an atmosphere of air (see Scheme 12) [44]. Complex **[Ir-27]** (2.5 mol%) was found to be the best catalyst in the presence of 1.5 equiv. of KOH at 120°C for 24 h. This synthetic methodology uses not just water as a green solvent instead of xylene, but also it requires a lower amount of base and a shorter reaction time under non-protective atmosphere in comparison with the methodology reported by Wan et al. Furthermore, both 1,2-propanediol and 2,3-butanediol were reacted with a wide range of vicinal aromatic diamines as well as with 2-nitro aniline derivatives. Good yields were obtained with a broad range of substrates (62–90%) [44].

2.6 Poly(Silyl Ethers) Via Iridium-Catalyzed Dehydrocoupling Polymerization

In the 2004 DOE report "Top Value-Added Chemicals from Biomass" it was proposed that polymeric materials from renewables should be one of the most desired targets [1]. Bio-based polymers, in fact, are derived from bio-based feedstock materials, and therefore, they allow a reduction of the total greenhouse gas emissions and of the overall carbon footprint of the final products. Bio-based carbon content has thus become an important added value for all companies, who want to improve the sustainability across their entire value chains. Polysilylethers (PSEs) show interesting properties like a low glass transition temperature (T_g), good thermal stability, biocompatibility, and high gas permeability but unfortunately, they are not derived from bio-based materials [48]. However, the most frequently used monomers for the synthesis of PSEs are dihydroxy compounds. The latter in turn can be diols which are indeed accessible from renewable feedstock. Driven by this

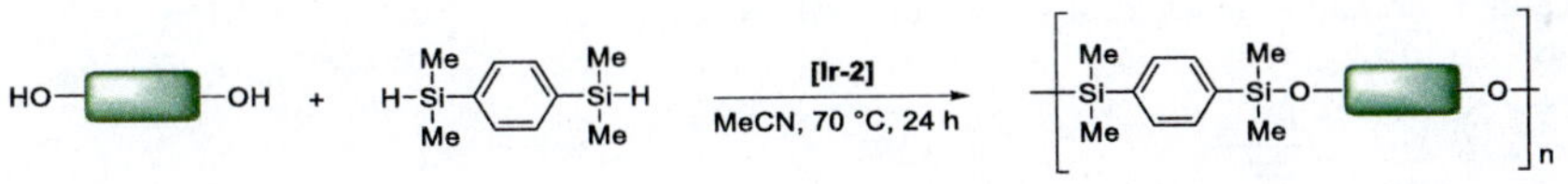

Scheme 13 Catalytic dehydrocoupling polymerization for the synthesis of partially bio-based PSEs

knowledge, Zhou and coworkers reported the iridium-catalyzed synthesis of partially bio-based PSEs by using bio-derived diols and a commercially available silane (see Scheme 13) [48].

In this protocol the bio-based diols 1,10-decanediol, 1,4-pentanediol, and 1,4-butanediol as BB monomers were reacted with 1,4-bis(dimethyl-silyl)benzene as AA type monomer. The dehydrocoupling polymerization is initiated by an Ir-H species which is presumably formed via a hydrolysis reaction of the hydrosilane with the starting iridium complex **[Ir-2]**. Subsequently, this newly formed Ir-H species can undergo a metathesis reaction with the bio-based diol that results in the formation of an Ir-alkoxy species which in turn reacts further with the silane to generate the dehydrocoupling product.

For all three diols, high molecular weight polymers were obtained (M_n = 24,100–43,800) in moderate to excellent yields (54–92%). The synthesized PSEs exhibited good thermal stability (T_5 = 418°C to 437°C) and low glass transition temperatures down to −49.6°C [48].

3 Iridium-Catalyzed Transformations of Bio-Based Furan Compounds

3.1 *Iridium-Catalyzed Synthesis of 1-Hydroxyhexane-2,5-Dione*

5-Hydroxymethylfurfural (5-HMF, see Table 1) is a furan compound with both an aldehyde and a hydroxymethyl group and it can be obtained from bio-based feedstock materials like cellulose, glucose, and fructose [54–57]. Although cellulose, starch, or preferably lignocellulose as readily available raw materials are considered more attractive than sugars, high yields of 5-HMF are currently only achieved by acidic treatment of fructose. Unfortunately, this in turn leads to relatively high production costs for 5-HMF. Additionally, the instability of 5-HMF limits its isolation in high yields and hence its current applications. To overcome this problem, the Dutch company Avantium produces the methyl ether of 5-HMF, more stable than 5-HMF itself [58]. Alternatively, Haworth and Jones reported a simple hydrolysis approach (90% yield) from 5-chloromethylfurfural [59], which can be also derived from cheap bulk cellulosic material as shown by Mascal and coworkers [60]. Currently, HMF is produced on relatively small scale by the Swiss company AVA-Biochem. Despite the problems which still need to be solved, the interest in

Table 1 Iridium-catalyzed conversion of HMF to HHD

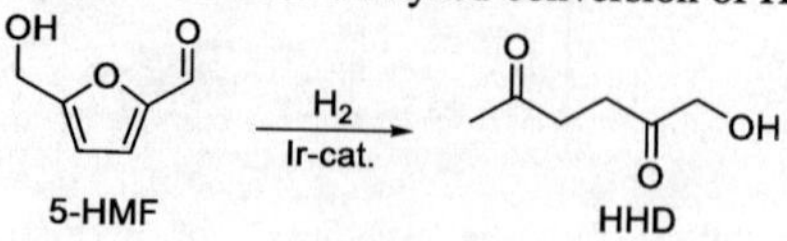

Entry	Catalyst (loading)	Conditions	Conv. [%]	Yield [%]	Ref.
1	**[Ir-31]** (0.26 mol %)	120°C, H_2O, 7 bar H_2, 2 h	100	86	[49]
2	**[Ir-32]** (0.01 mol %)	130°C, FBS (pH = 2.5), 2 h	100	92(85)[a]	[50]
3	**[Ir-32]** (0.10 mol %)	130°C, H_2O, 5 wt% $Al_2(SO_4)_3$, 30 bar H_2, 4 h	100	77	[51]
4	**[Ir-33]** (0.01 mol %)	130°C, FBS (pH = 2.5), 2 h	100	~ 95	[50]
5	**[Ir-34]** (0.0008 mol%)	120°C, H_2O (pH = 3.4), 35 bar H_2, 6 h	88	67	[52]
6	**[Ir-35]** (0.50 mol %)	120°C, H_2O, 10 bar H_2, 2 h	100	76(69)[a]	[53]
7	**[Ir-36]** (0.075 mol %)	140°C, PBS (pH = 2.5), 60 bar H_2, 1 h	100	71[a]	[53]

[a]Isolated yield

the commercialization of 5-HMF is immense, based on the enormous number of publications dealing with its synthesis and derived products [54]. One of the 5-HMF products which has recently received high attention in the field of homogenous iridium catalysis is 1-hydroxyhexane-2,5-dione (HHD, see Table 1), the hydrogenated ring-opened form of 5-HMF [61].

Only a few homogenous iridium catalysts are known for the synthesis of HHD, even though homogenous catalysts seem generally to perform better than heterogenous systems [61] (Fig. 8). The group of Zhang and coworkers was the first to report an iridium-catalyzed hydrogenation of 5-HMF in aqueous solution in order to synthesize HHD [49]. Bipyridine Cp*-Ir complexes were found to be efficient catalysts for the selective formation of HHD and **[Ir-31]** emerged as the best candidate under their optimized conditions (Table 1, Entry 1). Subsequently, Deng, Fu, and coworkers carried out an in-depth investigation on the different parameters of the hydrolytic transfer hydrogenation of HMF, by exploring the effect of electron-donating and electron-withdrawing groups on the bipyridyl ligand, the influence of the temperature and of the pH [50]. The obtained results revealed that temperature and pH are important parameters to maximize catalytic performance. Particularly, full conversions and high selectivities up to 95% were obtained at 130°C in aqueous formate buffer solutions (FBS, pH = 2.5) with catalysts **[Ir-32]** and **[Ir-33]** (Table 1, Entries 2 and 4). Moreover, the authors were the first to isolate HHD in 85% yield on a large scale [50].

Unfortunately, attempts to reuse the catalyst resulted in a decrease of catalytic activity and only 70% yield was achieved. The same catalyst **[Ir-32]** was used by

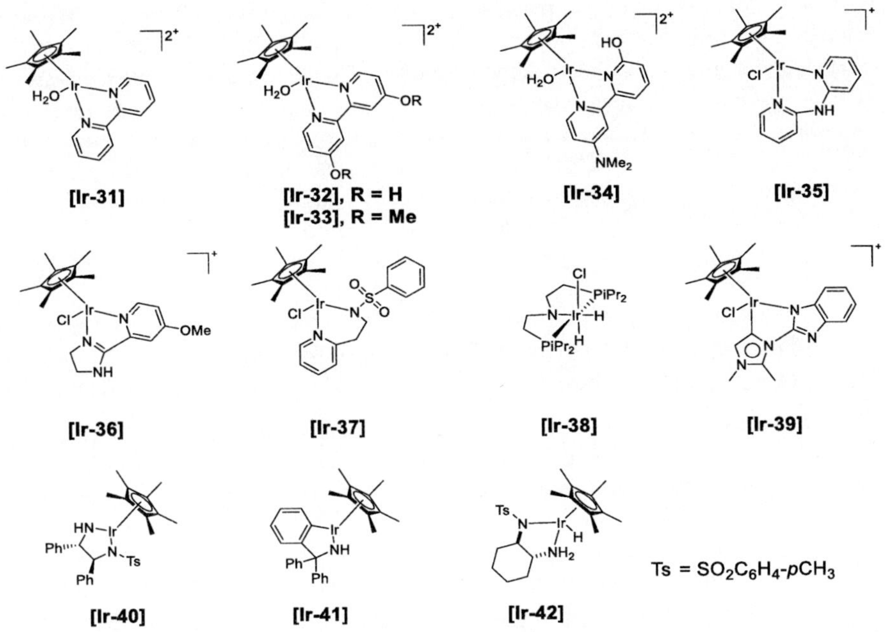

Fig. 8 Iridium complexes for the hydrogenation of furan compounds

them to synthesize HHD via hydrolytic hydrogenation (30 bar) with a selectivity of 77% in the presence of 5 wt% of $Al_2(SO_4)_3$ (Table 1, Entry 3) [51]. Zhang and coworkers tested a new series of Cp*-iridium(III) complexes bearing different substituted bipyridine ligands [52]. The best catalyst, **[Ir-34]**, has two substituents on the two rings of the bipyridine ligand: on one side it contains an *N*-dimethylamino group as wing-spike, and on the other side it contains a hydroxyl group in ortho-position with respect to the nitrogen. At a comparatively low catalyst loading (0.0008 mol% of **[Ir-34]**) at 120°C in acidic water (pH = 3.4) for 6 h under 35 bar of H_2, a good conversion (88%) and a reasonably good selectivity towards HHD (67%) were obtained (Table 1, Entry 5). Additionally, the authors compared the performance of the catalyst under transfer hydrogenation conditions with respect to the hydrogenation conditions, by using formic acid as the hydrogen source. In the latter case, the use of 0.005 mol% of **[Ir-34]** with 2 equiv. of formic acid at 120°C led to a yield of 60% in HHD after 2 h of reaction [52]. Two years later de Vries and coworkers investigated new structural motifs in the ligand sphere, finding complexes **[Ir-35]** and **[Ir-36]** to be the best catalysts for the hydrolytic ring-opening of 5-HMF resulting in a selectivity up to 76% in HHD (Table 1, Entries 6–7) [53]. Furthermore, the same authors were able to isolate 69% of HHD and to fully characterize it including by single crystal X-ray analysis. They could further improve the efficiency of **[Ir-36]** by adjusting the pH with a phosphate buffer solution (PBS, pH = 2.5). Under these conditions the catalyst loading was reduced to 0.075 mol% while maintaining the same selectivity towards the formation of HHD. The latter result was also used to scale up the reaction in an experiment where 25 g of 5-HMF was used and where HHD was obtained in a good isolated yield of 71% after 1 h [53].

In conclusion, it has been demonstrated that Cp* iridium complexes bearing additional bidentate ligands show a high activity and a good selectivity in the conversion of 5-HMF into HHD (see Table 1). However, the scalability of the HHD production from 5-HMF through homogenous iridium-catalyzed hydrogenations is still strongly limited by the formation of humins which may be reduced by using high dilutions, and by the poor reusability of the catalysts [61].

3.2 *Conversion of Furfural Compounds to Ketones*

Due to the promising results obtained in the synthesis of HHD from 5-HMF, an increasing interest in the further applications of HHD has emerged. Interestingly, several iridium-catalyzed reactions using 5-HMF as the starting material were reported where HHD was formed in situ and then it further cyclized to two different kinds of cyclopentanones (see Scheme 14) [61].

The formation of cyclopentanones is the result of an intramolecular aldol condensation of HHD which, depending on the acidity of the solution, leads to the formation of either 3-hydroxymethyl cyclopentanone (**7**) or 2-hydroxy-3-methylcyclopent-2-enone (**8**), respectively [55]. For the synthesis of **7** only heterogenous catalysts were known till 2017, when the group of Fu investigated for the first time half-sandwich iridium complexes coupled with metal oxides for this reaction [51]. The authors investigated the influence of different parameters like the kind of iridium complex and the kind of Brønsted or Lewis acid used in the reaction. As a conclusion of their study, the best results (100% conversion; 82% of **7**)

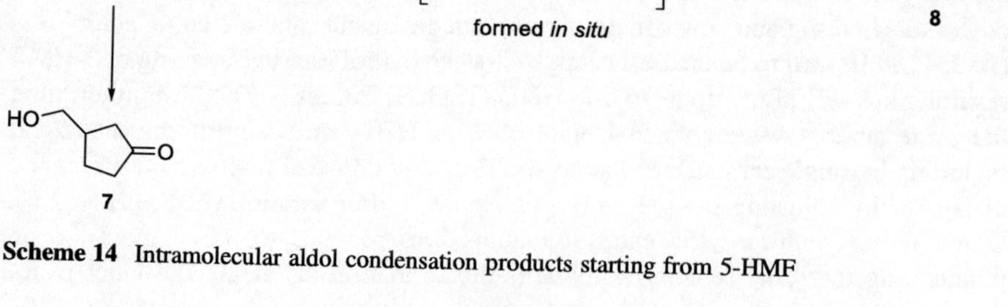

Scheme 14 Intramolecular aldol condensation products starting from 5-HMF

Table 2 Effect of Lewis acid or Brønsted acid on the iridium-catalyzed conversion of furfural compounds

Entry	Substrate	Acid	Conv. [%]	Yield [%]	Products
1	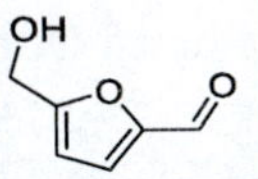	40 wt% γ-Al_2O_3	100 [a]	2/82 [a]	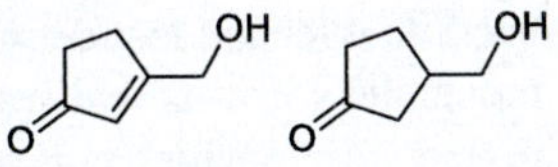
2	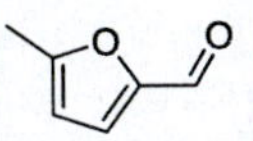	45 wt% γ-Al_2O_3	100 [a]	30/24 [a]	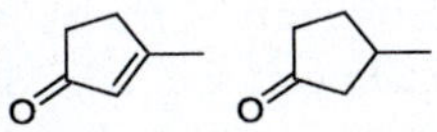
3	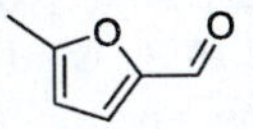	PBS (pH = 1.5)	100 [a]	39 [a]	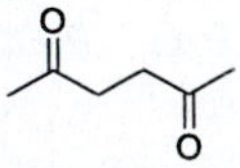
4	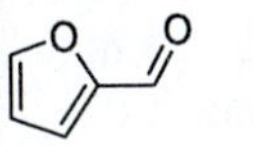	52 wt% γ-Al_2O_3	100 [a]	60/~2 [a]	
5	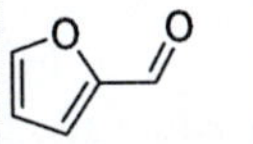	PBS (pH = 1.0)	100 [b]	41/9 [b]	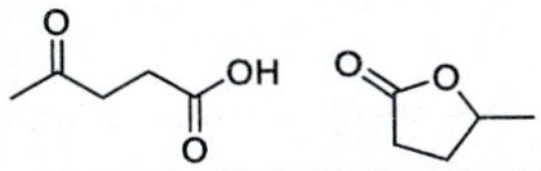

[a]5-HMF, 5-methyl furfural or furfural aq. (0.067 M, 3 mL), **[Ir-32]** (0.1 mol%), H_2 30 bar, 130°C, 4 h
[b]Furfural (0.60 mmol), 2 mL pH = 1.0 phosphate buffer solution (0.1 M), **[Ir-32]** (0.0083 mol%), 10 bar H_2, 120°C, 4 h

were obtained when 0.1 mol% of **[Ir-32]** was used in combination with 40 wt% of γ-Al_2O_3 as a Lewis acid cocatalyst at 130°C under 30 bar of H_2 for 4 h.

In addition to 5-HMF as a starting material (Table 2, Entry 1), other bio-derived furan compounds like 5-methyl furfural (Table 2, Entries 2–3) and furfural (Table 2, Entries 4–5) could be hydrogenated in the same manner to 3-methyl cyclopentenone and cyclopentenone in 30% and 60% yield, respectively (Table 2). Under Brønsted acid catalysis using a phosphate buffer solution (PBS) 5-methyl furfural (Entry 3) and furfural (Entry 5) were converted to the straight chain ketones 2,5-hexanedione and levulinic acid in moderate yields.

When the intramolecular aldol condensation of HHD is carried out under basic conditions, **8** is formed. The latter is a flavoring agent which is commonly synthesized from adipic acid in a multistep reaction that uses toxic and carcinogenic reactants [61]. For this reason, a catalytic protocol for the synthesis of **8** from benign bio-based materials would be highly desired. The base-promoted formation of **8** by using HHD as a starting material was firstly discovered by Deng, Fu, and coworkers [51]. Later, de Vries and coworkers carried out an in-depth study on the role of the base and of the solvent [53].

As a conclusion, an isolated yield of 72% of **8** in an aqueous KOH solution at 60°C after 15 min was achieved. Furthermore, the authors decided to combine their previous procedure for the synthesis of HHD mediated by **[Ir-35]** with the KOH-mediated aldol condensation reaction in order to synthesize **8** in a one-pot reaction. Based on this methodology, an overall yield of 55% of **8** was achieved.

3.3 Iridium-Catalyzed (Transfer)Hydrogenation of Furfurals

Furfural (FAL) was isolated for the first time by Johann Wolfgang Döbereiner in 1831 and, due to the variety of possible transformations which it can undergo, has become over the last century one of the most important biomass-based C_5-platform molecules [55]. Its wide range of applications can be found in the recently published review article which summarizes the synthesis of more than 80 chemicals directly or indirectly derived from FAL [62]. The production of FAL is currently carried out in lignocellulosic biorefineries by using agricultural and forestry waste as starting materials. Furfuryl alcohol (FOL) is by far the largest FAL derivative, with 65% of the overall FAL production volume used for its synthesis. FOL is in fact an important intermediate primarily used in the production of resins, pharmaceuticals, and in the manufacture of fragrances. Currently, the gas-phase Cu-catalyzed hydrogenation of FAL is the major route towards FOL, because the reaction in liquid phase is economically unattractive due to the high operating costs of batch reactors [62].

Efficient iridium catalysts for the (transfer)hydrogenation of FAL to FOL were reported by Deng, Fu, and coworkers. Their initial synthetic procedure required the use of semi-sandwich iridium catalysts **[Ir-32]** or **[Ir-33]** in 0.0083 mol% under 10 bar H_2 at 120°C for one hour [63]. Subsequently, the effect of pH (adjusted by a phosphate buffer solution (PBS) and H_3PO_4) on catalytic activity for catalysts **[Ir-33]** and **[Ir-32]** was examined. As a result, the highest TOF (1,600 h^{-1}) for **[Ir-33]** was found at pH = 3.5, while for **[Ir-32]** the highest TOF (7,340 h^{-1}) was achieved at pH = 5. The authors explained the higher activity of catalyst **[Ir-32]** with respect to **[Ir-33]** with its ability to generate an electro-donating aryloxy anion which is crucial for the performance of this type of catalyst. Transfer hydrogenation using formic acid was also examined with these catalysts (Table 3, Entries 1–3). With

Table 3 Recently reported iridium catalysts for the (transfer)hydrogenation of FAL to FOL

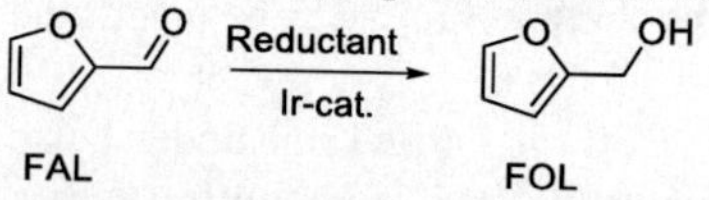

Entry	Catalyst (loading)	Conditions	Yield [%]	TON	TOF [h^{-1}]	Ref
1	**[Ir-32]** (0.0083 mol%)	H_2O, 120°C, 1 h, HCOOH	>99	12,067	12,067	[63]
2	**[Ir-32]** (0.0017 mol %)	H_2O, 120°C, 1 h, HCOOH	23	13,877	13,877	[63]
3	**[Ir-33]** (0.0083 mol %)	H_2O, 120°C, 1 h, HCOOH	30	3,668	3,668	[63]
4	**[Ir-37]** (1 mol%)	*i*PrOH, 85°C, 0.5 h	95	95	190	[64]
5	**[Ir-39]** (0.5 mol%)	EtOH/H_2O, 30°C, 2 h, 1 bar H_2	94	470	235	[65]

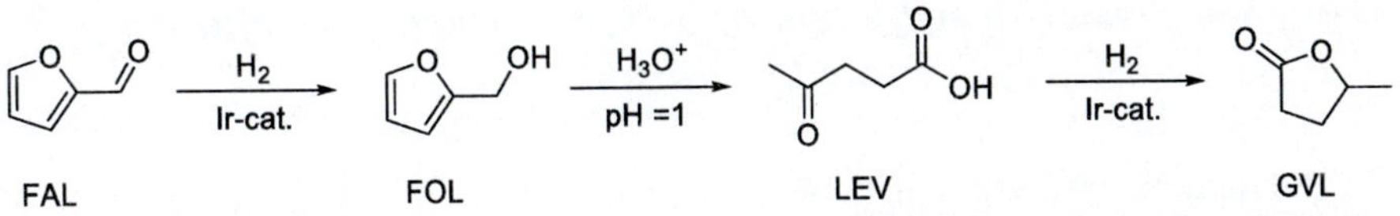

Scheme 15 One-pot conversion of FAL to GVL by employing the iridium complexes **[Ir-32]** and **[Ir-33]**

catalyst **[Ir-33]** 30.4% yield was achieved at pH 4 and with catalyst **[Ir-32]** a yield >99% was observed at pH 4.5. Finally, Fu and coworkers also reported the direct conversion of furfural into levulinic acid (LEV) and gamma-valerolactone (GVL) in a one-pot reaction (Scheme 15). Specifically, when furfural was reduced under 10 bar H_2 for 4 h in the presence of 0.0083 mol% of **[Ir-33]** at 120°C in PBS (pH =1) LEV was produced in 24% yield and GVL in 35% yield. In comparison, by using **[Ir-32]** under the same conditions levulinic acid was obtained in 41% yield and GVL in 9% yield [63].

Use of complex **[Ir-37]** was reported by O'Connor and coworkers, who converted FAL into FOL with an excellent 95% yield, by using 1 mol% **[Ir-37]** in combination with isopropanol as the hydrogen source at 85°C for 30 min [64]. Unfortunately, the authors did not investigate lower catalyst loadings, hence a relatively low TON (95) was reported; also, the TOF was only 190 h^{-1} (Table 3, Entry 4). Choudhury and coworkers reported the catalyst **[Ir-39]** which they used to hydrogenate FAL to FOL (yield: 94%) under atmospheric hydrogen pressure and under base-free and very mild conditions (Table 3, Entry 5) [65]. The group of Rauchfuss went a step back, to make furfuryl alcohol directly from xylose [66]. Thus in the first step, xylose underwent dehydration catalyzed by 0.1 equiv. of formic acid in DMSO/THF (1/10 v/v), followed by neutralization with 0.1 equiv. of NEt_3 and subsequent addition of 0.5 mol% of **[Ir-40]**. Thereafter, the transfer hydrogenation reaction took place in additional THF with the freshly made furfural and in the presence of replenished formic acid to obtain FOL in 63% isolated yield.

Nielsen and coworkers reported a highly active iridium complex for the hydrogenation of bio-derived 5-methyl furfural (MF) to 5-methyl furfuryl alcohol, a potential biodiesel precursor [67]. In the presence of 0.05 mol% **[Ir-38]** and 2 mol % NaOEt as base in an ethanol/water mixture under 30 bar H_2 pressure at 60°C excellent yields up to 99% were achieved after 150 min. Decreasing the catalyst loading to 0.005 mol% and 0.0005 mol% resulted in 91% yield (5 h) and 56% yield (48 h), respectively, at 120°C under 30 bar H_2 pressure. Finally, the authors performed a scale-up reaction with 7.9 mmol of MF with 0.01 mol% **[Ir-38]** under 30 bar H_2 at 120°C for 2 h which gave an isolated yield of 97%.

3.4 *Selective Hydrogenation of 5-Hydroxymethyl Furfural to 2,5-Bis(Hydroxymethyl)Furan*

2,5-Bis(hydroxymethyl)furan (BHMF) is a key molecule for the preparation of polymers, resins, crown ethers, and rayons and for this reason it is a potentially promising downstream product of biorefineries [68]. Although both homogenous and heterogenous catalysts are known for the selective hydrogenation of 5-HMF to BHMF, the literature about homogenously catalyzed hydrogenation to form BHMF is scarce [68]. The first example of homogenous iridium catalysts was reported by Rauchfuss and coworkers [66]. The authors tested a range of Cp* iridium complexes (**[Ir-40-42]**) (0.1–0.5 mol%) all of which hydrogenated 5-HMF quantitatively within a few hours at 40°C in THF (see Table 4, Entries 3–6). The same iridium catalysts were also used for the direct conversion of fructose to BHMF. By performing the reaction at 40°C in the presence of 0.5 mol% of **[Ir-40]** with 2 mmol of D-fructose, 2 mmol of formic acid and 0.2 mmol of base (NaOH or NEt_3) in THF, an excellent BHMF yield of 99% was obtained. The effect of the added base was immense: the reaction was completed within 2 h in contrast to 12 h without base (80% yield of BHMF). In order to find a more active system, Deng, Fu, and coworkers carried out an in-depth study of Cp* iridium complexes bearing a bipyridine moiety. The best results were obtained with **[Ir-32]**, a Cp^*Ir^{III} half-sandwich complex able to withstand the acidic reaction conditions (pH 4.0–6.5) (Table 4, Entry 1) [50]. By using an aqueous formate buffer solution (FBS) at 130°C for 2 h and 0.01 mol% of **[Ir-32]**, the formation of BHMF was observed at pH >3. At pH 5.2 BHMF was obtained almost quantitatively. The use of formic acid as the hydrogen source resulted in

Table 4 Iridium catalysts known for the selective transfer hydrogenation of HMF to BHMF

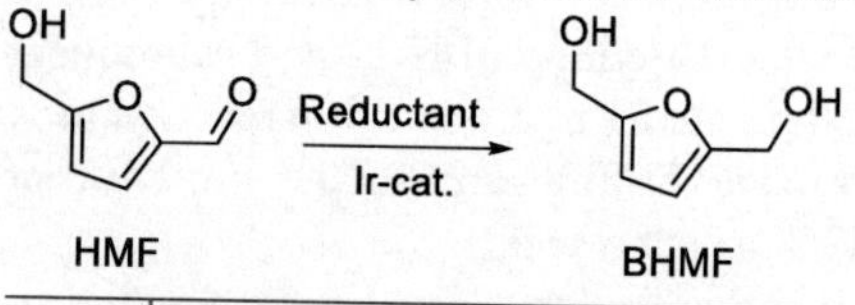

Entry	Catalyst (loading)	Conditions	Yield [%]	TON	TOF $[h^{-1}]$	Ref.
1	**[Ir-32]** (0.01 mol %)	H_2O, 130°C, 2 h, HCOOH, pH 5–5.5	~98	9,800	4,900	[50]
2	**[Ir-37]** (1.0 mol %)	*i*PrOH, 85°C, 0.5 h	99	100	200	[64]
3	**[Ir-40]** (0.5 mol %)	THF, 40°C, 2 h, HCOOH	99	500	250	[66]
4	**[Ir-40]** (0.1 mol %)	THF, 40°C, 4 h, HCOOH	70	700	175	[66]
5	**[Ir-41]** (0.5 M mol%)	THF, 40°C, 1 h, HCOOH	99	500	500	[66]
6	**[Ir-42]** (0.5 mol %)	THF, 40°C, 1 h, HCOOH	99	500	500	[66]

generally higher yields with respect to the use of H_2 (10 bar) in PBS. The latter observation was explained with the high extent of degradation of BHMF in PBS and with the poor solubility of H_2 in water. Additionally, the authors underlined the importance of balancing the acidity of the reaction system. In fact, the presence of FBS leads to a steadily growth in pH during the reaction and at a later stage of the reaction prevents BHMF from degradation. Alternatively, to formic acid, the group of O'Connor and coworkers used isopropanol as a transfer hydrogenation reagent [64]. Under the same conditions used for the transfer hydrogenation of furfural (see Table 3, Entry 4), BHMF was quantitatively formed from 5-HMF after 30 min at 85°C with 1 mol% of **[Ir-37]** (Table 4, Entry 2). The transfer hydrogenation with **[Ir-37]** follows an outer sphere mechanism. In contrast, according to Fu and coworkers, catalyst **[Ir-32]** forms an active hydride species which operates via an inner sphere mechanism.

4 Iridium-Catalyzed Conversions of Levulinic Acid

4.1 Hydrogenation to GVL

Levulinic acid (LEV) is a platform chemical which can be obtained in high yields from liognocellulose via a simple acid-catalyzed conversion [69]. Hydrogenation of LEV results in gamma-valerolactone (GVL), another very useful platform chemical, which has been investigated for use as a solvent, a fuel additive, as well as a precursor for bulk and fine chemicals (Scheme 16) [70].

To improve the efficiency of this process, a hydrogenation catalyst that can withstand acidic aqueous conditions from the biomass hydrolysis step would be a great advantage. For this reason, Fu and coworkers screened some half-sandwich Iridium catalysts, which were known for the hydrogenation of CO_2 in water [71]. Among these, catalyst **[Ir-33]** (Fig. 9), with electron-donating methoxy groups

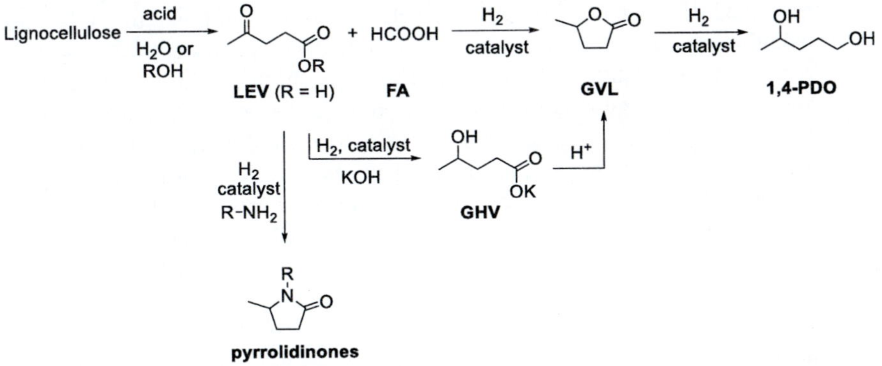

Scheme 16 The levulinic acid platform

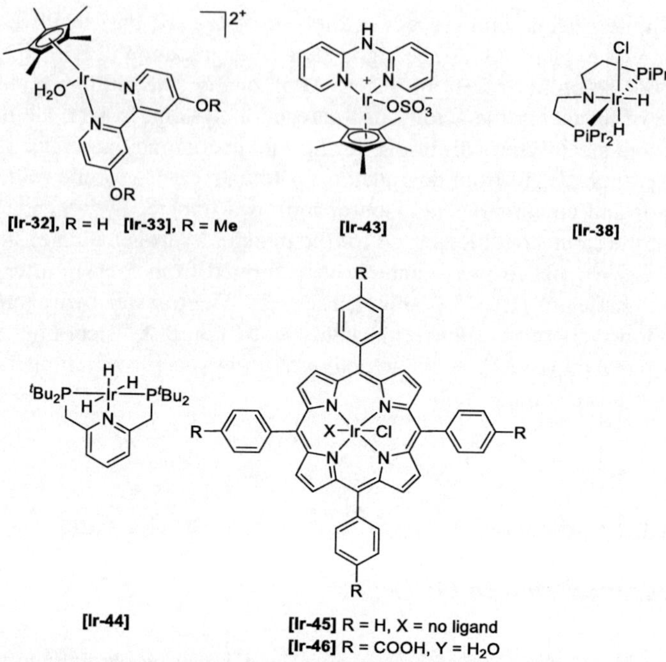

Fig. 9 Iridium catalysts for the hydrogenation of levulinic acid

on the bipyridine ligand was the most efficient and achieved 98% conversion of LEV to GVL in 4 h at 120°C and 10 bar hydrogen pressure (Table 5, Entry 1). The TON of this catalyst was determined to be 78,000, the highest reported figure for this reaction up to this point. The researchers also tested its activity in the transfer hydrogenation using formic acid (FA), as this compound is a commonly used hydrogen donor in these procedures but is also produced in roughly equal amounts to LEV in the synthesis from biomass. With two equivalents of FA they achieved a yield of 99% GVL under the same reaction conditions (Table 5, Entry 2). Recycling experiments were performed by extracting the GVL from the aqueous reaction mixture followed by the addition of more substrate. After five runs no decrease in yield was detected. After confirming the stability of the catalyst towards sulfuric acid, the researchers also applied their hydrogenation methodology to a mixture of LEV and FA generated by the acid-catalyzed hydrolysis of glucose. The yield of GVL from glucose was 45%. Using cellulose as the starting material, the GVL yield was 34%.

The same group developed a mixed hydrogenation/TH procedure by performing the reaction under 10 bar of hydrogen in a formate buffer (Table 5, Entry 3) [63]. The pH optimum was 4.5 and the contribution of formic acid to the hydrogenation reaction was indicated by comparison to other buffer solutions at the same pH. **[Ir-32]**, which contains 4,4′-hydroxy substituents on the bipyridine ligand, performed better than **[Ir-33]**. The TOF reached 12,200 h^{-1}. The authors attributed this to the higher electron-donating ability of the deprotonated hydroxyl groups, present at this pH.

Table 5 Reaction conditions for Ir-catalyzed hydrogenation of LEV to GVL[a]

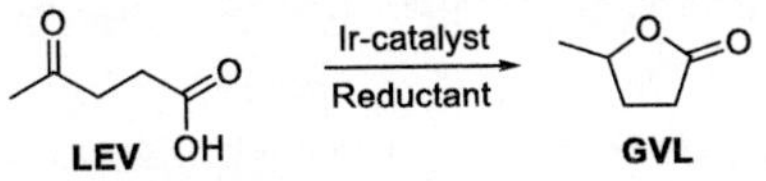

Entry	Catalyst (Loading)	Temp	H_2 (bar)	Additive	Yield (%)	TON (TOF)[b]	Ref.
1	**[Ir-33]** (0.01 mol %)	120°C	10	–	98	78,000 (2,166)	[71]
2	**[Ir-33]** (0.01 mol %)	120°C	–	2 equiv HCOOH	99	9,900 (2,478)	[71]
3	**[Ir-32]** (0.0083 mol%)	120°C	10	1M FBS	61	12,200 (12,200)	[63]
4	**[Ir-43]** (0.05 mol %)	130°C	5	–	99	174,000 (2,418)	[72]
5	**[Ir-43]** (0.10 mol %)	110°C	–	2 equiv HCOOH	99	9,000 (126)	[72]
6[c,d]	**[Ir-38]** (0.05 mol %)	60°C	20	NaOEt (5 mol%)	99	9,300 (390)	[73]
7[d]	**[Ir-44]** (0.05 mol %)	100°C	50	LiOH (1.2 equiv)	98	71,000 (1,482)	[74]
8[e]	**[Ir-45]** (0.20 mol %)	100°C	10	–	100	1,574 (132)	[75]

[a]Solvent is water except when noted otherwise
[b]TOFs were in most cases calculated by the authors of this chapter from the reported TON values and are given as TOF = h^{-1}. They only serve for a rough comparison as they were not calculated at the same degree of conversion
[c]Substrate is ethyl levulinate
[d]Solvent is EtOH
[e]Solvent is DMF

Fischmeister and coworkers developed the Cp*Ir(SO_4) complex **[Ir-43]** containing a bis 2-pyridylamine as ligand which they tested for the reduction of LEV under both hydrogenation and transfer hydrogenation (TH) conditions (Table 5, Entries 4–5) [72]. The researchers found that whereas FA as the hydrogen donor gave full conversion of LEV to GVL, there was no conversion with isopropanol. Control experiments confirmed that **[Ir-43]** is a very efficient FA dehydrogenation catalyst, causing full dehydrogenation within 1 h under the reaction conditions used for TH. This prompted the authors to investigate the mechanism and it was found that in fact FA is dehydrogenated to form H_2, which in turn hydrogenates the LEV. Experiments in open versus closed systems showed that there is a combined contribution of TH and dehydrogenation of FA followed by hydrogenation of LEV to the formation of GVL under these conditions. Under hydrogenation conditions full conversion to GVL could be achieved with only 5 bar hydrogen pressure and a maximum TON of 174,000 albeit at longer reaction times. The catalyst loading could be decreased to 0.001 mol% whilst still achieving 97% yield (after 32 h, at 130°C). Recycling experiments demonstrated that the catalyst could be reused three times without loss of activity. The authors also investigated the reaction starting from glucose via acid hydrolysis. The GVL yield over two steps was 57% using hydrogen gas and 44% under TH conditions using only the FA generated in the reaction as the hydrogen donor.

The Nielsen group investigated the hydrogenation of alkyl levulinates to GVL [73]. These ethyl or methyl esters of LEV are obtained when alcohols are used as the reaction medium for the acid-catalyzed conversion of carbohydrates instead of water and are also produced as a side stream in some processes [76]. Among the range of PNP and SNS pincer catalysts tested, **[Ir-38]** based on the MACHO ligand, earlier developed by Takasago, was the best, able to convert ethyl levulinate to GVL under very mild conditions (20 bar hydrogen and 60°C, neat with only small amounts of alcohol added) (Table 5, Entry 6). In contrast to the previous catalysts, this pincer catalyst requires 5 mol% of base (NaOEt) for activation. The highest TON reached was 9,300. Methyl levulinate reacted faster than the ethyl ester under the same conditions. In a small scale-up the authors tested the catalyst under very mild conditions – 25 bar hydrogen and 25°C. Here they could achieve full conversion with 0.05 mol% of catalyst after 72 h.

Zhou and coworkers screened a range of PNP, PNN, and NNN pincer ligands with iridium in the presence of base. The best catalyst was PNP complex **[Ir-44]**, with the nitrogen analogue also giving high yields [74]. In the optimization of the reaction conditions it was found that base was necessary for the reaction to occur; when it was lowered from 1.2 equiv. to 0.5 equiv., the yield decreased to below 10%. With optimized reaction conditions LEV could be converted to GVL in 98% yield (Table 5, Entry 7). A TON of 71,000 was measured, albeit at quite high hydrogen pressure of 100 bar.

Sakthivel and coworkers synthesized several iridium porphyrin complexes and tested them in the hydrogenation of LEV to GVL and 1,4-pentanediol under base-free conditions [75]. The catalyst with the best performance and selectivity for LEV was **[Ir-45]** (Table 5, Entry 8). The carboxylated version of the catalyst, **[Ir-46]**, was not as efficient as **[Ir-45]**. However, when the authors performed reuse experiments by adding more batches of LEV into the reaction mixture, **[Ir-46]** produced mostly 1,4-pentanediol (1,4-PDO), the further hydrogenation product of GVL, and the yields of 1,4-PDO increased with each cycle, reaching a selectivity of 81% after five runs. For **[Ir-45]** the yield of 1,4-PDO also increased with each run so that the selectivity for LEV was only 66% for run four. The authors attributed this to the buildup of water in the system, but did not perform any control experiments to confirm this or to exclude the formation of nanoparticles.

The group of Voutchkova-Kostal employed TH for the hydrogenation of LEV to the potassium salt of gamma-hydroxyvaleric acid (GHV) (Scheme 16) [77]. Under most reaction conditions, even in the presence of base, this compound immediately cyclizes to form the lactone GVL. Based on their efficient NHC iridium complex for the acceptorless dehydrogenation of glycerol (vide supra), glycerol was investigated as the hydrogen donor and a range of NHC-based complexes with pendant sulfonate groups for better water solubility were tested. In the screening, the same complex used in the previous publication, carbonyl complex **[Ir-10]** with two sulfonated NHC ligands, was found to be the most active. Two equivalents of KOH per mol substrate was found to be the optimum amount of additive, while the reaction was carried out in excess glycerol mixed with an equal volume of water. The base is needed to convert the dehydrogenated glycerol to LA, and to stabilize the formed (GHV) as its salt. Coupling these two hydrogenation/dehydrogenation reactions achieves the generation of two valuable products in one pot. The base concentration was adjusted to achieve full conversion of LEV to GHV. Under these conditions, the yield of LA was 2–3 equivalents due to additional acceptorless dehydrogenation of glycerol. With the optimized conditions, quantitative conversion of LEV to GHV as its potassium salt could be achieved at a very low catalyst loading of 1–10 ppm in 2 h at 150°C under microwave heating. The maximum TOF achieved with 1 ppm catalyst was 50,550 h^{-1}. As in their previous report [31], these really high turnovers appear to be due to the better heat and mass transfer due to microwave heating. When the experiment was repeated under conventional heating, 24 h was required to reach full conversion. Other hydrogen donors were screened but were not as efficient. Although slower, using (bio-)ethanol, full conversion of LEV could be achieved within 3 h. While this is an interesting report, it is not explained why in this case no GVL is formed, even though it apparently happens in basic conditions in other reports. Attempts were made to separate the two carboxylate products, which were not successful due to the similar properties of the two compounds. However, acidification of the reaction medium led to the cyclization of GHV to GVL which could then be separated from the other product.

4.2 *Reductive Amination to Pyrrolidinones*

The reductive amination of LEV to *N*-substituted 5-methyl pyrrolidinones was also investigated by several groups. These compounds are building blocks for inks, pharmaceuticals, fibers, or surfactants and can be used as solvents, similar to the aprotic polar solvent NMP. The catalysts employed for this transformation are summarized in Fig. 10 and the reaction conditions in Table 6. Building on their previous work for the base-free hydrogenation of LEV to GVL in water (*vide supra*),

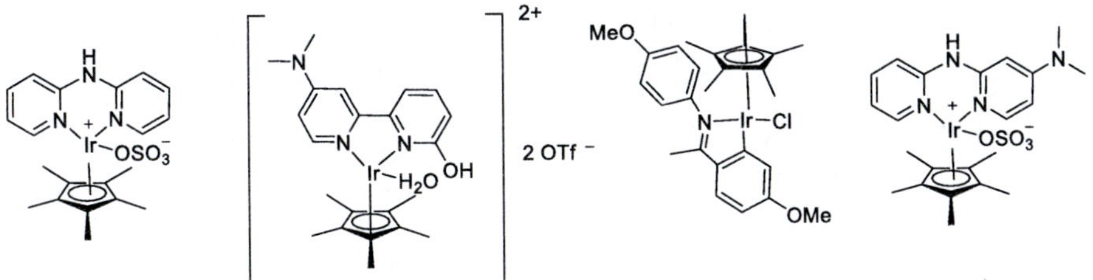

Fig. 10 Catalysts employed for the reductive amination of LA to pyrrolidinones

Table 6 Reaction conditions for the reductive amination of LEV to pyrrolidinones

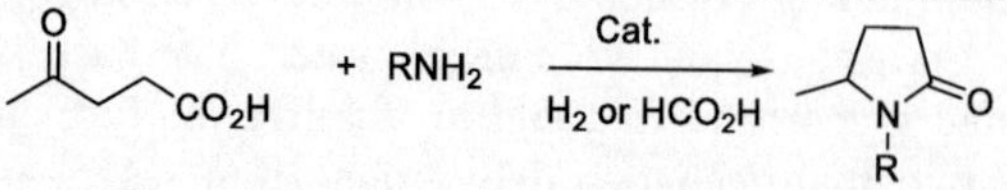

Entry	Catalyst (loading)	Temp.	H_2 (bar)	Base/additive	Yield[a]	Ref
1	**Ir-43** (0.05)	110°C	5	–	90%	[78]
2	**Ir-34** (0.05)	80°C	20	–	95%	[79]
3	**Ir-47** (0.05)	80°C	–	FA/formate buffer pH 3.5	91%	[80]
4	**Ir-48** (0.05)	60°C	–	2 equiv. FA	97%	[81]

[a]Yield is given for the reaction with aniline

Fischmeister and coworkers employed the same catalyst **[Ir-43]** for the reductive amination of LEV with a range of aliphatic and aromatic amines [78]. In neat conditions with 0.05 mol% catalyst loading and 5 bar hydrogen pressure they achieved excellent yields for unhindered substituted anilines, while sterically demanding *ortho*-substituents lowered the yields somewhat (Table 6, Entry 1). Doubly *ortho*-substituted anilines gave moderate yields which could be increased a lot by doubling the catalyst loading. At the reaction temperature of 110°C **[Ir-43]** did not hydrogenate the amido and cyano side chains, acetates however were partially reduced to the alcohol. A range of primary aliphatic and benzylic amines also gave yields of 90% and higher. In addition, the authors performed a process test in which glucose was first hydrolyzed to LEV, and after filtration of insoluble side products and neutralization of the acid, aniline and **[Ir-43]** were added to convert the crude LEV to the pyrrolidinone. The yield over these two steps from glucose was 31%. One of the synthesized compounds, N-Propyl-5-methyl-2-pyrrolidone was assessed as a bio-sourced replacement for the highly regulated and toxic solvent NMP using the Ru-catalyzed arylation of 2-phenylpyridine as a model reaction. While the conversion was slightly lower, the selectivity for the two possible products was different in each solvent, showing that there is scope for tailoring the solvent to a specific reaction. However, as the authors note, in terms of toxicity, there is probably no benefit in using the bio-derived version.

Zhang and coworkers employed catalyst **[Ir-34]** for the reductive amination of LEV (Table 6, Entry 2) [79]. Using 20 bar H_2 at 80°C they achieved good yields in the conversion of a range of electron-rich aryl amines as well as of an aliphatic and a benzylic amine at a catalyst loading of 0.05%.

Xiao and coworkers investigated the reductive amination to pyrrolidinones via transfer hydrogenation from formic acid [80]. They used iridacycle **[Ir-47]** with electron-donating methoxy groups on the ligand in water, but found that it only performed well in a narrow pH range, the optimum being pH 3.5 (Table 6, Entry 3). This necessitated the use of a FA/formate buffer system. With this, they achieved good to very good yields when using various substituted aryl amines with low catalyst loadings of 0.05 mol%. For more challenging substrates like longer chain aliphatic amines the amount of catalyst had to be increased to 0.2 mol%.

The Fischmeister group also recently demonstrated the application of base free transfer hydrogenation for the reductive amination of LEV [81]. Formic acid was used as the renewable hydrogen donor together with an updated version of their catalyst containing an electron-donating dimethylamine group in the ligand (**[Ir-48]**). At a catalyst loading of 0.05 mol% at 60°C using 2 equiv. of FA, a range of aromatic amines with electronically and sterically demanding substituents were converted in good to excellent yields, as well as some aliphatic and benzylic amines (Table 6, Entry 4). Again, the researchers investigated the reaction starting from glucose, using the FA formed as the side product as the hydrogen donor. The yield over these two steps was 23%, which is comparable to other literature reports but a bit lower than using their other hydrogenation procedure. The benefit here however is that no additional hydrogen or hydrogen donor needs to be added.

5 Iridium-Catalyzed Dehydrogenation of Biomass Compounds

Hydrogen has the potential to become a universal energy carrier in the future as it is light and its combustion only creates water. However, a lot of research remains to be done on the efficient generation and reversible storage of this gas.

Electrolysis of water is one way to generate renewable hydrogen; however, this is currently still a very energy intensive process, so that researchers have been looking into producing hydrogen from biomass sources via dehydrogenation reactions. In addition, several small molecules are investigated as potential hydrogen carriers. One such molecule is formic acid, which can be made via hydrogenation of CO_2 and is also a side product in the synthesis of LEV (Schemes 16 and 17) [82, 83].

As already mentioned in the introduction, the hydrogenation of CO_2 to generate small molecule hydrogen carriers such as formic acid, methanol, or bicarbonate is a very active area of research with a lot of potential for homogeneous catalysis, including iridium. In this chapter we will not discuss this topic and interested readers are referred to a recent review [1]. One thing to keep in mind however is that the hydrogenation of CO_2 is generally much more difficult than the dehydrogenation of molecules such as FA, due to the stability of the CO_2 molecule.

This section will discuss the recent literature on the dehydrogenation of biomass derived compounds. Depending on the research, the focus here either lies with the production of hydrogen or the synthesis of other compounds. The discussed catalysts are summarized in Fig. 11.

$$\text{HCOOH} \underset{\text{cat1 or cat2}}{\overset{\text{cat1}}{\rightleftharpoons}} CO_2 + H_2$$

Scheme 17 Dehydrogenation of FA/hydrogenation of CO_2

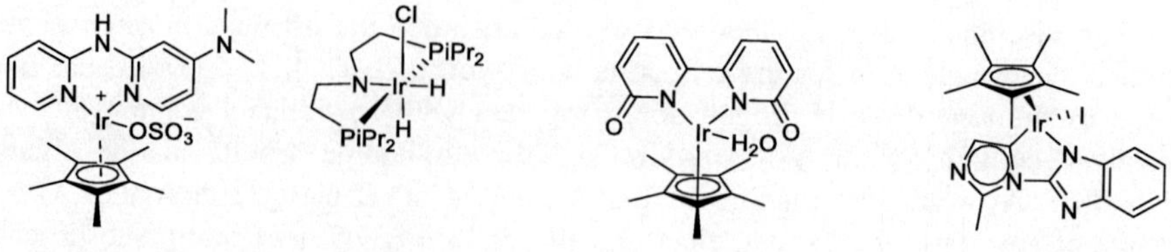

[Ir-48] [Ir-38] [Ir-49] [Ir-50]

Fig. 11 Catalysts used for the dehydrogenation of biomass compounds

5.1 Hydrogen Generation from Biomass

Based on their findings that the TH of LA by formic acid was for a large part due to a hydrogenation reaction after a very fast dehydrogenation of FA, the Fischmeister group investigated their **[Ir-43]** and **[Ir-48]** catalysts (*vide supra*) as well as some related complexes for the dehydrogenation of formic acid under neat and base-free conditions, simulating the conditions needed for fuel cells [84]. Initial ligand screening in aqueous formic acid found that catalyst **[Ir-48]** with an additional electron-donating dimethylamine group on the ligand compared to **[Ir-43]**, and an unsubstituted bridging N-H was the most efficient catalyst for this reaction with a TOF of 12,321 h^{-1} at a loading of 0.01 mol% at 80°C in water. pH screening revealed that the catalyst worked best without added base. Based on this result, the reaction was investigated in different concentrations of FA. While it worked in neat FA with a TOF of 4,800 h^{-1}, the highest reported value to date, the optimum concentration was 3 M, with a TOF of 13,058 h^{-1}. Storage tests of the catalyst in formic acid also revealed that it remained active with only slight decrease in TOF after 10 days.

Beller and coworkers demonstrated that the pincer complex **[Ir-38]** could be used in ppm amounts for hydrogen generation from a range of complex biomass sources including mono- and disaccharides, cellulose, lignocellulose, and even spent cigarette filters (cellulose acetate), which appears to be an important waste stream in Germany [85]. Several catalysts including the commercially available Ru-Macho catalysts were screened, initially using L-fructose as a model substrate. Here it was found that **[Ir-38]** outperformed the Ru catalysts. The optimum amount of base was 1.5 equiv. of NaOH. The role of the base is two-fold: on the one hand, it is needed to activate the pincer complex, and, on the other hand, it traps most of the formed CO_2 as bicarbonate. The protocol was then extended to cellulose and different types of lignocellulose. It was found that for efficient hydrogen generation the biomass first needed to be hydrolyzed into sugars, due to the very low solubility of (ligno)cellulose. Therefore, a three-step protocol was adopted: First, the biomass was depolymerized in NMP by acid hydrolysis using HCl for 2.5 h at 120°C after which time the reaction mixture was cooled to 95°C, neutralized, and 1.5 eq NaOH were added and in the second step reacted for another hour at 120°C. In the third step, **[Ir-38]** was added for hydrogen generation at the same temperature. With cellulose the catalyst reached at TON of 6,285 after 2 h at 20.3 ppm loading, and with bamboo lignocellulose a TON of 2,844 after 8 h at a loading of 20.9 ppm. Unfortunately, only a small percentage of the sugars was dehydrogenated. When calculated on the total theoretically available hydrogen from the sugar molecule (6), the yield of this process is only around 3.3%. This is much lower than other proposed methods for the generation of renewable hydrogen from biomass, such as aqueous phase reforming. Here a hydrogen yield from glucose of 51% can be achieved, albeit at much higher temperatures [86].

5.2 *Biomass Molecules as Hydrogen Donors for Transfer Hydrogenation*

Another area where dehydrogenation of biomass compounds might be valorized is their use as sacrificial hydrogen donors in transfer hydrogenation. The most commonly used hydrogen donor in TH reactions is isopropanol, which is currently not produced renewably; however, FA is also often employed, as discussed in the chapters above. Fujita and coworkers investigated the use of glucose as a hydrogen donor for the TH of carbonyls to alcohols with Ir catalysts containing different substituents on the bipyridyl ligands [87]. The best catalyst was **[Ir-49]**. Using two equiv. of glucose per mol of substrate and substoichiometric amounts of base in water at 100°C, several acetophenone derivatives could be reduced to the alcohols in good yields using 0.1 mol% of **[Ir-49]** (Scheme 18). However, the reaction required 1 mol% of catalyst for all other aromatic and aliphatic ketones. For acetophenones with methoxy or halide substituents however, ether formation from the resultant alcohols was the main reaction, necessitating a change of solvent. In dimethylacetamide (DMA), good to excellent yields of the desired alcohols could be achieved. Aromatic aldehydes were also reduced to the alcohols. The authors performed experiments to elucidate the mechanism and found that only the anomeric C1 hydroxy group of glucose was dehydrogenated resulting in the formation of gluconolactone; hence, the need for 2 equivalents in order to reach full conversion. Coordination of glucose to the catalyst is facilitated by the carbonyl groups on the ligand.

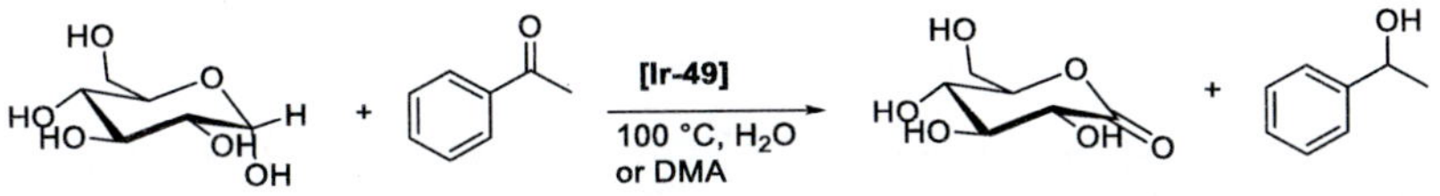

Scheme 18 Transfer hydrogenation of acetophenone using glucose as hydrogen donor

Choudhury and coworkers employed an iridium catalyst with a strong sigma donating abnormal imidazolydine NHC ligand, **[Ir-50]**, for the transfer hydrogenation of CO_2 to formate, using glycerol as the hydrogen donor [88]. The authors demonstrate that **[Ir-50]** is a very effective catalyst for the TH of CO_2 at ambient pressure. They had previously shown that this catalyst is also very efficient in the same reaction using dihydrogen gas [89]. Glycerol, which is generated as a side product in the transesterification of vegetable oils to biodiesel (*vide infra*), was determined to be the best hydrogen donor, giving better results than the common reductant isopropanol. K_2CO_3 was found to be a better base than KOH, and control experiments were performed to exclude the generation of formate from K_2CO_3. Within 12 h the catalyst reached a TOF of 90 h^{-1} at 150°C reaction temperature. In comparison, in their previous publication the same catalyst achieved a TOF of 58 h^{-1} under atmospheric pressure and 30°C, while the reverse reaction, the dehydrogenation of formic acid, proceeded with a TOF of 100,000 h^{-1} at 90°C. The use of glycerol as a hydrogen donor for TH in combination with iridium and ruthenium catalysts has recently been reviewed [31].

5.3 *Iridium-Catalyzed Dehydrogenation of Sugars to Sugar Acids*

With a focus on the other product, instead of hydrogen, Mata and coworkers employed Cp*NHC-Ir complex **[Ir-51]** for the selective dehydrogenation of glucose to gluconic acid in water (Scheme 19 top) [90]. This compound is currently produced via enzymatic oxidation in industry. They achieved a yield of 88% with no side

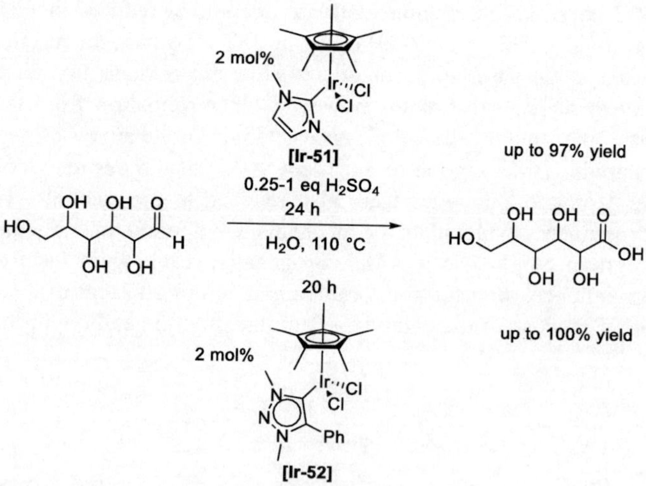

Scheme 19 Dehydrogenation of sugars to sugar acids

product formation. The reaction was accelerated by the addition of 25 mol% H_2SO_4. 1 equiv. of sulfuric acid could increase the yield to 97%. The researchers also applied the reaction to starch as a model carbohydrate. Here two reactions can be performed in one pot as starch hydrolysis is achieved by sulfuric acid in water. After 50 h the yield of gluconic acid was 50%, with 30% glucose and no side product formation. A mechanism was proposed on the basis of ESI-MS experiments combined with DFT calculations. Catalyst precursors containing sulfate or chlorides convert into the active aqua catalyst on dissolution in water. The two water ligands are displaced by glucose in its ring-opened form being bound as a chelate. A nucleophilic attack of H_2O on the aldehyde group of glucose leads to a gem-diol, which reacts via the release of a proton and beta hydride elimination leading to the release of gluconic acid and the generation of an iridium hydride complex, which takes up the proton and releases hydrogen gas to reform the initial complex. The observed acceleration when acid was added to the system is probably due to the faster protonation of the catalyst in this last step.

Following on from this initial study, the same group reported a triazole carbene catalyst **[Ir-52]** which could dehydrogenate a range of sugars to their corresponding acids with high selectivity (Schem 19, bottom) [91]. This catalyst did not require acid to work leading to a large improvement over the previous work. The authors tested different *N*-substituents on the triazole ligand (Me, Bu, and glucose) of which the methyl induced the highest yield: quantitative conversion of glucose to gluconic acid could be achieved at a catalyst loading of 2% in water. The glucose substituted catalyst gave high initial rates but then degraded at the optimum reaction temperature of 110°C. Several other sugars were then screened as substrates and most were converted to the corresponding acids in yields of >90%. Catalyst **[Ir-51]** from the group's previous work was screened alongside (with 0.25 eq H_2SO_4) and in some cases gave slightly higher yields than **[Ir-52]**. Additionally, the butyl version was also slightly better in some cases. Progress of the reaction was monitored by NMR and was determined to be first order in the catalyst. Again, the authors used ESI-MS for mechanistic studies. Based on these results they postulated that the catalyst forms a dicationic hydride bridged bimetallic complex, which they then synthesized and obtained the X-ray structure for. This dimer appears to be the resting state for the catalyst. The catalytic cycle proceeds as described in the previous publication above. One notable difference is that for the triazolylidene ligand no extra acid is necessary. This is rationalized by the fact that the proton which is released by the rearrangement of the gem-diol can be transported intramolecularly via the N2 of the ligand. This is not possible with NHC ligands and points to a reason for the necessity of acid in the previous work.

6 Iridium-Catalyzed Conversion of Lignin

Lignin is the most abundant source of bio-derived aromatics on earth (see Fig. 1), being on average 30% of lignocellulose by weight. However, due to its heterogeneity, it has so far been much more difficult to valorize it compared to the polysaccharide component of lignocellulose. Lignin contains several different linkages between several aromatic monomers, the composition and ratios of which differ a lot depending on the source. The most common linkage is the β-O-4 aryl ether linkage, which has therefore been the target of most research to date [92]. So far, the focus has been on the generation of (substituted) phenols, and currently the only compound that is produced selectively from lignin in about 3% yield is vanillin [93]. In addition to the variation in composition, the structure and properties of extracted lignins varies depending on the source, and the structure and properties of extracted lignins are very different depending on the method by which they are produced. Due to the mix of monomers and linkages in lignin, the result of lignin depolymerization is always a mixture of compounds. In the following, four recent examples of targeting different parts of the β-O-4 linkage with homogeneous iridium catalysts are discussed.

Fu and coworkers investigated the dehydrogenative oxidation of the alpha hydroxyl group to the ketone in a β-O-4 linked model compound (Scheme 20a) [94]. Oxidation to the ketone was reported to lower the bond dissociation energy for the β-O-4 bond and so should facilitate the depolymerization of lignin. A screen of different metal catalysts (Ru, Rh, Ir) revealed that **[Ir-49]** was by far the best catalyst for this reaction. This catalyst was tested on a range of lignin model substrates under optimized conditions (dioxane, 150°C, 1 mol% catalyst loading, 20 h). Most of these were dehydrogenated to the corresponding ketone in yields around 80%, with a few examples of lower yields. DFT calculations confirmed that the alpha keto product is

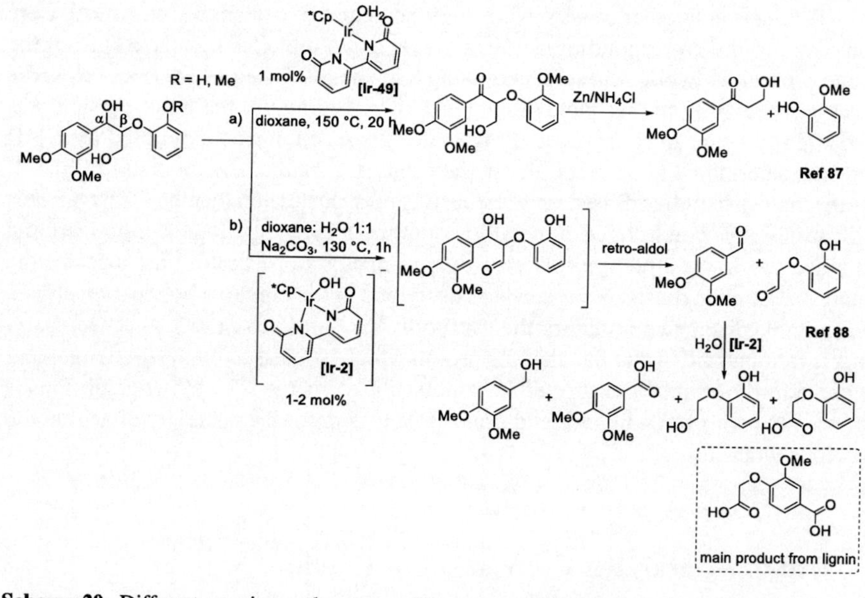

Scheme 20 Different reaction pathways on the common β-O-4 linkage in lignin (model compound): (**a**) selective dehydrogenation of the α-hydroxyl, (**b**) selective dehydrogenation of the γ-hydroxyl, followed by base-catalyzed retro-aldol reaction and Ir-catalyzed hydrogenation/dehydrogenation reactions to yield stable alcohols and acids

more stable than the gamma. Based on the calculations a mechanism was proposed which proceeds via abstraction of a hydrogen from the carbon by the iridium atom and abstraction of the hydrogen of the hydroxyl group by one of the carbonyl groups of the bipyridine ligand. Following the release of the product, the catalyst returns to its original state by the release of a molecule of H_2. The authors then applied the catalyst to real lignin, which they extracted with acidic dioxane from birch wood, a source high in β-O-4 linkages. To confirm that the gas released from the reaction is hydrogen, they channeled it into another reaction using a Ru catalyst for the hydrogenation of decene to decane, which was then detected.

In order to depolymerize the Ir-pretreated lignin, first the authors tested a 2-step protocol on the model compound, where the depolymerization was achieved efficiently by Zn/NH_4Cl. A formate buffer system was also tested, but due to the efficient dehydrogenation of formic acid by the Ir catalyst remaining in solution this was not usable. This procedure was then also applied to real lignin. GPC analysis of lignin samples before and after the treatment steps confirmed that lignin could not be depolymerized by the Zn system without the iridium pretreatment, while after the two steps it showed a decrease in molecular weight.

Two years later the group of Bruijnincx employed the anionic version of the bipyridone catalyst **[Ir-49], [Ir-2]**, to catalyze the conversion of lignin into aromatic diols and diacids via dehydrogenation of the primary (β) hydroxyl followed by retro-aldol Cα – Cβ bond cleavage (Scheme 20b) [95]. This catalyst, **[Ir-2]**, was chosen as it was known to be active in primary alcohol dehydrogenation (for example EG, see Sect. 2.1.2). In a basic water/dioxane mixture (pH 11) **[Ir-49]** converted spontaneously into its anionic form **[Ir-2]**, and this catalyst system was able to cleave several lignin model compounds, also including phenol-containing ones, efficiently within 1 h. Next, the authors applied their method to real lignin (softwood enzyme), which resulted in an oil in 82% yield, which was confirmed by GPC to have much lower molecular weights. The other 18% were higher molecular weight material. A 2D HSQC NMR analysis showed that there were hardly any β-O-4 linkages left in this material and also the β-5 (phenylcoumaran) linkages were highly reduced. Another type of common linkage in lignin, the resinol (β-β) unit was not touched. Analysis of the oil fraction by GC confirmed the presence of the monomers expected from the model compound experiments. The yield of identified monomers was 3.8 wt% which is about 40% of the theoretical yield. Dioxasolv lignin was also tested and gave a slightly higher yield, mainly due to vanillin derived from the end groups. To simplify the product mixture, the authors refluxed the reaction mixture in air for a further 16 h, which led to the partial conversion of the monomers into a diacid derivative (see Scheme 16) in 2.4% yield, demonstrating that it can be possible to tune the depolymerization reaction to give a few products.

De Vries, Barta, and coworkers investigated iridium-catalyzed decarbonylation as one of the several strategies to prevent recondensation reactions of the aldehyde fragments which are formed by acidolysis of the β-O-4 linkage (Scheme 21a) [96]. In a one-pot reaction they achieved acid-catalyzed cleavage of β-O-4 model compounds by triflic acid and decarbonylation of the resulting aldehydes to methyl aromatics by $[Ir(cod)Cl]_2$ with triphenylphosphine ligand **[Ir-53]**. They then applied

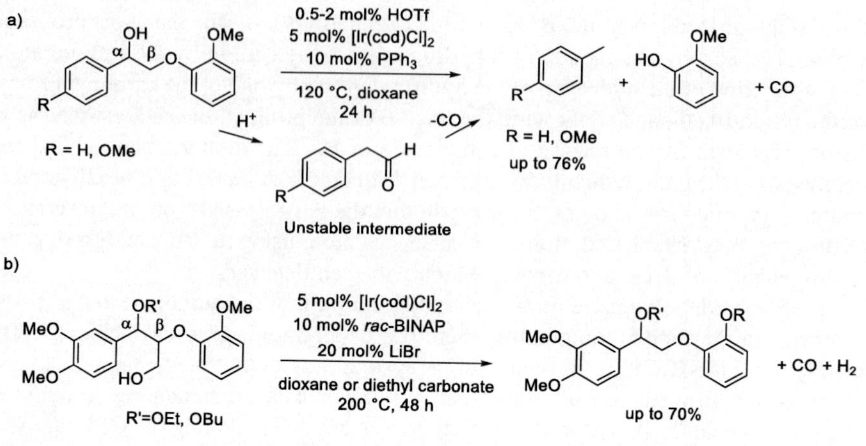

Scheme 21 Iridium-catalyzed decarbonylation of β-O-4 lignin model compounds

to methodology to walnut dioxasolv lignin. The content β-O-4 linkages in this type of lignin was determined to be 45% by 2D HMQC NMR. The theoretical maximum monomer yield from breaking β-O-4 bonds connected to monomeric units was 10%. The decarbonylation of lignin yielded 2 wt% of monomers, C1 and C0 phenolics, the latter are probably derived from vanillin-type aldehydes, which are known to form during lignin acidolysis.

Deuss and coworkers recently investigated the acceptorless dehydrogenative decarbonylation of the gamma-carbinol in the β-O-4 linkage (Scheme 21b) [97]. This transformation allows for access to different structures from lignin, whilst at the same time producing syngas (a mixture of CO and H_2), which is a valuable starting material for the synthesis of alkanes, and which can also be produced via the gasification of biomass. For this purpose the authors choose the catalyst system previously reported by Olsen and Madsen [98] consisting of $[Ir(cod)Cl]_2$ combined with *rac*-BINAP ligand (**[Ir-54]**) with LiCl (4 eq w.r.t. Ir) as an additive. Using simple model compounds the reaction conditions were optimized in 1,4-dioxane (or diethyl carbonate) as solvent at 190°C (142°C inside the reaction vessel). They found that the presence of phenolic residues even enhanced the yield. The generation of syngas was confirmed by GC and the ratio of CO:H_2 was circa 1.1:1. The authors then tested α-ethoxylated and -butoxylated β-O-4 model substrates, which would be present in lignin extracted with ethanol or butanol, respectively. The reason for using α-alkoxylated species is to prevent the retro-aldol splitting as in the case of Bruijnincx's work. These compounds required higher catalyst loadings and temperature as well as double the reaction time to achieve satisfactory yields. A phenylcoumaran model compound could also be dehydrogenated and decarbonylated under the same conditions. The authors then applied their catalyst system to different types of organosolv lignins and could demonstrate the defunctionalization of these substrates, alongside the production of syngas. However, when using real lignin as the substrate the ratio between CO and H_2 became a lot higher, indicating that the hydrogen released is reacting with the substrate at other sites under the reaction conditions.

7 Iridium-Catalyzed Transformations of Fatty Acids

7.1 *Vegetable Oil as a Feedstock*

Vegetable oils, and to a smaller degree animal fats, are a large source of renewable chemicals [99]. In contrast to lignocellulose, vegetable oils are also food products, however currently a lot of oil crops such as rapeseed are grown in large quantities for the production of biodiesel, so that the conversion of a small fraction of these into chemicals will not affect the food production. Additionally, oils can be harvested from algae, which avoids the use of arable land for their production.

For biodiesel production, the triglycerides from vegetable oils are transesterified with methanol to produce fatty acid methyl esters (FAMEs), leaving large amounts of glycerol as a side product [100]. Due to this, much research has been done on the valorization of glycerol, including a range of iridium-catalyzed reactions. These are discussed in Sect. 2.4. Additionally, as was already discussed in Sect. 5.2, glycerol is also interesting as a hydrogen donor.

The fatty acids from vegetable oils are mostly unsaturated or polyunsaturated, although saturated fatty acids also exist, for example in coconut and palm oil. Animal fat, on the other hand, contains mostly saturated fatty acids, such as stearic acid.

Oleic acid, a monounsaturated omega-9 fatty acid is the most common fatty acid in nature. It is present in up to 70% in canola oil, which is made from rapeseed.

7.2 *Iridium-Catalyzed Transformations of Fatty Acids*

7.2.1 Hydrogenation

A high content of polyunsaturated FAMEs in biodiesel reduces its performance: the stability and energy density of the fuel are lowered, while viscosity and emission of hydrocarbons are increased. In the USA the main feedstock for producing biodiesel are soybean and corn oil, which are high in polyunsaturated fatty acids, unlike in Europe where high oleic acid content canola oil is most commonly used. Fully saturated FAMEs, such as those produced from animal fat, on the other hand, have high melting points and are prone to precipitation in diesel blends [41]. The best performance is achieved by high oleate content.

Williams and coworkers investigated iridium complex **[Ir-9]** from their previous work on the acceptorless dehydrogenation of glycerol to LA (see Sect. 2.4.1) for the hydrogenation of polyunsaturated FAMEs using glycerol and methanol as hydrogen

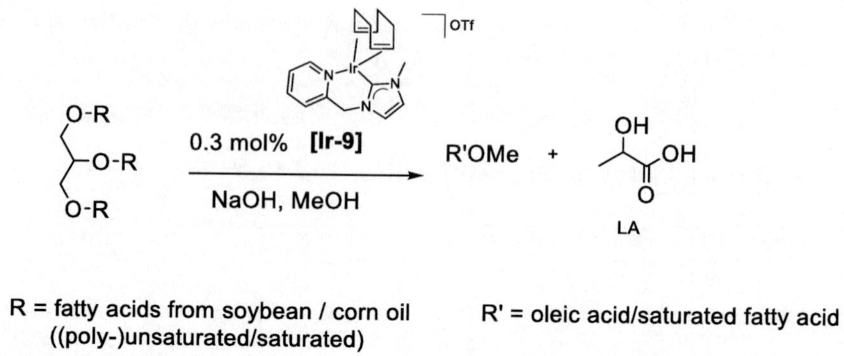

Scheme 22 Combined methanolysis of vegetable oil, hydrogenation of polyunsaturated fatty acids and conversion of glycerol to lactic acid

donor (Scheme 22) [101]. This reaction could be combined in one pot with the synthesis of FAMEs from corn oil and methanol. As one molecule of glycerol can only generate one equivalent of H_2, the additional hydrogen was produced by the dehydrogenation of methanol. Under the reaction conditions investigated, full and selective hydrogenation of the polyunsaturated fatty acids to oleate and other monounsaturated acids occurred when using 25 equiv. of methanol on top of the one equivalent of glycerol from the triglyceride. The glycerol was converted into lactate in >99%. Using 25 equivalents of glycerol, hydrogenation also proceeded in >99%. An isolated yield of monounsaturated FAMEs of 65% could be achieved. Control experiments demonstrated that a higher methanol loading led to lower conversions due to poisoning of the catalyst. The role of the base in the hydrogenation reaction was investigated and it was found that it was not required for methanol dehydrogenation. However, no lactate was formed from glycerol unless stoichiometric base was present. The reaction was normally performed in closed vessels. To test the contribution of TH, a reaction was performed in an open system. Here the hydrogenation yield was 25%, confirming that most of the hydrogenation was by dihydrogen liberated from the H donors.

Even though not desirable for biodiesel, full hydrogenation of fatty acids was also investigated. By raising the temperature from 120 to 150°C and increasing the catalyst loading substantially from 0.3 to 6 mol%, complete hydrogenation could be achieved in 95%. In an effort to reduce the iridium loading, the authors found that adding Fe-MACHO catalyst in 3 mol% to the original amount of iridium could achieve complete hydrogenation in 90% yield.

Darensbourg and coworkers investigated a water-soluble analogue of Vaska's complex with the ligand (*m*-sulfonatophenyl)diphenylphosphine, **[Ir-55]**, for the hydrogenation and *cis-trans* isomerization of unsaturated (small) carboxylic acids and further applied it to the hydrogenation of soybean lecithin liposomes, as a model for biological membrane modification [102].

Lecithin is a general term for a group of phospholipids. The fatty acid composition of this sample is listed in Fig. 12. Under fairly mild conditions (60°C, 1 bar

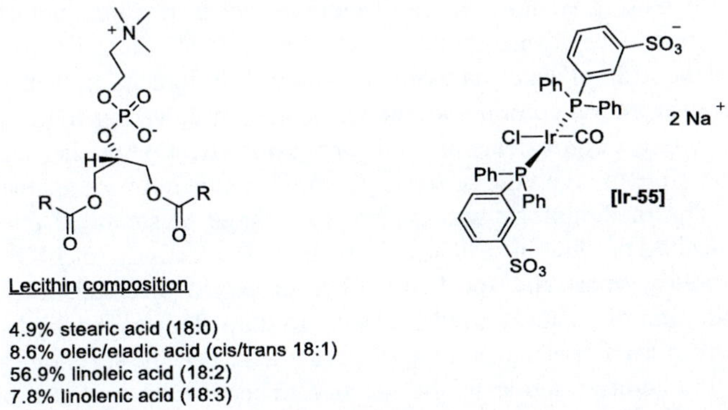

Fig. 12 Iridium catalyst used in the isomerization of lecithin and lecithin composition

hydrogen) the authors found that even though there was some hydrogenation of the multiple double bonds, the isomerization of the natural cis form to the trans isomer was much faster. Using 0.01 mmol catalyst/10 mg substrate, the ratio of *cis* to *trans* double bonds in 18:1 fatty acids went from 93.2:6.7 to 58.1:41.9 in just 30 min. At the same time there was only 9.7% of hydrogenation of double bonds. Hydrogenation of trans alkenes is much more difficult so that full hydrogenation could not be achieved. Additionally, it might be more difficult for the catalyst to reach the sites due to structural changes in the liposome due to the isomerization. This comparably low activity of the catalyst is however enough to make enough changes to the liposomes for biomedical applications.

7.2.2 Decarbonylation

Aliphatic linear alpha olefins (LAOs) are important industrial chemicals, produced on large scale for the synthesis of surfactants and lubricants as well as comonomers in HDPE and linear low density PE, depending on their chain length [103]. Linear internal olefins (LIOs), the isomerization product in this reaction, are also industrially relevant products, mostly used in drilling fluids and paper sizing [104].

Ryu and coworkers applied Vaska's catalyst, $IrCl(CO)(PPh_3)_2$, **[Ir-56]**, to the decarbonylation of long chain aliphatic carboxylic acids (Scheme 23) [105]. Using stearic acid from palm oil, they first screened several iridium precursors and

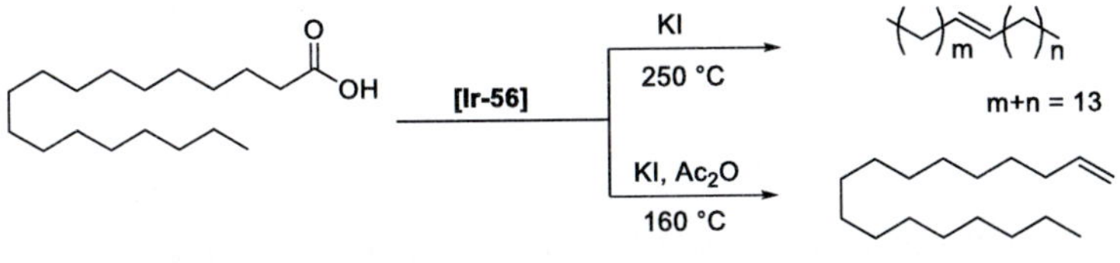

Scheme 23 Decarbonylation of saturated fatty acids to yield selectively internal or terminal olefins

additives. When using 20 mol% KI as an additive at a reaction temperature of 250°C, internal alkenes were formed with selectivities >98%. A test with the iodine analogue of the iridium catalysts worked equally well, indicating that the KI might activate the catalyst via chloride–iodide exchange. By lowering the temperature to 160°C and adding 2 equivalents of acetic anhydride (Ac_2O) together with 0.5 eq of KI, it was possible to invert the selectivity to produce the terminal alkenes with 98% selectivity. The methodology was applied to a range of saturated fatty acids, all giving the desired product in >80% yield with 99% selectivity for LIOs and >76% yield and >96% selectivity for LAOs. The proposed mechanism starts by the oxidative addition of an acid anhydride to the iridium. This acid anhydride is either generated from two molecules or carboxylic acid, or by reaction with Ac_2O, indicating this additive's role in the reaction. The formed acyl complex is then decarbonylated releasing CO. The alkene is released by β-hydride elimination, leaving an iridium hydride complex, which can either release a carboxylic acid and return to the initial state or catalyze the isomerization of the double bond.

Hapiot and coworkers later applied this catalytic system to the decarbonylation of unsaturated fatty acids to produce linear alpha olefins (Scheme 24) [104]. The authors performed extensive screening of reaction conditions with oleic acid as a model substrate. This was chosen for ease of analysis, as it only contains one double bond. In a phosphane ligand screening with **[Ir-56]** or $[Ir(cod)Cl]_2$ as the iridium source it was found that the best combination was 3:1 PPh_3: $[Ir(cod)Cl]_2$ at 5 mol% loading and 160°C for 5 h (**[Ir-57]**). Acetic anhydride and potassium iodide were again used as additives. Without Ac_2O there was hardly any conversion. Just like in Ryu's work, the optimum amounts of the additives for best conversion and selectivity to the terminal alkene were determined to be 2 equiv. Ac_2O and 0.5 equiv. KI. With **[Ir-56]**, on the other hand, there was good conversion without Ac_2O, however almost exclusively the internal alkene product was formed. The reaction system was then applied to other unsaturated fatty acids as well as α,ω-diacids. All substrates showed good to very good conversions with selectivities to the terminal alkene of >80%. The catalyst could be recycled by distilling off the product and other components. When fresh substrate was added the conversion and selectivity decreased somewhat. The selectivity could be restored by adding fresh triphenylphosphine; however, the yield was still 10% lower than with fresh catalyst.

A few years later the same group investigated different ligands for this transformation, as phosphane ligands are generally not desirable when not necessary due to

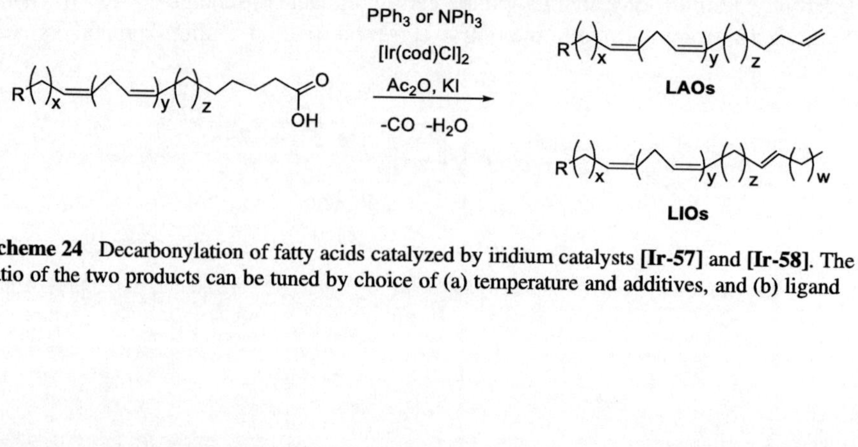

Scheme 24 Decarbonylation of fatty acids catalyzed by iridium catalysts **[Ir-57]** and **[Ir-58]**. The ratio of the two products can be tuned by choice of (a) temperature and additives, and (b) ligand

stability issues [106]. A range of different amine ligands chosen for their air stability were screened with $[Ir(cod)Cl]_2$ under the same reaction conditions as before. Aliphatic tertiary amines gave good conversions, while for aromatic amines and those with delocalized electrons the conversions were much lower, reflecting their weaker binding to the iridium. The best ligand in the screening was the triphenylphosphine analogue NPh_3, forming the amine analogue catalyst **[Ir-58]**. However, it could not reach the activity of the PPh_3 catalyst from the previous publication. While the selectivity for LAOs was much higher with the phosphane ligand, it was found that it was possible to tune the reaction conditions to reach very high selectivity for LIOs with the amine ligands.

7.2.3 Hydroformylation

Carpentier and coworkers investigated the use of iridium among other metals for the hydroformylation-isomerization reaction of fatty acid derivatives [107]. The study focused on 10-undecenenitrile as the substrate (Scheme 25). This compound can be readily prepared by the reaction of 10-undecenoic acid from castor oil with ammonia and is a precursor to polyamide-12 [108].

Testing the bis-phosphite ligand Biphephos, originally developed by Union Carbide, which they had employed in their previous study with rhodium [109], with other metals, the authors found that with iridium the TOF was only five times slower than that of the rhodium system (730 vs 3,320 h^{-1}). Other metals tested (Pd, Ru) were a lot slower. This is an interesting result, as currently most hydroformylation catalysts are based on rhodium, a very expensive and rare metal, which is considered vastly superior to all others for this reaction. Even though iridium is not cheap either, it is much cheaper than rhodium, making this a potentially viable process. One problem with hydroformylation is isomerization of the double bond of the alkene to internal positions, which then gives rise to undesired branched hydroformylation products. The selectivity to the terminal aldehyde with the iridium catalyst (**[Ir-59]**) in this case was 99:1. Some of the starting material was isomerized to internal alkenes; however, this was only hydroformylated to a minor extent. For 10-undecenenitrile, the amount of internal alkenes increased from 5 to 22% during the reaction. Some other fatty acid derivatives were screened and showed the same high selectivity to the linear products. Additionally, the iridium

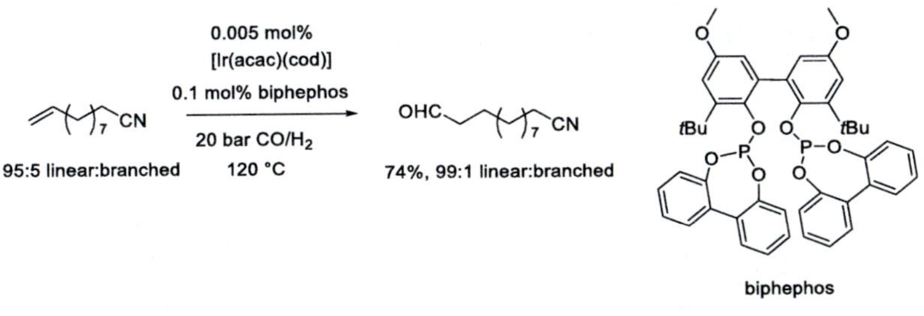

Scheme 25 Hydroformylation of fatty acid derivatives with **[Ir-59]**

could be reused three times by distilling off the reaction mixture and product and adding more ligand.

7.2.4 Trialkylsilylation

As we have already seen, terminal olefins are often more desired for further functionalization, giving rise to linear, end-functionalized compounds. Riepl and coworkers investigated the iridium-catalyzed isomerizing dehydrogenative silylation of methyl oleate, the most common FAME (Scheme 26) [110]. Silicon containing compounds like this are interesting precursors for industry in the fields of adhesives, lubricants, plasticizers, and polymers. Initially the authors tried to apply an existing methodology for isomerizing silylation to methyl oleate [111]. This consisted of 5 mol% $[Ir(OMe)Cl]_2$, 10 mol% 2,2′-bipyridine ligand combined with 3 equivalents of norbornene as a hydrogen acceptor, and 3 equivalents of triethylsilane. However, it turned out that the bipyridine ligand was inhibiting the isomerization step, so that very little desired product was formed even after prolonged reaction times. When the ligand was omitted, and $[Ir(OMe)(cod)]_2$, **[Ir-60]**, was used, the yield of the terminal silylated product rose to 69% (Scheme 26). Further changing of reaction conditions, such as increasing the temperature from 60°C led to a deactivation of the catalyst. Sadly, the procedure did not work with other bulkier trialkylsilanes.

7.2.5 Hydroboration

Angelici and coworkers employed $[Ir(coe)_2Cl]_2$ in combination with dppe ligand (**[Ir-61]**) for the isomerizing hydroboration of methyl oleate with pinacolborane [112]. In their protocol at room temperature, they achieved a yield of 45% of the desired borylated product, while the main product of the reaction was the hydrogenated FAME methyl stearate with 47% (Scheme 27). The fact that there is so much

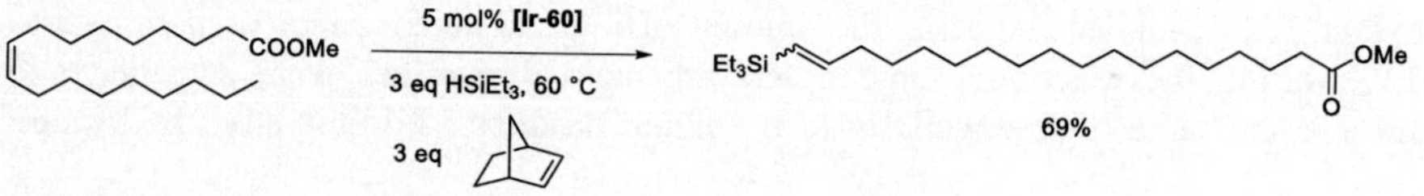

Scheme 26 Iridium-catalyzed trialkylsilylation of methyl oleate

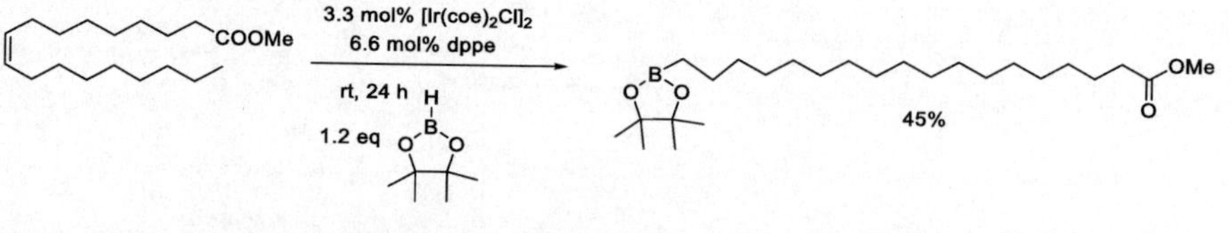

Scheme 27 Isomerizing hydroboration of methyl oleate catalyzed by **[Ir-61]**

of the hydrogenation product formed, which does not happen with shorter chain internal alkenes under the same conditions, led the authors to conclude that the isomerization step is much slower for this substrate. $[Ir(cod)Cl]_2$ gave no borylation product under the same conditions. When the reaction temperature was increased to 50°C, a low yield of 25% was achieved, with a large amount of the starting material still present after 24 h.

8 Conclusion

In this chapter we have summarized research in the field of iridium-catalyzed biomass conversion. This is a developing field, most of the work discussed here was published in the last 8 years. With the recent progress in the conversion of cellulosic biomass and the availability of several platform chemicals, there is now more and more scope for homogeneous catalysis in this area, due to its generally higher selectivity at milder conditions. We therefore expect to see more publications in this field in the future. One massive advantage of iridium over other metals, especially with regard to biomass, is its excellent stability towards water, air, and acidic conditions. We have seen that a range of useful fine chemicals and precursors, monomers, and even polymers can be synthesized from biomass using iridium catalysis. Mechanistically, many of the reactions discussed are hydrogenation/dehydrogenation reactions. This is an important reaction pathway for biomass, which is highly functionalized and oxygen-rich compared to the well-known oil-based building blocks. Other mechanisms such as decarbonylation/hydroformylation and others have also been applied to bio-based compounds. The valorization of lignin still lags behind the saccharide-based biomass but also here iridium catalysts have shown promising potential in the initial defunctionalization and the selective generation of a few monomers from the complex mixture. We expect to see more research in this direction in the coming years.

Finally, the dehydrogenation of biomass to generate hydrogen, a highly useful product with the potential to be a main energy carrier in the future is also being investigated. With the development of more efficient catalysts, the generation of renewable hydrogen from biomass should soon become cost-effective.

References

1. Werpy T, Petersen G (2004) Top value added chemicals from biomass volume I – results of screening for potential candidates from sugars and synthesis gas. https://doi.org/10.2172/15008859
2. Bozell JJ, Petersen GR (2010) Technology development for the production of biobased products from biorefinery carbohydrates-the US Department of Energy's "top 10" revisited. Green Chem 12:539–554

3. Deuss PJ, Barta K, de Vries JG (2014) Homogeneous catalysis for the conversion of biomass and biomass-derived platform chemicals. Cat Sci Technol 4:1174–1196
4. Sordakis K, Tang C, Vogt LK, Junge H, Dyson PJ, Beller M, Laurenczy G (2018) Homogeneous catalysis for sustainable hydrogen storage in formic acid and alcohols. Chem Rev 118:372–433
5. De Clercq R, Dusselier M, Sels BF (2017) Heterogeneous catalysis for bio-based polyester monomers from cellulosic biomass: advances, challenges and prospects. Green Chem 19:5012–5040
6. Zhang Y, Liu D, Chen Z (2017) Production of C2-C4 diols from renewable bioresources: new metabolic pathways and metabolic engineering strategies. Biotechnol Biofuels 10:1–20
7. Stadler BM, Wulf C, Werner T, Tin S, De Vries JG (2019) Catalytic approaches to monomers for polymers based on renewables. ACS Catal 9:8012–8067
8. Pang J, Zheng M, Sun R, Wang A, Wang X, Zhang T (2016) Synthesis of ethylene glycol and terephthalic acid from biomass for producing PET. Green Chem 18:342–359
9. van der Waal JC, Gruter GJM, Claassens-Dekker P (2016) Process for preparing ethylene glycol from a carbohydrate source via transition metal-catalyzed hydrogenolysis. WO2016114660A1
10. Wu J, Shen L, Chen ZN, Zheng Q, Xu X, Tu T (2020) Iridium-catalyzed selective cross-coupling of ethylene glycol and methanol to lactic acid. Angew Chem Int Ed 59:10421–10425
11. Zhan Y, Hou W, Li G, Shen Y, Zhang Y, Tang Y (2019) Oxidant-free transformation of ethylene glycol toward glycolic acid in water. ACS Sustain Chem Eng 7:17559–17564
12. Torbina VV, Vodyankin AA, Ten S, Mamontov GV, Salaev MA, Sobolev VI, Vodyankina OV (2018) Ag-based catalysts in heterogeneous selective oxidation of alcohols: a review. Catalysts 8:313
13. Wang YM, Lorenzini F, Rebros M, Saunders GC, Marr AC (2016) Combining bio- and chemo-catalysis for the conversion of bio-renewable alcohols: homogeneous iridium catalysed hydrogen transfer initiated dehydration of 1,3-propanediol to aldehydes. Green Chem 18:1751–1761
14. Lacroix SD, Pennycook A, Liu S, Eisenhart TT, Marr AC (2012) Amination and dehydration of 1,3-propanediol by hydrogen transfer: reactions of a bio-renewable platform chemical. Cat Sci Technol 2:288–290
15. Ma Y, Wang YM, Morgan PJ, Jackson RE, Liu X, Saunders GC, Lorenzini F, Marr AC (2018) Designing effective homogeneous catalysis for glycerol valorisation: selective synthesis of a value-added aldehyde from 1,3-propanediol via hydrogen transfer catalysed by a highly recyclable, fluorinated Cp*Ir(NHC) catalyst. Catal Today 307:248–259
16. Bothwell KM, Lorenzini F, Mathers E, Marr PC, Marr AC (2019) Basic ionic liquid gels for catalysis: application to the hydrogen borrowing mediated dehydration of 1,3-Propanediol. ACS Sustain Chem Eng 7:2686–2690
17. Bahé F, Grand L, Cartier E, Jacolot M, Moebs-Sanchez S, Portinha D, Fleury E, Popowycz F (2020) Direct amination of Isohexides via borrowing hydrogen methodology: Regio- and Stereoselective issues. Eur J Org Chem 2020:599–608
18. Jacolot M, Moebs-Sanchez S, Popowycz F (2018) Diastereoselective iridium-catalyzed amination of biosourced Isohexides through borrowing hydrogen methodology. J Org Chem 83:9456–9463
19. Froidevaux V, Negrell C, Caillol S, Pascault JP, Boutevin B (2016) Biobased amines: from synthesis to polymers; present and future. Chem Rev 116:14181–14224
20. Kadraoui M, Maunoury T, Derriche Z, Guillarme S, Saluzzo C (2015) Isohexides as versatile scaffolds for asymmetric catalysis. Eur J Org Chem 2015:441–457
21. Janvier M, Moebs-Sanchez S, Popowycz F (2016) Nitrogen-functionalized isohexides in asymmetric induction. Chimia 70:77–83
22. Imm S, Bähn S, Zhang M, Neubert L, Neumann H, Klasovsky F, Pfeffer J, Haas T, Beller M (2011) Improved ruthenium-catalyzed amination of alcohols with ammonia: synthesis of diamines and amino esters. Angew Chem Int Ed 50:7599–7603

23. Pingen D, Diebolt O, Vogt D (2013) Direct amination of bio-alcohols using ammonia. ChemCatChem 5:2905–2912
24. Zhang Y, Lim CS, Boon Sim DS, Pan HJ, Zhao Y (2014) Catalytic enantioselective amination of alcohols by the use of borrowing hydrogen methodology: cooperative catalysis by iridium and a chiral phosphoric acid. Angew Chem Int Ed 53:1399–1403
25. Gross J, Tauber K, Fuchs M, Schmidt NG, Rajagopalan A, Faber K, Fabian WMF, Pfeffer J, Haas T, Kroutil W (2014) Aerobic oxidation of isosorbide and isomannide employing TEMPO/laccase. Green Chem 16:2117–2121
26. Starr JN, Westhoff G (2014) Lactic acid, 7th edn. Ullmann's encyclopedia of industrial chemistry. https://onlinelibrary.wiley.com/doi/book/10.1002/14356007
27. Sharninghausen LS, Campos J, Manas MG, Crabtree RH (2014) Efficient selective and atom economic catalytic conversion of glycerol to lactic acid. Nat Commun 5:5084
28. Manas MG, Campos J, Sharninghausen LS, Lin E, Crabtree RH (2015) Selective catalytic oxidation of sugar alcohols to lactic acid. Green Chem 17:594–600
29. Lu Z, Demianets I, Hamze R, Terrile NJ, Williams TJ (2016) A prolific catalyst for selective conversion of neat glycerol to lactic acid. ACS Catal 6:2014–2017
30. Sun Z, Liu Y, Chen J, Huang C, Tu T (2015) Robust iridium coordination polymers: highly selective, efficient, and recyclable catalysts for oxidative conversion of glycerol to potassium lactate with dihydrogen liberation. ACS Catal 5:6573–6578
31. Finn M, Ridenour JA, Heltzel J, Cahill C, Voutchkova-Kostal A (2018) Next-generation water-soluble homogeneous catalysts for conversion of glycerol to lactic acid. Organometallics 37:1400–1409
32. Cheong Y-J, Sung K, Park S, Jung J, Jang H-Y (2020) Valorization of chemical wastes: Ir (biscarbene)-catalyzed transfer hydrogenation of inorganic carbonates using glycerol. ACS Sustain Chem Eng 8:6972–6978
33. Ahmed Foskey TJ, Heinekey DM, Goldberg KI (2012) Partial deoxygenation of 1,2-propanediol catalyzed by iridium pincer complexes. ACS Catal 2:1285–1289
34. Lao DB, Owens ACE, Heinekey DM, Goldberg KI (2013) Partial deoxygenation of glycerol catalyzed by iridium pincer complexes. ACS Catal 3:2391–2396
35. Haynes A, Maitlis PM, Morris GE, Sunley GJ, Adams H, Badger PW, Bowers CM, Cook DB, Elliott PIP, Ghaffar T, Green H, Griffin TR, Payne M, Pearson JM, Taylor MJ, Vickers PW, Watt RJ (2004) Promotion of iridium-catalyzed methanol Carbonylation: mechanistic studies of the Cativa process. J Am Chem Soc 126:2847–2861
36. Coskun T, Conifer CM, Stevenson LC, Britovsek GJP (2013) Carbodeoxygenation of biomass: the carbonylation of glycerol and higher polyols to monocarboxylic acids. Chem Eur J 19:6840–6844
37. Crabtree RH (2019) Transfer hydrogenation with glycerol as H-donor: catalyst activation, deactivation and homogeneity. ACS Sustain Chem Eng 7:15845–15853
38. Farnetti E, Kašpar J, Crotti C (2009) A novel glycerol valorization route: chemoselective dehydrogenation catalyzed by iridium derivatives. Green Chem 11:704–770
39. Crotti C, Kašpar J, Farnetti E (2010) Dehydrogenation of glycerol to dihydroxyacetone catalyzed by iridium complexes with P–N ligands. Green Chem 12:1295–1300
40. Crotti C, Farnetti E, Guidolin N (2010) Alternative intermediates for glycerol valorization: iridium-catalyzed formation of acetals and ketals. Green Chem 12:2225–2231
41. Silva PHR, Gonçalves VLC, Mota CJA (2010) Glycerol acetals as anti-freezing additives for biodiesel. Bioresour Technol 101:6225–6229
42. Deutsch J, Martin A, Lieske H (2007) Investigations on heterogeneously catalysed condensations of glycerol to cyclic acetals. J Catal 245:428–435
43. Cataldo M, Nieddu E, Gavagnin R, Pinna F, Strukul G (1999) Hydroxy complexes of palladium(II) and platinum(II) as catalysts for the acetalization of aldehydes and ketones. J Mol Catal A Chem 142:305–316
44. Chakrabarti K, Maji M, Kundu S (2019) Cooperative iridium complex-catalyzed synthesis of quinoxalines, benzimidazoles and quinazolines in water. Green Chem 21:1999–2004

45. Labed A, Jiang F, Labed I, Lator A, Peters M, Achard M, Kabouche A, Kabouche Z, Sharma GVM, Bruneau C (2015) Iridium-catalyzed sustainable access to functionalized Julolidines through hydrogen autotransfer. ChemCatChem 7:1090–1096
46. Minakawa M, Watanabe K, Toyoda S, Uozumi Y (2018) Iridium-catalyzed direct cyclization of aromatic amines with diols. Synlett 29:2385–2389
47. Lv D, Xie Z, Gu B, Wu H, Wan H (2016) Highly efficient synthesis of quinoxaline derivatives catalized by iridium complex. Russ J Gen Chem 86:2887–2890
48. Zhai XY, Wang XQ, Ding YX, Zhou YG (2019) Partially biobased polymers: the synthesis of polysilylethers via dehydrocoupling catalyzed by an anionic iridium complex. Chin Chem Lett:1–4
49. Xu Z, Yan P, Xu W, Liu X, Xia Z, Chung B, Jia S, Zhang ZC (2015) Hydrogenation/hydrolytic ring opening of 5-HMF by Cp∗-iridium(III) half-sandwich complexes for bioketones synthesis. ACS Catal 5:788–792
50. Wu WP, Xu YJ, Zhu R, Cui MS, Li XL, Deng J, Fu Y (2016) Selective conversion of 5-hydroxymethylfuraldehyde using Cp∗Ir catalysts in aqueous Formate buffer solution. ChemSusChem 9:1209–1215
51. Xu YJ, Shi J, Wu WP, Zhu R, Li XL, Deng J, Fu Y (2017) Effect of Cp*iridium(III) complex and acid co-catalyst on conversion of furfural compounds to cyclopentanones or straight chain ketones. Appl Catal A 543:266–273
52. Xu Z, Yan P, Li H, Liu K, Liu X, Jia S, Zhang ZC (2016) Active Cp∗iridium(III) complex with ortho-hydroxyl group functionalized bipyridine ligand containing an electron-donating group for the production of diketone from 5-HMF. ACS Catal 6:3784–3788
53. Wozniak B, Spannenberg A, Li Y, Hinze S, de Vries JG (2018) Cyclopentanone derivatives from 5-hydroxymethylfurfural via 1-hydroxyhexane-2,5-dione as intermediate. ChemSusChem 11:356–359
54. Van Putten RJ, Van Der Waal JC, De Jong E, Rasrendra CB, Heeres HJ, de Vries JG (2013) Hydroxymethylfurfural, a versatile platform chemical made from renewable resources. Chem Rev 113:1499–1597
55. Mika LT, Cséfalvay E, Németh Á (2018) Catalytic conversion of carbohydrates to initial platform chemicals: chemistry and sustainability. Chem Rev 118:505–613
56. Wang T, Nolte MW, Shanks BH (2014) Catalytic dehydration of C6 carbohydrates for the production of hydroxymethylfurfural (HMF) as a versatile platform chemical. Green Chem 16:548–572
57. Rosatella AA, Simeonov SP, Frade RFM, Afonso CAM (2011) 5-Hydroxymethylfurfural (HMF) as a building block platform: biological properties, synthesis and synthetic applications. Green Chem 13:754–793
58. de Jong E, Dam MA, Sipos L, Gruter GJM (2012) Biobased monomers, polymers, and materials. In: ACS symposium series Ch. 1, 1–13, vol 1105. American Chemical Society
59. Haworth WN, Jones WGM (1944) The conversion of sucrose into Furan compounds. Part 1. 5-hydroxymethylfurfural and some derivatives. J Chem Soc:667–670
60. Mascal M, Nikitin EB (2008) Direct, high-yield conversion of cellulose into biofuel. Angew Chem Int Ed 47:7924–7926
61. Wozniak B, Tin S, de Vries JG (2019) Bio-based building blocks from 5-hydroxymethylfurfural via 1-hydroxyhexane-2,5-dione as intermediate. Chem Sci 10:6024–6034
62. Mariscal R, Maireles-Torres P, Ojeda M, Sádaba I, López Granados M (2016) Furfural: a renewable and versatile platform molecule for the synthesis of chemicals and fuels. Energy Environ Sci 9:1144–1189
63. Wu WP, Xu YJ, Chang SW, Deng J, Fu Y (2016) pH-regulated aqueous catalytic hydrogenation of biomass carbohydrate derivatives by using semisandwich iridium complexes. ChemCatChem 8:3375–3380

64. Townsend TM, Kirby C, Ruff A, O'Connor AR (2017) Transfer hydrogenation of aromatic and linear aldehydes catalyzed using Cp*Ir(pyridinesulfonamide)Cl complexes under base-free conditions. J Organomet Chem 843:7–13
65. Garhwal S, Maji B, Semwal S, Choudhury J (2018) Ambient-pressure and base-free aldehyde hydrogenation catalyst supported by a bifunctional abnormal NHC ligand. Organometallics 37:4720–4725
66. Thananatthanachon T, Rauchfuss TB (2010) Efficient route to hydroxymethylfurans from sugars via transfer hydrogenation. ChemSusChem 3:1139–1141
67. Padilla R, Koranchalil S, Nielsen M (2020) Efficient and selective catalytic hydrogenation of furanic aldehydes using well defined Ru and Ir pincer complexes. Green Chem. https://doi.org/10.1039/D1030GC01543A
68. Long J, Xu Y, Zhao W, Li H, Yang S (2019) Heterogeneous catalytic upgrading of biofuranic aldehydes to alcohols. Front Chem 7:529
69. Hayes DJ, Fitzpatrick S, Hayes MHB, Ross JRH (2008) Kamm B, Gruber PR, Kamm M (eds) Biorefineries-industrial processes and products. Wiley-VCH, Weinheim, pp 139–164
70. Alonso DM, Wettstein SG, Dumesic JA (2013) Gamma-valerolactone, a sustainable platform molecule derived from lignocellulosic biomass. Green Chem 15:584–595
71. Deng J, Wang Y, Pan T, Xu Q, Guo QX, Fu Y (2013) Conversion of carbohydrate biomass to γ-valerolactone by using water-soluble and reusable iridium complexes in acidic aqueous media. ChemSusChem 6:1163–1167
72. Wang S, Huang H, Dorcet V, Roisnel T, Bruneau C, Fischmeister C (2017) Efficient iridium catalysts for base-free hydrogenation of Levulinic acid. Organometallics 36:3152–3162
73. Padilla R, Jørgensen MSB, Paixão MW, Nielsen M (2019) Efficient catalytic hydrogenation of alkyl levulinates to γ-valerolactone. Green Chem 21:5195–5200
74. Li W, Xie JH, Lin H, Zhou QL (2012) Highly efficient hydrogenation of biomass-derived levulinic acid to γ-valerolactone catalyzed by iridium pincer complexes. Green Chem 14:2388–2390
75. Anjali K, Aswini MS, Aswin P, Ganesh V, Sakthivel A (2019) Iridium tetra(4-carboxyphenyl) porphyrin, calix[4]pyrrole and Tetraphenyl porphyrin complexes as potential hydrogenation catalysts. Eur J Inorg Chem 2019:4087–4094
76. Démolis A, Essayem N, Rataboul F (2014) Synthesis and applications of alkyl levulinates. ACS Sustain Chem Eng 2:1338–1352
77. Wang K, Heltzel J, Sandefur E, Culley K, Lemcoff G, Voutchkova-Kostal A (2020) Transfer hydrogenation of levulinic acid from glycerol and ethanol using water-soluble iridium N-heterocyclic carbene complexes. J Organomet Chem 919:121310–121310
78. Wang S, Huang H, Bruneau C, Fischmeister C (2017) Selective and efficient iridium catalyst for the reductive amination of levulinic acid into pyrrolidones. ChemSusChem 10:4150–4154
79. Xu Z, Yan P, Jiang H, Liu K, Zhang ZC (2017) Iridium-catalyzed reductive amination of levulinic acid to pyrrolidinones under H2 in water. Chin J Chem 35:581–585
80. Wei Y, Wang C, Jiang X, Xue D, Li J, Xiao J (2013) Highly efficient transformation of levulinic acid into pyrrolidinones by iridium catalysed transfer hydrogenation. Chem Commun 49:5408–5410
81. Wang S, Huang H, Bruneau C, Fischmeister C (2019) Formic acid as a hydrogen source for the iridium-catalyzed reductive amination of levulinic acid and 2-formylbenzoic acid. Cat Sci Technol 9:4077–4082
82. Guan C, Pan YP, Zhang TH, Ajitha MJ, Huang KW (2020) An update on formic acid dehydrogenation by homogeneous catalysis. Chem Asian J 15:937–946
83. Onishi N, Laurenczy G, Beller M, Himeda Y (2018) Recent progress for reversible homogeneous catalytic hydrogen storage in formic acid and in methanol. Coord Chem Rev 373:317–332
84. Wang S, Huang H, Roisnel T, Bruneau C, Fischmeister C (2019) Base-free dehydrogenation of aqueous and neat formic acid with iridium(III) Cp*(dipyridylamine) catalysts. ChemSusChem 12:179–184

85. Li Y, Sponholz P, Nielsen M, Junge H, Beller M (2015) Iridium-catalyzed hydrogen production from monosaccharides, disaccharide, cellulose, and lignocellulose. ChemSusChem 8:804–808
86. Cortright RD, Davda RR, Dumesic JA (2002) Hydrogen from catalytic reforming of biomass-derived hydrocarbons in liquid water. Nature 418:964–967
87. Yoshida M, Hirahata R, Inoue T, Shimbayashi T, Fujita KI (2019) Iridium-catalyzed transfer hydrogenation of ketones and aldehydes using glucose as a sustainable hydrogen donor. Catalysts 9:503
88. Kumar A, Semwal S, Choudhury J (2019) Catalytic conversion of CO 2 to Formate with renewable hydrogen donors: an ambient-pressure and H 2 -independent strategy. ACS Catal 9:2164–2168
89. Semwal S, Kumar A, Choudhury J (2018) Iridium-NHC-based catalyst for ambient pressure storage and low temperature release of H2 via the CO2/HCO2H couple. Cat Sci Technol 8:6137–6142
90. Borja P, Vicent C, Baya M, García H, Mata JA (2018) Iridium complexes catalysed the selective dehydrogenation of glucose to gluconic acid in water. Green Chem 20:4094–4101
91. Mollar-Cuni A, Byrne JP, Borja P, Vicent C, Albrecht M, Mata JA (2020) Selective conversion of various monosaccharaides into sugar acids by additive-free dehydrogenation in water. ChemCatChem 12:3746–3752
92. Sun ZH, Fridrich B, de Santi A, Elangovan S, Barta K (2018) Bright side of lignin Depolymerization: toward new platform chemicals. Chem Rev 118:614–678
93. Fache M, Boutevin B, Caillol S (2016) Vanillin production from lignin and its use as a renewable chemical. ACS Sustain Chem Eng 4:35–46
94. Zhu R, Wang B, Cui M, Deng J, Li X, Ma Y, Fu Y (2016) Chemoselective oxidant-free dehydrogenation of alcohols in lignin using Cp∗Ir catalysts. Green Chem 18:2029–2036
95. Lancefield CS, Teunissen LW, Weckhuysen BM, Bruijnincx PCA (2018) Iridium-catalysed primary alcohol oxidation and hydrogen shuttling for the depolymerisation of lignin. Green Chem 20:3214–3221
96. Deuss PJ, Scott M, Tran F, Westwood NJ, de Vries JG, Barta K (2015) Aromatic monomers by in situ conversion of reactive intermediates in the acid-catalyzed depolymerization of lignin. J Am Chem Soc 137:7456–7467
97. Zhang Z, Ziijlstra DS, Lahive CW, Deuss PJ (2020) Combined lignin defunctionalisation and synthesis gas formation by acceptorless dehydrogenative decarbonylation. Green Chem:4778523
98. Olsen EPK, Madsen R (2012) Iridium-catalyzed dehydrogenative decarbonylation of primary alcohols with the liberation of syngas. Chem Eur J 18:16023–16029
99. Biermann U, Bornscheuer U, Meier MAR, Metzger JO, Schafer HJ (2011) Oils and fats as renewable raw materials in chemistry. Angew Chem Int Ed 50:3854–3871
100. Mishra VK, Goswami R (2018) A review of production, properties and advantages of biodiesel. Biofuels 9:273–289
101. Lu Z, Cherepakhin V, Kapenstein T, Williams TJ (2018) Upgrading biodiesel from vegetable oils by hydrogen transfer to its fatty esters. ACS Sustain Chem Eng 6:5749–5753
102. Kovács J, Todd TD, Reibenspies JH, Joó F, Darensbourg DJ (2000) Water-soluble organometallic compounds. 9. Catalytic hydrogenation and selective isomerization of olefins by water-soluble analogues of Vaska's complex. Organometallics 19:3963–3969
103. de Vries JG, Deuss PJ, Barta K (2017) Kamer PCJ, Vogt D, Thybaut J (eds) Contemporary catalysis: science, technology, and applications. Royal Society of Chemistry, London, pp 29–73
104. Ternel J, Lebarbé T, Monflier E, Hapiot F (2015) Catalytic decarbonylation of biosourced substrates. ChemSusChem 8:1585–1592
105. Maetani S, Fukuyama T, Suzuki N, Ishihara D, Ryu I (2011) Efficient iridium-catalyzed decarbonylation reaction of aliphatic carboxylic acids leading to internal or terminal alkenes. Organometallics 30:1389–1394

106. Ternel J, Léger B, Monflier E, Hapiot F (2018) Amines as effective ligands in iridium-catalyzed decarbonylative dehydration of biosourced substrates. Cat Sci Technol 8:3948–3953
107. Ternel J, Couturier JL, Dubois JL, Carpentier JF (2015) Rhodium versus iridium catalysts in the controlled tandem hydroformylation-isomerization of functionalized unsaturated fatty substrates. ChemCatChem 7:513–520
108. Miao X, Fischmeister C, Dixneuf PH, Bruneau C, Dubois JL, Couturier JL (2012) Polyamide precursors from renewable 10-undecenenitrile and methyl acrylate via olefin cross-metathesis. Green Chem 14:2179–2183
109. Ternel J, Couturier JL, Dubois JL, Carpentier JF (2013) Rhodium-catalyzed tandem isomerization/hydroformylation of the bio-sourced 10-undecenenitrile: selective and productive catalysts for production of polyamide-12 precursor. Adv Synth Catal 355:3191–3204
110. Huber T, Firlbeck D, Riepl HM (2013) Iridium-catalysed isomerising trialkylsilylation of methyl oleateq. J Organomet Chem 744:144–148
111. Lu B, Falck JR (2010) Iridium-catalyzed (Z)-trialkylsilylation of terminal olefins. J Org Chem 75:1701–1705
112. Ghebreyessus KY, Angelici RJ (2006) Isomerizing-hydroboration of the monounsaturated fatty acid ester methyl oleate. Organometallics 25:3040–3044

Top Organomet Chem (2021) 69: 397–454
https://doi.org/10.1007/3418_2020_60

Published online: 29 December 2020

Iridium Nanoparticles for Hydrogenation Reactions

Luis M. Martínez-Prieto, Israel Cano, and Piet W. N. M. van Leeuwen

Contents

Abstract The use of iridium nanoparticles (Ir NPs) as catalysts for hydrogenation reactions is reviewed with an emphasis on the recent advances in this area. Different types of Ir NPs are examined: NPs immobilised on supports, ligand-stabilised NPs,

L. M. Martínez-Prieto
ITQ, Instituto de Tecnología Química, CSIC-Universitat Politécnica de València, Valencia, Spain
e-mail: luismiguel.martinez@csic.es

I. Cano
Departamento de Física Aplicada, Facultad de Ciencias, Universidad de Cantabria, Santander, Spain
e-mail: israel.canorico@unican.es

P. W. N. M. van Leeuwen (✉)
LPCNO, Laboratoire de Physique et Chimie des Nano-Objets, Institut National des Sciences Appliquées-Toulouse, Toulouse, France
e-mail: vanleeuw@insa-toulouse.fr

confined NPs, NPs stabilised by ionic liquids and polymers and NPs generated in situ without stabilising agent. A key issue is the role of the stabiliser in the catalytic process (activity, selectivity and recyclability). General trends in the use of conditions, stabilisers, additives and co-catalysts were also observed. In spite of the advances achieved in the last decade, there is still a quest for Ir NP-based catalysts with sufficient selectivity to be industrially applied in fine chemistry.

Keywords Carbon materials · Cyclohexene hydrogenation · Hydrogenation reactions · Ionic liquids · Iridium nanoparticles (Ir NPs) · Ligand-stabilised Ir NPs · Metal oxide supports · Nanocatalysis · Polymer supports · α,β-unsaturated carbonyl compounds

1 Introduction

Catalysis is crucial for green (or sustainable) chemistry, which has the objective to reduce or eliminate the use and generation of hazardous substances. Apart from being potentially selective, catalytic processes permit the use of stoichiometric amounts of reagents, which reduces the amount of by-products and residues. Catalysis can be homogeneous or heterogeneous, depending on whether the substrates and catalysts are in the same phase or not. Each type of catalysis presents a series of advantages and disadvantages. For example, homogeneous catalysts tend to be more selective and active than heterogeneous ones, but often they are less stable and recyclable. In between, we can find metal nanoparticles (MNPs), which combine the benefits of heterogeneous and homogeneous catalysts [1–3]. MNPs, as well as heterogeneous catalysts, possess the possibility to be easily recovered and subsequently reused. Besides, MNPs present a large number of active sites due to their high surface/volume ratio, which makes them very active in catalysis approaching homogeneous ones [4–6]. Their high conversions and selectivities together with an easy product separation make MNPs ideal catalysts for catalytic processes. For all this, MNP catalysis is gaining a lot of interest in both academia and industry.

The main objective in MNP catalysis is to develop appropriate tools to produce well-defined nano-objects with controlled size, morphology and catalytic properties. To this end, stabilisers are the best way to control these parameters. Among all stabilising agents used for MNP stabilisation, we can find surfactants [7–10], polymers [11, 12], ionic liquids [13, 14], ligands [15, 16] and supports [17, 18], which substantially improve the MNP stability. In addition to protecting MNPs through electronic and/or steric stabilisations, they also are able to modify the catalytic properties. For example, through a synergistic effect with the nanoparticle surface, stabilising ligands are able to enhance the stability and activity of MNPs [19, 20].

Hydrogenation of unsaturated compounds using molecular H_2 is one of the most efficient and atom-economical way for the formation of high added value compounds. For example, the hydrogenation of aromatic compounds is of great relevance for industry and organic synthesis [21], since the corresponding cyclohexanes/cyclohexenes are precursors for industrial organic chemistry [22] or key intermediates in pharmaceuticals [23]. Recent works have demonstrated the capability of MNPs for the catalytic hydrogenation of olefins [24, 25], alkynes [26–28], arenes [29, 30], ketones [31–33], aldehydes [34–36], nitrocompounds [37, 38] and carboxylic acids [39, 40], with good levels of chemoselectivities. Among the MNPs commonly used in catalysis, iridium nanoparticles (Ir NPs) have been widely used due to their high activity and superior oxidation resistance when exposed to air. In addition, their large number of oxidation states and the ability to coordinate and activate a great number of substrates (hydrogen, olefins, carbonyl compounds, etc.) make them efficient catalyst for hydrogenation reactions [41]. In this chapter we intend to give an overview of the recent advances of Ir NPs in hydrogenation reactions. In principle, the Ir NP catalysts will be divided into (1) ligand-stabilised Ir NPs, (2) IL-stabilised Ir NPs, (3) polymer-stabilised Ir NPs, (4) "naked" Ir NPs and (5) supported Ir NPs, emphasising the influence of the stabilising agent in the hydrogenation activity.

2 Hydrogenation Reactions Catalysed by Ligand-Stabilised Iridium Nanoparticles

The use of organic ligands as stabilisers for the synthesis of metal nanoparticles (MNPs) with applications in catalysis has been widely explored. These ligands usually have a simple function as a stabilising agent. However, there are also many examples of ligand-stabilised MNPs in which the ligand acts both as stabiliser and as functional ligand playing a role in the catalytic process [42]. In this context, Claver, Roucoux and coworkers described the synthesis of Ir NPs by H_2 reduction of bis(1,5-cyclooctadiene)diiridium(I) dichloride ($[Ir(cod)Cl]_2$) in the presence of chiral diphosphite ligands derived from carbohydrates (Scheme 1a) [43]. These diphosphite-stabilised Ir NPs were employed as catalysts for methylanisole hydrogenation (Scheme 1b), leading to significant variations in activity depending on the ligand acting as stabilising agent (Table 1). However, no insights about the factors that produce these differences in activity were given. In addition, very low enantiomeric excess values (ees) were obtained, although a slight increase in the ees up to 6% was achieved with Rh and Ru NPs prepared this way, which suggest some influence of the ligand on the hydrogenation process. Finally, no hydrogenation reaction was observed with Ir NPs prepared from $[Ir(cod)_2]BF_4$ (Table 1). This was attributed to a poisoning effect caused by the coordination of the counteranion to the nanoparticle surface.

Claver and coworkers also employed chiral diphosphines for the synthesis of ligand-stabilised Ir NPs supported on SiO_2 [44]. Ir NPs of 1.6 nm in size ligated by (2*R*,4*R*)-2,4-bis(diphenylphosphino)pentane ((*R*,*R*)-BDPP, Scheme 2) were

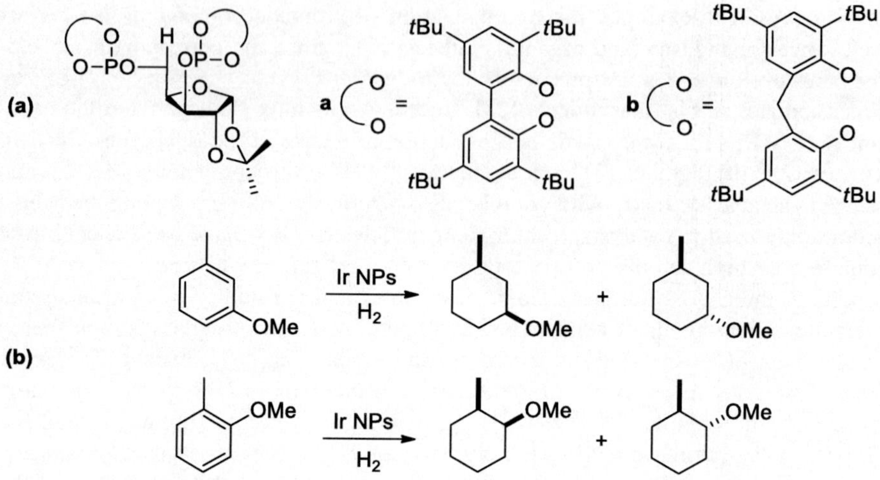

Scheme 1 (**a**) Diphosphite ligands used as stabilising agents for Ir NPs; (**b**) hydrogenation of *m*– and *o*–methylanisole by Ir NPs

Table 1 Hydrogenation of *m*– and *o*–methylanisole by Ir NPs stabilised with chiral diphosphite ligands derived from carbohydrates

L	Size (nm)	Substrate	Conversion (%)	Cis (%)	ee (%)	TON
a	1.3	*m*-methylanisole	87	81	–	70
b	1.5	*m*-methylanisole	14	75	–	12
a[a]	1.4	*m*-methylanisole	0	–	–	–
a	1.3	*o*-methylanisole	96	100	1	77
b	1.5	*o*-methylanisole	4	100	2	3
a[a]	1.4	*o*-methylanisole	0	–	–	–

Reaction conditions: Substrate/Ir = 40, methylanisole (1.24 mmol), pentane (10 mL), H_2 (40 bar), room temperature (RT), 24 h
[a] $[Ir(cod)_2]BF_4$ as precursor

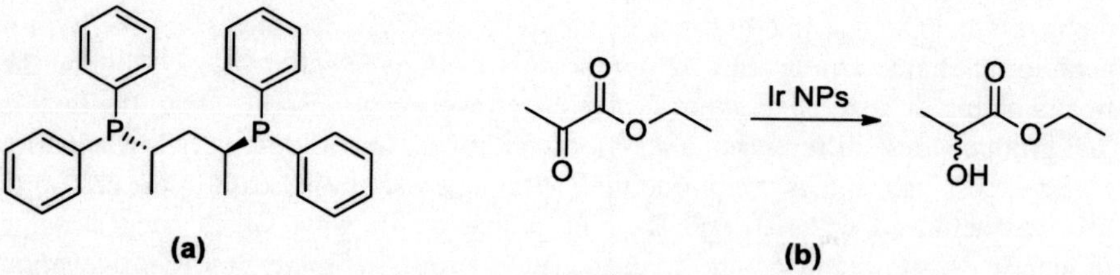

Scheme 2 (**a**) (*R,R*)-BDPP ligand employed for the synthesis of Ir NPs. (**b**) Asymmetric hydrogenation of ethyl pyruvate to ethyl lactate

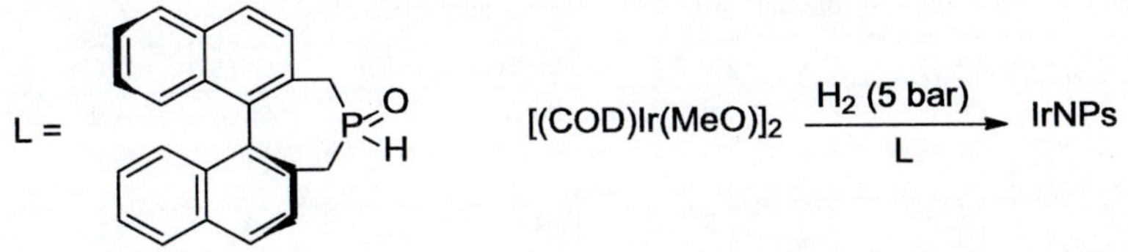

Scheme 3 Synthesis of Ir NPs by H_2 reduction of $[Ir(OMe)(COD)]_2$ in the presence of 3,5-dihydro-4*H*-dinaphtho[2,1-c:1′,2′-e] phosphepine-4-oxide

synthesised through $NaBH_4$ reduction of $IrCl_3$, which were subsequently immobilised in SiO_2 by impregnation (1 wt% Ir). These Ir NPs gave 90% conversion and 23% ee in the asymmetric hydrogenation of ethyl pyruvate (substrate/Ir = 100, 40 bar H_2, RT, 460 min), which shows a chiral induction from the ligand.

In the same way, Cano, van Leeuwen and coworkers described the synthesis of Ir NPs of 1.4(0.2) nm in size by H_2 reduction of $[Ir(OMe)(COD)]_2$ in the presence of a chiral secondary phosphine oxide (SPO), 3,5-dihydro-4*H*-dinaphtho[2,1-c:1′,2′-e] phosphepine-4-oxide (Scheme 3) [45]. The Ir NPs contain both Ir(0) and Ir(I) atoms, and a 1.75 Ir:SPO ratio was found, which entails that 90% surface iridium atoms are bound to SPO ligands. In addition, IR and ^{31}P NMR analyses suggested that the SPO molecules are bound to the nanoparticle surface as phosphoryl anions.

These chiral NPs were active catalysts for the asymmetric hydrogenation of prochiral ketones, leading to ees up to 56% (Table 2). In line with previous studies on hydrogenation by SPO-ligated Au NPs [46, 47], a heterolytic hydrogenation mechanism through a ligand-metal cooperative effect was proposed, in such a way that the oxygen atom of the SPO operates as a Lewis base and takes the proton (H^+), while a neighbouring iridium atom acts as a Lewis acid and captures the hydride (H^-).

Among the different coordination modes of SPOs, these ligands can coordinate to metals (Scheme 4) as a neutral phosphinous acid (**1**), as deprotonated phosphinito anion (**2**) and as monoanionic bidentate ligand (**3**). In this context, subsequent IR and ^{31}P MAS NMR investigations on Ir NPs ligated by different SPOs (Scheme 4) demonstrated that neutral acid coordination type increases with the SPO basicity [48]. In addition, these studies showed that dicyclohexylphosphine oxide (Scheme 4c) is mainly bound as a bidentate ligand. However, *tert*-butyl(phenyl)phosphine oxide and diphenylphosphine oxide (Scheme 4a and b, respectively) also coordinate as the phosphinous acid and the phosphinito anion on the NP surface. These three types of SPO-stabilised Ir NPs showed high selectivity towards the allylic alcohol product in the hydrogenation of cinnamaldehyde (substrate/Ir = 400, 10 bar H_2, THF, RT), for which a heterolytic H_2 cleavage was also suggested.

Finally, the Ir NPs ligated by *tert*-butyl(phenyl)phosphine oxide were also employed in the hydrogenation of several substituted aldehydes (Table 3) [49]. In addition to the high selectivity observed in cinnamaldehyde hydrogenation, these NPs of 1.33(0.28) nm in size were highly selective towards the unsaturated alcohol

Table 2 Asymmetric hydrogenation of ketones mediated by Ir NPs

Entry	Substrate	Ir NP	Conversion (%)	ee (%)	Configuration
1	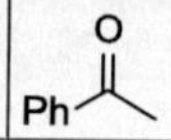	R	88	55	S
2		S	47	52	R
3	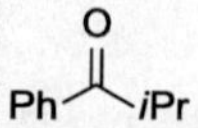	R	38	30	S
4		S	11	27	R
5	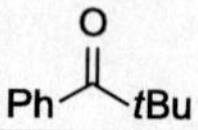	R	90	22	S
6		S	38	20	R
7	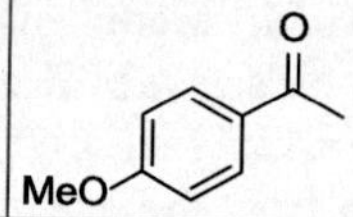	R	63	56	S
8		S	14	39	R
9	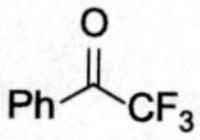	R	100	10	S
10		S	100	8	R
11	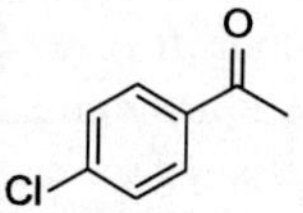	R	98	50	S
12		S	51	34	R
13	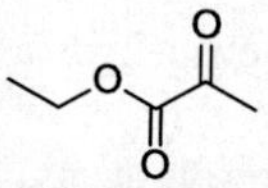	R	100	26	S
14		S	100	30	R

Reaction conditions: Substrate/Ir = 100, substrate (0.25 mmol), THF (0.75 mL), 40 bar H_2, RT, 18 h

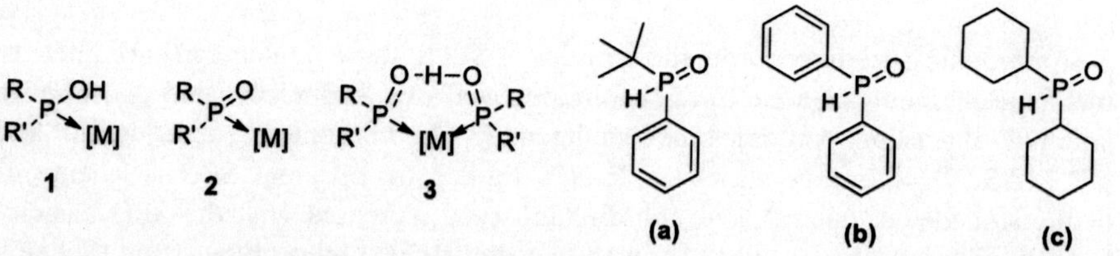

Scheme 4 Three possible coordination modes of SPOs to metals (**1–3**) and non-chiral secondary phosphine oxides employed for the synthesis of Ir NPs (**a-c**)

in the hydrogenation of *p*-cyanobenzaldehyde and 2-octynal, although that obtained in the reduction of *p*-nitrobenzaldehyde was lower.

A ligand effect was also reported in the hydrogenation of α,β-unsaturated carbonyl compounds by Ir NPs (2.1 (0.2) nm) stabilised with $Ph_2P(CH_2CH_2O)_{22}CH_3$ [50]. These NPs showed high selectivity towards the hydrogenation of the aldehyde group. By contrast, a preference for the C=C bond of unsaturated ketones was also observed (Table 4). The authors suggested different adsorption models for aldehydes and ketones on the NP surface. The C=O group would be preferentially adsorbed via

Table 3 Catalytic hydrogenation of substituted aldehydes by SPO-stabilised Ir NPs

Substrate	Product	P (bar)	Conv. (%)	Sel. (%)	TOF (h^{-1})
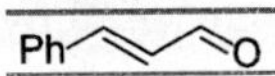	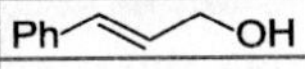	10	99	99	22
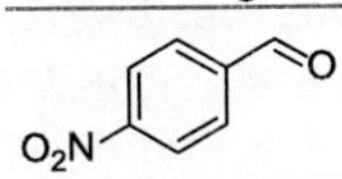	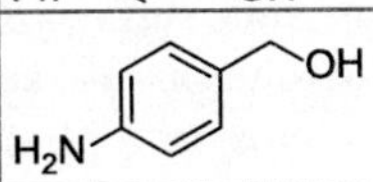	20	>99	95	22
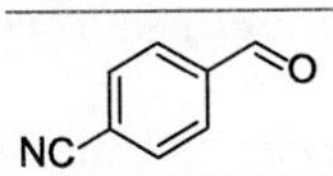	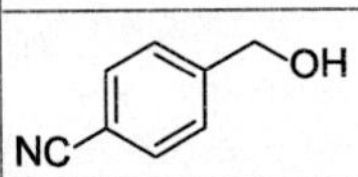	40	>99	>99	11
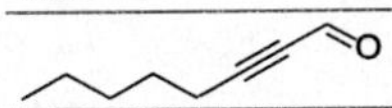	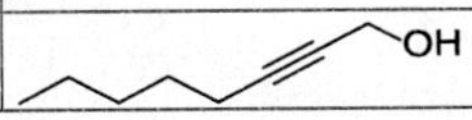	40	80	96	18

Reaction conditions: Substrate/Ir = 400, Ir NPs (0.0025 mmol Ir), THF (0.75 mL), RT. 18 h

Table 4 Hydrogenation of α,β-unsaturated carbonyl compounds mediated by Ir NPs stabilised by $Ph_2P(CH_2CH_2O)_{22}CH_3$

Substrate	Product	Time (h)	Conv. (%)	Selec. (%)[a]
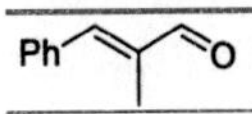	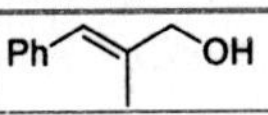	3	96	>99
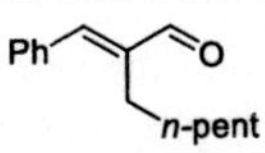	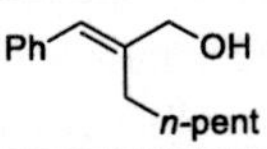	7	90	>99
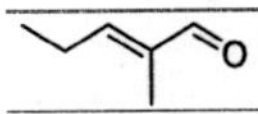	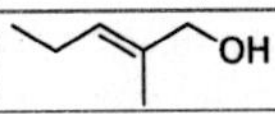	7	72	98
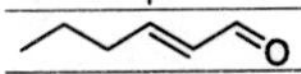	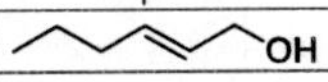	3	>99	96
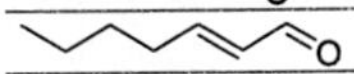	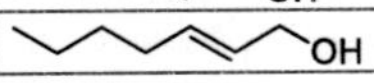	3.5	>99	94
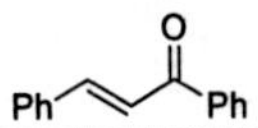	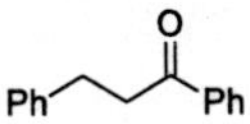	9	98	>99
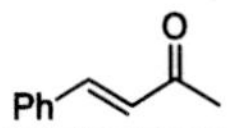	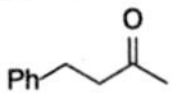	2.5	76	90
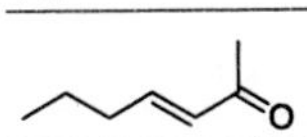	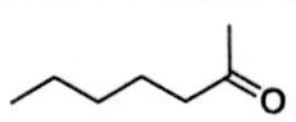	3.5	>99	>99
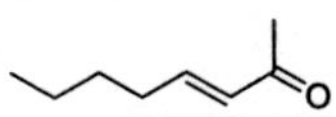	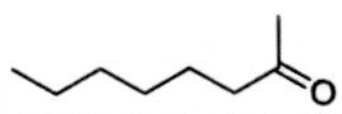	4	>99	>99
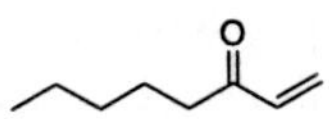	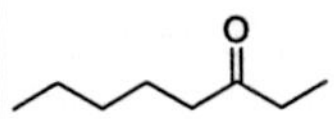	1	>99	>99

Reaction conditions: Substrate/Ir = 100, 1-pentanol (4 mL), H_2O (4 mL) containing 0.0067 mmol Ir NPs. 10 bar H_2, 70°C

[a]The remainder is the saturated alcohol

the aldehydes, giving high selectivities towards the unsaturated alcohol. However, preference for the C=C bond adsorption would occur when the substrate is a ketone. It was proposed that ligands bound to the NP surface constrain the surface site geometries which only allow certain adsorption models.

As commented above, there are also several examples on hydrogenation by Ir NP-based catalytic systems in which the ligand only acts as a stabilising agent.

Roucoux, Chaudret and coworkers employed an aqueous solution of Ir(0) NPs stabilised by N,N-dimethyl-N-cetyl-N-(2-hydroxyethyl)ammonium chloride for arene hydrogenation [51]. The Ir NPs of 1.9 (0.7) nm in size were synthesised by $NaBH_4$ reduction of $IrCl_3$ in the presence of this surfactant under microwave-assisted conditions. These NPs were stored in solution and subsequently used for the hydrogenation of several aromatic compounds in biphasic media (substrate/Ir = 100, water, 40 bar H_2, 20°C), showing good recycling behaviour and TOF values up to 400 h^{-1} for the hydrogenation of anisole.

In 2005, Korgel and coworkers described the use of oleic acid and oleylamine, tetraoctylammonium bromide (TOAB), trioctylphosphine (TOP) and tetraoctylphosphonium bromide (TOPB) to stabilise Ir NPs, which were subsequently employed as catalysts for 1-decene hydrogenation [52]. The NP size decreases in the order TOP (10–100 nm) > TOPB (~4 nm) > oleic–oleylamine (~2 and ~ 4 nm by selective precipitation) > TOAB (1.5–3 nm). Interestingly, TOPB-ligated Ir NPs showed higher activity than those stabilised by TOAB, while the use of TOP and oleic acid and oleylamine led to catalytically inactive NPs. It was proposed that the ligand desorption allows the access of 1-decene to metal surface sites. The smallest NPs (TOAB) would show stronger ligand binding and thus are less active (4 s^{-1} at 75°C, 0.21 bar H_2 and 1,000:1-decene/Ir mass ratio) than the bigger NPs stabilised by the presumably labile TOPB ligand (270 s^{-1}). Similarly, oleic acid and oleylamine would bind very strongly to the surface, and therefore the corresponding NPs are not active. Indeed, before decreasing in the fifth cycle, an increase in TOF values was observed for the first four cycles in the recycling experiments with TOAB-ligated Ir NPs (4 → 13 → 50 → 124 s^{-1}), which was attributed to this ligand desorption. Although the ligand desorption is key for the observed activity, this example does not describe a real role of the ligand in the catalytic process.

In this vein, Finke and coworkers described the use of Ir NPs stabilised only by H^+Cl^-, $Ir(0)_n{\cdot}(H^+Cl^-)_a$, as structural unit to prepare new Ir NPs by exchange with more strongly binding ligands such as tributylamine (Bu_3N), tridodecylamine ($Dodec_3N$) and trioctylphosphine ($P(Oct)_3$) [53]. The $Ir(0)_n{\cdot}(H^+Cl^-)_a$ NPs are generated by H_2 reduction of $[Ir(cod)Cl]_2$ in acetone, and H^+Cl^- can be easily replaced in situ by other ligands of interest that coordinate more strongly. Indeed, the analogous H_2 reduction of $[Ir(cod)Cl]_2$ in the presence of these ligands allows one to obtain new nanoparticles, which were employed as catalysts for cyclohexene hydrogenation (Table 5). This indicates that the initial $Ir(0)_n{\cdot}(H^+Cl^-)_a$ nanoparticles

Table 5 Hydrogenation of cyclohexene catalysed by Ir NPs stabilised with different ligands

Stabiliser	Ir NP	Size (nm)	TTOs[a]	Initial TOF (s^{-1})	TOF (s^{-1})
Bu_3N	$Ir(0)_n{\cdot}(Bu_3NH^+Cl^-)^a$	1.8 ± 0.6	25,000[b]	1.1	0.2
$Dodec_3N$	$Ir(0)_n{\cdot}(Dodec_3NH^+Cl^-)^a$	1.8 ± 0.6		~1.4	0.2

Reaction conditions: cyclohexene/Ir = 1,375, Ir (1.2 mM), cyclohexene (1.65 M), acetone, H_2 (2.76 bar), 22°C
[a]TTOs: total turnovers
[b]Cyclohexene/Ir = 61,000

Table 6 Hydrogenation of various unsaturated substrates mediated by bimetallic $NiIr_4$ and monometallic Ir NPs

Substrate	NP	Time (h)	Conv. (%)	Selectivity (%)	TOF (h^{-1})
1-octene	$NiIr_4$[a]	14	100	–	5,025
Diphenylacetylene	$NiIr_4$[b]	1	90	67:23 (*Z*:*E*)	35
			Yield (%)	COL:HCOL:HCAL:A	
Cinnamaldehyde	$NiIr_4$[c]	3	19	23:2:3:72	419
Cinnamaldehyde	Ir[c]	3	14	22:3:2:73	675
Cinnamaldehyde	$NiIr_4$[d]	22	9	60:7:33:0	10

Reaction conditions: 20 mg $NiIr_4$, hexane (1–5 mL), H_2 (10 bar), 80°C
CAL cinnamaldehyde, *COL* cinnamic alcohol, *HCOL* hydrocinnamic alcohol, *HCAL* hydrocinnamic aldehyde, *A* acetals
[a]30 mg $NiIr_4$, RT
[b]THF (30 mL)
[c]CAL (16 mmol), methanol (30 mL)
[d]CAL (16 mmol), isopropanol (30 mL)

may be used as building blocks for the production of catalytically active Ir NPs with a controlled ligand composition.

The use of organic compounds that serve as solvent, as reducing agent and as stabiliser is an interesting strategy to prepare ligand-stabilised MNPs. Following this approach, highly crystalline $NiIr_4$ NPs (Ir/Ni ratio: 4) were synthesised by heating of $IrCl_3$ in oleylamine at 300°C, which were subsequently isolated by centrifugation [54]. These bimetallic NPs of 9 nm in size were an active catalyst for the hydrogenation of 1-octene, diphenylacetylene and cinnamaldehyde (Table 6). Monometallic Ir NPs were also prepared through a similar procedure (ethylene glycol, 150°C) and used as a reference. High activities were observed in the hydrogenation of 1-octene and cinnamaldehyde. However, the latter gave a mixture of several hydrogenation products, including acetals. These acetal by-products were generated by reaction between cinnamaldehyde and the alcohol solvent employed in the catalytic transformation, although the process is reversible and their presence was minimised with long reaction times. Finally, the hydrogenation of diphenylacetylene proceeds with high selectivity to stilbene. Unfortunately, the $NiIr_4$ catalyst did not show advantages as compared to monometallic Ir NPs.

3 Hydrogenation Reactions Catalysed by Iridium Nanoparticles Stabilised in Ionic Liquids

Ionic liquids (ILs), salts of which the melting point is below 100°C [55], have received special attention in catalysis because they possess important features that make them very suitable as, for instance, good thermal stability, high conductivity, low vapour pressure, high catalytic activity, ease of catalyst recovery and product purification and easy fine-tuning basicity-nucleophilicity by simple modification of

IL structure [56, 57]. Interestingly, ILs can operate as electrostatic and/or electrosteric stabilisers for metallic nanoparticles preventing their agglomeration [58]. Therefore, MNPs can be generated and immobilised in ILs, allowing the straightforward separation of products and the recycling of the system [59, 60]. Since the pioneering work of Dupont et al. [61], in the last two decades, IL-stabilised MNPs have attracted a large interest by researchers working in catalysis. In this study, iridium nanoparticles (Ir NPs) of 2.0(0.4) nm in size were generated by H_2 reduction of $[Ir(cod)Cl]_2$ in 1-butyl-3-methylimidazolium hexafluorophosphate ($bmimPF_6$). The resulting Ir(0)/$bmimPF_6$ system was employed as catalyst for the hydrogenation of several alkenes under mild reaction conditions (4 bar H_2 and 75°C), showing high activities (TOF = 6,000 h^{-1}) and a good recycling behaviour. The reported nanocatalysts can be reused for up to at least seven cycles without any significant loss in activity. This publication was the starting point for a series of studies by this group on catalytic hydrogenation mediated by IL-stabilised Ir NPs. Indeed, the Ir(0) NPs stabilised by $bmimPF_6$ were used, both as a solid (referred to nanoparticles that were isolated through centrifugation and washed with organic solvents and subsequently redispersed in the neat substrate) or redispersed in the ionic liquid (biphasic conditions), for the hydrogenation of arenes [62] and ketones [63] (Table 7). Similarly, these Ir NPs were successfully recycled up to 7 and 15 times, respectively.

In a subsequent publication, competitive hydrogenation experiments of alkylbenzenes (benzene, toluene, ethylbenzene, isopropylbenzene, *sec*-butylbenzene and *tert*-butylbenzene) allowed the researchers to determine the selectivity constants (S) against toluene of this type of Ir NPs (Table 8) and other $bmimPF_6$-stabilised MNPs, which may be employed to predict the relative reactivity of new monoalkylbenzenes pairs [64]. The correlation of initial reaction rate constant for toluene/benzene and toluene/monoalkylbenzene hydrogenation experiments with the Taft equation indicates that the rates of hydrogenation of alkylbenzenes only depend on steric impediment of the alkyl substituents. Finally, it was suggested that bulky substituents slow the hydrogenation rate.

The kinetics for the hydrogenation of 1-decene catalysed by $[Ir(cod)Cl]_2$) in $bmimPF_6$ was also investigated, showing an autocatalytic mechanism in which iridium nanoclusters are formed [65]. The curves obtained in the hydrogenation of different alkenes were fitted to a model that implies (1) nucleation ($A \rightarrow B$, $k_1 = 2.04\ h^{-1}$) and (2) autocatalytic surface growth ($A + B \rightarrow 2B$, $k_2 = 5{,}125\ M^{-1}\ h^{-1}$). The small and monomodal size of NPs in the Ir/$bmimPF_6$ system (2.0(0.4) nm) indicates a strong stabilisation by the IL and that (1) is faster than (2). Indeed, XPS, EXAFS and SAXS studies proved the presence of a protective IL shell with a diameter of ca. 2.8–4.0 nm surrounding the Ir NPs for several Ir/bmimX systems (X = PF_6, BF_4, NTf_2) [66]. Additional kinetic experiments on the hydrogenation of 1-decene by the Ir/$bmimPF_6$ system demonstrated that the catalyst follows a monomolecular surface reaction mechanism and thus acts as a heterogeneous catalytic system for alkene hydrogenation [65], that is:

Table 7 Hydrogenation of arenes and ketones mediated by $bmimPF_6$-stabilised Ir NPs used as a solid (Ir NPs) or redispersed in 1 mL of $bmimPF_6$ (Ir/IL)

Catalyst	Substrate	Product	Time (h)	Sel. (%)	TOF (h^{-1})	Ref.
Ir NPs		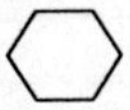	2	100	125	[62]
Ir/IL	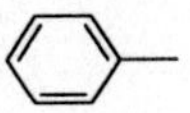	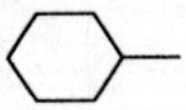	25	100	44[a]	[62]
Ir/IL		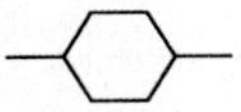	18	5:1[b]	24[c]	[62]
Ir NPs		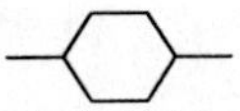	12	3:1[b]	21	[62]
Ir NPs	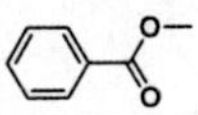	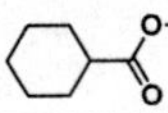	18	100	13	[62]
Ir NPs	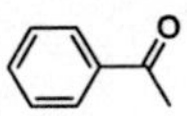	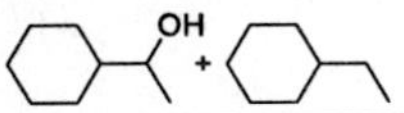	16	58/42	15	[62]
Ir NPs	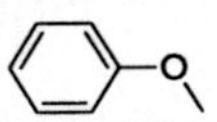	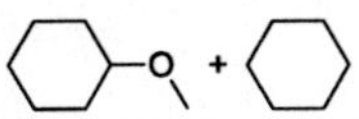	18	84/16	10	[62]
Ir NPs	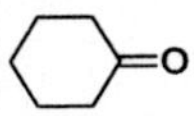	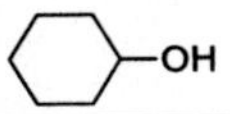	2	100	125	[63]
Ir NPs	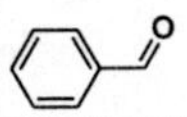	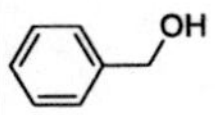	15	100	17	[63]
Ir NPs	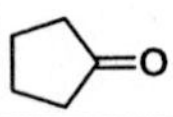	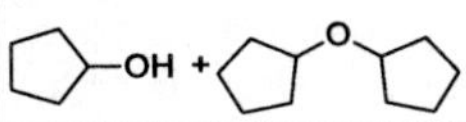	4	88/12	62.5	[63]
Ir NPs	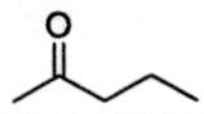	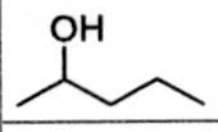	2.5	100	96	[63]
Ir NPs	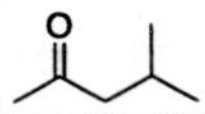	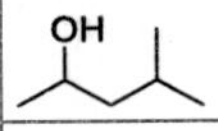	2.5	100	96	[63]
Ir NPs	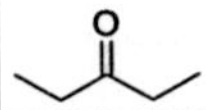	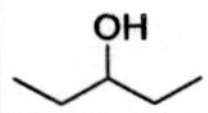	3.7	100	68	[63]
Ir NPs	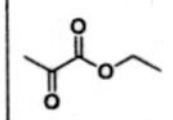	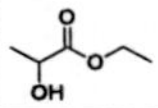	2.5	100	98	[63]
Ir NPs	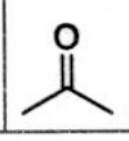	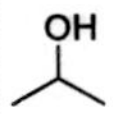	2.0	100	119	[63]

Reaction conditions: Substrate/Ir = 250, H_2 (4 bar), 75°C. Ir NPs used as a solid stands for NPs isolated by centrifugation and washed with organic solvents and redispersed in the neat substrate. Ir/IL: biphasic conditions

[a]Substrate/Ir = 1,200

[b]cis:trans

[c]Substrate/Ir = 500

Table 8 Constant selectivity (*S*) for competitive hydrogenation experiments of alkylbenzenes against toluene and relative initial reaction rate (k_{rel}) for the hydrogenation of toluene and several alkylbenzenes by bmimPF_6-stabilised Ir NPs

Alkyl	*S*	Charton (*ν*)	Taft (E_s)	log $k_{rel.}$
H	1.61	−0.004	1.24	0.301
Me	1	0	0	0
Et	0.86	0.12	−0.07	−0.12
i-Pr	0.35	0.20	−0.47	−0.48
s-Bu	0.21	0.46	−1.13	−0.60
t-Bu	0.13	0.68	−1.54	−0.98

Reaction conditions: Ir NPs (0.026 mmol), substrate/catalyst = 250, H_2 (4 bar), 75°C

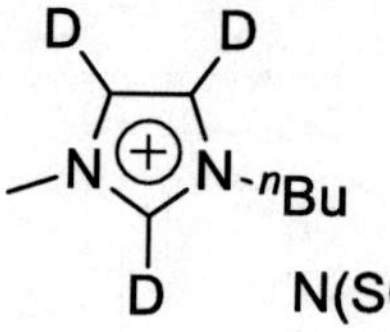

Fig. 1 Deuterated IL used in reference [67]

$$\left(v = \frac{k_c K[S]}{1 + K[S]}\right)$$

The kinetic studies also showed that the hydrogenation reaction is a mass transfer-controlled process at $[H_2] < 4$ bar. The catalytic kinetic constant (*kc*) at $[H_2] \geq 4$ bar and the adsorption constant (*K*) are independent of H_2 pressure. This reveals that the hydrogenation process shows no dependence on $[H_2]$ and thus is only determined by [1-decene] in IL.

Later on, the same group demonstrated the participation of N-heterocyclic carbenes (NHCs) in hydrogenation reactions catalysed by Ir NPs dissolved in imidazolium-based ionic liquids [67]. bmimNTf_2-stabilised Ir(0) NPs of 4.2(0.8) nm redispersed in deuterated bmimNTf_2 ([BMI]-d_3.NTf_2, Fig. 1) were employed as catalyst for cyclohexene hydrogenation. Interestingly, the Ir NPs also promoted D/H exchange reactions at C2, C4 and C5 positions of the imidazolium cation, which proves the formation of NHCs species. A preferential D/H exchange process at the least acidic C4 and C5 rather than at the more acidic C2 suggests that the C2–H bond is less available to generate carbenes. This steric impediment indicates that the imidazolium-based IL is structured as ion pairs, in which the most favourable interaction between cation and anion is through the C2–H. Consequently, the ionic liquid interacts with the NP surface preferentially as aggregates of $\{[(Im)_x(X)_{x-n}]^{n+}[(Im)_{x-n}(X)_x]^{n-}\}_n$ type.

The hydrogenation of cyclohexene was also described for Ir NPs stabilised in imidazolium-based ILs containing different alkyl chains [68]. These NPs were prepared by H_2 reduction of different iridium precursors in $C_1C_4ImBF_4$ or $C_1C_{10}ImBF_4$ (Scheme 5). The use of $[Ir(cod)Cl]_2$ provided spherical NPs of different sizes (2.5(0.5) nm $C_1C_4ImBF_4$ and 3.6(0.9) nm in $C_1C_{10}ImBF_4$), whereas Ir NPs

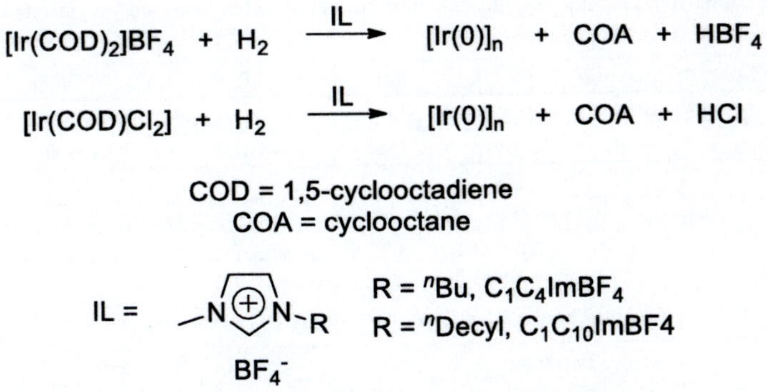

Scheme 5 Formation of Ir NPs in ILs using different Ir and ILs precursors

Table 9 Hydrogenation of 1-hexene, cyclohexene and benzene catalysed by bmimPF_6-stabilised Ir NPs, both isolated and supported on polymeric membranes

Catalyst	Substrate	Ir NPs (mg)	Time (h)[a]	TOF (h^{-1})[b]
Non-supported Ir NPs	1-hexene	5.0	0.04	638
CA/Ir	1-hexene	1.6	0.04	592
CA/bmimNTf_2/Ir	1-hexene	1.6	0.03	750
Non-supported Ir NPs	Cyclohexene	5.0	0.04	578
CA/Ir	Cyclohexene	1.6	0.09	40
CA/bmimNTf_2/Ir	Cyclohexene	1.6	0.09	270
Non-supported Ir NPs	Benzene	5.0	0.10	266
CA/Ir	Benzene	1.6	1.46	18
CA/bmimNTf_2/Ir	Benzene	1.6	0.31	78

Reaction conditions: substrate/Ir = 250, H_2 (4 bar), 75°C
[a]Time at 10% conversion
[b]TOF at 10% conversion. CA/Ir: Ir NPs isolated by centrifugation and supported on CA. CA/bmimNTf_2/Ir: Ir NPs isolated by centrifugation, redispersed in bmimNTf_2 and supported on CA

synthesised from $[Ir(cod)_2]BF_4$ are spherical or worm-like depending on the IL employed (spherical of 1.9(0.4) nm in $C_1C_4ImBF_4$ and worm-like in $C_1C_{10}ImBF_4$). The Ir NPs generated from $[Ir(cod)_2]BF_4$ show higher activity in the hydrogenation of cyclohexene, which is attributed to the smaller NP size or the surface poisoning effect of Cl^- species on NPs prepared from $[Ir(cod)Cl]_2$. Finally, almost all Ir/IL systems were successfully reused up to eight cycles without any significant loss in the catalytic activity.

Ir(0) NPs stabilised by bmimPF_6 (2.1 nm size) were synthesised using DuPont's method [61, 63] and subsequently immobilised on a cellulose acetate (CA)-based polymeric membrane [69]. The CA/bmimNTf_2/Ir catalytic system showed higher activity in the hydrogenation of 1-hexene than the non-supported Ir NPs, while the latter was much more active for cyclohexene and benzene hydrogenation (Table 9).

Table 10 Biphasic liquid-liquid hydrogenation of cyclohexene catalysed by various Ir/bmimBF_4 systems

Substrate/catalyst	P (bar)	T (°C)	Time (min)	Conv. (%)	TOF (h^{-1})	Ref.
2,000	4	4	75	97	1,940	[70][a]
2,000	4	4	75	100	800	[70][a]
2,108	10	10	90	36	222	[71][b]

[a]0.08 ml Ir/IL dispersion (0.2 metal wt%, c = 0.0128 mmol/mL) and cyclohexene (0.2 mL)
[b]0.09 mmol Ir in 1.5 mL IL (1 metal wt%, c = 0.06 mmol/mL) and cyclohexene (20 mL)

Table 11 Hydrogenation of benzene and cyclohexene by Ir/TRGO catalysts

	Benzene		Cyclohexene	
Catalyst	Conversion (%)	TOF (h^{-1})	Conversion (%)	TOF (h^{-1})
Ir@TRGO (MW)[a]	80.4	10,280	85.5	68,250
Ir@TRGO (e-beam)[b]	97.4	4,861		

Conditions: H_2 (10 bar), 100°C, 1 h
[a]Benzene: Ir@TRGO (5.3 mg, 3.6% Ir), benzene/Ir = 4,490. Cyclohexene: Ir@TRGO (14 mg, 3.6% Ir), cyclohexene/Ir = 3,991, 3 min
[b]Ir@TRGO (8.6 mg, 7.2% Ir), benzene/Ir = 3,943

The iridium catalysts were successfully recycled up to seven times, but an improved recyclability was observed for the polymeric catalytic systems.

In addition, it is worth noting the work of Janiak and coworkers, in which Ir NPs stabilised in bmimX (X = BF_4, OTf) and BtMANTf_2 (BtMA = n-butyl-tri-methyl-ammonium) were synthesised by thermal decomposition of $Ir_4(CO)_{12}$ [70]. Well-dispersed Ir NPs of 1.1–3.6 nm in size were obtained in bmimX, while the use of BtMANTf_2 as stabilising media led to highly agglomerated Ir NPs. The resulting Ir/bmimBF_4 system was a very active catalyst for the biphasic liquid-liquid hydrogenation of cyclohexene, with TOF values up to 1,940 h^{-1} (Table 10, entries 1 and 2). Following this approach, both microwave-assisted and photolytic conditions were employed to decompose $Ir_6(CO)_{16}$ in bmimBF_4 and thus generate Ir NPs of 0.8 and 1.4 nm in size, respectively [71]. The Ir/bmimBF_4 system obtained under microwave (MW) conditions was again used as cyclohexene hydrogenation catalyst, but in this case, the observed activity was lower (Table 10, entry 3).

The iridium-carbonyl precursor $Ir_4(CO)_{12}$ described above was also used to generate graphene-supported Ir NPs in ILs [72]. $Ir_4(CO)_{12}$ dispersed in bmimBF_4 was thermally decomposed under MW conditions (200 W, 250°C, 3 × 20 min) or by electron-beam (e-beam) irradiation (10 MeV, 3 × 1 s) in the presence of graphite oxide (TRGO), leading to the formation of Ir NPs with 2.7(0.7) and 3.6(1.0) nm in size, respectively. Such Ir/TRGO catalytic systems were highly active in the hydrogenation of benzene and cyclohexene (Table 11). TOF values up to ~10,300 h^{-1} were observed in the hydrogenation of benzene to cyclohexane. In addition, the catalysts were reused up to 10 (MW-synthesised NPs) and 5 (e-beam-generated NPs) times, showing good recycling behaviour. Based on the correlation between catalytic activity and NP size, the authors suggest that there could be an optimal NP diameter

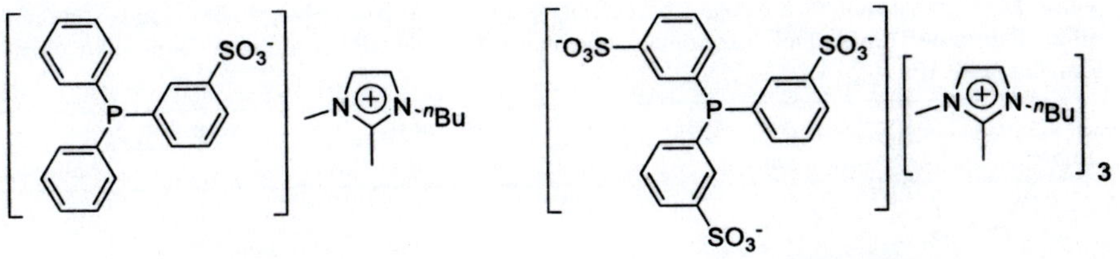

Fig. 2 Phosphine-functionalised imidazolium-based ionic liquids

or surface which does not necessarily correspond to the smallest NPs. However, no insights about this proposal were given.

The last example concerns the use of Ir NPs stabilised by phosphine-functionalised ILs as hydrogenation catalysts [73]. Worm-like nanoparticles of 12.2 nm in size were synthesised by H_2 reduction of [Ir(cod)Cl]$_2$ in [bmim][tppm] and [bmim][tppt]. (Fig. 2), which were subsequently isolated through several washing/centrifugation cycles.

These Ir(0) NPs dispersed in bmimPF_6 were found to be recyclable and selective catalysts for the hydrogenation of nitroarenes, aromatic carbonyl compounds and quinoline derivatives and their analogues (Table 12). In general, nitroarenes were reduced to anilines with very high selectivities. The Ir NPs were also very selective towards the carbonyl functionality in the hydrogenation of aldehydes, while the reduction of aromatic ketones yielded the corresponding alcohol products with moderate selectivities. Finally, complete selectivities towards the hydrogenation of the heteroaromatic group were observed in the hydrogenation of quinoline and its analogues. Although the selective hydrogenation of carbonyl and quinoline compounds through a heterolytic hydrogenation mechanism has been extensively described, the authors do not examine whether dihydrogen is cleaved via a homolytic or heterolytic mechanism.

4 Hydrogenation Reactions Catalysed by Iridium Nanoparticles Stabilised by Polymers

Different polymers can serve as stabilising agents for MNPs, which usually act as steric stabilisers but may also provide an electrostatic stabilisation. Among them, poly(N-vinyl-2-pyrrolidone) (PVP) is probably the most employed polymer for the synthesis of MNPs. There are many reports describing the use of this bulky, non-ionic and water-soluble polymer as a stabiliser, reducing agent, growth modifier or dispersant for metal nanoparticles [74].

In a first example, Ir NPs stabilised with PVP (MW = 40,000) were prepared by thermal decomposition of $IrCl_3$ in NaOH solution under conventional heating and MW-assisted conditions [75]. The Ir NPs of 1.1–1.2 nm in size were highly selective

Table 12 Chemoselective hydrogenation of nitroarenes, aromatic carbonyl compounds and quinoline derivatives and their analogues mediated by Ir NPs stabilised with phosphine-functionalised ILs

		Ir@[bmim][tppm]		Ir@[bmim][tppt]	
Substrate	Product	Conv. (%)	Sel. (%)	Conv. (%)	Sel. (%)
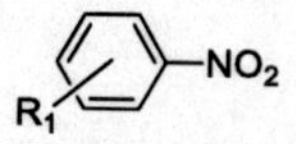	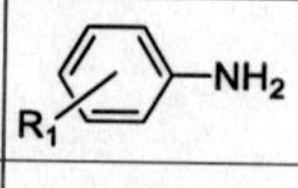	100	72–100	100	100
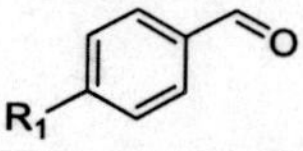	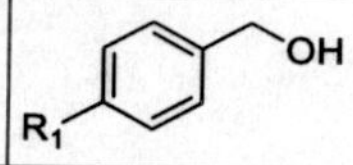	42–100	100		
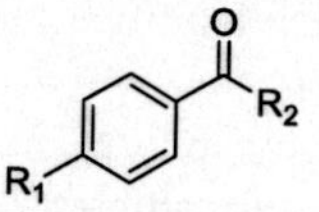	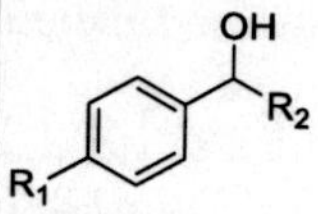	8–90	88–96	20	85
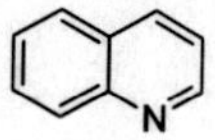	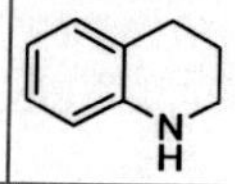	94	100	92	100
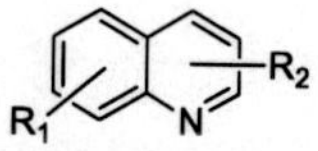	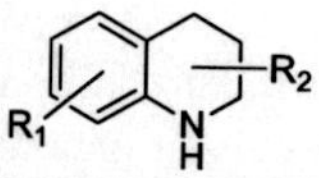	81–100	100		
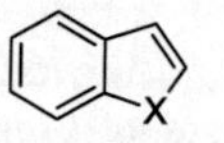 X = NH, O, S	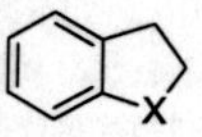	22–33	46–100		

Conditions: 1.8 M substrate in $bmimPF_6$ (1 mL), substrate/Ir = 100, H_2 (30 bar). Nitroarenes: 50°C, 5 h. Ketones: 75°C, 10 h. Aldehydes: 30°C, 10 h. Quinolines: 50°C, 10 h

Table 13 Hydrogenation of citronellal by PVP–stabilised Ir NPs

Catalyst	NP size (nm)	Conversion (%)	Selectivity (%)	TOF (h^{-1})
Ir NPs (MW)	1.1(0.19)	22.4	100	36
Ir NPs (heating)	1.2(0.26)	30.8	97	50

Reaction conditions: H_2 (10 bar), 60°C, 2 h

in the hydrogenation of citronellal to citronellol (Table 13), which is probably favoured by the steric impediment in the C=C double bond.

Later on, Tsukuda and coworkers employed the same polymer to stabilise Ir NPs [76]. The reduction of H_2IrCl_6 by $NaBH_4$ or ethylene glycol in the presence of PVP yielded Ir NPs of 1.8(0.3) and 2.2(0.5) nm in size, respectively. Through an extensive characterisation, it was concluded that the NPs contain mainly Ir(0) and a small amount of IrO_2 (<11%) and there are an electronic donation from PVP that makes the NP surface to present a slightly negative charge. These Ir NPs showed high selectivities towards the nitro group in the hydrogenation of substituted nitroarenes (Table 14), for which the authors suggested a preferential adsorption of the NO_2 groups on IrO_2 sites and the activation of H_2 at Ir(0) active sites, in a similar

Table 14 Chemoselective hydrogenation of substituted nitroarenes mediated by PVP-stabilised Ir NPs prepared through reduction of H_2IrCl_6 with ethylene glycol

Substrate	Product	Conversion (%)	Selectivity (%)
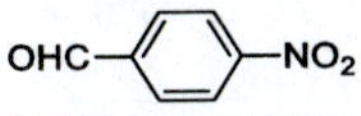	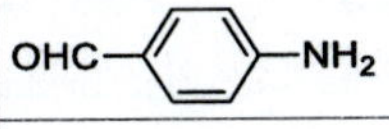	99	99
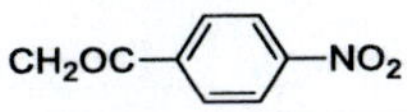	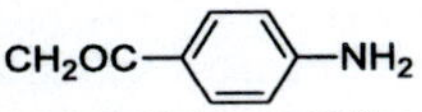	100	100
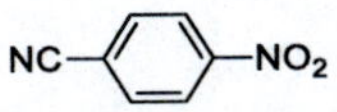	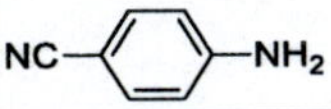	95	100
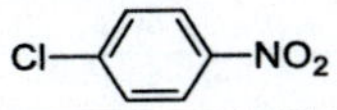	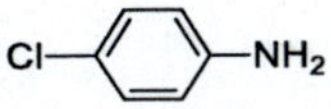	100	99

Reaction conditions: substrate/Ir = 50, solvent (ethyl acetate), H_2 (1 bar), 25°C, 2 h

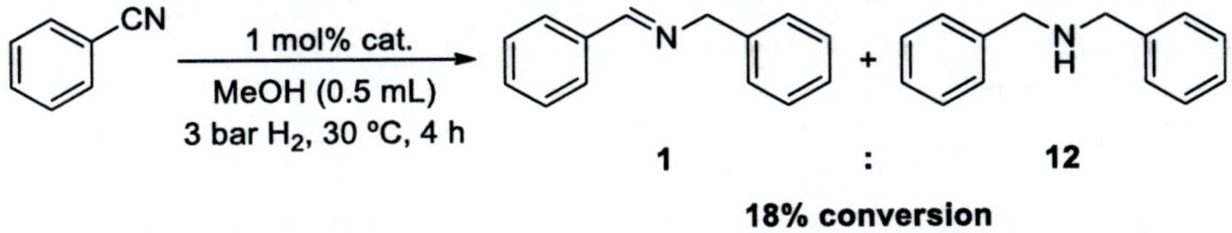

Scheme 6 Hydrogenation of benzonitrile by PVP-stabilised Ir NPs

way to previous reports on nitroarene reduction mediated by MNPs supported on metal oxides [77, 78].

Along this line, Ir NPs of 2.5(0.7) nm stabilised with PVP were synthesised by ethanol reduction of $IrCl_4$ under MW-assisted conditions (<400 W, 165°C, 15 min) [79]. Such Ir NPs were employed as catalysts for the hydrogenation of benzonitrile under mild conditions (Scheme 6), although the observed activity was lower than those shown by other transition metal nanoparticles (Rh, Pd, Pt), while the selectivity towards the imine product was very poor.

Other types of polymers have been investigated for the stabilisation of Ir NPs with applications as hydrogenation catalysts. For example, Nagashima and coworkers described the use of ammonium salts of hyperbranched polystyrene (HPS–$NR_3^+Cl^-$; R = ^{n}Bu or ^{n}Oct) to stabilise transition metal nanoparticles [80]. Ir NPs of 1.9(0.4) and 1.7(0.3) nm in size were prepared by $NaBH_4$ reduction of $IrCl_3$ in the presence of HPS– ^{n}Bu$R_3^+Cl^-$ and HPS– ^{n}Oct$R_3^+Cl^-$, respectively. These NPs were used as catalysts for cyclohexene hydrogenation, but, as in the aforementioned work, the activity was very low compared to those shown by nanoparticles of other metals such as Ru, Rh, Pd and Pt (≤21% conversion; 0.2 mol%, 1 bar H_2, 30°C, 1 h).

Finally, Jagirdar and coworkers employed iridium nanosponges as catalysts for alkene hydrogenation [81]. The solid-state reduction of H_2IrCl_6 with ammonia borane resulted in the formation of Ir(0) NPs immobilised in a BNH_x polymer. Then, water was added to this polymer, which led to a vigorous H_2 release and the generation of a nanosponge containing Ir NPs (2–3 nm size). The Ir nanosponge was

Table 15 Hydrogenation of alkenes mediated by Ir nanosponges

Substrate	Product	Time (h)	TOF (h^{-1})
Styrene	Ethylbenzene	4	500
4-Chlorostyrene	4-Chloro ethylbenzene	4	500
4-Methylstyrene	4-methyl ethylbenzene	4	500
1-Hexene	Hexane	4	500
1-Octene	Octane	5	400
1-Hexadecene	Hexadecane	6	333
1-Octadecene	Octadecene	6	333
Cyclohexene	Cyclohexane	4	500
Cyclooctene	Cyclooctane	5	400
1,5-Cyclooctadiene	Cyclooctane	9	222
Ethyl acrylate	Ethyl propionate	3.5	571
t-butyl acrylate	*t*-butyl propionate	4	500
Methyl methacrylate	Methyl-2-methyl propionate	3.75	533
Cyclohexenone	Cyclohexanone + cyclohexanol (89:11)	3.83	522
4-Vinylcyclohexene	Ethyl cyclohexane	9	222
R-limonene	*p*-Menth-2-ene[a]	4.5	444
R-limonene	*p*-Menth-2-ene[b]	0.5	4,000

Reaction conditions: substrate/Ir = 2,000, substrate (20 mmol), CH_2Cl_2, H_2 (4 bar), 30°C
[a]Substrate/Ir = 1,000
[b]Substrate/Ir = 1,000, 75°C, *n*-heptane

found to be a highly active hydrogenation catalyst for a wide range of alkenes (Table 15), where the steric hindrance on the C=C bond seems to play an important role. In addition, the catalyst was recovered via filtration and reused up to seven cycles without any significant loss in the activity.

5 Hydrogenation Reactions Catalysed by Iridium Nanoparticles Generated In Situ Without Stabilising Agent

There are several examples of hydrogenation by Ir NPs generated and stabilised in situ with the available species (solvents, precursor components, etc.) that do not require an extra stabilising agent. The majority of these works were carried out by Finke's research group, which has extensively studied the formation mechanism of Ir NPs and their applications in hydrogenation catalysis. This group reported the formation of Ir(0) NPs (1.8 nm) by H_2 reduction of $[Ir(cod)Cl]_2$ in acetone as solvent (2.76 bar H_2, 22°C; Scheme 7a) [82]. The reaction proceeds with the concomitant hydrogenation of acetone to isopropanol with 95% selectivity (TON = 16,400, initial TOF = 1.9 s^{-1}), the remainder being diisopropyl ether (Scheme 7b). Interestingly, the addition of molecular sieves increased the selectivity up to 100%

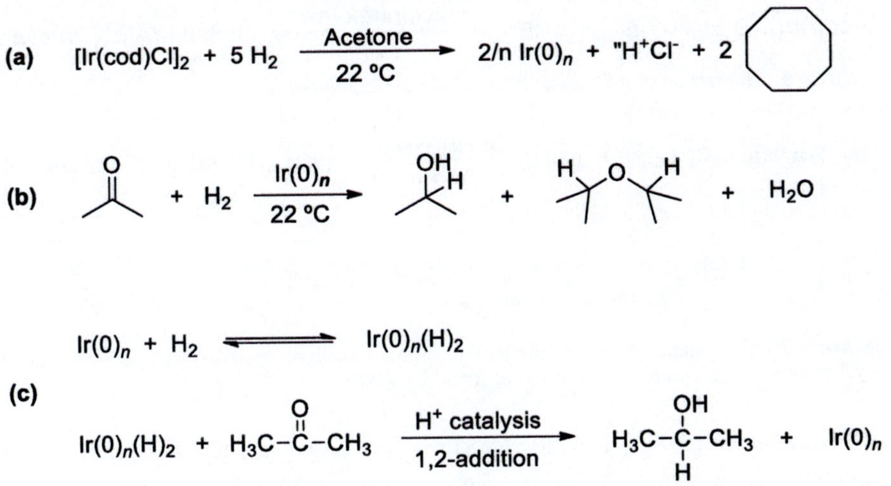

Scheme 7 (**a**) Formation of Ir(0) NPs by H_2 reduction of $[Ir(cod)Cl]_2$ in acetone; (**b**) concomitant hydrogenation of acetone; (**c**) proposed hydrogenation mechanism

(TON = 18,800), which points to an acid-assisted hydrogenation process. This was demonstrated by the use of a strong and non-coordinating base named Proton Sponge (1,8-bis-(dimethylamino)naphthalene) that scavenged the protons (H^+) from the reaction mixture and inhibited the hydrogenation process. An ionic hydrogenation mechanism was proposed through 1,2-addition to the acetone in which Ir–H and H^+ species are involved (Scheme 7c). This mechanism was supported by studies with deuterated acetone and kinetic experiments that showed a first-order dependence on [Ir] and [H^+].

In the same way, the formation of Ir(0) NPs by hydrogenation of $[Ir(cod)Cl]_2$ in benzene (2.76 bar H_2, 22°C) was described and the concomitant hydrogenation of the latter to cyclohexane [83]. Complete conversion was observed in 8.7 h, with TOF = 25 s^{-1} and TON =5,250. Ir NPs of <10 nm in size were observed at 4 h of reaction, and a size >20 nm was determined at the end of the process. Thus, agglomeration to larger NPs occurs as the process advances. The authors introduced the concept of weakly ligated, labile ligands (WLLs), which bind weakly to the NP surface and provide a minimum level of stabilisation but can be easily decoordinated. The only available surface ligands are benzene, H_2 (and H^-) and H^+Cl^-, the latter being a model example of WLL. Indeed, the use of $[Ir(cod)(CH_3N)_2]BF_4$ and $[Ir(cod)(CH_3N)_2]PF_6$ as precursors did not produce benzene hydrogenation. This indicates that BF_4^- and PF_6^- anions coordinate more strongly to the NP surface while H^+Cl^- provide a poor stabilisation, as can be seen by the formation of aggregates and bulk iridium, but it leads to more active Ir NPs catalysts.

This research group also described the synthesis and characterisation of an iridium Ziegler catalyst, which was prepared by reaction between $[(cod)Ir(\mu\text{-}O_2C_8H_{15})]_2$ and $AlEt_3$ in cyclohexane (Scheme 8) [84]. An extensive study of

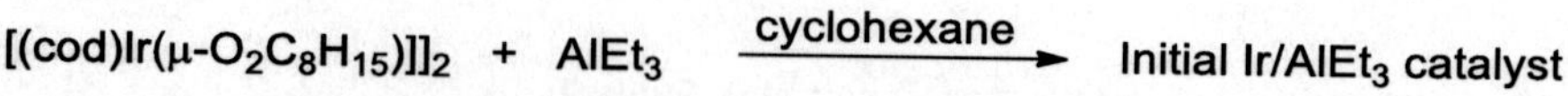

Scheme 8 Preparation of the Ziegler-type iridium catalytic system

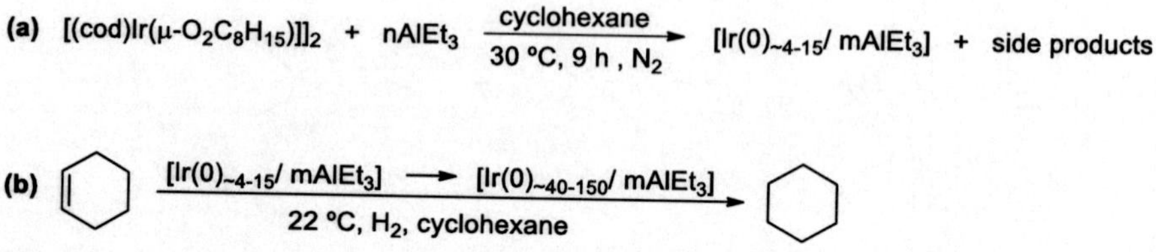

Scheme 9 (**a**) Preparation of the Ziegler-type iridium catalytic system; (**b**) hydrogenation of cyclohexene and concomitant formation of $Ir(0)_{\sim 40–150}$ NPs

the solution by Z-contrast STEM, XAFS and MALDI MS showed a mixture of monometallic Ir complexes and $Ir_{4–15}$ nanoclusters of 0.5–0.7 nm in size.

The use of this solution in cyclohexene hydrogenation led to the formation of larger $Ir(0)_{40–150}$ nanoclusters with 1.0–1.6 nm in size. Interestingly, the population of these nanoclusters increased during the hydrogenation process, for which a black precipitate and a concomitant increase in activity were also observed. Consequently, although activity by *homogeneous* iridium species was not discarded, the $Ir(0)_{40–150}$ nanoclusters are the most active catalyst in the system. Indeed, the precipitate displays comparable catalytic activity as the maximum activity obtained at the end of the hydrogenation process. This was further supported by mercury poisoning tests [85]. The addition of Hg(0) when the process is in the maximum rate regime completely inhibited the hydrogenation reaction, while some activity, attributed to non-poisoned homogeneous catalysts as Ir complexes and subnanometer $Ir_{4–15}$ clusters, is observed if the addition of Hg(0) is carried out at initial stages of hydrogenation process.

In a subsequent publication, this methodology was applied to prepare a series of Ir (0) NPs differing in the $[(cod)Ir(\mu\text{-}O_2C_8H_{15})]_2$:$AlEt_3$ ratio (Scheme 9) [86]. As in the previous work, the $[(cod)Ir(\mu\text{-}O_2C_8H_{15})]_2$ + $AlEt_3$ solution (Al/Ir = 2, 3, 5) generated highly active $Ir(0)_{40–150}$ NPs during the hydrogenation reaction of cyclohexene. It was found that the NPs obtained by addition of 3 or 5 equivalents of $AlEt_3$ are stable at 200°C (no agglomeration was observed after heating at this temperature for 30 min a solution of Ir NPs prepared in dodecane), while a Al/Ir $\geq$ 2 it is necessary to avoid nanoparticle aggregation and formation of bulk Ir(0) during the hydrogenation process at RT (Scheme 9b).

The different Ir NPs can be isolated as a solid and subsequently used as catalysts for cyclohexene hydrogenation, providing exceptionally high activities with TOF values up to 3,500 h^{-1} for Al/Ir = 2 (Table 16). From the catalytic results, it can be seen that not only the stability but also the catalytic activity and catalyst lifetime are affected by the Al/Ir ratio.

An in-depth study based on experimental tests provided strong evidence for nanoparticle stabilisation by $AlEt_3$ or some of its derivatives [87]. It was proposed

Table 16 Hydrogenation of cyclohexene by Ziegler-type iridium nanoparticles

Al/Ir ratio	Stability at 200°C	NP size (nm)	TTO	TOF (h^{-1})
2	No	1.4 ± 0.7	180,000 after 52 h	3,500
3	Yes	1.5 ± 0.5	155,000 after 144 h	1,100
4	Yes	1.7 ± 0.4	100,000 after 150 h	700

Reaction conditions: cyclohexene/Ir = 1,375, Ir (0.02 mM), cyclohexene (1.65 M), solvent (cyclohexane), H_2 (2.76 bar), 22°C
TTO total turnover

Fig. 3 Representation of possible stabilisers on the surface of Ir(0) NPs generated from [(cod)Ir(μ-$O_2C_8H_{15}$)]$_2$ + $AlEt_3$

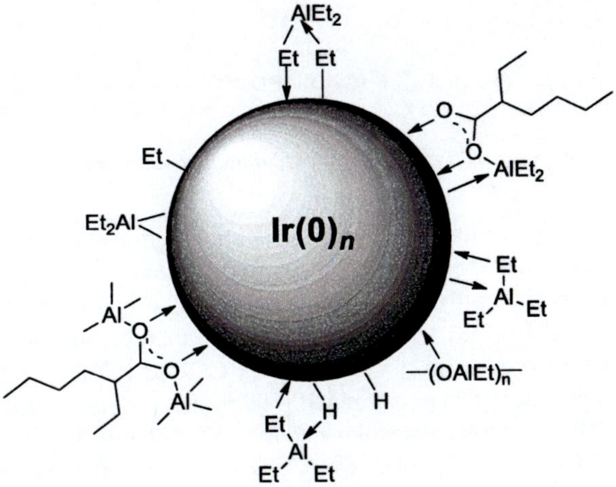

that $AlEt_3$ species are bound to the NP surface in the vicinity of active sites, and thus an increase in the $AlEt_3$ concentration hinders the access of substrate molecules. Indeed, the catalytic activity decrease at higher Al/Ir ratios (Table 16) and free $AlEt_3$ was found in the [(cod)Ir(μ-$O_2C_8H_{15}$)]$_2$ + $AlEt_3$ catalytic solution for Al/Ir $\geq$ 3. However, the results of this study also showed that the stabilisers may be non-charged aluminium alkyl carboxylates, $AlEt_2(O_2C_8H_{15})$ or Al–O–Al containing alkylalumoxanes generated by hydrolysis of $AlEt_3$ (Fig. 3).

Similarly, Finke and coworkers employed a polyoxometalate-containing organometallic precursor, $[Bu_4N]_5Na_3[(1,5\text{-COD})Ir{\cdot}P_2W_{15}Nb_3O_{62}]$, for the preparation of Ir NPs [88]. This precursor was reduced with H_2 in propylene carbonate as solvent, yielding $Ir(0)_n$ clusters that are highly sensitive of synthesis conditions. Isolable and redispersable nanoclusters with a monodisperse size of 2.1(0.3) nm were obtained at 22°C and 1.2 mM precursor concentration (Table 17). However, non-redispersable and polydisperse NPs of 5(2) nm in size were generated at 60°C and 0.24 mM concentration. It was found that under the latter conditions, a $(1,5\text{-COD})Ir(solvent)_2^+$ species is generated, which undergoes a fast and uncontrolled reduction, and thus leads to the formation of low-quality $Ir(0)_n$ clusters. As expected, these Ir NPs exhibit low activity in cyclohexene hydrogenation in comparison with that shown by the monodisperse nanoclusters (Table 17).

Table 17 Concomitant hydrogenation of cyclohexene by $Ir(0)_n$ nanoclusters prepared in situ through H_2 reduction of $[Bu_4N]_5Na_3[(1,5\text{-COD})Ir{\cdot}P_2W_{15}Nb_3O_{62}]$

[Precursor] (mM)	T (°C)	NP size (nm)	Redispersable	TTOs
1.2	22	2.1 ± 0.3	Yes	56,000
0.24	60	5.0 ± 2.0	No	7,300

Reaction conditions: cyclohexene/Ir = 1,400, solvent (propylene carbonate), H_2 (2.76 bar)
TTO total turnover

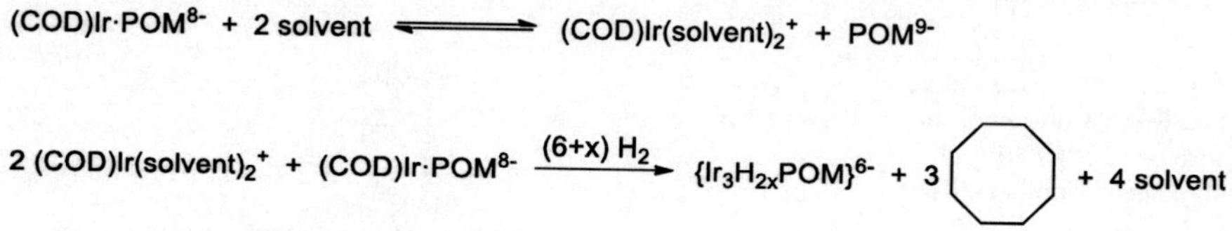

Scheme 10 Alternative termolecular nucleation mechanism

This cyclohexene hydrogenation reaction was subsequently employed to investigate the kinetics of the nucleation process that yields the Ir NPs at 22°C, $Ir(0)_{\sim 300}$, from this polyoxometalate-containing organometallic precursor (abbreviated as $(COD)Ir{\cdot}POM^{8-}$). The study showed that the nucleation is a second-order homogeneous process in the precatalyst ($(COD)Ir{\cdot}POM^{8-}$) [89]. Then different nucleation mechanisms were examined. The results provided evidence for a mechanism named "alternative termolecular nucleation" (Scheme 10) [90]. In such a mechanism, one $(COD)Ir{\cdot}POM^{8-}$ and two $(COD)Ir^+$ species interact to generate a kinetically effective nucleus (KEN) with $\{Ir_3H_{2x}POM\}^{8-}$ composition. This KEN contains hydrogen in the structure and is the key species that leads to the formation of $Ir(0)_{\sim 300}$ NPs.

Finally, it is worth mentioning the work of Zhang and coworkers on hydrogenation of quinoline derivatives by "naked" IrO_2 NPs [91]. The IrO_2 NPs (2 nm) were synthesised by a ball-milling reaction between $IrCl_3$ and NaOH at RT. These highly stable NPs were active catalysts for the hydrogenation of quinoline derivatives and other N-heterocyclic compounds (Table 18) under very mild conditions (H_2 balloon, RT). Very high selectivities towards the hydrogenation of the heterocyclic ring were observed. Interestingly, substrates with substituents located in the vicinity of the N-pyridine atom were reduced with very high yields. However, low yields were obtained with reactants substituted at remote positions (Table 18), and thus harsher conditions were required to afford good results (60 bar H_2, 60°C). In addition, the recyclability of the IrO_2 NPs was evaluated in the hydrogenation of 2-methylquinoline. The nanocatalyst showed an excellent recycling behaviour as was reused up to 30 cycles without any significant loss in activity and selectivity. Surprisingly, XPS showed no sign of reduction of Ir(IV).

The authors proposed that the hydrogenation process occurs by a mechanism where IrO_2 stabilises a homolytic hydrogen, which activates the N-heterocycle through the N atom. Simultaneously, this moiety interacts with the IrO_2 NP by a weak hydrogen bond between a neighbouring methyl H_2C–H and a surface oxygen

Table 18 Hydrogenation of quinoline derivatives and other N-heterocycles mediated by IrO_2 NPs. *Left* neighbouring substituents, *right* none or remote substituents

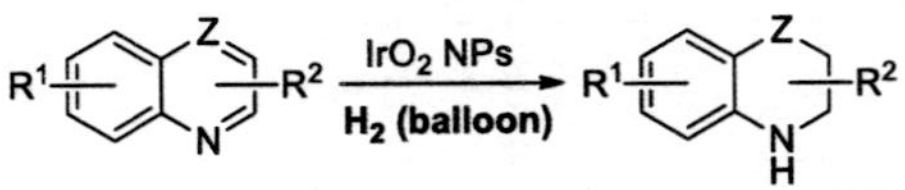

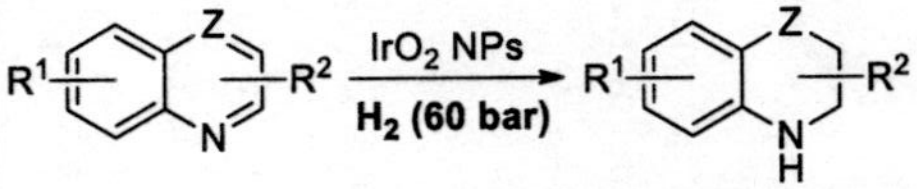

Z	R^1	R^2	Yield (%)[a]	Z	R^1	R^2	Yield (%)[b]
CH	H	2-Me	98	CH	H	H	98 (<5)
CH	H	3-Me	98	CH	H	2-Ph	97 (43)
CH	H	2-NH_2	96	CH	H	4-Me	42 (0)
CH	6-Me	2-Me	97	CH	H	6-Me	98 (<5)
CH	8-Me	H	98	CH	H	6-MeO	97 (<5)
CH	6-Br	2-Me	98	CH	6-NH_2	H	96 (0)
CH	8-Br	2-Me	97	CH	H	7-Me	97 (18)
N	H	2-Me	96	CH	8-Cl	H	95 (31)
N	H	2,3-dimethyl	97	N	H	H	98 (<5)
N	5-Me	H	91	N	H	2-Ph	96 (<5)

Reaction conditions: Substrate (0.2 mmol), IrO_2 NPs (1.2 mg, 0.04 mmol), MeOH (2 mL), 24 h. I

[a] H_2 balloon, RT. Isolated yields

[b] H_2 (60 bar), 60°C. Isolated yields (H_2 balloon and RT in brackets)

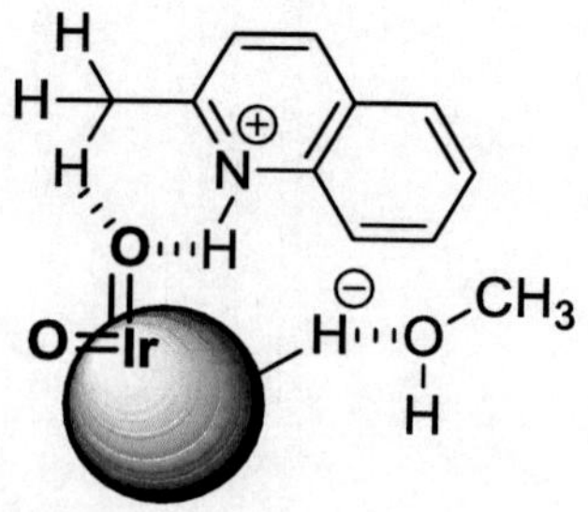

Fig. 4 Proposed interaction between the IrO_2 NP and the substrate through C–H···O (=Ir = O) hydrogen bonding [91]

atom (Fig. 4). Methanol interacts via oxygen with the surface hydride in the proposed mechanism, which seems as yet somewhat speculative.

6 Supported Ir NPs

Supported MNPs are one of the most suitable catalysts for industrial applications since the immobilisation of MNPs on solid supports increases their stability and recyclability [92]. In addition, MNPs supported on high surface area materials present a higher active metal surface than conventional heterogeneous catalysts, making them more active. However, supports not only stabilise MNPs but also are able to modify their catalytic properties. The following section collects the most relevant recent hydrogenation examples of reactions catalysed by supported Ir NPs, with some discussions towards the influence of stabilising supports on the catalysis.

6.1 Ir NPs Supported on Carbon Materials

Carbon materials are one of the most commonly used supports in catalysis. They present most of the requirements for a catalyst support, such as thermal stability, high surface area or chemical inactivity. All of this makes them ideal supports for the immobilisation of MNPs. Moreover, the possibility to modify them by introducing doping atoms or functional groups, which can modify the electronic properties of the supported metal, makes these materials attractive for heterogeneous catalysis. The most common carbon materials used for MNP stabilisation are activated carbon, fullerenes, carbon nanotubes (CNTs), multi-walled CNTs (MWCNTs), carbon nanofibers (CNFs), graphenes and reduced graphene oxides (rGO). Most of them, decorated with Ir NPs, have been used for hydrogenation reactions. In 2010, Faria, in collaboration with Serp, reported the use of Ir NPs supported on MWCNTs for the selective hydrogenation of cinnamaldehyde to the corresponding alcohol [93]. Two types of MWCNT-supported Ir NPs were used as catalysts (before and after a reduction treatment to remove the excess oxygenated surface groups of the MWCNTs), and interesting differences were observed in their selectivity (Fig. 5).

a)

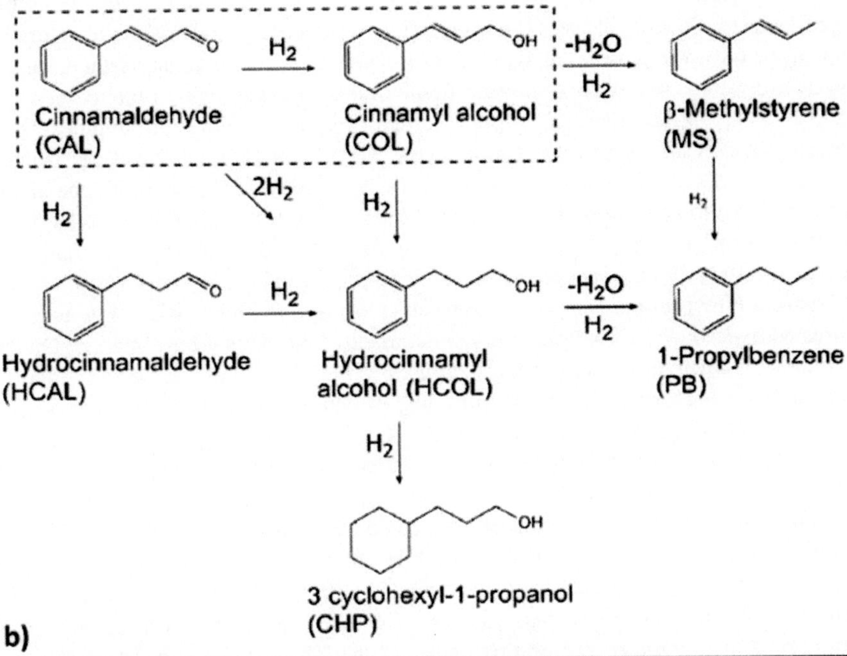

b)

Catalyst	Selectivity (%)			
	COL	HCAL	HCOL	Others
Ir/MWCNT	56	18	10	16
Ir/MWCNT973	68	17	11	4
Pt/MWCNT	8	51	13	28
Pt/MWCNT973	68	10	9	13

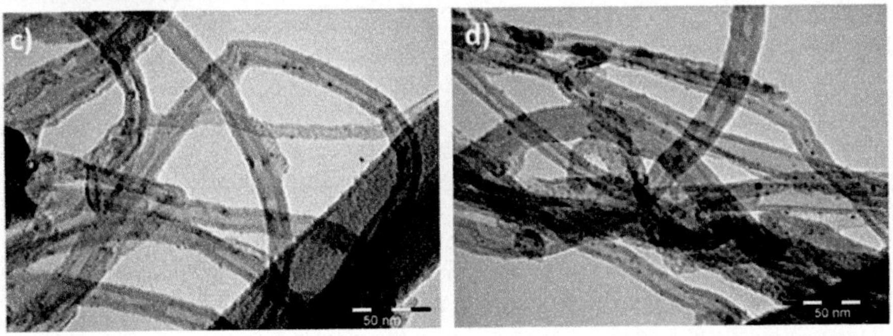

Fig. 5 (**a**) Catalytic hydrogenation of cinnamaldehyde. (**b**) Selectivity towards COL, HCAL, HCOL and other products obtained at 50% conversion with MWCNT-supported Ir and Pt NPs. Bottom: TEM micrographs of Ir/MWCNT catalysts before (**c**) and after (**d**) a post-reduction thermal treatment at 973 K. Reprinted with permission from Ref. [93]. Copyright 2017 Elsevier

After the reduction treatment, the MWCNT-supported Ir NPs (Ir/MWCNT973) showed a higher selectivity to cinnamyl alcohol than the same catalyst before (Ir/MWCNT). Specifically, at 50% conversion, a selectivity towards the unsaturated alcohol of 68% was observed. The higher selectivity was explained by the stronger interaction between the Ir and the carbon nanotubes after removal of the oxygenated surface groups. An identical support effect, but even more pronounced, was observed for Pt NPs supported on the same MWCNTs (Fig. 5). However, Mayer et al. observed that the oxidative treatment of CNTs only had a small influence on the catalytic activity of supported Ir NPs. Here, the CNT-supported Ir NPs were prepared through a hydrogenolytic approach and tested in the hydrogenation of quinaldine [94]. It was observed that the temperature of the hydrogenolysis had a substantial effect on the catalytic activity of the supported NPs. When low temperatures were used (−30°C), large and almost inactive particles were obtained. On the contrary, higher hydrogenolysis temperatures (60 or 150°C) assist the formation of smaller NPs (1–20 nm), which exhibit a higher catalytic activity in the hydrogenation of N-heterocycles. Shortly after, Du/Fan et al. employed Ir NPs supported on CNTs for the selective hydrogenation of biomass-derived levulinic acid (LA) into γ-valerolactone (GVL) [95]. Apart from carrying out the hydrogenation under mild aqueous conditions (50°C; 2 MPa of H2), these CNT-supported Ir NPs can operate in the presence of formic acid (FA), which is present in the LA obtained from the acid hydrolysis of lignocellulosic biomass. In 2016, Ma et al. applied Ir NPs immobilised in MWCNTs as catalyst for the selective hydrogenation of nitroaromatics into the corresponding anilines [96]. This Ir catalyst not only showed an excellent activity and selectivity for a wide range of nitroaromatics but also presented a remarkable stability, the catalyst being recycled and reused up to five times without any signal of deactivation. Ir NPs, but this time supported on CNFs, were also used by Motoyama et al. for the selective hydrogenation of functionalised nitroarenes and imines [97]. Here the Ir NPs were generated in three types of CNFs (platelet, CNFs-P; tubular, CNF-T; and herringbone, CNF-H) and activated carbon (AC) by thermal decomposition of an iridium carbonyl clusters, $Ir_4(CO)_{12}$. Among them, Ir/CNF-T demonstrated to be the most efficient catalyst for the hydrogenation of nitroarenes, being highly active and selective even in the presence of other functional groups such as nitriles, ketones or esters.

Murzin et al. used a mesoporous carbon material, namely, Sibunit, which combine the benefits of graphite (conductivity and stability) and active carbons (high adsorption capacity and surface area), to support Ir NPs [98]. The resulting catalysts, with different metal loadings (0.5, 1 and 3 wt%), were used for the hydrogenation of citral under atmospheric hydrogen pressure at 70°C. These catalysts were not selective at all, obtaining a mixture of geraniol, nerol, the isopropyl ethers of geraniol and nerol (catalysis was carried out in 2-propanol), citronellal and citronellal diisopropyl acetal (Fig. 6). Ir/Sibunit (1 wt%), which present the larger NPs (6 nm), produces less geraniol and nerol than the other catalysts, whereas Ir/Sibunit (3 wt%) forms mostly the ether products.

Ir NPs intercalated between graphite layers (Ir-GIC) were presented by Shirai et al. for the selective hydrogenation of cinnamaldehyde into cinnamyl alcohol in

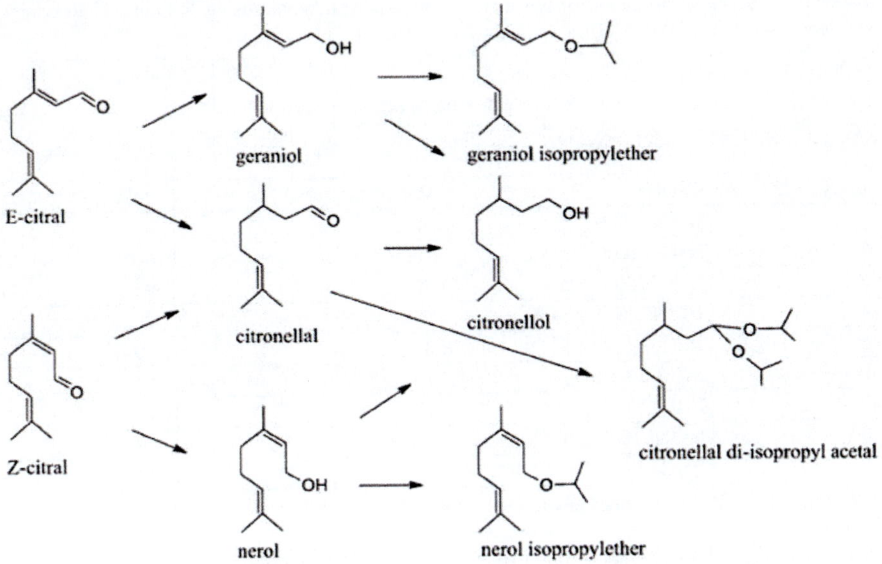

Fig. 6 Reaction scheme for the hydrogenation of citral. Reprinted with permission from Ref. [98]. Copyright 2012 Springer

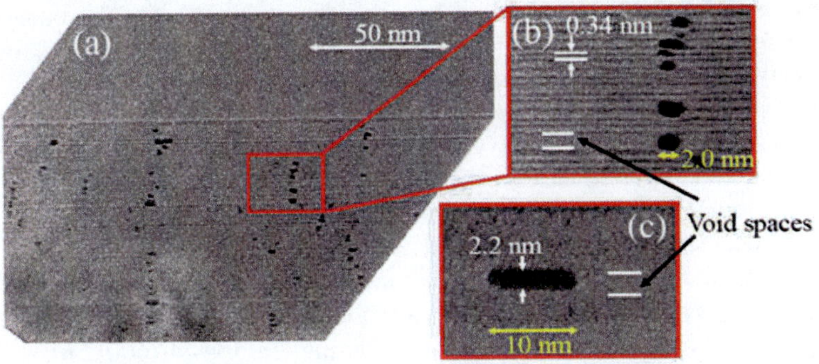

Fig. 7 TEM images of side view of Ir-GIC for iridium nanodisks of Ir-GIC. Reprinted with permission from Ref. [99]. Copyright 2019 Chemical Society of Japan

supercritical CO_2 (sc-CO_2) [99]. In particular, disk-shape MNPs (average diameter of 8 nm and average thickness of 2 nm; Fig. 7) intercalated between graphite layers were generated by reduction of $IrCl_4$. Comparing the activity of Ir-GIC with Ir NPs supported on graphite surface (Ir/Gmix), the selectivity towards the cinnamyl alcohol is much higher for Ir-GIC. The higher selectivity of Ir-GIC (> 90% at full conversion) is explained by the different substrate-metal interactions between the two catalytic systems. sc-CO_2 transports the cinnamaldehyde with the aromatic ring parallel to the graphite layers, and the substrate preferentially adsorbs by the

Table 19 Reductive amination of nitrobenzene with benzaldehyde using Ir catalysts. Reprinted with permission from Ref. [100]. Copyright 2017 Wiley-VCH

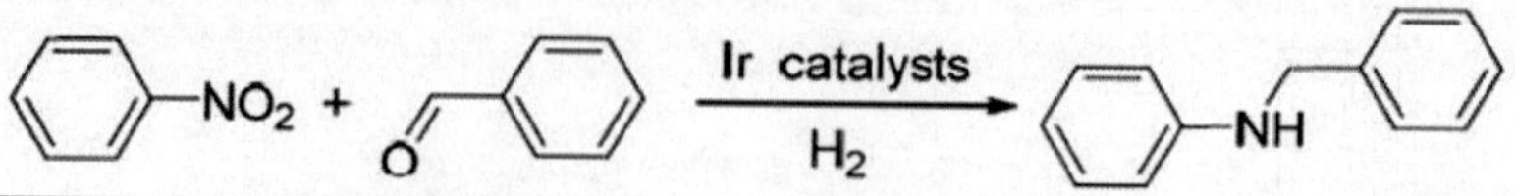

Entry	Catalyst	Conversion[a] (%)	Yield[a] (%)
1	Ir@NC(500-2 h)	95	91
2	Ir@NC(600-2 h)	100	100
3	Ir@NC(700-2 h)	96	93
4	Ir@NC(600-3 h)	99	98
5	IL/C(600-2 h)	0	0
6	Ir/C(600-2 h)	10	7
7	Ir-IL/C	12	10
8 [b]	Ir@NC(600-2 h)	Trace	0
9[c]	Pd/C	100	33

[a]Determined by GC, n-hexadecane was used as internal standard
[b]Without H_2
[c]0.1 mol% Pd (Pd/C) was added; IL = [MCNI]C1; C = active carbon
[d]Reaction conditions: 1.0 mmol nitrobenzene; 1.4 mmol benzaldehyde; Ir@NC(600-2 h) (Ir 0.1 mol %, 28 mg) added; toluene (2.0 mL); at 80°C; under 20 bar H_2; 18 h

carbonyl group with the intercalated Ir nanodisks of Ir-GIC. On the other hand, cinnamaldehyde adsorbs arbitrarily on the supported Ir NPs of Ir/Gmix, showing a low selectivity (44% at 90% of conversion).

The possibility to functionalise carbon materials by introducing heteroatoms, such as N, can influence the catalytic activity of the supported Ir NPs. In line with this, Huang et al. reported an Ir catalyst based on Ir Nps supported on an N-doped carbon material for the reductive amination of ketones and aldehydes with nitroaromatics and amines [100]. The Ir NPs supported on N-doped carbon were generated by pyrolysis of $IrCl_3$ in the presence of an ionic liquid [MCNI]Cl (1-methyl-3-cyanomethylimidazoulium chloride) and activated carbon. The best catalyst resulted from the pyrolysis at 600°C during 2 h, Ir@NC(600-2 h), showing total selectivity at full conversion in the reductive amination of nitrobenzene with benzaldehyde using H_2 gas (Table 19, entry 2). A clear effect of the N-doped support was observed, since Ir NPs generated from pyrolysis of iridium trichloride in activated carbon showed much less activity (Table 19, entry 6). They also observed that N-doped carbon facilitates the dispersion and the stability of the Ir NPs on Ir@NC(600-2 h).

Another efficient way to improve the selectivity of supported MNPs is the introduction of a second metal. Drelinkiewicz et al. reported monodisperse and well-distributed Pd-Ir alloy NPs supported on carbon (Pd-Ir/C) for the hydrogenation of cinnamaldehyde (Fig. 8, top) [102]. Monometallic Pd NPs are highly active and selective towards C=C hydrogenation, whereas Ir NPs are more selective to C=O bond hydrogenation, but at the same time they are less active. Thus, they found that

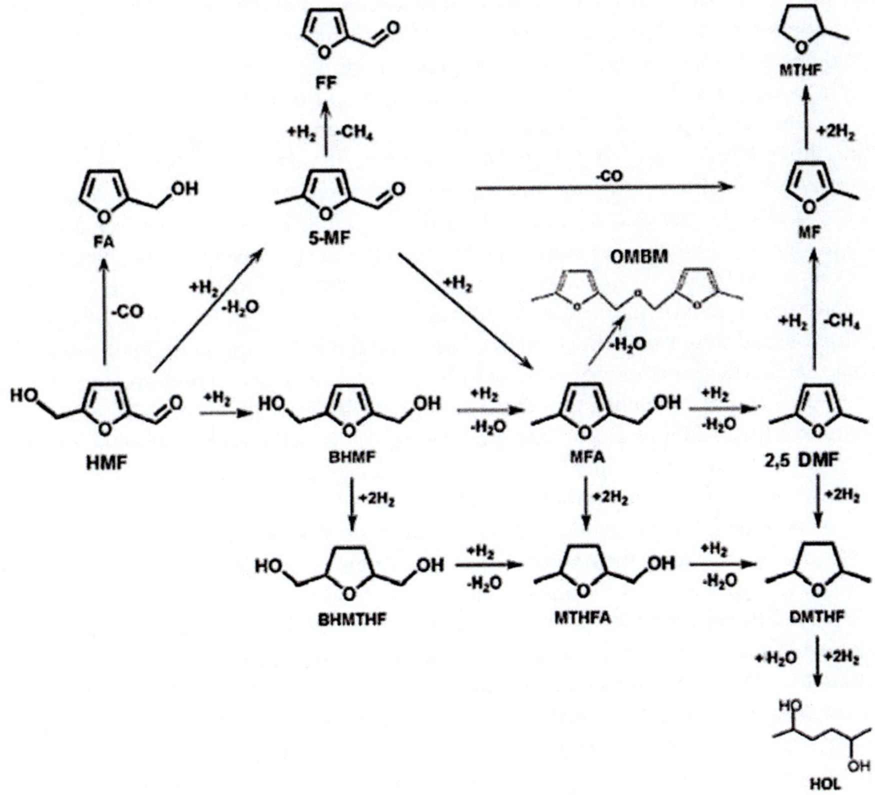

Fig. 8 Reaction scheme for the hydrogenation/hydrodeoxygenation of hydroxymethylfurfural (HMF). Reprinted with permission from Ref. [101]. Copyright 2019 Elsevier

the monometallic Pd/C catalyst was more active than the bimetallic one, while Pd-$Ir_{0.1}$/C was more selective towards the formation of the unsaturated alcohol. Therefore, the incorporation of a small amount of Ir (0.1 equiv.) on the Pd surface is able to modify the nanoparticle catalytic selectivity.

Another recent example is the catalytic transformation of 5-hydroxymethylfurfural (HMF) to 2,5-dimethylfuran (DMF) over bimetallic PtIr NPs supported on a mesoporous carbon material, CMK-3, described by Beltramone et al. (Fig. 8) [101]. While the monometallic catalyst (Pt-CMK-3) gave 50% selectivity towards DMF at 90% conversion, the bimetallic one (PtIr-CMK-3) showed a higher activity/selectivity (86% at 98% conversion). In conclusion, they observed that the PtIr alloy catalyst is more active and selective towards the formation of 2,5-DMF than the analogous Pt catalyst, due to a synergistic effect between Ir and Pt.

Two-dimensional carbon-based materials like graphene, graphene oxide or reduced graphene oxide are being widely used for the deposition of MNPs

[103]. In particular, there are a couple of recent examples about catalysts based on Ir NPs supported on graphene materials for hydrogenation reactions. In 2015, Janiak and coworkers presented Ir NPs supported on thermally reduced graphite oxide, or "graphene", (Ir@TRGO) as a recyclable catalyst for the hydrogenation of benzene or cyclohexene [72]. The Ir NPs were generated by thermal decomposition of $Ir_4(CO)_{12}$ in the presence of the graphene material in an ionic liquid (1-butyl-3-methylimidazolium tetrafluoroborate) by microwave or electron-beam irradiation, as was already introduced in Sect. 2 on IL supported Ir NPs. Ir@TRGO showed a high activity in the solvent-free hydrogenation of benzene to cyclohexane (TOF $= 10{,}000\ h^{-1}$) under relatively mild conditions (100°C, 10 bar H_2). This catalyst was reused more than ten times without any significant loss of activity. Moreover, through size effect studies, they found that the optimum NP diameter for benzene hydrogenation in these conditions and with this kind of catalysts is 3.6 ± 1.0 nm. Reduced graphene-supported iridium NPs (Ir/rGO) were also employed by Rong et al. for the hydrogenation of p-chloronitrobenzene (*p*-CNB) and cinnamaldehyde (CAL) [104]. Here, graphene-supported Ir NPs were prepared following a one-pot hydrothermal method, in which $H_2IrCl_6{\cdot}6H_2O$ was reduced together with graphene oxide under different H_2 pressures and temperatures. Interestingly, by adjusting the hydrothermal conditions, it was possible to modulate the catalytic properties of the obtained catalysts. Ir/rGO obtained at higher H_2 pressure displayed better activities in the hydrogenation reactions due to a higher reduction degree of the Ir NPs. On the other hand, when low hydrothermal temperatures were used, the reduction of the graphene oxide was not complete, and different reactivities were observed for *p*-CNB and CAL due to distinct support-substrate interactions. Graphene with more oxygenated surface groups (resulting from low hydrothermal temperatures) facilitate the interaction between nitro groups and the surface via H-bonds. However, graphene with less oxygenated surface groups (higher hydrothermal temperatures) favours the adsorption of CAL on the graphene surface through π-π stacking interactions (Fig. 9). Therefore, they demonstrated that substrate-support interactions strongly influence the catalytic activity of graphene-support Ir NPs.

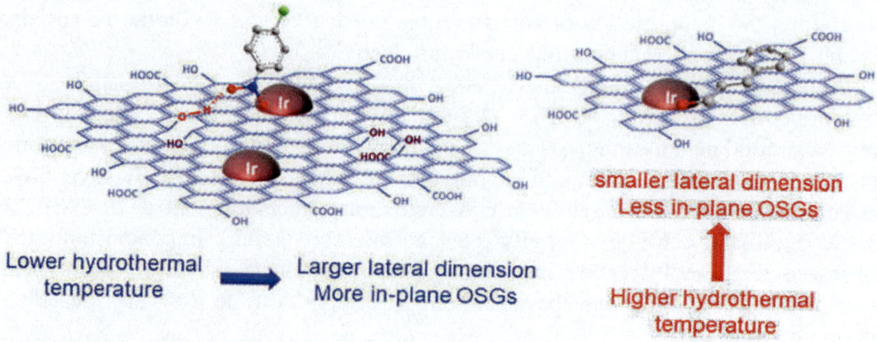

Fig. 9 Different adsorption modes of p-CNB (left) and CAL (right) on Ir/rGO depending on the amount of oxygenated surface groups. Reprinted with permission from Ref. [104]. Copyright 2019 Elsevier

6.2 *Metal Oxide-Supported Ir NPs*

6.2.1 Ir NPs Supported on TiO_2

During the last decade, catalysts based on MNPs supported on metal oxides have awakened a great interest in heterogeneous catalysis. One of the most often employed and studied metal oxide supports is titanium oxide or titania (TiO_2). The great interest for TiO_2-supported MNPs comes from their strong metal support interaction, chemical stability and high surface area, which is reflected in an improved catalytic activity and stability.

In 2012, Luo et al. reported a series of catalysts based on Ir NPs supported on TiO_2 (Ir/TiO_2) for the vapour-phase selective hydrogenation of crotonaldehyde to the corresponding unsaturated alcohol [105]. Ir/TiO_2 catalysts with an Ir content of 3 wt% were synthesised by the reduction of titania impregnated with H_2IrCl_6 in H_2 at different temperatures (100, 200, 300, 400 and 500°C). It was found that the reduction temperature has a great influence in the activity and selectivity of the resulting catalysts. While catalysts reduced at 100 and 500°C show poor activity and selectivity after 500 min under flow conditions (~25% selectivity at ~3% conversion), Ir/TiO_2 reduced at 300°C present the highest activity/selectivity (75% selectivity at 27% conversion). Ir/TiO_2 reduced at 100°C (Ir/TiO_2_100) contains a large number of Lewis acid sites ($Ir^{\delta+}$), which strongly interact with the reactant through the carbonyl oxygen (σ2), favouring the interaction between the double bond and Ir (0) species (π), and the subsequent formation of butanal (Fig. 10). On the other hand, Ir/TiO_2_300 with a lower content of $Ir^{\delta+}$ species, and a majority of Ir (0), exhibit a medium strength substrate-metal interaction between the carbonyl carbon and Ir (0) sites (σ1), which is reflected in a higher conversion and selectivity towards crotyl alcohol (Fig. 10). However, with Ir/TiO_2_500, there is a strong interaction between carbonyl carbon and Ir(0) species (σ1) that impedes the desorption of crotonaldehyde and supresses the activity. Therefore, the activity and selectivity of these titania-supported Ir NPs are highly dependent on the nature of substrate-metal interactions (π, σ1 and σ2). This adsorption model of crotonaldehyde on Ir NPs was previously proposed by Reyes et al. in 2003 [106].

A similar catalytic behaviour was observed by Rojas et al. in the hydrogenation of citral over Ir NPs supported on TiO_2 (Ir/TiO_2), prepared by deposition of $H_2IrCl_6 \cdot 6H_2O$ at high H_2 pressure [107]. Comparing the results obtained for Ir/TiO_2 before and after a reduction treatment [H_2 flow (30 mL/min) at 500°C for 2 h], the activity and selectivity towards the hydrogenation of the carbonyl group

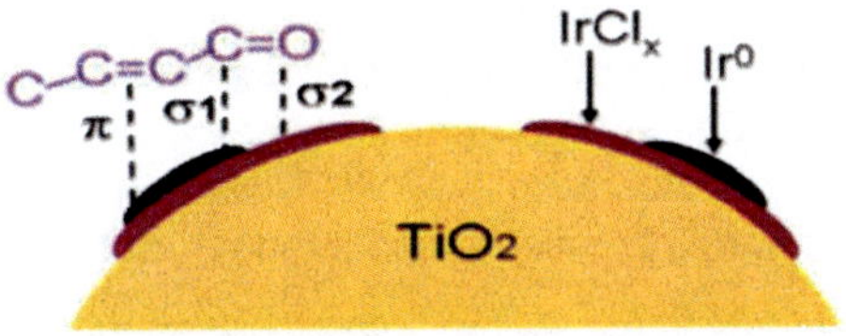

Fig. 10 Adsorption mode of crotonaldehyde on Ir/TiO_2 proposed by Luo et al. Reprinted with permission from Ref. [105]. Copyright 2012 Elsevier

were totally different. While as-synthesised Ir/TiO_2 showed a very low conversion and poor selectivity towards the corresponding alcohol (26% selectivity and 1% conversion after 3 h reaction), pre-reduced Ir/TiO_2 greatly enhanced the selective hydrogenation of the carbonyl group achieving a selectivity towards geraniol of around ca. 90% at a conversion of 26% (3 h reaction). The enhancement of both activity and selectivity is probably due to an increase in the ratio of Ir(0)/$Ir^{\delta+}$ after the reduction treatment.

In 2016, Hagelin-Weaver, Bowers et al. investigated the effects on metal-support interactions on the pairwise selective parahydrogen addition over Ir/TiO_2 and the subsequent hydrogenation of propene [108]. Comparing Ir/TiO_2 with Ir supported in other non-reducible oxides such as SiO_2 and Al_2O_3, they found that after a reduction treatment of Ir/TiO_2 at 500°C, the pairwise selectivity increases up to 20 times. The strong metal-support interaction (SMSI) of the partially reduced titania, together with the presence of flat-shaped Ir NPs and Cl^- ions, is responsible for this increase in the pairwise selectivity. Toledo-Antonio et al. also studied the influence of metal-support interactions on Ir NPs over TiO_2 but using this time the hydrogenation of cyclohexene as model reaction [109]. In particular, they generated Ir NPs on anatase TiO_2 nanoparticles and anatase TiO_2 nanotubes via wet impregnation of H_2IrCl_6 followed by a calcination (400°C under air, 4 h) and hydrogenation steps (400°C under H_2, 2 h). 1.3 nm half truncated cuboctahedral nanoparticles were generated on TiO_2 nanoparticles, while full 1.4 nm cuboctahedral particles were obtained on TiO_2 nanotubes. The differences in morphology and metal-support interactions resulted in different catalytic performances. At 0°C, Ir NPs on TiO_2 nanotubes showed a slightly higher activity than the truncated ones on TiO_2 nanoparticles (TOFs: 4.7 Vs 3.2 s^{-1}). The lower activity of Ir truncated NPs was explained by their poorer metallic character due to a different metal-support interaction. Therefore, the TiO_2 crystal planes that interact with Ir NPs strongly influence both the morphology and the catalytic activity of the resulting catalysts. The same group also generated bimetallic (IrPt) NPs on TiO_2 nanotubes (NT) via wet impregnation of H_2IrCl_6 and H_2PtCl_6. The catalytic activity of these bimetallic materials was checked in the cyclohexene disproportionation reaction [110]. In general, the bimetallic catalyst was more active in cyclohexene disproportionation reaction than the corresponding monometallic ones. More specifically, under 150°C cyclohexene is hydrogenated to cyclohexane, whereas above 200°C, cyclohexene is dehydrogenated to benzene (Fig. 11). However, Ir does not only promote Pt catalysts. In 2014 Li et al. reported how Ir is also able to promote Au/TiO_2 catalysts for the selective hydrogenation of cinnamaldehyde [111]. Monometallic Ir/TiO_2 is more active than Au/TiO_2 but much less selective towards the formation of cinnamyl alcohol (Fig. 12a). The bimetallic catalysts Au-Ir/TiO_2, however, showed a hydrogenation rate five times higher than Au-Ir/TiO_2 but a similar selectivity to the alcohol (84%). Therefore, the incorporation of Ir on Au-Ir/TiO_2 increases the activity while maintaining the selectivity of Au. This synergic effect was explained by substrate-metal interactions. They proposed that cinnamaldehyde is preferentially adsorbed on bimetallic NPs through the carbonyl groups, since carbonyl carbon is adsorbed on gold atoms (negatively

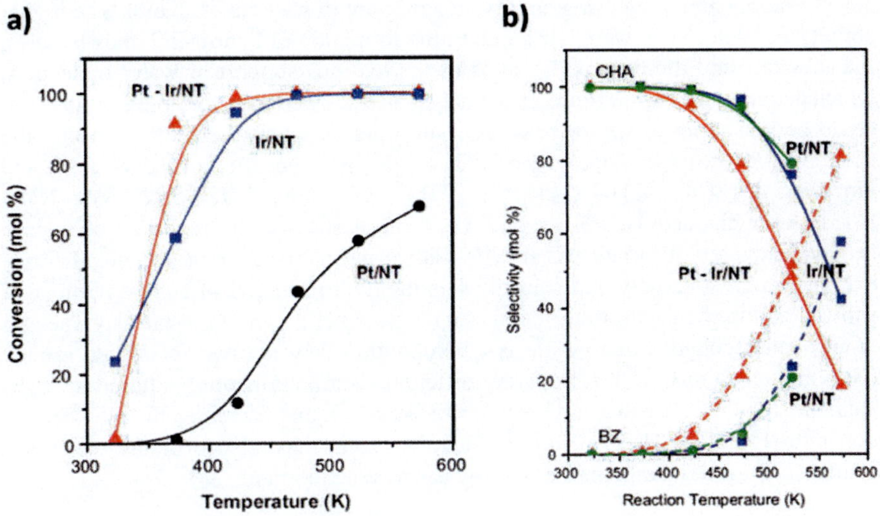

Fig. 11 (**a**) Cyclohexene hydrogenation to cyclohexane and dehydrogenation to benzene over Ir/NT, Pt/NT and PtIr/NT catalysts. (**b**) Selectivity to cyclohexane and benzene. Reprinted with permission from Ref. [110]. Copyright 2012 Elsevier

a)

Catalyst	Time (h)	Conv. (%)	Sel. (%)			Reaction rate (h^{-1})[b]
			COL	HCAL	HCOL	
Au/TiO_2	4	8.6	83.8	12.4	3.8	3.2
$Au\text{-}Ir/TiO_2$	1	12.0	83.4	11.2	5.4	17.7
Ir/TiO_2	1	10.9	55.6	38.0	6.4	15.9

[a] Reaction conditions: 0.3 g CALD, 150 mg catalyst and 20 mL 2-propanol/H_2O as solvent, 100 °C, 2 MPa P_{H2}.
[b] Reaction rate was calculated by the moles of CALD converted per mol Au and Ir per hour with TiO_2-supported Au, Ir, and Au–Ir catalysts.

b)

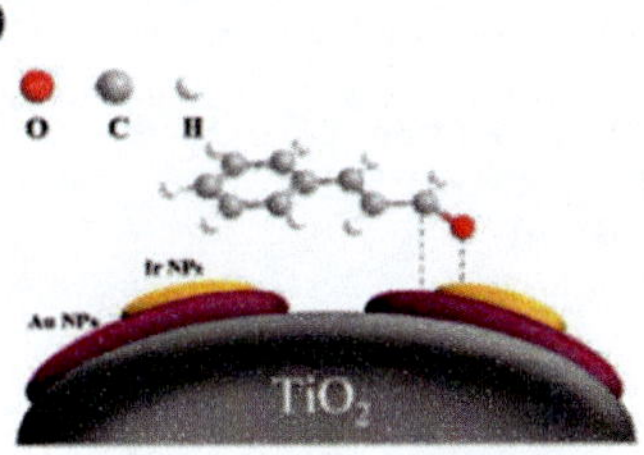

Fig. 12 (**a**) Hydrogenation of cinnamaldehyde over Au/TiO_2, Ir/TiO_2 and $Au\text{-}Ir/TiO_2$. (**b**) Adsorption mode of cinnamaldehyde on $Au\text{-}Ir/TiO_2$ proposed by Li et al. Reprinted with permission from Ref. [111]. Copyright 2014 Elsevier

charged), while carbonyl oxygen is adsorbed on Ir ones (positively charged; Fig. 12b).

6.2.2 Ir NPs Supported on ZrO_2

ZrO_2, or zirconia, is also a metal oxide support suitable for MNP stabilisation due to its high surface area and good thermal stability. For example, in 2010, Ir NPs supported on hydrous zirconia ($Ir/ZrO_2.xH_2O$) were straightforwardly prepared by a co-precipitation method by Li et al [112]. This catalyst, $Ir/ZrO_2.xH_2O$, compared with Ir/ZrO_2 prepared by impregnation method, showed a higher activity and selectivity in the hydrogenation of haloaromatic nitro compounds to the corresponding amine by using as solvent a mixture of water and ethanol. The higher activity and selectivity were attributed to the formation of hydrogen bonds between the substrate and the hydroxyl groups of the hydrous support or water molecules, which activates the nitro group. This catalyst also showed a remarkable stability as it could be reused up to six times without any loss of activity/selectivity. Along the same line, Reyes et al. employed Ir/ZrO_2 for the hydrogenation of *meta*-substituted nitrobenzenes (-H, -CHO, -OCH3, -CH3, -Cl, -CN, -CH=CH2 and $-NO_2$) [113]. After characterisation the Ir/ZrO_2 catalyst showed the presence of mainly zero valent, partially oxidised, Ir NPs with a mean diameter of 1.8 nm. Ir/ZrO_2 exhibited a great activity and selectivity in the hydrogenation of a great number of substituted nitrobenzenes to the corresponding amine, except for *m*-nitrostyrene and *m*-nitrobenzaldehyde where there is a competition between two functional groups (NO_2 and C=O or C=C). Comparing different electron donating/withdrawing substituents, they observe that the presence of donor groups decreases the NO_2 reduction rate (Fig. 13). These results point to a strong dependence of the electronic character of the substrate on the hydrogenation reaction rate.

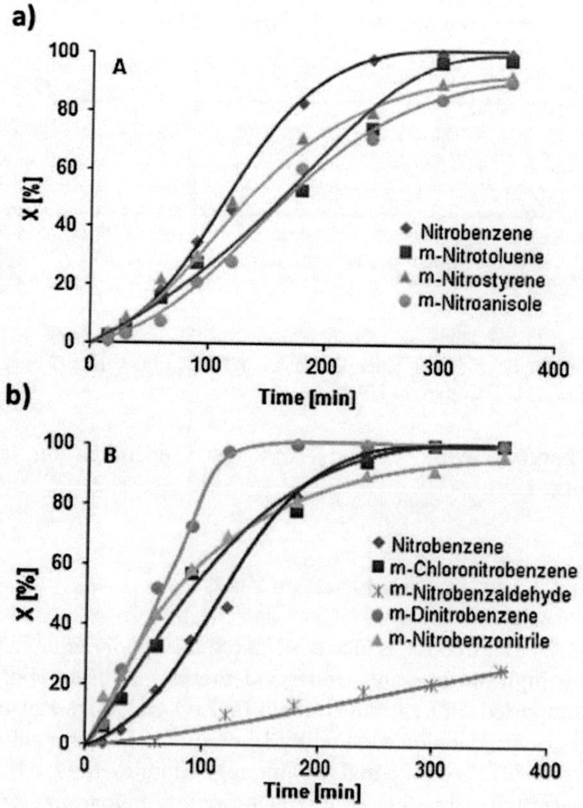

Fig. 13 (**a**) Hydrogenation of (**a**) electron-donor and (**b**) electron-withdrawing *meta*-substituted nitrobenzenes over Ir/ZrO_2. Reprinted with permission from Ref. [113]. Copyright 2013 Elsevier

6.2.3 Ir NPs Supported on MoO_x

Reducible metal oxides such as MoO_x, with variable oxidation states that can be easily reduced during catalysis (or catalyst preparation), have emerged as promising supports for MNPs. They normally present numerous oxygen vacancies and strong metal-support interactions, which results in efficient catalysis. For example, Gao et al. reported Ir NPs supported on hydrogenated MoO_x nanorods for the selective hydrogenation of α, β,-unsaturated aldehydes [114]. Following a one-pot synthetic method, they formed Ir NPs supported on hydrogenated nanorods of MoO_x (Ir/H-MoO_x), while simultaneously generating the Ir NPs and the hydrogenated support (H-MoO_x; Fig. 14a). More specifically, Ir NPs dissociates the H_2 to H atoms, which subsequently spill over to the MoO_3 surface as a proton and electron, doping the support with hydrogen. The obtained catalysts, Ir/H-MoO_x, showed a high chemoselectivity in the hydrogenation of cinnamaldehyde to cinnamyl alcohol (93% selectivity at full conversion). Ir/H-MoO_x also promotes the selective hydrogenation of a large number of other substrates with multiple functional groups, such as crotonaldehyde, citral, furfural, chloronitrobenzene or nitrobenzoic acid. The high selectivity was explained by the presence of $Ir^{\delta-}$ species on Ir/H-MoO_x which facilitates the adsorption of the cinnamaldehyde through the carbonyl group (Fig. 14b). The shelf life of the catalyst was investigated, showing good levels of

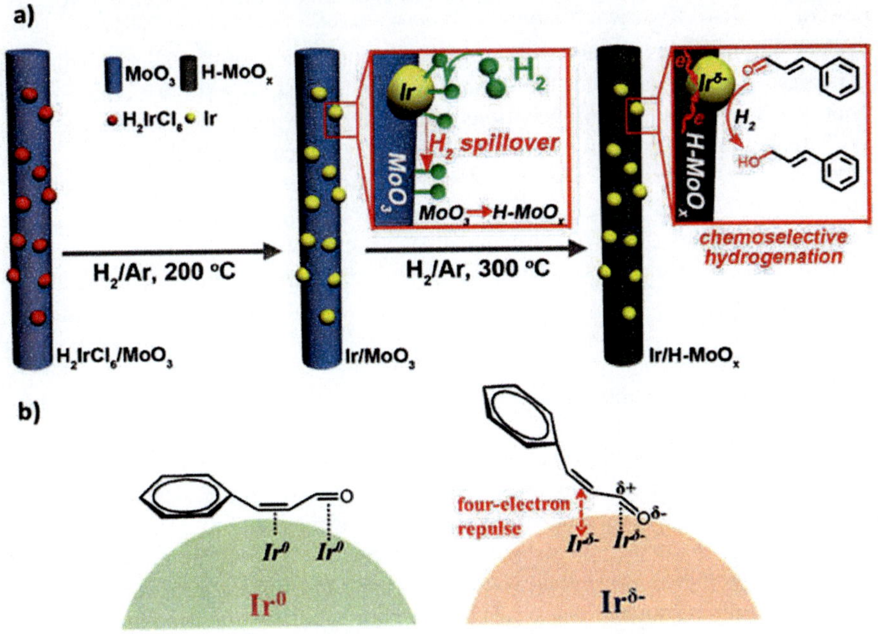

Fig. 14 Illustration of (**a**) one-pot synthesis of Ir/H-MoO_x for selective hydrogenation reactions and (**b**) different adsorption modes of cinnamaldehyde on Ir(0) and $Ir^{\delta-}$ surfaces. Reprinted with permission from ref. [114]. Copyright 2016 Elsevier

recyclability. After five successive catalytic runs, it presented similar values of activity and selectivity (>90%). Somewhat later, the same group studied how the hydrogen doping into MoO_3 nanorods modulate the metal-support interactions and enhance the chemoselectivity of supported Ir NPs in the hydrogenation of furfural to furfuryl alcohol [115]. Through a controlled hydrogen doping by varying the reduction temperature, the contents of low-valence Mo species (Mo5+ and Mo4+) and $Ir^0/Ir^{\delta+}$ ratios on the surface were adjusted to obtain the most efficient catalyst. In particular, Ir/H-MoO_x reduced at 400°C showed the best activity (TOF: 4.62 min^{-1}) and selectivity (> 99%) in the hydrogenation of furfural to furfuryl alcohol in mild conditions (30°C and 20 bar H_2). These works highlight the importance of supports in MNP catalysis, which can modify the electronic properties of the metal centres and therefore enhance the nanoparticle catalytic behaviour.

6.2.4 Ir NPs Supported on Magnetic NPs

The immobilisation of MNPs on magnetic responsive supports, such as Fe_3O_4, $MnFe_2O_4$ or $CoFe_2O_4$, facilitates the catalysts recovery and the separation of the products after catalysis. In 2012, Rossi et al. reported the generation of Ir NPs by H_2 reduction of $IrCl_3$ on a magnetic support based on Fe_3O_4 NPs covered by an amino-modified silica shell [116]. The resulting magnetically recoverable catalyst, Fe_3O_4@SiO_2-NH_2-Ir^0 (Fig. 15b), exhibited a remarkable activity in the hydrogenation of alkenes (e.g. TOF: 6,000 h^{-1} in the hydrogenation of cyclohexene to cyclohexane). It was observed that the functionalisation of the silica shell with amino groups was crucial for the stability of the catalyst. The surface amino group improves the metal-support interactions and enhance the recyclability of the catalyst. With the help of an external magnet, the catalyst was recovered and reused up to six times without any loss of activity. Moreover, the amino groups on SiO_2 surface promote the metal adsorption during the synthesis and minimises the metal leaching during the catalysis (Fig. 15a). Again, this work emphasised the importance of the support on MNP catalysis, which plays a crucial role in the stability and recyclability of the catalyst.

6.2.5 Ir NPs Supported on MgO

MgO is a well-known high surface area support widely used in heterogeneous catalysis. Like other basic supports, MgO is capable to modify the catalytic properties of MNPs. Another efficient way to modify the MNP reactivity is through the use of additives. For example, Tomishige et al. presented Ir NPs supported on MgO and modified with Fe cations as effective heterogeneous catalyst for the selective hydrogenation of α,β-unsaturated ketones [117]. In particular, Fe cation-modified Ir/MgO produced the corresponding α,β-unsaturated secondary alcohol with a high activity (initial TOF = 12 min^{-1}) and selectivity (up to 90%). The catalyst was obtained in situ by mixing Ir/MgO and the Fe cation precursor [Ir/$MgO^{+}Fe(NO_3)_3$]. Kinetic

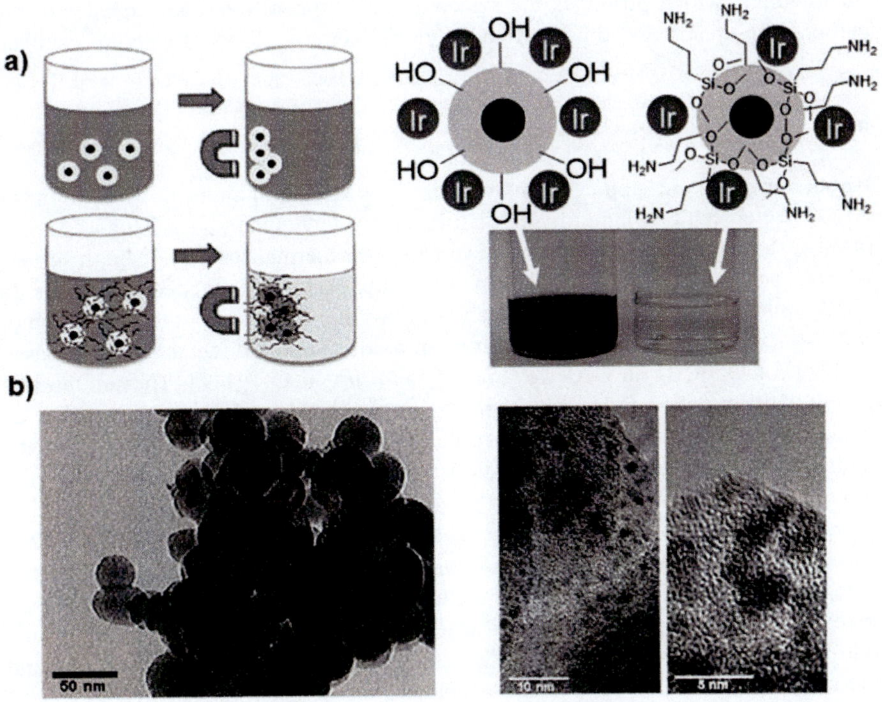

Fig. 15 (a) Supernatant solutions obtained after magnetic separation of Ir^0 catalysts immobilised on non-functionalised supports and amino-functionalised supports. (b) TEM micrographs of $Fe_3O_4@SiO_2$-NH_2-Ir^0. Reprinted with permission from Ref. [116]. Copyright 2013 Wiley-VCH

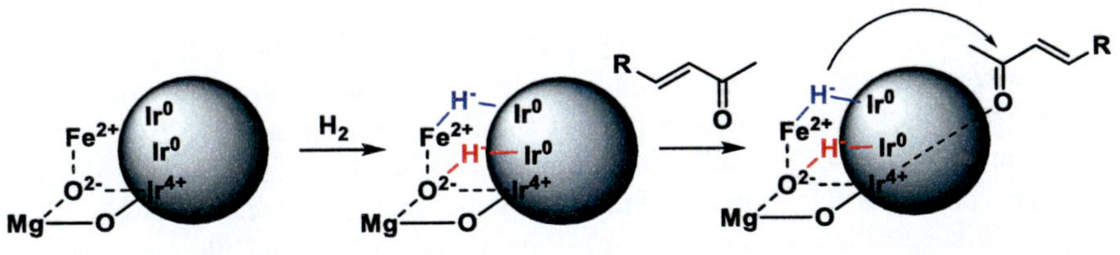

Fig. 16 Proposed mechanism for the hydrogenation of α,β-unsaturated ketones over Ir/MgO$^+$Fe $(NO_3)_3$. Adapted and reproduced with permission from ref. [117]. Copyright 2017 American Chemical Society

studies and catalyst characterisations proposed that the active sites are the interface among Ir^0, Ir^{4+} and Fe^{2+} species on the MgO support. More specifically, the cation Fe^{2+} next to the anion O^{2-} dissociate the H_2 in a heterolytic way and generate H^- and H^+ species (Fig. 16). Next to these hydride and proton species, C=O bonds adsorb on Ir^{4+} sites through a dipole-dipole interactions and form a six-membered intermediate, which promotes the selective, outer-sphere hydrogen transfer to the carbonyl group and explains the high activity/selectivity observed with this catalyst.

6.2.6 Ir NPs Supported on Al_2O_3

Another metal oxide support extensively used for MNP immobilisation is alumina or Al_2O_3, which can be found in various forms including alpha (α-Al_2O_3) and gamma (γ-Al_2O_3). In addition to a high chemical and hydrothermal stability, its high surface area and acid-base properties make it a useful support for heterogeneous catalysis. In 2010, Finke et al. carried out kinetic and mechanistic studies about the formation of Ir NPs on γ-Al_2O_3 (Ir(0)$_n$/γ-Al_2O_3) from an oxide-supported metal complex, based on [Ir(1,5-COD)Cl]$_2$ on γ-Al_2O_3 (Ir(1,5-COD)Cl/γ-Al_2O_3) [118]. They found that that the NP formation followed a two-step mechanism based on a slow continuous nucleation and a fast autocatalytic surface growth. In fact, *Finke* et al. used the hydrogenation of cyclohexene to cyclohexane as a reporter reaction, being able to monitor the formation of Ir NPs (2.9 ± 0.4 nm) during the synthesis (acetone, 22°C, 3 bar H_2 and 1.700 equiv. of cyclohexene). The resulting catalysts, Ir(0)n/γ-Al2O, showed a higher activity and longer lifetime than previous catalysts reported until the date based on Ir NPs supported on alumina. In the same year, Piccolo et al. investigated the decomposition of Ir(acac)$_3$ (acac = acetylacetonate) impregnated on silica-alumina (40 wt% silica, ASA) by TEM, in situ XRD and thermogravimetry-differential thermal analysis-mass spectrometry (TG-DTA-MS) [119]. Here, the heating treatments under oxidative (O_2), reductive (H_2) and inert conditions (Ar) were compared. After the calcination under air of Ir(acac)$_3$ on ASA, large and not very well-distributed IrO_2 NPs (8 ± 4 nm) were formed, the acac ligands having been burnt to CO_2 and H_2O. However, under reductive conditions small Ir NPs (1.4 ± 0.2 nm) are formed, while the acac ligand is decomposed (hydrogenolysis) and hydrogenated to alkanes and water. Lastly, thermal treatment under Ar was not able to completely remove acac ligands from the MNP surface. In a consecutive work, the same group employed Ir/ASA for the tetralin hydroconversion in the presence of H_2S as an active, robust and thiotolerant catalyst [120].

The behaviour of Ir NPs supported on alumina and silica in the enantioselective hydrogenation of ethyl pyruvate was studied by Bachiller-Baeza et al. in 2013 [121]. More specifically, they generated Ir NPs on silica and alumina by two different synthetic methods (incipient wetness impregnation and flame spray pyrolysis) and tested these catalysts in the ethyl pyruvate hydrogenation using cinchonidine as chiral modifier. Although these catalysts showed good catalytic activities, they only exhibit moderate enantioselectivities to (R)-ethyl lactate [enantiomeric excess: 26% (Ir/SiO_2) and 15% (Ir/Al_2O_3); Fig. 17]. However, interesting differences depending on the preparation method and the support were found. It was observed that the choice of support is more important than the preparation method, since the metal-support interactions strongly influence the surface structures of the Ir NPs (i.e. size, morphology and electron density). According to this, the best catalyst turned out to be Ir/SiO_2 prepared by the wetness impregnation method.

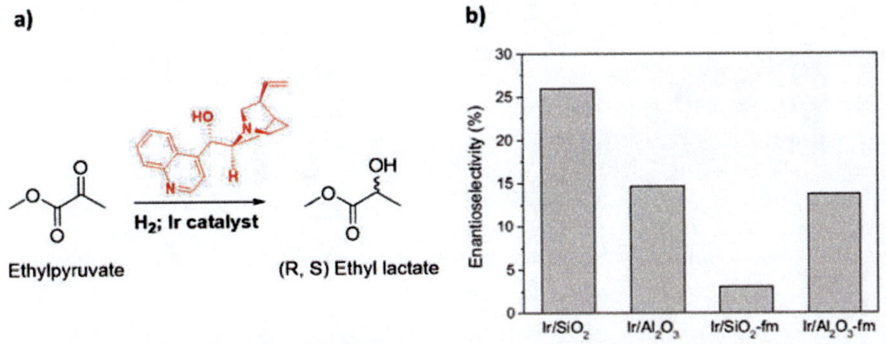

Fig. 17 (**a**) Enantioselective hydrogenation of ethyl pyruvate using cinchonidine (red) as chiral modifier and supported Ir NPs as catalysts. (**b**) Enantiomeric excess (ee) to (R)-ethyl lactate at full conversion over Ir catalysts. React. Cond.: EtOH, 298 K, 20 bar H2, 42 mg Ir catalyst (Ir content: 5 wt%). Adapted and reproduced from ref. [121]. Copyright 2013 Elsevier

Structural symmetric dendrimers have also been used to generate and stabilise Ir NPs. The strong dendrimer-metal precursor interactions facilitate and control the formation of the nanoparticles. For example, in 2008, Williams et al. obtained Ir NPs supported on γ-Al_2O_3 (Ir/γ-Al_2O_3) following a dendrimer metal nanocomposite approach (DMN) [122]. In particular, $IrCl_3 \cdot 3H_2O$ was first complexed with a four generation hydroxyl-terminated polyamidoamine [(G4OH) PAMAM] dendrimer, thanks to the large amount of amines sites available for Ir^{3+} coordination. Then, Ir/γ-Al_2O_3 was prepared through a wet impregnation of the DMN precursor on the alumina followed by an oxidation/reduction treatment. The resultant catalyst (1 wt %) was tested in the hydrogenation of benzonitrile and compared with Ir/γ-Al_2O prepared by a conventional method (wet impregnation of $IrCl_3 \cdot 3H_2O$ instead of the DMN precursor). Although the DMN approach produces more well-dispersed NPs overall with large Ir loadings, dendrimer-derived catalysts show similar activity and selectivity for benzonitrile hydrogenation compared to conventionally prepared catalyst.

Mixed oxides are also attracting attention as MNP support since they present a wide range of acidities. In 2012, Buriak et al. reported an interesting comparative study about mono- and multimetallic nanoparticles supported on different metal oxide supports for the hydrogenation of a series of aromatic compounds [123]. In particular, the catalytic activity of 72 different catalysts (with different MNP and support compositions) were compared in the hydrogenation of arenes (toluene, naphthalene) and heteroarenes (pyridine, quinoline, indole, thiophene and benzothiophene). Combination of 4 metals (Rh, RhPt, RhIr, RuPt, IrPt and RhPtIr) over 12 different supports (pure Al_2O_3, SiO_2 and TiO_2 and 9 combinations between them) were used to create the catalyst library and perform the screening study. The catalyst synthesis followed a one-pot process in which the metallic salts (NP precursors) were mixed with metal alkoxides (oxide support precursors) in a solution of EtOH/H_2O/HCl. 56 catalysts of the initial 72 demonstrated to be active in

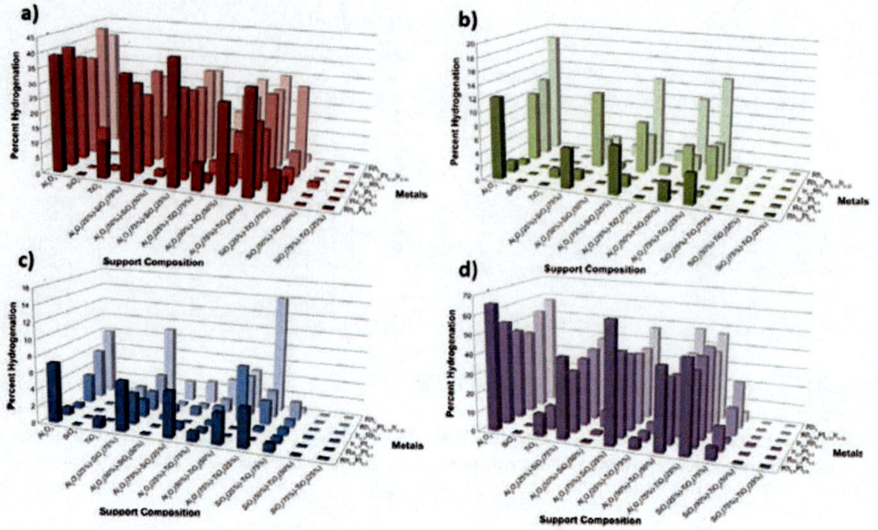

Fig. 18 Hydrogenation of toluene (**a**), indole (**b**), naphthalene (**c**) and quinoline (**d**) using mono- and multimetallic NPs supported on metal oxide supports. React. Cond.: isopropanol (20 ml); 24 h; 10 bar H_2 pressure; 9–16 mg of catalysts (metal content: 1 wt%). Adapted and reproduced from ref. [123]. Copyright 2012 with permission from American Chemical Society

the hydrogenation reactions. The inactive catalysts were those based on MNPs supported on SiO_2, SiO_2(25%)-TiO_2(75%) and SiO_2(50%)-TiO_2(50%). Among the substrates, only five [toluene, naphthalene, indole, quinoline and benzothiophene (< 4% conversion)] were hydrogenated under the applied conditions (Fig. 18). Among all the reactants studied, quinoline was the most active. A support effect is evident for toluene and quinolone, where the combination of metals is not as influential as the metal oxide chosen as support, which plays a more important role. For example, catalysts based MNPs supported on Al_2O_3 or Al_2O_3-containing supports are the most active, even when the MNP size and oxidation state were comparable with other catalysts. On the other hand, for naphthalene and indole hydrogenations, both support and metal compositions have a substantial effect in the catalyst activity. This combinatory study highlights the great influence of support and metal composition of MNPs for hydrogenation reactions, being crucial the accurate election of them to obtain efficient catalysts.

Although metal-support interactions strongly affect the catalytic activity of supported MNPs, a recent study of Wang et al. revealed that the dispersion of the support in the reaction media also plays an important role and even more pronounced than this MNP-support interaction [124]. Investigating the catalytic activity of Ir NPs supported on different supports, such as the mesoporous zeolite SBA-15, N-doped carbon, active carbon or various metal oxides (γ-Al_2O_3, ZrO_2, MgO, TiO_2 and CeO_2) in the hydrogenation of benzoic acid to cyclohexanecarboxylic acid in water under mild condition, they showed that the stability of support dispersions is

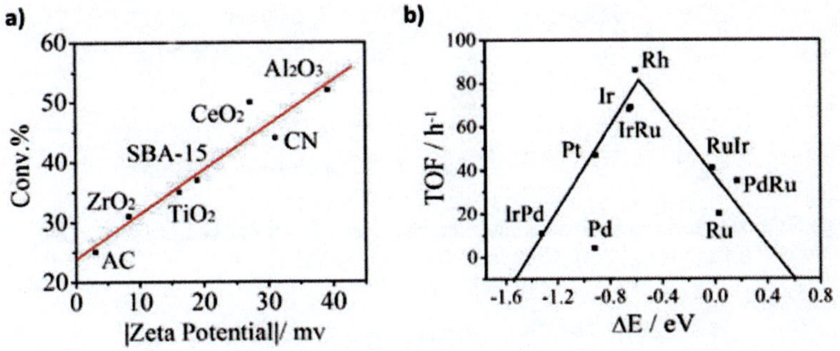

Fig. 19 (**a**) Relationship between the conversion of benzoic acid over various Ir catalysts and the zeta potential of their corresponding supports. (**b**) Catalytic activity (TOF/h^{-1}) as a function of the difference of adsorption energies between H_2 and H_2O (ΔE/eV) on different mono- and bimetallic catalysts. Reprinted with permission from Ref. [124]. Copyright 2017 the Royal Society of Chemistry

a key factor. Plotting the zeta potential of the different support used Vs the conversion of the corresponding Ir catalysts, a clear support dispersion effect can be observed (Fig. 19a). The catalytic activity of the catalysts linearly increases with the absolute value of zeta potential of the supports, Ir γ-Al_2O_3 being the fastest catalyst. Joint experimental/theoretical studies demonstrated a "volcano" curve between TOFs of benzoic acid hydrogenation and the difference of adsorption energies between H_2 and H_2O on different mono- and bimetallic catalysts. This curve presents an optimum energy value about −0.6 eV, which means that H_2 must adsorb 0.6 eV more efficiently than H_2O (Fig. 19b). This explains the high activity of Ir and Rh catalysts that are on the top of the volcano curve. Therefore, the activity of this aqueous hydrogenation reaction is guided by the adsorption energies of reactants and solvent.

6.2.7 Ir NPs Supported on AlO(OH)

Park et al. compared the activity of Ir and Rh NPs entrapped in aluminium oxyhydroxide [Rh/AlO(OH) and Ir/AlO(OH)] in the hydrogenation of aromatics and ketones [125]. In general, Ir/AlO(OH) showed a lower activity in the hydrogenation of arenes than the rhodium catalyst, which showed a TON over 20,000 in the hydrogenation of benzene and anisole under solvent-free conditions (75°C; 4 bar H_2). However, for the hydrogenation of ketones, catalysts were highly dependent on the reaction conditions. While Rh/AlO(OH) shows a higher conversion at room temperature (hydrogen balloon, n-hexane as solvent), the activity of Ir/AlO(OH) is much higher under solvent-free conditions (75°C; 4 bar H_2) (Fig. 20). The reuse and recyclability of these stable catalysts were investigated. For example, Rh/AlO (OH) retained its activity during ten cycles under mild conditions (n-hexane, room temperature and hydrogen balloon) and at least during five cycles under solvent-free conditions at 75°C and 4 bar H_2. Fan et al. used a similar catalytic system, Ir NPs

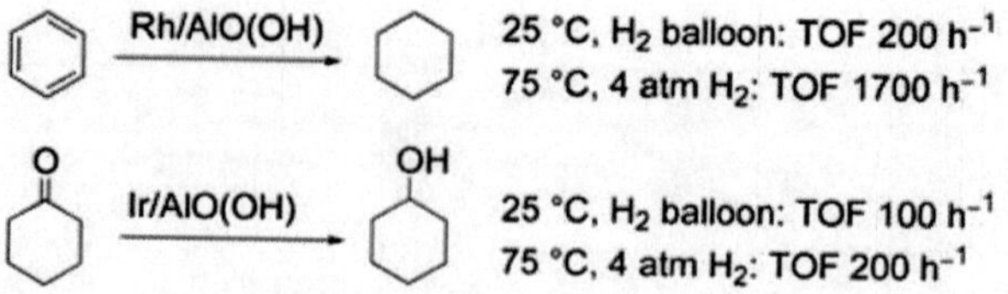

Fig. 20 Hydrogenation of aromatics and ketones over Rh/AlO(OH) and Ir/AlO(OH). Reprinted with permission from Ref. [124]. Copyright 2007 Wiley-VCH

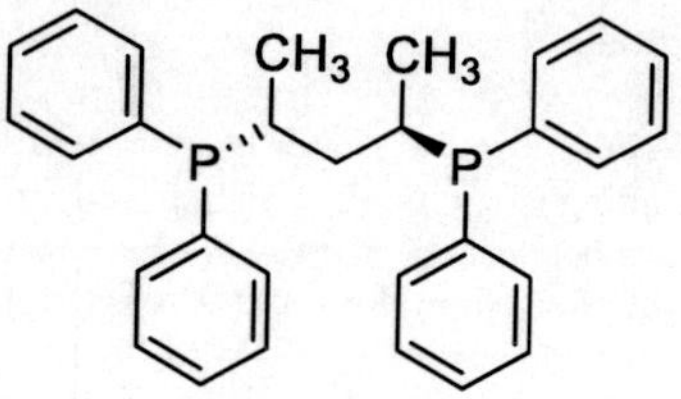

Fig. 21 (*R,R*)-BDPP chiral ligand used in ref. [44]

entrapped in AlO(OH), for the selective hydrogenation of *p*-chloronitrobenzene to *p*-chloroaniline under mild conditions (35°C, 1 bar H_2, 1 mmol substrate, solvent: 5 mL ethanol +2.0 mL H_2O, 2 h) [126]. The excellent activity and selectivity of this catalyst are attributed to the formation of hydrogen bonds between the nitro group of the substrate and the surface hydroxyl groups of the catalyst, which promotes the hydrogenation of $-NO_2$ groups. Moreover, the hydrogen bonds formed between the aniline and water facilitate the desorption of the products from the catalyst surface.

6.3 SiO_2-Supported Ir NPs

SiO_2, also called silica, has been widely used to stabilise MNPs. Unlike reducible metal oxides like TiO_2, which present strong metal-support interactions and influence in a higher degree the MNP properties, the irreducible oxides like silica are normally more inert supports. In a collaborative effort, Reyes and Claver generated Ir, Pt and Rh NPs on SiO_2 by reduction of the corresponding metallic salts in the presence of a chiral diphosphine, (2*R*,4*R*)-2,4-bis(diphenylphosphino)pentane [(*R*, *R*)-BDPP; Fig. 21] [44]. These chiral SiO_2-supported MNPs were tested in the asymmetric hydrogenation of numerous ketones (ethyl pyruvate, acetophenone, 2,3-butanedione, 1-phenylpropane-1,2-dione and 3,4-hexanedione). All catalysts showed good levels of conversion and enantioselectivity (up to 78% with Rh NPs) due to the functionalisation of the MNP surface with the chiral phosphine ligand. For more details see Sect. 1 on ligand-stabilised Ir NPs. This study was the first example of enantioselective hydrogenation of ketones induced by the presence of a chiral phosphine ligand on the metal surface of supported MNPs.

Tsukuda et al. reported a size-controlled synthesis of 15, 30 or 60 atoms Ir clusters within a OH-terminated poly(amidoamine) (PAMAM) dendrimer of generation 6 (Ir_x:G6; $x = 15$, 30 or 60) [127]. After deposition of Ir_x:G6 in SiO_2, these Ir catalysts selective hydrogenated the –NO_2 group of 2-nitrobenzaldehyde into 2-aminobenzaldehyde and anthranil under mild conditions (30°C, 1 bar of H_2). A clear size dependency in the selectivity of the reaction was observed; the smaller the iridium cluster is, the more selective the catalyst is towards the formation of 2-aminobenzaldehyde (Table 20).

As we have previously mentioned, an efficient way to improve the selectivity of supported MNPs in the hydrogenation of substrates with different functional groups is the use of a second metal. In 2013, Tomishige and coworkers reported silica-supported Ir NPs decorated with ReO_x (Ir-ReO_x/SiO_2) as active catalyst for the selective hydrogenation of unsaturated aldehydes towards the corresponding alcohols [128]. The catalysis was performed in water at low temperature (30°C) and moderate H_2 pressure (8 bar). Here, the Ir atoms, in cooperation with the basic oxide species, heterolytically dissociate H_2 into $H^{\delta+}$ and $H^{\delta-}$ (Fig. 22a). This heterolytic H_2 cleavage explains the high activity and selectivity exhibited by this catalyst in the gram-scale hydrogenation of crotonaldehyde (92% conversion and 90% selectivity; initial TOF = 12,600 h^{-1}). This synergy between Ir NPs and ReOx species significantly increased the selectivity, keeping the characteristic high activity of noble metals. Another interesting synergistic effect was observed by Liao et al. on bimetallic PdIr NPs supported on mesoporous silica nanoparticles (PdIr/MSN) prepared by an impregnation-hydrogenation method [129]. They observed that the incorporation of Ir on Pd NPs significantly enhances the activity and selectivity of the catalyst in the hydrogenation of nitrobenzene to aniline. Here, there is an evident promoting effect, since bimetallic NPs with a molar Ir/Pd ratio of 0.1 (PdIr0.1/MSN) present an activity up to 28 and 8 times higher than those containing monometallic Ir/MSN and Pd/MSN, respectively (Fig. 22b). The promotional effect of Ir was explained by two factors: (1) an increase in the number of surface-active sites, consequence of a small NP size, and (2) a modification of the electronic properties of the MNP surface, which intensifies the electronic interactions. Liang et al. also observed a significant promotion effect of Ir in the selective hydrogenation of phenol to cyclohexanone over bimetallic PtIr NPs supported on hollow mesoporous silica spheres (HMSs) [130]. They generated the bimetallic PtIr on HMSs with different metal molar ratios (Ir/Pt: 0.5, 0.2 or 0.1) following a simple impregnation-reduction method. All catalysts showed similar selectivities towards cyclohexanone (80–85%), but different conversions were obtained depending on the amount of Ir incorporated. The most active catalyst was the one with a Ir/Pt molar ratio of 0.1 (PtIr0.1), with 60% of conversion (8.6 times higher than the monometallic Pt/HMS; Fig. 22c, left) and good levels of stability (4 recycle times; Fig. 22c, right). Here, the positive effect of the Ir is attributed to the enhancement of the metal dispersion (higher number of active sites) and the modulation of the NP electronic state (influence the MNP catalytic behaviour). Another example about a synergistic effect was reported by Ding et al. [131], in which ultrasmall silica-supported bimetallic NPs (Pd-Ir/SiO_2, Pd-Pt/SiO_2, Pd-Au/SiO_2, Pt-Ir/SiO_2 and Ru-Pt/SiO2) were active in the selective

Table 20 Hydrogenation of aromatic compounds over Ir NPs/montmorillonite. Reprinted with permission from Ref. [127]. Copyright 2016 Wiley-VCH

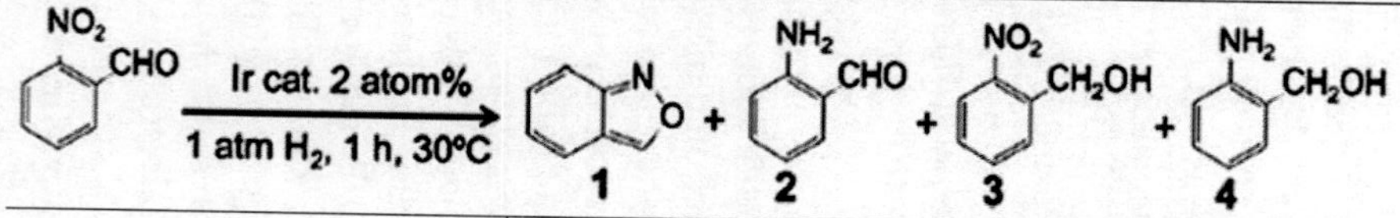

Catalyst	Conversion (%)	Selectivity (%)			
		1	2	3	4
Ir15: G6/SiO_2	98	17	83	<1	0
	98	17	83	<1	0
Ir30: G6/SiO_2	98	21	79	<1	0
	98	24	76	<1	0
Ir60: G6/SiO_2	98	42	58	<1	0
	98	40	60	<1	0

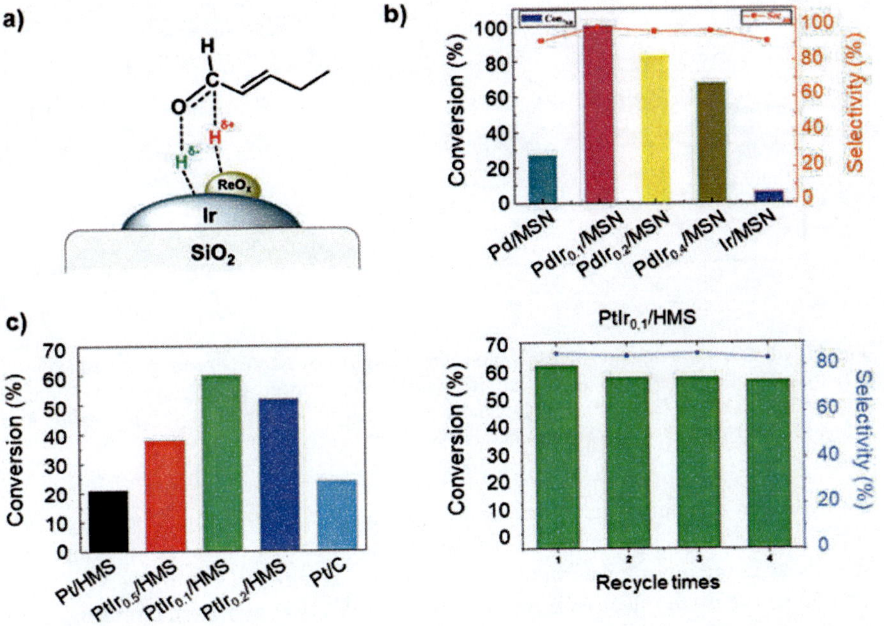

Fig. 22 (**a**) Proposed heterolytic H_2 cleavage mechanism for the hydrogenation of crotonaldehyde over Ir-ReO_x/SiO_2 of ref. [128]. (**b**) Conversion and selectivity to aniline in the hydrogenation of nitrobenzene with mono- (Pd/MSN and Ir/MSN) and bimetallic PdIr/MSN. Catal. Cond.: 10 mL ethanol; 0.04% cat.; 9 bar H_2; room temperature; 30 min. Reprinted with permission from ref. [129]. Copyright 2015 American Chemical Society. (**c**) *Left:* Phenol hydrogenation over mono- and bimetallic catalysts. *Right:* Recycling study with $PtIr_{0.1}$/HMS. Catal. Cond.: 50 mg catalyst (1.5% metal content); 0.5 g phenol; 10 ml acetate; 50°C; 0.5 bar H_2; 1 h. Reprinted with permission from ref. [130]. Copyright 2017 Royal Society of Chemistry

hydrogenation of acetylene. The enhanced activity compared to the monometallic parent catalysts was explained by a weaker coordination of acetylene and ethylene on the bimetallic surfaces.

In 2018, Henkelman and Humphrey reported for the first time bimetallic AgIr alloy NPs [132]. Naturally, Ag and Ir are immiscible metals, but using a microwave-assisted polyol synthesis method, small AgIr alloy NPs (2.5–5.5 nm) were successfully obtained. Bimetallic alloy NPs were prepared with different metal compositions, $Ag_xIr_{(100-x)}$ NPs ($x = 6$–31), from $Ag(NO_3)$ and $IrCl_3$ precursors. Ir-rich NPs, stabilised with PVP and dispersed in SiO_2, showed a higher activity in the gas-phase alkene hydrogenation (cyclohexene to cyclohexane) than the analogous pure Ir catalyst (Fig. 23a). More specifically, Ag_6Ir_{94} showed a TOF 75% higher than pure Ir NPs, even presenting a large particle size (i.e. Ag_6Ir_{94} NPs = 2.5 nm; Ir NPs = 1.7 nm). DFT calculations explain this difference in activity by the larger proportion of weakly bound H atoms on Ag_6Ir_{94} NPs, due to their large size and higher surface/edge sites ratio. The weaker H-binding sites of Ag_6Ir_{94} NPs facilitate

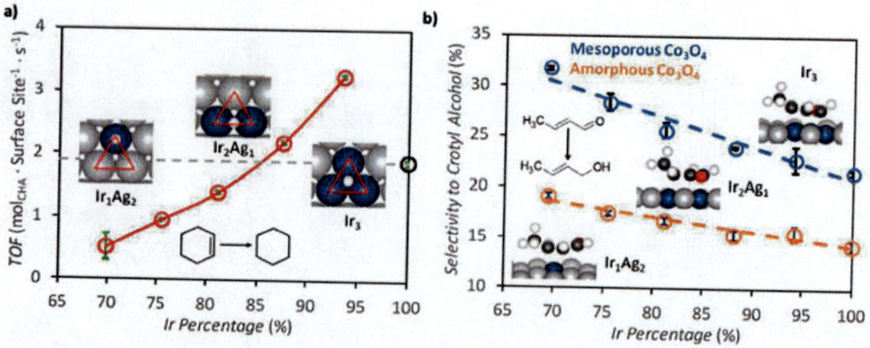

Fig. 23 (**a**) TOFs as a function of %Ir composition in AgIr alloy NPs (red) and compared to pure Ir NPs (black plot and grey line). (**b**) Selectivity to crotyl alcohol over $Ag_xIr_{(100-x)}$/a-Co_3O_4 (orange) and $Ag_xIr_{(100-x)}$/m-Co_3O_4 (blue) as a function of %Ir composition. Reprinted with permission from ref. [132]. Copyright 2018 American Chemical Society

the hydrogen transfer to the co-adsorbed cyclohexene. On the other hand, AgIr NPs supported on Co_3O_4 present a linear dependence between metal composition and selectivity in the hydrogenation of α,β-unsaturated aldehydes. The more Ag rich the catalyst is, the more selective is the catalyst towards C=O hydrogenation (Fig. 23b). Moreover, the Co_3O_4 support plays an important role in the hydrogenation mechanism, since the use of mesoporous Co_3O_4 considerably increases the selectivity. DFT calculations explain this trend through different synergic factors: (1) Ag-rich NPs present higher proportion of weak H-binding sites due to their larger size, which promote the hydrogen spillover to the Co_3O_4 support; (2) Ag-rich NPs present more Ag atoms which are weaker C=C adsorption sites; and (3) the larger Ag-rich NPs also present a higher weak surface/strong edge adsorption site ratio; thus C=C hydrogenation is more hindered in larger NPs.

In addition to carbon-based materials, metal oxides and silica supports, one can find another less used material for MNP stabilisation such as clays. For example, Dutta et al. generated Ir NPs (~4 nm) inside the nanopores of a modified montmorillonite clay by an incipient wetness impregnation method (Ir NPs/montmorillonite), by using $IrCl_3$ as metal precursor and ethylene glycol as reductant [133]. This Ir catalysts presented a high catalytic activity in the hydrogenation of a great number of aromatic compounds under solvent-free conditions (maximum TOF = 79 h^{-1}; Table 21). Moreover, the catalyst was easily recovered from the reaction media by filtration and was reused many times without any significant loss of catalytic activity.

6.4 Zeolite-Supported Ir NPs

Zeolites are rigid anionic frameworks of aluminosilicates with well-defined channels and cavities. Zeolites have numerous applications; one of the most important is the

Table 21 Hydrogenation of aromatic compounds over Ir NPs/montmorillonite[a]. Reprinted with permission from ref. [132]. Copyright 2016 Royal Society of Chemistry

Entry	Arene	Product	Conversion[b] (%)	TOF[c] (h^{-1})
1	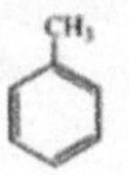		100[1] 96[2] 95[3]	79 76 75
2	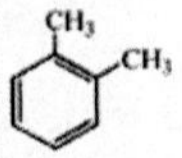	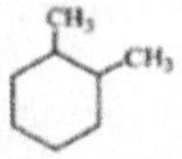	90	71
3	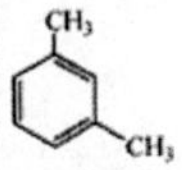	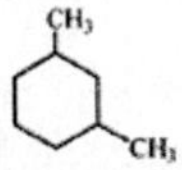	97	77
4	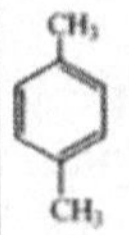	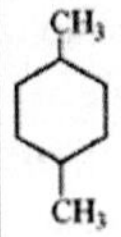	100	79
5	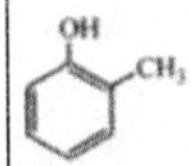	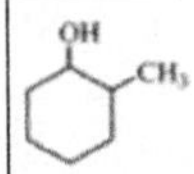	43	34
6	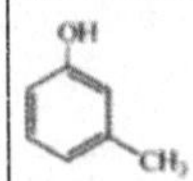	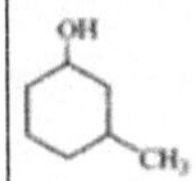	65	51
7	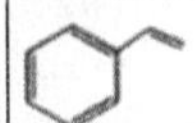	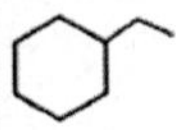	45	35
8	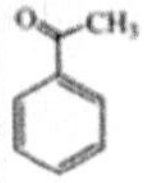	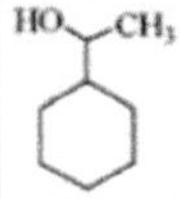	73	57

[a]Conditions: catalyst 50 mg, H_2 pressure: 5 bars temperature: 75°C, time: 6 h, stirring: 500 rpm
[b]Determined by GC analysis
[c]TOF with respect to product

separation of molecules (adsorbents, gas purification, etc.). In addition, zeolites have been widely used as catalyst by the petrochemical industry and lately as MNP support because of its higher surface-volume ratio and the possibility to modify the MNP catalytic activity. In 2014, Özkar et al. reported Ir NPs supported on a zeolite with faujasite (FAU) framework as a green nanocatalyst for the hydrogenation of aromatics [134]. The nanocatalyst, IrNPs@FAU, was obtained following a two-step procedure: (1) first, Ir^{3+} cations were inserted in FAU via an ion-exchange process; (2) then, the Ir^{3+}-exchanged FAU was chemically reduced by $NaBH_4$, generating Ir NPs (5.8 ± 2.1 nm) on the FAU surface and Ir nanoclusters inside the zeolite cavities. The resulting catalyst showed a high activity and stability in the hydrogenation of several neat aromatic substrates (benzene, toluene, mesitylene and

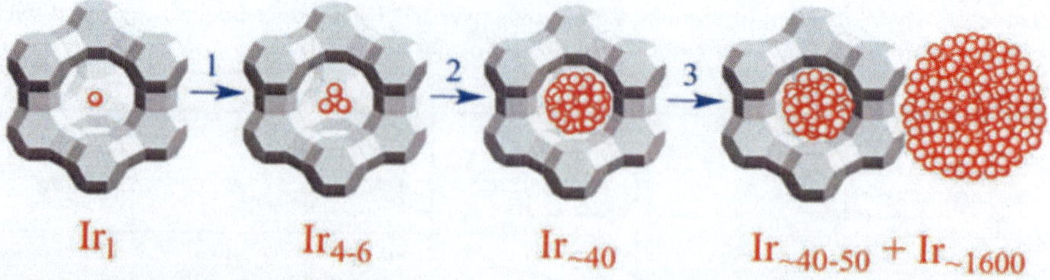

Fig. 24 Sequential agglomerative sintering of atomically dispersed Ir_1/zeolite Y during the cyclohexene hydrogenation at 72°C. Reprinted with permission from ref. [135]. Copyright 2015 American Chemical Society

o-xylene) under mild conditions (room temperature and 3 bar H_2). For example, benzene was completely hydrogenated to cyclohexane with a TOF = 3,215 h^{-1}. In addition, the nanocatalyst showed a high stability versus sintering and leaching, being reused up to five times preserving more than 50% of its initial activity during the last run. Most importantly, these zeolite-supported Ir NPs exhibited a catalytic lifetime without precedents in the neat benzene hydrogenation, with a total turnover number (TTON) of 197,000 over 92 h. This catalytic process was presented as green catalysis since it satisfies most of the green chemistry fundamentals, such as 100% selectivity, the transformation of all substrates into products, solvent-free, etc. A year later, Özkar, in collaboration with Gates and Finke, described the agglomerative sintering of atomically dispersed Ir_1/zeolite Y during the cyclohexene hydrogenation at 72°C [135]. In particular, they characterised the sintering of Ir_1/zeolite Y catalyst after 3 cycles of 3,800 total turnovers (TTOs) by HAADF/STEM and EXAFS spectroscopy. They observed that after the first 3,800 TTOs, the initial atomically dispersed Ir atoms (Ir_1) sinter into Ir clusters of 4–6 atoms ($Ir_{4–6}$). Later, after 2 cycles of 3,800 TTOs (7,600 TTOs), Ir clusters of 40 atoms approx. ($Ir_{\sim 40}$) are the dominant catalytic species. Finally, after the third cycle (11,400 TTOs), two populations of nano-objects were detected, Ir clusters of 40–50 atoms inside the cavities of the zeolite ($Ir_{\sim 40–50}$) and Ir nanoparticles with an average size of 3.5 ± 0.9 nm ($Ir_{\sim 1,600}$), which are generated on the zeolite surface (Fig. 24). This sequential agglomerative sintering considerably decreases the hydrogenation activity of the catalyst, reducing the initial hydrogenation rate up to eight times.

Catalysts based on supported MNPs are of great importance for the industrial synthesis of numerous chemicals. However, MNPs tend to sinter under reaction conditions, deteriorating the catalytic properties of the materials. Therefore, materials containing active MNPs inside their pores are of high interest for catalytic applications since their stability considerably increases. For example, Hölderich et al. synthesised materials containing noble MNPs inside the MCM-41 pores by direct synthesis (Fig. 25) [136]. In particular, Ir, Rh and Pd NPs inside MCM-41 ([Me]*x*-MCM-41) were prepared by using surfactant stabilised nanoparticles during the formation of MCM-41. The obtained [Me]*x*-MCM-41 were active catalysts in the hydrogenation of different cyclic alkenes (cyclohexene, norbornene, cyclooctene

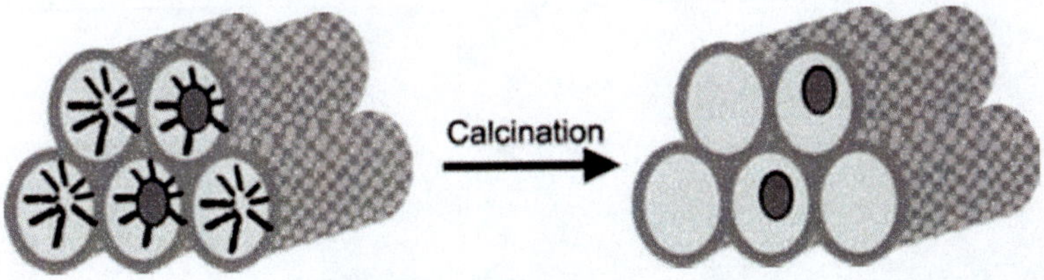

Fig. 25 Illustration of the synthesis of [Me]*x*-MCM-41 by using surfactant stabilised MNPs. Reprinted with permission from ref. [136]. Copyright 2002 Royal Society of Chemistry

and cyclododecene). Since the pore size of the zeolite is large enough (31 Å) to allow diffusion of all substrates without hindrance, the differences in activity were mainly caused by the alkene strain energy. For instance, the conversion of norbornene was much higher than that of the other substrates because the hydrogenation of the norbornene structure is energetically more favoured. Comparing the different metals, [Rh]*x*-MCM-41 showed the highest activity, followed by Ir and Pd.

6.5 *Ir NPs Supported on Metal-Organic Frameworks (MOFs)*

In addition to the aforementioned supports, metal-organic frameworks (MOFs) have emerged as a significant group of porous materials for MNP stabilisation. These hybrid crystalline materials based on 3D networks of metal ions coordinated to organic ligands are suitable host-guest systems for the stabilisation of ligand-free metal nanoparticles. Moreover, they present a pore structure easily tuneable by the selection of the proper organic linker. In this context, Zahmakiran generated Ir NPs on the structure of a zeolitic imidazole framework ZIF-8 (Zn(MeIM); MeIM = 2-methylimidazole) by hydrogenolysis of the organometallic precursor Ir(COD)(MeCp) which was previously impregnated into ZIF-8 [137]. The resultant material, IrNPs@ZIF-8, demonstrated to be an efficient catalyst in the hydrogenation of cyclohexene and phenylacetylene in terms of activity, selectivity and durability. Apart from showing a great stability (reused up to five times), it hydrogenates cyclohexene in solvent-free conditions (25°C, 3 bar H_2) with an initial TOF = 370 h^{-1} and phenylacetylene to styrene with high selectivity ($\geq$ 90% at full conversion under mild conditions (40°C, 3 bar H_2)). A similar gas-phase loading was employed by Kempe et al. to generate crystalline Ir NPs (1.7 nm) inside the pores of the chromium terephthalate metal-organic framework [MIL-101 (Cr); Fig. 26], which showed a high activity in the hydrogenation of cyclohexene (full conversion with only 10 ppm of Ir) [138].

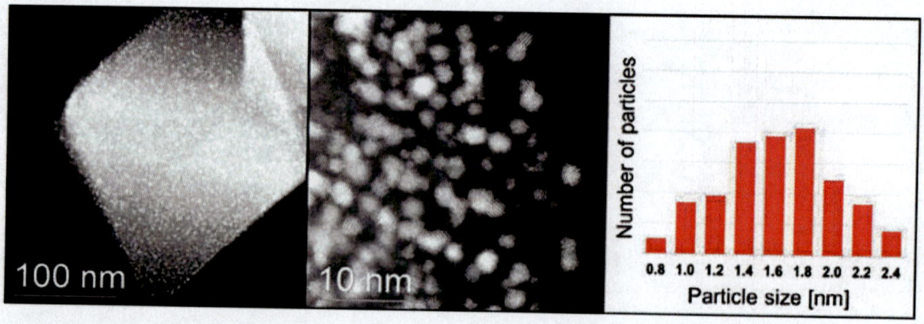

Fig. 26 (HR)-ADF-STEM micrographs and size histogram of Ir NPs inside the pores of MIL-101 (Cr). Reprinted with permission from ref. [138]. Copyright 2002 Elsevier

7 Concluding Remarks

As can be seen by the numerous works mentioned in the present chapter, iridium nanoparticles have attracted large interest as hydrogenation catalysts. NPs immobilised on supports, ligand-stabilised NPs, confined NPs and NPs stabilised by ionic liquids and polymers are examples of Ir NP-based catalytic systems employed to carry out the hydrogenation process. Also remarkable is the use of Ir NPs generated in situ that do not need stabilising agents. Although iridium is an expensive transition metal, Ir NPs generally exhibit high activity as hydrogenation catalysts. In addition, its low tendency to oxidation allows a good recyclability. These features can make Ir NPs worthwhile from an industrial point of view.

However, Ir NPs have shown a limited efficiency in terms of selectivity, and thus new approaches and strategies are required to make the reduction process sufficiently selective to be industrially relevant. Indeed, important advances have been achieved in recent years in the development of highly chemoselective Ir NP-based systems. These catalysts mainly consist of nanoparticles supported on oxides, oxidised carbon materials or oxygenated surfaces, for which the catalytic reaction sometimes proceeds through an interaction between metal and support. The use of Ir NPs stabilised by ligands that additionally play a role in the process is also a very promising strategy to carry out selective, partial hydrogenation of unsaturated substrates. Metal nanoparticles functionalised by organic ligands can be immobilised on one of the many supports mentioned, which merges the two approaches creating a plethora of interesting interactions between metals, support, ligands and substrates. The new research area can be sure of a tremendous industrial interest [92], and thus it can be expected that new and stimulating advances in this field will appear in the near future.

References

1. Heiz U, Landman U (2007) Nanocatalysis. Springer, Berlin
2. Astruc D (2008) Nanoparticles and catalysis. Wiley-VCH, Weinheim
3. Philippot K, Serp P (2013) Nanomaterials in catalysis. Wiley-VCH, Weinheim
4. Liu L, Corma A (2018) Metal catalysts for heterogeneous catalysis: from single atoms to nanoclusters and nanoparticles. Chem Rev 118:4981–5079
5. Ni B, Wang X (2015) Face the edges: catalytic active sites of nanomaterials. Adv Sci 2:1500085
6. Mostafa S, Behafarid F, Croy JR, Ono LK, Li L, Yang JC, Frenkel AI, Cuenya BR (2010) Shape-dependent catalytic properties of Pt nanoparticles. J Am Chem Soc 132:15714–15719
7. Schulz J, Roucoux A, Patin H (1999) Unprecedented efficient hydrogenation of arenes in biphasic liquid-liquid catalysis by re-usable aqueous colloidal suspensions of rhodium. Chem Commun:535–536
8. Stowell CA, Korgel BA (2005) Iridium nanocrystal synthesis and surface coating-dependent catalytic activity. Nano Lett 5:1203–1207
9. Roucoux A, Schulz J, Patin H (2003) Arene hydrogenation with a stabilised aqueous rhodium (0) suspension: a major effect of the surfactant counteranion. Adv Synth Catal 345:222–229
10. Iablokov V, Beaumont SK, Alayoglu S, Pushkarev VV, Specht C, Gao JH, Alivisatos AP, Kruse N, Somorjai GA (2012) Size-controlled model Co nanoparticle catalysts for CO2 hydrogenation: synthesis, characterization, and catalytic reactions. Nano Lett 12:3091–3096
11. Pan C, Pelzer K, Philippot K, Chaudret B, Dassenoy F, Lecante P, Casanove MJ (2001) Ligand-stabilized ruthenium nanoparticles: synthesis, organization, and dynamics. J Am Chem Soc 123:7584–7593
12. Duteil A, Queau R, Chaudret B, Mazel R, Roucau C, Bradley JS (1993) Preparation of organic solutions or solid films of small particles of ruthenium, palladium, and platinum from organometallic precursors in the presence of cellulose derivatives. Chem Mater 5:341–347
13. Dupont J, Fonseca GS, Umpierre AP, Fichtner PFP, Teixeira SR (2002) Transition-metal nanoparticles in imidazolium ionic liquids: recyclable catalysts for biphasic hydrogenation reactions. J Am Chem Soc 124:4228–4229
14. Ryu J, Sanchez L, Keul HA, Raj A, Bockstaller MR (2008) Imidazolium-based ionic liquids as efficient shape-regulating solvents for the synthesis of gold nanorods. Angew Chem Int Ed 47:7639–7643
15. Martínez-Prieto LM, Chaudret B (2018) Organometallic ruthenium nanoparticles: synthesis, surface chemistry, and insights into ligand coordination. Acc Chem Res 51:376–384
16. Martínez-Prieto LM, van Leeuwen PWNM (2020) van Leeuwen PWNM, Claver C (eds) Ligand effects in ruthenium nanoparticle catalysis. Recent advances in nanoparticle catalysis. Springer-Nature, Cham
17. Chinthaginjala JK, Villa A, Su DS, Mojet BL, Lefferts L (2012) Nitrite reduction over Pd supported CNFs: metal particle size effect on selectivity. Catal Today 183:119–123
18. Ismail AA, Hakki A, Bahnemann DW (2012) Mesostructure Au/TiO2 nanocomposites for highly efficient catalytic reduction of *p*-nitrophenol. J Mol Catal A Chem 358:45–151
19. Martinez-Prieto LM, Baquero EA, Pieters G, Flores JC, de Jesus E, Nayral C, Delpech F, van Leeuwen PWNM, Lippens G, Chaudret B (2017) Monitoring of nanoparticle reactivity in solution: interaction of l-lysine and Ru nanoparticles probed by chemical shift perturbation parallels regioselective H/D exchange. Chem Commun 53:5850–5853
20. Martinez-Prieto LM, Cano I, Marquez A, Baquero EA, Tricard S, Cusinato L, del Rosal I, Poteau R, Coppel Y, Philippot K, Chaudret B, Campora J, van Leeuwen PWNM (2017) Zwitterionic amidinates as effective ligands for platinum nanoparticle hydrogenation catalysts. Chem Sci 8:2931–2941
21. Sheldo RA, Arends I, Hanefeld U (2007) Green chemistry and catalysis. Wiley-VCH, Weinheim
22. Weissermel K, Arpe HJ (1993) Industrial organic chemistry. VCH, New York

23. Lovering F, Bikker J, Humblet C (2009) Escape from flatland: increasing saturation as an approach to improving clinical success. J Med Chem 52:6752–6756
24. García-Antón J, Axet MR, Jansat S, Philippot K, Chaudret B, Pery T, Buntkowsky G, Limbach HH (2008) Reactions of olefins with ruthenium hydride nanoparticles: NMR characterization, hydride titration, and room-temperature C–C bond activation. Angew Chem Int Ed 47:2074–2078
25. Hu Y, Yu Y, Hou Z, Li H, Zhao X, Feng B (2008) Biphasic hydrogenation of olefins by functionalized ionic liquid-stabilized palladium nanoparticles. Adv Synth Catal 350:2077–2085
26. Fiorio JL, López N, Rossi LM (2017) Gold–ligand-catalyzed selective hydrogenation of alkynes into cis-alkenes via H2 Heterolytic activation by frustrated Lewis Pairs. ACS Catal 7:2973–2980
27. Delgado JA, Benkirane O, Claver C, Curulla-Ferre D, Godard C (2017) Advances in the preparation of highly selective nanocatalysts for the semi-hydrogenation of alkynes using colloidal approaches. Dalton Trans 46:12381–12403
28. López-Vinasco AM, Martínez-Prieto LM, Asensio JM, Lecante P, Chaudret B, Cámpora J, van Leeuwen PWNM (2020) Novel nickel nanoparticles stabilized by imidazolium-amidinate ligands for selective hydrogenation of alkynes. Catal Sci Tech 10:342–350
29. Fang M, Sánchez-Delgado RA (2014) Ruthenium nanoparticles supported on magnesium oxide: a versatile and recyclable dual-site catalyst for hydrogenation of mono- and poly-cyclic arenes, N-heteroaromatics, and S-heteroaromatics. J Catal 311:357–368
30. Rakers L, Martínez-Prieto LM, López-Vinasco AM, Philippot K, van Leeuwen PWNM, Chaudret B, Glorius F (2018) Ruthenium nanoparticles ligated by cholesterol-derived NHCs and their application in the hydrogenation of arenes. Chem Commun 54:7070–7073
31. Cano I, Tschan MJL, Martinez-Prieto LM, Philippot K, Chaudret B, van Leeuwen PWNM (2016) Enantioselective hydrogenation of ketones by iridium nanoparticles ligated with chiral secondary phosphine oxides. Catal Sci Tech 6:3758–3766
32. Martinez-Prieto LM, Ferry A, Rakers L, Richter C, Lecante P, Philippot K, Chaudret B, Glorius F (2016) Long-chain NHC-stabilized RuNPs as versatile catalysts for one-pot oxidation/hydrogenation reactions. Chem Commun 52:4768–4771
33. Mitsudome T, Yamamoto M, Maeno Z, Mizugaki T, Jitsukawa K, Kaneda K (2015) One-step synthesis of Core-gold/Shell-ceria nanomaterial and its catalysis for highly selective Semihydrogenation of alkynes. J Am Chem Soc 137:13452–13455
34. Cano I, Martinez-Prieto LM, Fazzini PF, Coppel Y, Chaudret B, van Leeuwen PWNM (2017) Characterization of secondary phosphine oxide ligands on the surface of iridium nanoparticles. Phys Chem Chem Phys 19:21655–21662
35. Wan W, Nie X, Janik MJ, Song C, Guo X (2018) Adsorption, dissociation, and spillover of hydrogen over Au/TiO2 catalysts: the effects of cluster size and metal–support interaction from DFT. J Phys Chem C 122:17895–17916
36. Luza L, Rambor CP, Gual A, Alves Fernandes J, Eberhardt D, Dupont J (2017) Revealing hydrogenation reaction pathways on naked gold nanoparticles. ACS Catal 7:2791–2799
37. Corma A, Serna P (2006) Chemoselective hydrogenation of nitro compounds with supported gold catalysts. Science 313:332–334
38. Li G, Zeng C, Jin R (2015) Chemoselective hydrogenation of nitrobenzaldehyde to nitrobenzyl alcohol with unsupported Au nanorod catalysts in water. J Phys Chem C 119:11143–11147
39. Pritchard J, Filonenko GA, van Putten R, Hensen EJM, Pidko EA (2015) Heterogeneous and homogeneous catalysis for the hydrogenation of carboxylic acid derivatives: history, advances and future directions. Chem Soc Rev 44:3808–3833
40. Martínez-Prieto LM, Puche M, Cerezo-Navarrete C, Chaudret B (2019) Uniform Ru nanoparticles on N-doped graphene for selective hydrogenation of fatty acids to alcohols. J Catal 377:429–437

41. Scholten JD (2013) From soluble to supported iridium metal nanoparticles, active and recyclable catalysts for hydrogenation reactions. Curr Org Chem 17(4):348–363
42. Martínez-Prieto LM, van Leeuwen PWNM (2020) Ligand effects in ruthenium nanoparticle catalysis. In: Turner N, Claver C, van Leeuwen PWNM (eds) Recent advances in nanoparticle catalysis. Springer, Berlin
43. Gual A, Godard C, Philippot K, Chaudret B, Denicourt-Nowicki A, Roucoux A, Castillón S, Claver C (2009) Carbohydrate-derived 1,3-diphosphite ligands as chiral nanoparticle stabilizers: promising catalytic systems for asymmetric hydrogenation. ChemSusChem 2:769–779
44. Ruiz D, Oportus M, Godard C, Claver C, Fierro JLG, Reyes P (2012) Novel metal nanoparticles stabilized with (2*R*,4*R*)-2,4-bis(diphenylphosphino)pentane on SiO_2. Their use as catalysts in Enantioselective hydrogenation reactions. Curr Org Chem 16:2754–2762
45. Cano I, Tschan MJ-L, Martinez-Prieto LM, Philippot K, Chaudret B, van Leeuwen PWNM (2016) Enantioselective hydrogenation of ketones by iridium nanoparticles ligated with chiral secondary phosphine oxides. Cat Sci Technol 6:3758–3766
46. Cano I, Chapman AM, Urakawa A, van Leeuwen PWNM (2014) Air-stable gold nanoparticles ligated by secondary phosphine oxides for the chemoselective hydrogenation of aldehydes: crucial role of the ligand. J Am Chem Soc 136:2520–2528
47. Cano I, Huertos MA, Chapman AM, Buntkowsky G, Gutmann T, Groszewicz PB, van Leeuwen PWNM (2015) Air-stable gold nanoparticles ligated by secondary phosphine oxides as catalysts for the chemoselective hydrogenation of substituted aldehydes: a remarkable ligand effect. J Am Chem Soc 137:7718–7727
48. Cano I, Martínez-Prieto LM, Fazzini PF, Coppel Y, Chaudret B, van Leeuwen PWNM (2017) Characterization of secondary phosphine oxide ligands on the surface of iridium nanoparticles. Phys Chem Chem Phys 19:21655–21662
49. Cano I, Martínez-Prieto LM, Chaudret B, van Leeuwen PWNM (2017) Iridium versus iridium: nanocluster and monometallic catalysts carrying the same ligand behave differently. Chem Eur J 23:1444–1450
50. Li W, Wang Y, Chen P, Zeng M, Jiang J, Jin Z (2016) Thermoregulated phase-transfer iridium nanoparticle catalyst: highly selective hydrogenation of the C=O bond for α,β-unsaturated aldehydes and the C=C bond for α,β-unsaturated ketones. Cat Sci Technol 6:7386–7390
51. Mévellec V, Roucoux A, Ramirez E, Philippot K, Chaudret B (2004) Surfactant-stabilized aqueous iridium(0) colloidal suspension: an efficient reusable catalyst for hydrogenation of Arenes in biphasic media. Adv Synth Catal 346:72–76
52. Stowell CA, Korgel BA (2005) Iridium nanocrystal synthesis and surface coating dependent catalytic activity. Nano Lett 5:1203–1207
53. Mondloch JE, Özkar S, Finke RG (2018) "Weakly ligated, labile ligand" nanoparticles: the case of Ir(0)*n*·(H+Cl−)*m*. ACS Omega 3:14538–14550
54. Egeberg A, Dietrich C, Kind C, Popescu R, Gerthsen D, Behrens S, Feldmann C (2017) Bimetallic nickel-iridium and nickel-osmium alloy nanoparticles and their catalytic performance in hydrogenation reactions. ChemCatChem 9:3534–3543
55. Wasserscheid P, Keim W (2000) Ionic liquids-new "solutions" for transition metal catalysis. Angew Chem Int Ed 39:3772–3789
56. Aparicio S, Atilhan M, Karadas F (2010) Thermophysical properties of pure ionic liquids: review of present situation. Ind Eng Chem Res 49:9580–9595
57. Hulsbosch J, De Vos DE, Binnemans K, Ameloot R (2016) Biobased ionic liquids: solvents for a green processing industry? ACS Sustain Chem Eng 4:2917–2931
58. Dupont J, Scholten JD (2010) On the structural and surface properties of transition-metal nanoparticles in ionic liquids. Chem Soc Rev 39:1780–1804
59. Migowski P, Dupont J (2007) Catalytic applications of metal nanoparticles in Imidazolium ionic liquids. Chem Eur J 13:32–39
60. Scholten JD, Leal BC, Dupont J (2012) Transition metal nanoparticle catalysis in ionic liquids. ACS Catal 2:184–200

61. Dupont J, Fonseca GS, Umpierre AP, Fichtner PFP, Teixeira SR (2002) Transition-metal nanoparticles in Imidazolium ionic liquids: recyclable catalysts for biphasic hydrogenation reactions. J Am Chem Soc 124:4228–4229
62. Fonseca GS, Umpierre AP, Fichtner PFP, Teixeira SR, Dupont J (2003) The use of Imidazolium ionic liquids for the formation and stabilization of Ir0 and Rh0 nanoparticles: efficient catalysts for the hydrogenation of Arenes. Chem Eur J 9:3263–3269
63. Fonseca GS, Scholten JD, Dupont J (2004) Iridium nanoparticles prepared in ionic liquids: an efficient catalytic system for the hydrogenation of ketones. Synlett 9:1525–1528
64. Fonseca GS, Silveira ET, Gelesky MA, Dupont J (2005) Competitive hydrogenation of alkyl-substituted Arenes by transition-metal nanoparticles: correlation with the alkyl-steric effect. Adv Synth Catal 347:847–853
65. Fonseca GS, Domingos JB, Nome F, Dupont J (2006) On the kinetics of iridium nanoparticles formation in ionic liquids and olefin hydrogenation. J Mol Cat A-Chem 248:10–16
66. Fonseca GS, Machado G, Teixeira SR, Fecher GH, Morais J, Alves MCM, Dupont J (2006) Synthesis and characterization of catalytic iridium nanoparticles in imidazolium ionic liquids. J Colloid Interf Sci 301:193–204
67. Scholten JD, Ebeling G, Dupont J (2007) On the involvement of NHC carbenes in catalytic reactions by iridium complexes, nanoparticle and bulk metal dispersed in imidazolium ionic liquids. Dalton Trans:5554–5560
68. Migowski P, Zanchet D, Machado G, Gelesky MA, Teixeira SR, Dupont J (2010) Nanostructures in ionic liquids: correlation of iridium nanoparticles' size and shape with imidazolium salts' structural organization and catalytic properties. Phys Chem Chem Phys 12:6826–6833
69. Faria VW, Brunelli MF, Scheeren CW (2015) Iridium nanoparticles supported in polymeric membranes: a new material for hydrogenation reactions. RSC Adv 5:84920–84926
70. Redel E, Krämer J, Thomann R, Janiak C (2009) Synthesis of Co, Rh and Ir nanoparticles from metal carbonyls in ionic liquids and their use as biphasic liquid–liquid hydrogenation nanocatalysts for cyclohexene. J Organomet Chem 694:1069–1075
71. Vollmer C, Redel E, Abu-Shandi K, Thomann R, Manyar H, Hardacre C, Janiak C (2010) Microwave irradiation for the facile synthesis of transition-metal nanoparticles (NPs) in ionic liquids (ILs) from metal–carbonyl precursors and Ru-, Rh-, and Ir-NP/IL dispersions as biphasic liquid–liquid hydrogenation Nanocatalysts for cyclohexene. Chem Eur J 16:3849–3858
72. Esteban RM, Schütte K, Brandt P, Marquardt D, Meyer H, Beckert F, Mülhaupt R, Kölling H, Janiak C (2015) Iridium@graphene composite nanomaterials synthesized in ionic liquid as re-usable catalysts for solvent-free hydrogenation of benzene and cyclohexene. Nano-Struct Nano-Objects 2:11–18
73. Jiang H-Y, Xu J, Sun B (2018) Selective hydrogenation of aromatic compounds using modified iridium nanoparticles. Appl Organometal Chem:e4260
74. Koczkur KM, Mourdikoudis S, Polavarapu L, Skrabalak SE (2015) Polyvinylpyrrolidone (PVP) in nanoparticle synthesis. Dalton Trans 44:17883–17905
75. Tu W-x, He B-l, H-f L, Luo X-l, Liang X (2005) Catalytic properties of polymer-stabilized colloidal metal nanoparticles synthesized by microwave irradiation. Chin J Polym Sci 23:211–217
76. Sharif MJ, Maity P, Yamazoe S, Tsukuda T (2013) Selective hydrogenation of nitroaromatics by colloidal iridium nanoparticles. Chem Lett 42:1023–1025
77. Corma A, Serna P, Concepción P, Calvino JJ (2008) Transforming nonselective into chemoselective metal catalysts for the hydrogenation of substituted Nitroaromatics. J Am Chem Soc 130:8748–8753, and references therein
78. Makosch M, Lin W-I, Bumbálek V, Sá J, Medlin JW, Hungerbühler K, van Bokhoven JA (2012) Organic Thiol modified Pt/TiO2 catalysts to control chemoselective hydrogenation of substituted Nitroarenes. ACS Catal 2:2079–2081, and references therein

79. Nishida Y, Chaudhari C, Imatome H, Sato K, Nagaoka K (2018) Selective hydrogenation of nitriles to secondary imines over Rh-PVP catalyst under mild conditions. Chem Lett 47:938–940
80. Gao L, Kojima K, Nagashima H (2015) Transition metal nanoparticles stabilized by ammonium salts of hyperbranched polystyrene: effect of metals on catalysis of the biphasic hydrogenation of alkenes and arenes. Tetrahedron 71:6414–6423
81. Ghosh S, Jagirdar BR (2017) Synthesis of mesoporous iridium nanosponge: a highly active, thermally stable and efficient olefin hydrogenation catalyst. Dalton Trans 46:11431–11439
82. Özkar S, Finke RG (2005) Iridium(0) nanocluster, acid-assisted catalysis of neat acetone hydrogenation at room temperature: exceptional activity, catalyst lifetime, and selectivity at complete conversion. J Am Chem Soc 127:4800–4808
83. Bayram E, Zahmakıran M, Özkar S, Finke RG (2010) In situ formed "weakly ligated/labile ligand" iridium(0) nanoparticles and aggregates as catalysts for the complete hydrogenation of neat benzene at room temperature and mild pressures. Langmuir 26:12455–12464
84. Alley WM, Hamdemir IK, Wang Q, Frenkel AI, Li L, Yang JC, Menard LD, Nuzzo RG, Ozkar S, Johnson KA, Finke RG (2010) Iridium Ziegler-type hydrogenation catalysts made from [(1,5-COD)Ir(μ-O2C8H15)]2 and AlEt3: spectroscopic and kinetic evidence for the Ir*n* species present and for nanoparticles as the fastest catalyst. Inorg Chem 49:8131–8147
85. Widegren JA, Finke RG (2003) A review of the problem of distinguishing true homogeneous catalysis from soluble or other metal-particle heterogeneous catalysis under reducing conditions. J Mol Cat A Chem 198:317–341
86. Hamdemir IK, Özkar S, Yih K-H, Mondloch JE, Finke RG (2012) Hydrocarbon-soluble, isolable Ziegler-type Ir(0)*n* nanoparticle catalysts made from [(1,5-COD)Ir(μ-O2C8H15)]2 and 2–5 equivalents of AlEt3: their high catalytic activity, long lifetime, and AlEt3-dependent, exceptional, 200°C thermal stability. ACS Catal 2:632–641
87. Hamdemir IK, Özkar S, Finke RG (2013) Exceptionally thermally stable, hydrocarbon soluble Ziegler-type Ir(0)*n* nanoparticle catalysts made from [Ir(1,5-COD)(m-O2C8H15)]2 plus AlEt3: tests of key hypotheses for their unusual stabilization. J Mol Cat A-Chem 378:333–343
88. Ott LS, Finke RG (2008) Supersensitivity of transition-metal nanoparticle formation to initial precursor concentration and reaction temperature*: understanding its origins. J Nanosci Nanotechnol 8:1551–1556
89. Laxson WW, Finke RG (2014) Nucleation is second order: an apparent kinetically effective nucleus of two for Ir(0)n nanoparticle formation from [(1,5-COD)IrI·P2W15Nb3O62]8− plus hydrogen. J Am Chem Soc 136:17601–17615
90. Özkar S, Finke RG (2017) Nanoparticle nucleation is Termolecular in metal and involves hydrogen: evidence for a kinetically effective nucleus of three {Ir3H2x·P2W15Nb3O62}6− in Ir(0)n nanoparticle formation from [(1,5-COD)IrI·P2W15Nb3O62]8− plus hydrogen. J Am Chem Soc 139:5444–5457
91. Ji Y-G, Wei K, Liu T, Wu L, Zhang W-H (2017) "Naked" iridium(IV) oxide nanoparticles as expedient and robust catalysts for hydrogenation of nitrogen heterocycles: remarkable vicinal substitution effect and recyclability. Adv Synth Catal 359:933–940
92. Albani D, Li Q, Vilé G, Mitchell S, Almora-Barrios N, Witte PT, López N, Pérez-Ramírez J (2017) Interfacial acidity in ligand-modified ruthenium nanoparticles boosts the hydrogenation of levulinic acid to gamma-valerolactone. Green Chem 19:2361–2370
93. Machado BF, Gomes HT, Serp P, Kalck P, Faria JL (2010) Liquid-phase hydrogenation of unsaturated aldehydes: enhancing selectivity of multiwalled carbon nanotube-supported catalysts by thermal activation. ChemCatChem 2:190–197
94. Rueping M, Koenigs RM, Borrmann R, Zoller J, Weirich TE, Mayer J (2011) Size-selective, stabilizer-free, hydrogenolytic synthesis of iridium nanoparticles supported on carbon nanotubes. Chem Mater 23:2008–2010
95. Du X, Liu Y, Wang J, Cao Y, Fan K (2013) Catalytic conversion of biomass-derived levulinic acid into γ-valerolactone using iridium nanoparticles supported on carbon nanotubes. Cuihua Xuebao 34:993–1001

96. Li H-B, Liu L, Ma X-Y (2016) Effective hydrogenation of Haloaromatic nitro compounds catalysed by iridium nanoparticles deposited on multiwall carbon nanotubes. Synth React Inorg Met Org Nano Met Chem 46:1499–1505
97. Motoyama Y, Taguchi M, Desmira N, Yoon S-H, Mochida I, Nagashima H (2014) Chemoselective hydrogenation of functionalized nitroarenes and imines by using carbon nanofiber-supported iridium nanoparticles. Chem Asian J 9:71–74
98. Bernas H, Simakova I, Prosvirin IP, Maeki-Arvela P, Leino R, Murzin DY (2012) Hydrogenation of citral over carbon supported iridium catalysts. Catal Lett 142(6):690–697
99. Kato S, Nanao H, Shirai M (2019) Graphite intercalated iridium nanodisks for cinnamaldehyde hydrogenation. Chem Lett 48:1262–1265
100. Sui D, Mao F, Fan H, Qi Z, Huang J (2017) General reductive amination of aldehydes and ketones with amines and Nitroaromatics under H2 by recyclable iridium catalysts. Chin J Chem 35:1371–1377
101. Ledesma B, Juarez J, Mazario J, Domine M, Beltramone A (2019) Bimetallic platinum/iridium modified mesoporous catalysts applied in the hydrogenation of HMF. Catal Today. https://doi.org/10.1016/j.cattod.2019.06.037
102. Szumelda T, Drelinkiewicz A, Kosydar R, Goral-Kurbiel M, Gurgul J, Duraczynska D (2017) Formation of Pd-group VIII bimetallic nanoparticles by the "water-in-oil" microemulsion method. Colloids Surf A Physicochem Eng Asp 529:246–260
103. Navalon S, Dhakshinamoorthy A, Alvaro M, Garcia H (2016) Metal nanoparticles supported on two-dimensional graphenes as heterogeneous catalysts. Coord Chem Rev 312:99–148
104. Wang Y, Rong X, Wang T, Wu S, Rong Z, Wang Y, Qu J (2019) Influence of graphene surface chemistry on Ir-catalyzed hydrogenation of p-chloronitrobenzene and cinnamaldehyde: weak molecule-support interactions. J Catal 377:524–533
105. Chen P, Lu J-Q, Xie G-Q, Hu G-S, Zhu L, Luo L-F, Huang W-X, Luo M-F (2012) Effect of reduction temperature on selective hydrogenation of crotonaldehyde over Ir/TiO2 catalysts. Appl Catal A 433-434:236–242
106. Reyes P, Aguirre MC, Melián-Cabrera I, López Granados M, Fierro JLG (2002) Interfacial properties of an Ir/TiO2 system and their relevance in crotonaldehyde hydrogenation. J Catal 208:229–237
107. Rojas H, Martinez JJ, Mancipe S, Borda G, Reyes P (2012) Citral hydrogenation over novel niobia and titania supported Au, Ir-Au and Ir catalysts. React Kinet Mech Catal 106:445–455
108. Zhao EW, Zheng H, Ludden K, Xin Y, Hagelin-Weaver HE, Bowers CR (2016) Strong metal-support interactions enhance the pairwise selectivity of Parahydrogen addition over Ir/TiO2. ACS Catal 6:974–978
109. Toledo-Antonio JA, Angeles-Chavez C, Cortes-Jacome MA, Cuauhtemoc-Lopez I, Lopez-Salinas E, Mosqueira ML, Ferrat G (2016) Metal support interaction effects on the reducibility of Ir nanoparticles on Titania nanotubes. Top Catal 59:366–377
110. Toledo-Antonio JA, Angeles-Chavez C, Cortes-Jacome MA, Cuauhtemoc-Lopez I, Lopez-Salinas E, Perez-Luna M, Ferrat-Torres G (2012) Highly dispersed Pt-Ir nanoparticles on titania nanotubes. Appl Catal A 437-438:155–165
111. Zhao J, Ni J, Xu J, Xu J, Cen J, Li X (2014) Ir promotion of TiO2 supported Au catalysts for selective hydrogenation of cinnamaldehyde. Catal Commun 54:72–76
112. Fan G-Y, Zhang L, Fu H-Y, Yuan M-L, Li R-X, Chen H, Li X-J (2010) Hydrous zirconia supported iridium nanoparticles: an excellent catalyst for the hydrogenation of haloaromatic nitro compounds. Catal Commun 11:451–455
113. Campos C, Torres C, Oportus M, Pena MA, Fierro JLG, Reyes P (2013) Hydrogenation of substituted aromatic nitrobenzenes over 1% 1.0 wt.%Ir/ZrO2 catalyst: effect of meta position and catalytic performance. Catal Today 213:93–100
114. He S, Xie L, Che M, Chan HC, Yang L, Shi Z, Tang Y, Gao Q (2016) Chemoselective hydrogenation of α,β-unsaturated aldehydes on hydrogenated MoOx nanorods supported iridium nanoparticles. J Mol Catal A Chem 425:248–254

115. Xie L, Chen T, Chan HC, Shu Y, Gao Q (2018) Hydrogen doping into MoO3 supports toward modulated metal-support interactions and efficient furfural hydrogenation on iridium nanocatalysts. Chem Asian J 13:641–647
116. Xie L, Chen T, Chan HC, Shu Y, Gao Q (2018) Hydrogen doping into MoO3 supports toward modulated metal-support interactions and efficient furfural hydrogenation on iridium nanocatalysts. Chem Asian J 13:641–647
117. Tamura M, Yonezawa D, Oshino T, Nakagawa Y, Tomishige K (2017) In situ formed Fe cation modified Ir/MgO catalyst for selective hydrogenation of unsaturated carbonyl compounds. ACS Catal 7:5103–5111
118. Mondloch JE, Wang Q, Frenkel AI, Finke RG (2010) Development plus kinetic and mechanistic studies of a prototype supported-nanoparticle heterogeneous catalyst formation system in contact with solution: Ir(1,5-COD)Cl/γ-Al2O3 and its reduction by H2 to Ir(0)n/γ-Al2O3. J Am Chem Soc 132(28):9701–9714
119. Nassreddine S, Bergeret G, Jouguet B, Geantet C, Piccolo L (2010) Operando study of iridium acetylacetonate decomposition on amorphous silica-alumina for bifunctional catalyst preparation. Phys Chem Chem Phys 12:7812–7820
120. Nassreddine S, Casu S, Zotin JL, Geantet C, Piccolo L (2011) Thiotolerant Ir/SiO2-Al2O3 bifunctional catalysts: effect of support acidity on tetralin hydroconversion. Cat Sci Technol 1:408–412
121. Dongil AB, Bachiller-Baeza B, Rodriguez-Ramos I, Guerrero-Ruiz A, Mondelli C, Baiker A (2013) Structural properties of alumina- and silica-supported iridium catalysts and their behavior in the enantioselective hydrogenation of ethyl pyruvate. Appl Catal A 451:14–20
122. Lopez-De Jesus YM, Vicente A, Lafaye G, Marecot P, Williams CT (2008) Synthesis and characterization of dendrimer-derived supported iridium catalysts. J Phys Chem C 112:13837–13845
123. Beckers NA, Huynh S, Zhang X, Luber EJ, Buriak JM (2012) Screening of heterogeneous multimetallic nanoparticle catalysts supported on metal oxides for mono-, poly-, and heteroaromatic hydrogenation activity. ACS Catal 2:1524–1534
124. Tang M, Mao S, Li X, Chen C, Li M, Wang Y (2017) Highly effective Ir-based catalysts for benzoic acid hydrogenation: experiment- and theory-guided catalyst rational design. Green Chem 19:1766–1774
125. Park IS, Kwon MS, Kang KY, Lee JS, Park J (2007) Rhodium and iridium nanoparticles entrapped in aluminum oxyhydroxide nanofibers: catalysts for hydrogenations of arenes and ketones at room temperature with hydrogen balloon. Adv Synth Catal 349:2039–2047
126. Fan G, Zhang C (2014) Effective hydrogenation of p-chloronitrobenzene over iridium nanoparticles entrapped in aluminum oxy-hydroxide under mild conditions. Adv Mater Res 881-883:267–270
127. Higaki T, Kitazawa H, Yamazoe S, Tsukuda T (2016) Partially oxidized iridium clusters within dendrimers: size-controlled synthesis and selective hydrogenation of 2-nitrobenzaldehyde. Nanoscale 8:11371–11374
128. Tamura M, Tokonami K, Nakagawa Y, Tomishige K (2013) Rapid synthesis of unsaturated alcohols under mild conditions by highly selective hydrogenation. Chem Commun 49:7034–7036
129. Yang H, Huang C, Yang F, Yang X, Du L, Liao S (2015) Mesoporous silica nanoparticle supported PdIr bimetal catalyst for selective hydrogenation, and the significant promotional effect of Ir. Appl Surf Sci 357:558–563
130. Zheng Y, He P, Fang Y, Yang X, Liang H (2017) Hollow mesoporous silica supported PtIr bimetal catalysts for selective hydrogenation of phenol: significant promotion effect of iridium. RSC Adv 7:31582–315879
131. Ding K, Cullen DA, Zhang L, Cao Z, Roy AD, Ivanov IN, Cao D (2018) A general synthesis approach for supported bimetallic nanoparticles via surface inorganometallic chemistry. Science 362:560–564

132. Guo H, Li H, Jarvis K, Wan H, Kunal P, Dunning SG, Liu Y, Henkelman G, Humphrey SM (2018) Microwave-assisted synthesis of classically immiscible Ag-Ir alloy nanoparticle catalysts. ACS Catal 8:11386–11397
133. Das P, Sarmah PP, Borah BJ, Saikia L, Dutta DK (2016) Aromatic ring hydrogenation catalysed by nanoporous montmorillonite supported Ir(0)-nanoparticle composites under solvent free conditions. New J Chem 40:2850–2855
134. Tonbul Y, Zahmakiran M, Ozkar S (2014) Iridium(0) nanoparticles dispersed in zeolite framework: a highly active and long-lived green nanocatalyst for the hydrogenation of neat aromatics at room temperature. Appl Catal B 148-149:466–472
135. Bayram E, Lu J, Aydin C, Browning ND, Ozkar S, Finney E, Gates BC, Finke RG (2015) Agglomerative sintering of an atomically dispersed Ir1/zeolite Y catalyst: compelling evidence against Ostwald ripening but for bimolecular and autocatalytic agglomeration catalyst sintering steps. ACS Catal 5(6):3514–3527
136. Niederer JPM, Arnold ABJ, Holderich WF, Spliethof B, Tesche B, Reetz M, Bonnemann H (2002) Noble metal nanoparticles incorporated in mesoporous hosts. Top Catal 18:265–269
137. Zahmakiran M (2012) Iridium nanoparticles stabilized by metal organic frameworks (IrNPs@ZIF-8): synthesis, structural properties and catalytic performance. Dalton Trans 41:12690–12696
138. Friedrich M, Klarner M, Hermannsdoerfer J, Kempe R (2018) Nanometer-scaled iridium particles gas-phase-loaded into the pores of the metal-organic framework MIL-101. Polyhedron 155:441–446

Top Organomet Chem (2021) 69: 455–456
https://doi.org/10.1007/3418_2020_73

Published online: 28 January 2021

Correction to: Chapters

Francisco J. Fernández-Alvarez and Luis A. Oro

Correction to:

Chapter "Iridium-Catalyzed Homogeneous Hydrogenation and Hydrosilylation of Carbon Dioxide" in:

Francisco J. Fernández-Alvarez and Luis A. Oro, Top Organomet Chem,
https://doi.org/10.1007/3418_2020_52

The original version of the below chapter was inadvertently published with the incorrect copyright holder name. It has been now updated with the correct copyright holder name.

https://doi.org/10.1007/3418_2020_52

Correction To: Iridium-Catalyzed Undirected Homogeneous C–H Borylation Reaction

Elena Fernández

Correction to:

Chapter "Iridium-Catalyzed Undirected Homogeneous C–H Borylation Reaction" in:

Elena Fernández, Top Organomet Chem,
https://doi.org/10.1007/3418_2020_53

The original version of the below chapter was inadvertently published with the incorrect copyright holder name. It has been now updated with the correct copyright holder name.

https://doi.org/10.1007/3418_2020_53

The updated online version of these chapters can be found at
https://doi.org/10.1007/3418_2020_52
https://doi.org/10.1007/3418_2020_53
https://doi.org/10.1007/3418_2020_54
https://doi.org/10.1007/3418_2020_55
https://doi.org/10.1007/3418_2020_56
https://doi.org/10.1007/3418_2020_58

Correction To: Electroreduction of Carbon Dioxide by Homogeneous Iridium Catalysts

Ryoichi Kanega

Correction to:

Chapter "Electroreduction of Carbon Dioxide by Homogeneous Iridium Catalysts" in:

Ryoichi Kanega, Top Organomet Chem,

https://doi.org/10.1007/3418_2020_54

The original version of the below chapter was inadvertently published with the incorrect copyright holder name. It has been now updated with the correct copyright holder name.

https://doi.org/10.1007/3418_2020_54

Correction To: Iridium-Catalyzed Silylation

Manuel Iglesias and Luis A. Oro

Correction to:

Chapter "Iridium-Catalyzed Silylation" in:

Manuel Iglesias and Luis A. Oro, Top Organomet Chem,

https://doi.org/10.1007/3418_2020_55

The original version of the below chapter was inadvertently published with the incorrect copyright holder name. It has been now updated with the correct copyright holder name.

https://doi.org/10.1007/3418_2020_55

Correction To: Iridium-Catalyzed Dehydrogenative Reactions

Takuya Shimbayashi and Ken-ichi Fujita

Correction to:

Chapter "Iridium-Catalyzed Dehydrogenative Reactions" in:

Takuya Shimbayashi and Ken-ichi Fujita, Top Organomet Chem,

https://doi.org/10.1007/3418_2020_56

The original version of the below chapter was inadvertently published with the incorrect copyright holder name. It has been now updated with the correct copyright holder name.

https://doi.org/10.1007/3418_2020_56

Correction To: Iridium Catalysts for Hydrogen Isotope Exchange

Marc Reid

Correction to:

Chapter "Iridium Catalysts for Hydrogen Isotope Exchange" in:

Marc Reid, Top Organomet Chem,

https://doi.org/10.1007/3418_2020_58

The original version of the below chapter was inadvertently published with the incorrect copyright holder name. It has been now updated with the correct copyright holder name.

https://doi.org/10.1007/3418_2020_58

The original chapters were corrected.

Top Organomet Chem (2021) 69: C1–C1
https://doi.org/10.1007/978-3-030-69083-0_74

Correction to: Iridium Catalysts for Organic Reactions

Luis A. Oro and Carmen Claver

Correction to:
L. A. Oro, C. Claver (eds.),
Iridium Catalysts for Organic Reactions,
Topics in Organometallic Chemistry 69,
https://doi.org/10.1007/978-3-030-69083-0

This book was inadvertently published without updating the following corrections:

p. 229, the word "provably" changed to "probably"
p. 231, the letter "a" added before "6/1"
p. 232, deleted the words "at 11"
p. 239, the word "pathways" changed to "tools"
p. 245, the word "catalyst" changed to "catalysts"
p. 248, the equation [IrH(κ [9]-hqca)(COE)] changed to [IrH(κ^3-*O*,*N*,*O'*-hqca (COE)]
p. 255, deleted "(Scheme 28)" after the word "intermediate"
p. 255, the word "Scheme 28" added after the word coupling
p. 261, the size of Scheme 37 is reduced
p. 320, in reference 28, the name "Gunaseker" is changed to "Gunasekar"
p. 321, in reference 48, the year 2001 is changed to 2011
p. 322, in reference 75, the word "Sustain" is changed to "Sustainable"
p. 324, in reference 103, the word "WA" is changed to "WE"

These corrections had been updated.

The updated online version of these chapters can be found at
https://doi.org/10.1007/3418_2020_55
https://doi.org/10.1007/3418_2020_52